T0291026

Heat Transfer Basics

Heat Transfer Basics

A Concise Approach to Problem Solving

By

Jamil Ghojel

Published by John Wiley & Sons, Inc., Hoboken, New Jersey.

Published simultaneously in Canada.

Library of Congress Cataloging-in-Publication Data:
Names: Ghojel, J., author. | John Wiley & Sons, publisher.
Title: Heat transfer basics : a concise approach to problem solving / by Jamil Ghojel.
Description: Hoboken, New Jersey : John Wiley & Sons, [2024] | Includes bibliographical references and index.
Identifiers: LCCN 2023013359 (print) | LCCN 2023013360 (ebook) | ISBN 9781119840268 (hardback) |
ISBN 9781119840275 (adobe pdf) | ISBN 9781119840282 (epub)
Subjects: LCSH: Heat--Transmission--Textbooks. | Heat--Transmission--Problems, exercises, etc.
Classification: LCC QC320.34 .G46 2023 (print) | LCC QC320.34 (ebook) | DDC 621.402/2--dc23/eng/20230928
LC record available at https://lccn.loc.gov/2023013359
LC ebook record available at https://lccn.loc.gov/2023013360

Cover image(s): © Etienne Outram/Shutterstock
Cover design: Wiley

Set in 9.5/12.5pt STIXTwoText by Integra Software Services Pvt. Ltd., Pondicherry, India

SKY10060096_111723

Contents

Preface

As the title implies, the book covers basic but essential material for heat transfer courses at different levels. The book has been made self-contained so that certain parts can be used as introductory material and others as material for more advanced levels. The book is also intended for use by practicing professionals in various engineering fields. Basic also implies concise coverage of a wide range of topics at sufficient depth without blowing up the number of pages and cost to extremes, as has been the case with many excellent textbooks in recent years.

The book is primarily concerned with conduction, convection, and radiation heat transfer. The book consists of 14 chapters and 14 appendices. Chapter 1 is a general brief introduction to heat transfer concepts. Chapters 2 and 3 cover the topics of one-dimensional steady-state conduction and extended surfaces, which are presented in considerably more detail and depth than most other similar textbooks. Chapter 4 deals with two-dimensional steady-state conduction issues and Chapter 5 discusses one- and multi-dimensional non-steady (transient) conduction in some detail. Numerical methods of solving steady and transient conduction problems are presented with a number of illustrative examples and spreadsheet screenshots. Following an introduction to fundamentals of convection heat transfer in Chapter 6, forced convection in external and internal flows are discussed in Chapters 7 and 8, respectively. Natural (free) convection is covered in Chapter 9. Chapters 10 and 11 deal with fundamentals of radiation heat transfer, shape factors, and the exchange of thermal radiation between surfaces separated by transparent and non-transparent media. In dealing with radiation in participating gaseous media, the conventional Hottel charts have been replaced by recently developed charts by Alberti et al. (2016, 2018. 2020) on the basis of more accurate spectral data. The comprehensive set of charts include gas emissivity of water vapour, carbon dioxide, and carbon monoxide together with pressure and binary gas-overlap correction factors. Essentials of heat exchangers and the classical thermal analysis using the log-mean-temperature-difference and NTU-effectiveness approaches is presented in Chapter 12.

Heat transfer with phase change, and mass transfer are also covered in Chapters 13 and 14, respectively. The book has 14 appendices (A to N) four of which (B, C, D, and N) are included in the text and the remainder can be accessed online at "www.wiley.com/go/ghojel/heat_transfer."

Since the primary goal of the book is to provide students with the knowledge and skills needed for practical engineering practice, both analytical and graphical solution techniques are presented. For the latter approach, most charts in the text are duplicated with enlarged versions in the appendices. Also, since computer applications play an integral role in solving complex problems, "Solaria Thermal" software package (Harley Thermal) is used in the book, where applicable, to reinforce the relevance of computer-aided solutions of heat transfer problems.

Preface

Acknowledgements

The author wishes to acknowledge and thank Dave Rosato, president and founder of Harley Thermal, for kindly providing access to the "Solaria Thermal" computer modelling software package which has been used in this book to model selected conduction and convection examples. His insight and advice on the use of Solaria throughout the preparation of the manuscript is greatly appreciated.

Thanks are also due to Dr David Trujillo (TRUCOMP – Inverse Problems) for providing the latest version of INTEMP (Inverse heat conduction software package), which was used to update the simulation of steel-concrete contact resistance in cylindrical structural columns.

About Solaria Thermal

Solaria is a general-purpose thermal modelling software package incorporating model generation, solving and post processing functions. It incorporates Finite Difference and Conjugate Gradient solvers to solve steady-state and transient models of constant, temperature-dependent, or time-dependent thermophysical properties. It can also handle anisotropic (three-dimensional) material conductivities. The software is relatively easy to learn and model generation and post processing is mostly intuitive. A free demo version with a limited number of nodes can be obtained from https//www.solariathermal.com.

List of Symbols

A	Area, m^2
A_b	Area of unfinned surface (prime area), m^2
A_c	Cross-sectional area, m^2
A_m	Fin profile area, m^2
A_p	Fin profile area, m^2
A_s	Surface area, m^2
C	Molar concentration, $kmol/m^3$; gas emissivity pressure correction factor
C_D	Drag coefficient
c_p	Specific heat at constant pressure, $J/kg.K$
C_r	Capacitance ratio
C_t	Thermal capacitance, J/K
c_v	Specific heat at constant volume, $J/kg.K$
D	Diameter, m
D_{AB}	Binary mass diffusivity, m^2/s
D_h	Hydraulic diameter, m
E	Emissive power, W/m^2
E_b	Blackbody emissive power, W/m^2
$E_{b\lambda}$	Blackbody emissive power per unit wave length (monochromatic), $W/m^2.\mu m$
Ec	Eckert number
$\dot{E}_g$	Rate of energy generation, W
F	Shape factor; drag force, N; correction factor for heat exchangers
Eu	Euler number
F_f	Fouling factor, $m^2.K/W$
Fo	Fourier number
Fr	Froud number
f	Friction factor, coefficient of friction or friction coefficient
G	Irradiation, W/m^2
Gr	Grashof number
g	Gravitational acceleration, m/s^2
h	Heat transfer coefficient, $W/m^2.K$; conductance, $W/m^2.K$
$\overline{h}$	Surface-average heat transfer coefficient
h_{fg}	Latent heat of vaporization, J/kg

h'_{fg}	Modified heat of vaporization, J/kg
h_m	Convection mass transfer coefficient, m/s
h_{rad}	Radiation heat transfer coefficient, $W/m^2.K$
I	Electric current, A
i	Radiation intensity, W/m^2
J	Radiosity, W/m^2; mass diffusion flux, $kg/m^2.s$
$\bar{J}$	Molar diffusion flux, $kmol/m^2.s$
Ja	Jacob number
k	Thermal conductivity, $W/m.K$
k_{eff}	Effective conductivity, $W/m.K$
L	Length, m
$LMTD$	Log mean temperature difference
M	Molar mass, $kmol$
$\dot{M}$	Molar flow rate, $kmol/s$
m	Mass, kg
$\dot{m}$	Mass flow rate, kg/s
N	Number of moles of substance, number of tubes
Nu	Nusselt number
NTU	Number of transfer units
$\dot{N}$	species molar transfer rate, $kmol/s$
P	Perimeter, m
Pe	Peclet number
Pr	Prandtl number
p	Pressure, N/m^2, atm, bar
Q	Energy transfer, J
$\dot{Q}$	Rate of energy transfer, W
q	Heat flux, W/m^2
$\dot{q}$	Rate of energy generation per unit volume, W/m^3
R	Cylinder radius, m; gas constant, $J/kg.K$; electric resistance, Ω
$\bar{R}$	Universal gas constant, $J/kmol.K$
Ra	Rayleigh number
Re	Reynolds number
R_c	Thermal contact resistance,
R_t	Thermal resistance, K/W
r	Radius, m
r_i	Inner radius, m
r_o	Outer radius, m
S	Conduction shape factor, m
S_D	Diagonal pitch of a tube bank, m
S_L	Longitudinal pitch of a tube bank, m
S_T	Traverse pitch of a tube bank, m
Sc	Schmidt number

Sh	Sherwood number
St	Stanton number
T	Temperature, K
t	Temperature, °C; time, sec
U	Fluid velocity, m/s; overall heat transfer coefficient, $W/m^2.K$; internal energy, J
V	Volume, m^3
v	Specific volume, m^3/kg
W	Width, m
We	Weber number
w_i	Species mass fraction
X	Vapour quality
y_i	Mole fraction of a species

Greek letters

α	Thermal diffusivity, m^2/s; absorptivity
β	Coefficient of volume thermal expansion, $1/K$
δ	Hydrodynamic boundary layer thickness, m; thickness (slab, plate), m
δ_t	Thermal boundary layer thickness, m
ε	Emissivity; heat exchanger effectiveness; average roughness height, m
ε_f	Fin effectiveness
ϵ	Correction factor (heat exchangers)
η	Efficiency
η_f	Fin efficiency
η_o	Overall efficiency
θ	Temperature difference, K
φ	Relative humidity
λ	Wavelength, $m, \mu m$
μ	Dynamic viscosity, kg/sm; molecular (molar) mass, $kg/kmol$
ϑ	Kinematic viscosity, m^2/s; frequency of radiation, $1/s$
ρ	Mass density, kg/m^3; reflectivity
σ	Stefan-Boltzmann constant, $W/m^2.K^4$; surface tension, N/m
τ	Shear stress, N/m^2; transmissivity; time step
ω	Solid angle, sr (steradian)

Subscripts

A	Species in binary mixture
atm	Atmosphere
B	Species in binary mixture
b	Blackbody; bulk temperature, K

c	Cross-sectional; cold fluid; characteristic; corrected; convective
cr	Critical
$cond$	Conduction
$conv$	Convection
D	Drag
e	Effective;
f	Fluid properties; fin condition; film (temperature)
g	Generation
h	Hydrodynamic; hot fluid
ins	Insulation
iso	Isothermal
L	Length, m
l	Saturated liquid conditions
m	Mean value
m	Maximum
o	Centre condition; outlet; outer
rad	Radiation
r	Root
s	Surface conditions, storage
sat	Saturated conditions
t	Thermal, tip
v	Saturated vapour conditions
x	Local conditions
w	Wall
Σ	Total
∞	Free stream conditions

Superscripts

o	Standard emissivity

About the Companion Website

Heat Transfer Basics: A Concise Approach to Problem Solving is accompanied by a companion website:

www.wiley.com/go/ghojel/heat_transfer

This website includes:

- Instructor's Manual
- Solution Manual
- Appendices A, E to M

1

Basic Concepts of Heat and Mass Transfer

1.1 Heat Transfer and Its Relationship With Thermodynamics

Heat transfer can be looked at as a branch of classical thermodynamics with specific distinguishing features. This can best be explained by considering the processes taking place in the cylinder of the internal combustion engine schematically shown in Figure 1.1a.

The thermodynamic processes taking place in the cylinder are elegantly and succinctly stated by the first law of thermodynamics (principle of conservation of energy) in the form

$$Q - W = \Delta U \tag{1.1}$$

where

- Q is the heat generated by the combustion of the injected fuel in the cylinder at the end of the compression process, resulting in high-pressure, high temperature gaseous products
- W is the work done by the piston as the combustion products expand with the piston moving downwards
- ΔU is the balance of the heat left from the combustion process to raise the internal energy and temperature of the gases inside the cylinder.

The first law of thermodynamics states that energy can be converted (transformed) from one form (heat) to another (work) but cannot be created or destroyed and that the process takes place between two states at equilibrium. The first law does not provide information on the amount of energy that can be transformed, energy transformation direction, and its effectiveness. This gap is filled by the second law of thermodynamics which stipulates that only part of the heat will be converted to work, heat will flow spontaneously from a high temperature source to a lower temperature sink, and the flow will be accompanied by irreversible changes that degrade the resultant energy making it difficult to be utilized further. Applied to the example under consideration, indicative heat balance in internal combustion engines is given in Table 1.1.

The disposal of the heat component shown under "losses" in Table 1.1 is the domain of the science of heat transfer. This process of disposal is explained below and some new terms are introduced without explanation at this stage to be defined in greater detail later.

The system shown in Figure 1.1b is essentially heat exchange between the high-temperature combustion gases and the circulating cooling water in the water jacket. Heat is first transferred from the hot gases at temperature T_{com} to the inner cylinder wall at temperature T_1 by convection

Heat Transfer Basics: A Concise Approach to Problem Solving, First Edition. Jamil Ghojel.

Companion website: www.wiley.com/go/ghojel/heat_transfer

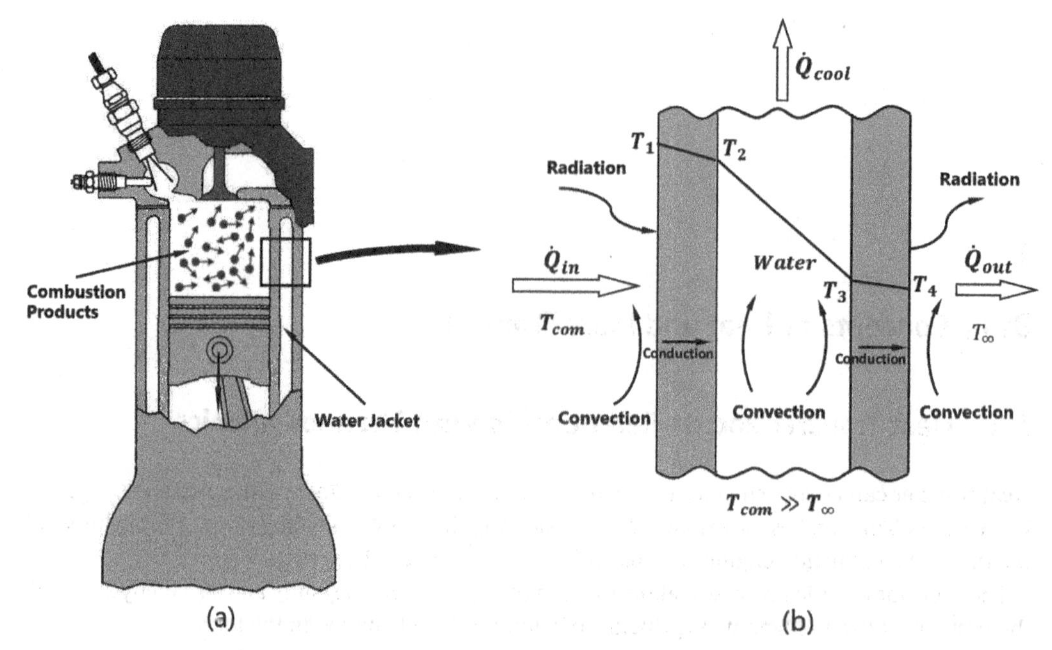

Figure 1.1 Heat transfer modes in the cylinder of internal combustion engine: (a) engine cylinder schematic; (b) control volume.

Table 1.1 Energy balance in internal combustion engines.

	% Energy supplied	
Component	**Petrol engine**	**Diesel engine**
Useful work	32	45
Exhaust system	38	28
Cooling system and surface losses	30	27

and radiation followed by conduction through the inner wall of the jacket with the temperature dropping to T_2. Heat is then transferred to the circulating water at temperature $T_w \approx (T_1 + T_2)/2$ by convection, followed by transfer by convection to the inner surface of the outer jacket wall at temperature T_3. The final conduction process takes place through the outer water jacket wall with a temperature drop to T_4 followed by heat losses from the outer cylinder surface by both convection and radiation to the atmosphere. A parallel heat exchange process takes place in the engine cooling system (not shown here), whereby the cooling water is circulated by a pump in a heat exchanger (radiator) to be cooled and returned to the engine. The flow rate and corresponding dimensions of the components of the cooling system must be selected to maintain optimum operating conditions of engine components.

From the above description of the processes in the cylinder, it is apparent that heat flow involves time-dependent factors and the fluids are in temperature nonequilibrium states. Therefore, the basic laws of thermodynamics are not adequate to assess the heat transfer processes described and proper analysis requires the use of tools that are provided by other branches of science, such as fluid dynamics, physics, and mathematics.

In heat transfer, the first law of thermodynamics is widely used, albeit without the work component and is generally known as the principle of conservation of energy (in short, energy or heat balance). Applied to the control volume in Figure 1.1, the heat balance can be expressed as

Rate of heat transfer into the control volume, $\dot{Q}_{in}$ =
Rate of heat removed by the cooling water, $\dot{Q}_{cool}$ +
Rate of heat loss to the surrounding media, $\dot{Q}_{out}$

$$\dot{Q}_{in} = \dot{Q}_{cool} + \dot{Q}_{out} \tag{1.2}$$

More generally, taking into account that heat transfer processes may be accompanied by internal energy generation $\dot{Q}_g$ and/or energy storage (accumulation) within the control volume $\dot{Q}_s$, the energy balance equation for a control volume can be written as

$$\dot{Q}_{in} - \dot{Q}_{out} + \dot{Q}_g = \dot{Q}_s \tag{1.3}$$

Internal heat generation could be due to electrical heating, chemical processes, or nuclear reactions.

The field of applications of heat transfer is wide and diverse covering, for example, energy generation, transport, process industry, building industry, agriculture, medical industry, and aviation and space industries. In the following sections, the three modes of heat transfer will be discussed separately but it should be kept in mind that they continually interact and rarely occur independently.

1.2 Heat Conduction

Consider a solid layer or static liquid layer contained between two long vertical plates at two different temperatures T_1 and T_2 (Figure 1.2). The molecules of the medium between the plates are depicted as small circles, with the molecules at higher temperature in black. The molecules exist in continuous agitated random motion indicated by the small arrows, with the degree of agitation being directly proportional to the temperature. According to the second law of thermodynamics, there is a net molecular energy transfer from the hotter side of the layer to the colder side and, consequently, net thermal energy (heat) transfer and this process is known as conduction.

The heat conduction process is governed by Fourier's law, stated in terms of the rate of heat transfer

$$\dot{Q} = -kA\frac{dT}{dx}\ J/s \text{ or } W \tag{1.4}$$

where

A – is the area through which heat flows, m^2

dT/dx – is the temperature gradient in the x direction, K/m, and

k – is the constant of proportionality known as the thermal conductivity, which is a physical property of the conduction medium and could vary significantly with temperature, W/mK

Often, Eq. (1.4) is written in terms of the heat flux q, defined as $\dot{Q}/A$

$$q = -k\frac{dT}{dx}\ W/m^2 \tag{1.5}$$

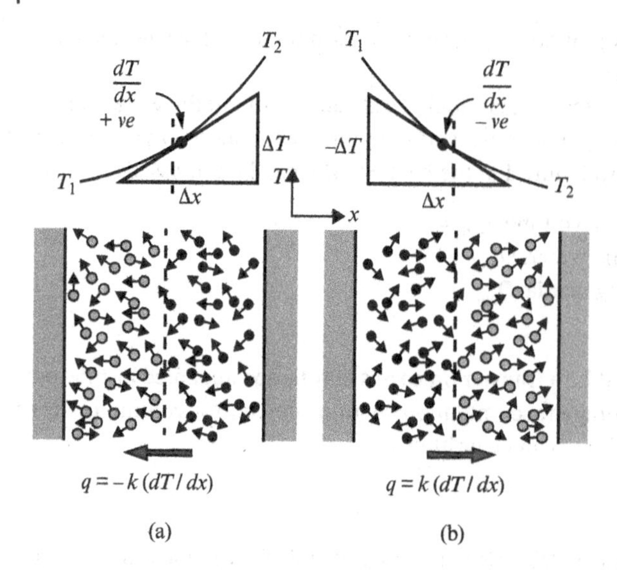

Figure 1.2 Conduction heat transfer and sign convention: (a) positive temperature gradient; (b) negative temperature gradient.

The negative sign in Fourier's law is deliberately introduced to obtain a positive heat flow when the temperature gradient is negative (decreasing temperature in the x-direction) (Figure 1.2b) and a negative heat flow when the temperature gradient is positive (increasing temperature in the x-direction) (Figure 1.2a).

For a slab of thickness δ and finite temperature gradient of ΔT, Eq. (1.4) can be rearranged as

$$\Delta T = \frac{\delta}{Ak}\dot{Q} \tag{1.6}$$

This equation is analogous to the equation relating voltage potential to resistance and current in electrical circuits (Figure 1.3)

$$\Delta V = RI \tag{1.7}$$

The thermal resistance for the conduction process is then

$$R = \frac{\delta}{Ak} K/W \tag{1.8}$$

Similar to electrical resistances, thermal resistances can be arranged in series or in parallel, depending on the physical model of the heat transfer problem.

For steady-state one-dimensional problems, in which the thickness of a plate is much smaller than its height, the temperature gradient in Eqs (1.4) and (1.5) can be approximated by the temperature difference divided by the thickness of the one-dimensional layer

Figure 1.3 Analogy between: (a) electrical and (b) thermal conduction circuits.

Table 1.2 Thermal conductivity for selected materials.

	Thermal conductivity
Material	***W/m.K***
Air	0.0625
Aluminium alloy	168
Carbon steel	64
Copper	386
Fibreglass	0.028
Water	0.611

$$\dot{Q} = -kA\frac{T_2 - T_1}{x_2 - x_1} = -kA\frac{\Delta T}{\Delta x} \tag{1.9}$$

Table 1.2 shows some indicative values of the thermal conductivity for some materials. More comprehensive data on thermal conductivity is provided in Appendix A.

Example 1.1 In the model shown in Figure 1.2, the distance between the two plates is 100 mm, the left plate is maintained at 50°C and the right plate at 20°C. Calculate the heat flux through the plate if the space between the plates is filled with air, water, fibreglass, and copper.

Solution

Since the temperature gradient is constant, the flux is directly proportional to the thermal conductivity

$$q = -k\frac{T_2 - T_1}{x_2 - x_1} = -k\frac{20 - 50}{0.1} = 300\ kW$$

The results of the calculations are shown in Table E1.1.

It is apparent that the higher the thermal conductivity, the less resistant the material is to heat flow. Most resistance (least heat flux) is exhibited by insulating materials. Air and water can be used as insulating materials under certain circumstances to reduce heat losses.

Table E1.1 Calculated heat flux for air, water, and fibreglass.

	Thermal conductivity	*q*
Material	***W/m.K***	***kW/m²***
Air	0.0625	0.01875
Water	0.611	0.1833
Fibreglass (insulating material)	0.028	0.084
Copper	386	115.8

1.3 Heat Convection

Convection is the process of heat transfer between fluids and solid boundaries. Referring to Figure 1.1, we observe the following heat transfer processes that come under this definition:

- From high-temperature gases to the inner walls of the cylinder
- From heated cylinder wall to the circulating water in the water jacket
- From the heated circulating water to the cooler outer cylinder wall
- From heated outer wall to the surrounding air at ambient temperature.

In all these processes, heat transfer is due to the combined effect of molecular motion, as illustrated in Figure 1.2, and macroscopic motion of the fluid portions. The macroscopic motion in the cylinder is caused by the turbulent motion of the combustion products as the piston moves downwards at high speed and in the water jacket by the action of the water pump. Macroscopic motion is the motion of fluid portions (parcels) from one position to another, resulting in enthalpy exchange between high- and low-temperature zones in the fluid, as a result of which the rate of heat transfer in the fluid as a whole and to the solid wall increases and the process is referred to as forced convection. The rate of heat transfer can be increased at will simply by increasing the rapidity of motion of fluid parcels by external means such as pumps and blowers. In some cases, moderate fluid motion may be caused by the density gradient in the fluid, in which case heat transfer is by free convection. In our example of the engine cylinder, this can occur at the exterior walls if the engine is stationary. If the engine is powering a motor vehicle, heat will be dissipated to the atmosphere by forced convection.

Convection heat transfer is more complex than conduction due to the interaction of fluid motion with surface effects and the formation of hydrodynamic and thermal boundary layers and the presence of velocity and temperature profiles. Figure 1.4 shows velocity and temperature profiles for forced and free convection.

Despite the aforementioned complications, the convective heat transfer rate $\dot{Q}_c$ and heat flux q_c can be calculated by a simple equation known as Newton's law of cooling

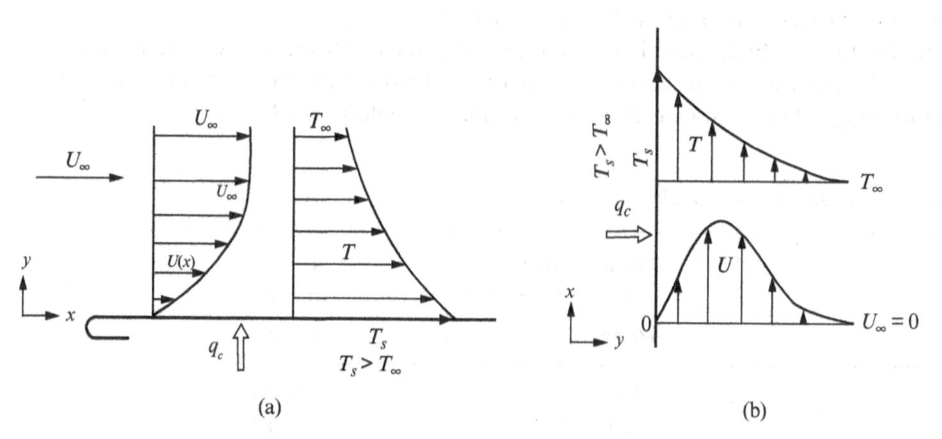

Figure 1.4 Schematic diagrams of velocity and temperature profiles over heated plates: (a) forced convection from horizontal plate; (b) free convection from vertical plate.

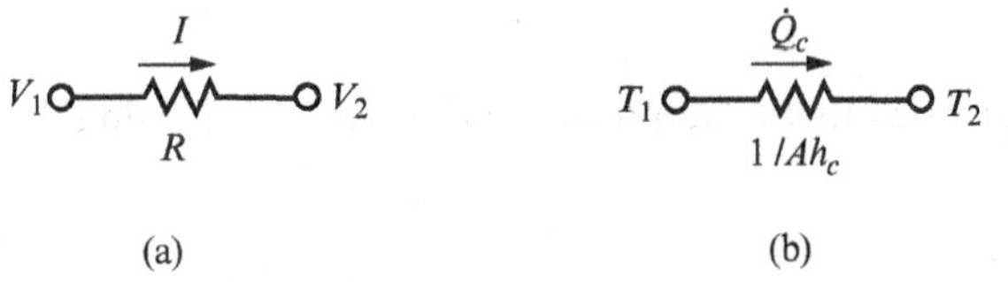

Figure 1.5 Analogy between electrical (a) and thermal convection (b) circuits.

Table 1.3 Heat transfer coefficient for selected fluids.

Fluid	Heat transfer coefficient $W/m^2.K$
Air (free convection)	$6-30$
Air (forced convection)	$10-500$
Water (free convection)	$200-600$
Water (forced convection)	$300-1{,}800$
Oil (forced convection)	$60-2000$
Superheated steam (forced convection)	$30-300$

$$\dot{Q}_c = h_c A \Delta T \qquad W \tag{1.10}$$

$$q_c = h_c \Delta T \; W/m^2 \tag{1.11}$$

where

$\dot{Q}_c$ – rate of convective heat transfer, W
q_c – heat flux, W/m^2
A – area through which heat flows, m^2
ΔT – difference between the surface temperature T_s and the free stream temperature T_∞, K or °C
h_c – proportionality factor known as the heat transfer coefficient, W/m^2K

The thermal circuit for a simple heat transfer by convection is shown in Figure 1.5. The convection resistance is

$$R = \frac{1}{Ah_c} K/W \tag{1.12}$$

The heat transfer coefficient is determined from experimental results and, strictly speaking, h_c in Eqs (1.10) and (1.11) should be an average value $\bar{h}_c$ over area A. The direction of the heat flux q_c (Figure 1.4) depends on whether the plate temperature T_s is higher or lower than the free stream temperature T_∞.

Values of heat transfer coefficient for selected fluids in contact with a solid surface are shown in Table 1.3.

Example 1.2 A solid cylindrical element 2 m long and 2.5 cm diameter is heated electrically by passing 1.5 A current at 80 V voltage until it reaches a steady surface temperature $T_s = 140°C$. If the temperature of the surrounding air is 16°C, estimate the average heat transfer coefficient for the heat transfer process between the tube surface and the surrounding air.

Solution

We assume that there is no energy storage and no other heat transfer modes. Hence, applying the energy equation

$$\dot{Q} = \dot{Q}_g = VI = 80 \times 1.5 = 120\ W$$

The surface area of the element is

$$A = \pi DL = \pi \times 0.025 \times 2 = 0.157\ m^2$$

From Newton's law of cooling

$$h = \frac{\dot{Q}}{A\Delta T} = \frac{120}{0.157 \times (140 - 16)} = 6.16\ W/m^2.K$$

1.4 Thermal Radiation

Radiation heat transfer in the example of Figure 1.1 is shown by the curly arrows inside the cylinder and at the exterior surface of the cylinder. Compared to the two heat transfer models discussed so far, radiation heat transfer has the following distinctive features:

- All surfaces above absolute zero temperature emit electromagnetic radiation energy E in all directions (Figure 1.6). The higher the temperature, the more energy is emitted
- The emitted energy is attributed to changes in the electron configurations of the constituent atoms and molecules at the atomic scale
- Radiation energy is transported by electromagnetic waves travelling at the speed of light
- When radiation emitted by one surface falls on a recipient surface and is absorbed, the random molecular energy increases causing the temperature of the recipient to increase
- The presence of a material medium is not a requirement for radiation energy transfer and can occur efficiently in vacuum
- Radiation energy is exchanged between emitting surfaces, resulting in net energy direction from the high-temperature source to the low-temperature target.

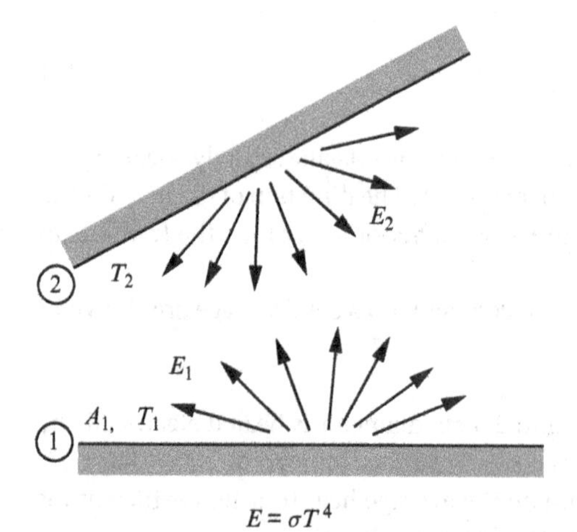

Figure 1.6 Emissive power of heated surfaces and radiation energy exchange.

The upper limit to the emitted energy is given by the Stefan–Boltzmann law, which states that any blackbody surface above a temperature of absolute zero radiates heat at a rate proportional to the fourth power of the absolute temperature

$$E_b = \sigma T_s^4 \qquad W/m^2 \tag{1.13}$$

The constant of proportionality σ is the Stefan–Boltzmann constant ($\sigma = 5.67 \times 10^{-8} W/m^2.K^4$), subscript b (for black) denotes idealized blackbody or black surface, and subscript s denotes surface. A net transfer of radiant heat can take place between any two bodies if a temperature gradient exists. If a convex small black body 1 at temperature T_1 exchanges radiation energy with a large black enclosure 2 at temperature T_2 ($T_2 > T_1$), as shown in Figure 1.7, all the radiant energy incident upon surface 1 is fully absorbed and the net rate of radiant heat transfer to that surface is given by

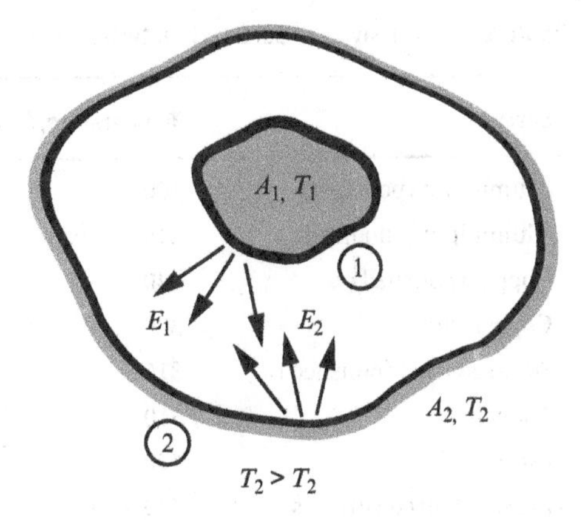

Figure 1.7 Convex black object in black enclosure.

$$\dot{Q}_b = \sigma A_1 (T_2^4 - T_1^4) \qquad W \tag{1.14}$$

Real surfaces, or grey surfaces, are not blackbody surfaces and the emitted energy (heat flux) by a real surface is less than that of a blackbody emission at the same temperature, and Eq. (1.13) acquires the form

$$E_g = \varepsilon \sigma T_s^4 \qquad W/m^2 \tag{1.15}$$

where ε is a property termed *emissivity* or *emittance* that has a value less than unity. If the surfaces of objects 1 and 2 in Figure 1.7 are assumed grey, all the radiation emitted from object 1 is absorbed by the surrounding enclosure (object 2) and the latter behave like a black body at T_2. The net radiative heat transfer to object 1 is given by

$$\dot{Q} = \varepsilon_1 \sigma A_1 (T_2^4 - T_1^4) \qquad W \tag{1.16}$$

If the two objects exchanging radiation have a specific geometry and finite size (Figure 1.6) and different emissivities, the radiation heat flow rate between them is

$$\dot{Q}_{12} = \sigma A_1 F_{12} (T_2^4 - T_1^4) \qquad W \tag{1.17}$$

F_{12} is known as the shape factor, view factor, configuration factor, or transfer factor and it accounts for the emissivities and relative geometries of actual bodies. The term "shape factor" will be used in this book. Determination of shape factors for some geometries could be mathematically challenging, and some examples will be presented in Chapter 10. Table 1.4 shows selected values of emissivity.

Table 1.4 Emissivity of selected materials and surfaces.

Surface	Temperature, K	Emissivity, ε
Aluminium (polished)	500	0.04
Aluminium (anodized)	310	0.94
Copper (polished)	340	0.041
Copper (dull)	320	0.15
Stainless steel (polished)	310	0.25
Asphalt	310	0.93
Glass	420	0.88
Light-coloured surfaces	530–800	0.25–0.5
Dark-coloured surfaces	530–800	0.4–0.8

Example 1.3 A spherical object of diameter $D = 25\,mm$ and temperature $T_1 = 800\,K$ is placed centrally inside an evacuated spherical shell with a nearly black wall that is maintained at temperature $T_2 = 350\,K$. Calculate the net rate of radiation heat transfer from the small sphere to the inner wall of the spherical shell for surface emissivities 0.03 and 0.85.

Solution

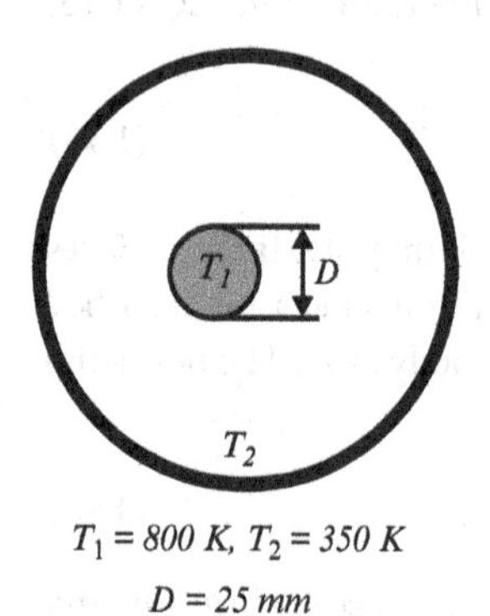

$T_1 = 800\ K,\ T_2 = 350\ K$

$D = 25\ mm$

$$A_1 = \pi D^2 = \pi(0.025)^2 = 1.963 \times 10^{-3}\ m^2$$

Applying Eq. (1.12), we obtain for $\varepsilon_1 = 0.03$

$$\dot{Q} = \varepsilon_1 \sigma A_1 (T_2^4 - T_1^4) = 0.03 \times 5.67 \times 10^{-8} \times 1.963 \times 10^{-3} (800^4 - 350^4) = 1.315\,W$$

For $\varepsilon_1 = 0.85$

$$\dot{Q} = 37.26\ W$$

1.5 Mass Transfer

Convective heat transfer is accompanied by transfer (movement) of the fluid, either on a macroscopic scale with a pump or a blower as a motive force in forced convection, or on a microscopic scale at the level of molecules moving under the effect of buoyancy forces within the fluid in free convection. However, there is another mode of mass transfer that is encountered in processes such as condensation, evaporation, distillation, drying, humidification, and so on. This mode is known as mass transfer. Heat and mass transfer are usually studied together, as they occur simultaneously in many practical processes.

Mass transfer occurs at the molecular level as a result of diffusion of matter from a region of high concentration to a region of low concentration, as shown schematically for a single species in Figure 1.8.

If the molecular concentration of a gaseous or liquid species is greater on the left side of the container than on the right side, more molecules will move from left to right than in the opposite direction with a net mass transfer. The diffusion rate for this process is governed by Fick's law of diffusion, which states that the molar mass flow rate of a constituent per unit area is proportional to the concentration gradient

$$\frac{\dot{M}}{A} = -D\frac{dC}{dx}\ kmole/m^2.s \tag{1.18}$$

and the molar flow rate

$$\dot{M} = -DA\frac{dC}{dx}\ kmole/s \tag{1.19}$$

where

$\dot{M}$ – molar flow rate, $kmole/s$

D – diffusion coefficient, m^2/s

A – area normal to flow direction, m^2

C – molar concentration of the species per unit volume, $kmole/m^3$

The negative sign in Flick's law is introduced to obtain a positive mass flux when the concentration gradient is negative (decreasing concentration in the x direction). Fick's law describes the diffusion of mass down a species concentration gradient in a manner similar to Fourier's law, which describes heat conduction down a temperature gradient.

Mass transfer occurs only in mixtures of two substances (species) or more, with at least one species within the mixture moving from a region of high concentration to a region of low concentration. Consider the binary system shown in Figure 1.9 comprising species A and B on both sides of a partition in the middle of a chamber of width L (Figure 1.9a). If the partition is removed, both fluids will tend to diffuse with the net transfer of species A being to the right (black arrow) and that of species B to the left (grey arrow) (Figure 1.9b). The process continues until an equilibrium state is reached with no further diffusion of either species.

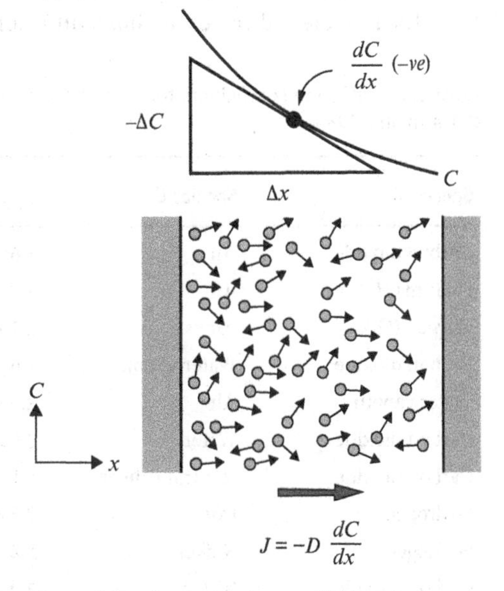

Figure 1.8 Concentration profile in mass transfer process.

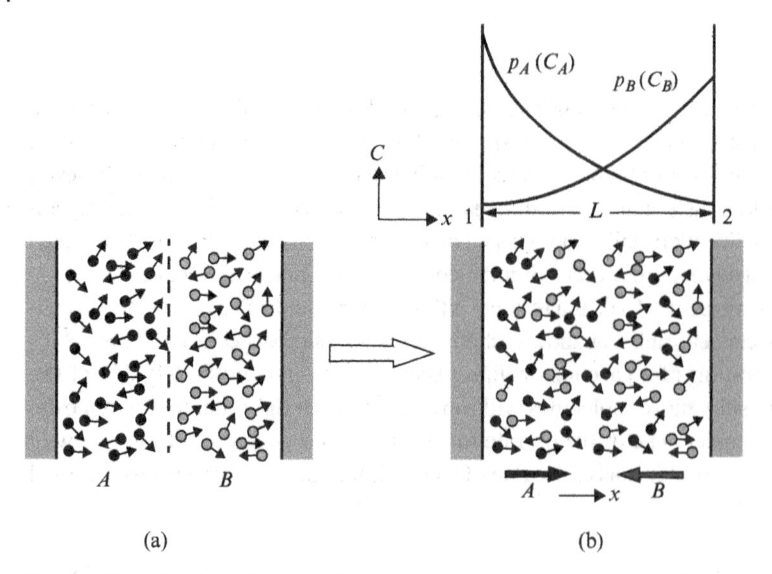

Figure 1.9 Schematic diagram of mass transfer in a binary gaseous mixture.

The molar flux of species A in the mixture according to Fick's law is

$$\frac{\dot{M}_A}{A} = -D_{AB}\frac{dC_A}{dx} \tag{1.20}$$

and that of species B in the opposite direction

$$\frac{\dot{M}_B}{A} = -D_{BA}\frac{dC_B}{dx} \tag{1.21}$$

The condition for the process to take place without density gradient is that D_{AB} be equal to D_{BA}. Table 1.5 lists selected values of binary diffusion coefficients.

Table 1.5 Diffusion coefficients for selected binary mixtures at 1 atm and 298 K.

Species *A*	**Species *B***	**D_{BA}, m^2/s**
Carbon dioxide	Air	1.6×10^{-5}
Helium (*He*)	Air	7.2×10^{-5}
Oxygen (O_2)	Air	2.1×10^{-5}
Carbon dioxide	Water vapour	1.6×10^{-5}
Water vapour	Air	2.5×10^{-5}
Carbon dioxide	Water	2.0×10^{-9}
Carbon dioxide	Natural rubber	1.1×10^{-10}
Hydrogen	Iron	2.6×10^{-13}
Hydrogen	Nickel	1.2×10^{-12}
Carbon dioxide	Water	2.0×10^{-9}

The molar concentration C_A of species A is defined as the number of moles of A per unit volume of the mixture and that for species B as the number of moles of B per unit volume of the mixture. For an ideal gas

$$C_A = \frac{N_A}{V} = \frac{p_A}{\bar{R}T},\ C_B = \frac{N_B}{V} = \frac{p_B}{\bar{R}T}\ kmole/m^3$$

where p_A and p_B are the partial pressures of species A and B.

Fick's law can also be written as a mass flow rate $\dot{m}_A$ in terms of mass concentration

$$\dot{m}_A = -\rho D_{AB}\frac{d\rho_A/\rho}{dx} \quad kg/s \tag{1.22}$$

where ρ_A is the density of species A and ρ is the density of the mixture.

Referring to Figure 1.9, the molar diffusion rate ($kmol/s$) of species A through a nonreacting plane wall (gaseous medium $A+B$) can be found by integrating for a chamber width L

$$\dot{M}_A = A\frac{D_{AB}}{\bar{R}T}\frac{P_{A1}-P_{A2}}{x_2-x_1} \tag{1.23}$$

where P_{A1} and P_{A2} are partial pressures of species A at x_1 and x_2, $\bar{R}$ is the universal gas constant ($J/kmol.K$), and T is the temperature of the mixture (K). Distances are in metres (m).

The process depicted in Figure 1.9 is comparable to a gas, such as helium, diffusing through a plane plastic membrane or leaking from a spherical silica container. Leakage of a gas from a chamber into the atmosphere is approximated by a process of diffusion between two large chambers connected by a passage. The expressions for a fluid in contact with non-reacting cylindrical and spherical walls are

Cylindrical tube: length L, internal radius r_1, external radius r_2

$$\dot{M}_A = 2\pi L\frac{D_{AB}}{\bar{R}T}\frac{P_{A1}-P_{A2}}{ln(r_2/r_1)} \tag{1.24}$$

Spherical shell: internal radius r_1, external radius r_2

$$\dot{M}_A = 4\pi r_1 r_2\frac{D_{AB}}{\bar{R}T}\frac{P_{A1}-P_{A2}}{r_2-r_1} \tag{1.25}$$

Example 1.4 Hydrogen gas is stored in a spherical shell made of nickel and has an outer diameter of 4.0 m and thickness of 50 mm. The molar concentration of hydrogen at the inner surface is 0.087 $kmol/m^3$ and at the outer surface is negligible. Determine the mass flow rate of hydrogen by diffusion through the nickel container wall. Take the diffusion coefficient for hydrogen as $1.2\times10^{-12}\ m^2/s$.

Solution

Equation (1.21) can be rewritten in terms of concentrations as

$$\dot{M}_A = 4\pi r_1 r_2 D_{AB}\frac{C_{A1}-C_{A2}}{r_2-r_1}$$

$$\dot{M}_A = 4\pi \times 1.95 \times 2.0 \times 1.2 \times 10^{-12} \frac{0.078 - 0.0}{0.05} = 9.174 \times 10^{-11} \; kmol/s$$

The mass rate of hydrogen diffusion (molecular mass of hydrogen ($\mu_H = 2$)).

$$\dot{m}_A = \mu_H \dot{M}_A = 2 \times 9.174 \times 10^{-11} = 1.8348 \times 10^{-10} \; kg/s$$

Mass transfer can also occur in a fluid flow system in a manner similar to natural and forced convection heat transfer and the process is known as convective mass transfer. Often, both mass and convective heat transfer can occur simultaneously. A species concentration gradient can cause a rise in the temperature gradient and a consequent heat transfer in addition to mass transfer driven by the temperature gradient between the surface of the fluid and the atmosphere. These topics will be covered in more detail in Chapter 14.

Problems

1.1 The brick wall of a house having a width of 7 m and a height of 7 m is subjected to a temperature gradient, as shown in Figure P1.1. The left surface 1 is at 6°C and the right surface 2 is at 16°C. If the brick is 30 cm thick and its thermal conductivity is $0.6 W/m.K$, determine the direction and rate of the conduction heat transfer.

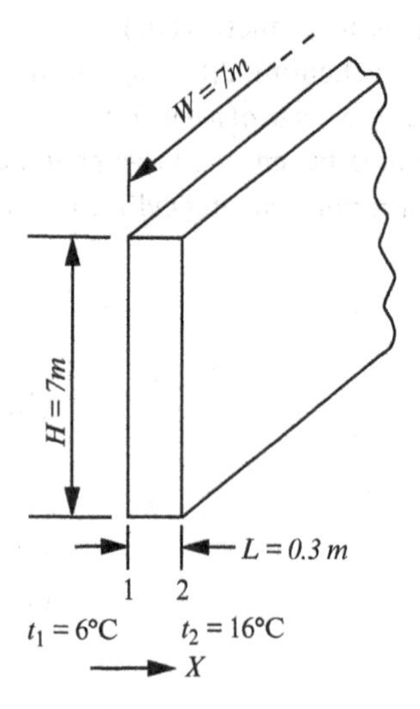

Figure P1.1

1.2 A temperature difference of 85°C is measured across an insulating material of thickness $L = 0.15m$ and surface area $A = 4 \; m^2$. The material is fibreglass having thermal conductivity of $k = 0.028 W/m.K$. Determine the heat lost in an hour.

1.3 A glass plate ($k_g = 1.019 W/m.K$) at outside surface temperature $T_1 = 40°C$ is layered with an insulation material ($k_l = 0.19 W/m.K$) at outside surface temperature $T_3 = 20°C$, as shown in Figure P1.3. The thickness of the glass plate is $\delta_g = 5mm$ and that for the insulation material is $\delta_l = 15mm$. The surface area of the glass is $0.05 \; m^2$. Determine the rate of heat transferred by conduction and the temperature of the interface T_2.

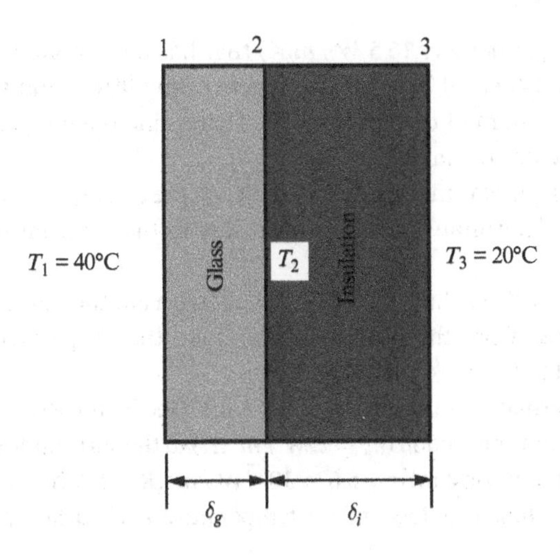

Figure P1.3

1.4 During a cold winter's night, a 1-cm ice sheet has formed on the outer surface of a cubical water tank of 3-cm wall thickness and measuring $0.6m \times 0.6m \times 0.6m$ on the outside. If the temperature of the water in the tank is increased to 30°C and maintained at that level, how long would it take for the ice to melt completely? Take the thermal conductivity of the tank material as 14.5 $W/m.K$, density of ice as $920 kg/m^3$, and the enthalpy of melting of ice (enthalpy of fusion) as $333.6 kJ/kg$

1.5 An electrical component 4 mm in diameter and 15 mm long dissipates 0.6 W of power as shown in Figure P1.5. Determine the heat energy dissipated into the surroundings in 24 hours and the heat flux from the surface of the component.

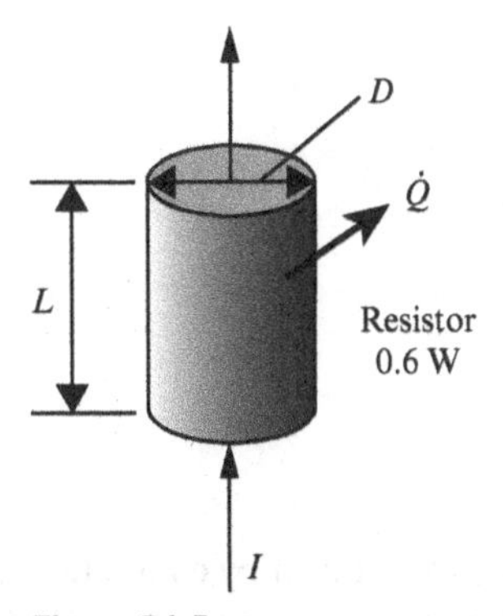

Figure P1.5

1.6 A resistance heater is used to raise the air temperature in a well-insulated room at atmospheric pressure ($4 \times 5 \times 6\ m^3$) from 7 to 25°C within 20 min. Calculate the required power generated by the heater.

1.7 1.2 kg of water is brought to the boil from an initial temperature of 15°C in a well-insulated electric kettle of 1.35 kW rating. The kettle is 0.5 kg and has an average specific heat of 0.75 $kJ / kg.K$. If the specific heat of water is 4.18 $kJ / kg.K$, determine the time required for the water to start boiling.

1.8 Water is boiled in a steel pot (thermal conductivity 15.5 $W/m.K$) that has a flat bottom of $5mm$ thick and $23cm$ diameter. The pot is heated by a gas flame generating $950W$ and the temperature of the inner surface of the bottom of the pot is 104°C. Determine the temperature of the surface of the pot in contact with the flame.

1.9 Air at 20°C flows over a hot plate of 1.5 m^2 that is maintained at 300°C. If the convective heat transfer coefficient is $h_c = 20W/m^2.K$, determine the rate of heat loss to the air from the surface of the plate.

1.10 A surface of 1.5 m^2 at a temperature of 300°C exchanges radiative heat with another surface at 40°C. Determine the rate of heat loss from the hotter surface. Take the shape factor $F_{12} = 0.5$. The two surfaces are assumed to be black bodies.

1.11 A steel plate $(0.6 \times 0.7 \times 0.05m)$ having thermal conductivity $k = 45W/m.K$ (see Figure P1.11) is losing heat from one surface by radiation and convection ($h_c = 22W/m^2.K$) to the surroundings at temperature $T_\infty = 288K$. Stefan–Boltzmann constant $\sigma = 5.67 \times 10^{-8} W/m^2.K^4$ and the emissivity of the steel plate is 0.3. Determine the total heat loss and the temperature T_2 of surface 2.

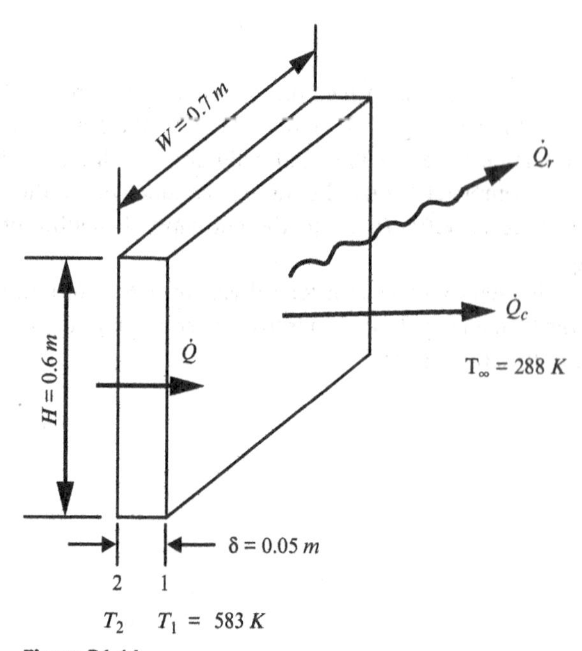

Figure P1.11

1.12 An electronic component of 0.001 m^2 surface area dissipates $0.4W$ of heat by convection and radiation to the surroundings at $293K$. If the convection heat transfer coefficient is $6W/m^2.K$, what is the temperature of the component surface?

1.13 A plastic membrane 0.3 mm thick at 25°C is exposed to a permeable gas with a pressure gradient between opposite sides of the membrane. The diffusion coefficient of the gas in the plastic is $D_{AB} = 8.7 \times 10^{-8}\ m^2/s$ and the molar concentrations of the gas on the two sides of the membrane are $0.0036 kmol/m^3$ and $0.0015 kmol/m^3$. Calculate the molar diffusion flux of the gas through the membrane.

1.14 A 2-m rubber tube is used to transport hydrogen gas at 2.0 bar and 34°C. The inner and outer diameters of the tube are $25mm$ and $40mm$, respectively. Diffusion coefficient of hydrogen in rubber is $D_{AB} = 2.2 \times 10^{-8}\ m^2/s$ and the molar concentration of hydrogen at the inner surface is 0.00625 $kmol/m^3$. Calculate the mass diffusion flux of hydrogen assuming concentration of hydrogen at the outer surface of the tube to be negligible.

1.15 A spherical shell of 3.5 m inner diameter and 3.58 m outer diameter is made from nickel and contains hydrogen at $350K$. The molar concentration of hydrogen at the inner surface is $0.085 kmol/m^3$ and zero at the outer surface. Calculate the mass diffusion rate of hydrogen through the spherical nickel wall. The molecular mass and the diffusion coefficient of hydrogen are respectively $D_{H_2-Ni} = 1.2 \times 10^{-12}\ m^2/s$ and $\mu_{H_2} = 2$.

2

One-Dimensional Steady-State Heat Conduction

Conduction heat transfer occurs mainly in solids, but it can also occur in liquids and gases to a lesser extent. In conduction, heat flows from molecule to molecule in the solid and the main prerequisite for the process to occur is the need for a continuous temperature gradient through the body. If the flow of heat is along parallel straight lines normal to the surface, the process is called steady-state conduction, which is the subject of this chapter. Transient or unsteady conduction will be considered in a later chapter.

2.1 General Heat Conduction Equation

An elementary volume dv of a heat conducting material is considered for analysis to determine the governing equations of the conduction process in three coordinate systems. It is assumed that:

1) The material is homogeneous and isotropic (a material is said to be isotropic if its properties are independent of the direction along which they are measured)
2) The deformation of the elementary volume with temperature is negligible
3) The internal heat source and storage (if applicable) are uniformly distributed within the volume dv
4) Temperature change is multi-dimensional.

2.1.1 Cartesian Coordinate System

The cubic element of volume $dv = dx\,dy\,dz$ in Cartesian coordinates is shown in Figure 2.1.

The rates of heat inflow through surfaces $dy\,dz, dx\,dz,$ and $dx\,dy$ are $\dot{Q}_x, d\dot{Q}_y$ and $d\dot{Q}_z$ respectively. The rate of heat outflow through the equal and opposite surfaces are $d\dot{Q}_{x+dx}, d\dot{Q}_{y+dy},$ and $d\dot{Q}_{z+dz}$ respectively. The heat rates are in J/s or W.

The functions $d\dot{Q}_{x+dx}, \dot{Q}_{y+dy}$ and $d\dot{Q}_{z+dz}$ are continuous in the incremental changes $dx, dy,$ and dz and can be expressed as Taylor's series

$$d\dot{Q}_{x+dx} = d\dot{Q}_x + \frac{\partial}{\partial x}\left(d\dot{Q}_x\right)dx + \frac{\partial^2}{\partial x^2}\left(d\dot{Q}_x\right)\frac{dx^2}{2!} + \ldots$$

$$d\dot{Q}_{y+dy} = d\dot{Q}_y + \frac{\partial}{\partial y}\left(d\dot{Q}_y\right)dy + \frac{\partial^2}{\partial y^2}\left(d\dot{Q}_y\right)\frac{dy^2}{2!} + \ldots$$

$$d\dot{Q}_{z+dz} = d\dot{Q}_z + \frac{\partial}{\partial z}\left(d\dot{Q}_z\right)dz + \frac{\partial^2}{\partial z^2}\left(d\dot{Q}_z\right)\frac{dz^2}{2!} + \ldots$$

Heat Transfer Basics: A Concise Approach to Problem Solving, First Edition. Jamil Ghojel.

Companion website: www.wiley.com/go/ghojel/heat_transfer

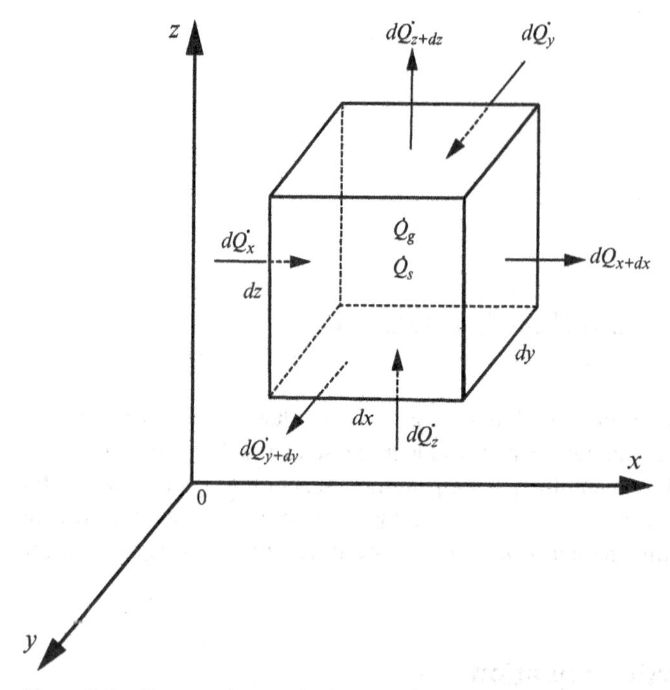

Figure 2.1 Elemental control volume in Cartesian coordinates.

The terms higher than the first two in the Taylor's series are usually ignored, leading to the final form of the heat rates leaving the control volume

$$d\dot{Q}_{x+dx} = d\dot{Q}_x + \frac{\partial}{\partial x}\left(d\dot{Q}_x\right)dx$$

$$d\dot{Q}_{y+dy} = d\dot{Q}_y + \frac{\partial}{\partial y}\left(d\dot{Q}_y\right)dy$$

$$d\dot{Q}_{z+dz} = d\dot{Q}_z + \frac{\partial}{\partial z}\left(d\dot{Q}_z\right)dz$$

Applying the law of conservation of energy (first law of thermodynamics) to the control volume dv, we can write

$$d\dot{Q}_{in} + \dot{Q}_g - d\dot{Q}_{out} = \dot{Q}_s \tag{2.1}$$

where

$$d\dot{Q}_{in} = d\dot{Q}_x + d\dot{Q}_y + d\dot{Q}_z$$

$$d\dot{Q}_{out} = d\dot{Q}_{x+dx} + d\dot{Q}_{y+dy} + d\dot{Q}_{z+dz}$$

If the heat generated per unit volume is denoted as $\dot{q}$, the total internally generated heat $\dot{Q}$ is

$$\dot{Q}_g = \dot{q}dx\,dy\,dz \qquad (\dot{q} \text{ is in } W/m^3)$$

The rate of internally stored energy $\dot{Q}_s$ is

$$\dot{Q}_s = \rho c_p \frac{\partial T}{\partial t} dx\, dy\, dz$$

where ρ is the density, c_p is the specific heat of the material, and T is the instantaneous temperature of the material during the conduction process.

Equation (2.1) can now be written as

$$d\dot{Q}_x + d\dot{Q}_y + d\dot{Q}_z - \left[d\dot{Q}_x + \frac{\partial}{\partial x}\left(d\dot{Q}_x\right)dx + d\dot{Q}_y + \frac{\partial}{\partial y}\left(d\dot{Q}_y\right)dy + d\dot{Q}_z + \frac{\partial}{\partial z}\left(d\dot{Q}_z\right)dz \right]$$
$$+\dot{q}dx\, dy\, dz = \rho c_p \frac{\partial T}{\partial t} dx\, dy\, dz$$

or

$$-\frac{\partial}{\partial x}\left(d\dot{Q}_x\right)dx - \frac{\partial}{\partial y}\left(d\dot{Q}_y\right)dy - \frac{\partial}{\partial z}\left(d\dot{Q}_z\right)dz + \dot{q}dx\, dy\, dz = \rho c_p \frac{\partial T}{\partial t} dx\, dy\, dz \quad (2.2)$$

If q_x, q_y, and q_z are the heat fluxes in the x, y- and z-directions, then

$$d\dot{Q}_x = q_x dy\, dz$$

$$d\dot{Q}_y = q_y dx\ dz$$

$$d\dot{Q}_z = q_z dx\ dy$$

Equation (2.2) becomes

$$-\frac{\partial}{\partial x}(q_x dy\, dz)dx - \frac{\partial}{\partial y}(q_y dx\ dz)dy - \frac{\partial}{\partial z}(q_z dx\ dy)dz + \dot{q}dx\, dy\, dz = \rho c_p \frac{\partial T}{\partial t} dx\, dy\, dz \quad (2.3)$$

Dividing both sides of Eq. (2.3) by $dv(= dx\, dy\, dz)$, we finally obtain

$$-\frac{\partial q_x}{\partial x} - \frac{\partial q_y}{\partial y} - \frac{\partial q_z}{\partial z} + \dot{q} = \rho c_p \frac{\partial T}{\partial t} \quad (2.4)$$

Fourier's law of heat conduction rate, written in terms of the heat fluxes, along the x, y, and z-directions are

$$q_x = -k\frac{\partial T}{\partial x}$$

$$q_y = -k\frac{\partial T}{\partial y}$$

$$q_z = -k\frac{\partial T}{\partial z}$$

Equation (2.4) can now be rewritten as

$$\frac{\partial}{\partial x}\left(k\frac{\partial T}{\partial x}\right) + \frac{\partial}{\partial y}\left(k\frac{\partial T}{\partial y}\right) + \frac{\partial}{\partial z}\left(k\frac{\partial T}{\partial z}\right) + \dot{q} = \rho c_p \frac{\partial T}{\partial t} \quad (2.5)$$

Equation (2.5) is the general differential conduction equation in Cartesian coordinates.

2.1.2 Cylindrical Coordinate System

The conduction equation in terms of cylindrical coordinates is better suited to solving conduction in bodies of cylindrical shape. An analysis similar to the previous case can be conducted by applying the energy equation to the cylindrical control volume $dv = r d\varphi\, dr\, dz$, as shown in Figure 2.2.

$$d\dot{Q}_{r+dr} = d\dot{Q}_r + \frac{\partial}{\partial r}\left(d\dot{Q}_r\right)dr$$

$$d\dot{Q}_{\varphi+d\varphi} = d\dot{Q}_\varphi + \frac{\partial}{\partial \varphi}\left(d\dot{Q}_\varphi\right)d\varphi$$

$$d\dot{Q}_{z+dz} = d\dot{Q}_z + \frac{\partial}{\partial z}\left(d\dot{Q}_z\right)dz$$

The relationships between the inflow and outflow heat rates are, respectively

$$d\dot{Q}_{r+dr} = d\dot{Q}_r + \frac{\partial}{\partial r}\left(d\dot{Q}_r\right)dr$$

$$d\dot{Q}_{\varphi+d\varphi} = d\dot{Q}_\varphi + \frac{\partial}{\partial \varphi}\left(d\dot{Q}_\varphi\right)d\varphi$$

$$d\dot{Q}_{z+dz} = d\dot{Q}_z + \frac{\partial}{\partial z}\left(d\dot{Q}_z\right)dz$$

Application of the energy equation to the control volume yields

$$d\dot{Q}_r + d\dot{Q}_\varphi + d\dot{Q}_z - \left[d\dot{Q}_r + \frac{\partial}{\partial r}\left(d\dot{Q}_r\right)dr + d\dot{Q}_\varphi + \frac{\partial}{\partial \varphi}\left(d\dot{Q}_\varphi\right)d\varphi + d\dot{Q}_z + \frac{\partial}{\partial z}\left(d\dot{Q}_z\right)dz\right]$$
$$+\dot{q}\, r d\varphi\, dr\, dz = \rho c_p \frac{\partial T}{\partial t} r\, d\varphi\, dr\, dz$$

or

$$-\frac{\partial}{\partial r}\left(d\dot{Q}_r\right)dr - \frac{\partial}{\partial \varphi}\left(d\dot{Q}_\varphi\right)d\varphi - \frac{\partial}{\partial z}\left(d\dot{Q}_z\right)dz + \dot{q}\, r\, d\varphi\, dr\, dz = \rho c_p \frac{\partial T}{\partial t} r\, d\varphi\, dr\, dz \tag{2.6}$$

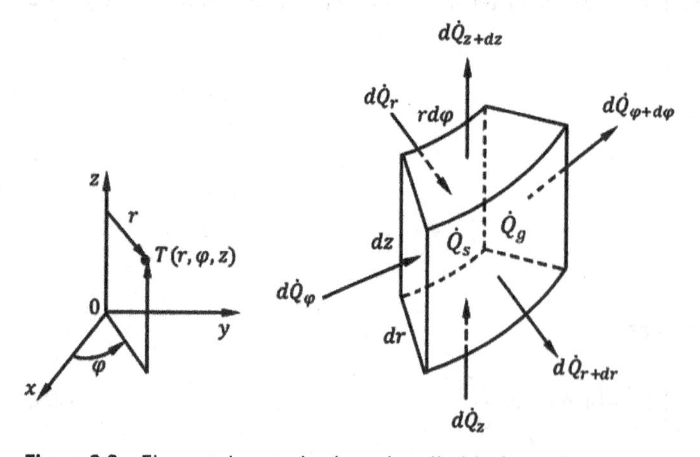

Figure 2.2 Elemental control volume in cylindrical coordinates.

If q_r, q_φ, and q_z are the heat fluxes in the r,φ and z-directions, then

$$d\dot{Q}_r = q_r r d\varphi\, dz$$

$$d\dot{Q}_\varphi = q_\varphi dr\, dz$$

$$d\dot{Q}_z = q_z r d\varphi\, dr$$

Equation (2.6) can be written as

$$-\frac{\partial}{\partial r}(q_r r\, dr\, d\varphi\, dz)-\frac{\partial}{\partial \varphi}(q_\varphi d\varphi\, dr\, dz)-\frac{\partial}{\partial z}(q_z r\, d\varphi\, dr\, dz)+\dot{q} r d\varphi\, dr\, dz = \rho c_p \frac{\partial T}{\partial t} r\, d\varphi\, dr\, dz \tag{2.7}$$

Dividing both sides of Eq. (2.7) by $dv(=r\,d\varphi\,dr\,dz)$ yields

$$-\frac{1}{r}\frac{\partial}{\partial r}(rq_r)-\frac{1}{r}\frac{\partial q_\varphi}{\partial \varphi}-\frac{\partial q_z}{\partial z}+\dot{q}=\rho c_p\frac{\partial T}{\partial t} \tag{2.8}$$

Since the area of the element $r\,d\varphi\,dz$ is a function of variable r, the latter is lumped together with q_r in the first term, as shown in Eq. (2.8). Fourier's law of heat conduction rate, written in terms of heat flux, along r,φ, and z-directions are

$$q_r = -k\frac{\partial T}{\partial r}$$

$$q_\varphi = -\frac{k}{r}\frac{\partial T}{\partial \varphi}$$

$$q_z = -k\frac{\partial T}{\partial z}$$

Substitution of the heat fluxes in Eq. (2.8) yields

$$\frac{1}{r}\frac{\partial}{\partial r}\left(kr\frac{\partial T}{\partial r}\right)+\frac{1}{r^2}\frac{\partial}{\partial \varphi}\left(k\frac{\partial T}{\partial \varphi}\right)+\frac{\partial}{\partial z}\left(k\frac{\partial T}{\partial z}\right)+\dot{q}=\rho c_p\frac{\partial T}{\partial t} \tag{2.9}$$

Equation (2.9) is the general conduction equation in cylindrical coordinates.

2.1.3 Spherical Coordinate System

The conduction equation in terms of spherical coordinates is used to solve conduction problems in bodies of spherical shape. An analysis similar to the previous case can be conducted by applying the energy equation to the spherical control volume $dv = r^2 \sin\theta d\theta\, d\varphi\, dr$, as shown in Figure 2.3.

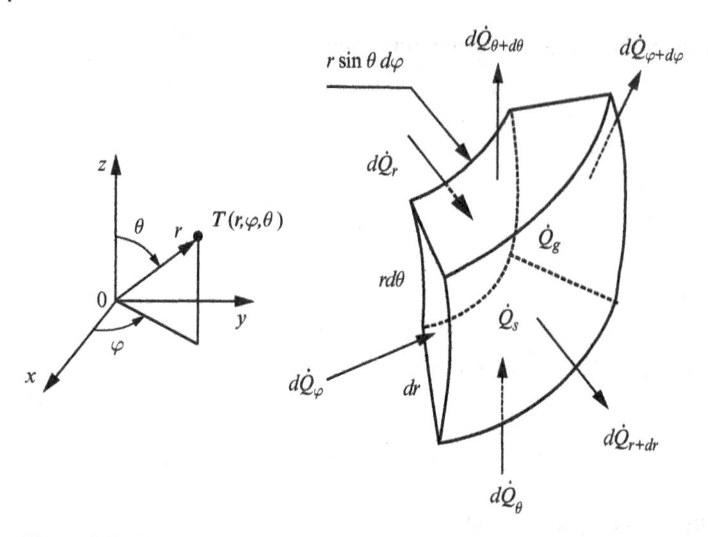

Figure 2.3 Elemental control volume in spherical coordinates.

The relationships between the inflow and outflow heat rates are

$$d\dot{Q}_{r+dr} = d\dot{Q}_r + \frac{\partial}{\partial r}\left(d\dot{Q}_r\right)dr$$

$$d\dot{Q}_{\varphi+d\varphi} = d\dot{Q}_\varphi + \frac{\partial}{\partial \varphi}\left(d\dot{Q}_\varphi\right)d\varphi$$

$$d\dot{Q}_{\theta+d\theta} = d\dot{Q}_\theta + \frac{\partial}{\partial \theta}\left(d\dot{Q}_\varphi\right)d\theta$$

Application of the energy equation to the control volume yields

$$\left(d\dot{Q}_r + d\dot{Q}_\varphi + d\dot{Q}_\theta\right)$$
$$-\left[d\dot{Q}_r + \frac{\partial}{\partial r}\left(d\dot{Q}_r\right)dr + d\dot{Q}_\varphi + \frac{\partial}{\partial \varphi}\left(d\dot{Q}_\varphi\right)d\varphi + d\dot{Q}_\theta + \frac{\partial}{\partial \theta}\left(d\dot{Q}_\theta\right)d\theta\right]$$
$$+\dot{q}r^2 \sin\theta d\theta\, d\varphi\, dr = \rho c_p \frac{\partial T}{\partial t}\, r^2 \sin\theta d\theta\, d\varphi\, dr$$

or

$$-\frac{\partial}{\partial r}\left(d\dot{Q}_r\right)dr - \frac{\partial}{\partial \varphi}\left(d\dot{Q}_\varphi\right)d\varphi - \frac{\partial}{\partial \theta}\left(d\dot{Q}_\theta\right)d\theta + \dot{q}r^2 \sin\theta\, d\theta\, d\varphi\, dr\, dz = \rho c_p \frac{\partial T}{\partial t}\, r^2 \sin\theta\, d\theta\, d\varphi\, dr \quad (2.10)$$

If q_r, q_φ, and q_θ are the heat fluxes in the r,φ and θ-directions, then

$$d\dot{Q}_r = q_r r^2 \sin\theta\, d\theta\, d\varphi$$

$$d\dot{Q}_\varphi = q_\varphi r d\theta dr$$

$$d\dot{Q}_\theta = q_\theta r \sin\theta\; d\varphi\, dr$$

Equation (2.10) can now be written as

$$\frac{\partial}{\partial r}\left(r^2 q_r\right)\sin\theta d\theta\, d\varphi\, dr + \frac{\partial}{\partial \varphi}(q_\varphi)r\, d\theta\, d\varphi\, dr + \frac{\partial}{\partial \theta}(q_\theta \sin\theta) r\, d\varphi\, d\theta\, dr + \dot{q} r^2 \sin\theta d\theta\, d\varphi\, dr = \rho c_p \frac{\partial T}{\partial t}\, r^2 \sin\theta d\theta\, d\varphi\, dr \tag{2.11}$$

Dividing both sides of Eq. (2.11) by $dv(= r^2 \sin\theta d\theta\, d\varphi\, dr)$, we obtain

$$-\frac{1}{r^2}\frac{\partial}{\partial r}\left(r^2 q_r\right) - \frac{1}{r\sin\theta}\frac{\partial q_\varphi}{\partial \varphi} - \frac{1}{r\sin\theta}\frac{\partial}{\partial\theta}(q_\theta \sin\theta) + \dot{q} = \rho c_p \frac{\partial T}{\partial t} \tag{2.12}$$

Area ($r^2 \sin\theta\ d\theta\ d\varphi$) is a function of r and area ($r\sin\theta\ d\varphi dr$) is a function of θ (or $\sin\theta$); hence, the variables r^2 and $\sin\theta$ in brackets in the first and third terms are not cancelled out and the two terms are written as shown in Eq. (2.12). Fourier's law of heat conduction rate, written in terms of heat flux, along r, φ, and θ-directions are

$$q_r = -k\frac{\partial T}{\partial r}$$

$$q_\varphi = -\frac{k}{r\sin\theta}\frac{\partial T}{\partial \varphi}$$

$$q_\theta = -\frac{k}{r}\frac{\partial T}{\partial \theta}$$

hence, Eq. (2.21) can be rewritten as

$$\frac{1}{r^2}\frac{\partial}{\partial r}\left(kr^2\frac{\partial T}{\partial r}\right) + \frac{1}{r^2\sin^2\theta}\frac{\partial}{\partial\varphi}\left(k\frac{\partial T}{\partial\varphi}\right) + \frac{1}{r^2\sin\theta}\frac{\partial}{\partial\theta}\left(k\sin\theta\frac{\partial T}{\partial\theta}\right) + \dot{q} = \rho c_p\frac{\partial T}{\partial t} \tag{2.13}$$

Equation (2.13) is the general conduction equation in spherical coordinates.

2.2 Special Conditions of the General Conduction Equation

Equations can be derived from the three variants of the general conduction equations for special cases by specifying different thermal conductivities and modes of energy generation and storage.

2.2.1 Constant Thermal Conductivity *k* With Energy Storage and Generation

A property known as thermal diffusivity α is introduced, defined as $\alpha = k / \rho c_p$ in units of m^2 / s.

Equations (2.5), (2.9), and (2.13) are reduced, respectively, to

$$\frac{\partial^2 T}{\partial x^2} + \frac{\partial^2 T}{\partial y^2} + \frac{\partial^2 T}{\partial z^2} + \frac{\dot{q}}{k} = \frac{1}{\alpha}\frac{\partial T}{\partial t} \tag{2.14}$$

$$\frac{1}{r}\frac{\partial}{\partial r}\left(r\frac{\partial T}{\partial r}\right) + \frac{1}{r^2}\frac{\partial^2 T}{\partial \varphi^2} + \frac{\partial^2 T}{\partial z^2} + \frac{\dot{q}}{k} = \frac{1}{\alpha}\frac{\partial T}{\partial t} \tag{2.15}$$

$$\frac{1}{r^2}\frac{\partial}{\partial r}\left(r^2\frac{\partial T}{\partial r}\right) + \frac{1}{r^2\sin^2\theta}\frac{\partial^2 T}{\partial\varphi^2} + \frac{1}{r^2\sin\theta}\frac{\partial}{\partial\theta}\left(\sin\theta\frac{\partial T}{\partial\theta}\right) + \frac{\dot{q}}{k} = \frac{1}{\alpha}\frac{\partial T}{\partial t} \tag{2.16}$$

2.2.2 Variable Thermal Conductivity and No Internal Energy Storage and Generation

Equations (2.5), (2.9), and (2.13) are reduced, respectively, to

$$\frac{\partial}{\partial x}\left(k\frac{\partial T}{\partial x}\right)+\frac{\partial}{\partial y}\left(k\frac{\partial T}{\partial y}\right)+\frac{\partial}{\partial z}\left(k\frac{\partial T}{\partial z}\right)=0 \tag{2.17}$$

$$\frac{1}{r}\frac{\partial}{\partial r}\left(kr\frac{\partial T}{\partial r}\right)+\frac{1}{r^2}\frac{\partial}{\partial \varphi}\left(k\frac{\partial T}{\partial \varphi}\right)+\frac{\partial}{\partial z}\left(k\frac{\partial T}{\partial z}\right)=0 \tag{2.18}$$

$$\frac{1}{r^2}\frac{\partial}{\partial r}\left(kr^2\frac{\partial T}{\partial r}\right)+\frac{1}{r^2\sin^2\theta}\frac{\partial}{\partial \varphi}\left(k\frac{\partial T}{\partial \varphi}\right)+\frac{1}{r^2\sin\theta}\frac{\partial}{\partial \theta}\left(k\sin\theta\frac{\partial T}{\partial \theta}\right)=0 \tag{2.19}$$

2.2.3 Variable Thermal Conductivity With Internal Energy Generation and No Energy Storage

Equations (2.5), (2.9), and (2.13) are reduced, respectively, to

$$\frac{\partial}{\partial x}\left(k\frac{\partial T}{\partial x}\right)+\frac{\partial}{\partial y}\left(k\frac{\partial T}{\partial y}\right)+\frac{\partial}{\partial z}\left(k\frac{\partial T}{\partial z}\right)+\dot{q}=0 \tag{2.20}$$

$$\frac{1}{r}\frac{\partial}{\partial r}\left(kr\frac{\partial T}{\partial r}\right)+\frac{1}{r^2}\frac{\partial}{\partial \varphi}\left(k\frac{\partial T}{\partial \varphi}\right)+\frac{\partial}{\partial z}\left(k\frac{\partial T}{\partial z}\right)+\dot{q}=0 \tag{2.21}$$

$$\frac{1}{r^2}\frac{\partial}{\partial r}\left(kr^2\frac{\partial T}{\partial r}\right)+\frac{1}{r^2\sin^2\theta}\frac{\partial}{\partial \varphi}\left(k\frac{\partial T}{\partial \varphi}\right)+\frac{1}{r^2\sin\theta}\frac{\partial}{\partial \theta}\left(k\sin\theta\frac{\partial T}{\partial \theta}\right)+\dot{q}=0 \tag{2.22}$$

2.3 One-Dimensional Steady-State Conduction

Heat conduction in simple geometries such as plane walls (plates), hollow cylinders, hollow spheres, and their composite variants will be considered. Initially, internal heat generation and storage will be ignored in these cases together with heat dissipation to the surroundings by other modes of heat transfer. In later sections, internal heat generation and heat dissipation from outer surfaces will be analysed. The effect of both constant and temperature-dependent thermal conductivity on the temperature profiles during the conduction process will also be analysed.

2.3.1 Plane Wall (or Plate) Without Heat Generation and Storage

Three cases will be considered in this section: conduction with constant thermal conductivity, conduction with temperature-dependent thermal conductivity (both in single layer plate), and conduction in a multi-layer (composite) plate with conductivities that are not temperature-dependent.

2.3.1.1 Constant Thermal Conductivity

Consider a homogeneous wall of thickness δ with heat flowing in the x-direction and the temperature changing from T_1 to T_2 in that direction only (Figure 2.4). If the wall is very large in the y- and z-directions, then the temperatures in these directions can be assumed constant ($\partial T/\partial y=\partial T/\partial z=0$) and Eq. (2.17) will be reduced to a one-dimensional conduction equation

$$\frac{\partial}{\partial x}\left(k\frac{\partial T}{\partial x}\right)=0 \tag{2.23}$$

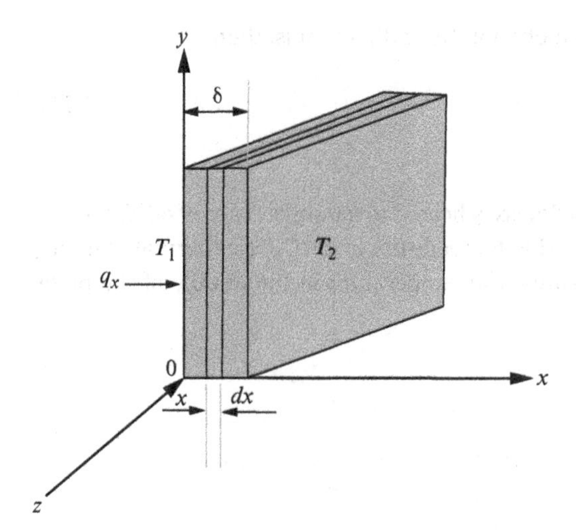

Figure 2.4 Heat conduction through plane wall.

If we further assume constant thermal conductivity ($k = const$), the conduction equation is transformed to

$$\frac{\partial^2 T}{\partial x^2} = 0 \tag{2.24}$$

Integrating Eq. (2.24) twice yields

$$\frac{dT}{dx} = C_1$$

$$T = C_1 x + C_2 \tag{2.25}$$

where C_1, C_2 are constants that can be determined from the boundary conditions:

At $x = 0$, $T = T_1$, and $C_2 = T_1$
At $x = \delta$, $T = T_2$, and $C_1 = -(T_1 - T_2)/\delta$

Substituting for C_1 and C_2 in Eq. (2.25) yields the equation for temperature distribution profile in the plane wall

$$T = T_1 - \frac{(T_1 - T_2)}{\delta} x \tag{2.26}$$

The heat flux in the x-direction using Fourier's law is

$$q_x = -k\frac{\partial T}{dx} = -kC_1$$

Substituting for C_1, we obtain

$$q_x = \frac{k}{\delta}(T_1 - T_2)$$

If the area of the wall is A, the rate of heat conduction in the x-direction is, then

$$\dot{Q}_x = \frac{kA}{\delta}(T_1 - T_2)W \tag{2.27}$$

Example 2.1 A semi-infinite plate of 12 mm thickness is heated to a temperature of 40°C on one side and exposed on the other side to the surrounding temperature of 25°C. The thermal conductivity of the plate's material is 3.5 $W/m.K$. Determine the temperature in the middle of the plate and the heat flux.

Solution

From Eq. (2.26)

$$T = T_1 - \frac{(T_1 - T_2)}{\delta}x$$

At $x = 0.006$

$$T = 40 - \frac{(40\quad 20)}{0.01}0.006 = 30°\text{C}$$

From Eq. (2.27)

$$q = \frac{\dot{Q}_x}{A} = \frac{3.5 \times 20}{0.012} = 5833 \quad W/m^2$$

2.3.1.2 Temperature-Dependent Thermal Conductivity

If the temperature gradient in the wall is large, it may be necessary to take into account the variability of thermal conductivity with temperature. Thermal conductivity varies with temperature for almost all metals and metal alloys, as can be seen from the data in Appendix A, accessible at "www.wiley.com/go/Ghojel/heat transfer." To simplify the analysis of the problem with variable k, the temperature dependence can be approximated by a linear relationship of the following form

$$k = k_0(1 + bT) \tag{2.28}$$

where k_0 is a constant of proportionality linked to the conductivity at 0°C, and b is a material-specific constant that can be positive or negative, depending on whether k is increasing or decreasing with temperature.

From Eq. (2.23), the one-dimensional conduction equation with variable thermal conductivity and with no heat generation and heat storage is transformed to

$$\frac{d}{dx}\left[k_0(1 + bT)\frac{dT}{dx}\right] = 0$$

Integrating yields

$$k_0(1 + bT)\frac{dT}{dx} = C_1$$

Separating the variables and again integrating yields

$$T+\frac{b}{2}T^2=\frac{C_1}{k_0}x+C_2$$

$$\frac{b}{2}T^2+T-\left(\frac{C_1}{k_0}x+C_2\right)=0 \tag{2.29}$$

The boundary conditions are

At $x=0, T=T_1$, at $x=\delta, T=T_2$; hence,

$$C_2=T_1+\frac{b}{2}T_1^2$$

and

$$C_1=\frac{k_0}{\delta}\left[\left(T_2+\frac{b}{2}T_2^2\right)-\left(T_1+\frac{b}{2}T_1^2\right)\right]$$

The solution to the quadratic Eq. (2.29) is then

$$T=-\frac{1}{b}\pm\sqrt{\frac{1}{b^2}+\frac{2}{b}\left(\frac{C_1}{k_0}x+C_2\right)} \tag{2.30}$$

The conduction heat rate is

$$\dot{Q}=-Ak_0(1+bT)\frac{dT}{dx}$$

and

$$\frac{dT}{dx}=\frac{C_1}{k_0(1+bT)}$$

hence, the heat flux

$$q_x=\frac{\dot{Q}}{A}=C_1$$

$$q_x=\frac{k_0}{\delta}\left[1+\frac{b}{2}(T_1+T_2)\right](T_1-T_2)=k_m\frac{(T_1-T_2)}{\delta} \tag{2.31}$$

In this equation

$$k_m=k_0\left[1+\frac{b}{2}(T_1+T_2)\right]=k_0[1+bT_m]$$

where the mean temperature is defined as $T_m=(T_1+T_2)/2$.

The temperature profile given by Eq. (2.30) can be rewritten to include the heat flux q_x

$$T=-\frac{1}{b}\pm\sqrt{\left(\frac{1}{b}+T_1\right)^2-\frac{2q_x}{bk_0}x} \tag{2.32}$$

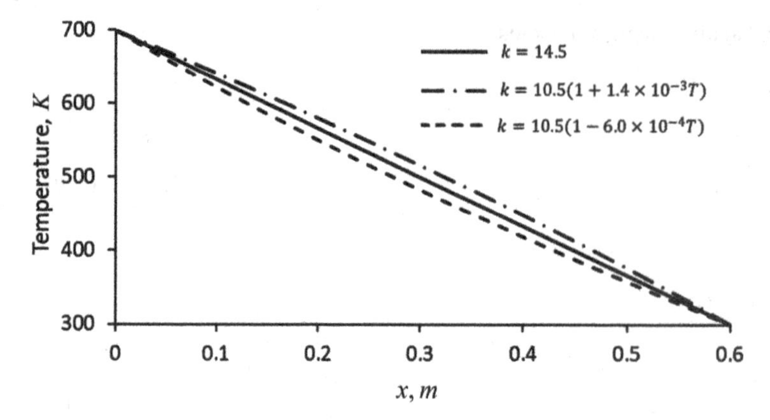

Figure 2.5 Effect of variability of the thermal conductivity on the temperature profiles in a plane wall.

The temperature profiles for constant and variable thermal conductivities in a plane wall for $\delta = 0.4m$, $T_1 = 700K$, and $T_2 = 300K$ are shown in Figure 2.5. For $k = const$, the profile is linear, and for $k = f(T)$, the profile is non-linear and the shape and magnitude depend on whether b is positive or negative. When b is negative, the $\pm$ sign in Eq. (2.32) should be replaced with the negative sign to get the correct answer.

Example 2.2 A plate 2.5 cm thick has one side maintained at 95°C and the other at 35°C. The thermal conductivity is temperature-dependent given by $k = 16.58(1 - 0.00234T)$ where T is in degrees K. Determine the temperature at the centre plane of the plate and the heat flux through the plate.

Solution

$$T = -\frac{1}{b} \pm \sqrt{\frac{1}{b^2} + \frac{2}{b}\left(\frac{C_1}{k_0}x + C_2\right)}$$

$$C_1 = \frac{k_0}{\delta}\left[\left(T_2 + \frac{b}{2}T_2^2\right) - \left(T_1 + \frac{b}{2}T_1^2\right)\right]$$

$$C_1 = \frac{16.58}{0.025}\left[\left(368 - \frac{0.00234}{2}368^2\right) - \left(308 + \frac{0.00234}{2}308^2\right)\right] = -8319.7$$

$$C_2 = T_1 + \frac{b}{2}T_1^2 = 368 - \frac{0.00234}{2}368^2 = 209.5$$

$$T = -\frac{1}{b} - \sqrt{\frac{1}{b^2} + \frac{2}{b}\left(\frac{C_1}{k_0}x + C_2\right)}$$

The plus sign before the square root symbol in Equation (2.30) results in an unrealistic temperature; hence, the selection of the negative sign.

$$T_{x=0.0125} = \frac{1}{0.00234} - \sqrt{\frac{1}{0.00234^2} + \frac{2}{0.00234}\left(\frac{-8319.7}{16.58}0.0125 + 209.5\right)} = 333\ K\,(60°\text{C})$$

$$q_x = k_m\frac{(T_1 - T_2)}{\delta}$$

$$k_m = k_0\left[1+\frac{b}{2}(T_1+T_2)\right] = 16.58\left[1-\frac{0.00234}{2}(368+308)\right] = 3.47 \quad W/m.K$$

$$q_x = 3.47\frac{(368-308)}{0.025} = 8328\,W$$

2.3.1.3 Composite Plane Wall

Consider the composite plane wall (or plate) shown in Figure 2.6 but with n layers of different materials of the same height and width having thicknesses δ_1, δ_2, ... δ_n and surface temperatures of the layers T_1, T_2, ... T_n, respectively. The heat fluxes through the layers are then

$$q_x = \frac{k_1}{\delta_1}(T_1 - T_2)$$

$$q_x = \frac{k_2}{\delta_2}(T_2 - T_3)$$

$$\vdots$$

$$q_x = \frac{k_n}{\delta_n}(T_n - T_{n+1})$$

The temperature gradients between each pair will then be

$$T_1 - T_2 = q_x\frac{\delta_1}{k_1}$$

$$T_2 - T_3 = q_x\frac{\delta_2}{k_2}$$

$$\vdots$$

$$T_n - T_{n+1} = q_x\frac{\delta_n}{k_n}$$

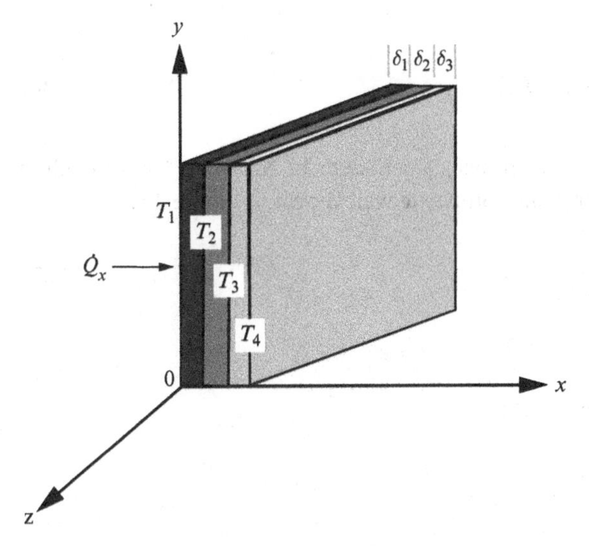

Figure 2.6 Three-layer composite plane wall.

Summation of the left and right sides of temperature gradient equations yields

$$T_1 - T_{n+1} = q_x\left(\frac{\delta_1}{k_1} + \frac{\delta_2}{k_2} + \ldots + \frac{\delta_n}{k_n}\right)$$

from which the heat flux is

$$q_x = \frac{T_1 - T_{n+1}}{\frac{\delta_1}{k_1} + \frac{\delta_2}{k_2} + \ldots + \frac{\delta_n}{k_n}}$$

The rate of heat flow across surface area A

$$\dot{Q}_x = \frac{T_1 - T_{n+1}}{\frac{\delta_1}{Ak_1} + \frac{\delta_2}{Ak_2} + \ldots + \frac{\delta_n}{Ak_n}} = \frac{T_1 - T_{n+1}}{R_1 + R_2 + \ldots + R_n} \tag{2.33}$$

From electrical analogy $\dot{Q}_x$ could be thought of as the current, $(T_1 - T_{n+1})$ as the potential difference, and the denominator on the right-hand side of Eq. (2.33) as the overall resistance

$$R = R_1 + R_2 + \ldots + R_n = \frac{1}{A}\sum_1^n \frac{\delta_i}{k_i} \quad K/W \tag{2.34}$$

The rate of heat flow can be expressed in terms of an overall heat transfer coefficient U as follows

$$\dot{Q}_x = UA\Delta T = \frac{\Delta T}{\frac{1}{UA}} \tag{2.35}$$

Equating Eqs (2.33) and (2.35) results in

$$\frac{1}{UA} = \frac{\delta_1}{Ak_1} + \frac{\delta_2}{Ak_2} + \ldots + \frac{\delta_n}{Ak_n} = R_1 + R_2 + \ldots + R_n$$

For a constant heat conduction area

$$\frac{1}{U} = \frac{\delta_1}{k_1} + \frac{\delta_2}{k_2} + \ldots + \frac{\delta_n}{k_n} = A(R_1 + R_2 + \ldots + R_n) \tag{2.36}$$

The units for U are the same as for the heat transfer coefficient in Newton's law of cooling $(W/m^2.K)$. The heat rate for the three-layer plane composite wall shown in Figure 2.6, is

$$\dot{Q}_x = \frac{T_1 - T_4}{\frac{\delta_1}{Ak_1} + \frac{\delta_2}{Ak_2} + \frac{\delta_3}{Ak_3}} = \frac{\Delta T}{R_1 + R_2 + R_3} \tag{2.37}$$

or

$$\dot{Q}_x = UA\Delta T$$

where

$$\frac{1}{U} = \frac{\delta_1}{k_1} + \frac{\delta_2}{k_2} + \frac{\delta_3}{k_3} = A(R_1 + R_2 + R_3)$$

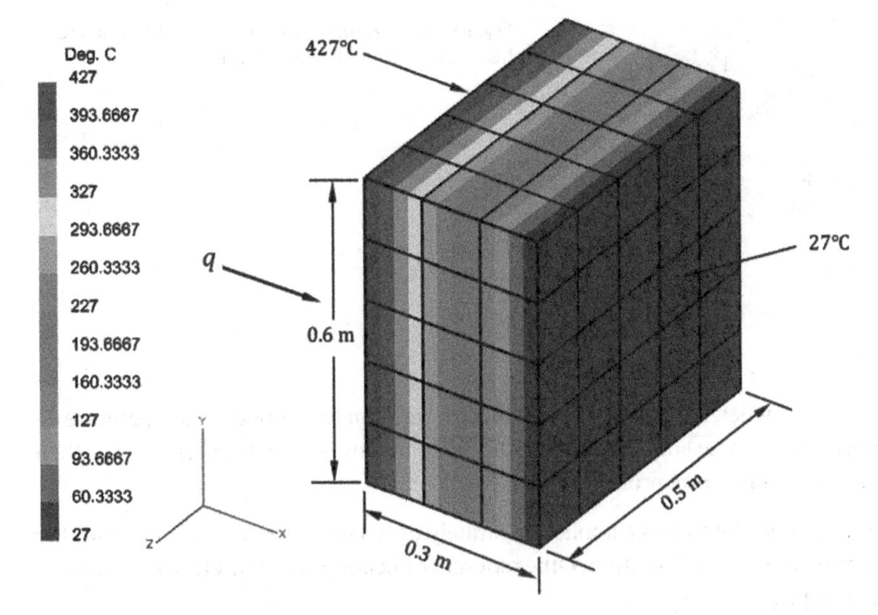

Figure 2.7 Computer model of three-layer composite plane wall of same material.

The temperature gradient across the plate can be written in terms of the thermal resistances as

$$\Delta T = T_1 - T_4 = \frac{\dot{Q}_x}{\dfrac{\delta_1}{Ak_1} + \dfrac{\delta_2}{Ak_2} + \dfrac{\delta_3}{Ak_3}} = \frac{\dot{Q}_x}{R_1 + R_2 + R_3}$$

Figure 2.7 shows the results of computer modelling of heat conduction in three-layer composite wall of the same material ($k = const$).

2.3.2 Boundary Conditions

To solve the conduction problem in the plate in Figure 2.4, Eq. (2.27) was derived to calculate the heat rate using the temperatures of the parallel surfaces of the plate surfaces T_1 and T_2 as the boundary conditions without reference to the heating sources of these temperatures. There are other boundary conditions that are encountered in practice and shown in Figure 2.8 for heat conduction direction from left to right. It should be noted that for surface temperatures, symbols T_1 and T_2 or T_{w1} and T_{w2} will be used interchangeably throughout the text.

- Heat transfer by convection (free or forced) from fluid 1 at temperature $T_{\infty 1}$ to the left surface of the solid plate at temperature T_{w1} in accordance with Newton's law of cooling $\dot{Q}_{c1} = h_{c1}(T_{\infty 1} - T_{w1})$ and from the right surface at temperature T_{w2} to fluid 2 at temperature $T_{\infty 2}$ in accordance with $\dot{Q}_{c2} = h_{c2}(T_{w2} - T_{\infty 2})$. The ambient temperatures of the fluids $T_{\infty 1}$, and $T_{\infty 2}$ need to be defined in addition to the heat transfer coefficients h_{c1} and h_{c2}.
- Heat transfer by radiation to and from the two surfaces in accordance with the equations $\dot{Q}_{r1} = \sigma(T_\infty^4 - T_{w1}^4)$ and $\dot{Q}_{r2} = \sigma(T_{w2}^4 - T_\infty^4)$, respectively, where σ is the Stefan–Boltzmann constant. The radiation source temperature T_∞ needs to be defined in addition to the radiation heat rate to find the surface temperatures. Temperature T_∞ could be assumed equal to the temperatures $T_{\infty 1}$ and $T_{\infty 2}$ that are typically the temperatures of surrounding enclosures such as the walls of a room.

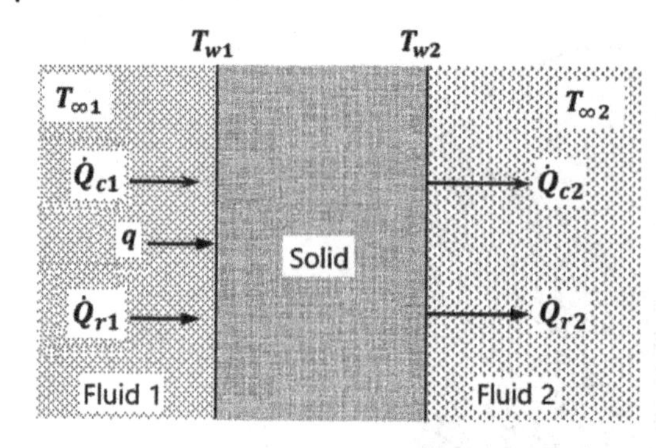

Figure 2.8 Examples of boundary conditions for heat conduction in a solid plate.

- Heat flux q (W/m^2) on the left surface can be specified in addition to a boundary temperature T_2 on the right surface. Under these boundary conditions, Fourier's law is used to determine T_{w1}, then solve Eq. (2.30) for the temperature profile.

The above boundary conditions can be applied separately or in combination, depending on the specifics of the problem under consideration. Other sources of heating could be electrical, electromagnetic, or radioisotopic.

Sometimes it is convenient to use an electrical analogy to assess complex conduction schemes using thermal resistances. In electricity, Ohm's law relates the voltage difference (potential difference) $V_1 - V_2$, electrical resistance R, and flow of electrical current I as $V_1 - V_2 = IR$ (Figure 2.9a)

If we rewrite Fourier's law in Eq. (2.27) as

$$\Delta T = T_1 - T_2 = \dot{Q}_x \left(\frac{\delta}{kA}\right)$$

we can think of $T_1 - T_2$ as the potential difference causing thermal current $\dot{Q}_x$ to flow through a thermal resistance δ / kA (Figure 2.9b).

The conduction thermal resistance will then be

$$R = \frac{\delta}{kA} \quad K/W \tag{2.38}$$

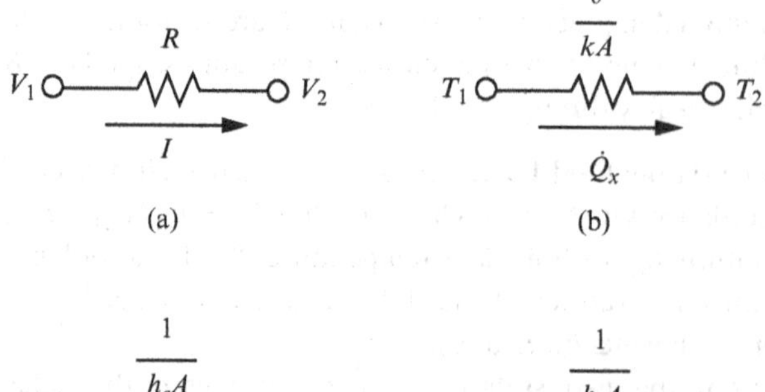

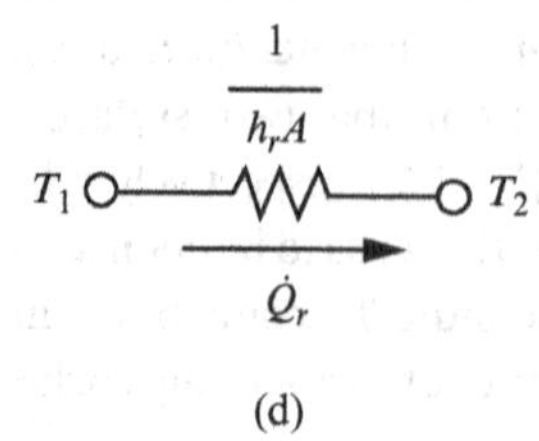

Figure 2.9 Comparison between electrical and thermal circuits: (a) electrical circuit; (b) conduction circuit; (c) convection circuit; and (d) radiation circuit.

A similar thermal circuit can be obtained for convection heat transfer by considering Newton's law of cooling for the temperature gradient $T_1 - T_2$

$$\dot{Q}_c = h_c A(T_1 - T_2)$$

where h_c is the heat transfer coefficient $(W/m^2.K)$, A is the area of the surface to which heat is being transferred by convection, T_1 is the fluid temperature, and T_2 is the temperature of the solid surface. Newton's law can be rewritten as

$$\Delta T_c = T_1 - T_2 = \dot{Q}_c\left(\frac{1}{h_c A}\right)$$

from which the convection thermal resistance, as shown in Figure 2.9c, is

$$R_{c1} = \frac{1}{h_c A} \quad K/W \tag{2.39}$$

The radiation heat transfer for the temperature gradient $T_1 - T_2$

$$\dot{Q}_r = \sigma A F_{12}(T_1^4 - T_2^4) = \sigma A F_{12}(T_1^2 - T_2^2)(T_1^2 + T_2^2)$$

$$\dot{Q}_r = \sigma A F_{12}(T_1 - T_2)(T_1 + T_2)(T_1^2 + T_2^2)$$

The radiation heat transfer can also be written in a similar way to Newton's law

$$\dot{Q}_r = h_r A(T_1 - T_2)$$

where h_r is the radiation conductance given by

$$h_r = \sigma F_{12}(T_1 + T_2)(T_1^2 + T_2^2) \quad W/m^2.K$$

In this equation, σ is the Stefan–Boltzmann constant $(5.67 \times 10^{-8}\ W/m^2.K^4)$, A is surface area, and F_{12} is the shape factor (fraction of radiation energy leaving surface 1 and falling on surface 2). The radiation thermal resistance is then

$$R_r = \frac{T_1 - T_2}{\dot{Q}_r} = \frac{1}{h_r A} \quad K/W \tag{2.40}$$

For a small temperature difference between the two surfaces $(T_1 \simeq T_2)$, we can write

$$h_r = \sigma F_{12}(T_1 + T_2)(T_1^2 + T_2^2) \approx 4\sigma F_{12} T_m^3 \quad W/m^2.K$$

and

$$R_r \approx \frac{1}{4A\sigma F_{12} T_m^3} \quad K/W \tag{2.41}$$

$T_m = (T_1 + T_2)/2$ is the mean of temperatures T_1 and T_2. The radiation thermal resistance is shown in Figure 2.9d.

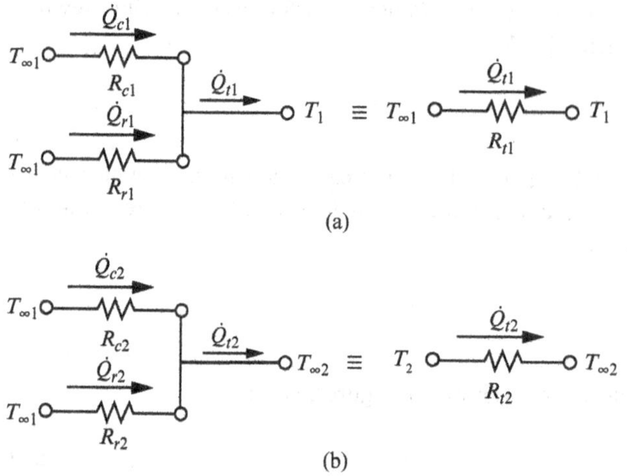

Figure 2.10 Thermal circuits for combined convection and radiation heat transfer for Figure 2.8: (a) surroundings to surface 1; (b) surface 2 to surroundings.

The thermal circuits for combined convection and radiation heat transfer at the boundaries of the solid body in Figure 2.8 are shown in Figure 2.10. It is assumed here that irradiation to surface 1 is from a radiation source of temperature $T_{\infty 1}$, and irradiation from surface 2 at temperature T_2 is to a radiation sink of temperature $T_{\infty 2}$.

The total heat transfer rate $\dot{Q}_t$ is

$$\dot{Q}_t = \frac{\Delta T}{R_t}$$

where

$$\Delta T = T_{\infty 1} - T_1 \text{ for heat transfer to surface 1}$$
$$\Delta T = T_2 - T_{\infty 2} \text{ for heat transfer from surface 2}$$

From analogy to two electric resistances in parallel, the total thermal resistances at the two walls are, respectively

$$R_{t1} = \frac{1}{\dfrac{1}{h_{c1}A} + \dfrac{1}{h_{r1}A}}, \quad R_{t2} = \frac{1}{\dfrac{1}{h_{c2}A} + \dfrac{1}{h_{r2}A}}$$

2.3.3 Hollow Cylinder (Tube) Without Heat Generation and Storage

2.3.3.1 Constant Thermal Conductivity

Consider the hollow cylinder of length L shown in Figure 2.11. The inner diameter is r_1 and the outer diameter is r_2. This case can also be considered as cylindrical wall of thickness $t = r_2 - r_1$. For one-dimensional and steady-state conduction (no heat generation or storage) and constant thermal conductivity, taking into account that the z-axis is along the cylinder axis, the differential conduction Eq. (2.13) is reduced to

$$\frac{\partial}{\partial r}\left(r\frac{\partial T}{\partial r}\right) = 0 \tag{2.42}$$

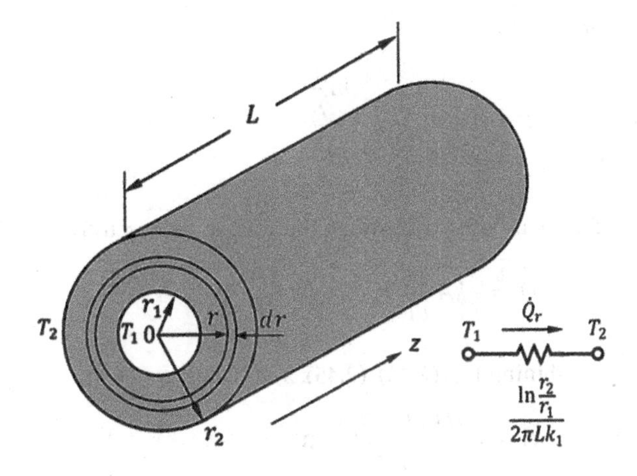

Figure 2.11 Hollow cylinder and its thermal circuit for one-dimensional heat conduction.

Integrating Eq. (2.42) yields

$$\frac{dT}{dr} = \frac{C_1}{r} \tag{2.43}$$

Integrating Eq. (2.43) we obtain

$$T = C_1 \ln r + C_2 \tag{2.44}$$

Substituting the boundary conditions $T_{(r=r_1)} = T_1$ and $T_{(r=r_2)} = T_2$ in Eq. (2.44) results in two simultaneous equations

$$T_1 = C_1 \ln r_1 + C_2$$

$$T_2 = C_1 \ln r_2 + C_2$$

Solving the resulting simultaneous equations gives the constants C_1 and C_2

$$C_1 = \frac{T_1 - T_2}{\ln \frac{r_1}{r_2}} \tag{2.45}$$

$$C_2 = T_1 - (T_1 - T_2)\frac{\ln r_1}{\ln \frac{r_1}{r_2}}$$

C_2 can also be written as

$$C_2 = T_2 - (T_1 - T_2)\frac{\ln r_2}{\ln \frac{r_1}{r_2}}$$

Equation (2.44) can now be written as

$$T = T_1 + (T_1 - T_2)\frac{\ln \frac{r}{r_1}}{\ln \frac{r_1}{r_2}} \tag{2.46}$$

or

$$T = T_2 + (T_1 - T_2)\frac{\ln\frac{r}{r_2}}{\ln\frac{r_1}{r_2}} \tag{2.47}$$

The heat flow rate through the wall according to Fourier law is

$$\dot{Q}_r = -kA\frac{dT}{dr} = -2\pi rLk\frac{dT}{dr} \quad W \tag{2.48}$$

Combining Eqs (2.43), (2.45), and (2.48), we obtain

$$\dot{Q}_r = \frac{2\pi Lk\left(T_1 - T_2\right)}{\ln\frac{r_2}{r_1}} \quad W \tag{2.49}$$

or

$$T_1 - T_2 = \dot{Q}_r\frac{\ln\frac{r_2}{r_1}}{2\pi Lk}$$

Thermal resistance is therefore

$$R = \frac{\ln\frac{r_2}{r_1}}{2\pi Lk}$$

The thermal circuit for the cylindrical wall is shown in Figure 2.11.

2.3.3.2 Temperature-Dependent Thermal Conductivity

The thermal conductivity of a material can vary significantly with temperature, and for most metals and metal alloys, it can be approximated by a linear function of the form

$$k = k_0(1 + bT)$$

From Eq. (2.18), the equation for one-dimensional conduction with variable conductivity without heat generation and heat storage can be written as

$$\frac{1}{r}\frac{\partial}{\partial r}\left(kr\frac{\partial T}{\partial r}\right) = 0$$

or

$$\frac{d}{dr}\left[k_0 r(1 + bT)\frac{dT}{dr}\right] = 0$$

Integrating the above equation yields

$$k_0(1 + bT)\frac{dT}{dr} = \frac{C_1}{r} \tag{2.50}$$

Separating the variables and integrating again yields

$$k_0\left(T+\frac{b}{2}T^2\right)=C_1\ln r+C_2$$

or

$$\frac{b}{2}T^2+T-\frac{1}{k_o}(C_1\ln r+C_2)=0 \tag{2.51}$$

Boundary conditions: $T_{(r=r_1)}=T_1$ and $T_{(r=r_2)}=T_2$, hence

$$k_0\left(T_1+\frac{b}{2}T_1^2\right)=C_1\ln r_1+C_2$$

$$k_0\left(T_2+\frac{b}{2}T_2^2\right)=C_1\ln r_2+C_2$$

Constants C_1 and C_2 are obtained from the above simultaneous equations

$$C_1=\frac{k_0\left(T_1+\frac{b}{2}T_1^2\right)-k_0\left(T_2+\frac{b}{2}T_2^2\right)}{\ln\left(\frac{r_1}{r_2}\right)}$$

$$C_2=k_0\left(T_1+\frac{b}{2}T_1^2\right)-C_1\ln r_1$$

The solution to the quadratic Eq. (2.51) is

$$T=-\frac{1}{b}+\sqrt{\frac{1}{b^2}+\frac{2}{b}\Phi} \tag{2.52}$$

where

$$\Phi=\frac{1}{k_0}(C_1\ln r+C_2)=\left(T_1+\frac{b}{2}T_1^2\right)+\frac{\left(T_1+\frac{b}{2}T_1^2\right)-\left(T_2+\frac{b}{2}T_2^2\right)}{\ln\left(\frac{r_1}{r_2}\right)}\ln\left(\frac{r}{r_1}\right)$$

The heat rate is

$$\dot{Q}=-k_0(1+bT)A\frac{dT}{dr}=-2\pi rLk_0(1+bT)\frac{dT}{dr}$$

Combining with Eq. (2.50) results in

$$\dot{Q}=-2\pi LC_1=2\pi L\frac{k_0\left(T_1+\frac{b}{2}T_1^2\right)-k_0\left(T_2+\frac{b}{2}T_2^2\right)}{\ln\left(\frac{r_2}{r_1}\right)}$$

or

$$\dot{Q} = \frac{2\pi L k_m (T_1 - T_2)}{\ln\left(\frac{r_2}{r_1}\right)} \quad W \tag{2.53}$$

where k_m is the thermal conductivity at the mean temperature $T_m = (T_1 + T_2)/2$

$$k_m = k_0\left[1 + b\left(\frac{T_1 + T_2}{2}\right)\right] \tag{2.54}$$

The heat rate per unit length

$$\frac{\dot{Q}}{L} = \frac{2\pi k_m (T_1 - T_2)}{\ln\left(\frac{r_2}{r_1}\right)} \quad W/m \tag{2.55}$$

The heat flux through the inner surface

$$q_1 = \frac{\dot{Q}}{2\pi r_1 L} = \frac{k_m (T_1 - T_2)}{r_1 \ln\left(\frac{r_2}{r_1}\right)} \quad W/m^2 \tag{2.56}$$

The heat flux through the outer surface

$$q_2 = \frac{\dot{Q}}{2\pi r_2 L} = \frac{k_m (T_1 - T_2)}{r_2 \ln\left(\frac{r_2}{r_1}\right)} \quad W/m^2 \tag{2.57}$$

The temperature profile given by Eq. (2.52) can be formulated to include the conduction heat rate

$$T = -\frac{1}{b} \pm \sqrt{\left(\frac{1}{b} + T_1\right)^2 - \frac{\dot{Q}}{b\pi L k_0}\ln\left(\frac{r}{r_1}\right)} \tag{2.58}$$

Three temperature profiles for conduction in the wall of a hollow cylinder are shown in Figure 2.12. The data used: $r_1 = 0.2m$, $r_2 = 0.7m$, $T_1 = 700K$, and $T_2 = 300K$. The thermal conductivities used are

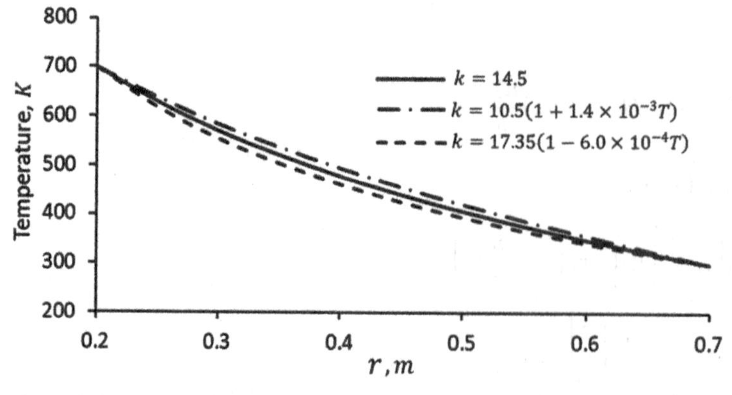

Figure 2.12 Effect of variability of the thermal conductivity on the temperature profiles in a hollow cylinder.

$$k = \text{const.} = 14.5\ W/m.K$$

$$k = 10.5(1 + 1.4 \times 10^{-3} T)\ b > 0$$

$$k = 17.3(1 - 6.0 \times 10^{-3} T)\ b < 0$$

The shapes of the temperature profiles for the three cases are non-linear and similar, with the temperatures in the cylindrical wall being higher or lower than those for constant k, depending on the sign of b.

2.3.3.3 Composite Cylinder

Consider a hollow cylindrical wall comprising n layers of different materials of the same length having surface temperatures of the layers T_1, T_2, ... T_n, respectively. Since the conduction process is one-dimensional, $d\dot{Q}_z / dr$ is constant over the entire length of the cylindrical layers. Since the heat fluxes are variable in the r-direction, Eq. (2.49) can be used to describe the heat conduction through each layer

$$\dot{Q}_r = \frac{2\pi L k_1 (T_1 - T_2)}{\ln \frac{r_2}{r_1}}$$

$$\dot{Q}_r = \frac{2\pi L k_2 (T_2 - T_3)}{\ln \frac{r_3}{r_2}}$$

$$\vdots$$

$$\dot{Q}_r = \frac{2\pi L k_n (T_n - T_{n+1})}{\ln \frac{r_{n+1}}{r_n}}$$

These equations can be rewritten in terms of temperature gradient across each layer as

$$T_1 - T_2 = \frac{\ln \frac{r_2}{r_1}}{2\pi L k_1} \dot{Q}_r$$

$$T_2 - T_3 = \frac{\ln \frac{r_3}{r_2}}{2\pi L k_2} \dot{Q}_r$$

$$\vdots$$

$$T_n - T_{n+1} = \frac{\ln \frac{r_{n+1}}{r_n}}{2\pi L k_n} \dot{Q}_r$$

Summing both sides of theses equations yield

$$T_1 - T_{n+1} = \frac{\ln \frac{r_2}{r_1}}{2\pi L k_1} \dot{Q}_r + \frac{\ln \frac{r_3}{r_2}}{2\pi L k_2} \dot{Q}_r + \ldots + \frac{\ln \frac{r_{n+1}}{r_n}}{2\pi L k_n} \dot{Q}_r$$

or

$$\dot{Q}_r = \frac{T_1 - T_{n+1}}{\frac{\ln \frac{r_2}{r_1}}{2\pi L k_1} + \frac{\ln \frac{r_3}{r_2}}{2\pi L k_2} + \ldots + \frac{\ln \frac{r_{n+1}}{r_n}}{2\pi L k_n}} = \frac{T_1 - T_{n+1}}{R_1 + R_2 + \ldots + R_n} \tag{2.59}$$

From an electrical analogy, where $\dot{Q}_r$ is the current and $\Delta T(=T_1 - T_{n+1})$ is the potential difference, the denominator on the right side of Eq. (2.59) will be the thermal resistance

$$R = R_1 + R_2 + \ldots + R_n = \sum_{i=1}^{n} \frac{\ln \frac{r_{i+1}}{r_i}}{2\pi L k_i} \quad K/W \tag{2.60}$$

The heat rate in terms of the overall heat transfer coefficient

$$\dot{Q}_r = UA\Delta T = \frac{\Delta T}{\frac{1}{UA}} \tag{2.61}$$

where

$$\frac{1}{UA} = \frac{\ln \frac{r_2}{r_1}}{2\pi L k_1} + \frac{\ln \frac{r_3}{r_2}}{2\pi L k_2} + \ldots + \frac{\ln \frac{r_{n+1}}{r_n}}{2\pi L k_n} = R_1 + R_2 + \ldots + R_n$$

Surface area A could be the inside or outside area of the composite cylinder. The conduction heat rate for the two-layer hollow cylinder, shown in Figure 2.13, is

$$\dot{Q}_r = \frac{T_1 - T_{n+1}}{\frac{\ln \frac{r_2}{r_1}}{2\pi L k_1} + \frac{\ln \frac{r_3}{r_2}}{2\pi L k_2}}$$

and the overall thermal resistance and heat transfer coefficient

$$R = \frac{\ln \frac{r_2}{r_1}}{2\pi L k_1} + \frac{\ln \frac{r_3}{r_2}}{2\pi L k_2} \tag{2.62}$$

$$\frac{1}{UA} = \frac{\ln \frac{r_2}{r_1}}{2\pi L k_1} + \frac{\ln \frac{r_3}{r_2}}{2\pi L k_2} \tag{2.63}$$

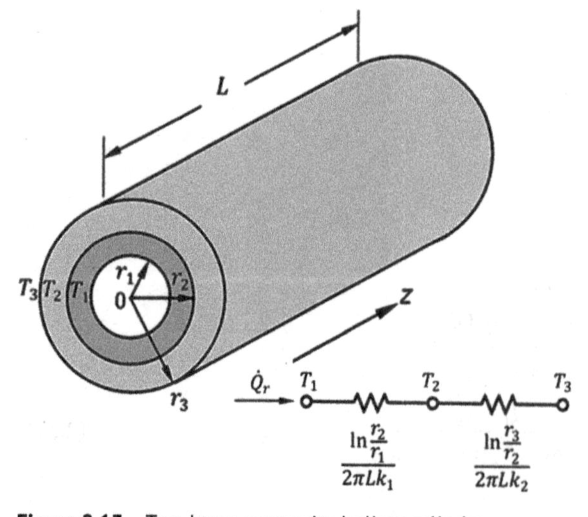

Figure 2.13 Two-layer composite hollow cylinder.

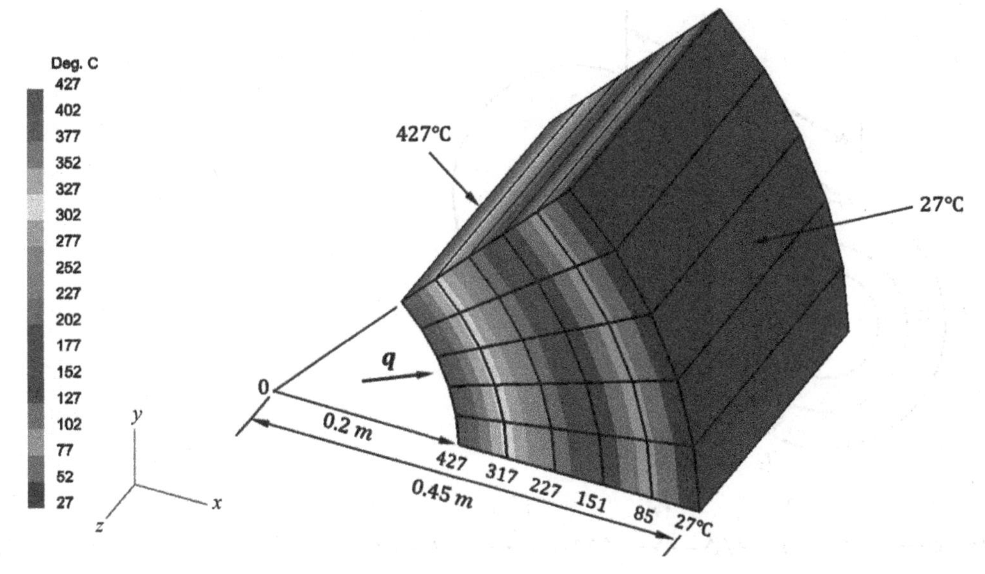

Figure 2.14 Results of computer model of five-layer composite cylinder of the same material (45° segment of a cylinder is shown with radii in metres).

The thermal circuit for the two-layer hollow cylinder is also shown in Figure 2.13.

Figure 2.14 shows the results of computer modelling of conduction heat transfer in a five-layer composite cylinder of the same material.

2.3.3.4 Critical Thickness of Cylinder Insulation

Insulation of cylindrical objects such as tubes and electric wires are often used as a means to control the heat rate to the surroundings. To evaluate the required insulation thickness, consider the hollow two-layer composite tube shown in Figure 2.15, in which the inner cylinder is a metal tube through which a heated fluid at temperature T_i is flowing and the outer layer is the insulating material exposed to heat transfer by convection to the surroundings at temperature T_∞.

In the thermal circuit, k_m, k_{ins} are, respectively, the thermal conductivities of the metal tube and insulation material; h_{ci}, h_{co} are, respectively, the inner and outer heat transfer coefficients. The total thermal resistance is

$$R_t = R_{ci} + R_m + R_{ins} + R_{co} = \frac{1}{2\pi r_1 L h_{ci}} + \frac{\ln\left(\frac{r_2}{r_1}\right)}{2\pi L k_m} + \frac{\ln\left(\frac{r_3}{r_2}\right)}{2\pi L k_{ins}} + \frac{1}{2\pi r_3 L h_{co}} \tag{2.64}$$

Equation (2.64) can be written as

$$R_t = C_1 + C_2 \ln\left(\frac{r_3}{r_2}\right) + \frac{C_3}{r_3} \tag{2.65}$$

where

$$C_1 = \left(\frac{1}{2\pi r_1 L h_{ci}} + \frac{\ln\left(\frac{r_2}{r_1}\right)}{2\pi L k_m}\right),\ C_2 = \frac{1}{2\pi L k_{ins}},\ C_3 = \frac{1}{2\pi L h_{co}}$$

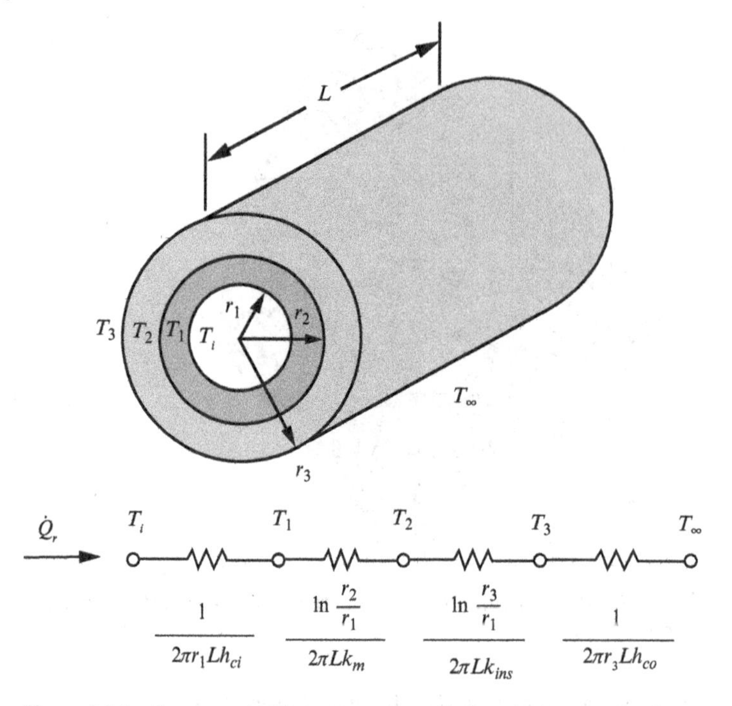

Figure 2.15 Two-layer hollow composite cylinder with convection heat transfer and its thermal circuit.

The heat rate in the radial direction is then

$$\dot{Q}_r = \frac{T_i - T_\infty}{R_t} = \frac{T_i - T_\infty}{C_1 + C_2 \ln\left(\frac{r_3}{r_2}\right) + \frac{C_3}{r_3}} \tag{2.66}$$

Figure 2.16 shows the heat rate and resistances plotted versus the outer radius r_3 with all other parameters in Eq. (2.64) being kept constant.

The total thermal resistance R_t exhibits a minimum at the critical radius r_{cr}, which can be found by differentiating Eq. (2.65) with respect to r_3 and equating the result to zero.

$$\frac{dR_t}{dr_3} = \frac{C_2}{r_3} - \frac{C_3}{r_3^2} = 0$$

from which

$$r_3 = r_{cr} = \frac{k_{ins}}{h_{co}} \tag{2.67}$$

For the data given in Figure 2.16, the critical radius $r_{cr} \cong 2.125/8.5 \cong 0.25m$.

The critical radius obtained from Eq. (2.67) is only an approximate estimate, since it assumes that Newton's law of colling is universally applicable to all convection heat transfer scenarios. However. it is accurate enough for use in engineering calculations.

When r_3 is increased, the conduction resistance R_{ci} and R_m remain unchanged, insulation resistance R_{ins} increases, and the outer convection resistance R_{co} decreases. Due to the different rates of

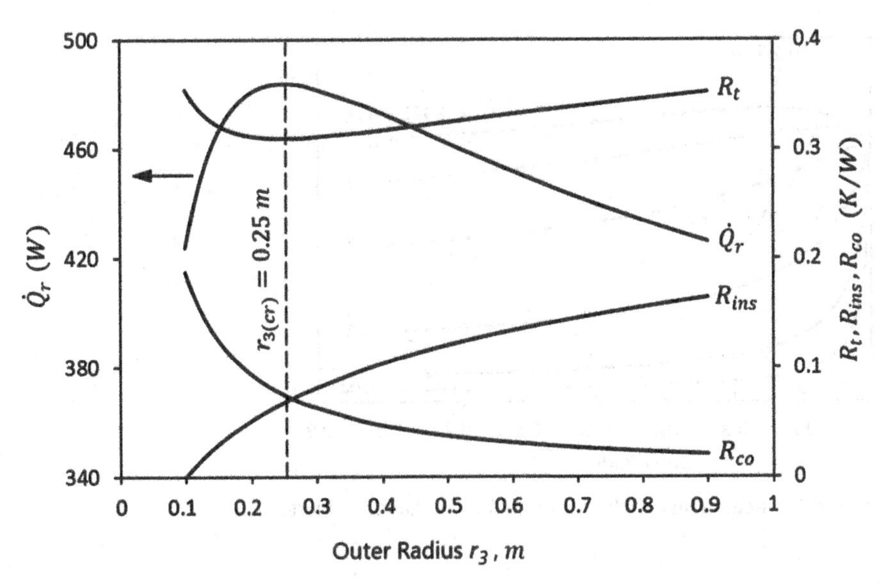

Figure 2.16 Thermal resistance versus outside radius of an insulated cylindrical wall $L = 1.0\,m$, $r_2 = 0.1\,m$, $k_m = 45\,W/m.K$, $k_{ins} = 2.125\,W/m.K$, $h_{ci} = 12\,W/m^2K$, $h_{co} = 8.5\,W/m^2.K$.

change of R_{ins} and R_{co} with r_3, the total resistance R_t initially decreases to its minimum value at the critical radius r_{cr} then starts increasing gradually. The heat rate increases sharply up to the critical radius, tapering off as r_3 increases further.

Figure 2.16 indicates that for $r_2 < r_3 < r_{cr}$ adding a layer of insulation to the steel cylinder will increase the heat rate $\dot{Q}_r$ (i.e., increase the heat loss). Figure 2.17 shows the effect of varying the critical radius on the heat rate by varying the insulation thermal conductivity, with all other parameters in Figure 2.16 remaining unchanged. It can be concluded from this figure that for reduction of heat loss by insulation to be effective, the outer radius must be greater than the critical radius ($r_3 > r_{cr}$). Having satisfied this condition, heat rate loss can be decreased progressively by reducing the thermal conductivity of the insulating material and increasing the insulation thickness. The latter measure becomes almost ineffective beyond a certain value depending on k_{ins}.

The feature of increasing heat rate $\dot{Q}_r$ with increasing insulation thickness up to the critical radius is usually utilized for protecting electrical components with ohmic dissipation (I^2R) from overheating by regulating the heat loss. Figure 2.15 can be used as a model for this case by assuming the inner circle of radius r_1 to be a solid electrical conductor sheathed by a special thin insulator (conductivity k_{ins1}) of radius r_2 which, in its turn, is covered by a second layer of insulating material of radius r_3 and conductivity k_{ins2} to regulate heat loss and serve as an additional electrical insulator. Assuming that T_i is now the isothermal temperature of the conductor, the thermal circuit in Figure 2.15 is reduced to the one shown in Figure 2.18.

The second insulation layer is usually selected so that it can enhance the heat loss from the conductor. From the upper two heat rate curves in Figure 2.17, it can be surmised that the optimum insulation radius for maximum heat transfer rate to the surroundings will be $r_3 = r_{cr}$.

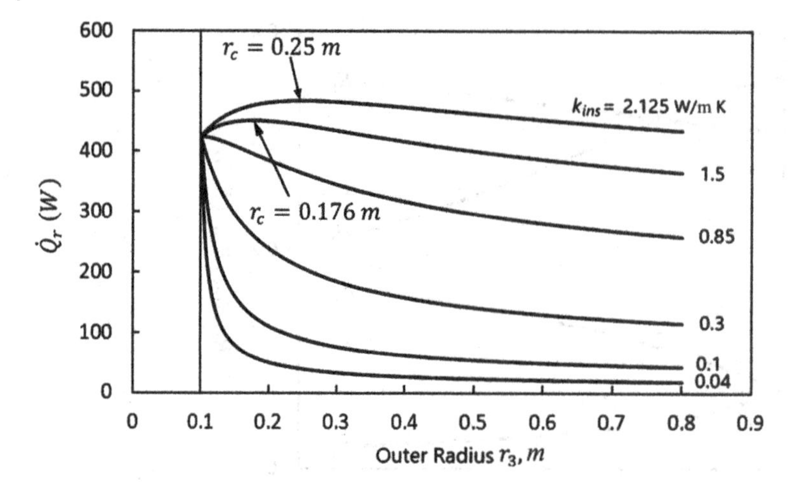

Figure 2.17 Effect of the critical radius on the heat rate for the shape in Figure 2.15.

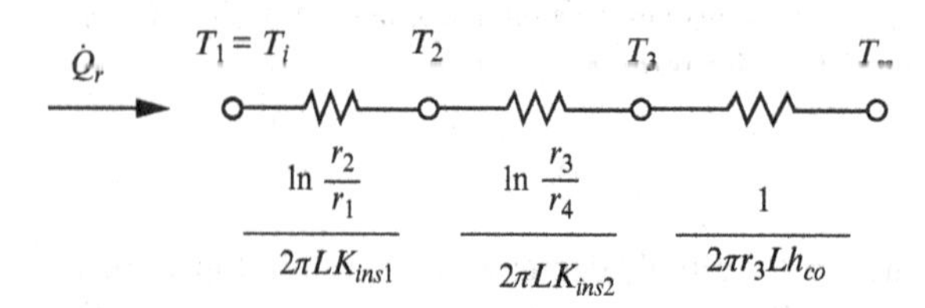

Figure 2.18 Thermal circuit for an electrical conductor with two insulation layers.

Example 2.3 An electrical conductor of radius r_1 (1 mm) at an isothermal temperature of 360 K is sheathed in a thin silica aerogel tube of radius r_2 (1.05 mm), which is covered with a thick layer of neoprene insulating material of radius r_3. The outside of the neoprene layer is exposed to convection heat transfer to the surroundings at 298 K with coefficient of heat transfer $h_{co} = 18\ W/m^2.K$. If the thermal conductivities of the silica gel and neoprene are, respectively, 0.024 and 0.17 $W/m.K$, determine

1) The optimum neoprene thickness for maximum heat transfer to the surroundings
2) The total thermal resistance and rate of heat loss to the surroundings per metre length of the conductor at optimum neoprene thickness
3) The temperatures of all surfaces at maximum heat transfer.

Solution

We refer to the cylindrical shape in Figure 2.15 and the thermal circuit in Figure 2.18

$$R_t = R_{ins1} + R_{ins2} + R_{co} = \frac{\ln\left(\frac{r_2}{r_1}\right)}{2\pi L k_{ins1}} + \frac{\ln\left(\frac{r_3}{r_2}\right)}{2\pi L k_{ins2}} + \frac{1}{2\pi r_3 L h_{co}}$$

Optimal thickness of neoprene for maximum heat transfer to the surroundings:

$$r_3 = r_{cr} = k_{ins2} / h_{co} = 0.17 / 18 = 0.00944\, m (9.44\, mm)$$

Total thermal resistance:

$$R_t = \frac{1}{L}\left[\frac{\ln\left(\frac{1.05}{1}\right)}{2\pi \times 0.024} + \frac{\ln\left(\frac{9.44}{1.05}\right)}{2\pi \times 0.17} + \frac{1}{2\pi \times 0.00944 \times 18}\right] = \frac{1}{L}(0.3235 + 2.056 + 0.9366)$$

$$R_t = \frac{3.32}{L} \quad K / W.m$$

hence,

$$\dot{Q}_r = \frac{T_i - T_o}{R_t} = \frac{(360 - 300)L}{3.32} = 18.07L \ W$$

$$\frac{\dot{Q}_r}{L} = 18.07\ W / m$$

Surface temperatures:

$$T_1 = T_i = 360K$$

$$T_2 = T_1 - R_{ins1}\dot{Q}_r = 360 - 0.3235 \times 18.07 = 354.2K$$

$$T_3 = T_2 - R_{ins2}\dot{Q}_r = 354.2 - 2.056 \times 18.07 = 317K$$

2.3.3.5 Effect of Order of Insulation Material

Consider a steel pipe carrying heated fluid with two layers of insulation of equal thickness but of different materials. To minimize the heat loss from the pipe surface, the material with the lower thermal conductivity should be in contact with the pipe surface. Reversing the order of the materials will cause an increase in the heat loss and the higher the ratio of thermal conductivities of the insulation materials, the greater the heat loss.

If the pipe is encapsulated in a three-layer insulation of equal thicknesses but of different materials, the heat loss can be minimized by arranging the materials so that their thermal conductivities are increasing outwards. Reversing the order will again cause the heat loss to increase. As long as the material with the lowest thermal conductivity is in contact with the pipe surface, the order of the other two layers has little effect.

Figure 2.19 shows the temperature profiles of the radial heat flow in a steel pipe with three layers of insulating material of equal thickness. Heat is transferred by convection from a heated fluid at $T_i = 400°C$ to the inner surface of the pipe, by conduction through the steel pipe wall and insulation layers, followed by heat transfer by convection to the surroundings at $T_o = 30°C$. The thermal conductivities of the insulating materials are increasing outwards in Figure 2.19a and decreasing in Figure 2.19b. The lower surface temperature of the outer surface in Figure 2.19a (49°C) is an indication of lower heat loss by convection.

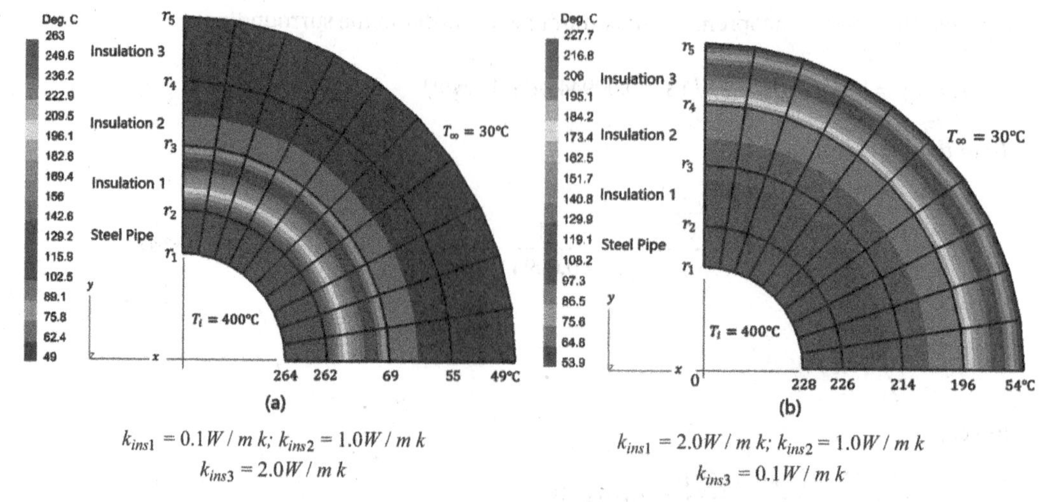

Figure 2.19 Computer modelling of the effect of insulation layer order on the temperature profiles for steel pipe $k_s = 14.5\ W/m.K, h_{ci} = 8\ W/m^2.K,\ h_{co} = 18, r_1 = 0.05\ m,\ r_2 = 0.07\ m,\ r_3 = 0.1\ m;\ r_4 = 0.13\ m,\ r_5 = 0.16\ m.$

2.3.4 Hollow Spherical Shell Without Heat Generation and Storage

2.3.4.1 Constant Thermal Conductivity

A hollow sphere with inner radius r_1 and external radius r_2 is shown in Figure 2.20. Uniform temperatures are maintained on the inner and external surfaces.

For a one-dimensional heat conduction process through the spherical wall of constant thermal conductivity in the r-direction, $\partial T/\partial \varphi$ and $\partial T/\partial \theta$ are equal to zero and Eq. (2.19) is reduced to

$$\frac{\partial}{\partial r}\left(r^2 \frac{\partial T}{\partial r}\right) = 0 \tag{2.68}$$

Integrating Eq. (2.68) twice yields

$$T = -\frac{C_1}{r} + C_2 \tag{2.69}$$

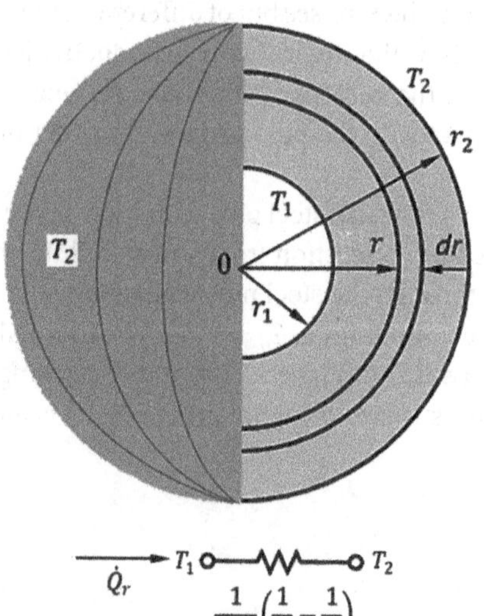

Figure 2.20 Spherical shell and its thermal circuit for one-dimensional heat conduction.

Boundary conditions: $T_{r=r_1} = T_1$, $T_{r=r_2} = T_2$

Substituting the boundary conditions into Eq. (2.69) results in

$$T_1 = -\frac{C_1}{r_1} + C_2$$

$$T_2 = -\frac{C_1}{r_2} + C_2$$

Solving the two simultaneous equations yields

$$C_1 = \frac{T_1 - T_2}{\dfrac{1}{r_2} - \dfrac{1}{r_1}}, \qquad C_2 = T_1 + \frac{1}{r_1}\frac{T_1 - T_2}{\dfrac{1}{r_2} - \dfrac{1}{r_1}}$$

Substituting for C_1 and C_2 into Eq. (2.69) and rearranging results in the final equation for the determination of the temperature profile in the spherical shell

$$T = T_1 - \frac{T_1 - T_2}{\dfrac{1}{r_1} - \dfrac{1}{r_2}}\left(\frac{1}{r_1} - \frac{1}{r}\right) \tag{2.70}$$

The heat flow rate through the wall according to Fourier law is

$$\dot{Q}_r = -kA\frac{dT}{dr}\,W$$

From Eq. (2.69), $dT/dr = C_1/r^2$; hence,

$$\dot{Q}_r = \frac{4\pi k(T_1 - T_2)}{\left(\dfrac{1}{r_1} - \dfrac{1}{r_2}\right)} = \frac{T_1 - T_2}{\dfrac{1}{4\pi k}\left(\dfrac{1}{r_1} - \dfrac{1}{r_2}\right)} = \frac{T_1 - T_2}{R} \tag{2.71}$$

where R is the thermal resistance in a spherical shell.

2.3.4.2 Temperature-Dependent Thermal Conductivity

The thermal conductivity of a material could vary significantly with temperature, and for most metals and metal alloys, it can be approximated by a linear function of the form

$$k = k_0(1 + bT)$$

From Eq. (2.19), the equation for one-dimensional conduction with variable conductivity without heat generation and heat storage can be written as

$$\frac{d}{dr}\left[r^2 k_0(1 + bT)\frac{dT}{dr}\right] = 0$$

Integrating yields

$$k_0(1 + bT)\frac{dT}{dr} = \frac{C_1}{r^2}$$

Separating the variables and integrating again yields the following equation:

$$k_0\left(T+\frac{bT^2}{2}\right)=-\frac{C_1}{r}+C_2 \tag{2.72}$$

Boundary conditions: $T_{r=r_1}=T_1$, $T_{r=r_2}=T_2$

Using these boundary conditions and following the same procedure for the hollow cylinder, we obtain a quadratic equation whose solution is

$$T=-\frac{1}{b}\pm\sqrt{\frac{1}{b^2}+\frac{2}{b}\Theta} \tag{2.73}$$

where

$$\Theta=\left(T_1+\frac{bT_1^2}{2}\right)+\frac{(1/r_1-1/r)}{(1/r_2-1/r_1)}\left[(T_1-T_2)+\frac{b}{2}(T_1^2-T_2^2)\right]$$

The heat flow rate for a spherical shell with averaged thermal conductivity k_m is

$$\dot{Q}_r=\frac{4\pi k_m(T_1-T_2)}{\left(\frac{1}{r_1}-\frac{1}{r_2}\right)} \tag{2.74}$$

where k_m can be taken as the thermal conductivity at the mean temperature $T_m=(T_1+T_2)/2$

$$k_m=k_0\left[1+b\left(\frac{T_1+T_2}{2}\right)\right]$$

Combining Eqs (2.73) and (2.74), we obtain a different formulation of the temperature profile

$$T=-\frac{1}{b}\pm\sqrt{\left(\frac{1}{b}+T_1\right)^2-\frac{\dot{Q}}{2\pi bk_0}\left(\frac{1}{r_1}-\frac{1}{r}\right)} \tag{2.75}$$

The temperature distribution in the wall of a hollow cylinder for $r_1=0.2m$, $r_2=0.45m$, $T_1=700K$, and $T_2=300K$ is shown in Figure 2.21. Three cases are compared:

$$k=const=14.5\ W/m.K$$

$$k=10.5(1+1.4\times10^{-3}T) \qquad b>0$$

$$k=17.3(1-6.0\times10^{-3}T) \qquad b<0$$

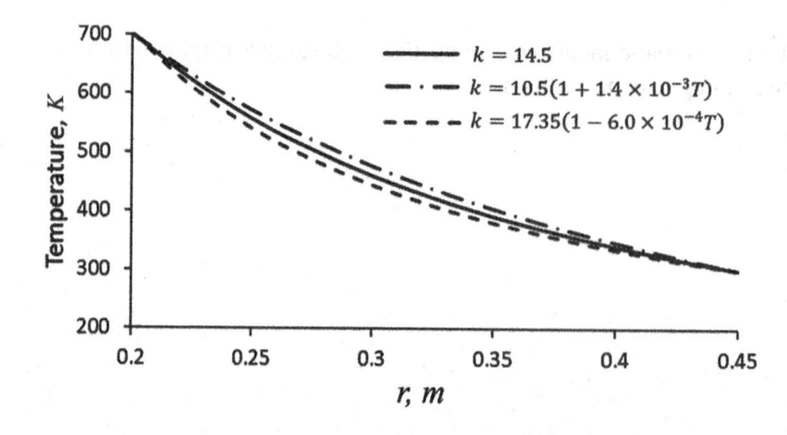

Figure 2.21 Effect of variability of the thermal conductivity on the temperature profiles in a spherical shell.

The shapes of the temperature profiles for the three cases are non-linear and are essentially the same with the temperatures in the wall of the spherical shell being higher or lower, depending on the sign of b.

2.3.4.3 Composite Spherical Shell

For a composite spherical shell with n layers of different materials, it can be shown that

$$\dot{Q}_r = \frac{T_1 - T_{n+1}}{\frac{1}{4\pi k_1}\left(\frac{1}{r_1} - \frac{1}{r_2}\right) + \frac{1}{4\pi k_2}\left(\frac{1}{r_2} - \frac{1}{r_3}\right) + \ldots + \frac{1}{4\pi k_n}\left(\frac{1}{r_n} - \frac{1}{r_{n+1}}\right)} \tag{2.76}$$

From an electrical analogy where $\dot{Q}_r$ is the current and $(\Delta T = T_1 - T_{n+1})$ is the potential difference, the denominator on the right side of Eq. (2.76) will be the total thermal resistance

$$R = R_1 + R_2 + \ldots + R_n = \sum_{i=1}^{n}\left[\frac{1}{4\pi k_i}\left(\frac{1}{r_i} - \frac{1}{r_{i+1}}\right)\right] \quad K/W$$

hence,

$$\dot{Q}_r = \frac{T_1 - T_{n+1}}{\sum_{i=1}^{n}\left[\frac{1}{4\pi k_i}\left(\frac{1}{r_i} - \frac{1}{r_{i+1}}\right)\right]} = \frac{T_1 - T_{n+1}}{R}$$

The heat rate in terms of the overall heat transfer coefficient

$$\dot{Q}_r = UA\Delta T = \frac{\Delta T}{\frac{1}{UA}} \tag{2.77}$$

where

$$\frac{1}{UA} = \frac{1}{4\pi k_1}\left(\frac{1}{r_1} - \frac{1}{r_2}\right) + \frac{1}{4\pi k_2}\left(\frac{1}{r_2} - \frac{1}{r_3}\right) + \ldots + \frac{1}{4\pi k_{i=n}}\left(\frac{1}{r_n} - \frac{1}{r_{n+1}}\right) = R_1 + R_2 + \ldots + R_n$$

Surface area A could be the inside or outside area of the composite sphere.

The thermal circuit for conduction heat transfer in a two-layer spherical shell is shown in Figure 2.22.

The results of computer modelling of heat transfer by conduction in a five-layer hollow composite sphere with inner radius $r_1 = 0.2\ m$ and outer radius $r_1 = 0.45\ m$ are shown in Figure 2.23. Heat is transferred by convection to the inner surface at $T_i = 400°\text{C}$ and $h_{ci} = 8\ W/m^2.K$ and from the outer surface to the surroundings at $T_\infty = 30°\text{C}$ and $h_{co} = 18\ W/m^2.K$. Figure 2.23a is for different materials of different thermal conductivities (1, 2, 3, 4, and 5 $W/m.K$ starting from the inner layer towards the outer layer) and Figure 2.23b is for the same material.

T_1 T_2 T_3

$\dot{Q}_r$

$\frac{1}{4\pi k_1}\left(\frac{1}{r_1} - \frac{1}{r_2}\right)$ $\frac{1}{4\pi k_2}\left(\frac{1}{r_2} - \frac{1}{r_3}\right)$

Figure 2.22 Thermal circuit for conduction in two-layer composite hollow cylinder.

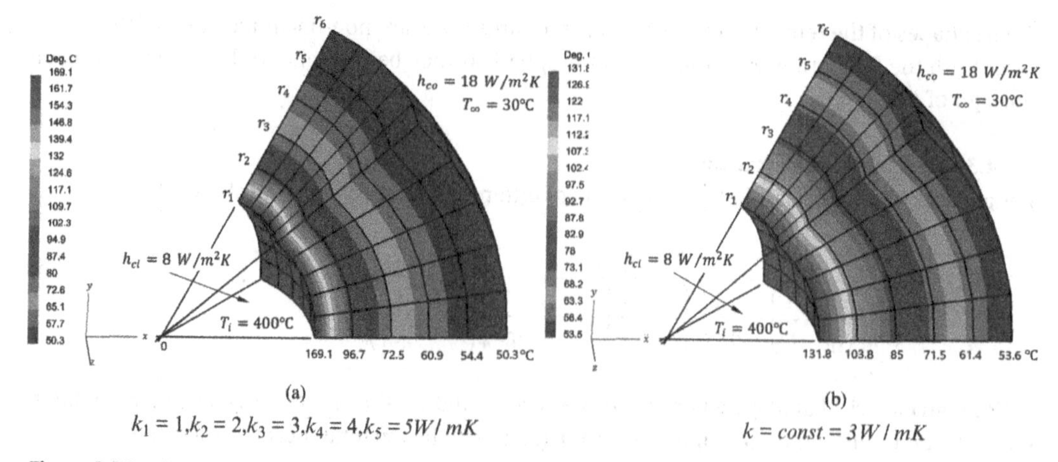

Figure 2.23 Computer modelling of five-layer composite sphere (a) different layer materials; and (b) same layer material.

Example 2.4 Consider the two spherical shells shown in Figure E2.4.1 with inner radius $r_i = 0.2m$, outer radius $r_o = 0.45m$, inner surface temperature $T_i = 700K$, and outer surface temperature $T_o = 300K$.

a) Solid spherical shell (hollow sphere) of homogeneous material of constant thermal conductivity
b) Spherical shell of five layers numbered from 1 to 5 of different materials and of equal thickness (0.05 m). The thermal conductivities for the materials are, respectively, 1, 2, 3, 4, and 5 $W/m.K$.

If there is no internal heat generation and storage, determine the temperature profiles for (a) and (b).

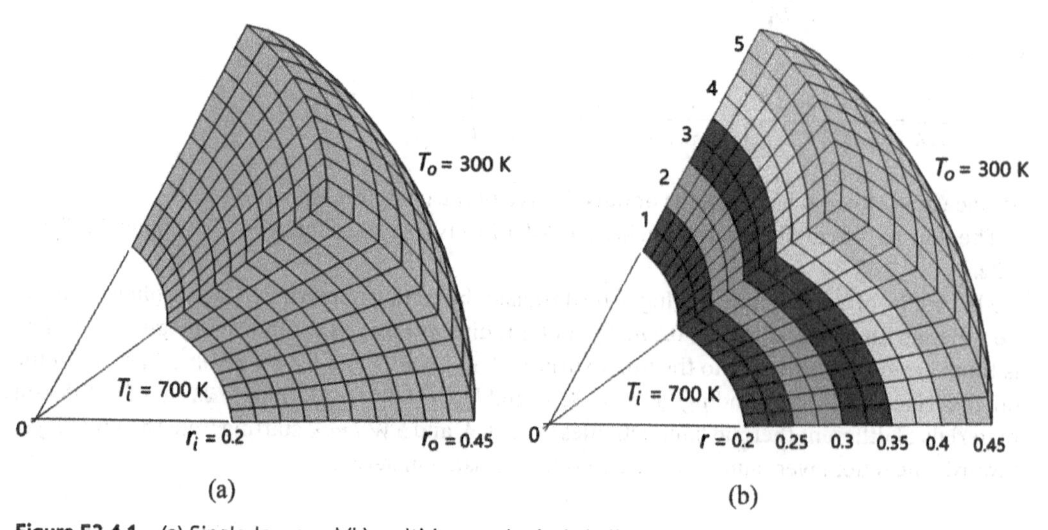

Figure E2.4.1 (a) Single-layer and (b) multi-layer spherical shells.

Solution

(a)

The temperature profile in the wall of a spherical shell is given by Eq. (2.70)

$$T = T_i - \frac{T_i - T_o}{\frac{1}{r_i} - \frac{1}{r_o}}\left(\frac{1}{r_i} - \frac{1}{r}\right)$$

The results of calculations at increments of 0.05 m are shown in Table E2.4:

Table E2.4 Results of calculations for the spherical shell in (a).

r*, *m	0.2	0.25	0.3	0.35	0.4	0.45
T*, *K	700	556	460	391.4	340	300
***T*,°C**	427	283	168	118.4	67	27

(b)

The total resistance

$$R = R_1 + R_2 + R_3 + R_4 + R_5$$

$$R = \frac{1}{4\pi k_1}\left(\frac{1}{r_1} - \frac{1}{r_2}\right) + \frac{1}{4\pi k_2}\left(\frac{1}{r_2} - \frac{1}{r_3}\right) + \frac{1}{4\pi k_3}\left(\frac{1}{r_3} - \frac{1}{r_4}\right) + \frac{1}{4\pi k_4}\left(\frac{1}{r_4} - \frac{1}{r_5}\right) + \frac{1}{4\pi k_5}\left(\frac{1}{r_5} - \frac{1}{r_6}\right)$$

$$= \frac{1}{4\pi \times 1}\left(\frac{1}{0.2} - \frac{1}{0.25}\right) + \frac{1}{4\pi \times 2}\left(\frac{1}{0.25} - \frac{1}{0.3}\right) + \frac{1}{4\pi \times 3}\left(\frac{1}{0.3} - \frac{1}{0.35}\right) + \frac{1}{4\pi \times 4}\left(\frac{1}{0.35} - \frac{1}{0.4}\right)$$

$$+ \frac{1}{4\pi \times 5}\left(\frac{1}{0.4} - \frac{1}{0.45}\right)$$

$$R = 0.079577 + 0.026526 + 0.012631 + 0.007105 + 0.004421 = 0.130261\ K/W$$

The rate of heat flow across the wall is constant and given by

$$\dot{Q}_r = \frac{T_1 - T_{n+1}}{R} = \frac{700 - 300}{0.13026} = 3071\,W$$

The temperatures of the layer surfaces:

$$r = 0.2\,m, T_1 = T_i = 700\ K\ (427°C)$$

$$r = 0.25m, T_2 = T_1 - \dot{Q}_r R_1 = 700 - 3071 \times 0.079577 = 455.6K\,(182.6°C)$$

$$r = 0.3m, T_3 = T_2 - \dot{Q}_r R_2 = 455.6 - 3071 \times 0.026526 = 374K\,(101°C)$$

$$r = 0.35m, T_4 = T_3 - \dot{Q}_r R_3 = 374 - 3071 \times 0.012631 = 335.2K\,(62.2°C)$$

$$r = 0.4m, T_5 = T_4 - \dot{Q}_r R_4 = 335.2 - 3071 \times 0.007105 = 313.4K\,(40.4°C)$$

$$r = 0.45m, T_6 = T_o = 300K\,(27°C)$$

The temperature plots for cases (a) and (b) are shown in Figure E2.4.2.

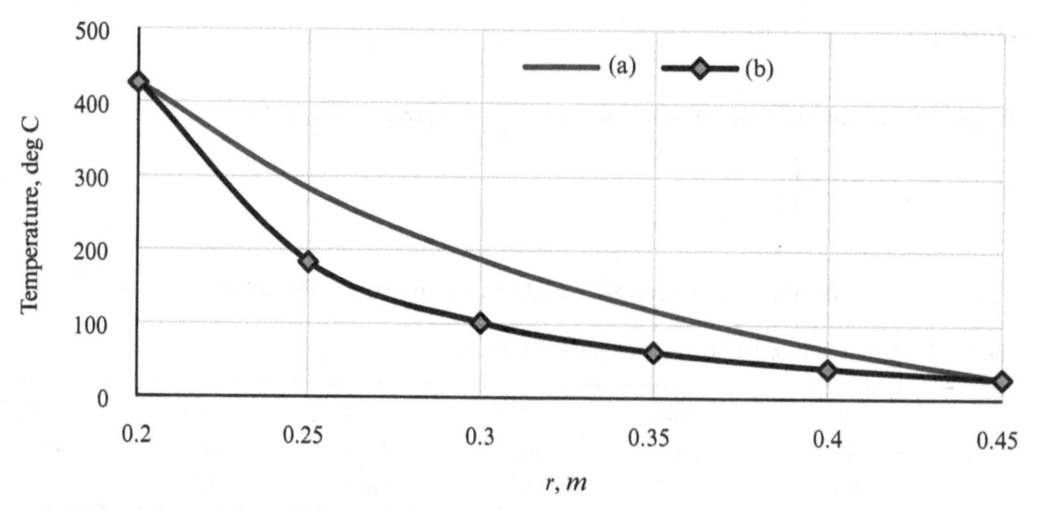

Figure E2.4.2 Temperature plots for cases (a) and (b) in Example 2.4.

2.3.5 Plate With Internal Heat Generation, No Heat Storage, and Uniform Heat Dissipation By Convection

In the previous sections, internal heat generation (or absorption) and heat storage in the general energy equation were ignored to simplify the analyses. However, situations may arise in which heat is generated or absorbed inside an object as a result of the Joule heating effect resulting from a current flowing in a conductor, heat liberation from fuel rods in a nuclear reactor, or heat resulting from the combustion of hydrocarbon fuels.

Ignoring heat storage and assuming constant thermal conductivity, the general law of conservation of energy in a Cartesian coordinate system for a flat plate or wall with internal heat generation is given by Eq. (2.20), which can be reduced for one-dimensional conduction to

$$\frac{\partial}{\partial x}\left(k\frac{\partial T}{\partial x}\right)+\dot{q}=0 \tag{2.78}$$

2.3.5.1 Constant Thermal Conductivity

Consider the plate shown in Figure 2.24 with thickness 2δ and the origin of x at the symmetry axis of the plate. If the dimensions of the plate in the other directions are sufficiently large, it can be assumed that the heat flow is one-dimensional across the plate thickness in the x-direction. Heat is generated internally at the rate of $\dot{q}$ (W/m^3) and the heat transferred through the plate is dissipated by convection from the two walls to the surroundings at temperature T_∞ with the heat transfer coefficient h. It is further assumed that both walls of the plate are uniformly cooled to a constant wall temperature T_w.

Integrating Eq. (2.80) twice with k being constant, we obtain

$$\frac{dT}{dx}+\frac{\dot{q}}{k}x=C_1 \tag{2.79}$$

$$T=-\frac{\dot{q}}{2k}x^2+C_1x+C_2 \tag{2.80}$$

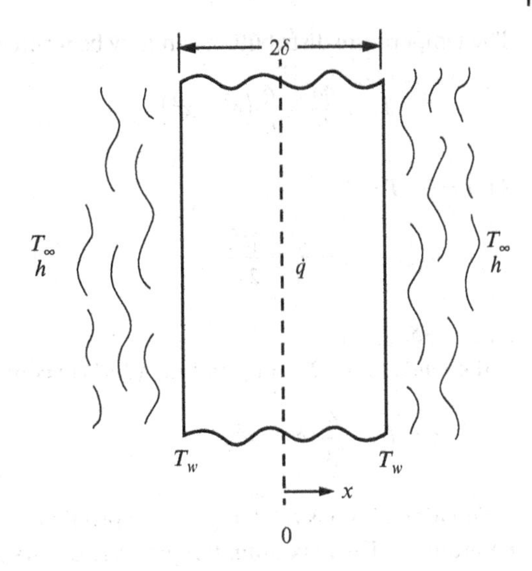

Figure 2.24 Plate with internal heat generation and uniformly cooled walls by convection.

For equal wall temperatures, the temperature profile exhibits a maximum at $x = 0$;

$$\left(\frac{dT}{dx}\right)_{x=0} = 0, T = T_0 = T_{max}, \; C_1 = 0$$

Substituting for C_1 into Eq. (2.79)

$$\frac{dT}{dx} = -\frac{\dot{q}}{k}x$$

and the heat flux q can be written as

$$q = -k\frac{dT}{dx} = \dot{q}x \quad W/m^2$$

At $x = \delta, T = T_w$; hence, from Eq. (2.80),

$$C_2 = T_w + \delta^2 \dot{q}/2k$$

Substituting C_1 and C_2 into Eq. (2.80), we obtain the final equation for the temperature distribution across the plate

$$T = T_w + \frac{\dot{q}}{2k}\left(\delta^2 - x^2\right) \tag{2.81}$$

The energy balance at the plate walls yields

$\dot{q}\delta = h(T_w - T_\infty)$, from which

$$T_w = T_\infty + \frac{\dot{q}\delta}{h} \tag{2.82}$$

The temperature distribution can now be written as

$$T = T_\infty + \frac{\dot{q}\delta}{h} + \frac{\dot{q}}{2k}\left(\delta^2 - x^2\right) \tag{2.83}$$

At $x = 0,\ \ T = T_{max}$

$$T_{max} = T_\infty + \frac{\dot{q}}{h} + \frac{\dot{q}\delta^2}{2k} \tag{2.84}$$

At $x = \pm\delta,\ T = T_w$

If h tends to ∞, $T_w = T_\infty$ and Eq. (2.83) becomes

$$T = T_w + \frac{\dot{q}}{2k}\left(\delta^2 - x^2\right) \tag{2.85}$$

Equation (2.85) is the temperature profile for heat conduction in a plate with known wall temperature T_w. The maximum temperature occurs at $x = 0$ and is given by

$$T_{max} = T_w + \frac{\dot{q}\delta^2}{2k}$$

2.3.5.2 Temperature-Dependent Thermal Conductivity

If we account for temperature dependence of the thermal conductivity as a linear function $k = k_o(1 + bT)$, Eq. (2.78) can be used without further simplification

$$\frac{d}{dx}\left(k\frac{dT}{dx}\right) + \dot{q} = 0$$

Using the same boundary conditions and integrating, it can be shown that the temperature distribution is

$$T = -\frac{1}{b} \pm \sqrt{\left(\frac{1}{b} + T_{max}\right)^2 - \frac{\dot{q}x^2}{k_o b}} \tag{2.86}$$

On the basis of Eq. (2.84), T_{max} is

$$T_{max} = T_\infty + \frac{\dot{q}\delta}{h} + \frac{\dot{q}\delta^2}{2k_m} \tag{2.87}$$

where

$$k_m = k_0\left[1 + \frac{b}{2}(T_w + T_{max})\right] \tag{2.88}$$

Equation (2.86) is solved iteratively by assuming values of T_{max} and calculating T_w (at $x = \mp\delta$) from Eqs (2.82) and (2.86) until the two results converge.

The calculated value of T_{max} can be used to calculate k_m from Eq. (2.88) and then use Eq. (2.87) to calculate T_{max} and compare the result with the iterated value. Once T_{max} is finally determined, the temperature profile can be plotted as a function of x using Eq. (2.86).

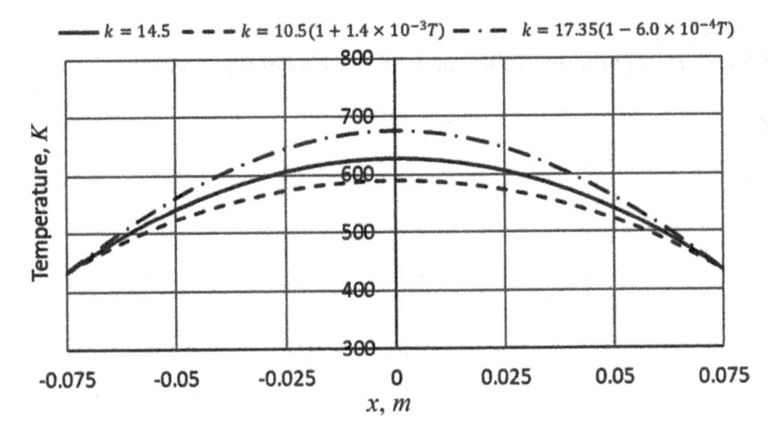

Figure 2.25 Temperature distribution in a plate with internal heat generation $T_\infty = 283K$, $h_1 = h_2 = 500\ W/m^2.K$, $\dot{q} = 10^6\ W/m^3$, $\delta = 0.075\ m$.

The temperature distribution in the wall of a plate of thickness $2\delta = 0.15m$ with origin of x at the symmetry of the plate is shown in Figure 2.25. Three cases are compared

$$k = \text{const.} = 14.5\,W/m.K$$

$$k = 10.5(1 + 1.4 \times 10^{-3} T) \quad b > 0$$

$$k = 17.3(1 - 6.0 \times 10^{-3} T) \quad b < 0$$

The shapes of the temperature profiles for the three cases are similar, with the temperatures in the cylindrical wall being higher or lower than the base case ($k = const$), depending on the sign of b.

2.3.6 Plate With Internal Heat Generation and Non-Uniform Heat Dissipation By Convection

Consider a plate with thickness 2δ (origin of x at the left side of the plate) and sufficiently large dimensions in the other directions to assume the heat flow to be one-dimensional across the plate thickness in the x-direction, as shown in Figure 2.26. Heat is generated internally at the rate of $\dot{q}$ (W/m^3) and the heat transferred through the plate is dissipated by convection from the two walls of unequal temperatures to the surroundings with heat transfer coefficient h_1 and h_2.

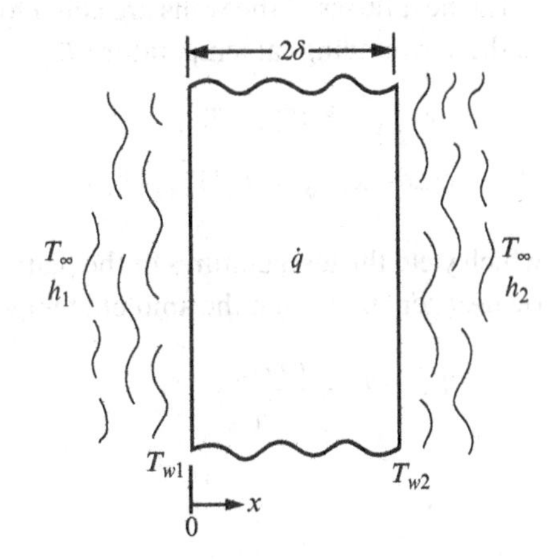

Figure 2.26 Plate with internal heat generation and non-uniformly cooled walls by convection.

2.3.6.1 Constant Thermal Conductivity

As in the case in Section 2.3.5.1, the temperature profile can be written as in Eq. (2.80)

$$T=-\frac{\dot{q}}{2k}x^2+C_1x+C_2$$

The boundary conditions

at $x=0, T=T_{w1}$

at $x=2\delta, T=T_{w2}$

hence,

$$C_2=T_{w1}$$

$$C_1=\frac{\dot{q}\delta}{k}+\frac{T_{w2}-T_{w1}}{2\delta}$$

The temperature distribution is then given by

$$T=-\frac{\dot{q}}{2k}x^2+\left(\frac{\dot{q}\delta}{k}+\frac{T_{w2}-T_{w1}}{2\delta}\right)x+T_{w1} \tag{2.89}$$

The unknown wall temperatures can be determined as follows:

- Find the location of maximum temperature by differentiating Eq. (2.89) and equating the result to zero

$$\frac{dT}{dx}=-\frac{\dot{q}}{k}x+\left(\frac{\dot{q}\delta}{k}+\frac{T_{w2}-T_{w1}}{2\delta}\right)=0$$

from which

$$x_{Tmax}=\frac{k}{2\delta\dot{q}}(T_{w2}-T_{w1})+\delta \tag{2.90}$$

- The heat flux along the x-axis is

$$q=\dot{q}x \quad W/m^2$$

where x is the distance measured from x_{Tmax} (where $dT/dx=0$) to the plate walls.

The heat fluxes at the walls are equal to the heat transferred by convection from the plate walls to the surroundings at temperature T_∞

$$\dot{q}x_{Tmax}=h_1(T_{w1}-T_\infty)$$

$$\dot{q}(2\delta-x_{Tmax})=h_2(T_{w2}-T_\infty)$$

which yield the temperatures of the plate walls in terms of the heat transfer coefficients, volumetric heat generated, and the ambient temperature

$$T_{w1}=T_\infty+\frac{\dot{q}x_{Tmax}}{h_1} \tag{2.91}$$

$$T_{w2} = T_\infty + \frac{\dot{q}(2\delta - x_{Tmax})}{h_2} \tag{2.92}$$

- From Eqs (2.91) and (2.92) we obtain

$$T_{w2} - T_{w1} = \frac{\dot{q}(2\delta - x_{Tmax})}{h_2} - \frac{\dot{q}x_{Tmax}}{h_1} \tag{2.93}$$

- x_{Tmax} can be determined from Eqs (2.90) and (2.93)

$$x_{Tmax} = \frac{2\delta\left(\dfrac{1}{h_2} + \dfrac{\delta}{k}\right)}{\left(\dfrac{1}{h_1} + \dfrac{1}{h_2} + \dfrac{2\delta}{k}\right)} \tag{2.94}$$

- Temperatures T_{w1} and T_{w2} can now be determined from Eqs (2.91) and (2.92) and substituted back into Eq. (2.89) to obtain the temperature as a function of x.

If the origin of x is in the centre of the plate, as in Figure 2.24, it can be shown that the temperature distribution in the plate will be given by

$$T = \frac{\dot{q}}{2k}\left(\delta^2 - x^2\right) + \left(\frac{T_{w2} - T_{w1}}{2\delta}\right)x + \left(\frac{T_{w2} + T_{w1}}{2}\right) \tag{2.95}$$

2.3.6.2 Temperature-Dependent Thermal Conductivity

If we account for temperature dependence of the thermal conductivity as a linear function $k = k_o(1 + bT)$, Eq. (2.78) can be used without further simplification

$$\frac{d}{dx}\left(k\frac{dT}{dx}\right) + \dot{q} = 0$$

Using the same boundary conditions and integrating, it can be shown that the temperature distribution can be written as

$$T = -\frac{1}{b} \pm \sqrt{\left(\frac{1}{b} + T_{w1}\right)^2 - \frac{\dot{q}}{bk_0}x^2 + \frac{2\dot{q}x_{Tmax}}{bk_0}x} \tag{2.96}$$

where, on the basis of Eq. (2.94)

$$x_{Tmax} = \frac{2\delta\left(\dfrac{1}{h_2} + \dfrac{\delta}{k_m}\right)}{\left(\dfrac{1}{h_1} + \dfrac{1}{h_2} + \dfrac{2\delta}{k_m}\right)}$$

$$T_{w1} = T_\infty + \frac{\dot{q}x_{Tmax}}{h_1}$$

$$T_{w2} = T_\infty + \frac{\dot{q}(2\delta - x_{Tmax})}{h_2}$$

$$k_m = k_0\left[1 + \frac{b}{2}(T_{w1} + T_{w2})\right]$$

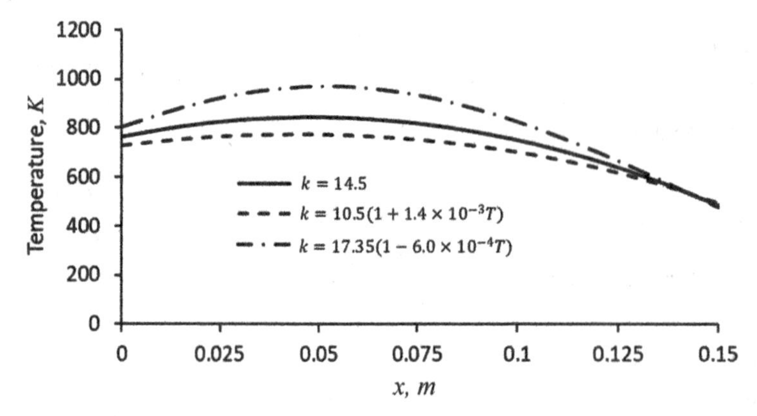

Figure 2.27 Temperature profiles in plate with internal heat generation and heat dissipation by convection $T_\infty = 10°C$, $h_1 = 100\ W/m^2.K, h_2 = 500\ W/m^2.K, \dot{q} = 10^6\ W/m^3$.

Table 2.1 Maximum and wall temperatures in an example of a plate with internal heat generation and non-uniformly cooled walls by convection.

	$k = 14.5\ W/m.K$	$k = 10.5(1+0.0014T)$	$k = 10.5(1-0.00066T)$
x_{Tmax}, m	0.0481	0.0445	0.052
T_{max}, K	844	779	929
T_{w1}, K	764	728	494
T_{w2}, K	487	494	479

The solution is carried out by assuming a value of k_m, calculating x_{Tmax}, wall temperatures T_{w1} and T_{w2}, and iterating until the assumed k_m and the calculated values from the last equation converge. The temperature profile can now be determined by substituting the final values of x_{Tmax} and T_{w1} into Eq. (2.96). The temperature distributions in a plate of non-uniformly cooled walls and of thickness $2\delta = 0.15m$ with origin of x at the left wall of the plate (as per Figure 2.26) are shown in Figure 2.27 for three cases and calculation results summarised in Table 2.1.

The shapes of the temperature profiles for the three cases are similar, with the temperatures in the plane wall being higher or lower than the base case ($k = const$) depending on the sign of b. The temperatures are the highest when $b < 0$, and lowest when $b > 0$. The difference in wall temperatures is due to different values of the heat transfer coefficient at the plate walls.

2.3.7 Solid Cylinder With Internal Heat Generation and Heat Dissipation By Convection

2.3.7.1 Constant Thermal Conductivity

Consider the solid cylinder shown in Figure 2.28 with constant thermal conductivity, internal heat generation at the rate of $\dot{q}$ (W/m^3), and no energy storage. The generated heat is uniformly dissipated by convection to the surroundings at temperature T_∞.

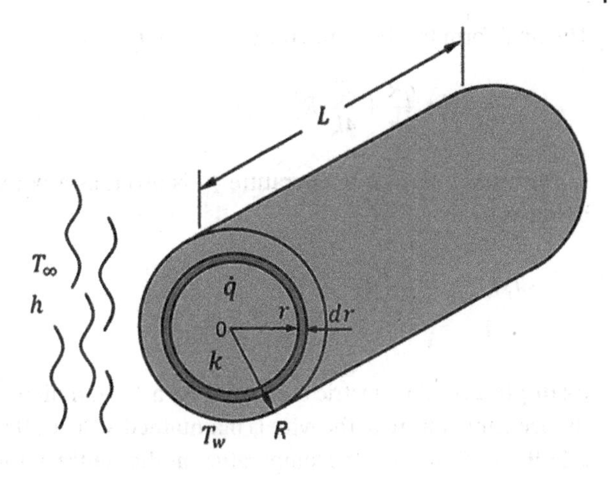

Figure 2.28 Solid cylinder with internal heat generation and convection heat dissipation.

If the length to radius ratio L/R is sufficiently large, the heat flow by conduction can be assumed one-dimensional in the r-direction and Eq. (2.18) is reduced to

$$\frac{\partial}{\partial r}\left(r\frac{\partial T}{\partial r}\right)+\frac{\dot{q}r}{k}=0$$

Integrating this equation twice yields

$$\frac{dT}{dr}+\frac{\dot{q}r}{2k}=\frac{C_1}{r} \tag{2.97}$$

$$T=-\frac{\dot{q}r^2}{4k}+C_1\ln r+C_2 \tag{2.98}$$

where C_1 and C_2 are the constants of integration.

Since the temperature profile is symmetrical relative to the centre,
$(dT/dr)_{r=0}=0$, and $C_1=0$
at $r=R$, from Eq. (2.97)

$$\left(\frac{dT}{dr}\right)_{r=R}=-\frac{\dot{q}R}{2k}$$

The rate of heat conduction at $r=R$ should be equal to the heat dissipated by convection from the outer surface; hence,

$$\dot{Q}=-kA\left(\frac{dT}{dr}\right)_{r=R}=\frac{\dot{q}RA}{2}=Ah(T_w-T_\infty)$$

from which the temperature at the wall is

$$T_w=T_\infty+\frac{\dot{q}R}{2h}$$

At $r=R$, $T=T_w$; hence, C_2 can be determined from Eq. (2.98)

$$C_2=T_\infty+\frac{\dot{q}R}{2h}+\frac{\dot{q}R^2}{4k}$$

The final form for the temperature distribution is

$$T_r = T_\infty + \frac{\dot{q}R}{2h} + \frac{\dot{q}}{4k}\left(R^2 - r^2\right) \tag{2.99}$$

If the outer surface temperature T_w is given as a boundary condition, the temperature profile becomes

$$T_r = T_w + \frac{\dot{q}}{4k}\left(R^2 - r^2\right) \tag{2.100}$$

Example 2.5 An electric current heats a 10-mm diameter wire at a rate of $2.4\times10^7\,W/m^3$. The surface temperature of the wire is maintained at 30°C. If the thermal conductivity of the conductor is $15\,W/m.K$, what is the temperature at the centre of the wire?

Solution

From Eq. (2.100)

$$T_r = T_w + \frac{\dot{q}}{4k}\left(R^2 - r^2\right)$$

At $r = 0$

$$T_r = 303 + \frac{2.4\times10^7}{4\times15}0.005^2 = 313K(40°C)$$

2.3.7.2 Temperature-Dependent Thermal Conductivity

If temperature dependence of the thermal conductivity is taken as $k = k_0(1 + bT)$, the temperature profile can then be written as

$$T_r = -\frac{1}{b} \pm \sqrt{\left(\frac{1}{b} + T_w\right)^2 + \frac{\dot{q}}{2bk_0}\left(R^2 - r^2\right)} \tag{2.101}$$

The square root is positive for $b > 0$, and negative for $b < 0$.

Figure 2.29 shows the calculated temperature profiles in a solid cylinder of radius $r_o = 0.075m$ in the radial direction for different thermal conductivities. The maximum temperature is at the centre for the three cases, being higher when $b < 0$ and lower when $b > 0$, compared with the value when $k = \text{const.}(b = 0)$.

2.3.8 Hollow Cylinder With Internal Heat Generation and Heat Dissipation By Convection From the Outer Surface

Figure 2.30 is a schematic diagram of a hollow cylinder of length L with the inner surface at temperature T_{w2} and convection heat transfer from the outer surface (heat transfer coefficient h_2) to the surroundings at temperature $T_{\infty 2}$. The internally generated energy is $\dot{q}\,W/m^3$.

2.3.8.1 Constant Thermal Conductivity

Equation (2.98) and its derivative also apply in this case

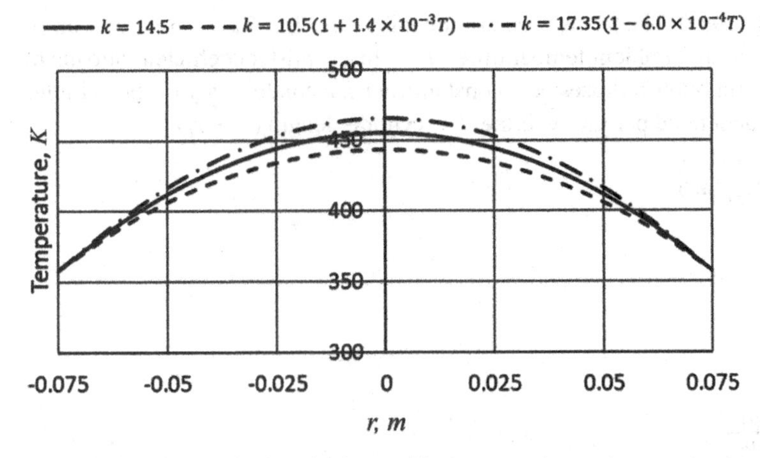

Figure 2.29 Solid cylinder with internal heat generation and convection heat dissipation to the surroundings $T_\infty = 10°C$, $h = 500\ W/m^2.K$, $\dot{q} = 1.0 \times 10^6\ W/m^3$.

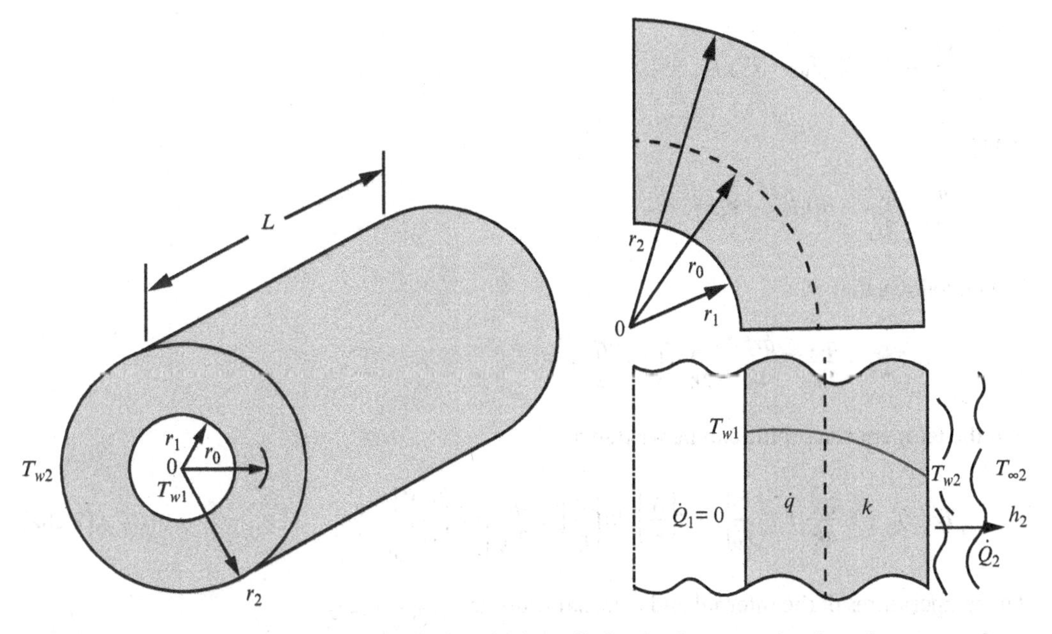

Figure 2.30 Heat transfer in a hollow cylinder with heat dissipation by convection from the outer surface.

$$T = -\frac{\dot{q}}{4k} r^2 + C_1 \ln r + C_2$$

$$\frac{dT}{dr} = -\frac{\dot{q}}{2k} r + \frac{C_1}{r}$$

The values of the constants C_1 and C_2 are dependent on the boundary conditions of the problem. The approach to the solution is similar to the solid cylinder, and the solutions for the case depicted in Figure 2.30 and the ones depicted in Figures 3.32 and 3.34 are broadly based on the methodology outlined in Isachenko et al. (1969). The terms used are the following: T_{w1}, T_{w2} – inner and outer surface temperatures; $T_{\infty 2}$ – outer ambient temperatures, h_2 – heat transfer coefficient outside of the cylinder; r_1, r_2 – inner and outer radiuses; k – constant thermal conductivity of the cylinder material; $\dot{q}$ – internal heat generated per unit volume, L – cylinder length ($L \gg r_2$).

The boundary conditions are:

at $r = r_1, \dot{Q}_1 = 0$, $(dT/dr)_{r=r_1} = 0$

hence,

$$C_1 = \frac{\dot{q}r_1^2}{2k}$$

At $r = r_2, T = T_{w2}$, and

$$\left(\frac{dT}{dr}\right)_{r=r_2} = -\frac{\dot{q}r_2}{2k} + \frac{\dot{q}r_1^2}{2kr_2}$$

$$\frac{\dot{Q}_2}{L} = -2\pi r_2 k\left(\frac{dT}{dr}\right)_{r=r_2}$$

$$\frac{\dot{Q}_2}{L} = 2\pi r_2 h_2 \left(T_{w2} - T_{\infty 2}\right)$$

hence,

$$\frac{\dot{q}r_2}{2} - \frac{\dot{q}r_1^2}{2r_2} = h_2 \left(T_{w2} - T_{\infty 2}\right)$$

It then follows that

$$C_2 = T_{\infty 2} + \frac{\dot{q}r_2}{2h_2} + \frac{\dot{q}r_2^2}{4k} - \frac{\dot{q}}{2h_2}\frac{r_1^2}{r_2} - \frac{\dot{q}r_1^2}{2k}\ln r_2$$

and the temperature profile can be written as

$$T_r = T_{\infty 2} + \frac{\dot{q}r_2^2}{4k}\left[1 - \left(\frac{r}{r_2}\right)^2 + 2\left(\frac{r_1}{r_2}\right)^2 \ln\left(\frac{r}{r_2}\right)\right] + \frac{\dot{q}r_2}{2h_2}\left[1 - \left(\frac{r_1}{r_2}\right)^2\right] \tag{2.102}$$

The temperatures of the internal and external walls are, respectively

$$T_{w1} = T_{\infty 2} + \frac{\dot{q}r_2^2}{4k}\left[1 - \left(\frac{r_1}{r_2}\right)^2 + 2\left(\frac{r_1}{r_2}\right)^2 \ln\left(\frac{r_1}{r_2}\right)\right] + \frac{\dot{q}r_2}{2h_2}\left[1 - \left(\frac{r_1}{r_2}\right)^2\right] \tag{2.103}$$

$$T_{w2} = T_{\infty 2} + \frac{\dot{q}r_2}{2h_2}\left[1 - \left(\frac{r_1}{r_2}\right)^2\right] \tag{2.104}$$

If the temperature of the outer wall T_{w2} is specified as a boundary condition instead of convective heat dissipation, the heat transfer coefficient for the outer surface can be equated to infinity ($h_2 = \infty$), which yields $T_{w2} = T_{\infty 2}$ and the temperature profile acquires the form

$$T_r = T_{w2} + \frac{\dot{q}r_2^2}{4k}\left[1 - \left(\frac{r}{r_2}\right)^2 + 2\left(\frac{r_1}{r_2}\right)^2 \ln\left(\frac{r}{r_2}\right)\right] \tag{2.105}$$

The heat conduction driving potential across the cylinder wall is the difference between the wall temperatures ($T_{w1} - T_{w2}$), which can be determined by substituting $r = r_1$ into Eq. (2.105), multiplying and dividing the contents in the square bracket by r_1^2, and rearranging

$$T_{w1} - T_{w2} = \frac{\dot{q}r_1^2}{4k}\left[\left(\frac{r_2}{r_1}\right)^2 - 2\ln\left(\frac{r_2}{r_1}\right) - 1\right] \tag{2.106}$$

2.3.8.2 Temperature-Dependent Thermal Conductivity

If the thermal conductivity is temperature-dependent and is defined as $k = k_0(1 + bT)$, the temperature profile will then be given by

$$T_r = -\frac{1}{b} \pm \sqrt{\left(\frac{1}{b} + T_{w2}\right)^2 + \frac{\dot{q}r_2^2}{2bk_0}\left[1 - \left(\frac{r}{r_2}\right)^2 + 2\left(\frac{r_1}{r_2}\right)^2 \ln\left(\frac{r}{r_2}\right)\right]} \tag{2.107}$$

The temperature T_{w2} can be specified as a boundary condition, or can be determined from Eq. (2.104) if heat is dissipated from the outer surface by convection.

Figure 2.31 shows the temperature profiles in the wall of a hollow cylinder with heat dissipation by convection from the outer surface for constant and temperature-dependent thermal conductivities calculated using Eqs. (2.102) and (2.107), respectively.

2.3.9 Hollow Cylinder With Internal Heat Generation and Heat Dissipation By Convection From the Inner Surface

Figure 2.32 is a schematic diagram of a hollow cylinder of length L with outer surface temperature T_{w2} and convection heat transfer from the inner surface at temperature T_{w1} and heat transfer coefficient h_1 to the surroundings at temperature $T_{\infty 1}$. The internally generated energy is $\dot{q}W/m^3$.

2.3.9.1 Constant Thermal Conductivity

As before, the temperature profile is given by

$$T_r = -\frac{\dot{q}}{4k}r^2 + C_1 \ln r + C_2$$

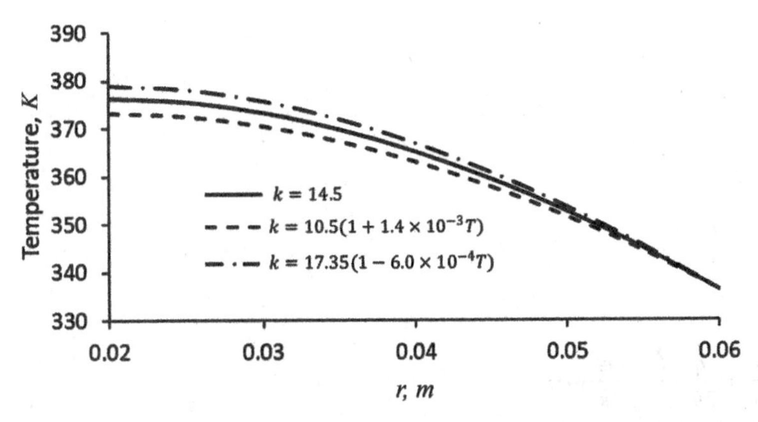

Figure 2.31 Temperature profiles in a hollow cylinder with heat dissipation by convection from the outer surface $r_1 = 0.02\ m$, $r_2 = 0.06\ m$, $\dot{q} = 10^6\ W/m^3$, $T_{\infty 2} = 10°C$, $h_2 = 500\ W/m^2.K$.

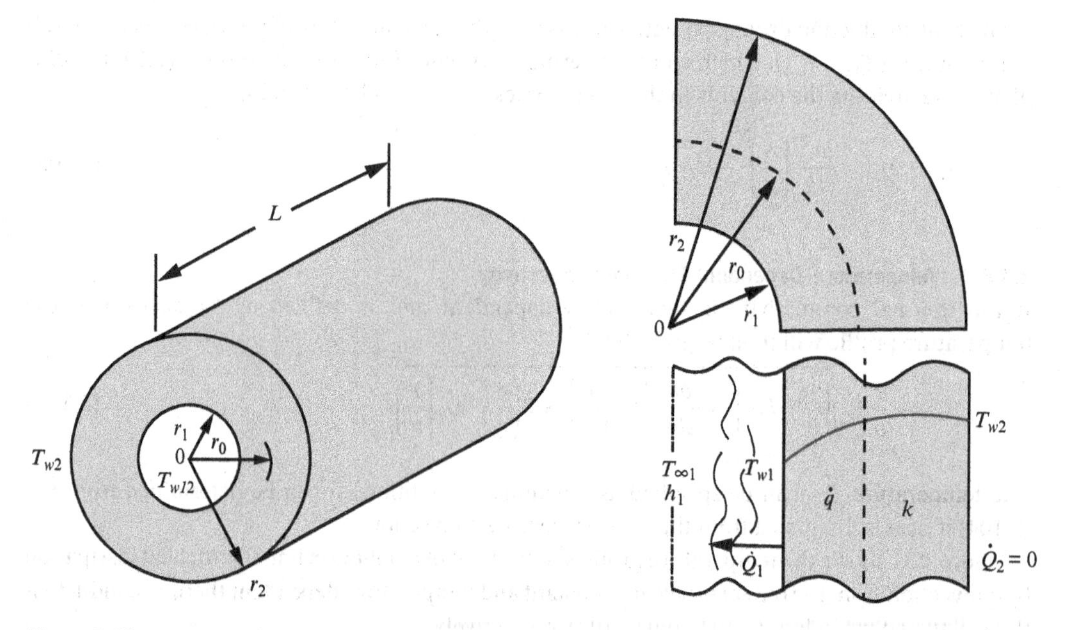

Figure 2.32 Heat transfer in a hollow cylinder with heat dissipation by convection from the inner surface.

The boundary conditions in this case are

At $r = r_2, \dot{Q}_2 = 0,\ (dT/dr)_{r=r_2} = 0$

hence,

$$C_1 = \frac{\dot{q}r_2^2}{2k}$$

At $r = r_1, T = T_{w1}$, and

$$\left(\frac{dT}{dr}\right)_{r=r_1} = -\frac{\dot{q}r_1}{2k} + \frac{\dot{q}r_2^2}{2kr_1}$$

$$\frac{\dot{Q}_1}{L} = 2\pi r_1 k\left(\frac{dT}{dr}\right)_{r=r_1}$$

$$\frac{\dot{Q}_1}{L} = 2\pi r_1 h_1 (T_{w1} - T_{\infty 1})$$

hence,

$$-\frac{\dot{q}r_1}{2} + \frac{\dot{q}r_2^2}{2r_1} = h_1 (T_{w1} - T_{\infty 1})$$

It then follows that

$$C_2 = T_{\infty 1} - \frac{\dot{q}r_1}{2h_1} + \frac{\dot{q}r_1^2}{4k} + \frac{\dot{q}}{2h_1}\frac{r_2^2}{r_1} - \frac{\dot{q}r_2^2}{2k}\ln r_1$$

and the temperature profile

$$T_r = T_{\infty 1} + \frac{\dot{q} r_1^2}{4k}\left[1 - \left(\frac{r}{r_1}\right)^2 + 2\left(\frac{r_2}{r_1}\right)^2 \ln\left(\frac{r}{r_1}\right)\right] + \frac{\dot{q} r_1}{2h_1}\left[\left(\frac{r_2}{r_1}\right)^2 - 1\right] \quad (2.108)$$

The temperatures of the inner and outer surfaces are, respectively

$$T_{w1} = T_{\infty 1} + \frac{\dot{q} r_1}{2h_1}\left[\left(\frac{r_2}{r_1}\right)^2 - 1\right] \quad (2.109a)$$

$$T_{w2} = T_{\infty 1} + \frac{\dot{q} r_1^2}{4k}\left[1 - \left(\frac{r_2}{r_1}\right)^2 + 2\left(\frac{r_2}{r_1}\right)^2 \ln\left(\frac{r_2}{r_1}\right)\right] + \frac{\dot{q} r_1}{2h_1}\left[\left(\frac{r_2}{r_1}\right)^2 - 1\right] \quad (2.109b)$$

If the temperature of the inner wall T_{w1} is specified as a boundary condition instead of convective heat dissipation, the heat transfer coefficient for the inner surface can be equated to infinity ($h_1 = \infty$), which yields $T_{w1} = T_{\infty 1}$, and the temperature profile acquires the form

$$T_r = T_{w1} + \frac{\dot{q} r_1^2}{4k}\left[1 - \left(\frac{r}{r_1}\right)^2 + 2\left(\frac{r_2}{r_1}\right)^2 \ln\left(\frac{r}{r_1}\right)\right] \quad (2.110)$$

The heat conduction driving potential across the cylinder wall is the difference between the wall temperatures ($T_{w2} - T_{w1}$), which can be determined by substituting $r = r_2$ in Eq. (2.110), multiplying and dividing the content in the square brackets by r_2^2, and rearranging

$$T_{w2} - T_{w1} = \frac{\dot{q} r_2^2}{4k}\left[\left(\frac{r_2}{r_1}\right)^2 + 2\ln\left(\frac{r_2}{r_1}\right) - 1\right] \quad (2.111)$$

2.3.9.2 Temperature-Dependent Thermal Conductivity

If the thermal conductivity is temperature-dependent and is defined as $k = k_0(1 + bT)$, the temperature profile will then be given by

$$T_r = -\frac{1}{b} \pm \sqrt{\left(\frac{1}{b} + T_{w1}\right)^2 + \frac{\dot{q} r_1^2}{2bk_0}\left[1 - \left(\frac{r}{r_1}\right)^2 + 2\left(\frac{r_2}{r_1}\right)^2 \ln\left(\frac{r}{r_1}\right)\right]} \quad (2.112)$$

The temperature T_{w1} can be a specified boundary condition, or can be determined from Eq. (2.109a) if heat is dissipated from the inner surface by convection.

Figure 2.33 shows the temperature profiles in the wall of a hollow cylinder with heat dissipation by convection from the inner surface for constant and temperature-dependent thermal conductivities calculated from Eqs (2.108) and (2.112), respectively.

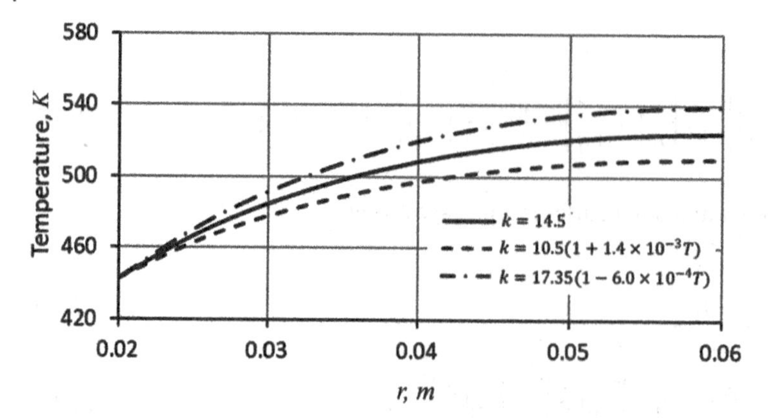

Figure 2.33 Temperature profiles in a hollow cylinder with heat dissipation by convection from the inner surface. $r_1 = 0.02\,m$, $r_2 = 0.06\,m$, $\dot{q} = 10^6\,W/m^3$, $T_{\infty 1} = 10°C$, $h_1 = 500\,W/m^2.K$.

2.3.10 Hollow Cylinder With Internal Heat Generation and Heat Dissipation By Convection From Both Inner and Outer Surfaces

Figure 2.34 is a schematic diagram of a hollow cylinder of length L, inner radius r_1 and outer radius r_2. Heat is dissipated by convection heat transfer from, respectively, the inner and outer surfaces (heat transfer coefficients h_1, h_2) to the surroundings at temperatures $T_{\infty 1}$ and $T_{\infty 2}$. The internally generated energy is $\dot{q}\ W/m^3$.

2.3.10.1 Constant Thermal Conductivity

The schematic diagram for this case in shown in Figure 2.34. Because of heat losses $\dot{Q}_1$, $\dot{Q}_2$ to the surroundings from the inner and outer surfaces of the hollow cylinder, the temperature profile exhibits a maximum value of T_{max} in the solid at $r = r_0$. Conceptually, the isothermal surface at $r = r_0$ separates the cylinder wall into two layers with two separate heat flows:

1) Heat flow towards the outer surface at $r = r_2$ with the driving potential $(T_{max} - T_{w2})$
2) Heat flow towards the inner surface at $r = r_1$ with the driving potential $(T_{max} - T_{w1})$.

From Eqs (2.106) and (2.111)

$$T_{max} - T_{w2} = \frac{\dot{q} r_0^2}{4k}\left[\left(\frac{r_2}{r_0}\right)^2 - 2\ln\left(\frac{r_2}{r_0}\right) - 1\right] \tag{2.113}$$

$$T_{max} - T_{w1} = \frac{\dot{q} r_0^2}{4k}\left[\left(\frac{r_1}{r_0}\right)^2 + 2\ln\left(\frac{r_0}{r_1}\right) - 1\right] \tag{2.114}$$

As before, the general equation of the temperature distribution and its derivative are, respectively

$$T = -\frac{\dot{q}}{4k} r^2 + C_1 \ln r + C_2 \tag{2.115}$$

$$\frac{dT}{dr} = -\frac{\dot{q}}{2k} r + \frac{C_1}{r}$$

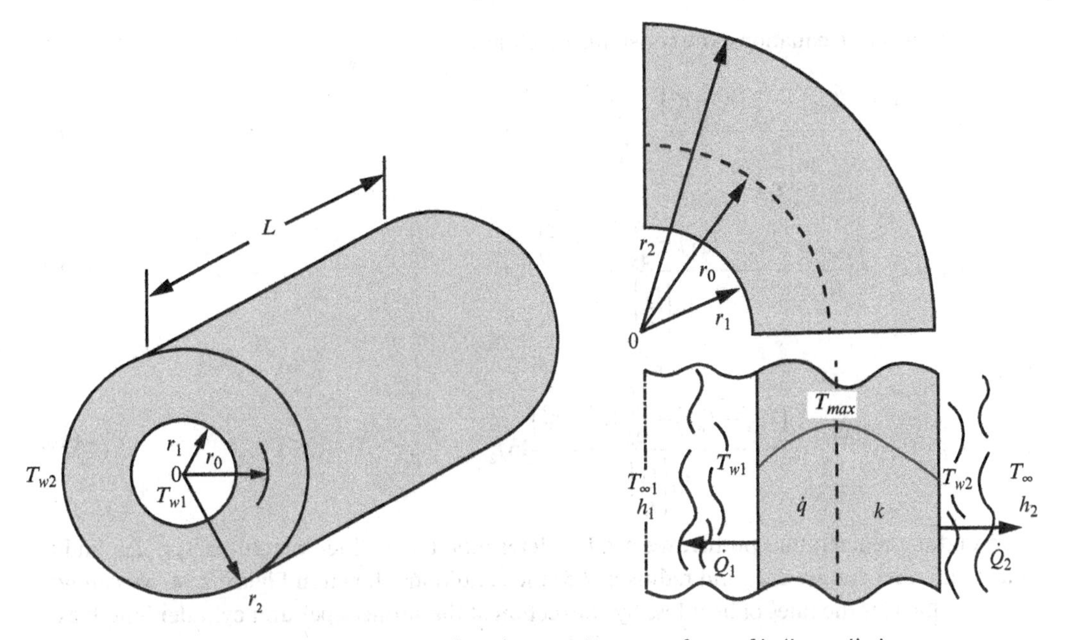

Figure 2.34 Heat dissipation by convection from inner and outer surfaces of hollow cylinder.

At $r = r_0$, $(dT/dr)_{r=r_0} = 0$
hence,

$$C_1 = \frac{\dot{q}r_0^2}{2k}$$

At $r = r_1, T = T_{w1}$, and

$$\left(\frac{dT}{dr}\right)_{r=r_1} = -\frac{\dot{q}r_1}{2k} + \frac{\dot{q}r_0^2}{2kr_1}$$

$$\frac{\dot{Q}_1}{L} = 2\pi r_1 k\left(\frac{dT}{dr}\right)_{r=r_1} = \pi\dot{q}\left(r_0^2 - r_1^2\right)$$

$$\frac{\dot{Q}_1}{L} = 2\pi r_1 h_1 (T_{w1} - T_{\infty 1})$$

At $r = r_2, T = T_{w2}$, and

$$\left(\frac{dT}{dr}\right)_{r=r_2} = -\frac{\dot{q}r_2}{2k} + \frac{\dot{q}r_1^2}{2kr_2}$$

$$\frac{\dot{Q}_2}{L} = -2\pi r_2 k\left(\frac{dT}{dr}\right)_{r=r_2} = \pi\dot{q}\left(r_2^2 - r_0^2\right)$$

$$\frac{\dot{Q}_2}{L} = 2\pi r_2 h_2 (T_{w2} - T_{\infty 2})$$

Based on the above equations, the constants C_1, C_2 are

$$C_1 = \frac{(T_{w2} - T_{w1}) + \frac{\dot{q}}{4k}\left(r_2^2 - r_1^2\right)}{\ln\left(\frac{r_2}{r_1}\right)} \tag{2.116}$$

$$C_2 = T_{w1} + \frac{\dot{q}r_1^2}{4k} - \frac{(T_{w2} - T_{w1}) + \frac{\dot{q}}{4k}\left(r_2^2 - r_1^2\right)}{\ln\left(\frac{r_2}{r_1}\right)} \ln r_1 \tag{2.117a}$$

or

$$C_2 = T_{w2} + \frac{\dot{q}r_2^2}{4k} - \frac{(T_{w2} - T_{w1}) + \frac{\dot{q}}{4k}\left(r_2^2 - r_1^2\right)}{\ln\left(\frac{r_2}{r_1}\right)} \ln r_2 \tag{2.117b}$$

To find the temperature profile, we need to determine the wall temperatures T_{w1}, T_{w2} and the maximum temperature T_{max} and radius r_o. For the conditions shown in Figure 2.34, we can write the equations for the rates of heat loss by convection at the surfaces per unit cylinder length as

$$\frac{\dot{Q}_1}{L} = 2\pi r_1 h_1 (T_{w1} - T_{\infty 1}) = \dot{q}\pi\left(r_0^2 - r_1^2\right) \tag{2.118}$$

$$\frac{\dot{Q}_2}{L} = 2\pi r_1 h_2 (T_{w2} - T_{\infty 2}) = \dot{q}\pi\left(r_2^2 - r_0^2\right) \tag{2.119}$$

from which

$$T_{w1} - T_{w2} = (T_{\infty 1} - T_{\infty 2}) + \frac{\dot{q}}{2}\left[\frac{\left(r_o^2 - r_1^2\right)}{r_1 h_1} - \frac{\left(r_2^2 - r_o^2\right)}{r_2 h_2}\right] \tag{2.120}$$

Also, subtracting Eq. (2.114) from Eq. (2.113), we obtain

$$T_{w1} - T_{w2} = \frac{\dot{q}r_0^2}{4k}\left[\left(\frac{r_2}{r_0}\right)^2 - \left(\frac{r_1}{r_0}\right)^2 + 2\ln\left(\frac{r_0}{r_2}\right) - 2\ln\left(\frac{r_0}{r_1}\right)\right] \tag{2.121}$$

The solution procedure is as follows:

1) Equating the right-hand sides of Eqs (2.120) and (2.121) yields r_0 (this can be easily done graphically in a spreadsheet)
2) Knowing r_0, temperatures T_{w1}, T_{w2} can be calculated from Eqs (2.118) and (2.119)
3) The maximum temperature is calculated from Eq. (2.113) or Eq. (2.114)
4) Constant C_1 is calculated from Eq. (2.116) and constants C_2 from Eq. (2.117a) or Eq. (2.117b)
5) For $r_0 < r < r_2$, the temperature profile is plotted using Eq. (2.102)
6) For $r_1 < r < r_0$, the temperature profile is plotted using Eq. (2.108).

Figure 2.35 shows the temperature profiles and computer modelling data for two distinct cases for a hollow cylinder with heat dissipation by convection from the inner and outer surfaces. Depending on the input data, the position and magnitude of the maximum temperature could be closer to the inner surface (Figure 2.35a) or to the outer surface (Figure 2.35b).

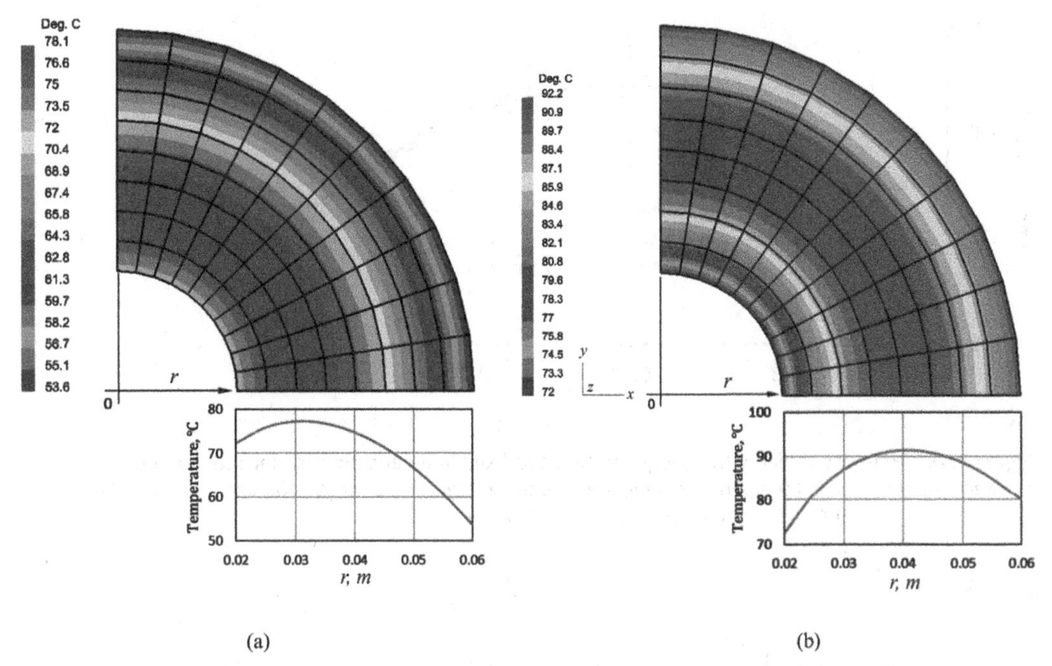

Figure 2.35 Hollow cylinder with heat dissipation by convection from inner and outer surfaces. $r_1 = 0.02\ m$, $r_2 = 0.06\,m$, $\dot{q} = 10^6\ W, k = 10\ W/m.K$; (a) $T_{\infty 1} = 15°C$, $T_{\infty 2} = 10°C, h_1 = 250\ W/m^2.K$, $h_2 = 500\ W/m^2.K$; (b) $T_{\infty 1} = 10°C$, $T_{\infty 2} = 15°C, h_1 = 500\ W/m^2.K$, $h_2 = 100\ W/m^2.K$.

The methodology outlined for a hollow cylinder in this section can also be used if the temperatures of the inner and outer walls are specified as boundary conditions instead of convection heat dissipation. This is done by making the heat transfer coefficients h_1 and h_2 equal to infinity, replacing the ambient temperatures $T_{\infty 1}$ and $T_{\infty 2}$ by the specified wall temperatures, calculating r_0 from Eq. (2.121), and calculating T_{max} from Eq. (2.113) or Eq. (2.114). The final temperature profiles are then plotted as follows:

For $r_0 < r < r_2$, the temperature profile is plotted using Eq. (2.102).
For $r_1 < r < r_0$, the temperature profile is plotted using Eq. (2.108).

2.3.10.2 Temperature-Dependent Thermal Conductivity

If the thermal conductivity is given as $k = k_0(1 + bT)$, the temperature profile can be worked out from the one-dimensional conduction equation in the cylindrical coordinate system and written as

$$\frac{b}{2}T^2 + T + \left(\frac{\dot{q}r^2}{4k_0} - \frac{C_1}{k_0}\ln r - \frac{C_2}{k_0}\right) = 0 \tag{2.122}$$

The constants C_1 and C_2 can be determined from the boundary conditions $T = T_{w1}$ at $r = r_1$ and $T = T_{w2}$ at $r = r_2$.

If

$$X = k_0\left(T_{w1} + \frac{b}{2}T_{w1}^2\right)$$

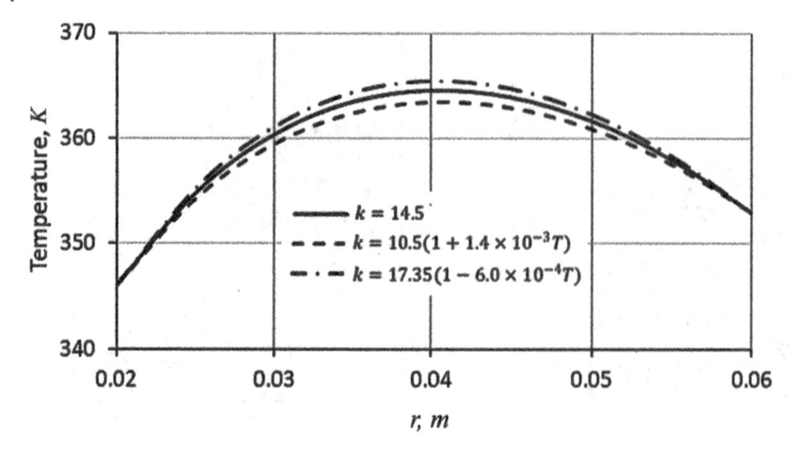

Figure 2.36 Hollow cylinder with heat generation and fixed inner and outer surface temperatures: constant conductivity and temperature dependent conductivity $r_1 = 0.02\,m$, $r_2 = 0.06\,m$, $\dot{q} = 10^6\,W/m^3$, $T_{\infty1} = 15°C$, $T_{\infty2} = 10°C$, $h_1 = 500\ W/m^2.K$, $h_2 = 250\ W/m^2.K$.

$$Y = k_0\left(T_{w2} + \frac{b}{2}T_{w2}^2\right)$$

the constants C_1 and C_2 can then be written as

$$C_1 = \frac{(X - Y) + \frac{\dot{q}}{4}\left(r_1^2 - r_2^2\right)}{\ln\left(\frac{r_1}{r_2}\right)}$$

$$C_2 = X + \frac{\dot{q}r_1^2}{4} - \frac{(X - Y) + \frac{\dot{q}}{4}\left(r_1^2 - r_2^2\right)}{\ln\left(\frac{r_1}{r_2}\right)}\ln r_1$$

The solution to Eq. (2.122) is

$$T_r = -\frac{1}{b} - \sqrt{\frac{1}{b^2} - \frac{2}{bk_0}\left(\frac{\dot{q}r^2}{4} - C_1 \ln r - C_2\right)} \tag{2.123}$$

Figure 2.36 compares the temperature profiles for the hollow cylinder with heat generation, constant conductivity, and temperature-dependent conductivity plotted using Eqs (2.115) and (2.123). Heat is dissipated by convection from the inner and outer surfaces to the surroundings at different temperatures.

2.3.11 Solid Sphere With Internal Heat Generation and Heat Dissipation By Convection and No Heat Storage

Figure 2.37 shows the schematic diagram for this case. The sphere of radius R is internally heated at the rate of $\dot{q}\ W/m^3$ and the heat generated is dissipated at the surface by convection (heat transfer coefficient h) to the surroundings at temperature T_∞.

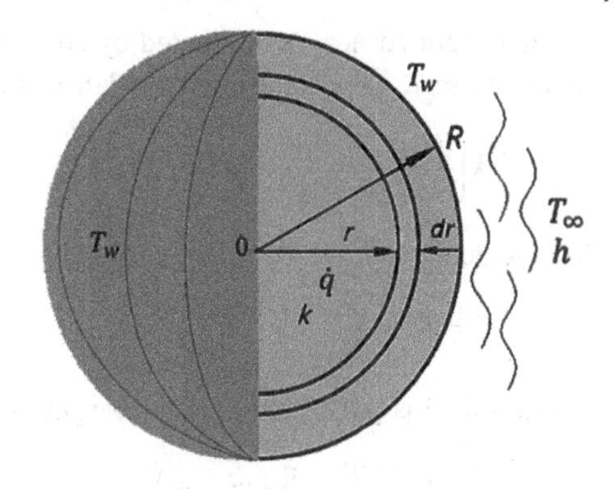

Figure 2.37 Conduction heat transfer in solid sphere with internal heat generation with convection on the outside.

2.3.11.1 Constant Thermal Conductivity

One-dimensional conduction equation with heat generation and constant conductivity in spherical coordinate system is

$$\frac{1}{r^2}\frac{d}{dr}\left(r^2\frac{dT}{dr}\right)+\frac{\dot{q}}{k}=0 \tag{2.124}$$

or

$$\frac{d}{dr}\left(r^2\frac{dT}{dr}\right)+\frac{\dot{q}r^2}{k}=0$$

Integrating yields

$$\frac{dT}{dr}+\frac{\dot{q}r}{3k}=\frac{C_1}{r^2} \tag{2.125}$$

Second integration finally yields

$$T=-\frac{\dot{q}r^2}{6k}-\frac{C_1}{r}+C_2 \tag{2.126}$$

C_1 and C_2 are the constants of integration.

At $r=0$, $dT/dr=0$

At $r=R$, $T=T_w$

hence, from Eqs (2.125) and (2.126)

$$C_1=0$$

$$C_2=T_w+\frac{\dot{q}R^2}{6k}$$

$$\left(\frac{dT}{dr}\right)_{r=R}=-\frac{\dot{q}R}{3k}$$

The equation for the temperature distribution in a solid sphere with internal heat generation is therefore

$$T=T_w+\frac{\dot{q}}{6k}\left(R^2-r^2\right) \tag{2.127}$$

The generated heat is dissipated by convection from the outer surface to the surroundings at temperature T_∞. From the energy balance at the surface ($r = R$),

$$-k\left(\frac{dT}{dr}\right)_{r=R} = \frac{\dot{q}R}{3} = h(T_w - T_\infty)$$

from which

$$T_w = T_\infty + \frac{\dot{q}R}{3h} \tag{2.128}$$

Combining Eqs (2.127) and (2.128), we get the temperature distribution equation

$$T = T_\infty + \frac{\dot{q}R}{3h} + \frac{\dot{q}}{6k}\left(R^2 - r^2\right) \tag{2.129}$$

Figure 2.38 shows the computer modelling results obtained for this case.

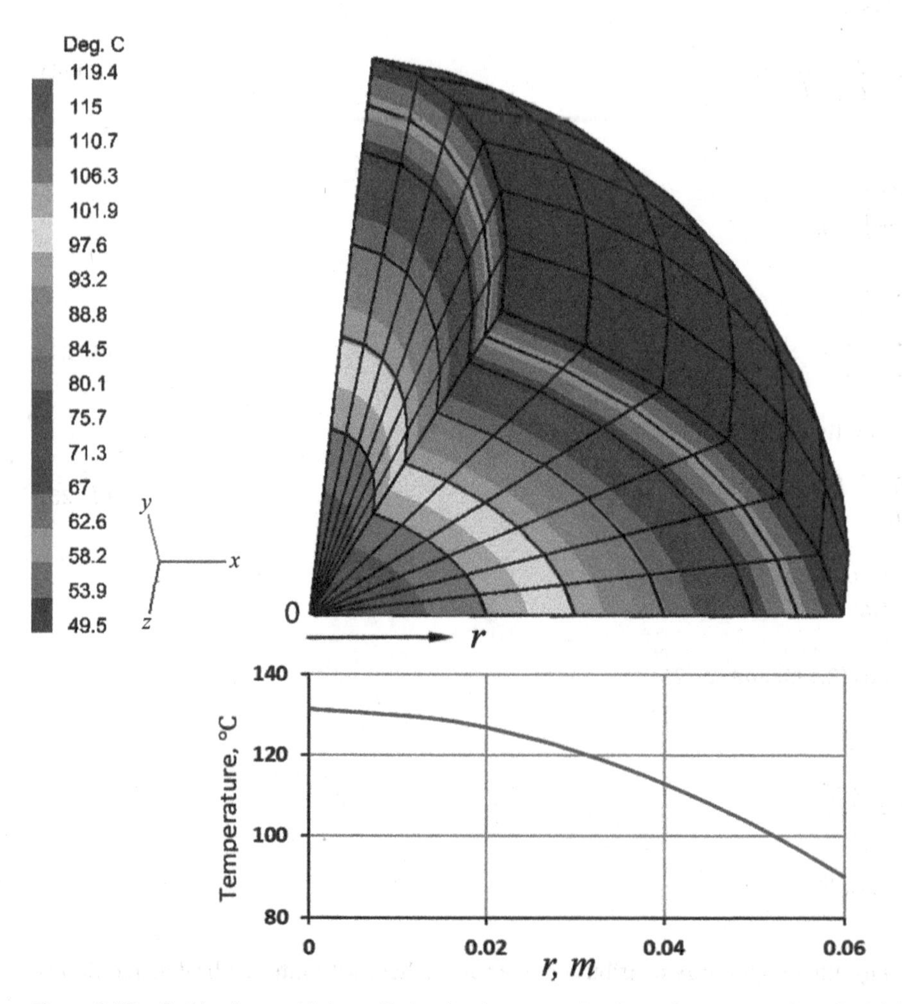

Figure 2.38 Solid sphere with heat dissipation by convection from the outer surface R=0.06 m, $\dot{q} = 10^6\, W/m^3, k = 14.5\, W/m.K, h = 500\, W/m^2.K,\ T_\infty = 10°C$.

Example 2.6 A solid steel sphere of 0.3 m diameter is cooled by convection ($h = 550\ W/m^2.K$) in air at $288\,K$. An internal heater generates $2\,MW/m^3$ in the sphere. What is the temperature at the centre and at a radial distance of 0.075 m from the centre? Take the thermal conductivity of steel as $14.5\,W/m.K$.

Solution

From Eq. (2.129) at $r = 0$

$$T_{r=0} = T_\infty + \frac{\dot{q}R}{3h} + \frac{\dot{q}}{6k}\left(R^2 - 0\right) = 288 + \frac{2\times10^6\times0.15}{3\times550} + \frac{2\times10^6}{6\times14.5}0.15^2 = 987K$$

$$T_{r=0.075} = T_\infty + \frac{\dot{q}R}{3h} + \frac{\dot{q}}{6k}\left(R^2 - r^2\right) = 288 + \frac{2\times10^6\times0.15}{3\times550} + \frac{2\times10^6}{6\times14.5}(0.15^2 - 0.075^2)$$

$$T_{r=0.075} = 858K$$

2.3.11.2 Temperature-Dependent Thermal Conductivity

If the thermal conductivity is temperature dependent and is defined as $k = k_0(1 + bT)$, the temperature profile can be worked out from the one-dimensional conduction equation in spherical coordinates

$$\frac{d}{dr}\left(kr^2\frac{dT}{dr}\right) + r^2\dot{q} = 0$$

Integrating, replacing k with $k = k_0(1 + bT)$, and rearranging yields

$$k_0(1 + bT)\frac{dT}{dr} + \frac{\dot{q}r}{3} = \frac{C_1}{r^2}$$

Separating the variables, integrating and rearranging yields

$$k_0\left(T + \frac{bT^2}{2}\right) + \frac{\dot{q}r^2}{6} + \frac{C_1}{r} - C_2 = 0$$

The constants C_1 and C_2 can be determined from the boundary conditions $dT/dr = 0$ at $r = 0$ and $T = T_w$ at $r = R$, leading to the equation for temperature profile in the radial direction

$$T_r = -\frac{1}{b} \pm \sqrt{\left(\frac{1}{b} + T_w\right)^2 + \frac{\dot{q}}{3bk_0}\left(R^2 - r^2\right)} \tag{2.130}$$

The temperature T_w can be a specified as a boundary condition, or can be determined from Eq. (2.128) if heat is dissipated from the outer surface by convection.

Figure 2.39 shows the temperature profiles in a solid sphere with heat dissipation by convection from the outer surface for constant and temperature-dependent thermal conductivities calculated from Eqs (2.129) and (2.130), respectively.

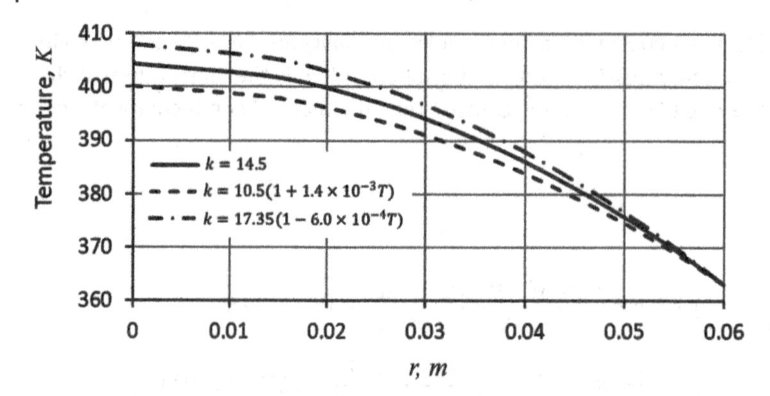

Figure 2.39 Temperature profiles in a solid sphere with heat dissipation by convection from the outer surface: constant and temperature-dependent conductivities in $W/m.K$) $r_1 = 0.02\,m$, $r_2 = 0.06\,m$, $\dot{q} = 10^6\,W/m^3$, $k_c = 14.5\ W/m.K, T_{\infty 1} = 10°C$, $h_1 = 250\ W/m^2.K$.

2.3.12 Hollow Sphere With Internal Heat Generation and Heat Dissipation By Convection From the Outer Surface and No Heat Storage

Figure 2.40 is a schematic diagram of a hollow sphere of internal radius r_1, external radius r_2, and convection heat transfer from the outer surface (heat transfer coefficient h_2) to the surroundings at temperature $T_{\infty 2}$. The internally generated energy per unit volume is $\dot{q}\ W/m^3$.

2.3.12.1 Constant Thermal Conductivity

The general conduction equations for a sphere are, as before

$$T = -\frac{\dot{q}r^2}{6k} - \frac{C_1}{r} + C_2 \tag{2.131}$$

and

$$\frac{dT}{dr} = -\frac{\dot{q}r}{3k} + \frac{C_1}{r^2}$$

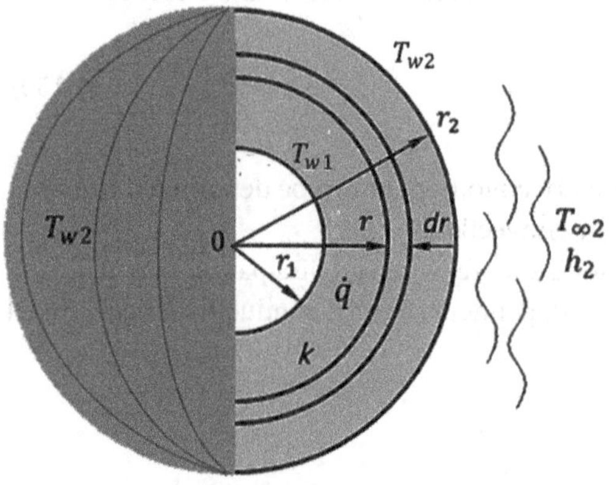

Figure 2.40 Conduction heat transfer in hollow sphere with internal heat generation and convection on the outside surface.

Boundary conditions

At $r = r_1, \dot{Q}_1 = 0, (dT/dr)_{r=r_1} = 0$

$$\left(\frac{dT}{dr}\right)_{r=r_1} = -\frac{\dot{q}r_1}{3k} + \frac{C_1}{r_1^2} = 0$$

from which

$$C_1 = \frac{\dot{q}r_1^3}{3k} \tag{2.132}$$

$$\frac{dT}{dr} = -\frac{\dot{q}r}{3k} + \frac{\dot{q}r_1^3}{3kr^2} \tag{2.133}$$

At $r = r_2, T = T_{w2}$

$$T_{w2} = -\frac{\dot{q}r_2^2}{6k} - \frac{\dot{q}r_1^3}{3kr_2} + C_2$$

from which

$$C_2 = T_{w2} + \frac{\dot{q}r_2^2}{6k} + \frac{\dot{q}r_1^3}{3kr_2} \tag{2.134}$$

Substitution for C_1 and C_2 into Eq. (2.131) results in

$$T_r = T_{w2} - \frac{\dot{q}r^2}{6k} - \frac{\dot{q}r_1^3}{3kr} + \frac{\dot{q}r_2^2}{6k} + \frac{\dot{q}r_1^3}{3kr_2}$$

or

$$T_r = T_{w2} - \frac{\dot{q}}{6k}\left(r^2 - r_2^2\right) - \frac{\dot{q}r_1^3}{3k}\left(\frac{1}{r} - \frac{1}{r_2}\right) \tag{2.135}$$

The wall temperature T_{w2} in terms of the temperature of the surroundings $T_{\infty 2}$ can be determined from the energy balance at the outer surface of the sphere

$$-k\left(\frac{dT}{dr}\right)_{r=r_2} = \frac{\dot{q}r_2}{3} - \frac{\dot{q}r_1^3}{3r_2^2} = h_2\left(T_{w2} - T_{\infty 2}\right)$$

hence,

$$T_{w2} = T_{\infty 2} + \frac{\dot{q}r_2}{3h_2} - \frac{\dot{q}r_1^3}{3h_2 r_2^2} \tag{2.136}$$

Equation (2.135) can now be rewritten as

$$T_r = T_{\infty 2} + \frac{\dot{q}}{3h_2 r_2^2}\left(r_2^3 - r_1^3\right) - \frac{\dot{q}}{6k}\left(r^2 - r_2^2\right) - \frac{\dot{q}r_1^3}{3k}\left(\frac{1}{r} - \frac{1}{r_2}\right) \tag{2.137}$$

The temperatures of the inner surface, therefore, is

$$T_{w1} = T_{\infty 2} + \frac{\dot{q}}{3h_2 r_2^2}\left(r_2^3 - r_1^3\right) - \frac{\dot{q}}{6k}(r_1^2 - r_2^2) - \frac{\dot{q}r_1^3}{3k}\left(\frac{1}{r_1} - \frac{1}{r_2}\right) \tag{2.138}$$

If the inner and outer wall temperatures are taken as boundary conditions, the conduction driving potential across the spherical wall will be the difference between the wall temperatures $(T_{w1} - T_{w2})$, which can be determined from Eqs (2.136) and (2.138) by equating the heat transfer coefficient h_2 in Eq. (2.138) to infinity $(h_2 = \infty)$

$$T_{w1} - T_{w2} = \frac{\dot{q}}{6k}\left(r_2^2 - r_1^2\right) - \frac{\dot{q}r_1^3}{3k}\left(\frac{1}{r_1} - \frac{1}{r_2}\right) \tag{2.139}$$

Figure 2.41 shows the results of computer modelling for this case.

2.3.12.2 Temperature-Dependent Thermal Conductivity

If the thermal conductivity is temperature-dependent and is defined as $k = k_0(1 + bT)$, the temperature profile can be worked out from the one-dimensional conduction equation in spherical coordinate system

$$\frac{d}{dr}\left(kr^2\frac{dT}{dr}\right) + r^2\dot{q} = 0$$

Integrating, replacing k with $k = k_0(1 + bT)$, and rearranging yields

$$k_0(1 + bT)\frac{dT}{dr} + \frac{\dot{q}r}{3} = \frac{C_1}{r^2}$$

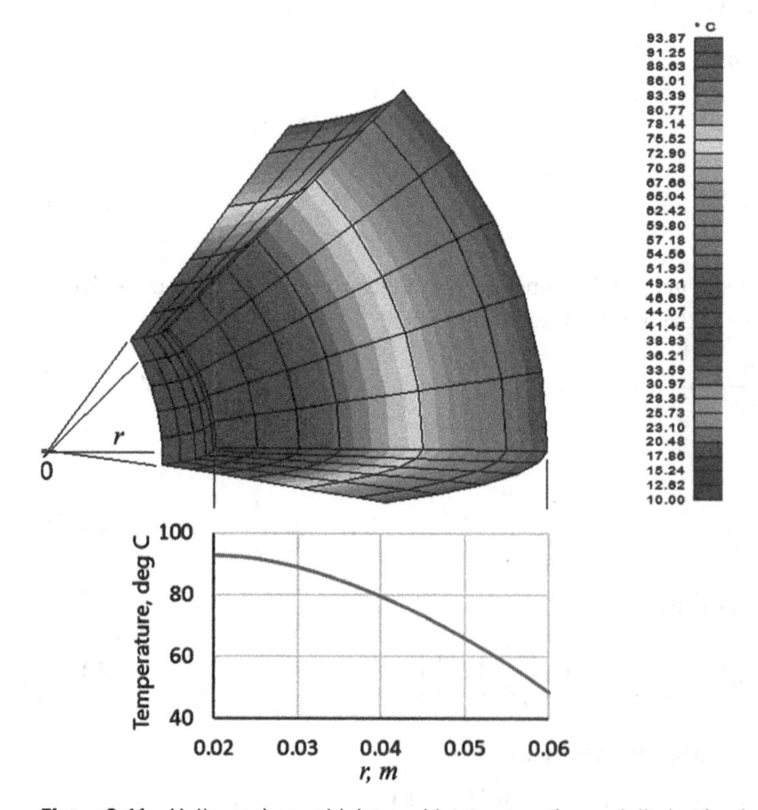

Figure 2.41 Hollow sphere with internal heat generation and dissipation by convection from the outer surface $\dot{q} = 10^6$ $W, k = 10$ $W/m.K$, $h_2 = 500$ $W/m^2.K$, $T_{\infty 2} = 10°C$.

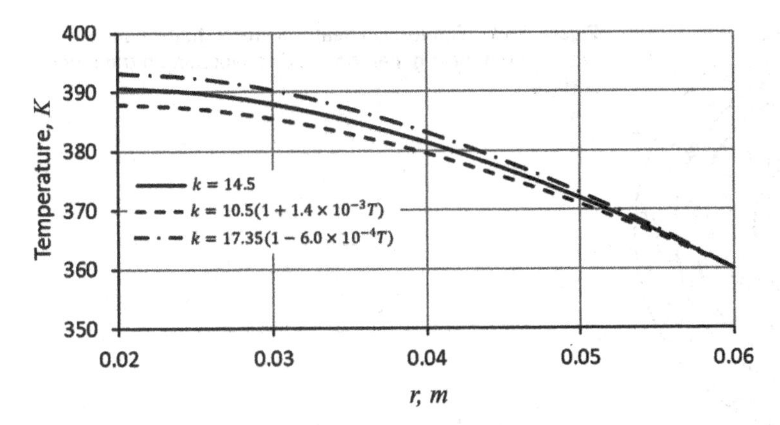

Figure 2.42 Temperature profiles in a hollow sphere with internal heat generation and heat dissipation by convection from the outer surface: constant and temperature-dependent conductivities $r_1 = 0.02\,m$, $r_2 = 0.06\,m$, $\dot{q} = 10^6\,W/m^3$, $T_{\infty 1} = 10°C$, $h_1 = 750\,W/m^2.K$.

Separating the variables, integrating, and rearranging yields

$$k_0\left(T + \frac{bT^2}{2}\right) + \frac{\dot{q}r^2}{6} + \frac{C_1}{r} - C_2 = 0$$

The constants C_1 and C_2 can be determined from the boundary conditions $dT/dr = 0$ at $r = r_1$ and $T = T_{w2}$ at $r = r_2$, leading to the following equation for the temperature profile in the radial direction

$$T_r = -\frac{1}{b} \pm \sqrt{\left(\frac{1}{b} + T_{w2}\right)^2 + \frac{\dot{q}}{3bk_0}\left(r_2^2 - r^2\right) + \frac{2\dot{q}r_1^3}{3bk_0}\left(\frac{1}{r_2} - \frac{1}{r}\right)} \tag{2.140}$$

The temperature of the outer wall T_{w2} can be a specified as a boundary condition, or can be determined from Eq. (2.136) if heat is dissipated from the outer surface by convection.

Figure 2.42 shows the temperature profiles in a hollow sphere with heat dissipation by convection from the outer surface for constant and temperature-dependent thermal conductivities calculated, respectively, from Eqs (2.137) and (2.140).

The temperature difference between the two curves in Figure 2.42 increases with the decreasing heat transfer coefficient h_2.

2.3.13 Hollow Sphere With Internal Heat Generation and Heat Dissipation By Convection From the Inner Surface and No Heat Storage

Figure 2.43 is a schematic diagram of a hollow sphere of internal radius r_1, external radius r_2, and convection heat transfer from the inner surface (heat transfer coefficient h_1) to the surroundings at temperature $T_{\infty 1}$. The internally generated energy per unit volume is $\dot{q}\ W/m^3$.

2.3.13.1 Constant Thermal Conductivity

The conduction equations are

$$T = -\frac{\dot{q}r^2}{6k} - \frac{C_1}{r} + C_2 \tag{2.141}$$

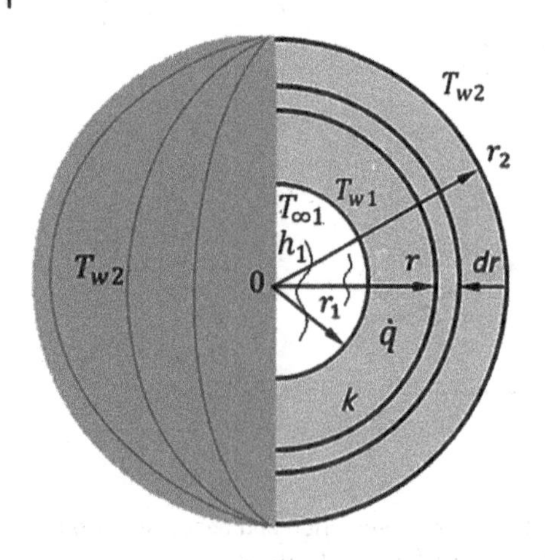

Figure 2.43 Conduction heat transfer in hollow sphere with internal heat generation and convection on the inner surface.

and

$$\frac{dT}{dr} = -\frac{\dot{q}r}{3k} + \frac{C_1}{r^2}$$

Boundary conditions

At $r = r_2$, $\dot{Q}_2 = 0$, $(dT/dr)_{r=r_2} = 0$

$$\left(\frac{dT}{dr}\right)_{r=r_1} = -\frac{\dot{q}r_2}{3k} + \frac{C_1}{r_2^2} = 0$$

from which

$$C_1 = \frac{\dot{q}r_2^3}{3k} \tag{2.142}$$

At $r = r_1, T = T_{w1}$

$$T_{w1} = -\frac{\dot{q}r_1^2}{6k} - \frac{\dot{q}r_2^3}{3kr_1} + C_2$$

from which

$$C_2 = T_{w1} + \frac{\dot{q}r_1^2}{6k} + \frac{\dot{q}r_2^3}{3kr_1} \tag{2.143}$$

Substituting for C_1 and C_2 into Eq. (2.141)

$$T_r = -\frac{\dot{q}r^2}{6k} - \frac{\dot{q}r_2^3}{3kr} + T_{w1} + \frac{\dot{q}r_1^2}{6k} + \frac{\dot{q}r_2^3}{3kr_1}$$

or

$$T_r = T_{w1} - \frac{\dot{q}}{6k}\left(r^2 - r_1^2\right) - \frac{\dot{q}r_2^3}{3k}\left(\frac{1}{r} - \frac{1}{r_1}\right) \tag{2.144}$$

The wall temperature T_{w1} in terms of the temperature of the surroundings $T_{\infty 2}$ can be determined from the energy balance at the inner surface of the sphere

$$k\left(\frac{dT}{dr}\right)_{r=r_1} = \frac{\dot{q}r_1}{3} - \frac{\dot{q}r_2^{\,3}}{3r_1^{\,2}} = h_1\left(T_{w1} - T_{\infty 1}\right)$$

from which

$$T_{w1} = T_{\infty 1} - \frac{\dot{q}r_1}{3h_1} + \frac{\dot{q}r_2^{\,3}}{3h_1\, r_1^{\,2}} = T_{\infty 1} + \frac{\dot{q}}{3h_1 r_1^{\,2}}\left(r_2^{\,3} - r_1^{\,3}\right) \tag{2.145}$$

Equation (2.144) can now be rewritten as

$$T_r = T_{\infty 1} + \frac{\dot{q}}{3h_1 r_1^{\,2}}\left(r_2^{\,3} - r_1^{\,3}\right) - \frac{\dot{q}}{6k}\left(r^2 - r_1^{\,2}\right) - \frac{\dot{q}r_2^{\,3}}{3k}\left(\frac{1}{r} - \frac{1}{r_1}\right) \tag{2.146}$$

The temperature of the outer surface is determined from Eq. (2.146) by putting $r = r_2$

$$T_{w2} = T_{\infty 1} + \frac{\dot{q}}{3h_1 r_1^{\,2}}\left(r_2^{\,3} - r_1^{\,3}\right) - \frac{\dot{q}}{6k}\left(r_2^{\,2} - r_1^{\,2}\right) - \frac{\dot{q}r_2^{\,3}}{3k}\left(\frac{1}{r_2} - \frac{1}{r_1}\right) \tag{2.147}$$

If the inner and outer wall temperatures are taken as boundary conditions, the conduction driving potential across the spherical wall will be the difference between the wall temperatures $(T_{w2} - T_{w1})$, which can be determined from Eqs (2.145) and (2.147) by equating the heat transfer coefficient for the inner surface in Eq. (2.147) to infinity $(h_1 = \infty)$

$$T_{w2} - T_{w1} = \frac{\dot{q}}{6k}\left(r_1^{\,2} - r_2^{\,2}\right) - \frac{\dot{q}r_2^{\,3}}{3k}\left(\frac{1}{r_2} - \frac{1}{r_1}\right) \tag{2.148}$$

Figure 2.44 shows the results of computer modelling for this case.

2.3.13.2 Temperature-Dependent Thermal Conductivity

If the thermal conductivity is temperature-dependent and is defined as $k = k_0(1 + bT)$, the temperature profile can be worked out from the one-dimensional conduction equation in spherical coordinate system

$$\frac{d}{dr}\left(kr^2\frac{dT}{dr}\right) + r^2\dot{q} = 0$$

Integrating, replacing k with $k = k_0(1 + bT)$, and rearranging yields

$$k_0(1 + bT)\frac{dT}{dr} + \frac{\dot{q}r}{3} = \frac{C_1}{r^2}$$

Separating the variables, integrating, and rearranging yields

$$k_0\left(T + \frac{bT^2}{2}\right) + \frac{\dot{q}r^2}{6} + \frac{C_1}{r} - C_2 = 0$$

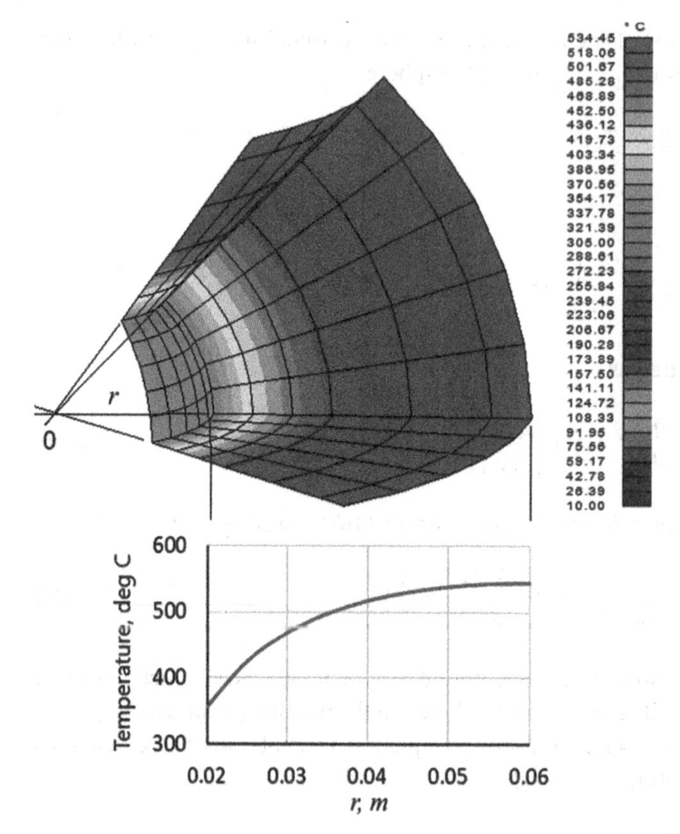

Figure 2.44 Hollow sphere with heat generation and heat dissipation by convection from the inner surface $\dot{q} = 10^6\ W, k = 10\ W/m.K,\ h_1 = 500\ W/m^2.K,\ T_{\infty 1} = 10°C$.

The constants C_1 and C_2 can be determined from the boundary conditions $dT/dr = 0$ at $r = r_2$ and $T = T_{w1}$ at $r = r_1$, leading to the following equation for the temperature profile in the radial direction:

$$T_r = -\frac{1}{b} \pm \sqrt{\left(\frac{1}{b} + T_{w1}\right)^2 + \frac{\dot{q}}{3bk_0}\left(r_1^2 - r^2\right) + \frac{2\dot{q}r_2^3}{3bk_0}\left(\frac{1}{r_1} - \frac{1}{r}\right)} \tag{2.149}$$

The temperature of the inner wall T_{w1} can be a specified as a boundary condition, or can be determined from Eq. (2.145) if heat is dissipated from the inner surface by convection.

Figure 2.45 shows the temperature profiles in a hollow sphere with internal heat generation and heat dissipation by convection from the inner surface for constant and temperature-dependent thermal conductivities calculated, respectively, from Eqs (2.146) and (2.149).

2.3.14 Hollow Sphere With Internal Heat Generation and Heat Dissipation By Convection From Both the Inner and Outer Surfaces and No Heat Storage

Figure 2.46 is a schematic diagram of a spherical shell (hollow sphere) of inner radius r_1 and outer radius r_2 with convection heat transfer from, respectively, the inner and outer surfaces (heat transfer coefficients h_1, h_2) to the surroundings at temperature $T_{\infty 1}$ and $T_{\infty 2}$. The internally generated energy is $\dot{q}\ W/m^3$.

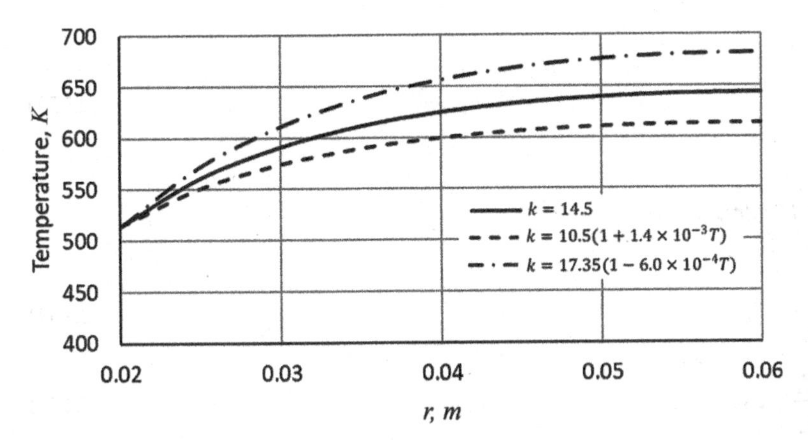

Figure 2.45 Temperature profiles in a hollow sphere with heat dissipation by convection from the inner surface: constant and temperature-dependent conductivities $r_1 = 0.02m$, $r_2 = 0.06m$, $\dot{q} = 10^6\ W/m^3$, $T_{\infty 1} = 10°C$, $h_1 = 750\ W/m^2.K$.

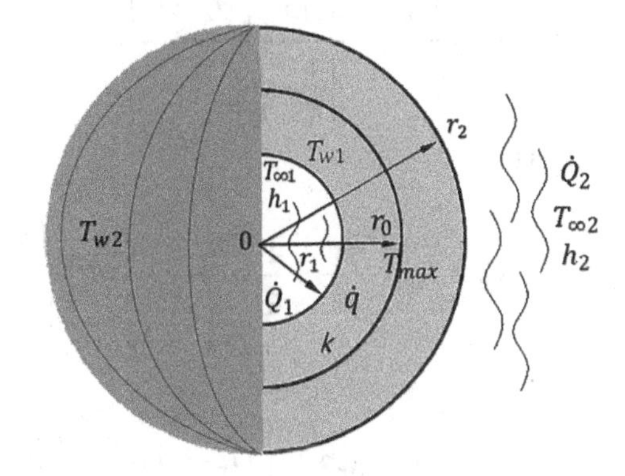

Figure 2.46 Hollow sphere with heat generation and heat dissipation by convection from both the inner and outer surfaces.

2.3.14.1 Constant Thermal Conductivity

Because of the heat losses $\dot{Q}_1$ and $\dot{Q}_2$ from the surroundings from the inner and outer surfaces of the hollow sphere, the temperature profile exhibits a maximum value of T_{max} in the solid at $r = r_0$. Conceptually, it can be argued that an isothermal surface at $r = r_0$ separates the wall of the sphere into two layers with two separate heat flows:

- Heat flow towards the outer surface at $r = r_2$ with the driving potential $(T_{max} - T_{w2})$
- Heat flow towards the inner surface at $r = r_1$ with the driving potential $(T_{max} - T_{w1})$.

From Eqs (2.139) and (2.148)

$$T_{max} - T_{w2} = \frac{\dot{q}}{6k}\left(r_2^2 - r_0^2\right) - \frac{\dot{q}r_0^3}{3k}\left(\frac{1}{r_0} - \frac{1}{r_2}\right) \tag{2.150}$$

$$T_{max} - T_{w1} = \frac{\dot{q}}{6k}(r_1^2 - r_0^2) - \frac{\dot{q}r_0^3}{3k}\left(\frac{1}{r_0} - \frac{1}{r_1}\right) \tag{2.151}$$

As before, the general equation of the temperature distribution and its derivative for a sphere are, respectively

$$T = -\frac{\dot{q}r^2}{6k} - \frac{C_1}{r} + C_2 \tag{2.152}$$

and

$$\frac{dT}{dr} = -\frac{\dot{q}r}{3k} + \frac{C_1}{r^2} \tag{2.153}$$

Boundary conditions

At $r = r_0, T = T_{max}$, $(dT/dr)_{r=r_0} = 0$

At $r = r_1, T = T_{w1}$, $\dot{Q} = \dot{Q}_1$

$$T_{w1} = -\frac{\dot{q}r_1^2}{6k} - \frac{C_1}{r_1} + C_2 \tag{2.154}$$

$$T_{w2} = -\frac{\dot{q}r_2^2}{6k} - \frac{C_1}{r_2} + C_2 \tag{2.155}$$

At $r = r_2, T = T_{w2}$, $\dot{Q} = \dot{Q}_2$

hence, from Eqs (2.154) and (2.155), we determine C_1

$$C_1 = \frac{\left(T_{w2} - T_{w1}\right) + \frac{\dot{q}}{6k}\left(r_2^2 - r_1^2\right)}{\left(\frac{1}{r_1} - \frac{1}{r_2}\right)} \tag{2.156}$$

C_2 is then

$$C_2 = T_{w1} + \frac{\dot{q}r_1^2}{6k} + \frac{1}{r_1}\frac{\left(T_{w2} - T_{w1}\right) + \frac{\dot{q}}{6k}\left(r_2^2 - r_1^2\right)}{\left(\frac{1}{r_1} - \frac{1}{r_2}\right)} \tag{2.157}$$

or

$$C_2 = T_{w2} + \frac{\dot{q}r_2^2}{6k} + \frac{1}{r_2}\frac{\left(T_{w2} - T_{w1}\right) + \frac{\dot{q}}{6k}\left(r_2^2 - r_1^2\right)}{\left(\frac{1}{r_1} - \frac{1}{r_2}\right)} \tag{2.158}$$

To find the temperature profile we need to determine the wall temperatures T_{w1}, T_{w2} and the maximum temperature T_{max} and radius r_o. For the conditions shown in Figure 2.46, we can write the equations for the rates of heat loss by convection at the surfaces as

$$\dot{Q}_1 = 4\pi r_1^2 h_1\left(T_{w1} - T_{\infty 1}\right) = \frac{4\pi\dot{q}}{3}\left(r_0^3 - r_1^3\right) \tag{2.159}$$

$$\dot{Q}_2 = 4\pi r_2^2 h_2\left(T_{w2} - T_{\infty 2}\right) = \frac{4\pi\dot{q}}{3r_2^2 h_2}\left(r_2^3 - r_0^3\right) \tag{2.160}$$

from which

$$T_{w1} = T_{\infty 1} + \frac{\dot{q}}{3r_1^2 h_1}\left(r_0^3 - r_1^3\right) \tag{2.161}$$

$$T_{w2} = T_{\infty 2} + \frac{\dot{q}}{3r_2^2 h_2}\left(r_2^3 - r_0^3\right) \tag{2.162}$$

$$T_{w1} - T_{w2} = (T_{\infty 1} - T_{\infty 2}) + \frac{\dot{q}}{3}\left[\frac{\left(r_o^3 - r_1^3\right)}{r_1^2 h_1} - \frac{\left(r_2^3 - r_o^3\right)}{r_2^2 h_2}\right] \tag{2.163}$$

Also, subtracting Eq. (2.151) from Eq. (2.150) yields

$$T_{w1} - T_{w2} = \frac{\dot{q}}{6k}\left(r_2^2 - r_1^2\right) + \frac{\dot{q} r_o^3}{3k}\left(\frac{1}{r_2} - \frac{1}{r_1}\right) \tag{2.164}$$

The solution procedure is as follows:

1) Equating the right-hand sides of Eqs (2.163) and (2.164) yields r_0 (this can easily be done graphically in a spreadsheet)
2) Knowing r_0, temperatures T_{w1}, T_{w2} can be calculated from Eqs (2.161) and (2.162)
3) The maximum temperature is calculated from Eq. (2.150) or Eq. (2.151)
4) The constant C_1 is calculated from Eq. (2.156) and C_2 from Eq. (2.157) or Eq. (2.158)
5) The temperature profile can now be plotted as function of the radius r using Eq. (2.152).

Figure 2.47 shows some results of computer modelling for this case. As for the hollow cylinder with convection on both the inner and outer surfaces, the temperature profile exhibits a maximum, with its location being dependent on the boundary conditions.

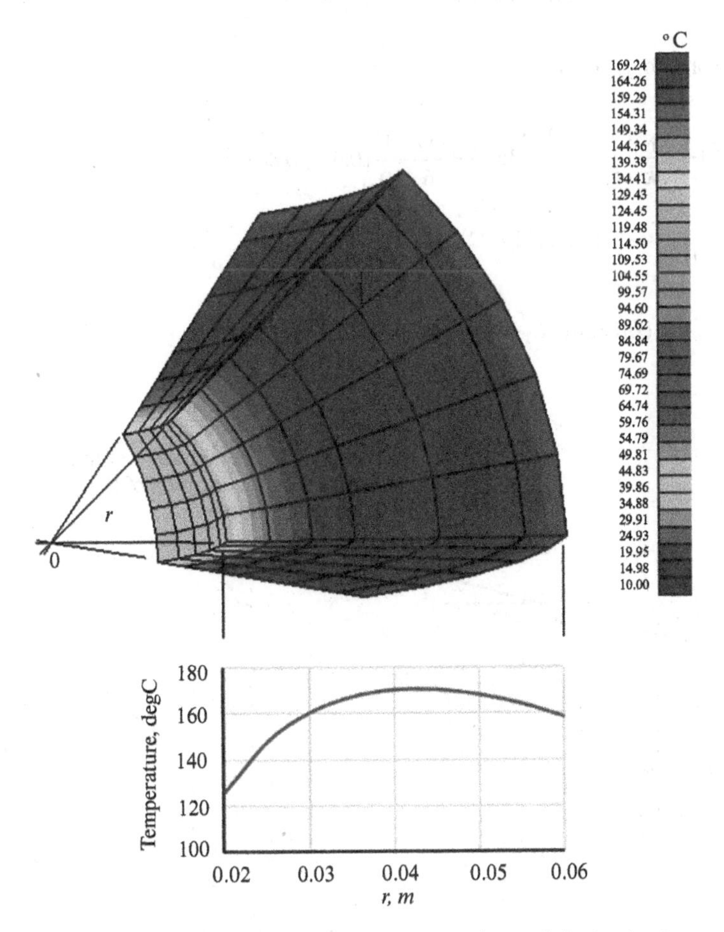

Figure 2.47 Hollow sphere with heat generation and dissipation by convection from the inner and outer surfaces $\dot{q} = 10^6\ W, k = 10\ W/m.K,\ h_1 = 500\ W/m^2.K\ \ h_2 = 100\ W/m^2.K, T_{\infty 1} = 10°C,\ T_{\infty 2} = 30°C$.

Example 2.7 The pressure of liquefied petroleum gas (LPG) is close to 0 kPa at −43°C. The gas is to be stored at these conditions in a spherical tank of $1.8m$ diameter and $0.1m$ wall thickness. Calculations show that to maintain these conditions, the spherical shell needs to be heated at the rate of $5.0\times10^5\,W/m^3$. If the heat transfer coefficient and ambient conditions are, respectively, $h_2 = 100\,W/m^2.K$ and 10°C, determine the wall temperatures, magnitude, and location of the maximum temperature in the shell, and plot the temperature profile. Take the thermal conductivity of the shell metal as $k = 10\,W/m.K$ and the heat transfer coefficient to the gas inside of the tank as $h_1 = 250\,W/m^2.K$.

Solution

The point of intersection of the plots of Eqs (2.163) and (2.164) yields $r_0 = 0.861m$ as shown in Figure E2.7.1.

The temperatures of the walls T_{w1}, T_{w2} can be calculated from Eqs (2.161) and (2.162)

$$T_{w1} = T_{\infty 1} + \frac{\dot{q}}{3r_1^2 h_1}\left(r_0^3 - r_1^3\right) = 230 + \frac{5\times10^5}{3\times0.8^2\times500}(0.861^3 - 0.8^3) = 295.8K(22.8°C)$$

$$T_{w2} = T_{\infty 2} + \frac{\dot{q}}{3r_2^2 h_2}\left(r_2^3 - r_0^3\right) = 283 + \frac{5\times10^5}{3\times0.8^2\times250}(0.9^3 - 0.861^3) = 357.7K(84.7°C)$$

The maximum temperature from Eq. (2.150)

$$T_{max} = T_{w2} + \frac{\dot{q}}{6k}\left(r_2^2 - r_0^2\right) - \frac{\dot{q}r_0^3}{3k}\left(\frac{1}{r_0} - \frac{1}{r_2}\right) = 357.7 + \frac{5\times10^5}{6\times10}\left(0.9^2 - 0.861^2\right)$$

$$-\frac{5\times10^5\times0.861^3}{3\times10}\left(\frac{1}{0.861} - \frac{1}{0.9}\right) = 394.6K\,(121.6°C)$$

To plot the temperature profile, the constants C_1 and C_2 are calculated

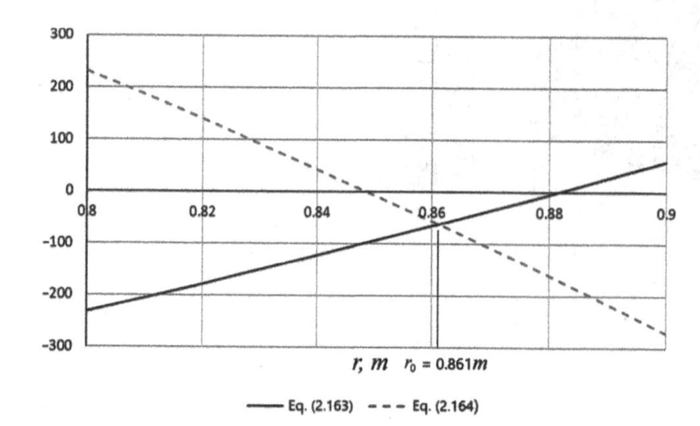

Figure E2.7.1 Graphical method for determining the location of the maximum temperature.

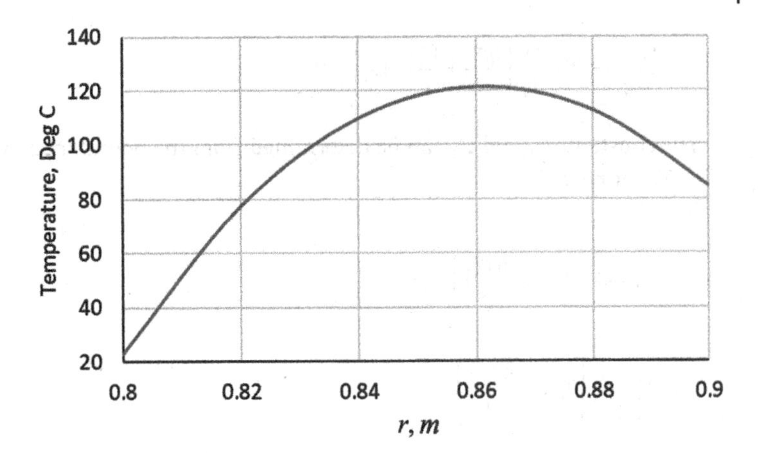

Figure E2.7.2 Temperature profile for Example 2.7.

$$C_1 = \frac{(T_{w2} - T_{w1}) + \frac{\dot{q}}{6k}(r_2^2 - r_1^2)}{\left(\frac{1}{r_1} - \frac{1}{r_2}\right)} = \frac{(357.7 - 295.8) + \frac{5\times10^5}{6\times10}(0.9^2 - 0.8^2)}{\left(\frac{1}{0.8} - \frac{1}{0.9}\right)} = 1.06\times10^4$$

$$C_2 = T_{w1} + \frac{\dot{q}r_1^2}{6k} + \frac{1}{r_1}\frac{(T_{w2} - T_{w1}) + \frac{\dot{q}}{6k}(r_2^2 - r_1^2)}{\left(\frac{1}{r_1} - \frac{1}{r_2}\right)} = 295.8 + \frac{5\times10^5\times0.8^2}{6\times10} + \frac{1.06\times10^4}{0.8} = 1.98\times10^4$$

The temperature profile is calculated from

$$T_r = -\frac{5\times10^5}{60}r^2 - \frac{1.06\times10^4}{r} + 1.98\times10^4$$

and plotted as shown in Figure E2.7.2.

The methodology outlined for a spherical shell in this section can be used if the temperatures of the inner and outer walls are specified as boundary conditions instead of convection heat dissipation. This is done by making the heat transfer coefficients h_1 and h_2 equal to infinity and replacing the ambient temperatures $T_{\infty1}$ and $T_{\infty2}$ by the specified temperatures.

2.3.14.2 Temperature-Dependent Thermal Conductivity With Specified Inner and Outer Surface Temperature

If the thermal conductivity is given as $k = k_0(1 + bT)$, the temperature profile can be worked out from the one-dimensional conduction equation in the spherical coordinate system yielding the quadratic equation

$$\frac{bT^2}{2}+T+\frac{1}{k_0}\left(\frac{\dot{q}r^2}{6}+\frac{C_1}{r}-C_2\right)=0 \tag{2.165}$$

The constants C_1 and C_2 can be determined from the boundary conditions $T=T_{w1}$ at $r=r_1$ and $T=T_{w2}$ at $r=r_2$.

If

$$X=k_0\left(T_{w1}+\frac{bT_{w1}^2}{2}\right)$$

$$Y=k_0\left(T_{w2}+\frac{bT_{w2}^2}{2}\right)$$

constants C_1 and C_2 can be written as

$$C_1=\frac{(X-Y)+\frac{\dot{q}}{6}\left(r_1^2-r_2^2\right)}{\left(\frac{1}{r_2}-\frac{1}{r_1}\right)}$$

$$C_2=X+\frac{\dot{q}r_1^2}{6}+\frac{1}{r_1}\frac{(X-Y)+\frac{\dot{q}}{6}\left(r_1^2-r_2^2\right)}{\left(\frac{1}{r_2}-\frac{1}{r_1}\right)}$$

The solution to Eq. (2.165) is

$$T_r=-\frac{1}{b}\pm\sqrt{\frac{1}{b^2}-\frac{2}{bk_0}\left(\frac{\dot{q}r^2}{6}+\frac{C_1}{r_1}-C_2\right)} \tag{2.166}$$

Figure 2.48 compares the temperature profiles for the hollow sphere with constant conductivity and temperature-dependent thermal conductivity plotted using Eqs (2.152) and (2.166). The wall temperatures are taken as $T_{w1}=369°C$, $T_{w2}=355°C$.

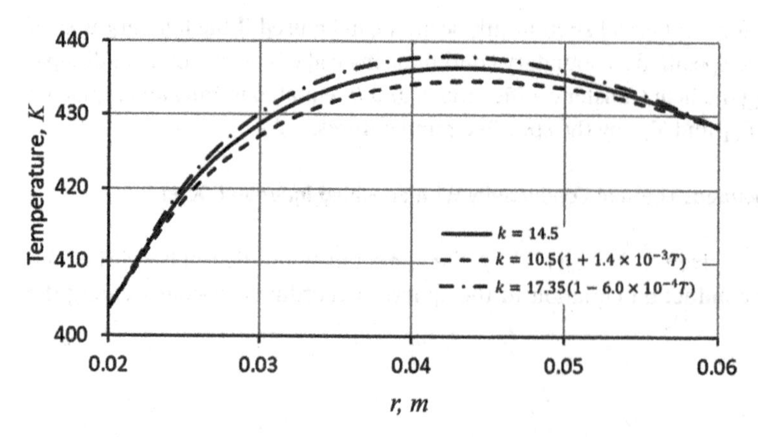

Figure 2.48 Hollow sphere with heat generation and specified inner and outer surface temperatures: constant thermal conductivity k_c and temperature-dependent conductivity k_t $r_1=0.02\,m$, $r_2=0.06\,m$, $\dot{q}=10^6\,W/m^3$, $h_1=500\,W/m^2.K$ $h_2=100\,W/m^2.K$, $T_{\infty1}=10°C$, $T_{\infty2}=30°C$.

2.4 Interface Contact Resistance

In the analysis of composite plane, circular, and spherical walls, perfect contact was assumed between adjoining surfaces with no contact resistance to heat flow. In engineering applications, material surfaces are hardly ever perfectly smooth and contact between pairs of materials occur at few places and voids are formed that can be filled with air, oil, water, or grease. Figure 2.49a depicts perfect contact and Figure 2.49b contact of two rough surfaces. An example of an extreme case of rough surfaces in contact is shown in Figure 2.50 for a steel–concrete interface in a cylindrical steel column filled with high-strength concrete at elevated temperature. Heat transfer at the interface occurs by conduction through the contacting asperities, through the gaps filled with liquid or gas, and through radiation through the gaps filled with a substance transparent to radiation or in a vacuum. The total heat transfer rate across the interface $\dot{Q}_i$ can be approximated as

$$\dot{Q}_i = \dot{Q}_c + \dot{Q}_r \tag{2.167}$$

where $\dot{Q}_c$ and $\dot{Q}_r$ are respectively the contact and radiation heat transfer rates. Strictly speaking, not all the gaps will be filled by a gas that are transparent to radiation and there could be a presence of opaque substances which will transmit heat by conduction. We will limit the discussion here to the two terms given in Eq. (2.167).

The total heat transfer rate across the interface $\dot{Q}_i$ is given by Newton's law of cooling

$$\dot{Q}_i = h_i A \Delta T_i = \frac{\Delta T_i}{\dfrac{1}{h_i A}} = \frac{\Delta T_i}{R_i} \tag{2.168}$$

where h_i is the interfacial conductance $(W/m^2.K)$, A is the nominal contact area, ΔT_i is the temperature gradient between the two surfaces, and R_i is the contact resistance $(R_i = 1/h_i A \; K/W)$.

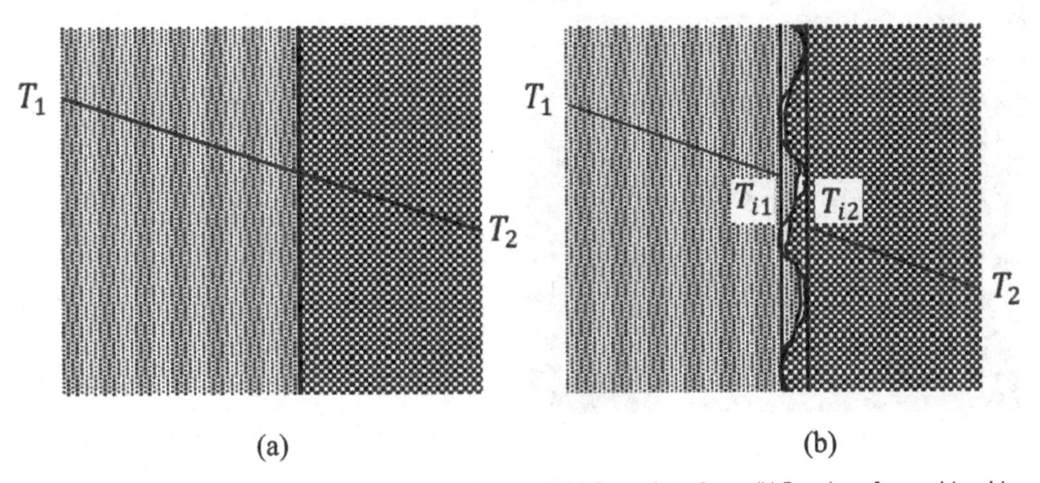

Figure 2.49 Interfacial contacts in a composite plane wall (a) Smooth surfaces; (b) Rough surfaces with voids.

The temperature gradient across the interface (Figure 2.49b) is

$$\Delta T = T_{i1} - T_{i2}$$

$$\dot{Q}_r = \sigma A F_{12}(T_{i1}^4 - T_{i2}^4) \tag{2.169}$$

where σ is the Stefan–Boltzmann constant ($5.67 \times 10^{-8}\, W/m^2K^4$) and F_{12} is the shape factor. The radiative resistance is then

$$R_r = \frac{T_{i1} - T_{i2}}{\dot{Q}_r} = \frac{T_{i1} - T_{i2}}{\sigma A F_{12}\left(T_{i1}^4 - T_{i2}^4\right)} \quad K/W$$

and the radiative conductance

$$h_r = \frac{\dot{Q}_r}{A\left(T_{i1} - T_{i2}\right)} = \frac{\sigma F_{12}\left(T_{i1}^4 - T_{i2}^4\right)}{T_{i1} - T_{i2}} \quad W/m^2.K \tag{2.170}$$

The radiative conductance can be simplified by introducing a mean interface temperature T_{im} (Bejan and Kraus 2003)

$$h_r \approx 4\sigma F_{12} T_{im}^3 \quad W/m^2.K$$

where

Figure 2.50 Steel–concrete rough interface in an unloaded circular column at 600°C (Ghojel 2004 with permission by Elsevier).

$$T_{im} = \frac{1}{2}(T_{i1} + T_{i2})$$

The process of heat transfer across an interface is complex, as the resistance is dependent on many variables such as contact geometry, ratio of actual contact area to the nominal area, type of fluid filling the voids, and contact pressure. Theoretical models have been developed to predict interface conductance encompassing both contact conductance and gap (void) conductance and used successfully to calculate the thermal joint conductance at interfaces formed by metals and mould compounds (Peterson and Fletcher 1987) and aluminium–ceramic interfaces in microelectronic applications (Yovanovitchet al. 1997). Table 2.2 shows selected data for interfacial conductance from heat transfer references.

Table 2.2 Selected values of metal-to-metal interfacial conductance.

Interface	Interfacial conductance ($W/m^2.K$)
Stainless steel/stainless steel	1900 – 3700
Stainless steel/aluminium	2200 – 12000
Copper/copper	10000 – 25000
Iron/aluminium	4000 – 40000

The current author combined experimental temperature measurements and the inverse heat analysis software INTEMP (2020), to estimate the interfacial conductance for the steel–concrete interface in a cylindrical steel column filled with concrete (Ghojel 2004). The two-dimensional finite-element model and three temperature profiles are shown in Figure 2.51.

The numbers in Figure 2.51a identify the nodes at which temperature measurements were taken. The results of this research lead to the following correlation for the average interfacial conductance between the steel tube and the concrete filling:

$$\bar{h}_i = a - b\, exp\left(cT^d\right) \quad W/m^2.K \tag{2.171}$$

where $a = 1926, b = 765.8, c = 339.9, d = -1.4$, and T is the temperature of the steel shell in °C. To account for the effect of contact resistance in composite walls, the contact resistance R_i in Eq. (2.168) can be added to the total resistance in plane, cylindrical, and spherical composite walls.

Figure 2.52 shows results of computer modelling of contact resistance between the flat surfaces of two blocks of identical material ($k = 20\,W/m.K$) and dimensions $0.2m \times 0.4m \times 0.4m$. The left and right surfaces are maintained at constant temperatures of 700 K and 300 K, respectively. The interfacial conductance is taken as $h_i = 500\,W/m^2.K$, which represents a thermal contact resistance of $R_c = 0.025\,K/W$ (radiation contact resistance is ignored). The resulting temperature drop due to contact resistance is $14K$.

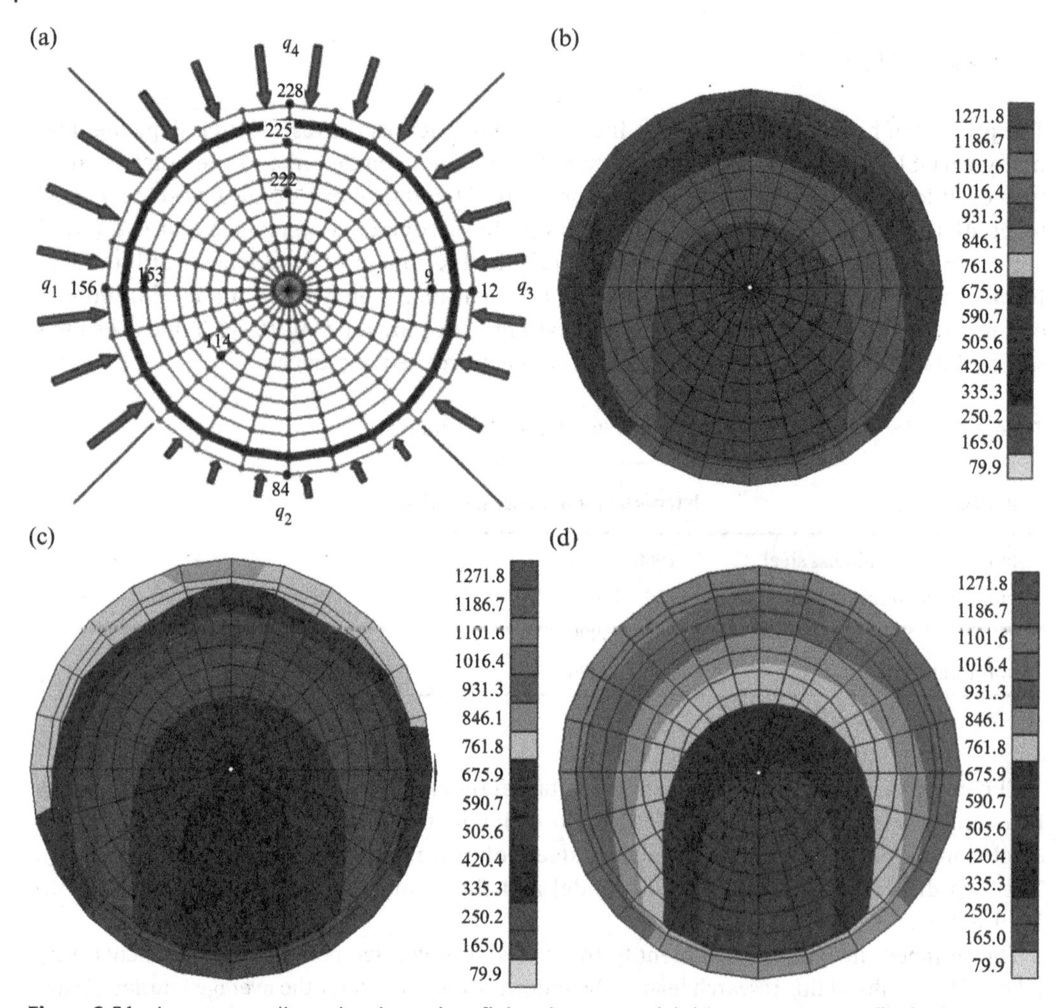

Figure 2.51 Inverse two-dimensional transient finite-element model: (a) temperature profile in the cross-section of the specimen after 30 min; (b) 60 min (c); and 90 min (d).

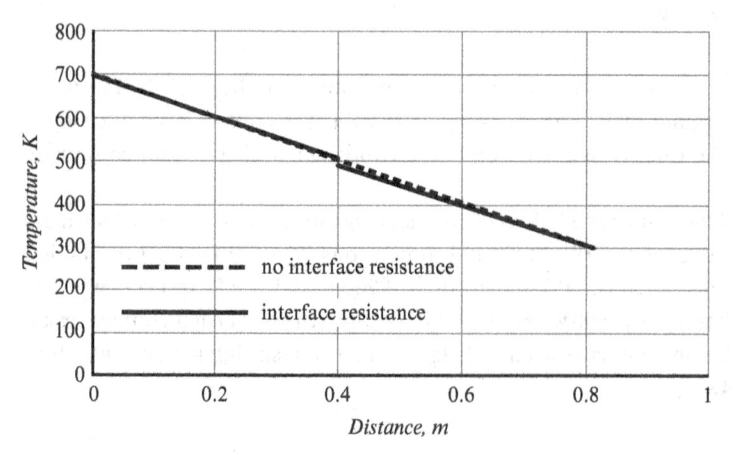

Figure 2.52 Effect of contact resistance on temperature profile between two solid blocks of identical thermal properties $k = 20\ W/m.K$, $h_i = 500\ W/m^2.K$.

Problems

2.1 Semi-infinite plate of $20mm$ thickness is heated to a temperature of 40°C on one side and exposed on the other side to the surroundings at a temperature of 25°C. The thermal conductivity of the plate's material is $4.5 W/m.K$. Determine the temperature at a point $15mm$ from the left side of the plate and the heat flux.

2.2 A refractory wall $0.5m$ thick has one side maintained at 1000°C and the other at 0°C. The thermal conductivity is temperature-dependent given by $k = 1.01(1 + 0.0014t)$, where t is in 0°C. Determine the heat flux through the plate.

2.3 A steel plate of $20mm$ thickness and $50 W/m.K$ thermal conductivity is layered with insulating material of 2 mm thickness and $1.0 W/m.K$ conductivity. The temperature of the outer surface of the steel plate is 250°C and that of the insulation outer surface 200°C. What is the heat flux through the composite wall?

2.4 For the Problem 2.3, what is the temperature of the interface between the steel plate and the insulating material?

2.5 Sketch the thermal network for the composite wall shown in Figure P2.5 and determine the heat flux assuming one-dimensional heat flow.

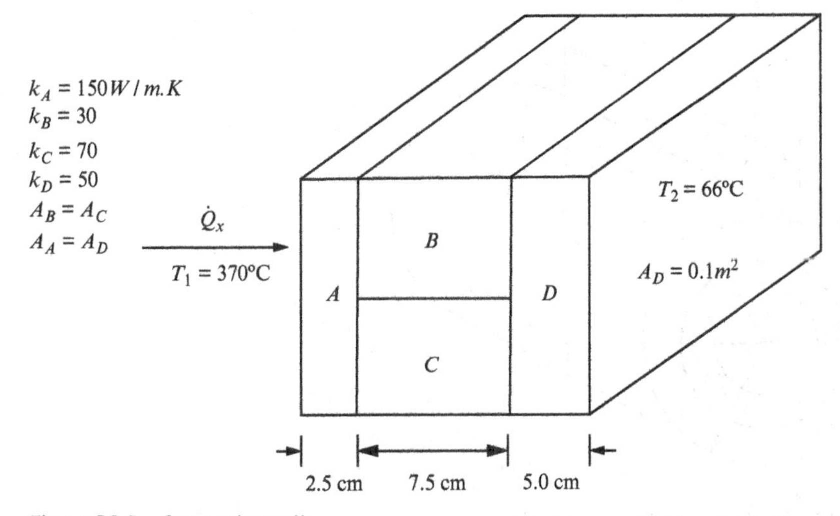

Figure P2.5 Composite wall.

2.6 For Problem 2.5, determine the temperatures at the inner surfaces of blocks A and D. Determine also the rate of heat transfer through blocks B and C.

2.7 A furnace wall is made up of an inside layer of silica brick $120mm$ thick (conductivity 1.7 $W/m.K$) covered with a layer of magnesite brick $240mm$ thick (conductivity 5.8 $W/m.K$). Due to imperfect contact at the interface between the two layers, there is an interfacial contact thermal resistance $R_i = 1/A_i h_i$, where interfacial conductance $h_i = 286\ W/m^2.K$. The temperatures at the inner and outer surfaces of the furnace are, respectively, 725°C and 110°C. Determine the heat flux through the furnace wall.

2.8 A solid metallic cylinder of 0.06 m diameter is cooled by convection ($h = 500\,W/m^2.K$) in air at $283K$. The cylinder is heated internally by a heater generating $1M\,W/m^3$. Plot the temperature profile for thermal conductivity $k = 14.5\,W/m.K$.

2.9 For the data in Problem 2.8, plot the temperature profiles for $k = 10, 14.5$, and $25\,W/m.K$.

2.10 Electric current heats a conductor of 12 mm diameter ($k = 15\,W/m.K$) at a rate of $2.2\times10^7\,W/m^3$. The heat generated is dissipated at the surface of the conductor by convection ($h = 300\,W/m^2.K$) to the surroundings at $T_\infty = 288K$. Calculate the temperatures at the centre and surface of the conductor.

2.11 An electric current heats a conductor of 10 mm diameter ($k = 15\,W/m.K$) at a rate of $2.4\times10^6\,W/m^3$. The heat generated is dissipated at the surface of the conductor by convection ($h = 300\,W/m^2.K$) to the surroundings at $T_\infty = 293°C$. Calculate the rate of heat dissipation per unit length to the surroundings.

2.12 A steel tube of $0.2m$ internal diameter and $0.216m$ external diameter ($k_s = 40\,W/m^2.K$) is covered with an insulation layer of $0.12m$ thickness ($k_i = 0.1\,W/m^2.K$), as shown in Figure P2.12. The tube is transporting steam at 300°C and the temperature of the surroundings is 25°C. The coefficients of convection heat transfer at the inner surface of the tube and outer surface of the insulator are, respectively, $100\,W/m^2.K$ and $8\,W/m^2.K$. Determine the total thermal resistance and the heat transfer rate per unit tube length.

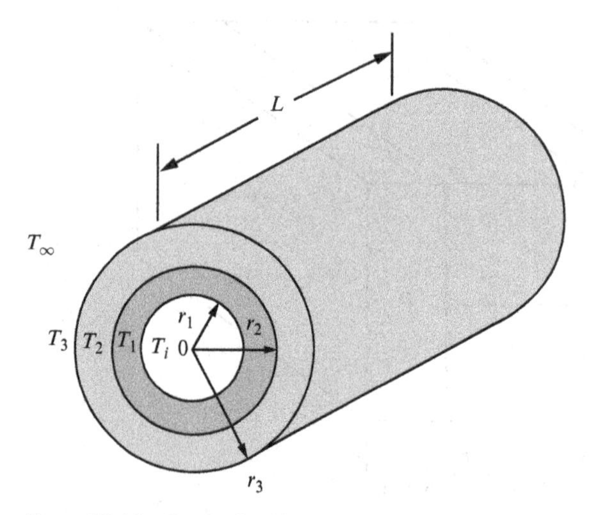

Figure P2.12 Steel tube with insulation layer.

2.13 For Problem 2.12, determine the following temperatures:

(a) Inner tube surface T_1

(b) Interface between outer steel–tube surface and insulation T_2

(c) Outer insulation surface T_3.

2.14 A 100-m steel tube of $0.094m$ internal diameter and $0.1m$ external diameter ($k_s = 16\,W/m^2.K$) is used to transport hot water at 80°C and is exposed to atmospheric air at 20°C. The heat transfer coefficient on the water side is $2000\,W/m^2.K$ and on the air side $200\,W/m^2.K$. Determine the rate of heat loss to the air.

2.15 A layer of insulation material of 50-mm thick is added to the tube in Problem 12.14. If the thermal conductivity of the insulation is $0.1\,W/m.K$, what is the percentage reduction in the rate of heat loss?

2.16 A compressed air pipe has inner radius of $0.2m$ and outer radius of $0.7m$. The inner and outer surface temperature are, respectively, 200°C and 27°C. For one-dimensional heat transfer in the radial r-direction,

(a) Plot variation of temperature in the pipe if the thermal conductivity is constant $k = 14.5 W/m.K$

(b) Plot variation of temperature in the pipe if the thermal conductivity changes with temperature $k = 10.5(1 + 0.0014T) W/m.K$, where T is in degrees K.

2.17 For Problem 2.16, determine the rate of heat conduction for (a) and (b).

2.18 A solid steel sphere of 0.2 m diameter is cooled by convection ($h = 400 W/m^2.K$) in air at $288K$. An internal heater generates $1M\ W/m^3$ in the sphere. What is the temperature at the centre and surface of the sphere? Take the thermal conductivity of steel as $14.5 W/m.K$, (213°C, 98°C).

2.19 A solid metallic sphere of 0.06 m diameter is cooled by convection ($h = 500 W/m^2.K$) in air at $283K$. An internal heater generates $1M\ W/m^3$ in the sphere. Plot the temperature profile for thermal conductivity $k = 14.5 W/m.K$.

2.20 For the data in Problem 2.19, plot the temperature profiles for $k = 10, 14.5$, and $25 W/m.K$.

2.21 A high-pressure sphere has inner radius of $0.2m$ and outer radius of $0.45m$. The inner and outer surface temperature are, respectively, 427°C and 27°C. For one-dimensional heat transfer in the radial r-direction:

(a) Plot variation of temperature in the pipe if the thermal conductivity is constant $k = 14.5 W/m.K$

(b) Plot variation of temperature in the pipe if the thermal conductivity changes with temperature as $k = 10.5(1 + 0.0014T) W/m.K$, where T is in degrees K.

2.22 For Problem 2.21, determine the rate of heat conduction for (a) and (b).

2.23 Consider a semi-infinite plate with a thickness of $150mm$ and with the origin of x at the symmetry axis of the plate. Heat $\dot{q} = 1.2 \times 10^6 W/m^3$ is internally generated and the heat transferred through the plate is dissipated by convection from the two walls to the surroundings at temperature 15°C with heat transfer coefficient of $500 W/m^2.K$. If both walls of the plate are uniformly cooled to a constant wall temperature, calculate the maximum temperature and the temperature at a distance of $0.025m$ from the centre. The thermal conductivity can be taken as equal to $14.5 W/m.K$.

2.24 A semi-infinite plate with a thickness of $0.15m$ and with the origin of x at the symmetry axis of the plate. Heat $\dot{q} = 1.0 \times 10^6 W/m^3$ is internally generated and the heat transferred through the plate is dissipated by convection from the two walls to the surroundings at temperature 10°C with heat transfer coefficient $500 W/m^2.K$. If both walls of the plate are uniformly cooled to a constant wall temperature, calculate the maximum temperature and the temperature at a distance of $0.025m$ from the centre of the plate. The thermal conductivity is $k = 10.5(1 + 0.0014T) W/m.K$, where T is in degrees K.

2.25 An electric kettle has a 2-kW resistance heater with a surface temperature of 120°C. The $5mm$ resistance wire is $1.0m$ long and has a thermal conductivity of $k = 20 W/m.K$. Assuming heat generation in the heater is uniform and the heat transfer is one-dimensional, what is the temperature at the centre of the wire?

2.26 Convection conditions are specified as the boundary condition at the surface of the wire in Problem 2.25 instead of surface temperature. The heat transfer coefficient is $40 W/m^2.K$ and the temperature of the water is 100°C. Determine the temperature at the centre and surface of the wire.

2.27 Heat is generated uniformly at the rate of $3.5\times10^7\ W/m^3$ in a spherical radioactive material ($d=10\,cm$) with a specified surface temperature of 95°C. For a steady one-dimensional heat conduction process, determine the temperature at the centre of the sphere if the thermal conductivity of the material is 18 $W/m.K$.

2.28 A plane wall ($k=45\,W/m.K$), 100mm thick, has heat generation at a uniform rate of $4.5\times10^6\ W/m^3$. The two sides of the wall are maintained at 180°C and 120°C. Assuming one-dimensional conduction, plot the temperature profile across the plate and determine the position and magnitude of maximum temperature.

2.29 A long rod of radius 50 cm with thermal conductivity of 10 $W/m.K$ contains radioactive material, which generates heat uniformly within the cylinder at a rate of $3.0\times10^4\ W/m^3$. The rod is cooled by convection from its cylindrical surface at $T_\infty=50°C$ with a heat transfer coefficient of $60\,W/m^2.K$. Determine the temperature at the centre and outer surface of the cylindrical rod.

2.30 Determine the temperature at the outer surface and centre of the cylindrical rod in Problem 2.29, assuming variable thermal conductivity $k=5(1+3.96\times10^{-3}T)$.

2.31 A long cylindrical tube has an inner radius of 50mm and an outer radius of 150mm. Heat is generated in the shell at the rate of $1.0\times10^3\ W/m^3$ and the thermal conductivity of the cylinder material as $0.5\,W/m.K$. It is observed that the maximum temperature occurs at a radius of 100mm and the temperature of the outer surface is 50°C:

(a) Deduce the relationship for the temperature profile for this specific problem from first principles

(b) Determine the maximum temperature and the temperature at the inner surface in the cylinder.

2.32 The walls of a tube transporting compressed gas is heated uniformly at the rate of $1.0\times10^6\ W/m^3$. The inside and outside radii of the tube are, respectively, 0.02m and 0.06m. Heat is dissipated by convection from the outer surface at $h=500\,W/m^2.K$, $T_\infty=10°C$. Plot the temperature variation in the cylinder wall and determine the temperature at the midpoint of the cylinder wall. Take the thermal conductivity of the tube material as $14.5\,W/m.K$.

2.33 Repeat Problem 2.32 for convection applied to the inner surface of the tube.

2.34 A thick-walled spherical tank, used to store high-pressure gas, is heated uniformly at the rate of $1.0\times10^6\ W/m^3$. The inside and outside radii of the spherical shell are, respectively, 0.02m and 0.06m. Heat is dissipated by convection from the outer surface at $h=250\,W/m^2.K$, $T_\infty=10°C$. Plot the temperature variation in the cylinder wall and determine the temperature at the midpoint of the cylinder wall. Take the thermal conductivity of the tube material as $14.5\,W/m.K$.

2.35 Repeat Problem 2.34 for convection applied to the inner surface of the sphere at $h=750\,W/m^2.K$ and $T_\infty=10°C$.

2.36 A thick-walled spherical shell is heated uniformly at the rate of $1.0\times10^6\ W/m^3$. The inside and outside radii of the spherical shell are, respectively, 0.02m and 0.06m. Heat is dissipated by convection from the inner and outer surfaces. The following data are provided:

- Inner surface $h_1=500\,W/m^2.K$, $T_{\infty1}=10°C$
- Outer surface $h_2=100\,W/m^2.K$, $T_{\infty2}=30°C$.

Determine the temperatures at both surfaces and at a depth of 0.01 m from the surface

3

Heat Transfer From Extended Surfaces

Extended surfaces or fins are small protrusions from a larger body into a surrounding fluid medium with the purpose of enhancing the rate of heating or cooling of the surface of the body.

The heat transfer rates for a single layer with convection heat transfer applied on the two outside surfaces of a plate, cylindrical shell, and spherical shell are, respectively

$$\dot{Q}_x = \frac{T_{\infty 1} - T_{\infty 2}}{\dfrac{1}{A_1 h_{c1}} + \dfrac{\delta}{Ak} + \dfrac{1}{A_2 h_{c2}}} \qquad (A_1 = A_2)$$

$$\dot{Q}_r = \frac{T_{\infty 1} - T_{\infty 2}}{\dfrac{1}{A_1 h_{c1}} + \dfrac{\ln \dfrac{r_2}{r_1}}{2\pi L k} + \dfrac{1}{A_2 h_{c2}}}$$

$$\dot{Q}_r = \frac{T_{\infty 1} - T_{\infty 2}}{\dfrac{1}{A_1 h_{c1}} + \dfrac{1}{4\pi k}\left(\dfrac{1}{r_1} - \dfrac{1}{r_2}\right) + \dfrac{1}{A_2 h_{c2}}}$$

In these equations, $T_{\infty 1}$, $T_{\infty 2}$ are the ambient temperatures; k is the thermal conductivity of the solid objects; h_{c1}, h_{c2} are the heat transfer coefficients; A_1, A_2 are the surface areas of the opposing sides of the objects; r_1 and r_2 are the inner and outer radii of the cylinder and sphere; and L is the length of the cylinder δ is the thickness of an infinite plate.

The denominator in each of these equations represent the thermal resistance to heat flow which can be reduced in order to increase the heat rate while maintaining the temperature gradient constant or can be adjusted to maintain constant heat rate while decreasing the temperature gradient. In either case, the thermal resistance can be reduced by increasing the area of one or both opposing surfaces. For a fixed geometry, the area can be increased by mounting fins of various shapes and sizes, examples of which are shown in Figures 3.1 and 3.2. If both surfaces are finned, the area on the side with the lower heat transfer coefficient is increased until $A_{f1} h_{c1} \cong A_{f2} h_{c2}$, where A_{f1}, A_{f2} are the areas of the finned surfaces on the two sides.

Heat Transfer Basics: A Concise Approach to Problem Solving, First Edition. Jamil Ghojel.

Companion website: www.wiley.com/go/ghojel/heat_transfer

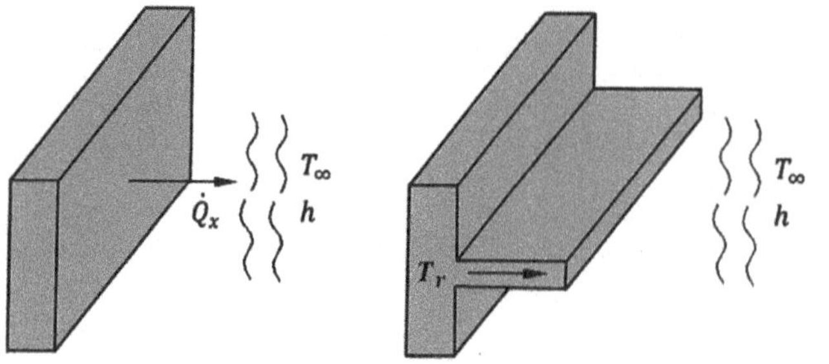

Figure 3.1 Heat transfer enhancement of a plate by means of a straight fin of rectangular profile (uniform thickness).

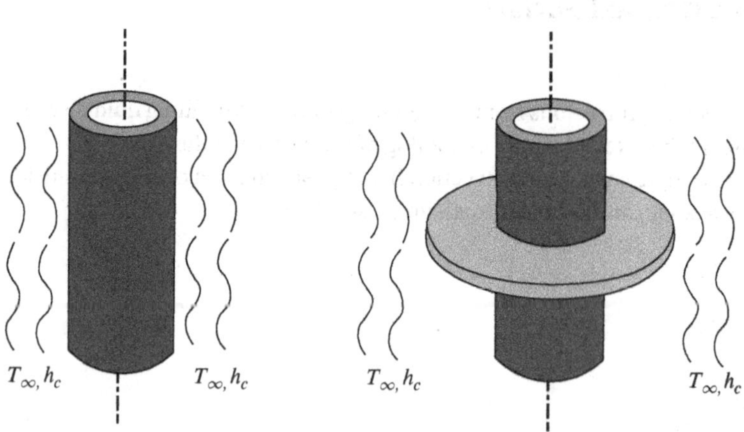

Figure 3.2 Heat transfer enhancement of hollow cylinder by means of an annular fin with uniform thickness.

3.1 Pin Fin of Rectangular Profile and Circular Cross-Section

3.1.1 Pin Fin of Finite Length and Un-Insulated Tip

Consider the mixed-mode heat transfer process in the pin fin shown in Figure 3.3. The pin has the shape of a slender solid cylinder forming a part of a plain plate. The analysis of the heat process will be based on the following assumptions:

- The pin has finite length L, diameter d, perimeter P_c, and cross-sectional area A_c
- The ratio of the length to diameter is large enough to assume negligible radial temperature variations in the cross-section
- Pin temperature at its root (base) T_r is constant
- The temperature changes only along the length of the pin in the x-direction
- The thermal conductivity is constant
- Heat is dissipated from the pin's circumferential surface ($A_f = \pi dL$) by convection with heat transfer coefficients h_f to the surroundings at temperature T_∞
- Heat is dissipated from the pin surface at the tip ($A_c = \pi d^2/4$) by convection with heat transfer coefficients h_t to the surroundings at temperature T_∞

Applying the law of conservation of energy (first law of thermodynamics) to the cylindrical element of the pin at temperature T, we can write:

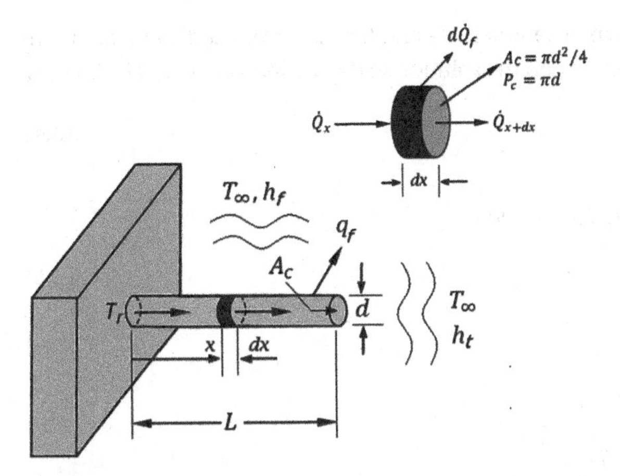

Figure 3.3 Mixed-mode heat transfer in a straight pin fin.

Rate of heat inflow − Rates of heat outflow = Rate of heat lost by convection

$$\dot{Q}_x - \dot{Q}_{x+dx} = d\dot{Q}_f \tag{3.1}$$

where

$$\dot{Q}_x = -kA_c \frac{dT}{dx}$$

$$\dot{Q}_{x+dx} = \dot{Q}_x + \frac{\partial}{\partial x}\left(\dot{Q}_x\right)dx = -kA_c \frac{dT}{dx} - kA_c \frac{d^2T}{dx^2}dx$$

$$d\dot{Q}_f = \pi d\, dx\, h_f\left(T - T_\infty\right) = P_c\, dx\, h_f\left(T - T_\infty\right)$$

where P_c and A_c are, respectively, the perimeter and area of the pin cross-section.

Substituting these values into Eq. (3.1) and dividing both sides by dx, we obtain the following second-order differential equation:

$$kA_c \frac{d^2T}{dx^2} = P_c\, h_f\left(T - T_\infty\right) \tag{3.2}$$

Since the ambient temperature T_∞ is constant, we can define the variable $\theta = T - T_\infty$ and use it to rewrite Eq. (3.2) as

$$kA_c \frac{d^2\theta}{dx^2} = P_c\, h_f \theta$$

or, for mathematical convenience, as

$$\frac{d^2\theta}{dx^2} - m^2\theta = 0 \tag{3.3}$$

where $m = +\sqrt{h_f P_c / kA_c} = \sqrt{4h_f / kd}$

For a given pin geometry and constant thermal conductivity and heat transfer coefficient, m is constant and has the units of inverse of length (m^{-1}) and the solution to the second-order Eq. (3.3) will be

$$\theta = C_1 e^{mx} + C_2 e^{-mx} \tag{3.4}$$

The boundary conditions are

At the root of the pin $x = 0, \theta = \theta_r$ and Eq. (3.4) yield

$$\theta_r = C_1 + C_2 \tag{3.5}$$

where $\theta_r = T_r - T_\infty$

At the tip of the pin $x = L$,

$$-kA_c\left(\frac{d\theta}{dx}\right)_{x=L} = h_t A_c (T_t - T_\infty) = h_t A_c \theta_t$$

or

$$\left(\frac{d\theta}{dx}\right)_{x=L} = -\frac{h_t \theta_t}{k} \tag{3.6}$$

The heat transfer coefficient at the tip of the pin h_t will typically be different from the value around the fin h_f.

For the boundary conditions at the tip of the pin, we obtain θ_t and $(d\theta/dx)_{x=L}$ from Eq. (3.4)

$$\theta_t = C_1 e^{mL} + C_2 e^{-mL} \tag{3.7}$$

$$\left(\frac{d\theta}{dx}\right)_{x=L} = mC_1 e^{mL} - mC_2 e^{-mL} \tag{3.8}$$

Combining Eqs (3.6), (3.7), and (3.8) yields

$$mC_1 e^{mL} - mC_2 e^{-mL} = -\frac{h_t}{k}\left(C_1 e^{mL} + C_2 e^{-mL}\right) \tag{3.9}$$

Simultaneous solution of Eqs (3.5) and (3.9) yields the constants C_1 and C_2

$$C_1 = \theta_r \frac{\left(1 - \frac{h_t}{mk}\right) e^{-mL}}{\left(e^{mL} + e^{-mL}\right) + \frac{h_t}{mk}\left(e^{mL} - e^{-mL}\right)}$$

$$C_2 = \theta_r \frac{\left(1 + \frac{h_t}{mk}\right) e^{mL}}{\left(e^{mL} + e^{-mL}\right) + \frac{h_t}{mk}\left(e^{mL} - e^{-mL}\right)}$$

Substituting for the values of C_1 and C_2 into Eq. (3.4), we obtain

$$\theta = \theta_r \frac{\left(1-\frac{h_t}{mk}\right)e^{-mL}e^{mx} + \left(1+\frac{h_t}{mk}\right)e^{mL}e^{-mx}}{\left(e^{mL}+e^{-mL}\right)+\frac{h_t}{mk}\left(e^{mL}-e^{-mL}\right)}$$

The equation for the temperature profile in the pin is usually written as

$$\frac{\theta}{\theta_r} = \frac{T-T_\infty}{T_r-T_\infty} = \frac{\left(e^{m(L-x)}+e^{-m(L-x)}\right)+\frac{h_t}{mk}\left(e^{m(L-x)}-e^{-m(L-x)}\right)}{\left(e^{mL}+e^{-mL}\right)+\frac{h_t}{mk}\left(e^{mL}-e^{-mL}\right)} \tag{3.10}$$

Making use of the hyperbolic functions

$$\left(e^{m(L-x)}-e^{-m(L-x)}\right) = 2\sinh m(L-x)$$

$$\left(e^{m(L-x)}+e^{-m(L-x)}\right) = 2\cosh m(L-x)$$

$$\left(e^{mL}-e^{-mL}\right) = 2\sinh mL$$

$$\left(e^{mL}+e^{-mL}\right) = 2\cosh mL$$

Equation (3.10) can be rewritten in a more compact form as

$$\frac{\theta}{\theta_r} = \frac{T-T_\infty}{T_r-T_\infty} = \frac{\cosh m(L-x)+\left(\frac{h_t}{mk}\right)\sinh m(L-x)}{\cosh mL+\left(\frac{h_t}{mk}\right)\sinh mL} \tag{3.11}$$

The temperature at the pin tip can be found from Eq. (3.10) by putting $x = L$

$$\frac{\theta_t}{\theta_r} = \frac{T_t-T_\infty}{T_r-T_\infty} = \frac{2}{\left(e^{mL}+e^{-mL}\right)+\frac{h_t}{mk}\left(e^{mL}-e^{-mL}\right)}$$

or, from Eq. (3.11)

$$\frac{\theta_t}{\theta_r} = \frac{T_t-T_\infty}{T_r-T_\infty} = \frac{1}{\cosh mL+\left(\frac{h_t}{mk}\right)\sinh mL}$$

The rate of heat transfer from the fin surface is equal to the rate of heat conducted through the pin at the root, i.e.,

$$\dot{Q}_x = -kA_c\left(\frac{d\theta}{dx}\right)_{x=0}$$

From Eq. (3.4), at $x = 0$

$$\left(\frac{d\theta}{dx}\right)_{x=0} = m(C_1 - C_2)$$

hence,

$$\dot{Q}_x = -kA_c m(C_1 - C_2)$$

Substituting for the values of C_1 and C_2 and rearranging, we obtain

$$\dot{Q}_x = kA_c m\theta_r \frac{\sinh mL + \left(\frac{h_t}{mk}\right)\cosh mL}{\cosh mL + \left(\frac{h_t}{mk}\right)\sinh mL} \quad W \tag{3.12}$$

3.1.2 Pin Fin of Finite Length and Insulated Tip

The boundary conditions for a fin with an insulated tip (no heat dissipation at the tip) are at $x = 0$, $\theta = \theta_r$ and at $x = L$, $(d\theta / dx)_{x=L} = 0$.

The equation for the temperature profile can be determined as before by finding the constants C_1 and C_2 and substituting them into Eq. (3.4). However, a simpler approach is to equate the heat transfer coefficient h_t in Eq. (3.11) to zero to obtain

$$\frac{\theta}{\theta_r} = \frac{T - T_\infty}{T_r - T_\infty} = \frac{\cosh m(L - x)}{\cosh mL} \tag{3.13}$$

The resulting temperature at the pin tip and rate of heat transfer from the fin surface are then, respectively

$$\frac{\theta_t}{\theta_r} = \frac{T_t - T_\infty}{T_r - T_\infty} = \frac{1}{\cosh mL}$$

$$\dot{Q}_x = kA_c m\theta_r \tanh mL \quad W \tag{3.14}$$

The equations for the insulated tip are simpler to handle and they can be used for fin pins without tip insulation by correcting the fin length as follows:

Area of the tip = area of the fin that needs to be added to account for heat loss from the tip

$$\frac{\pi}{4} d^2 = \pi d \Delta L$$

hence, $\Delta L = d / 4$, and the fin length in Eqs (3.13) and (3.14) will then be replaced by $L_c = L + \Delta L$.

Example 3.1 A steel rod 20 mm in diameter and 120 mm long is mounted on a wall at 250°C. The rod is exposed to the surroundings at 15°C. If the convection heat transfer coefficient is $12\,W/m^2.K$, determine the heat lost by the rod. Take the thermal conductivity of steel as $45\,W/m.K$.

Solution

$$L_c = L + \Delta L = L + d/4 = 120 + 20/4 = 125 mm$$

$$\dot{Q}_x = kA_c m\theta_r \tanh mL_c \quad W$$

$$A_c = \frac{\pi d^2}{4} = \frac{\pi \times 0.02^2}{4} = 3.142 \times 10^{-4}\ m^2$$

$$m=\sqrt{\frac{hP_c}{kA_c}}=\sqrt{\frac{12\times\pi\times0.02\times4}{45\times\pi\times0.02^2}}=7.3$$

$$mL_c=0.125\times7.3=0.913$$

$$\dot{Q}_x=45\times3.142\times10^{-4}\times7.3\times(250-15)\times\tanh 0.913=17W$$

3.1.3 Pin Fin of Infinite Length

The boundary conditions in this case are

At $x=0$ $\quad\theta=\theta_r=T_r-T_\infty$
At $x=\infty$ $\quad\theta=0$

From Eq. (3.4) at $x=0$

$$\theta_r=C_1+C_2$$

At $x=\infty$
$e^{-m\infty}=0$ and $\theta=0$; hence, $C_1e^{m\infty}=0$, which can be true only if $C_1=0$
Hence,

$$C_2=\theta_r$$

Substituting for the values of C_1 and C_2 in Eq. (3.4) we obtain

$$\frac{\theta}{\theta_r}=\frac{T-T_\infty}{T_r-T_\infty}=e^{-mx} \tag{3.15}$$

The rate of heat transfer from the fin surface is equal to the conduction heat transfer rate in the pin at the root, i.e.,

$$\dot{Q}_x=-kA_c\left(\frac{d\theta}{dx}\right)_{x=0}=-kA_c(-m\theta_r)$$

hence,

$$\dot{Q}_x=kA_cm\theta_r \quad W \tag{3.16a}$$

or

$$\dot{Q}_x=\sqrt{hP_ckA_c}\,\theta_r \tag{3.16b}$$

Figure 3.4 shows the effect of the type of material (thermal conductivity) on the temperature profile for a 0.2-m pin fin with no tip insulation. The temperatures for all materials decrease with increasing distance from the fin root (base), and the lower the thermal conductivity, the steeper the rate of decrease of temperature. If copper is used as fin material, the temperature of the fin becomes almost uniform along its entire length.

Figure 3.5 shows the effect of the coefficient of heat transfer h (assumed the same for the entire fin, including tip surface area) on the temperature profile for a 0.2-m pin fin with no tip insulation. The temperatures for all values of h decrease with increasing distance from the fin root (base), and the higher the heat transfer coefficient, the steeper the rate of decrease of temperature. Increasing h significantly can cause a sharp decrease in the dissipated heat rate, as can be seen in Figure 3.6.

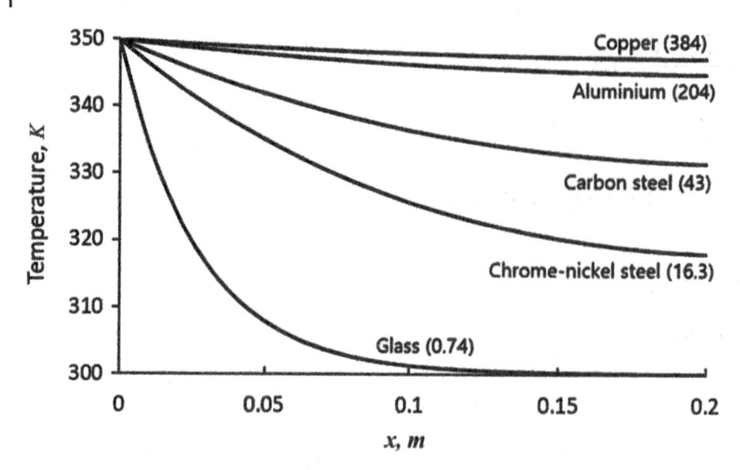

Figure 3.4 Effect of fin material on the temperature profile of a pin fin with no tip insulation $T_r = 350K$, $T_\infty = 300K$, $h_f = 15\,W/m^2.K$, $h_t = 15\,W/m^2.K$, $L = 0.2\,m$, $d = 0.06\,m$ (Values in brackets are thermal conductivities in $W/m.K$).

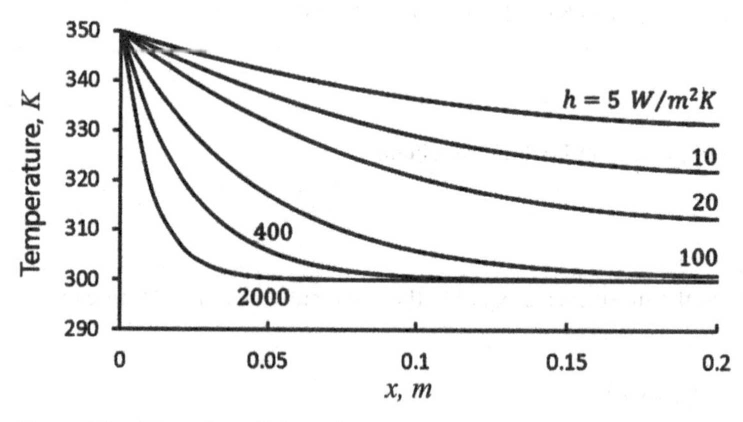

Figure 3.5 Effect of coefficient of heat transfer on the temperature profile of a pin fin with no tip insulation $h_f = h_f = h$, $k = 14.5\,W/m.K$, $T_r = 350K$, $T_\infty = 300K$, $L = 0.2\,m$, $d = 0.06\,m$.

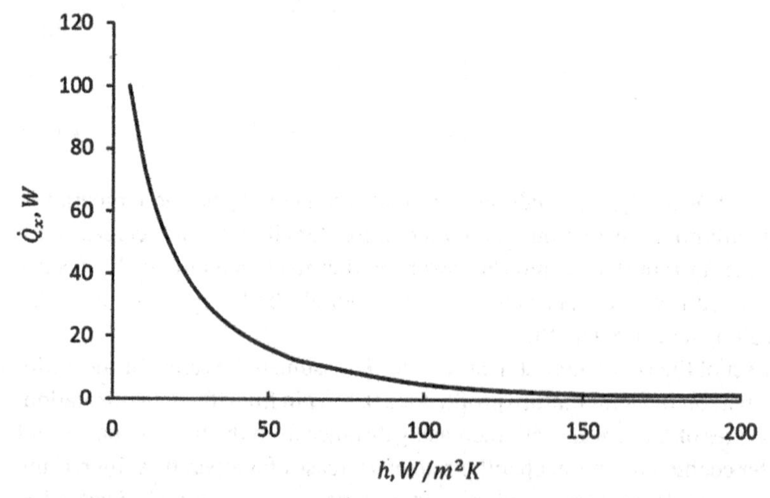

Figure 3.6 Effect of increasing the coefficient of heat transfer on the rate of heat dissipation by convection in a pin fin with no tip insulation $k = 14.5W/m.K$, $T_r = 350K$, $T_\infty = 300K$, $L = 0.2\,m$, $d = 0.06\,m$.

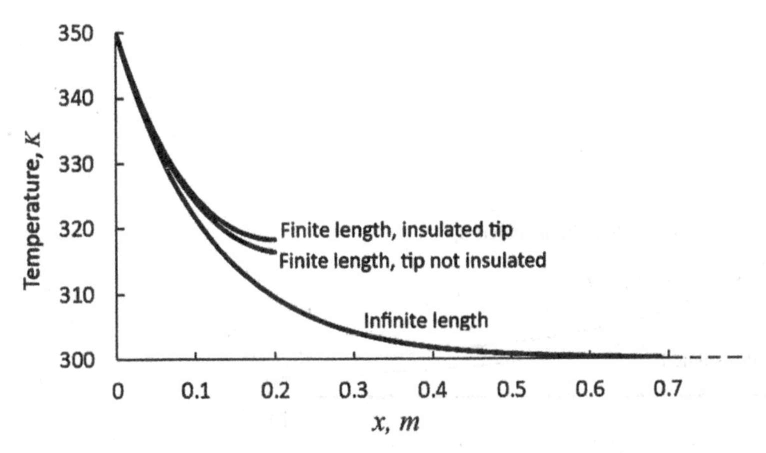

Figure 3.7 Temperature profiles in pin fins of different configurations $T_r = 350K$, $T_\infty = 300K$, $k = 14.5\ W/m.K$, $h_t = h_f = 15\ W/m^2.K$, $L = 0.2\ m$, $d = 0.06\ m$.

Very high values of h are encountered in high-velocity flows of fluids and boiling liquids and the rates of heat dissipation by convection in these cases are almost negligible.

Figure 3.7 shows the temperature profiles for 0.2-m long pin fin with and without tip insulation and for an infinite-length fin. The root (base) temperature is 350 K for all cases.

The temperatures profiles diverge as they decrease continuously towards the tip, with the values being highest for the fin with insulated tip.

3.1.4 Fin Efficiency

The effectiveness of the heat transfer process of a fin can be defined by a parameter known as the fin efficiency

$$\eta_f = \frac{\dot{Q}_{x(fin)}}{\dot{Q}_{x(iso)}} \tag{3.17}$$

where $\dot{Q}_{x(fin)}$ is given by Eq. (3.12) and $\dot{Q}_{x(iso)}$ is the heat transferred if the temperature of the entire fin is at the root temperature T_r (isothermal fin); i.e., $\theta = \theta_r$.

For a pin fin of finite length without tip insulation

$$\dot{Q}_{x(iso)} = \left(P_c L h_f + \frac{\pi}{4} d^2 h_t\right)(T_r - T_\infty)$$

and the fin efficiency is then

$$\eta_f = \frac{kA_c m\theta_r \dfrac{\sinh mL + \left(\dfrac{h_t}{mk}\right)\cosh mL}{\cosh mL + \left(\dfrac{h_t}{mk}\right)\sinh mL}}{\left(P_c L h_f + \dfrac{\pi}{4} d^2 h_t\right)} \tag{3.18}$$

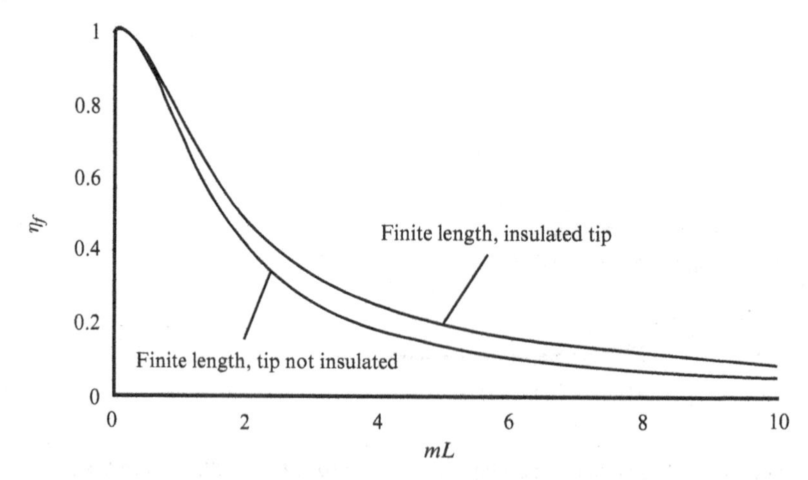

Figure 3.8 Fin efficiency η_f versus fin parameter mL: $T_r = 350K$, $T_\infty = 300K$, $k = 15\,W/m.K$, $h_t = h_f = var.$, $L = 0.2\,m$, $d = 0.06\,m$.

If the fin is insulated at the tip
$\dot{Q}_{x(fin)}$ is given by Eq. (3.14) and $\dot{Q}_{x(iso)} = P_c L h_f (T_r - T_\infty)$, and the fin efficiency is

$$\eta_f = \frac{kA_c m\theta_r \tanh mL}{P_c L h_f} = \frac{\tanh mL}{mL} \tag{3.19}$$

where $m = \sqrt{4h_f / kd} = \sqrt{2}\sqrt{2h_f / kd}$.

Figure 3.8 shows the pin fin efficiencies plotted versus the non-dimensional parameter mL, known as the fin parameter, for a finite-length pin fin with and without insulated tip. Equations (3.18) and (3.19) are used for the plots. The fin parameter is varied by varying the heat transfer coefficients h_f between 0.002 and 500 $W/m^2.K$ at a constant thermal conductivity $k = 15\,W/m.K$.

3.2 Straight Fin of Rectangular Profile and Uniform Thickness

Consider the straight fin shown in Figure 3.9. The fin profile is rectangular in shape, of constant thickness t, length L, and width w. Heat is conducted through the fin along its length in the x-direction only and is dissipated by convection. For a fin with large length-to-thickness ratio, the fin surface is $A_f = 2(Lw) + 2(Lt) \cong 2Lw$. The area at the tip is the cross-sectional area $A_t = wt$. The temperature profile for this pin can be determined using the same methodology used for the pin fin of rectangular profile by considering the elemental rectangular prism shown in Figure 3.9 and the equations will be identical to the ones derived for the pin fin, albeit having different magnitudes for the areas and perimeter of the cross-section.

The temperature profile

$$\frac{\theta}{\theta_r} = \frac{T - T_\infty}{T_r - T_\infty} = \frac{\cosh m(L-x) + \left(\frac{h_t}{mk}\right)\sinh m(L-x)}{\cosh mL + \left(\frac{h_t}{mk}\right)\sinh mL}. \tag{3.20}$$

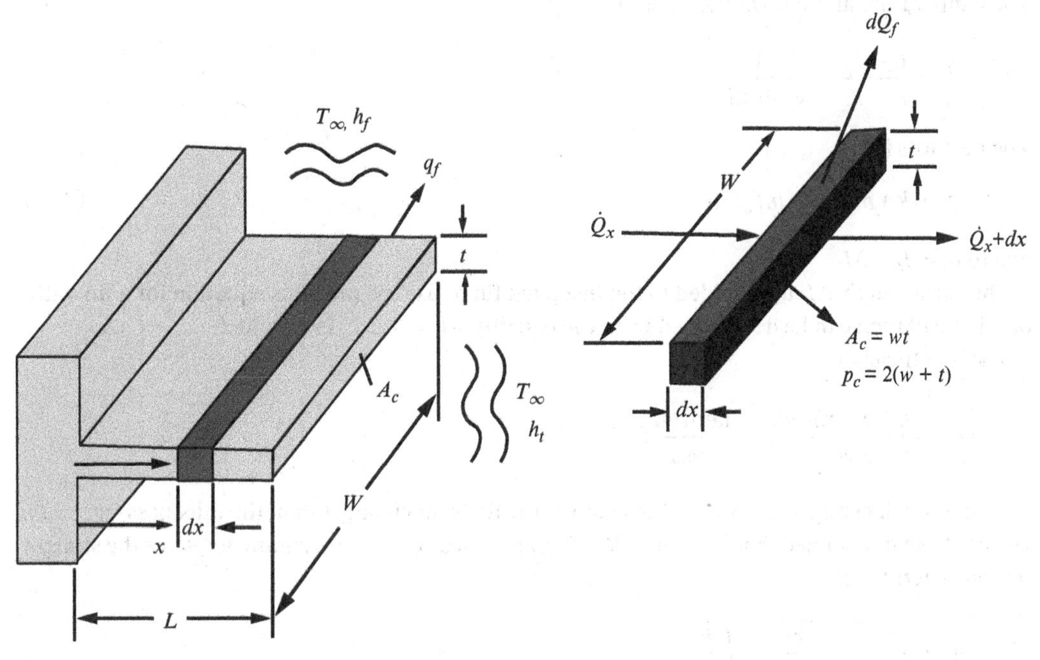

Figure 3.9 Straight fin with constant thickness and finite length.

The temperature at the fin tip

$$\frac{\theta_t}{\theta_r} = \frac{T_t - T_\infty}{T_r - T_\infty} = \frac{1}{\cosh mL + \left(\frac{h_t}{mk}\right)\sinh mL} \tag{3.21}$$

The rate of heat transfer from the fin surface (heat loss)

$$\dot{Q}_x = kA_c m\theta_r \frac{\sinh mL + \left(\frac{h_t}{mk}\right)\cosh mL}{\cosh mL + \left(\frac{h_t}{mk}\right)\sinh mL} \quad W \tag{3.22}$$

The fin efficiency

$$\eta_f = \frac{kA_c m}{\left(2wLh_f + wth_t\right)} \frac{\sinh mL + \left(\frac{h_t}{mk}\right)\cosh mL}{\cosh mL + \left(\frac{h_t}{mk}\right)\sinh mL} \tag{3.23}$$

It is assumed here that heat is dissipated by convection from the top, bottom surfaces with heat transfer coefficient h_f, and from the tip with heat transfer coefficient h_t.

If the tip of the fin is insulated, the temperature profile is

$$\frac{\theta}{\theta_r} = \frac{T - T_\infty}{T_r - T_\infty} = \frac{\cosh m(L_c - x)}{\cosh mL_c}$$

The temperature at the tip of the fin is

$$\frac{\theta_t}{\theta_r} = \frac{T_t - T_\infty}{T_r - T_\infty} = \frac{1}{\cosh mL_c}$$

The heat loss is

$$\dot{Q}_x = kA_c m\theta_r \tanh mL_c \quad W \tag{3.24}$$

where $L_c = L + \Delta L$

The extra length ΔL to be added to the insulated fin to use the previous equation for a fin without tip insulation can be determined from the equality $wt = 2w\Delta L$, from which $\Delta L = t/2$.

The fin efficiency

$$\eta_f = \frac{kA_c m \tanh mL_c}{2wL_c h_f} = \frac{\tanh mL_c}{mL_c}$$

In most engineering applications, the width of the fin is much larger than the thickness ($w >> t$). Hence, it can be assumed that $P_c = 2w + 2t \approx 2w$, and since $A_c = wt$, fin parameter m for the straight fin can be written as

$$m = \sqrt{\frac{hP_c}{kA_c}} = \sqrt{\frac{2hw}{kwt}} = \sqrt{\frac{2h}{kt}}$$

Figure 3.10 is an example of the temperature profile obtained by calculation and by computer modelling. The temperature at the root of the fin is kept constant at 60°C and the heat transfer

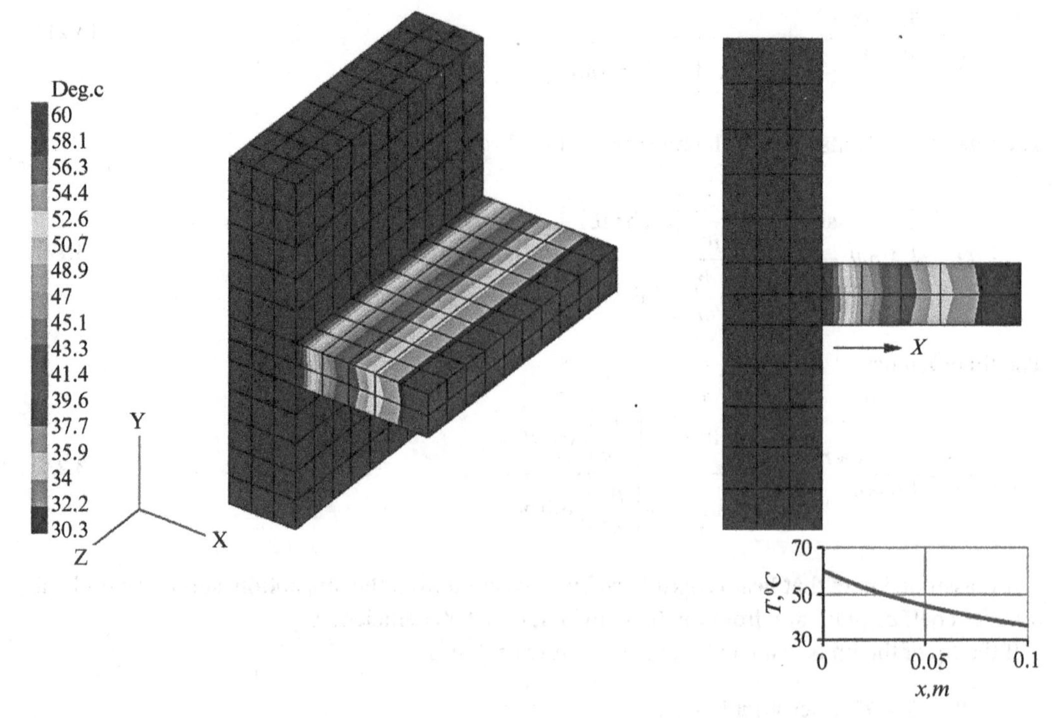

Figure 3.10 Temperature profile of a straight fin with the following data: $L = 0.1\ m$, $w = 0.2m$, $t = 0.03m$, $T_r = 60\ °C$, $T_\infty = 27°C$, $k = 5W/m.K$, $h_f = 10\ W/m^2.K$, $h_t = 75\ W/m^2.K$.

coefficients of convection from the fin surface and tip are, respectively, 10 and 75 $W/m^2.K$. The heat losses from the sides of the fin are ignored.

Example 3.2 A straight fin of rectangular profile has a thickness of 2.0 mm, length of 23 mm, and thermal conductivity of 14 $W/m.K$. The base of the fin is maintained at a temperature of 220°C, while the fin is exposed to a convection environment at 23°C with heat transfer coefficient 25 $W/m^2.K$. Calculate the heat lost per metre of fin width.

Solution

Referring to the fin in Figure 3.9,

$$L_c = L + \Delta L = 23 + \frac{2}{2} = 24mm$$

$$\dot{Q}_x = kA_c m\theta_r \tanh mL_c \quad W$$

$$A_c = wt = 0.002\ w = 0.048\ w$$

$$m = \sqrt{\frac{2h}{kt}} = \sqrt{\frac{50}{14 \times 0.002}} = 42.26$$

$$mL_c = 42.26 \times 0.024 = 1.014$$

$$\frac{\dot{Q}_x}{w} = 14 \times 0.002 \times 42.26 \times (220 - 23) \times \tanh 1.014 = 179 W/m$$

3.3 Pin Fin of Triangular Profile and Circular Cross-Section (Conical Pin Fin)

Figure 3.11 shows a pin fin of triangular (conical) profile with root diameter d_r and length L.

The governing differential equation for the fin is known as the modified Bessel equation

$$x^2 \frac{d^2\theta}{dx^2} + 2x \frac{d\theta}{dx} - m^2 x\theta = 0 \tag{3.25}$$

Solving this equation yields:

Temperature excess profile:

$$\frac{\theta}{\theta_r} = \frac{T - T_\infty}{T_r - T_\infty} = \left(\frac{L}{x}\right)^2 \frac{I_1\left(2\sqrt{2}m\sqrt{xL}\right)}{I_1\left(2\sqrt{2}mL\right)} \tag{3.26}$$

where $m = \sqrt{2h_f / kd_r}$

Heat flow rate:

$$\dot{Q}_f = \frac{\sqrt{2}\pi d_r^2 km\theta_r}{4} \frac{I_2\left(2\sqrt{2}mL\right)}{I_1\left(2\sqrt{2}mL\right)} \tag{3.27}$$

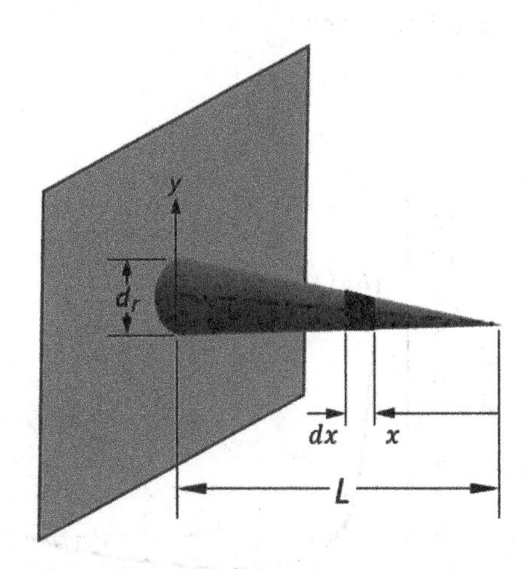

Figure 3.11 Pin fin of triangular profile and circular cross-section.

Fin efficiency:

$$\eta_f = \frac{4}{2\sqrt{2mL}} \frac{I_2\left(2\sqrt{2mL}\right)}{I_1\left(2\sqrt{2mL}\right)} \tag{3.28}$$

Equations (3.27) and (3.28) feature second-order modified Bessel functions of the first kind $I_2(x)$, where $x = 2\sqrt{2mL}$. Values of this function can be found in Table B.2 in Appendix B or calculated using the correlation in Table B.4. Bessel functions of higher order, such as $I_2(x)$ and $K_2(x)$, can be expressed by functions of lower orders for all integers by the following recurrence formulas

$$I_{n+1}(x) = I_{n-1}(x) - \frac{2n}{x} I_n(x),\ K_{n+1}(x) = K_{n-1}(x) + \frac{2n}{x} K_n(x)$$

For example, for $n = 1$,

$$I_2(x) = I_0(x) - \frac{2}{x} I_1(x),$$

$$K_2(x) - K_0(x) + \frac{2}{x} K_1(x)$$

Figure 3.12 is a plot of $I_2(x)$ and $K_2(x)$ for $x = 0 - 5$. More detailed tables of modified Bessel functions can be found in Abramowitz and Stegan (1972).

Figure 3.13 shows the efficiency of the conical pin fin plotted versus the non-dimensional fin parameter mL. Equation (3.28) is valid for values $mL = 0 - 3.5$.

3.4 Straight Fins of Variable Cross-Sectional Area

Practice shows that at a given heat transfer rate, straight fins with decreasing thickness are best suited for applications where reduction of the weight of fin is essential. Two types of fins will be considered below:

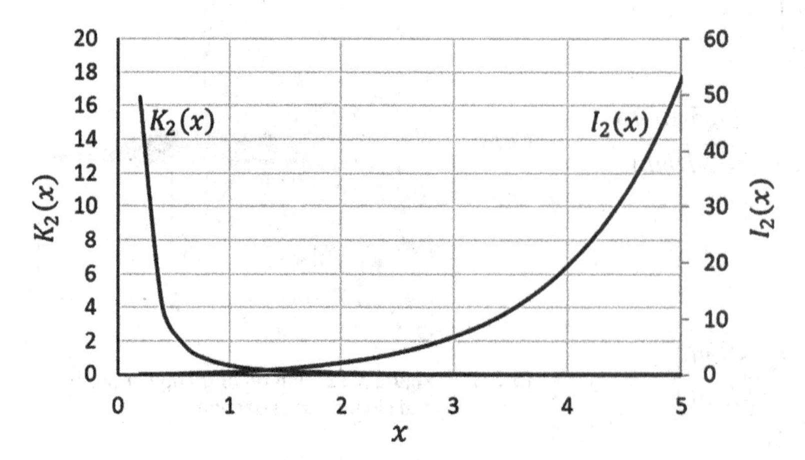

Figure 3.12 Second-order modified Bessel functions of the first and second kinds.

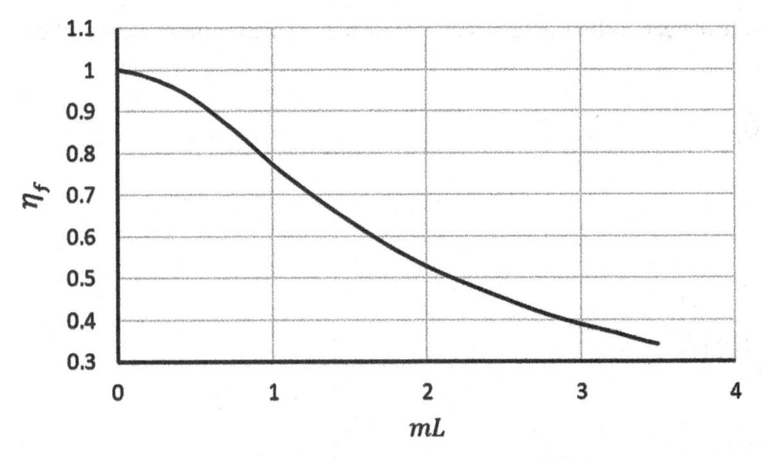

Figure 3.13 Fin efficiency of conical pin fin. $m = \sqrt{2h_f / kd_r}$.

1) Fin of triangular profile
2) Fin of trapezoidal profile.

3.4.1 Fin of Trapezoidal Profile

Consider the straight fin with trapezoidal profile shown in Figure 3.14. The fin length is L, width is w, thickness at the root is t_r, and thickness at the tip is t_t.

For mathematical convenience, the origin of the Cartesian coordinate system is located at the tip of a fictitious fin with a triangular profile of length x_r with the positive direction opposite to the direction of heat flow. As before, at a distance x from the origin 0, the energy equation for the element of width dx is:

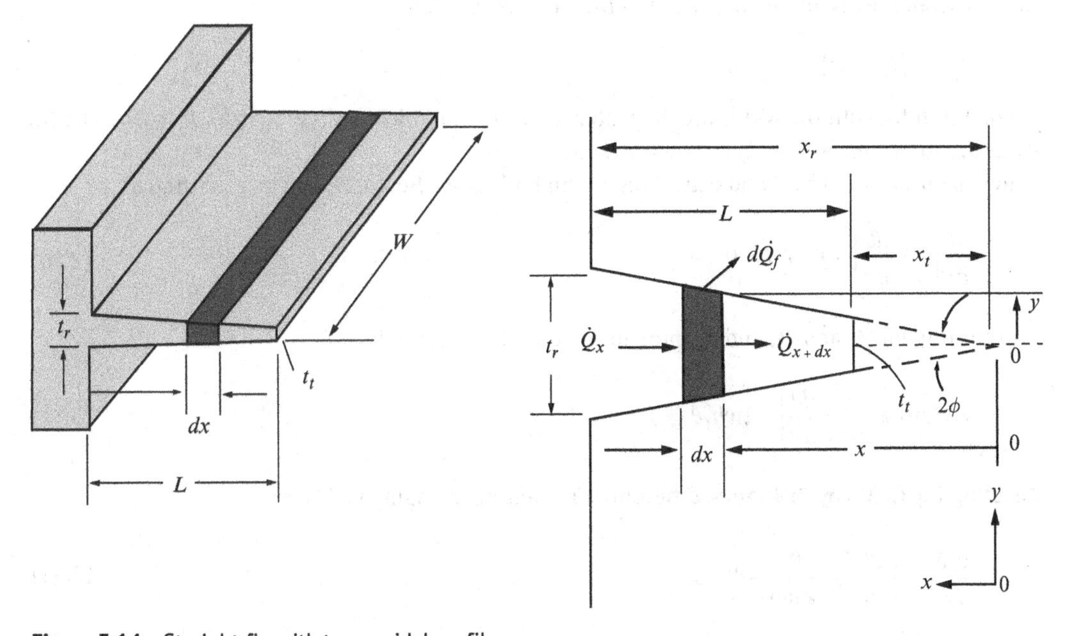

Figure 3.14 Straight fin with trapezoidal profile.

Rate of heat inflow at x – Rate of heat outflow at $(x+dx)=$ Rate of heat dissipation by convection from element surface

$$-\dot{Q}_{x+dx}-\left(-\dot{Q}_x\right)=d\dot{Q}_f$$

where

$$\dot{Q}_x=kA_c\frac{dT}{dx}$$

$$-\dot{Q}_{x+dx}=-\left[\dot{Q}_x+\frac{d}{dx}\left(\dot{Q}_x\right)dx\right]$$

$$d\dot{Q}_f=P_c\,dx\,h_f\left(T-T_\infty\right)$$

hence,

$$-\dot{Q}_x-\frac{d}{dx}\left(\dot{Q}_x\right)dx+\dot{Q}_x=d\dot{Q}_f$$

$$-\frac{d}{dx}\left(\dot{Q}_x\right)dx=P_c\,dx\,h_f\left(T-T_\infty\right)$$

or

$$\frac{d}{dx}\left(kA_c\frac{dT}{dx}\right)dx=P_c\,dx\,h_f\left(T-T_\infty\right) \tag{3.29}$$

P_c and A_c are, respectively, the perimeter and area of the fin cross-section at distance x from the origin. The highlighted element has thickness $2y$, where $y=2x\tan\phi$ and $\tan\phi=t_r/2x_r$. At a distance x from the origin, the perimeter of the fin cross-section is

$$P_c=2\left(2y+w\right)\cdot$$

For a thin fin with the width much greater than the root thickness, $(w>>2b)$, $P_c\cong 2w$. The fin cross-sectional area at x is $A_c=2yw=2xw\tan\phi$.

Replacing $\left(T-T_\infty\right)$ by θ and cancelling dx on both sides, Eq. (3.29) can be rewritten as

$$\frac{d}{dx}\left(kA_c\frac{d\theta}{dx}\right)-P_c\,h_f\theta=0 \tag{3.30}$$

Substituting for A_c and P_c, and keeping in mind that x is variable

$$2w\tan\phi k\frac{d}{dx}\left(x\frac{d\theta}{dx}\right)-2wh_f\theta=0 \tag{3.31}$$

Dividing Eq. (3.31) by $2wk\tan\phi$, differentiating, and rearranging yields

$$\frac{d^2\theta}{dx^2}+\frac{1}{x}\frac{d\theta}{dx}-\frac{h_f}{xk\tan\phi}\theta=0 \tag{3.32}$$

To solve Eq. (3.32), a new variable z is introduced and the equation is rewritten as

$$\frac{d^2\theta}{dz^2}+\frac{1}{z}\frac{d\theta}{dz}-\frac{1}{z}\theta=0 \tag{3.33}$$

where

$$z=\frac{h_f x}{k\tan\phi}=\beta x\left(dx=\frac{dz}{\beta},\beta=\frac{h_f}{k\tan\phi},\ \tan\phi=\frac{t_t}{2x_t}=\frac{t_r}{2x_r}=\frac{t_r-t_t}{2L}\right)$$

Equation (3.33) is a modified Bessel equation and has the general solution

$$\theta=C_1 I_0\left(2\sqrt{z}\right)+C_2 K_0\left(2\sqrt{z}\right) \tag{3.34}$$

where I_0 and K_0 are, respectively, zero-order modified Bessel functions of the first and second kinds.

The integration constants C_1 and C_2 are found from the boundary conditions:

- At the fin root $x=x_r$, $\theta=\theta_r=T_r-T_\infty$
- At the fin tip $x=x_t$, $\theta=\theta_t=T_t-T_\infty$ and, ignoring heat loss at the tip, $(d\theta/dx)_{x=x_t}=0$.

T_r and T_t are the temperatures at the root and tip of the fin, respectively. T_∞ is the temperature of the surroundings.

The resulting governing equations, using subscripts r and t for fin root and fin tip, respectively, are then:

Temperature profile:

$$\frac{\theta}{\theta_r}=\frac{T-T_\infty}{T_r-T_\infty}=\frac{I_0\left(2\sqrt{z}\right)K_1\left(2\sqrt{z_t}\right)+I_1\left(2\sqrt{z_t}\right)K_0\left(2\sqrt{z}\right)}{I_0\left(2\sqrt{z_r}\right)K_1\left(2\sqrt{z_t}\right)+I_1\left(2\sqrt{z_t}\right)K_0\left(2\sqrt{z_r}\right)} \tag{3.35}$$

Temperature at the tip:

$$\frac{\theta_t}{\theta_r}=\frac{T_t-T_\infty}{T_r-T_\infty}=\frac{I_0\left(2\sqrt{z_t}\right)K_1\left(2\sqrt{z_t}\right)+I_1\left(2\sqrt{z_t}\right)K_0\left(2\sqrt{z_t}\right)}{I_0\left(2\sqrt{z_r}\right)K_1\left(2\sqrt{z_t}\right)+I_1\left(2\sqrt{z_t}\right)K_0\left(2\sqrt{z_r}\right)} \tag{3.36}$$

Rate of heat dissipation by convection is equal to the rate of heat conduction at the root of the fin determined from Fourier's law

$$\dot{Q}_x=-kA_r\left(\frac{d\theta}{dx}\right)_{x=x_r}=-kt_r w\left(\frac{d\theta}{dx}\right)_{x=x_r}$$

From the differentiation rules of the modified Bessel functions

$$\frac{d}{dx}\left[I_0(ax)\right]=aI_1(ax);\ \frac{d}{dx}\left[K_0(ax)\right]=-aK_1(ax)$$

and noting that $z=\beta x$, we can write

$$\frac{d}{dx}\left[I_0\left(2\sqrt{z}\right)\right]=\frac{d}{dx}\left[I_0\left(2\sqrt{\beta x}\right)\right]=\sqrt{\frac{\beta}{x}}I_1\left(2\sqrt{\beta x}\right)=\sqrt{\frac{\beta}{x}}I_1\left(2\sqrt{z}\right)$$

$$\frac{d}{dx}\left[K_0\left(2\sqrt{z}\right)\right]=\frac{d}{dx}\left[K_0\left(2\sqrt{\beta x}\right)\right]=-\sqrt{\frac{\beta}{x}}K_1\left(2\sqrt{\beta x}\right)=-\sqrt{\frac{\beta}{x}}K_1\left(2\sqrt{z}\right)$$

Differentiating Eq. (3.35), substituting $z = z_r$ in the functions above, and rearranging yields

$$\left(\frac{d\theta}{dx}\right)_{x=x_r}=\theta_r\sqrt{\frac{\beta}{x_r}}\frac{I_1\left(2\sqrt{z_r}\right)K_1\left(2\sqrt{z_t}\right)-I_1\left(2\sqrt{z_t}\right)K_1\left(2\sqrt{z_r}\right)}{I_0\left(2\sqrt{z_r}\right)K_1\left(2\sqrt{z_t}\right)+I_1\left(2\sqrt{z_t}\right)K_0\left(2\sqrt{z_r}\right)}$$

$x_r = z_r / \beta$ and $\beta = h_f / k \tan\phi$

hence

$$\sqrt{\beta / x_r} = \beta / \sqrt{z_r} = h_f / k \tan\phi \sqrt{z_r}$$

The rate of heat transfer in the direction from the root to the tip is then

$$Q_x=\frac{h_f t_r w \theta_r}{\tan\phi\sqrt{z_r}}\frac{I_1\left(2\sqrt{z_r}\right)K_1\left(2\sqrt{z_t}\right)-I_1\left(2\sqrt{z_t}\right)K_1\left(2\sqrt{z_r}\right)}{I_0\left(2\sqrt{z_r}\right)K_1\left(2\sqrt{z_t}\right)+I_1\left(2\sqrt{z_t}\right)K_0\left(2\sqrt{z_r}\right)} \tag{3.37}$$

Fin efficiency is

$$\eta_f=\frac{t_r}{2L\tan\phi\sqrt{z_r}}\frac{I_1\left(2\sqrt{z_r}\right)K_1\left(2\sqrt{z_t}\right)-I_1\left(2\sqrt{z_t}\right)K_1\left(2\sqrt{z_r}\right)}{I_0\left(2\sqrt{z_r}\right)K_1\left(2\sqrt{z_t}\right)+I_1\left(2\sqrt{z_t}\right)K_0\left(2\sqrt{z_r}\right)} \tag{3.38}$$

Equations (3.35) to (3.38) can be evaluated using the modified Bessel functions from Table B.2 in Appendix B. Alternatively, the mathematical correlations in Table B.2 can be used. These correlations are mathematical representation of the tabular data for use in computer codes or in spreadsheet analysis. The correlations are valid for the values of x within the range $0-10$.

Plots of the zero- and first-order Modified Bessel functions of the first and second kinds are shown in Figure 3.15. These functions are of integer order $n = 0$, and $n = 1$.

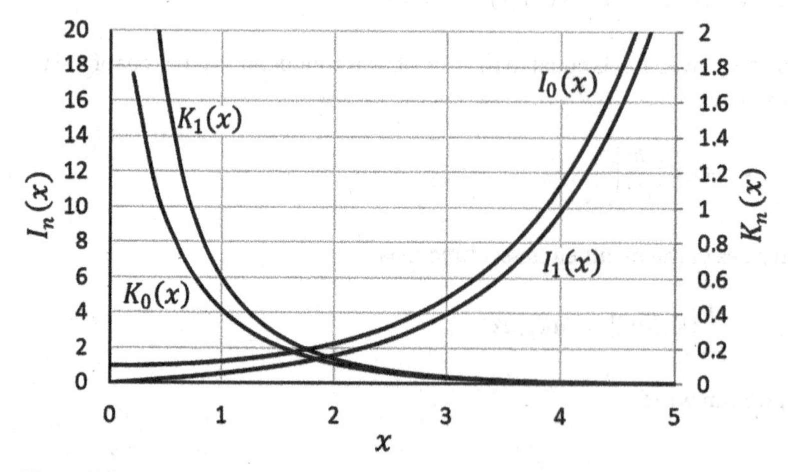

Figure 3.15 Modified Bessel functions of the first and second kinds.

3.4.2 Direct Solution of the Straight Fin of Trapezoidal Profile

Repeated solutions of Eqs (3.35) to (3.38) using modified Bessel functions could be cumbersome, and a more direct solution method for quick determination of the fin efficiency and heat rate can be obtained as follows:

- The efficiency given by Eq. (3.38) can be rewritten as a function of two independent variables (βL) and τ

$$\eta_f = \frac{\tau}{(\tau-1)} \frac{1}{\sqrt{\left(\frac{\tau}{\tau-1}\right)}\beta L} \times$$

$$\times \frac{I_1\left(2\sqrt{\left(\frac{\tau}{\tau-1}\right)}\beta L\right)K_1\left(2\sqrt{\left(\frac{1}{\tau-1}\right)}\beta L\right) - I_1\left(2\sqrt{\left(\frac{1}{\tau-1}\right)}\beta L\right)K_1\left(2\sqrt{\left(\frac{\tau}{\tau-1}\right)}\beta L\right)}{I_0\left(2\sqrt{\left(\frac{\tau}{\tau-1}\right)}\beta L\right)K_1\left(2\sqrt{\left(\frac{1}{\tau-1}\right)}\beta L\right) + I_1\left(2\sqrt{\left(\frac{1}{\tau-1}\right)}\beta L\right)K_0\left(2\sqrt{\left(\frac{\tau}{\tau-1}\right)}\beta L\right)} \quad (3.39)$$

where $\tau = t_r / t_t\,(\tau > 1.0)$, $\beta = h_f / k \tan\phi$, and $\tan\phi = (t_r - t_t)/2L$ (Figure 3.14).

- Equation (3.39) is solved repeatedly and the efficiency is plotted versus βL for a range of values of τ, as shown in Figure 3.16. A more detailed enlarged version of this figure is given in Figures B.1 and B.2 in Appendix B comprising 10 curves for the ranges $\beta L = 0.2 - 20$ and $\tau = 1.35 - 6$, which can be used for quick approximate determination of fin efficiency.

The fin length L used in all equations should be replaced by the corrected length that accounts for the heat loss at the fin tip $L_c = L + t/2$. Once the efficiency is determined, the heat flow rate can be determined from

$$\dot{Q}_x = 2L_c w h_f \theta_r \eta_f \quad (3.40)$$

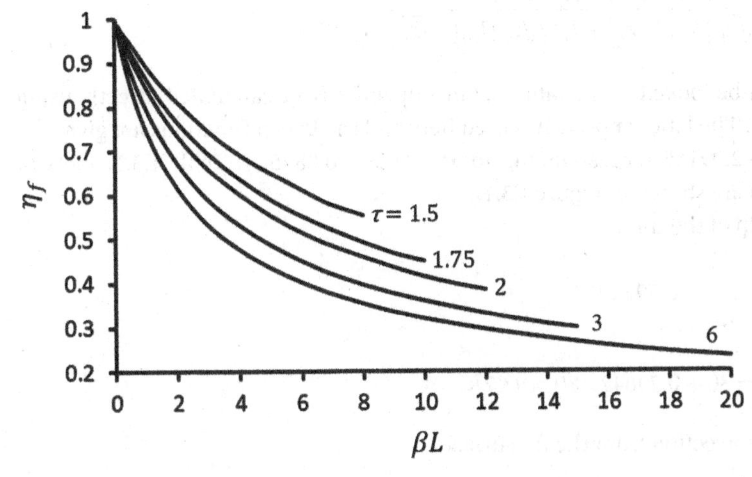

Figure 3.16 Efficiency of straight fin of trapezoidal profile versus βL at various values of τ.

Example 3.3 The following data are for the straight trapezoidal fin in Figure 3.14: $L=0.05m$, $w=1.0m$, $t_r=7mm$, $t_t=3mm$, $\theta_r=80°C$, $k=40W/mK$, $h_f=20W/m^2K$. Determine the temperature profile along the fin length and the heat loss rate if the ambient temperature is $T_\infty=30°C$.

Solution

$$\tan\phi=\frac{t_r-t_t}{2L}=\frac{7-3}{2\times1000\times0.05}=0.04$$

$$\beta=\frac{h_f}{k\tan\phi}=\frac{20}{40\times0.04}=12.5$$

$$x_r=\frac{t_r}{2\tan\phi}=\frac{7}{2\times1000\times0.04}=0.0875$$

$$L_c=L+t_t/2=0.05+0.003/2=0.0515m$$

$$x_t=x_r-L_c=0.0875-0.0515=0.036m$$

$$z=\beta x=12.5x$$

$$z_r=\beta x_r=12.5\times0.0875=1.09375$$

$$2\sqrt{z_r}=2.09165$$

$$z_t=\beta x_t=12.5\times0.036=0.44$$

$$2\sqrt{z_t}=1.341641$$

The equation for the temperature profile

$$\frac{\theta}{\theta_r}=\frac{T-T_\infty}{T_r-T_\infty}=\frac{I_0(2\sqrt{z})K_1(2\sqrt{z_t})+I_1(2\sqrt{z_t})K_0(2\sqrt{z})}{I_0(2\sqrt{z_r})K_1(2\sqrt{z_t})+I_1(2\sqrt{z_t})K_0(2\sqrt{z_r})}$$

The Bessel functions can be looked up in Table B.2 in Appendix B, or calculated directly using the correlation in Table B.2. The latter approach is used here and the Bessel functions are given for the range $2\sqrt{z}=1.34164-2.09165$ corresponding to $x=0.036-0.0875$ in Table E3.1. Plots of $T=f(X)$ and $\theta/\theta_r=f(X)$ are shown in Figure E3.1.

The temperature at the tip of the fin

$$\theta_t=\left(\frac{\theta}{\theta_r}\right)_{x=x_t}=\left(\frac{\theta}{\theta_r}\right)_{X=0.05}=0.79642$$

$$T_t=T_\infty+0.79642\theta_r=30+0.79642\times80=93.7°C$$

c) Heat rate dissipated by convection from the fin surface

$$\dot{Q}_x=\frac{h_f t_r w\theta_r}{\tan\phi\sqrt{z_r}}\frac{I_1(2\sqrt{z_r})K_1(2\sqrt{z_t})-I_1(2\sqrt{z_t})K_1(2\sqrt{z_r})}{I_0(2\sqrt{z_r})K_1(2\sqrt{z_t})+I_1(2\sqrt{z_t})K_0(2\sqrt{z_r})}$$

Table E3.1 Modified Bessel functions and calculated temperatures for Example 3.1.

X, m	x, m	$2\sqrt{z}$	$I_0(2\sqrt{z})$	$K_0(2\sqrt{z})$	$I_1(2\sqrt{z})$	$K_1(2\sqrt{z})$	θ/θ_r	T,°C
0	0.0875	2.09165	2.43220	0.10182	1.73161	0.12419	1.00000	110.00
0.005	0.0825	2.03101	2.33026	0.10965	1.63665	0.13441	0.96885	107.51
0.01	0.0775	1.96850	2.23101	0.11839	1.54374	0.14592	0.93953	105.16
0.015	0.0725	1.90394	2.13439	0.12822	1.45277	0.15897	0.91215	102.97
0.02	0.0675	1.83712	2.04034	0.13932	1.36363	0.17386	0.88686	100.95
0.025	0.0625	1.76777	1.94880	0.15195	1.27618	0.19097	0.86387	99.11
0.03	0.0575	1.69558	1.85973	0.16643	1.19028	0.21083	0.84345	97.48
0.035	0.0525	1.62019	1.77306	0.18317	1.10572	0.23411	0.82595	96.08
0.04	0.0475	1.54110	1.68876	0.20274	1.02231	0.26175	0.81184	94.95
0.045	0.0425	1.45774	1.60678	0.22590	0.93976	0.29505	0.80179	94.14
0.0515	0.0360	1.34164	1.50359	0.26322	0.83312	0.35005	0.79642	93.71

X – Distance measured from root to tip

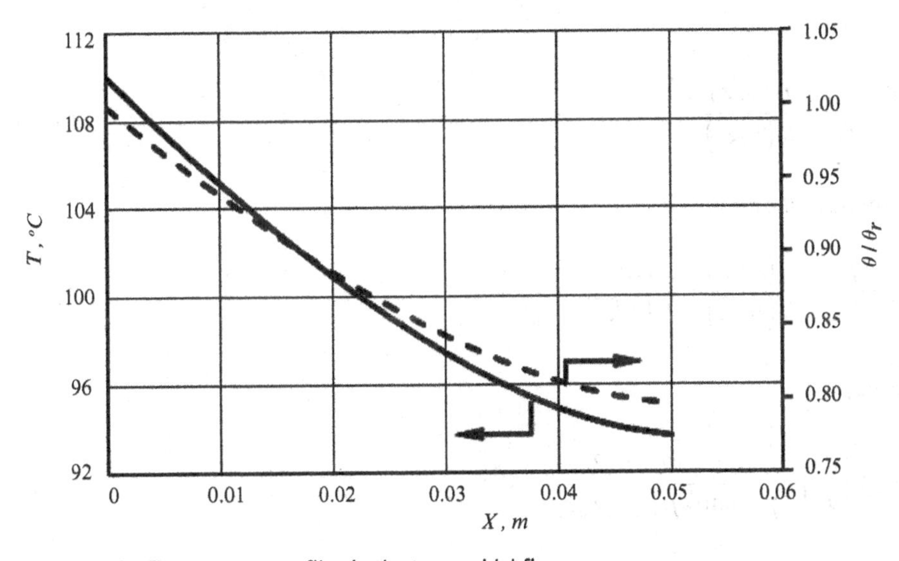

Figure E3.1 Temperature profiles in the trapezoidal fin.

$$\dot{Q}_x = \frac{20 \times 0.007 \times 1.0 \times 80}{0.04 \times \sqrt{1.09375}} \frac{1.73161 \times 0.33592 - 0.85771 \times 0.12419}{2.43220 \times 0.33592 + 0.85771 \times 0.10182} = 143.7\ W$$

3.4.3 Straight Fin of Triangular Profile

The fin of triangular profile shown in Figure 3.17 can be considered as a special case of the trapezoidal fin in Figure 3.14, with $x_r = L$, $x_t = 0$, $z_t = \beta L$, $\beta = h_f / k \tan\phi$, $z_t = 0$, $I_1(0) = 0$.

Equations (3.35) to (3.38) are then reduced to

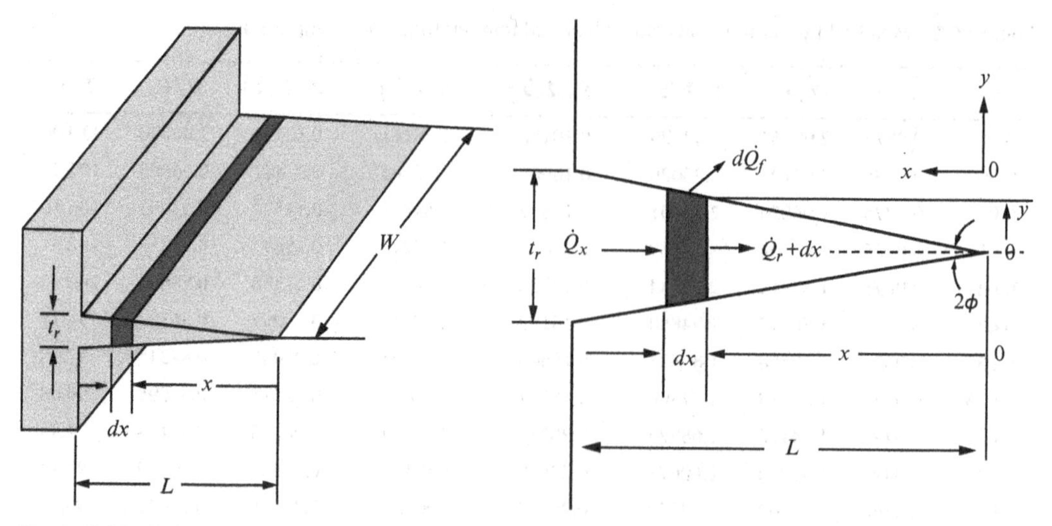

Figure 3.17 Straight fin with triangular profile.

Temperature profile:

$$\frac{\theta}{\theta_r} = \frac{T - T_\infty}{T_r - T_\infty} = \frac{I_0\left(2\sqrt{z}\right)}{I_0\left(2\sqrt{z_r}\right)} \tag{3.41}$$

For the temperature at the fin tip:

$$\frac{\theta_t}{\theta_r} = \frac{T_t - T_\infty}{T_r - T_\infty} = \frac{I_0\left(2\sqrt{z_t}\right)}{I_0\left(2\sqrt{z_r}\right)} = \frac{I_0(0)}{I_0\left(2\sqrt{z_r}\right)} = \frac{1}{I_0\left(2\sqrt{z_r}\right)} \tag{3.42}$$

Rate of heat transfer:

$$\dot{Q}_x = -kA_r\left(\frac{d\theta}{dx}\right)_{x=L} = \frac{h_f t_r w \theta_r}{\tan\phi\sqrt{z_r}} \frac{I_1\left(2\sqrt{z_r}\right)}{I_0\left(2\sqrt{z_r}\right)} \quad W \tag{3.43}$$

Fin efficiency:

$$\eta_f = \frac{2}{\left(2\sqrt{z_r}\right)} \frac{I_1\left(2\sqrt{z_r}\right)}{I_0\left(2\sqrt{z_r}\right)} \tag{3.44}$$

Knowing that $z_r = \beta L$, $\beta = h_f / k\tan\phi = 2h_f L / kt_r$, $m = \sqrt{2h_f / kt_r}$, Eq. (3.44) can also be expressed as

$$\eta_f = \frac{2}{(2mL)} \frac{I_1(2mL)}{I_0(2mL)} \tag{3.45}$$

Equations (3.41) to (3.45) can be evaluated using modified Bessel functions.

3.4.4 Correction Factor Solution Method for Straight Fins of Variable Cross-Sectional Area

The governing equations of straight fins of variable cross-sectional area, particularly trapezoidal fins, are cumbersome to handle and if a quick estimate of the heat loss rate is required and efficiency charts are unavailable, a correction factor method outlined by Mikheyev (1964) can be used. According to this method, the fin with a trapezoidal profile is replaced by an equivalent fin of rectangular profile, as shown in Figure 3.18, and the rate of heat dissipated from the fin surface is estimated from the following equation instead of Eq. (3.37)

$$\dot{Q}_x = \epsilon \dot{Q}_{eq} \left(\frac{A_f}{A_{eq}} \right) \tag{3.46}$$

The solution procedure is as follows:

- The heat loss $\dot{Q}_{eq}$ is calculated from Eq. (3.25) for a straight fin of rectangular profile with insulated tip having the same width, mean thickness $t = t_m = (t_r + t_t)/2$ and corrected length $L_c = L + t/2$ as the trapezoidal fin
- The surface area A_f of the trapezoidal fin, including the tip area, is calculated from $A_f = 2wL + wt_t = 2wL_c$ (Figure 3.18a). For a triangular fin $A_f = 2wL$
- The corrected surface area A_{eq} of the equivalent rectangular fin with insulated tip is calculated from $A_{eq} = 2w(L + t_m/2)$ (Figure 3.18b)

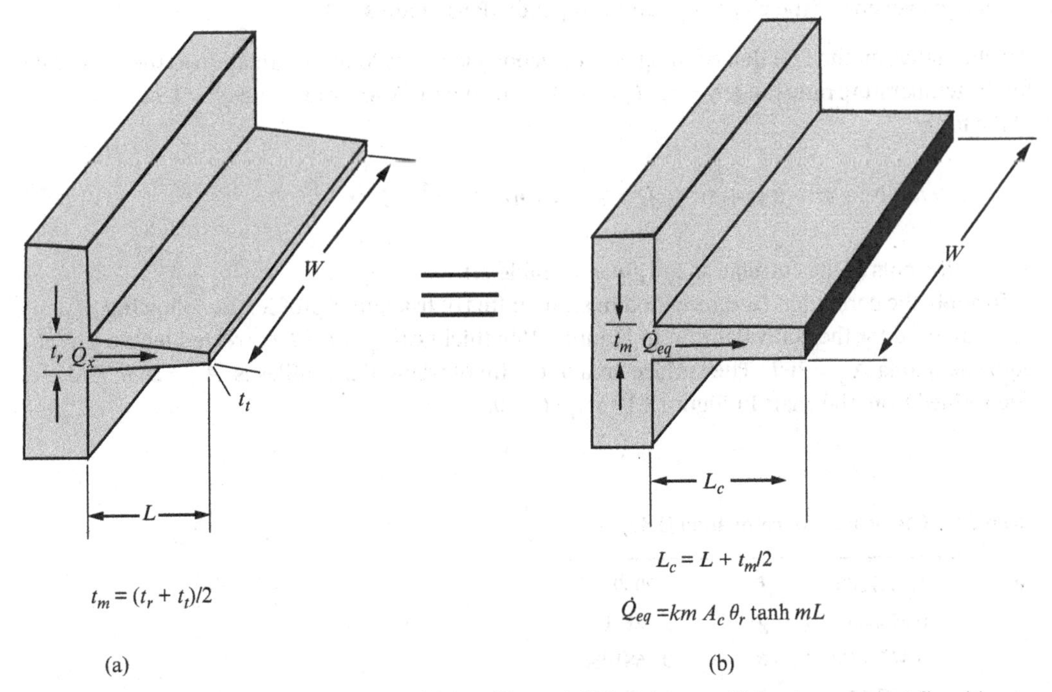

Figure 3.18 Straight fin with: (a) trapezoidal profile; and (b) its constant thickness equivalent fin with insulated tip.

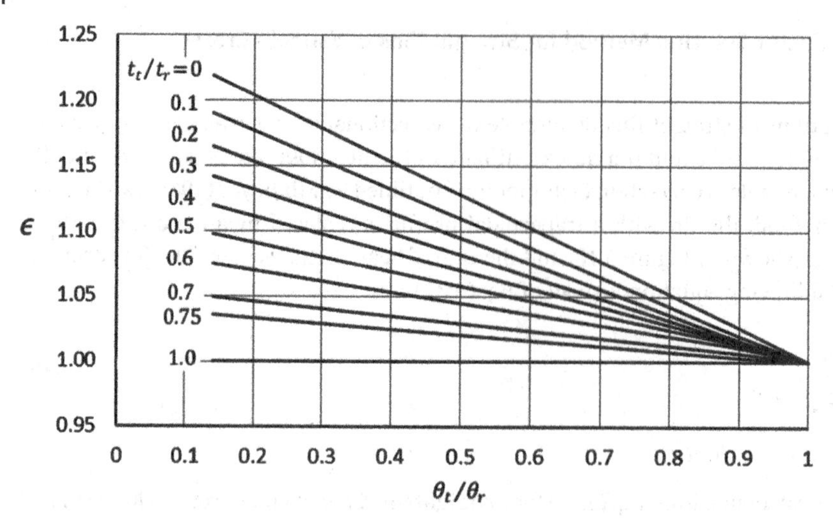

Figure 3.19 Correction factor chart for simplified calculation of heat loss from straight fins of trapezoidal profile.

- The correction factor ϵ, which is a function of both the temperature ratio θ_t/θ_r and thickness ratio t_t/t_r, is determined from the modified Mikheyev chart shown in Figure 3.19. The line corresponding to thickness ratio $t_t/t_r = 1$ in the chart represents a straight fin of rectangular profile (constant thickness) and the line $t_t/t_r = 0$ represents a straight fin with triangular profile. An enlarged version of the chart is given in Appendix B as Figures B.3.

An alternative method for determining the correction factor ϵ is to fit a correlation to the chart data for the temperature ratio range $\gamma = \theta_t/\theta_r = 0.14 - 1.0$, and thickness ratio range $\tau = t_t/t_r = 0 - 0.75$ of the form

$$\epsilon = a + b\gamma + c\tau + d\gamma^2 + e\tau^2 + f\gamma\tau + g\gamma^3 + h\tau^3 + i\gamma\tau^2 + j\gamma^2\tau \tag{3.47}$$

The coefficients of the correlation are given in Table 3.1.

To apply the correction factor method in straight fins of triangular profile, the following parameters are used for the equivalent fin in Figure 3.18b: thickness $t_m = t_r/2$; corrected length $L_c = L$; equivalent area $A_{eq} = 2wL$. The surface area of the fin of triangular profiles is $A_f = 2wL$, and ϵ is determined from the chart in Figure 3.19 at $t_t/t_r = 0$.

Table 3.1 Coefficients for correlation (3.47).

a	1.25472801	*f*	0.29394
b	-0.253865	*g*	1.02E-14
c	-0.34502196	*h*	-0.1681088
d	-1.66E-14	*i*	-1.64E-02
e	0.2075224	*j*	-8.74E-16

Example 3.4 Use the correction factor method to estimate the heat loss and thermal efficiency from the fin in Example 3.3.

Solution

Data for the straight fin of uniform thickness and insulated tip:
Mean thickness: $t_m = 0.5(0.007 + 0.003) = 0.005\ m$
Equivalent length: $L_c = L + 0.5t_m = 0.05 + 0.005/2 = 0.0525\ m$
Cross-sectional perimeter: $2(w + t_m) = 2(1 + 0.005) = 2.01\ m$
Cross-sectional area: $A_c = wt_m = 1 \times 0.005 = 0.005\ m^2$
Parameter m: $m = \sqrt{h_f P_c / kA_c} = \sqrt{(20 \times 2.01)/(40 \times 0.005)} = 14.1771/m$
Ratio of tip thickness to root thickness: $t_t/t_r = 0.003/0.007 = 0.4285$

Temperature ratio:

$$\frac{\theta_t}{\theta_r} = \frac{T_t - T_\infty}{T_r - T_\infty} = \frac{1}{\cosh mL} = \frac{1}{\cosh(14.17 \times 0.0525)} = 0.7751$$

Knowing θ_t/θ_r and t_t/t_r, the correction factor ϵ can be determined from Eq. (3.47) or from Figure 3.19. The value obtained is $\epsilon \cong 1.030277$.

The heat loss rate is

$$\dot{Q}_{eq} = \sqrt{kh_f P_c A_c}\theta_r \tanh mL_c = kmA_c\theta_r \tanh mL_c$$
$$= 40 \times 14.177 \times 0.005 \times 80 \times \tan\text{h}(14.177 \times 0.0525) = 143.3\ \ W$$

Equivalent heat transfer area $A_{eq} = 2wL_c = 2 \times 1.0 \times 0.0525 = 0.105\ \ m^2$

Heat transfer area for the trapezoidal fin:

$$A_f = 2wL + wt_t = 2 \times 1.0 \times 0.05 + 1.0 \times 0.003 = 0.103\ \ m^2$$

The heat loss rate from the trapezoidal fin:

$$\dot{Q}_x = \epsilon \dot{Q}_{eq}\left(\frac{A_f}{A_{eq}}\right) = 1.030277 \times 143.3 \times \left(\frac{0.103}{0.105}\right)$$
$$= 144.8W(\text{differenc of about}0.75\%\text{ compared with the exact solution})$$

Fin efficiency is given by:

$$\eta_f = \frac{\dot{Q}_x}{2wLh_f\theta_r} = \frac{144.8}{2 \times 1.0 \times 0.05 \times 20 \times 80} = 0.905$$

3.4.5 Straight Fin of Convex Parabolic Profile

The fin is shown schematically in Figure 3.20 and the governing differential equation is

$$\sqrt{x}\frac{d^2\theta}{dx^2} + \frac{1}{2\sqrt{x}}\frac{d\theta}{dx} - m^2\theta\sqrt{L} = 0 \tag{3.48}$$

where $\theta = T_x - T_\infty$, $m = \sqrt{2h_f / kt}$

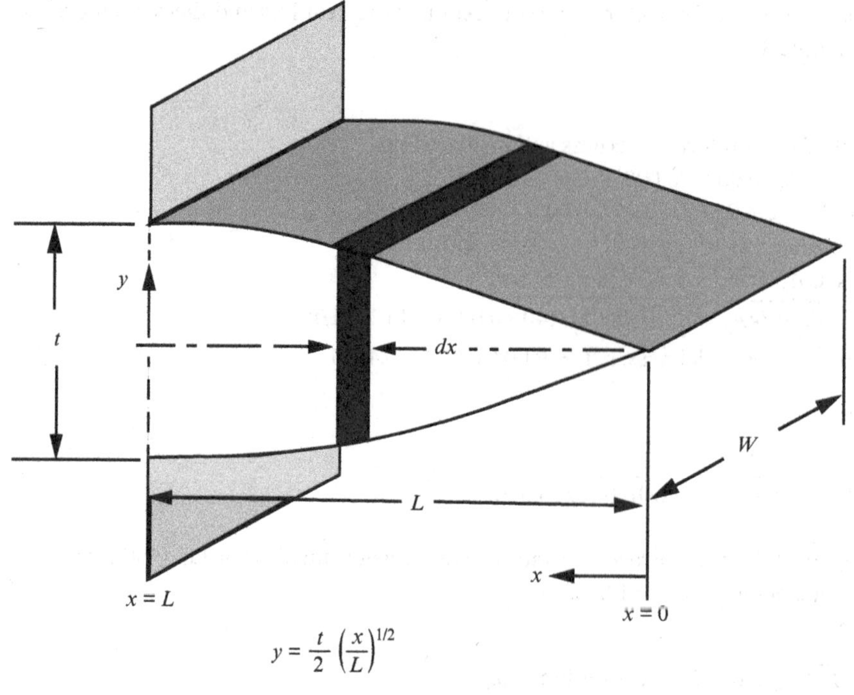

Figure 3.20 Straight fin of convex parabolic profile.

The solution to Eq. (3.48) leads to the following relations:

Excess temperature:

$$\frac{\theta}{\theta_r} = \left(\frac{x}{L}\right)^{1/4} \frac{I_{-1/3}\left(\frac{4}{3}mL^{1/4}x^{3/4}\right)}{I_{-1/3}\left(\frac{4}{3}mL\right)} \tag{3.49}$$

where $\theta_r = T_r - T_\infty$

Heat flow rate through fin cross-section at the root of the fin:

$$\dot{Q}_x = ktwm\theta_r \frac{I_{2/3}\left(\frac{4}{3}mL\right)}{I_{-1/3}\left(\frac{4}{3}mL\right)} \tag{3.50}$$

Fin efficiency:

$$\eta_f = \frac{\dot{Q}_x}{A_f h_f \theta_r} = \frac{ktwm}{A_f h_f} \frac{I_{2/3}\left(\frac{4}{3}mL\right)}{I_{-1/3}\left(\frac{4}{3}mL\right)} = \frac{1}{mL} \frac{I_{2/3}\left(\frac{4}{3}mL\right)}{I_{-1/3}\left(\frac{4}{3}mL\right)}$$

For $w >> t$, $A_f \approx 2wL$; hence,

$$\eta_f = \frac{\frac{4}{3}}{\left(\frac{4}{3}mL\right)} \frac{I_{2/3}\left(\frac{4}{3}mL\right)}{I_{-1/3}\left(\frac{4}{3}mL\right)} \tag{3.51}$$

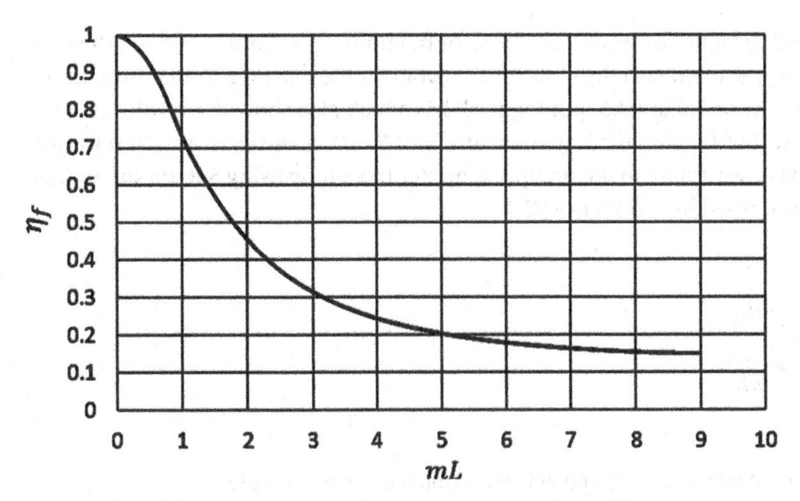

Figure 3.21 Efficiency of straight fin of convex parabolic profile.

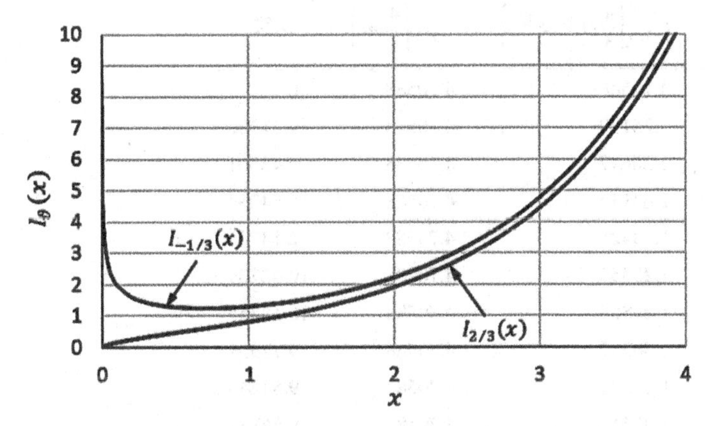

Figure 3.22 Plots of modified fractional-order Bessel functions.

The solution to Eqs (3.49) to (3.51) requires the knowledge of modified Bessel functions of fractional order $I_{2/3}$ and $I_{-1/3}$, which can be found in Table B.2 in Appendix B or calculated using the correlations in Table B.5. Figure 3.21 shows the efficiency of the straight fin of convex parabolic profile plotted as a function of the fin parameter mL.

Figure 3.22 is a plot of the modified fractional-order Bessel functions adapted from data published by the National Bureau of Standards, USA (1949). The curve of $I_{-1/3}(x)$ tends to infinity at $x = 0$ and displays a minimum at $x = 0.7$.

Example 3.5 Plot the temperature profile along a fin of convex parabolic profile for the following data:

$L = 0.1m$, $w = 1.0m$, $t = 6mm$, $k = 40 W/m.K$, $h_f = 60 W/m^2.K$, $\theta_r = 60°C$. Also, determine the fin efficiency and heat flow rate.

Solution

$$m = \sqrt{\frac{2h_f}{kt}} = \sqrt{\frac{2 \times 60}{0.006 \times 40}} = 22.36$$

The results of substituting values of Bessel functions of fractions $-1/3$ and $2/3$ from the correlations in Table B.5 in Appendix B into Eq. (3.49) and solving for the distance X measured from the root of the fin are shown in Table E3.5 and Figure E3.5, which also show the results of computer modelling. As a result of $I_{-1/3}(x)$ tending to ∞ at $x=0(X=0.1)$, the exact solution for the fin using Eq. (3.49) fails at or very close to the fin tip. Computer modelling using Solaria shows that $\theta_t \cong 8°C$, as shown by the dashed line in Figure E3.5.

Fin efficiency is

$$\eta_f = \frac{\frac{4}{3}}{\left(\frac{4}{3}mL\right)} \frac{I_{2/3}\left(\frac{4}{3}mL\right)}{I_{-1/3}\left(\frac{4}{3}mL\right)}$$

Table E3.5 Bessel functions of fractions - 1/3 and 2/3 and calculated temperatures for Example 3.5.

$X = L - x$	x	$\theta_r\left(\frac{x}{L}\right)^{1/4}$	$I_{-1/3}\left(\frac{4}{3}mL^{1/4}x^{3/4}\right)$	$I_{-1/3}\left(\frac{4}{3}mL\right)$	θ,°C
0	0.1	60	4.70588	4.70588	60
0.02	0.08	56.74449	3.25199	4.70588	39.21326
0.04	0.06	52.80670	2.26647	4.70588	25.43301
0.06	0.04	47.71624	1.63146	4.70588	16.54254
0.074	0.026	42.84445	1.33421	4.70588	12.14726
0.082	0.018	39.08133	1.25555	4.70588	10.42709
0.086	0.014	36.70145	1.29868	4.70588	10.12855
0.09	0.01	33.74047	1.36397	4.70588	9.77951
0.094	0.006	29.69539	1.51221	4.70588	9.54245
0.098	0.002	22.56361	1.99411	4.70588	9.56131
0.099	0.001	18.97366	2.40152	4.70588	9.68272

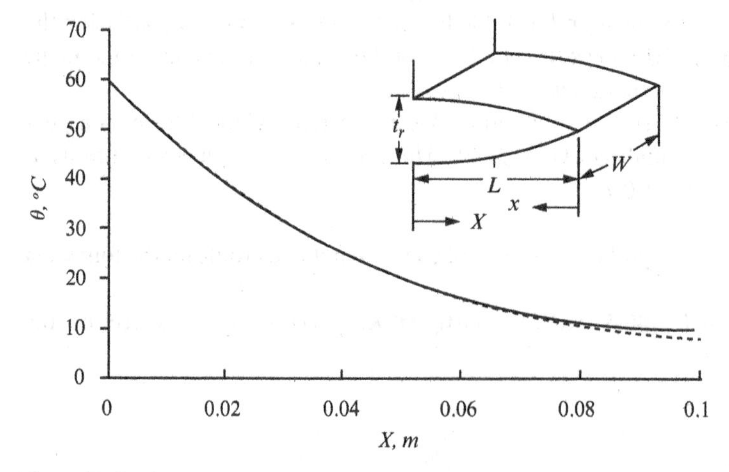

Figure E3.5 Temperature profile along the fin length.

$$mL = 2.236,\ \frac{4}{3}mL = 2.9813,\ I_{2/3}\left(\frac{4}{3}mL\right) = 4.37658,\ I_{-1/3}\left(\frac{4}{3}mL\right) = 4.71531$$

$$\eta_f = \frac{1}{2.236}\frac{4.32191}{4.70588} = 0.41$$

$$\eta_f = \frac{\dot{Q}_x}{A_f h_f \theta_r}$$

$$\dot{Q} = \eta_f A_f h_f \theta_r = 0.41 \times 0.2 \times 60 \times 60 = 297.5 W$$

3.4.6 Straight Fin of Concave Parabolic Profile

The fin is shown schematically in Figure 3.23 and the governing differential equation is

$$x^2 \frac{d^2\theta}{dx^2} + 2x\frac{d\theta}{dx} - m^2L^2\theta = 0 \tag{3.52}$$

The solution of this differential equation leads to the following relations:

Excess temperature:

$$\frac{\theta}{\theta_r} = \left(\frac{x}{L}\right)^{\alpha} \tag{3.53}$$

where $\theta_r = T_r - T_\infty$, $\alpha = -1/2 + (1/2)\left(1 + 4m^2L^2\right)^{1/2}$

Heat flow rate through fin cross-section at the root of the fin:

$$\dot{Q}_x = \frac{ktw\theta_r}{2L}\left[-1 + \sqrt{1 + 4(mL)^2}\right] \tag{3.54}$$

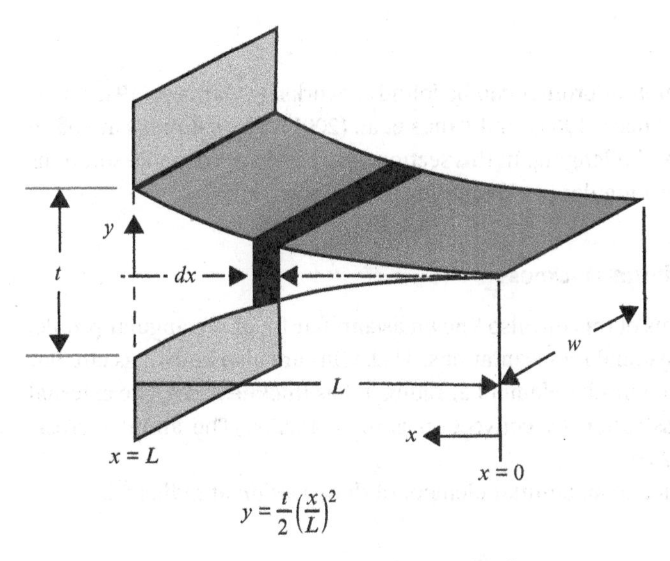

Figure 3.23 Straight fin of concave parabolic profile.

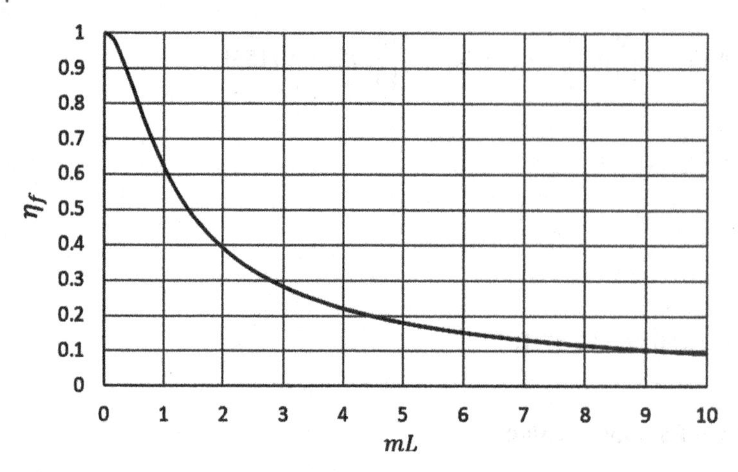

Figure 3.24 Efficiency of straight fin of concave parabolic profile.

Fin efficiency:

$$\eta_f = \frac{\dot{Q}_x}{A_f h_f \theta_r} = \frac{ktw\theta_r}{2L\left(2Lwh_f\theta_r\right)}\left[-1+\sqrt{1+4\left(mL\right)^2}\right] \tag{3.55}$$

For $w >> t$, $A_f \approx 2wL$, and by multiplying the denominator and numerator by $\left[-1-\sqrt{1+(2mL)^2}\right]$, Eq. (3.55) is reduced to

$$\eta_f = \frac{2}{1+\sqrt{4\left(mL\right)^2+1}} \tag{3.56}$$

The plot of Eq. (3.56) is shown in Figure 3.24.

3.5 Annular Fins

Exact solutions of annular fins of different profiles can be found in works by Gardner (1945), Bert (1963), Mikheyev (1964), Smith and Sucec (1969), and Kraus et al. (2001). The solutions are often lengthy and could be mathematically challenging. In this section, exact and approximate solutions will be given for the annular fin of rectangular profile (uniform thickness) only.

3.5.1 Straight Annular Fin of Uniform Thickness

Figure 3.25 shows schematic diagrams of this fin, also known as annular fin of rectangular profile, in two-dimensional and three-dimensional representations. These fins are also known as circular or annular fins. The dark highlighted annular element at radius r has thickness Δr. The external elemental area exposed to heat dissipation by convection is $A_f = 4\pi r\Delta r$. The annular cross-sectional area of the fin at r is $A_c = 2\pi rt$.

The energy balance equation applied to an annular element of thickness dr at radius r is

$$\dot{Q}_r - \dot{Q}_{r+dr} = d\dot{Q}_f$$

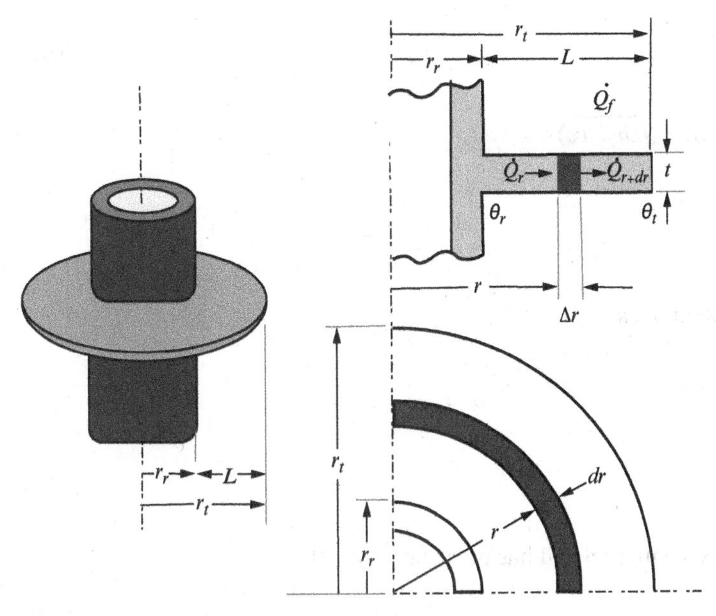

Figure 3.25 Straight annular fin of uniform thickness.

where

$$\dot{Q}_r = -kA_c \frac{d\theta}{dr}$$

$$\dot{Q}_{r+dr} = \dot{Q}_r + \frac{d\dot{Q}_r}{dr} dr = -kA_c \frac{d\theta}{dr} - \frac{d}{dr}\left(kA_c \frac{d\theta}{dr}\right)$$

$$d\dot{Q}_f = A_f h_f \theta$$

hence,

$$\frac{d}{dr}\left(kA_c \frac{d\theta}{dr}\right) = A_f h_f \theta$$

From the geometry, the perimeter at distance r from the centre of the annular fin is $P_c = 2\pi r$, the external elemental area exposed to heat dissipation by convection is $A_f = 2P_c dr = 4\pi r dr$, and the cross-sectional area of the fin at r is $A_c = 2\pi rt$. Considering that the fin cross-sectional area varies with r, the energy balance equation can now be written as

$$2\pi tk \frac{d}{dr}\left(r \frac{d\theta}{dr}\right) = 4\pi r \Delta r h_f \theta$$

$$2\pi tk \left[r \frac{d^2\theta}{dr^2} + \frac{d\theta}{dr}\right] = 4\pi r \Delta r h_f \theta$$

Letting Δr tend to zero, dividing both sides of the equation by $(2\pi rtk)$, and rearranging, we obtain

$$\frac{d^2\theta}{dr^2}+\frac{1}{r}\frac{d\theta}{dr}=\frac{2h_f}{tk}\theta \tag{3.57}$$

Let $z=\left(\sqrt{2h_f/tkr}\right)$, or $z=mr$ $(m=\sqrt{2h_f/tk})$
hence,

$$dr=\frac{1}{m}dz$$

Equation (3.57) can now be rewritten as

$$m^2\frac{d^2\theta}{dz^2}+\frac{m^2}{z}\frac{d\theta}{dx}-m^2\theta=0 \text{ or}$$

$$\frac{d^2\theta}{dz^2}+\frac{1}{z}\frac{d\theta}{dx}-\theta=0 \tag{3.58}$$

Equation (3.58) is modified Bessel equation and has the general solution

$$\theta=C_1I_0(z)+C_2K_0(z) \tag{3.59}$$

where $I_0(z)=I_0(mr)$ and $K_0(z)=K_0(mr)$ are, respectively, zero-order modified Bessel functions of the first and second kinds.

The integration constants C_1 and C_2 are found from the boundary conditions.

At the root of the fin $r=r_r$, $\theta=\theta_r=T_r-T_\infty$.

At the fin tip $r=r_t$, $\theta=\theta_t=T_t-T_\infty$, and ignoring heat loss at the tip, $(d\theta/dx)_{r=r_t}=0$.

Solving for C_1 and C_2 and substituting the result into Eq. (3.59), we obtain:

Temperature profile:

$$\frac{\theta}{\theta_r}=\frac{T-T_\infty}{T_r-T_\infty}=\frac{I_0(mr)K_1(mr_t)+I_1(mr_t)K_0(mr)}{I_0(mr_r)K_1(mr_t)+I_1(mr_t)K_0(mr_r)} \tag{3.60}$$

Temperature at the tip:

$$\frac{\theta_t}{\theta_r}=\frac{T_t-T_\infty}{T_r-T_\infty}=\frac{I_0(mr_t)K_1(mr_t)+I_1(mr_t)K_0(mr_t)}{I_0(mr_r)K_1(mr_t)+I_1(mr_t)K_0(mr_r)} \tag{3.61}$$

Rate of heat flow can be determined from Fourier's law:

$$\dot{Q}_r=-kA_c\left(\frac{d\theta}{dr}\right)_{r=r_r}=-k(2\pi r_r t)\theta_r\left(\frac{d\theta}{dr}\right)_{r=r_r}$$

From the differentiation rules of the modified Bessel functions

$$\frac{d}{dx}\left[I_0(ax)\right]=aI_1(ax);\ \frac{d}{dx}\left[K_0(ax)\right]=-aK_1(ax)$$

we can write

$$\frac{d}{dr}[I_0(mr)] = mI_1(mr)$$

$$\frac{d}{dr}[K_0(mr)] = -mK_1(mr)$$

Differentiating Eq. (3.61) and replacing r with r_r yields

$$\left(\frac{d\theta}{dr}\right)_{r=r_r} = \theta_r \frac{m[I_1(mr_r)K_1(mr_t) - I_1(mr_t)K_1(mr_r)]}{I_0(mr_r)K_1(mr_t) + I_1(mr_t)K_0(mr_r)}$$

hence,

$$\dot{Q}_r = 2\pi r_r t k \theta_r m \frac{I_1(mr_t)K_1(mr_r) - I_1(mr_r)K_1(mr_t)}{I_0(mr_r)K_1(mr_t) + I_1(mr_t)K_0(mr_r)} \qquad (3.62)$$

Fin efficiency:

$$\eta_f = \frac{\dot{Q}_r}{2\left[\pi\left(r_t^2 - r_r^2\right)\right]h_f\theta_r}$$

$$\eta_f = \frac{2r_r}{m\left(r_t^2 - r_r^2\right)} \frac{I_1(mr_t)K_1(mr_r) - I_1(mr_r)K_1(mr_t)}{I_0(mr_r)K_1(mr_t) + I_1(mr_t)K_0(mr_r)} \qquad (3.63)$$

Equations (3.60), (3.61), (3.62), and (3.63) can be evaluated using the Bessel functions in Table B.2 or correlations in Table B.3 in Appendix B. Plots of fin efficiency versus mL_c and $\rho(=t_t/t_r)$ can be found in Figures B.4 and B.5 in Appendix B.

3.5.2 Direct Solution of the Straight Annular Fin of Uniform Thickness

Repeating solutions of Eqs (3.61) to (3.63) using modified Bessel functions can be cumbersome and a more direct solution method for quick determination of the fin efficiency and heat rate can be obtained as follows:

- Equation (3.63) is rewritten as a function of two independent variables (mL) and ρ, where $m = \sqrt{2h_f/tk}$, $L = r_t - r_r$, and $\rho = r_t/r_r$

$$\eta_f = \frac{2}{(\rho+1)}\frac{1}{(mL)} \times$$

$$\times \frac{I_1\left(\frac{\rho}{\rho-1}mL\right)K_1\left(\frac{1}{\rho-1}mL\right) - I_1\left(\frac{1}{\rho-1}mL\right)K_1\left(\frac{\rho}{\rho-1}mL\right)}{I_0\left(\frac{1}{\rho-1}mL\right)K_1\left(\frac{\rho}{\rho-1}mL\right) + I_1\left(\frac{\rho}{\rho-1}mL\right)K_0\left(\frac{1}{\rho-1}mL\right)} \qquad (3.64)$$

- Equation (3.64) is solved repeatedly and the fin efficiency is plotted versus mL for different values of ρ to obtain the efficiency chart shown in Figure 3.26. Enlarged versions of this chart given in Figures B.4 and B.5 in Appendix B can be used for quick estimation of the efficiency of a

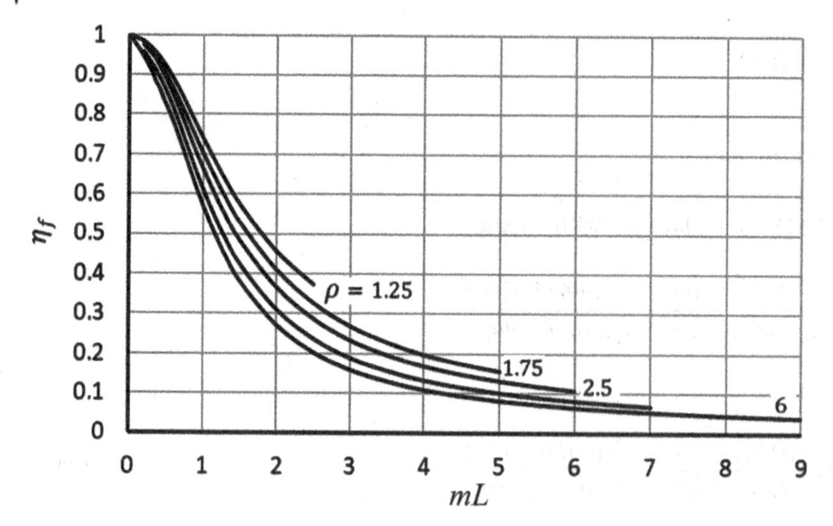

Figure 3.26 Efficiency chart of straight annular fin of uniform thickness for selected values of $\rho(L = r_{tc} - r_r)$.

straight annular fin of rectangular profile within a wide range of values of ρ and mL. Two charts are drawn for the ranges $mL = 0.5 - 9.0$, $\rho = 1.25 - 6.0$

The tip radius r_t in Eqs (3.61) to (3.64) should be replaced by the corrected radius r_{tc} to account for the heat loss from the fin edge ($r_{tc} = r_t + t/2$).

Example 3.6 The annular fin of rectangular profile, shown in Figure (3.25), has thickness $t = 3.6mm$, root radius $r_r = 60mm$, and tip radius $r_t = 120mm$. The fin is heated at the root to $\theta_r = 80°C$. The thermal conductivity of the fin material is $k = 30\,W/m.K$ and the heat transfer coefficient is $h_f = 30\,W/m^2.K$. Determine:

a) the temperature profile in the radial direction
b) fin heat loss rate
c) fin efficiency.

Solution

$$m = \sqrt{\frac{2h_f}{tk}} = \sqrt{\frac{2\times 30}{30\times 0.0036}} = 23.57$$

The corrected radius is

$$r_{tc} = r_t + t/2 = 0.12 + \frac{0.0036}{2} = 0.1218m$$

$$z = mr = 23.57r$$

$$z_r = mr_r = 23.57\times 0.06 = 1.4142$$

$$z_t = mr_{tc} = 23.57\times 0.1218 = 2.870854$$

The equation for the temperature profile is

$$\frac{\theta}{\theta_r}=\frac{T-T_\infty}{T_r-T_\infty}=\frac{I_0(mr)K_1(mr_t)+I_1(mr_t)K_0(mr)}{I_0(mr_r)K_1(mr_t)+I_1(mr_t)K_0(mr_r)}$$

The Bessel functions can be determined from Table B.2 or correlations in Table B.3 in Appendix B. The latter approach is used here and the Bessel functions are given for the range $r=0.06-0.12$ in Table E3.6. Ratios of excess temperatures θ/θ_r and fin temperature T are also given in the table. The plot $T=f(r)$ is shown in Figure E3.6.

Fin heat loss rate can be calculated from

$$\dot{Q}_r=2\pi r_r tkm\theta_r\times\frac{I_1(mr_t)K_1(mr_r)-I_1(mr_r)K_1(mr_t)}{I_0(mr_r)K_1(mr_t)+I_1(mr_t)K_0(mr_r)}$$

$$2\pi r_r tkm\theta_r=2\times\pi\times0.06\times0.0036\times30\times23.57\times80=76.773$$

Table E3.6 Bessel functions and calculated temperatures for Example 3.6.

r, m	mr	$I_0(mr)$	$K_0(mr)$	$I_1(mr)$	$K_1(mr)$	θ/θ_r	T, °C
0.06	1.414214	1.566639	0.239146	0.898694	0.314385	1.000000	110
0.07	1.649916	1.806482	0.176357	1.138457	0.224600	0.773663	91.89
0.08	1.885618	2.108010	0.131167	1.427835	0.162905	0.617933	79.43
0.09	2.121320	2.484189	0.098204	1.779890	0.119499	0.513030	71.04
0.10	2.357023	2.951619	0.073915	2.210430	0.088426	0.446313	65.71
0.11	2.592725	3.531306	0.055875	2.738791	0.065891	0.409694	62.78
0.1218	2.870854	4.395980	0.040349	3.520987	0.046904	0.380398	60.43

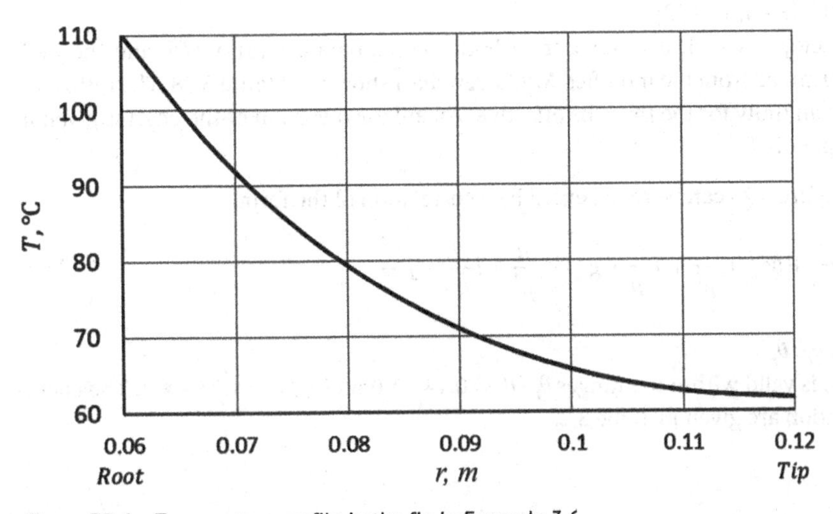

Figure E3.6 Temperature profile in the fin in Example 3.6.

$$\frac{I_1(mr_t)K_1(mr_r)-I_1(mr_r)K_1(mr_t)}{I_0(mr_r)K_1(mr_t)+I_1(mr_t)K_0(mr_r)}=\frac{3.520987\times0.314385-0.898694\times0.046904}{1.566639\times0.046904+3.520987\times0.239146}=1.16305$$

hence,

$$\dot{Q}_r=76.773\times1.16305=89.3W$$

Fin efficiency is given by

$$\eta_f=\frac{2r_r}{m\left(r_t^2-r_r^2\right)}\frac{I_1(mr_t)K_1(mr_r)-I_1(mr_r)K_1(mr_t)}{I_0(mr_r)K_1(mr_t)+I_1(mr_t)K_0(mr_r)}$$

$$\eta_f=\frac{2\times0.06}{23.57\left(0.1218^2-0.06^2\right)}\times1.16305=0.527$$

3.5.3 Correction Factor Solution Method for Annular Fins of Uniform Thickness

Mikheyev (1964) also proposed a correction factor method for the straight annular fin of uniform thickness, in which the heat loss from the annular fin can be calculated from

$$\dot{Q}_r=\epsilon\dot{Q}_{eq}\left(\frac{A_f}{A_{eq}}\right) \tag{3.65}$$

The solution procedure is as follows:

- The heat loss $\dot{Q}_{eq}$ is calculated from Eq. (3.24) for a straight fin of rectangular profile with insulated tip having a width of 1.0 m, the same thickness of the annular fin, and corrected length $L_c=(r_t-r_r)+t/2$ (Figure 3.27b)
- The corrected surface area A_f of the annular fin from which heat is dissipated by convection is calculated from $A_f=2\pi\left(r_{tc}^2-r_r^2\right)$, where $r_{tc}=r_t+t/2$
- The corrected surface area A_{eq} of the equivalent rectangular fin with insulated tip is calculated from $A_{eq}=2\times1.0\times[(r_t-r_r)+t/2]$
- The correction factor ϵ, which is a function of both the temperature ratio θ/θ_r and the radii ratio r_t/r_r, is determined from the modified Mikheyev chart shown in Figure 3.28. The ratio r_t/r_r must be higher than unity for the fin to be effective. An enlarged version of the chart is given in Figure B.6 in Appendix B.

The chart data of Figure 3.28 can be represented by a correlation of the form

$$\epsilon=a+b\gamma+\frac{c}{\rho}+d\gamma^2+\frac{e}{\rho^2}+f\frac{\gamma}{\rho}+g\gamma^3+\frac{h}{\rho^3}+i\frac{\gamma}{\rho^2}+j\frac{\gamma^2}{\rho} \tag{3.66}$$

where $\rho=r_t/r_r\gamma=\ \theta_t/\theta_r$

Correlation (3.66) is valid within the ranges $\theta_t/\theta_r=0.14-1.0$ and $r_t/r_r=1.25-4.0$. The coefficients of the correlation are given in Table 3.2.

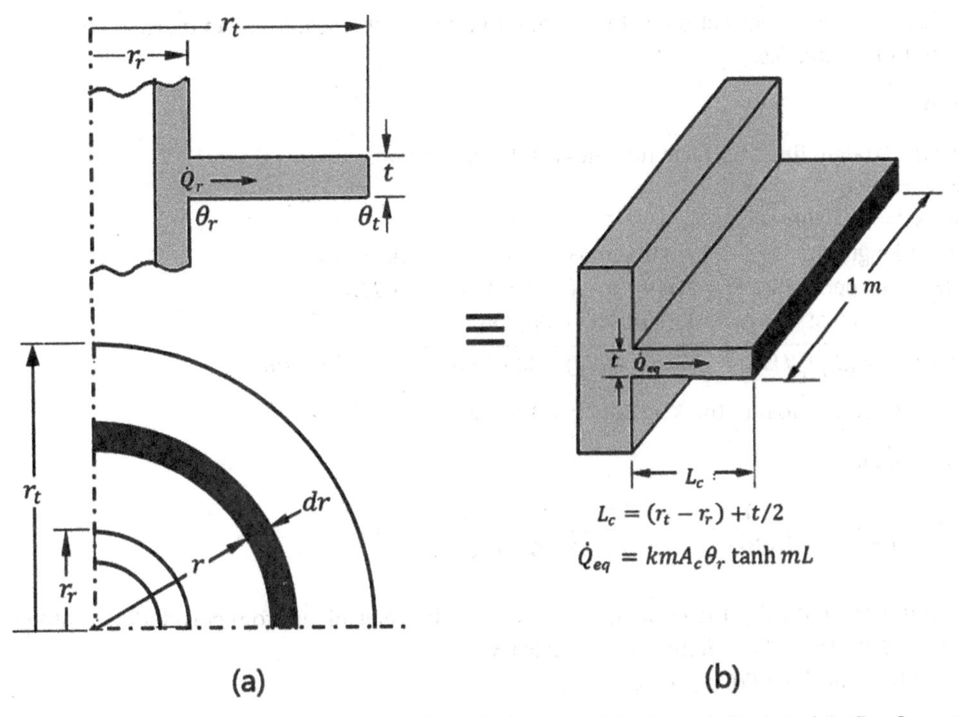

Figure 3.27 Straight annular fin: (a) of uniform thickness; and (b) its equivalent straight fin of constant thickness with insulated tip.

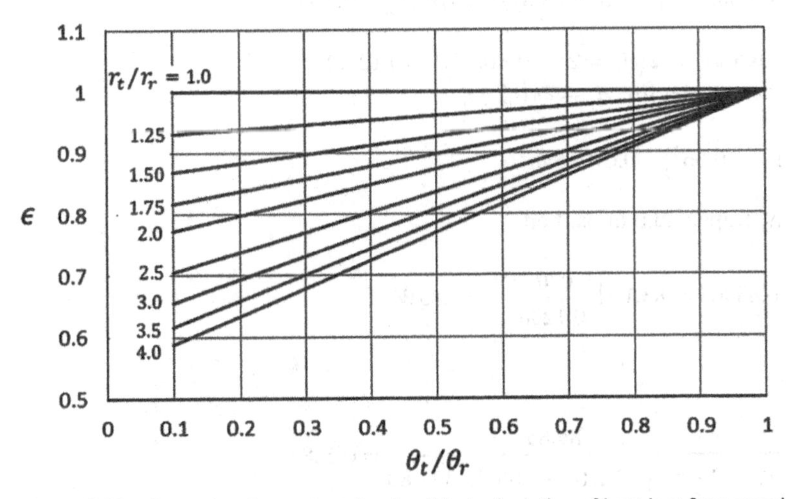

Figure 3.28 Correction factor chart for simplified calculation of heat loss from annular fins of constant thickness.

Table 3.2 Coefficients for correlation (3.66).

a	0.273736	f	-1.13816
b	0.716	g	7.87E-16
c	1.202486	h	6.31E-02
d	1.25E-15	i	0.422279
e	-0.53928	j	-4.27E-15

Example 3.7 Use the correction factor method to estimate the heat loss and thermal efficiency for the fin in Example 3.6.

Solution

Data for the straight fin of uniform thickness and insulated tip:
Width: $w = 1.0m$
Thickness: $t = 0.0036m$
Equivalent length: $L_c = (r_t - r_r) + 0.5t = 0.06 + 0.0036/2 = 0.0618m$
Cross-sectional perimeter: $P_c = 2(w+t) = 2(1+0.0036) = 2.0072m$
Cross-sectional area: $A_c = wt = 1 \times 0.0036 = 0.0036m^2$
Parameter: $m = \sqrt{h_f P_c / kA_c} = \sqrt{(30 \times 2.0072)/(30 \times 0.0036)} = 23.6126\ 1/m$

Ratio of tip thickness to root thickness: $r_t / r_r = 120/60 = 2.0$

Temperature ratio:

$$\frac{\theta_t}{\theta_r} = \frac{T_t - T_\infty}{T_r - T_\infty} = \frac{1}{\cosh mL_c} = \frac{1}{\cosh(23.6126 \times 0.0618)} = 0.441$$

Knowing θ_t / θ_r and r_t / r_r, the correction factor ϵ can be determined from correlation (3.66) or from Figure 3.28. The value obtained is $\epsilon = 0.85965$.
The heat loss rate from Eq. (3.24) is

$$\dot{Q}_{eq} = \sqrt{kh_f P_c A_c}\,\theta_r \tanh mL_c = kmA_c\theta_r \tanh mL_c$$
$$= 30 \times 23.6126 \times 0.0036 \times 80 \times \tanh(23.6126 \times 0.0618) = 183.1W$$

Equivalent heat transfer area: $A_{eq} = 2wL_c = 2 \times 1.0 \times 0.0618 = 0.1236m^2$
Heat transfer area for the annular fin: $A_f = 2\pi\left(r_{tc}^2 - r_r^2\right)$

$$A_f = 2 \times \pi \times \left(0.1218^2 - 0.06^2\right) = 0.07056\ m^2$$

The heat loss rate from the trapezoidal fin is then

$$\dot{Q}_x = \epsilon \dot{Q}_{eq}\left(\frac{A_f}{A_{eq}}\right) = 0.85965 \times 183.1 \times \left(\frac{0.07056}{0.1236}\right) = 89.85W$$

Fin efficiency is given by

$$\eta_f = \frac{\dot{Q}_x}{2\pi\left(r_t^2 - r_r^2\right)h_f\theta_r} = \frac{89.85}{2 \times \pi \times \left(0.1218^2 - 0.06^2\right) \times 30 \times 80} = 0.528$$

3.5.4 Circular (Annular) Fin of Triangular Profile

Exact solution of the differential equation for the temperature distribution for this configuration was provided by Smith and Sucec (1969), which showed that the fin efficiency is a function of two variables

$$\eta_f = f(\rho, \phi)$$

where

$$\rho = r_t / r_r \text{ and } \phi = r_t (r_t - r_r)\left(\frac{2h_f}{kt}\right).$$

The data by Smith and Sucec (1969) of the efficiency of the straight annular fin of triangular profile as a function of both parameter ρ and ϕ is plotted, as shown in Figure 3.29, using the following correlation fitted to the data

$$\eta_f = \boldsymbol{a} + \boldsymbol{b}\ln\phi + \frac{\boldsymbol{c}}{\rho} + \boldsymbol{d}\ln\phi^2 + \frac{\boldsymbol{e}}{\rho^2} + \frac{\boldsymbol{f}\ln\phi}{\rho} + \boldsymbol{g}\ln\phi^3 + \frac{\boldsymbol{h}}{\rho^3} + \frac{\boldsymbol{i}\ln\phi}{\rho^2} + \frac{\boldsymbol{j}\ln\phi^2}{\rho} \tag{3.67}$$

The coefficients of correlation (3.67) are given in Table 3.3.

Correlation (3.67) is valid within the ranges $\phi = 1.0 - 16.0$ and$\rho = 1.25 - 5.0$.

A more detailed and enlarged version of the chart in Figure 3.29 is given in Figure B.7 in Appendix B.

Comparison of calculations of fin efficiencies of three annular fins of triangular profiles using correlation (3.67) with results of exact solution by Smith and Sucec (1969) are presented in Table 3.4. It is evident that correlation (3.67) can reproduce the exact solution to a good degree of accuracy for the examples in the table

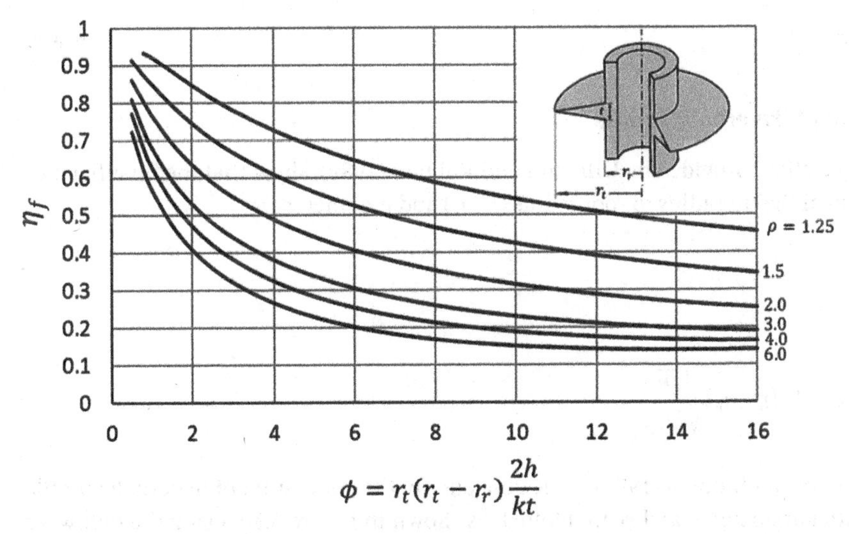

Figure 3.29 $\eta_f = f(\phi)$ plots for the efficiency of annular fin of triangular profile at constant ρ.

Table 3.3 Coefficients for correlation (3.67).

a	0.40965702	***f***	0.02842023
b	-0.24515984	***g***	0.01261382
c	1.12821633	***h***	0.71258346
d	0.01089931	***i***	0.24477565
e	-1.17528797	***j***	-0.10392467

Table 3.4 Examples of calculation of the efficiency of annular fin of triangular profile.

Data	Examples		
	Fin 1	Fin 2	Fin 3
t, m	0.00254	0.00508	0.000254
r_r, m	0.0508	0.058	0.0254
r_t, m	0.1016	0.0106	0.3175
$k, W/mK$	43.3	207.68	51.91
$h_f, W/m^2K$	22.7	204.42	85.92
r_t/r_r	2	2	1.25
$\phi = r_t(r_t - r_r)\left(\frac{2h_f}{kt}\right)$	2.13055	2.0	2.60419
η_f, eq (3.67)	0.612	0.626	0.778
η_f (Smith and Sucec)	0.607	0.620	0.788

The heat loss rate can now be calculated from

$$\dot{Q}_r = A_f h_f \theta_r \eta_f \quad W \tag{3.68}$$

3.5.5 Annular Fin of Hyperbolic Profile

Data for this configuration, provided by Ullmann and Kalman (1989), show that the fin efficiency is a function of ratio of the tip radius to root radius (r_t / r_r) and parameter ψ.

$$\eta_f = f(\rho, \psi)$$

where

$$\rho = r_t / r_r \text{ and } \psi = (r_t - r_r)\sqrt{\frac{2h}{kt_r}}$$

The data by Ullmann and Kalman (1989) of the efficiency of the annular fin of hyperbolic profile as a function of both parameter ρ and ϕ are plotted, as shown in Figure 3.30, using the following correlation fitted to the data:

$$\eta_f = \mathbf{a} + \mathbf{b}\ln\psi + \frac{\mathbf{c}}{\rho} + \mathbf{d}\ln\psi^2 + \frac{\mathbf{e}}{\rho^2} + \frac{\mathbf{f}\ln\psi}{\rho} + \mathbf{g}\ln\psi^3 + \frac{\mathbf{h}}{\rho^3} + \frac{\mathbf{i}\ln\psi}{\rho^2} + \frac{\mathbf{j}\ln\psi^2}{\rho} \tag{3.69}$$

The coefficients of the correlation (3.69) are given in Table 3.5.

Correlation (3.69) is valid within the ranges $\psi = 0.5 - 5.0$ and $\rho = 1.5 - 5.0$.

A more detailed and enlarged version of the chart in Figure 3.30 is given in Figure B.8 in Appendix B.

Having determined the fin efficiency, the heat loss rate can be calculated from

$$\dot{Q}_r = A_f h_f \theta_r \eta_f \quad W \tag{3.70}$$

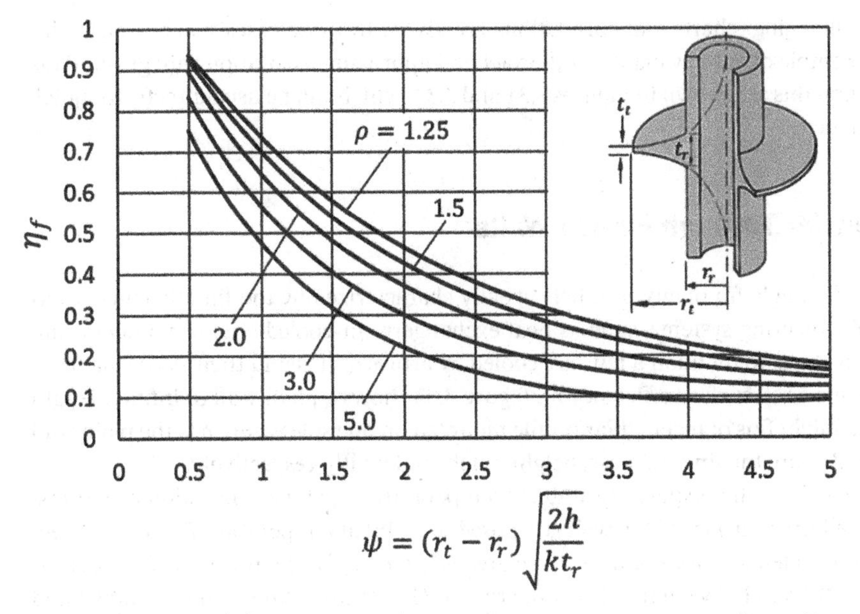

Figure 3.30 $\eta_f = f(\psi)$ plots for the efficiency of an annular fin of hyperbolic profile at constant ρ.

Table 3.5 Coefficients for correlation (3.69).

a	0.286578842	*f*	-0.121251598
b	-0.389611585	*g*	0.052924815
c	1.152866980	*h*	0.413931937
d	0.050146442	*i*	0.192314838
e	-1.070057653	*j*	-0.156215163

The convective surface can be taken as $A_f \cong 2\pi\left(r_t^2 - r_r^2\right)$ with an error less than 1% compared to the exact equation:

$$A_f = 2\pi r_r \left\{C - B + \left(t_r/4\right)\ln\left[\frac{\left(C - t_r/2\right)\left(B + t_r/2\right)}{\left(C + t_r/2\right)\left(B - t_r/2\right)}\right]\right\}$$

where

$$B = \sqrt{r_r^2 + \left(t_r/2\right)^2} \qquad C = \sqrt{\left(r_t^2/r_r\right)^2 + \left(t_r/2\right)^2}$$

3.6 Other Fin Shapes

Table 3.6 is a summary of the methodologies for the calculation of the efficiency of various types of pins and fins. The fins are of different configurations, including the ones discussed earlier. The methodologies are either equations of exact solutions or correlations and/or graphs for approximate calculations. In some cases, only equations are given, and in others, where exact solutions are mathematically challenging, charts and correlations are given. In those cases where exact solutions of moderate complexity are available, both exact and approximate solutions are given. Plots of efficiency of several fins are shown in Figures 3.31 and 3.32, which can be used directly for quick estimation of efficiencies.

3.7 Heat Transfer Through Finned Walls

The performance of a single fin of any profile is usually characterized by the fin efficiency η_f, as discussed so far. Engineering systems, such as heat exchangers, air-cooled engine cylinders, and heat sinks of silicon chips, to mention a few, are cooled by arrays of fins and their performance is characterized by the overall surface efficiency η_o. Figure 3.33 shows a plain wall of infinite length with an array of n straight fins of rectangular profile mounted on one side where δ is the plain wall thickness; t, L, p, and w are the fin thickness, height, pitch, and width, respectively.

The plain surface of the wall is exposed to ambient temperature $T_{\infty 1}$ and a convection heat transfer coefficient h_1; the finned surface of the wall is exposed to ambient temperature $T_{\infty 2}$ and convection heat transfer coefficients h_f for the fin surface area nA_f and h_p for the prime surface area A_p (the surface area of the wall between the fins at the roots). The temperatures of the unfinned and finned surfaces of the plate are designated T_{w1} and T_{w2}, respectively. The excess temperature at the fin roots is $\theta_r = T_{w2} - T_{\infty 2}$.

The total surface area of the finned side of the wall is

$$A_t = nA_f + A_p \tag{3.71}$$

The total rate of heat transfer from the finned side of the wall by convection is

$$\dot{Q}_t = \dot{Q}_f + \dot{Q}_p = \eta_f nA_f h_f \theta_r + A_p h_p \theta_r = \eta_f nA_f h_f \theta_r + \left(A_t - nA_f\right) h_p \theta_r$$

or, rearranging

$$\dot{Q}_t = A_t \theta_r h_f \left[\frac{h_p}{h_f} - \frac{nA_f}{A_t}\left(\frac{h_p}{h_f} - \eta_f\right)\right] \tag{3.72}$$

Fin overall surface efficiency for an array of straight fins of rectangular profile can be written as

$$\eta_o = \frac{\dot{Q}_t}{\dot{Q}_{max}}$$

where $\dot{Q}_{max}$ is the heat rate obtained if the total surface area on the finned side of the wall is maintained at the root temperature θ_r

hence,

$$\eta_o = \frac{A_t h_f \left[\frac{h_p}{h_f} - \frac{nA_f}{A_t}\left(\frac{h_p}{h_f} - \eta_f\right)\right]}{\left(nA_f h_f + A_p h_p\right)} \tag{3.73}$$

If a uniform heat transfer coefficient is assumed on the finned side ($h_f = h_p = h$), Eqs (3.72) and (3.73) are simplified to

Table 3.6 Methodologies for determining the efficiency of a range of pin, straight, and radial fins.

Pin (Spine) Fins of circular cross-section		
Profile	**Exact solution**	**Approximate solution**
Rectangular (cylindrical)	$\eta_f = \dfrac{\tanh\left(\sqrt{2}mL_c\right)}{\left(\sqrt{2}mL_c\right)}$ $L_c = L + r/2m = \sqrt{2h/kd}$	—
Triangular (conical)	$\eta_f = \dfrac{4}{\left(2\sqrt{2}mL\right)} \dfrac{I_2\left(2\sqrt{2}mL\right)}{I_1\left(2\sqrt{2}mL\right)}$ $m = \sqrt{2h_f/kd}$	Correlation in Table 3.7
Convex parabolic	$\eta_f = \dfrac{2}{\left(\dfrac{4\sqrt{2}}{3}mL\right)} \dfrac{I_1\left(\dfrac{4\sqrt{2}}{3}mL\right)}{I_0\left(\dfrac{4\sqrt{2}}{3}mL\right)}$ $m = \sqrt{2h_f/kd}$	Correlation in Table 3.7
Concave parabolic 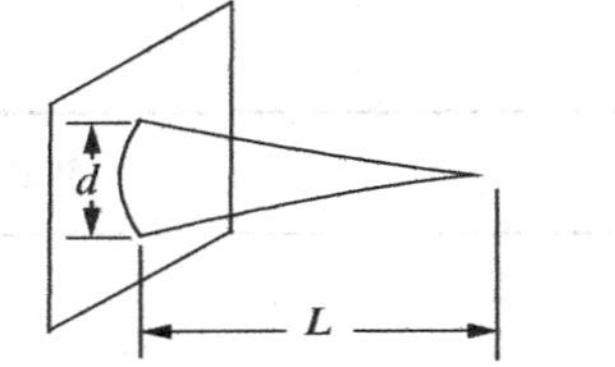 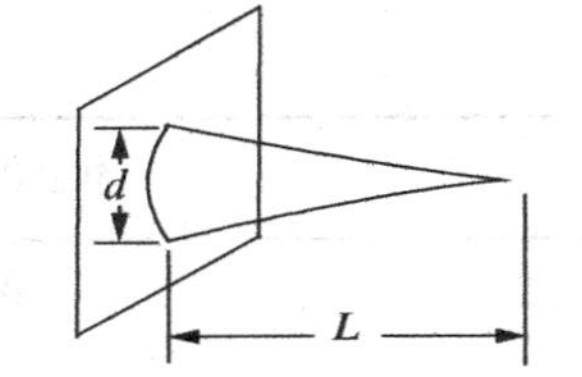	$\eta_f = \dfrac{2}{1 + \sqrt{(8/9)(mL)^2 + 1}}$ $m = \sqrt{2h_f/kd}$	—

(Continued)

Table 3.6 (Continued)

Straight (Longitudinal) Fins		
Profile	**Exact solution**	**Approximate solution**
Rectangular	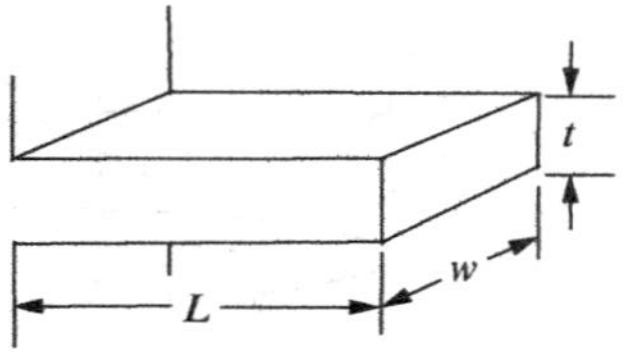 $\eta_f = \frac{1}{mL_c} \tanh mL_c$ $L_c = L + t/2$ $m = \sqrt{2h_f / kt_r}$	—
Trapezoidal	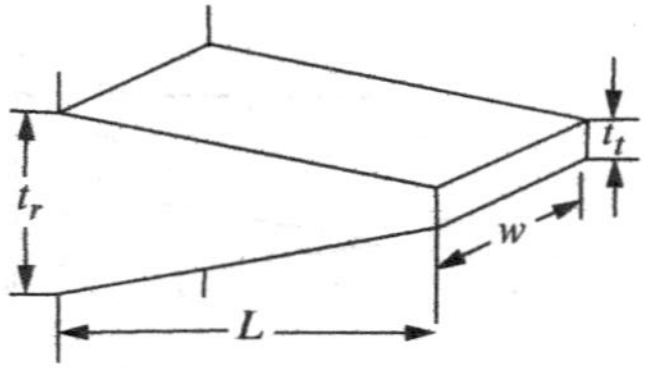 $\eta_f = \frac{t_r}{2L \tan\phi \sqrt{z_r}} \times$ $\frac{I_1(2\sqrt{z_r})K_1(2\sqrt{z_t}) - I_1(2\sqrt{z_t})K_1(2\sqrt{z_r})}{I_0(2\sqrt{z_r})K_1(2\sqrt{z_t}) + I_1(2\sqrt{z_t})K_0(2\sqrt{z_r})}$	Figure B.1, B.2 Section 3.4.4
Triangular	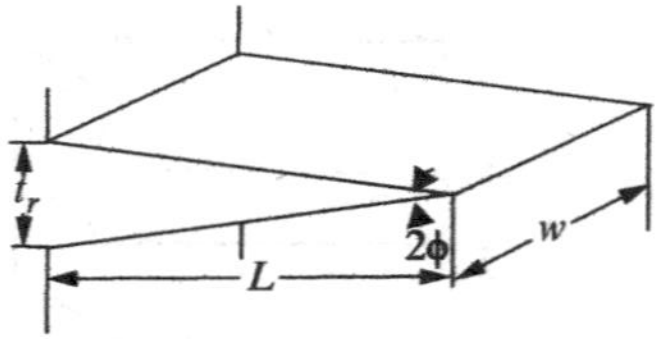 $\eta_f = \frac{1}{(mL)} \frac{I_1(2mL)}{I_0(2mL)}$ $m = \sqrt{2h_f / kt_r}$	Correlation in Table 3.7

Profile		Exact solution	Approximate solution
Convex parabolic	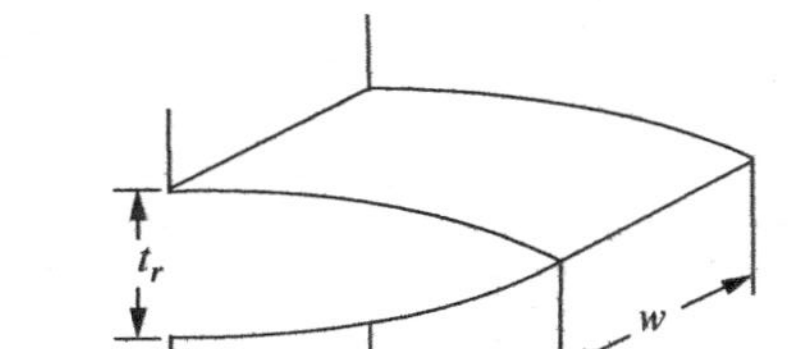	$\eta_f = \frac{1}{(mL)} \frac{I_{2/3}\left(\frac{4}{3}mL\right)}{I_{-1/3}\left(\frac{4}{3}mL\right)}$ $m = \sqrt{2h_f / kt_r}$	Correlation in Table 3.7
Concave parabolic		$\eta_f = \frac{2}{1+\sqrt{4(mL)^2+1}}$ $m = \sqrt{2h_f / kt_r}$	—

Annular (Radial) Fins

Profile		Exact solution	Approximate solution
Rectangular	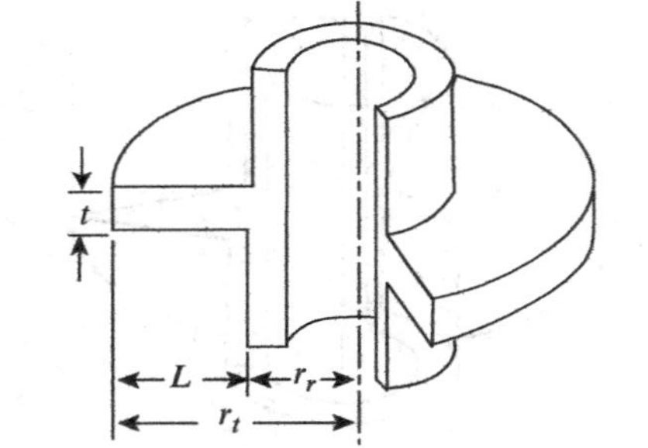	$\eta_f = \frac{2r_r}{m\left(r_t^2 - r_r^2\right)} \times$ $\frac{I_1(mr_t)K_1(mr_r) - I_1(mr_r)K_1(mr_t)}{I_0(mr_r)K_1(mr_t) + I_1(mr_t)K_0(mr_r)}$ $m = \sqrt{2h_f / tk}$	Appendix B: Figures B.4, B.5 $\rho = t_t / t_r$ Section 3.5.3

(Continued)

Table 3.6 (Continued)

Triangular	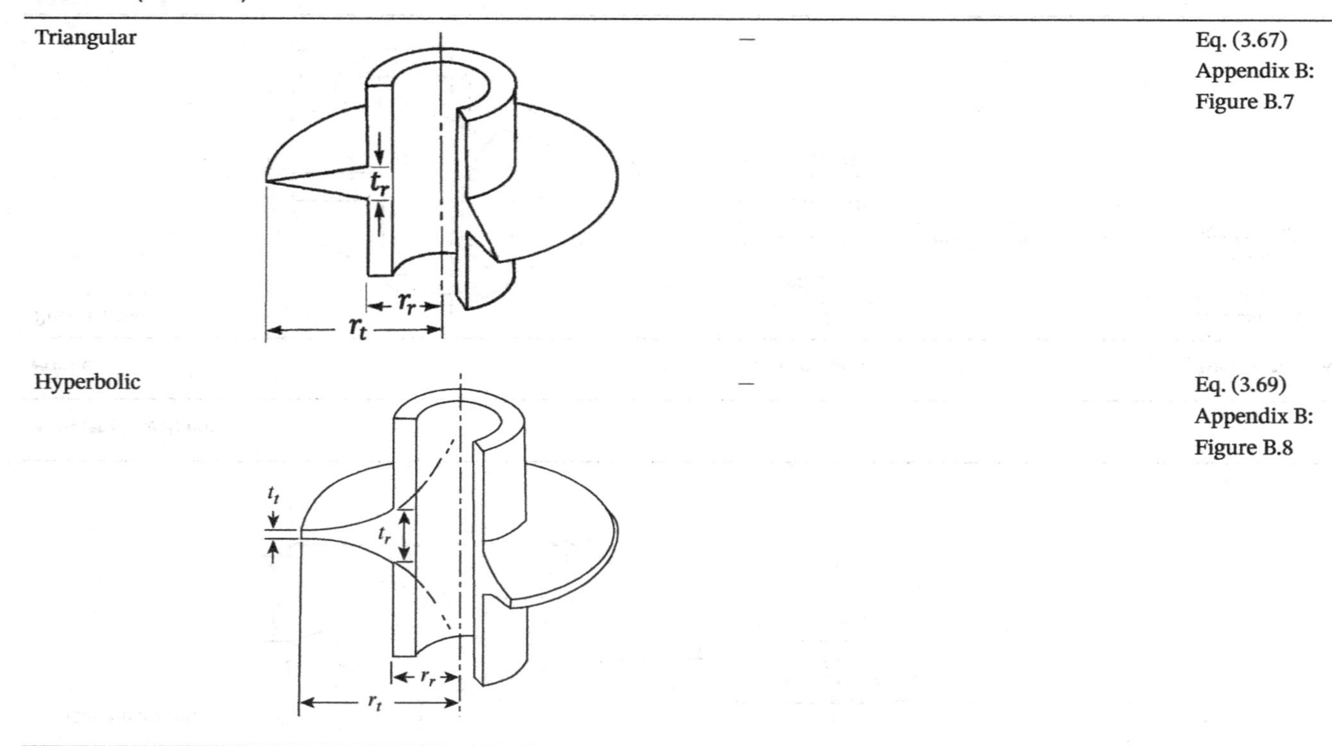	—	Eq. (3.67) Appendix B: Figure B.7
Hyperbolic		—	Eq. (3.69) Appendix B: Figure B.8

Table 3.7 Efficiency correlations for some pin and straight fins in Table 3.6.

Fin type	Triangular (conical) pin	Convex parabolic pin	Straight triangular fin	Straight convex parabolic fin
Correlation	$\eta_f = \sum_{n=0}^{6} a_n (\Phi)^n$ $\Phi = mL$			$\eta_f = \sum_{n=0}^{10} a_n (\Phi)^n$ $\Phi = mL$
	$m = \sqrt{2h_f / kd}$	$m = \sqrt{2h_f / kd}$	$m = \sqrt{2h_f / kt_r}$	$m = \sqrt{2h_f / kt_r}$
a_0	0.998756457	1.001468695	1.001762174	-0.000000098
a_1	0.001681818	-0.014448677	-0.020099956	0.000005544
a_2	-0.383730383	-0.532705612	-0.587075121	-0.000136523
a_3	0.174524201	0.360630593	0.419031085	0.001913584
a_4	-0.002829745	-0.106616159	-0.130228359	-0.016799480
a_5	-0.013482982	0.015070228	0.019323659	0.095658230
a_6	0.002241079	-0.000828586	-0.001114209	-0.352489460
a_7				0.803563246
a_8				-0.973610614
a_9				0.183616175
a_{10}				0.994491279

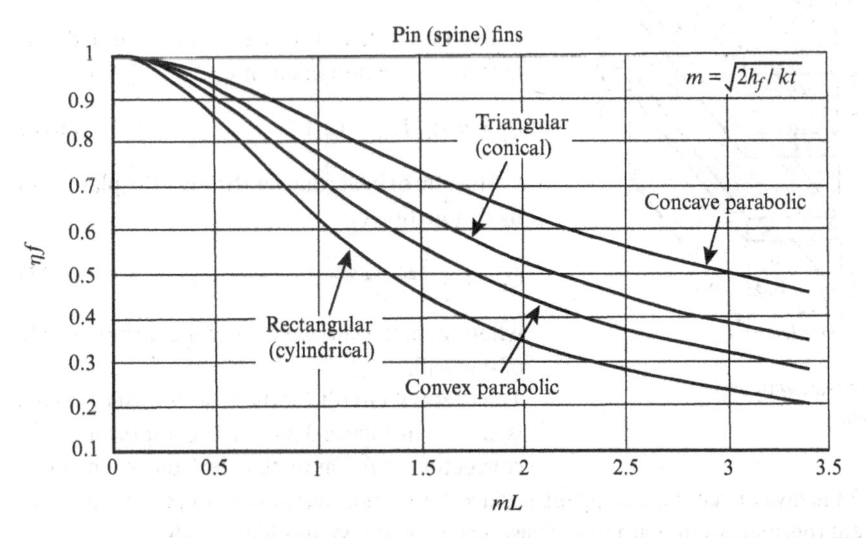

Figure 3.31 Efficiency of pin (spine) fins with circular cross-section and different profiles. Thickness at the root t_r, $m = \sqrt{2h_f / kt_r}$.

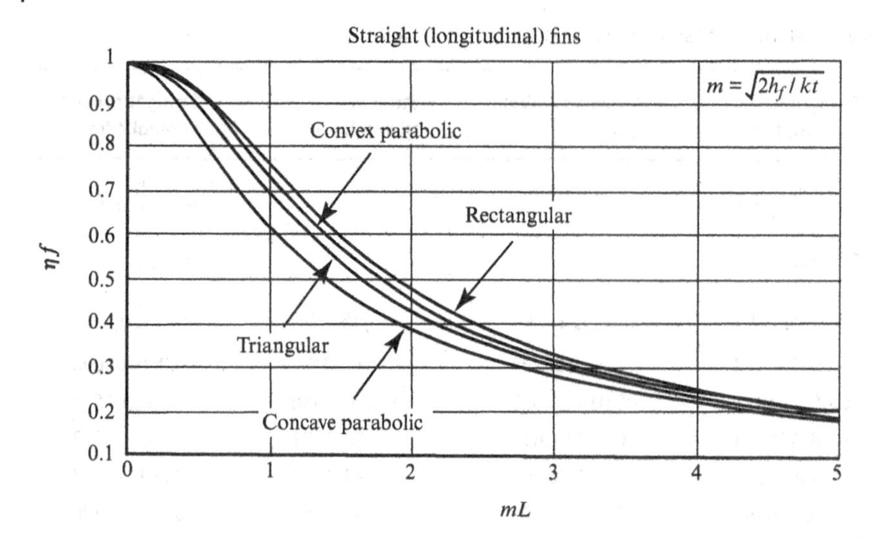

Figure 3.32 Efficiency of straight (longitudinal) fins of different profiles. Thickness at the root t_r, $m=\sqrt{2h_f/kt_r}$.

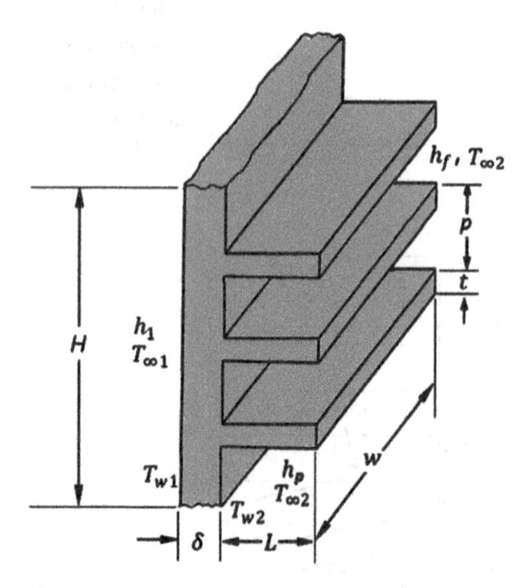

Figure 3.33 Plane wall with an array of straight fins.

$$\dot{Q}_t = A_t\theta_r h\left[1-\frac{nA_f}{A_t}(1-\eta_f)\right] \tag{3.74}$$

$$\eta_o = 1-\frac{nA_f}{A_t}(1-\eta_f) \tag{3.75}$$

The rate of heat transfer to the unfinned surface by convection (assuming $T_{\infty1} > T_{w1}$) is

$$\dot{Q}_1 = h_1A_1(T_{\infty1}-T_{w1}) \tag{3.76}$$

The rate of heat transfer through the plain wall by conduction is

$$\dot{Q}_c = kA_1\frac{T_{w1}-T_{w2}}{\delta} \tag{3.77}$$

where A_1 is the surface area of the unfinned side of the wall.

The thermal circuit for the heat transfer process is shown in Figure 3.34a. It is comprised of the convection and conduction resistances in series ($\dot{Q}_1$ and $\dot{Q}_c$) and fin array convection and prime surface convection resistances in parallel ($\dot{Q}_p$ and $\dot{Q}_f$). The equivalent thermal circuit with total resistance R_Σ is shown in Figure 3.34b.

$$\dot{Q}_\Sigma = \frac{(T_{\infty1}-T_{w1})}{R_\Sigma} \tag{3.78}$$

where

$$R_\Sigma = \frac{1}{h_1A_1}+\frac{\delta}{kA_1}+\frac{1}{\eta_f nA_f h_f+(A_t-nA_f)h_p}$$

$$R_\Sigma = \frac{1}{h_1 A_1} + \frac{\delta}{kA_1} + \frac{1}{A_t h_f \left[\frac{h_p}{h_f} - \frac{nA_f}{A_t} \left(\frac{h_p}{h_f} - \eta_f \right) \right]} \tag{3.79}$$

The overall heat transfer coefficient is

$$U = \frac{1}{R_\Sigma} = \frac{1}{\frac{1}{h_1 A_1} + \frac{\delta}{kA_1} + \frac{1}{A_t h_f \left[\frac{h_p}{h_f} - \frac{nA_f}{A_t} \left(\frac{h_p}{h_f} - \eta_f \right) \right]}} \tag{3.80}$$

$$\dot{Q}_\Sigma = U \left(T_{\infty 1} - T_{w1} \right)$$

If $h_f = h_p = h$, then

$$R_\Sigma = \frac{1}{h_1 A_1} + \frac{\delta}{kA_1} + \frac{1}{A_t h \left[1 - \frac{nA_f}{A_t} \left(1 - \eta_f \right) \right]} \tag{3.81}$$

or, combining with Eq. (3.73),

$$R_\Sigma = \frac{1}{h_1 A_1} + \frac{\delta}{kA_1} + \frac{1}{A_t h \eta_o} \tag{3.82}$$

The overall heat transfer coefficient is

$$U = \frac{1}{R_\Sigma} = \frac{1}{\frac{1}{h_1 A_1} + \frac{\delta}{kA_1} + \frac{1}{A_t h \eta_o}} \tag{3.83}$$

The fin efficiency η_f for a straight rectangular fin as before

$$\eta_f = \frac{\tanh mL_c}{mL_c} \tag{3.84}$$

where $m = \sqrt{2h_f / tk}$

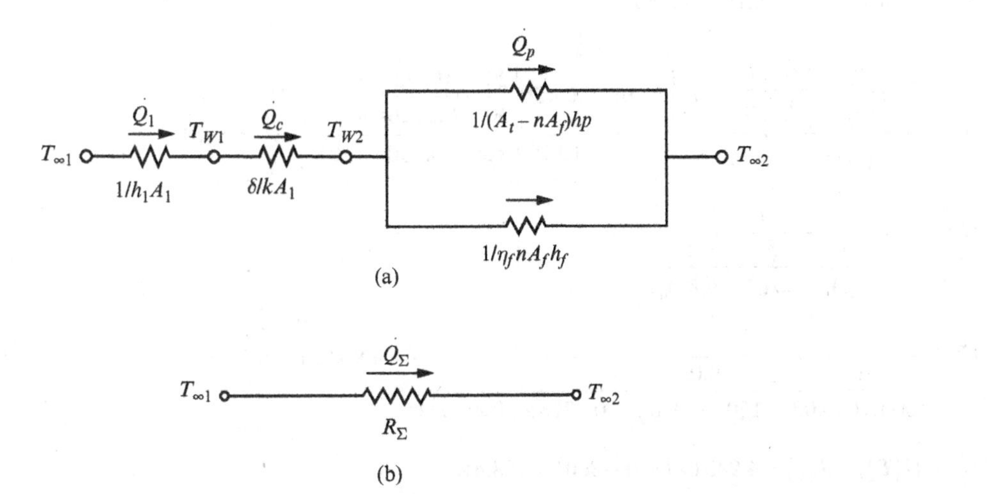

Figure 3.34 Thermal circuit of the heat transfer process in the fin array system in Figure 3.33.

The overall heat transfer coefficient in the absence of fins is

$$U=\frac{1}{R_\Sigma}=\frac{1}{\frac{1}{h_1A_1}+\frac{\delta}{kA_1}+\frac{1}{A_1h}}$$

Example 3.8 Determine the overall surface efficiency and heat transfer rate for the plain wall in Figure 3.33 with eight fins for the following data: $H=201mm$, $L=60mm$, $w=200mm$, $t=6mm$, $\delta=50mm$, $T_{\infty1}=500\,K$, $h_1=1000\,W/m^2.K$, $T_{\infty2}=300\,K$, $h_p=20\,W/m^2.K$, $h_f=30\,W/m^2.K$, $k=150\,W/m.K$.

Solution

The overall efficiency of the array is given by Eq. (3.73)

$$\eta_o=\frac{A_th_f\left[\frac{h_p}{h_f}-\frac{nA_f}{A_t}\left(\frac{h_p}{h_f}-\eta_f\right)\right]}{\left(nA_fh_f+A_ph_p\right)}$$

$$A_1=wH=0.2\times0.201=0.0402\ m^2$$

$$A_r=\left(H-nt\right)w=\left(0.201-8\times0.006\right)0.2=0.0306\ m^2$$

$$nA_f=8\times2\times wL=16\times0.2\times0.06=0.192\ m^2$$

$$A_t=A_r+nA_f=0.0306+0.192=0.2226\ m^2$$

$$m=\sqrt{2h_f/tk}=\sqrt{\frac{2\times30}{0.006\times150}}=8.165$$

$$L_c=L+t/2=0.06+0.003=0.063m$$

$$\eta_f=\frac{\tanh mL_c}{mL_c}=\frac{\tanh\left(8.165\times0.063\right)}{8.165\times0.063}=0.92$$

$$\eta_o=\frac{A_th_f\left[\frac{h_p}{h_f}-\frac{nA_f}{A_t}\left(\frac{h_r}{h_f}-\eta_f\right)\right]}{\left(nA_fh_f+A_rh_r\right)}=\frac{0.2226\times30\left[\frac{20}{30}-\frac{0.192}{0.2226}\left(\frac{20}{30}-0.92\right)\right]}{\left(0.192\times30+0.0306\times20\right)}=0.931$$

$$U=\frac{1}{R_\Sigma}=\frac{1}{\frac{1}{h_1A_1}+\frac{\delta}{kA_1}+\frac{1}{A_th_r\eta_o}}$$

$$U=\frac{1}{\frac{1}{1000\times0.0402}+\frac{0.05}{150\times0.0402}+\frac{1}{0.2226\times20\times0.931}}=4.9431W/K$$

$$\dot{Q}_\Sigma=U\left(T_{\infty1}-T_{w1}\right)=4.9431\times\left(500-300\right)=988.6W$$

The overall heat transfer coefficient for the finless plain wall is

$$U = \frac{1}{R_\Sigma} = \frac{1}{\frac{1}{h_1 A_1} + \frac{\delta}{k A_1} + \frac{1}{A_1 h}}$$

$$U = \frac{1}{\frac{1}{1000 \times 0.0402} + \frac{0.05}{150 \times 0.0402} + \frac{1}{0.0402 \times 20}} = 0.783 W / K$$

$$\dot{Q}_\Sigma = U\left(T_{\infty 1} - T_{w1}\right) = 0.783 \times \left(500 - 300\right) = 160.8W$$

The presence of fins increases the rate of heat dissipation from the wall more than five-fold.

Figure 3.35 shows computer modelling results of the finned surface in Example 3.8 with a plot of the temperature along a single fin. The rate of heat dissipated by convection predicted by the model is $\dot{Q}_\Sigma = 994W$, which is about 0.54% higher than the exact solution.

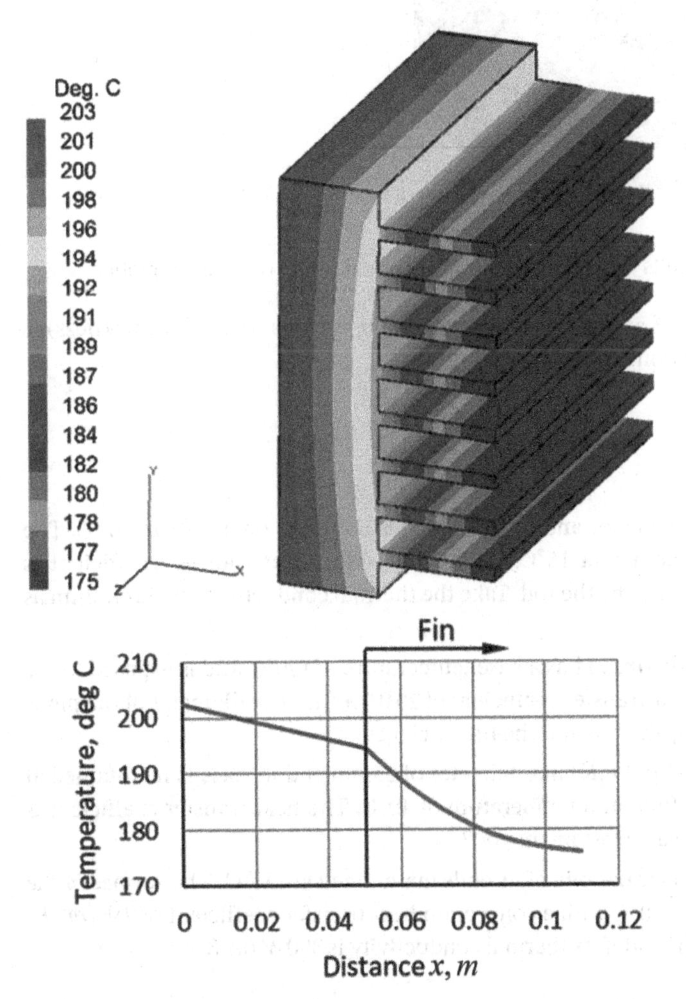

Figure 3.35 Computer modelling results of plain wall with eight straight fins of rectangular profile.

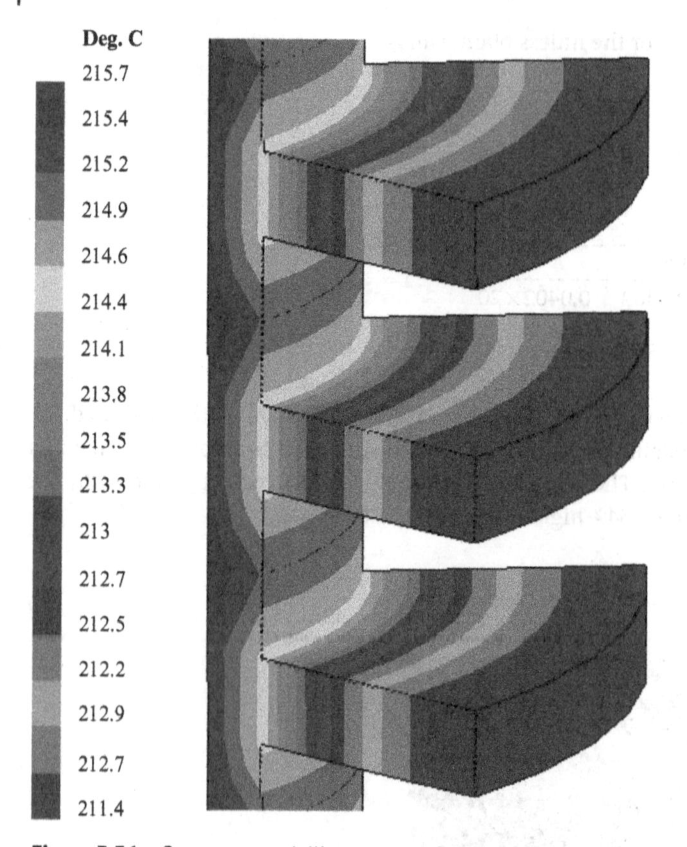

Figure 3.36 Computer modelling results of cylindrical wall with annular fins of rectangular profile.

A computer model of an array of annular fins of rectangular profiles (constant thickness) is shown in Figure 3.36 as an illustration.

Problems

3.1 An aluminium rod 20 mm in diameter and 120 mm long is mounted on a wall at 250°C. The rod is exposed to the surroundings at 15°C. If the convection heat transfer coefficient is $12 W/m^2.K$, determine the heat lost by the rod. Take the thermal conductivity of aluminium as $204 W/m.K$.

3.2 A long copper pin fin ($k = 375 W/m.K$) has a base temperature of 120°C and is exposed to the surroundings at 20°C with a heat-transfer coefficient of $20 W/m^2.K$. The diameter of the pin is $5 mm$. Calculate the heat lost by the rod and the fin efficiency.

3.3 A very long copper rod ($k = 375 W/m.K$) has diameter of $25 mm$ and its base is maintained at 90°C. The rod is exposed to a fluid at a temperature of 40°C. The heat transfer coefficient is $h = 4.5 W/m^2.K$. How much heat is lost by the rod?

3.4 A very long rod of $50 mm$ diameter has one of its ends maintained at 130°C. The surface of the rod is exposed to ambient air at 20°C with a convection heat transfer coefficient of $9 W/m^2.K$. Calculate the heat loss from the rod if its thermal conductivity is $390 W/m.K$.

3.5 A very long 2-mm diameter copper rod ($k = 380\,W/m.K$) extends from a surface at 120°C. The temperature of surrounding air is 25°C and the heat transfer coefficient over the rod is $10\,W/m^2.K$. Calculate the heat loss from the rod.

3.6 An aluminium fin 1.5 mm thick is placed on a circular tube with 27-mm outer diameter. The fin is 6 mm long. The tube wall is maintained at 150°C, the environment temperature is 15°C, and the convection heat-transfer coefficient is $20\,W/m^2.K$. Take the thermal conductivity of aluminium as $k = 210\,W/m.K$. Calculate the heat lost by the fin.

3.7 One end of a long rod 3 cm in diameter is inserted into a furnace with the outer end projecting into the outside air. Once the steady state is reached, the temperature of the rod is measured at two points, 15 cm apart and found to be 140°C and 100°C, when the atmospheric air is at 30°C with convection coefficient of $20\,W/m^2.K$. Calculate the thermal conductivity of the rod material.

3.8 A straight trapezoidal fin is $0.05m$ long and $1.0m$ wide. The thicknesses at the root and tip are, respectively, $7\,mm$ and $3\,mm$. The temperature difference at the root is $\theta_r = 80°C$. The heat transmitted by conduction at the root is dissipated to the surroundings at $h_f = 20\,W/m^2.K$. If the thermal conductivity of the fin material is $k = 40\,W/m.K$, determine the heat loss rate from the fin. You can use the fin efficiency charts.

3.9 The following data is provided for a fin of convex parabolic profile: $L = 0.1m$, $w = 1.0m$, $t = 6mm$, $k = 40\,W/m.K$, $h_f = 60\,W/m^2.K$, $\theta_r = 60°C$. Determine the fin efficiency and heat flow rate from the fin.

3.10 A straight fin has a concave parabolic profile $y = t(x/L)^2/2$, with root thickness $t = 6mm$ and length $L = 20mm$. Determine the rate of heat dissipation by the fin and fin efficiency if its base temperature is 200°C and it is exposed to fluid at 27°C with a heat transfer coefficient of $2800\,W/m^2.K$. Take the thermal conductivity of the fin material as $190\,W/m.K$.

3.11 If the fin in Problem 3.11 is replaced with a fin of convex parabolic profile and all other data kept unchanged, what are the efficiency and rate of heat loss?

3.12 The annular fin of rectangular profile has thickness $t = 3.6mm$, root radius $r_r = 60mm$, and tip radius $r_t = 120mm$. The fin is heated at the root to $\theta_r = 80°C$. The thermal conductivity of the fin material is $k = 30\,W/m.K$ and the heat transfer coefficient is $h_f = 30\,W/m^2.K$. Use the efficiency chart to estimate the rate of heat loss from the fin.

3.13 A radial fin of rectangular profile is exposed to surroundings at a temperature of 35°C and a heat transfer coefficient of $h = 40\,W/m^2.K$. The temperature at the fin base is 110°C and the fin is made from steel with $k = 40\,W/m.K$. The outer and inner diameters of the fin are 25 and $10\,cm$, respectively and the fin thickness is $0.25\,cm$. Determine the fin efficiency using analytical solution.

3.14 Calculate the rate of heat dissipation from the fin in Problem 3.13 using analytical solution.

3.15 Calculate the tip temperature of the fin in Problem 3.13 using analytical solution.

3.16 Straight fins of different profiles are exposed to surroundings at a temperature of 20°C and a heat transfer coefficient of $h = 40\,W/m^2.K$. In all cases, the temperature at the fin base is 90°C and the fins are made from steel with $k = 30\,W/m.K$. All fins are $10\,cm$ high with bases $0.80\,cm$ thick. Compare the fin efficiencies if the profiles are: (a) rectangular; (b) triangular; (c) concave parabolic; or (d) convex parabolic.

3.17 Straight fins of different profiles are exposed to surroundings at a temperature of 20°C and a heat transfer coefficient of $h = 40\,W/m^2.K$. In all cases, the temperature at the fin base is 90°C and the fins are made from steel with $k = 30\,W/m.K$. All fins are $10\,cm$ high with bases

0.80 cm thick. Compare the rates of heat dissipation from the fins per unit length if the profiles are: (a) rectangular; (b) triangular; (c) concave parabolic; and (d) convex parabolic.

3.18 Straight fins of different profiles are exposed to surroundings at a temperature of 20°C and a heat transfer coefficient of $h = 40\,W/m^2.K$. In all cases, the temperature at the fin base is 90°C and the fins are made from steel with $k = 30\,W/m.K$. All fins are 10 cm high with base 0.80 cm thick. Compare the fin tip temperatures if the profiles are: (a) rectangular; (b) triangular; (c) concave parabolic; and (d) convex parabolic.

3.19 Determine the overall surface efficiency and heat transfer rate for the plain wall in Figure 3.33 with six fins for the following data: $H = 201mm$, $L = 60mm$, $w = 200mm$, $t = 6mm$, $\delta = 50mm$, $T_{\infty 1} = 500\,K$, $h_1 = 1000\,W/m^2.K$, $T_{\infty 2} = 300\,K$, $h_p = 20\,W/m^2.K$, $h_f = 30\,W/m^2.K$, $k = 150\,W/m.K$.

3.20 The cylinder of a small air-cooled engine is made from aluminium alloy with thermal conductivity $k = 186\,W/m.K$. The cooling is provided by an array of eight radial fins of rectangular profile (uniform thickness), as shown in Figure P3.20. The temperature at the root of the fins is 525 K and heat is dissipated by convection at temperature $T_\infty = 320\,K$ and heat transfer coefficient $h = 75\,W/m^2.K$. Determine the rate of heat loss from the cylinder with and without fins.

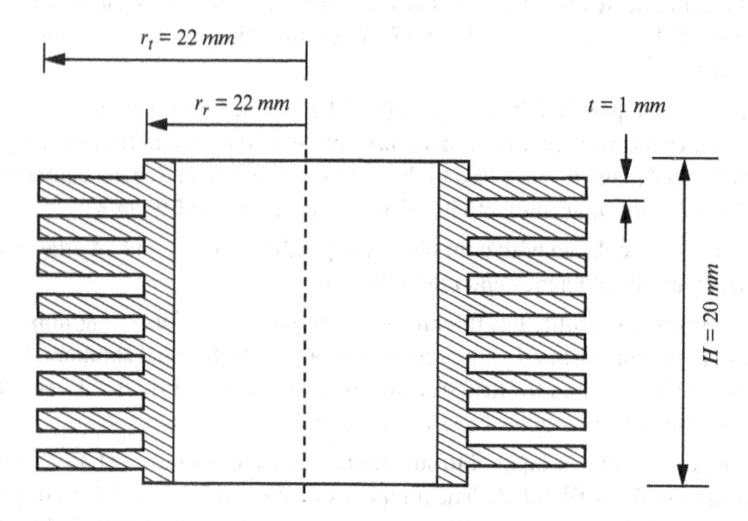

Figure P3.20 Schematic diagram of air-cooled engine cylinder.

4

Two-Dimensional Steady-State Heat Conduction

Several one-dimensional heat conduction processes under different boundary conditions were discussed in detail in Chapters 2 and 3. The main assumption underlying these processes was that heat flows only in one direction. This assumption is generally true for a limited number of cases and real heat transfer processes are almost always multi-dimensional. In this chapter, two-dimensional heat conduction processes will be analysed and solutions presented. The assumptions needed in order to simplify the analysis are:

- Steady-state flow
- No internal energy storage or generation
- The thermal conductivity is constant in value and does not vary with direction.

Three methods of evaluating two-dimensional conduction problems will be discussed:

1) Analytical method based on solving differential equations
2) Conduction shape factor method
3) Numerical method.

4.1 Analytical Method

Analytical solutions of two-dimensional conduction problems are useful where the model can be represented by a simple geometry and relatively simple boundary conditions. The method of separation of variables is widely used in solving two-dimensional steady-state conduction problems. The use of this method requires that the differential equation and all but one of the boundary conditions be homogeneous. A differential equation or boundary condition is homogeneous if each nonzero term contains the dependent variable or its derivatives (in heat conduction problems T, dT/dx, or d^2T/dT^2) and is not altered when the dependent variable in the equation is multiplied by a constant. Equation (2.17) in Chapter 2 meets the criteria and is homogeneous

$$\frac{\partial}{\partial x}\left(k\frac{\partial T}{\partial x}\right)+\frac{\partial}{\partial y}\left(k\frac{\partial T}{\partial y}\right)+\frac{\partial}{\partial z}\left(k\frac{\partial T}{\partial z}\right)=0$$

Equation (2.20) in Chapter 2, for conduction with heat generation

$$\frac{\partial}{\partial x}\left(k\frac{\partial T}{\partial x}\right)+\frac{\partial}{\partial y}\left(k\frac{\partial T}{\partial y}\right)+\frac{\partial}{\partial z}\left(k\frac{\partial T}{\partial z}\right)+\dot{q}=0$$

Heat Transfer Basics: A Concise Approach to Problem Solving, First Edition. Jamil Ghojel.

Companion website: www.wiley.com/go/ghojel/heat_transfer

is non-homogeneous because of the presence of the heat generation term $\dot{q}$.

The following are boundary conditions that are typically applicable in conduction heat transfer problems:

Non-homogeneous:
Constant temperature: $T_s = T_0$
Temperature as a function of position and time: $T_s = f(x)$ or $T_s = f(y)$
Heat flux: $-k\partial T / \partial x = q_x$, or $-k\partial T / \partial y = q_y$ (q_x and q_y) could be constant values or functions of position and time.
Convection heat transfer to a fluid: $-k\partial T / \partial x = h(T - T_\infty)$

Homogeneous:
Zero temperature: $T_s = 0$
Adiabatic surface (perfectly insulated): $\partial T / \partial x = 0$ or $\partial T / \partial y = 0$
Convection heat transfer to a fluid at zero temperature ($T_\infty = 0$): $-k\partial T / \partial x = hT$

The implementation of the method separation of variables also requires that the geometry of the region be described by orthogonal coordinate system (examples include rectangles, squares, cylinders, and spheres). Other shapes, such as triangles, trapezoids, and irregular shapes, require different solution techniques.

Comprehensive analytical solutions of two-dimensional conduction problems are well documented and can be found in publications such as Carslaw and Jaeger (1959), Arpaci (1966), Ozisik (1993), and Latif (2009). Such analysis is beyond the scope of this introductory text and only brief outlines of the solutions for some two-dimensional plates will be presented with examples highlighting the applications of the resulting equations.

The simplest example of a two-dimensional conduction model is the rectangular bar shown in Figure 4.1a. If we assume that the depth of the bar D in the z-direction is much greater than W and L, only the highlighted cross-section needs to be considered, as shown in Figure 4.1b.

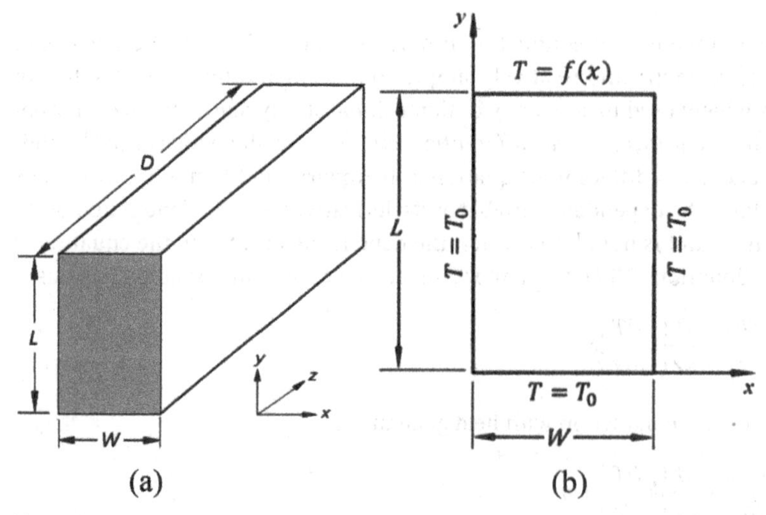

Figure 4.1 Schematics of heat conduction model: (a) Solid bar of rectangular cross-section; (b) Two-dimensional rectangle (grey shaded in Figure 4.1a).

The differential equation in Cartesian coordinates for a steady-state two-dimensional conduction with constant thermal conductivity, no energy storage, and no internal energy generation, can be obtained from Eq. (2.14)

$$\frac{\partial^2 T}{\partial x^2}+\frac{\partial^2 T}{\partial y^2}=0 \tag{4.1}$$

Four boundary conditions are applied at the outer edges of the rectangle in Figure 4.1b:

$$x=0,\ 0<y<L, T=T_0$$

$$y=0,\ 0<x<W, T=T_0$$

$$x=W,\ 0<y<L, T=T_0$$

$$y=L, 0<x<W, T=f(x)$$

The boundary conditions $T=T_0$ on the bottom and two sides of the rectangle are nonzero values and non-homogeneous and the method of separation of variables cannot be applied to solve the problem. One way around this is to define the temperature $\phi=T-T_0$, which transforms these boundary conditions to zero-value homogeneous boundary conditions $(\phi=0)$.

Equation (4.1) can now be rewritten as

$$\frac{\partial^2 \phi}{\partial x^2}+\frac{\partial^2 \phi}{\partial y^2}=0 \tag{4.2}$$

with the transformed boundary conditions

$$x=0,\quad 0<y<L,\ \phi=0$$

$$y=0,\quad 0<x<W, \phi=0$$

$$x=W,\ 0<y<L,\ \phi=0$$

$$y=L,\quad 0<x<W, T=f(x)$$

It is normally assumed that a solution $\phi(x,y)$ that satisfies Eq. (4.2) and the four boundary conditions can be expressed as a product of two functions

$X(x)$ and $Y(y)$

$$\phi(x,y)=X(x)Y(y) \tag{4.3}$$

from which

$$\frac{d^2\phi}{dx^2}=\frac{d^2X}{dx^2}Y$$

$$\frac{d^2\phi}{dy^2}=\frac{d^2Y}{dy^2}X$$

Substitution of the derivatives into Eq. (4.2) and rearranging yields

$$-\frac{1}{X}\frac{d^2X}{dx^2}=\frac{1}{Y}\frac{d^2Y}{dy^2} \tag{4.4}$$

Both sides of Eq. (4.4) must be equal to a constant for the equality to hold. If the constant is taken as equal to λ^2, we obtain two independent ordinary differential equations

$$\frac{d^2X}{dx^2}+\lambda^2X=0 \tag{4.5a}$$

$$\frac{d^2Y}{dy^2}-\lambda^2Y=0 \tag{4.5b}$$

The general solutions to the differential Eqs (4.5a) and (4.5b) are, respectively

$$X=C_1\cos\lambda x+C_2\sin\lambda x$$

$$X=C_3e^{-\lambda y}+C_4e^{\lambda y}$$

from which, based on Eq. (4.3)

$$\phi(x,y)=(C_1\cos\lambda x+C_2\sin\lambda x)(C_3e^{-\lambda y}+C_4e^{\lambda y})$$

Imposing the boundary conditions finally yields the general solution to the temperature field

$$\phi(x,y)=T-T_0=\sum_{n=1}^{\infty}c_n\sin\left(\frac{n\pi x}{W}\right)\frac{\sinh(n\pi y/W)}{\sinh(n\pi L/W)} \tag{4.6}$$

where

$$c_n=\frac{2}{W}\int_0^W f(x)\sin\left(\frac{n\pi x}{W}\right)dx \tag{4.7}$$

The arbitrary function $f(x)$ represents the temperature distribution along the top boundary in Figure 4.1b ($y=L$), which could be a constant value or sinusoidal function.

4.1.1 Two-Dimensional Plate With Finite Length and Width and Constant Boundary Conditions

4.1.1.1 Temperature Distribution

The model is shown schematically in Figure 4.2a in which the boundary condition at the top edge of the model is a constant temperature, i.e., $f(x)=T_1$, or $F(x)=T_1-T_0$.

Substituting T_1-T_0 into Eq. (4.7) yields

$$c_n=\frac{2}{W}\int_0^W (T_1-T_0)\sin\left(\frac{n\pi x}{W}\right)dx=-\frac{2}{W}(T_1-T_0)\frac{W}{n\pi}\left[\cos\left(\frac{n\pi x}{W}\right)\right]_0^W$$

$$c_n=-\frac{2}{n\pi}(T_1-T_0)[\cos(n\pi)-1] \tag{4.8}$$

Solving Eq. (4.8) for n, we obtain

$$c_n=\frac{2\left[1-(-1)^n\right]}{n\pi}(T_1-T_0)\quad \text{for}\quad n=1,3,5,\ldots$$

$$c_n=0\quad \text{for}\quad n=2,4,6,\ldots$$

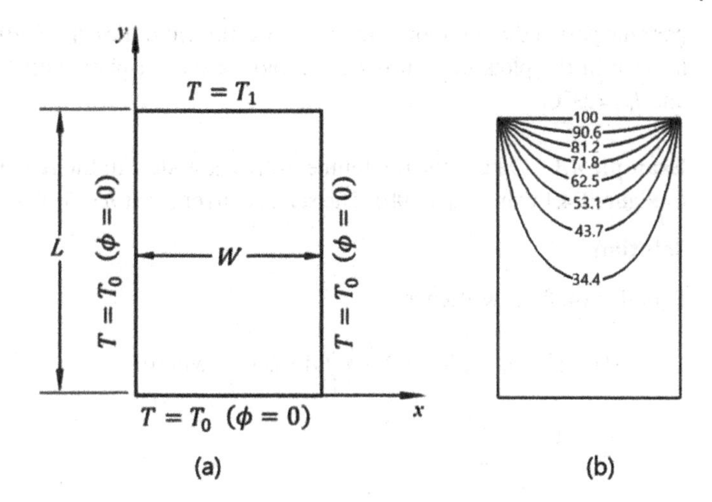

Figure 4.2 Schematic diagram of the two-dimensional plate (a), and plots of constant-temperature contours (b): $W = 0.4\,m$, $L = 0.5\,m$, $T_1 = 100°C$, $T_0 = 25°C$.

Substitution of c_n into Eq. (4.6) results in the equations for the temperature distribution in the $x-y$ plane

$$\frac{T-T_0}{T_1-T_0}=\frac{2}{\pi}\sum_{n=1}^{\infty}\frac{\left[1-(-1)^n\right]}{n}\sin\left(\frac{n\pi x}{W}\right)\frac{\sinh(n\pi y/W)}{\sinh(n\pi L/W)} \tag{4.9a}$$

or

$$T=T_0+(T_1-T_0)\frac{2}{\pi}\sum_{n=1}^{\infty}\frac{\left[1-(-1)^n\right]}{n}\sin\left(\frac{n\pi x}{W}\right)\frac{\sinh(n\pi y/W)}{\sinh(n\pi L/W)} \tag{4.9b}$$

Equation (4.9b) converges quickly to return the temperature at any point with coordinates (x, y) within the shape and can readily be solved in a spreadsheet.

Equation (4.9b) can be used to plot the temperature field $T(x, y)$ in the form of constant-temperature contours (Figure 4.2b), or the temperature profiles in the x- and y-directions (Figure 4.3). The

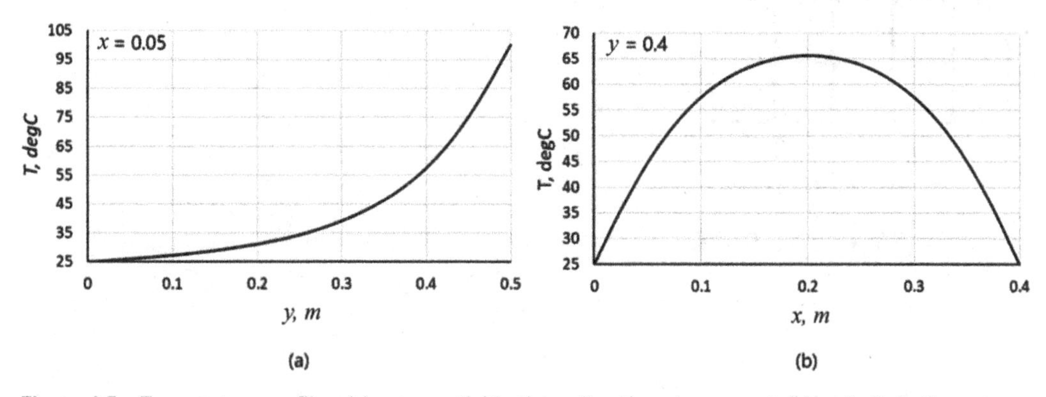

Figure 4.3 Temperature profiles: (a) exponential in the y-direction at $x = \text{const}$; (b) periodic in the x-directions at $y = \text{const}$.

periodic part of the equation is in the x-direction, and the exponential part in the y-direction as can be seen in the plots in Figure 4.3. All plots are for a plate with $L = 0.5m, W = 0.4m$, $T_1 = 100°C$, and $T_0 = 25°C$.

Example 4.1 Determine the temperatures at node 4 in the rectangular rod of unit depth shown in Figure E4.1 using Eq. (4.9b). The series converges at $n = 5$. Take $T_1 = 90°C$ and $T_0 = 25°C$.

Solution

Equation (4.9b) is written as

$$T(x,y) = T_0 + (T_1 - T_0)\frac{2}{\pi}\sum_{n=1}^{\infty}(A \times B \times C) \text{ where}$$

$$A = \frac{\left[1-(-1)^n\right]}{n}$$

$$B = \sin\left(\frac{n\pi x}{W}\right)$$

$$C = \frac{\sinh(n\pi y / W)}{\sinh(n\pi L / W)}$$

The results of calculations for $T(0.05, 0.2)$ are shown in Table E4.1.

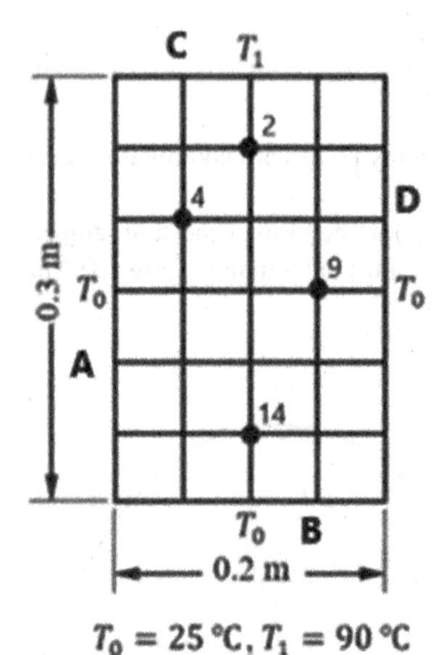

Figure E4.1 Two-dimensional model for Example 4.1.

Table E4.1 Calculation methodology of the two-dimensional model in Example 4.1.

n	A	B	C	$A\times B\times C$	$\frac{2}{\pi}\sum_{n=1}^{\infty}(A\times B\times C)$	$T(0.05,0.2)\ T_4$, °C
0	–	0	–	–	–	–
1	2	0.70710678	0.20750811	0.29346079	0.18682	37.14
2	0	1	0.04321376	0	0.18682	37.14
3	0.666666667	0.70710678	0.00898329	0.00423476	0.18952	37.32
4	0	1.2251E-16	0.00186744	0	0.18952	37.32
5	0.4	-0.7071067	0.00038820	-0.0001098	0.18945	37.31
6	0	-1	8.0699E-05	0	0.18945	37.31
7	0.285714286	-0.7071067	1.6775E-05	-3.3892E-06	0.18945	37.31
8	0	-2.4503E-16	3.4873E-06	0	0.18945	37.31

4.1.1.2 Rate of Heat Transfer

Consider the model shown in Figure 4.4 with two-dimensional heat conduction in the $x-y$-plane ($D>>L, D>>W$).

The incremental rate of heat transfer $d\dot{Q}_y$ in the y-direction through any surface Ddx in the $x-z$-plane between $y=0$ and $y=L$ can be calculated from Fourier's equation

$$d\dot{Q}_y = -k\frac{\partial T}{\partial y}Ddx \tag{4.10}$$

The total heat rate through surface WD is then

$$\dot{Q}_y = -\int_{x=0}^{W} k\frac{\partial T}{\partial y}Ddx \tag{4.11}$$

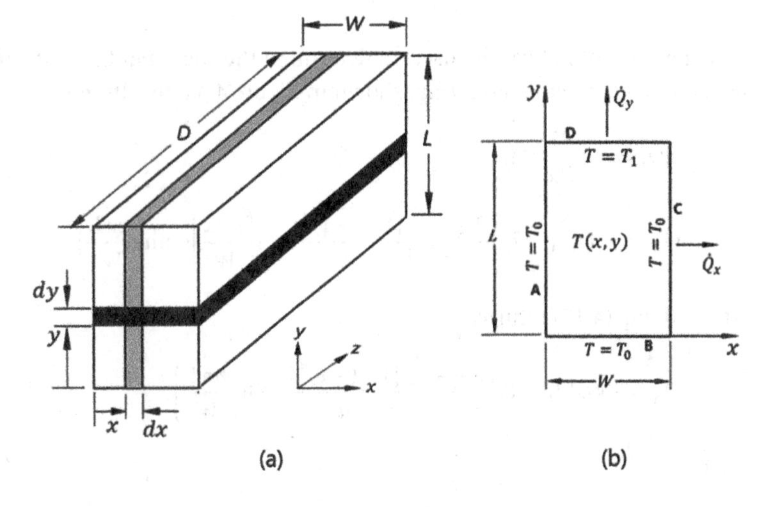

Figure 4.4 Rate of heat transfer in a rectangular rod in the y and x directions.

Differentiating Eq. (4.9b) with respect to y at fixed x and substituting in Eq. (4.11) results in

$$d\dot{Q}_y = -k(T_1 - T_0)\left\{\frac{2}{W}\sum_{n=1}^{\infty}\left[1-(-1)^n\right]\sin\left(\frac{n\pi x}{W}\right)\frac{\cosh(n\pi y/W)}{\sinh(n\pi L/W)}\right\}Ddx$$

and

$$\dot{Q}_y = -kD(T_1 - T_0)\frac{2}{W}\int_{x=0}^{W}\sum_{n=1}^{\infty}\left[1-(-1)^n\right]\frac{\cosh(n\pi y/W)}{\sinh(n\pi L/W)}\sin\left(\frac{n\pi x}{W}\right)dx$$

$$\dot{Q}_y = -kD(T_1 - T_0)\frac{2}{W}\sum_{n=1}^{\infty}\left[1-(-1)^n\right]\frac{\cosh(n\pi y/W)}{\sinh(n\pi L/W)}\frac{W}{n\pi}\left[-\frac{W}{n\pi}\cos\left(\frac{n\pi x}{W}\right)\right]_0^W$$

$$\dot{Q}_y = kD(T_1 - T_0)\frac{2}{\pi}\sum_{n=1}^{\infty}\frac{\left[1-(-1)^n\right]}{n}\frac{\cosh(n\pi y/W)}{\sinh(n\pi L/W)}\left[\cos(n\pi)-1\right] \tag{4.12}$$

$\left[\cos(n\pi)-1\right]=0$ for $n=2,4,6\ldots$ and -2 for $n=1,3,5\ldots$; hence, Eq. (4.12) is written as

$$\dot{Q}_y = -kD(T_1 - T_0)\frac{4}{\pi}\sum_{n=1}^{\infty}\frac{\left[1-(-1)^n\right]}{n}\frac{\cosh(n\pi y/W)}{\sinh(n\pi L/W)} \tag{4.13}$$

At $y=0$, Eq. (4.13) yields the heat rate through the lower surface.

Designating outer surfaces as A, B, C, and D, as shown in Figure 4.4b, we obtain

$$\dot{Q}_B = -kD(T_1 - T_0)\frac{4}{\pi}\sum_{n=1}^{\infty}\frac{\left[1-(-1)^n\right]}{n}\frac{1}{\sinh(n\pi L/W)} \tag{4.14}$$

At $y=L$, we obtain the heat rate through the upper surface

$$\dot{Q}_D = -kD(T_1 - T_0)\frac{4}{\pi}\sum_{n=1}^{\infty}\frac{\left[1-(-1)^n\right]}{n}\coth(n\pi L/W) \tag{4.15}$$

A similar method can be used to determine the incremental heat rate $d\dot{Q}_x$ in the x-direction across the incremental area Ddy, shown in Figure 4.4a, resulting in

$$d\dot{Q}_x = -k\frac{\partial T}{\partial x}Ddy \tag{4.16}$$

$$\dot{Q}_x = -kD(T_1 - T_0)\frac{2}{\pi}\sum_{n=1}^{\infty}\frac{[1-(-1)^n]}{n}\cos\left(\frac{n\pi x}{W}\right)\coth\left(\frac{n\pi L}{W}\right) \tag{4.17}$$

At $x=0$, Eq. (4.17) returns

$$\dot{Q}_A = -kD(T_1 - T_0)\frac{2}{\pi}\sum_{n=1}^{\infty}\frac{[1-(-1)^n]}{n}\coth\left(\frac{n\pi L}{W}\right) \tag{4.18}$$

At $x=W$

$$\dot{Q}_C = -kD(T_1 - T_0)\frac{2}{\pi}\sum_{n=1}^{\infty}\frac{[1-(-1)^n]}{n}\cos(n\pi)\coth\left(\frac{n\pi L}{W}\right) \tag{4.19}$$

Example 4.2 Find the heat rates per unit depth at the four surfaces of the rectangular rod in Example 4.1 if the thermal conductivity is $40 W/m.K$.

Solution

Applying Eqs (4.14), (4.15), (4.18), and (4.19), we find that the infinite series in Eq. (4.14) is the only one that converges rapidly giving $\dot{Q}_B = -118.96 W/m$. All other equations fail to converge, indicating heat rates tending to infinity, which is meaningless. This results from the stepwise temperature discontinuities at the top right and top left corners (the temperature changes from 100°C to 25°C).

4.1.2 Two-Dimensional Plate With Finite Length and Nonconstant Boundary Conditions

4.1.2.1 Temperature Distribution

Figure 4.5a shows a rectangular plate with non-homogeneous boundary conditions, one of which is spatially varying at the upper edge $T = f(x)$. The boundary conditions are then

$$x = 0,\quad 0 < y < L,\ T = T_0$$
$$y = 0,\quad 0 < x < W,\ T = T_0$$
$$x = W,\ 0 < y < L,\ \ T = T_0$$
$$y = L,\quad 0 < x < W,\ T = f(x)$$

Let $f(x)$ be a simple sinusoidal function of the form

$$f(x) = T = T_0 + T_m \sin\left(\frac{\pi x}{W}\right),\ \text{or}\ F(x) = T - T_0 = T_m \sin\left(\frac{\pi x}{W}\right)$$

Substituting $F(x)$ in Eq. (4.6) at $y = L$ gives

$$T_m \sin\left(\frac{\pi x}{W}\right) = \sum_{n=1}^{\infty} c_n \sin\left(\frac{n\pi x}{W}\right) \tag{4.20}$$

The equality in Eq. (4.20) is true if $c_n = T_m$ for $n = 1$.

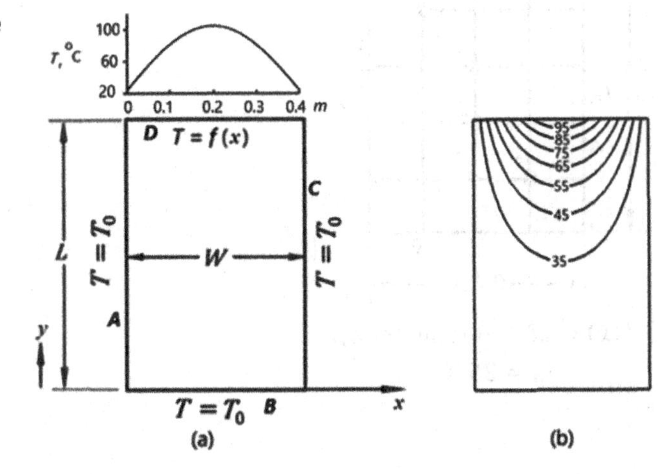

Figure 4.5 Schematic diagram of the model (a), and temperature contours (b) for two-dimensional conduction with spatially varying temperature at the top edge: $T = f(x)$, $W = 0.4\,m$, $L = 0.5\,m$, $T_0 = 25°C$, $T = f(x) = 25 + 80\sin(\pi x/W)$.

Equation. (4.6) is now transformed to

$$\phi(x,y) = T - T_0 = T_m \sin\left(\frac{\pi x}{W}\right)\frac{\sinh(\pi y / W)}{\sinh(\pi L / W)} \tag{4.21a}$$

$$T = T_0 + T_m \sin\left(\frac{\pi x}{W}\right)\frac{\sinh(\pi y / W)}{\sinh(\pi L / W)} \tag{4.21b}$$

Equation (4.21b) can be used to plot the temperature field $T(x, y)$ in the form of constant-temperature contours, as shown in Figure 4.5b.

Example 4.3 Determine the temperatures at nodes 1, 2, 3, and 4 in the rectangular plate shown in Figure E4.3 with two-dimensional conduction. A sinusoidal temperature function $f(x)$ is applied at the top edge of the plate.

Solution

Using Eq. (4.21b), we obtain

$$T(0.1, 0.1) = 25 + 80\sin\left(\frac{0.1\pi}{0.4}\right)\frac{\sinh(0.1\pi / 0.4)}{\sinh(0.5\pi / 0.4)} = 26.9°\text{C}$$

$$T(0.2, 0.2) = 25 + 80\sin\left(\frac{0.2\pi}{0.4}\right)\frac{\sinh(0.2\pi / 0.4)}{\sinh(0.5\pi / 0.4)} = 32.2°\text{C}$$

$$T(0.2, 0.3) = 25 + 80\sin\left(\frac{0.2\pi}{0.4}\right)\frac{\sinh(0.3\pi / 0.4)}{\sinh(0.5\pi / 0.4)} = 41.5°\text{C}$$

$$T(0.3, 0.4) = 25 + 80\sin\left(\frac{0.3\pi}{0.4}\right)\frac{\sinh(0.4\pi / 0.4)}{\sinh(0.5\pi / 0.4)} = 50.7°\text{C}$$

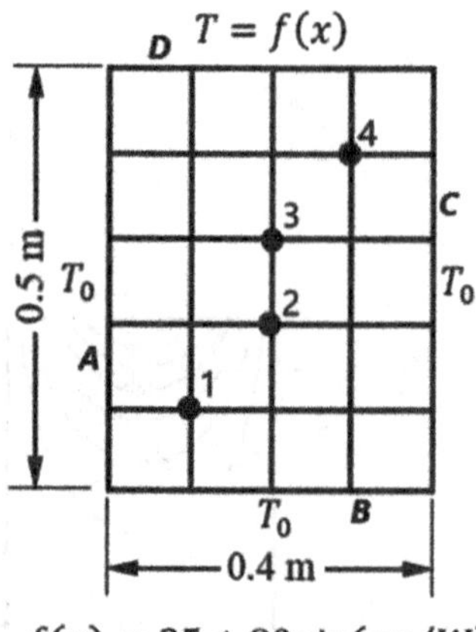

Figure E4.3 Two-dimensional model for Example 4.3.

4.1.2.2 Rate of Heat Transfer

Following the same method used in Section 4.1.1.2, the following expressions for the rates of heat transfer for this model can be obtained:

Heat transfer rate in the y-direction is

$$d\dot{Q}_y = -k\frac{\partial T}{\partial y}Ddx$$

$$\dot{Q}_y = -2kDT_m\frac{\cosh(\pi y/W)}{\sinh(\pi L/W)} \tag{4.22}$$

$y = 0$

$$\dot{Q}_B = -2kDT_m\frac{1}{\sinh(\pi L/W)} \tag{4.23}$$

$y = L$

$$\dot{Q}_D = -2kDT_m\coth(\pi L/W) \tag{4.24}$$

Heat transfer rate in the x-direction

$$d\dot{Q}_x = -k\frac{\partial T}{\partial x}Ddy$$

$$\dot{Q}_x = -kDT_m\cos\left(\frac{\pi x}{W}\right)\frac{1}{\sinh(\pi L/W)}\left[\cosh\left(\frac{\pi L}{W}\right) - 1\right] \tag{4.25}$$

$x = 0$

$$\dot{Q}_A = -kDT_m\frac{1}{\sinh(\pi L/W)}\left[\cosh\left(\frac{\pi L}{W}\right) - 1\right] \tag{4.26}$$

$x = W$

$$\dot{Q}_C = kDT_m\frac{1}{\sinh(\pi L/W)}\left[\cosh\left(\frac{\pi L}{W}\right) - 1\right] \tag{4.27}$$

Example 4.4 Calculate the heat rates at the four edges of the model in Example 4.3. Take $k = 40\,W/m.K$.

Solution

From Eqs (4.23), (4.24), (4.26), and (4.27) respectively, we find the heat rates per unit depth

$$\dot{Q}_B = 0.254\,kW/m$$

$$\dot{Q}_D = -6.17\,kW/m$$

$$\dot{Q}_A = -3.08\,kW/m$$

$$\dot{Q}_C = 3.08\,kW/m$$

4.1.3 Two-Dimensional Plate With Semi-Infinite Length

Figure 4.6 shows a two-dimensional plate with width W and semi-infinite length in the y-direction. The boundary conditions for this case are:

$$x = 0,\quad 0 \le y \le \infty,\ \phi = 0$$
$$y = 0,\quad 0 < x < W,\ \phi = 0$$
$$x = W, 0 \le y \le \infty,\ \phi = 0$$
$$y = \infty, 0 < x < W,\ \phi = 0$$

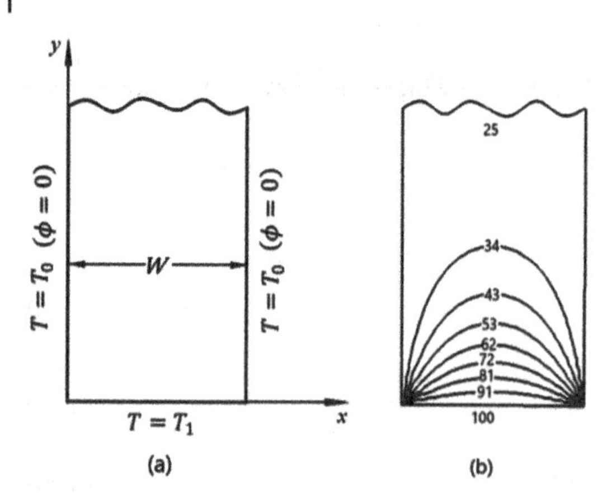

Figure 4.6 Schematic diagram of the model (a), and temperature contours (b) for two-dimensional semi-infinite plate: $W = 0.4\,m, T_1 = 100°C,\ T_0 = 25°C$.

The analytical solution in Welty et al. (2013) of this model for these boundary conditions results in Eq. (4.28):

$$T(x,y) = T_0 + \frac{4(T_1 - T_0)}{\pi} \sum_{n=0}^{\infty} \frac{\exp\left[\frac{-(2n+1)\pi y}{W}\right]}{2n+1} \sin\left[\frac{(2n+1)\pi x}{W}\right] \tag{4.28}$$

Equation (4.28) can be used to determine the temperature field in the form of contour plots as shown in Figure 4.6b.

Example 4.5 Consider the rectangular plate in Figure E4.5 with 0.4 m width and semi-infinite length and two-dimensional heat conduction. Determine

a) the temperature at nodes1, 2, 3, and 4
b) the height y at which $T(0.2, y) = 25°C$.

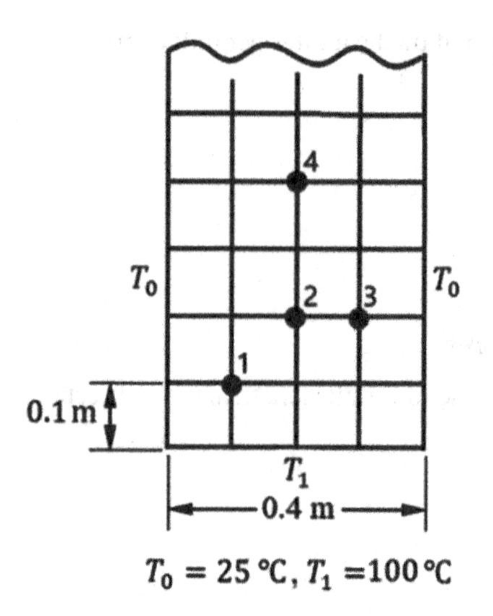

Figure E4.5 Two-dimensional model for Example 4.5.

Table E4.5 Calculation methodology of the two-dimensional model in Example 4.5.

n	A	B	C	$A \times B \times C$	$\frac{4}{\pi}\sum_{n=0}^{\infty} A \times B \times C$	$T(0.1,01)$
0	1	0.455938	0.707107	0.322397	0.410489	55.78664
1	0.333333	0.09478	0.707107	0.02234	0.438933	57.91994
2	0.2	0.019703	−0.70711	−0.00279	0.435385	57.65386
3	0.142857	0.004096	−0.70711	−0.00041	0.434858	57.61435

Equation (4.28) can be written as

$$T(x,y) = T_0 + (T_1 - T_0)\frac{4}{\pi}\sum_{n=0}^{\infty} A \times B \times C$$

where

$$A = \frac{1}{2n+1}$$

$$B = \frac{\exp\left[\frac{-(2n+1)\pi y}{W}\right]}{2n+1}$$

$$C = \sin\left[\frac{(2n+1)\pi x}{W}\right]$$

a) The results for $T(0.1,0.1)$ are shown in Table E4.5. Strictly speaking, convergence of the series occurs at $n = 8$. However, if the resulting temperature is approximated to one decimal place, $n = 3$ is a good place to stop.
The remaining temperatures are: $T(0.2,0.2) = 44.6°C$, $T(0.3,0.2) = 39.2°C$, $T(0.2,04) = 29.1°C$.
b) By trial and error, we find that $T(0.2,y) = 25°C$ at $y \cong 2.2m$.

4.1.4 Other Boundary Conditions

Several combinations of non-constant boundary conditions can be applied to the two-dimensional conduction model under consideration. The two temperature profiles $T = f(x)$ and $T = f(y)$, used here as examples, are given by Eqs (4.29a) and (4.29b) and plotted in Figure 4.7. Three examples are presented in Figure 4.8 with the boundary conditions displayed on the accompanying schematics of the models. In the absence of analytical techniques for solving such complex problems, computer modelling was used to obtain the temperature contours shown next to the schematics of the models.

$$T = f(x) = T_0 + 80\sin(\pi x/W) \tag{4.29a}$$

$$T = f(y) = T_0 + 50\left[2(y/L) - (y/L)^2\right] \tag{4.29b}$$

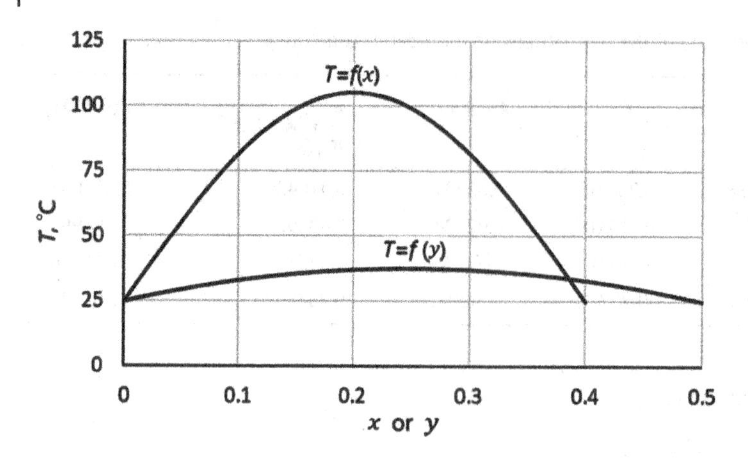

Figure 4.7 Temperature profiles applied on the edges of the two-dimensional conduction models.

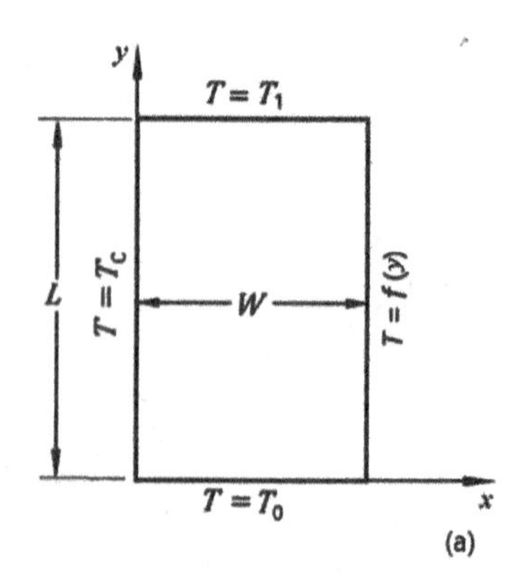

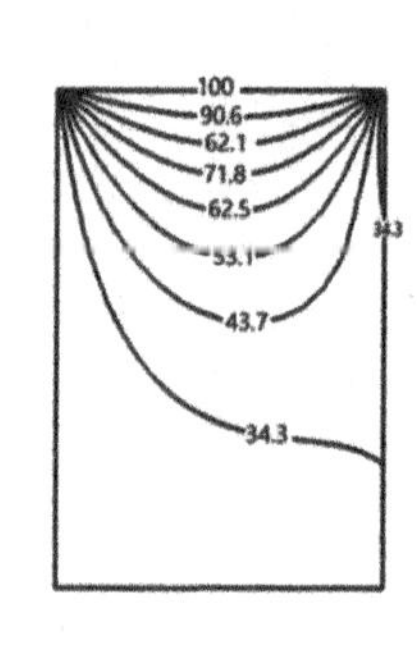

(a)

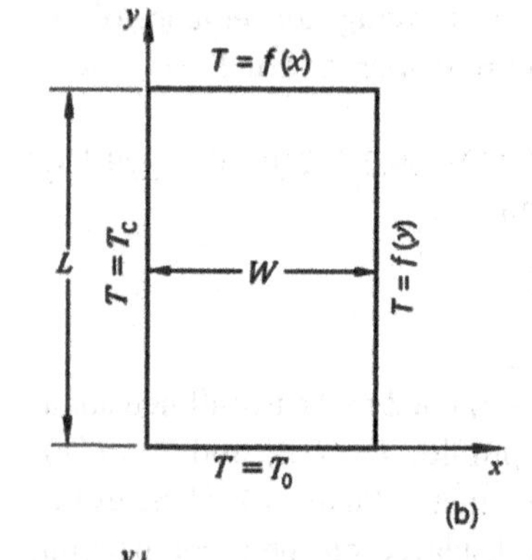

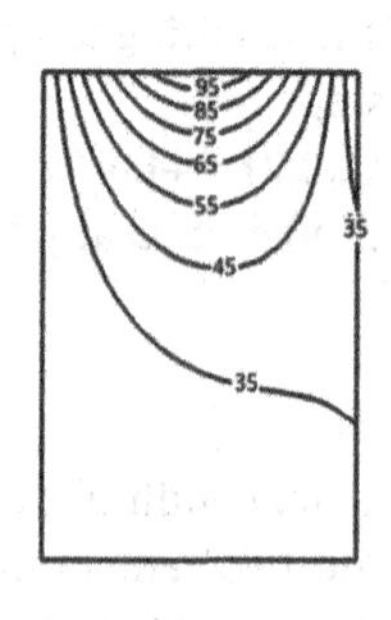

(b)

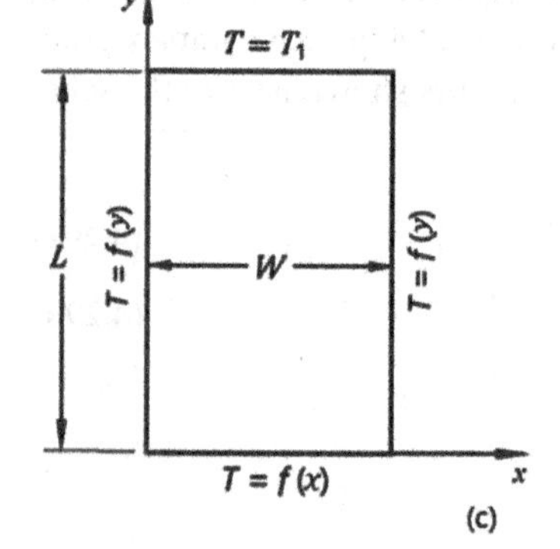

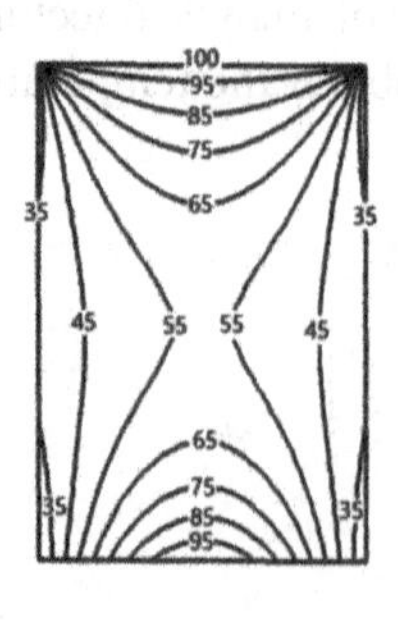

(c)

Figure 4.8 Computer simulation results of the effect of various boundary conditions on the temperature contours of a two-dimensional conduction model $T_0 = 25°C$, $T_1 = 100°C$.

4.1.5 Two-Dimensional Semi-Circular Plate (Or Cylinder) With Prescribed Boundary Conditions

Consider the simple example shown in Figure 4.9 of the steady-state two-dimensional conduction problem of a semi-circular plate of thickness δ or long cylinder of length L ($L >> R$). Generally, cylindrical coordinates are better suited to solving conduction problems in bodies of cylindrical shape, and for conduction in a system without heat generation and storage, Eq. (2.15) can be written as

$$\frac{\partial^2 T}{\partial r^2}+\frac{1}{r}\frac{\partial T}{\partial r}+\frac{1}{r^2}\frac{\partial^2 T}{\partial \varphi^2}=0 \tag{4.30}$$

The boundary conditions for this model are

$$0 \le r \le R,\ 0 \le \varphi \le \pi$$

$$\varphi = 0,\ T = T_0;\ \varphi = \pi,\ T = T_0;\ r = 0,\ T = T_0$$

$$r = R,\ T = f(\varphi)$$

If we define the temperatures $\phi = T - T_0$ and $f(\varphi) - T_0 = F(\varphi)$, the new boundary conditions are then

$$\varphi = 0,\ \phi = 0;\ \varphi = \pi,\ \phi = 0;\ r = 0,\ \phi = 0$$

$$r = R,\ \phi = F(\varphi)$$

Equation (4.20) is now written as

$$\frac{\partial^2 \phi}{\partial r^2}+\frac{1}{r}\frac{\partial \phi}{\partial r}+\frac{1}{r^2}\frac{\partial^2 \phi}{\partial \varphi^2}=0 \tag{4.31}$$

The method of separation of variables can be applied to solve the differential Eq. (4.31). If a separation of the form $\phi = \mathcal{R}(r)\Phi(\varphi)$ is assumed, we obtain

$$\frac{d^2\Phi}{d\varphi^2}+\lambda\Phi=0 \tag{4.32}$$

$$r^2\frac{d^2\mathcal{R}}{dr^2}+r\frac{d\mathcal{R}}{dr}-\lambda^2\mathcal{R}=0 \tag{4.33}$$

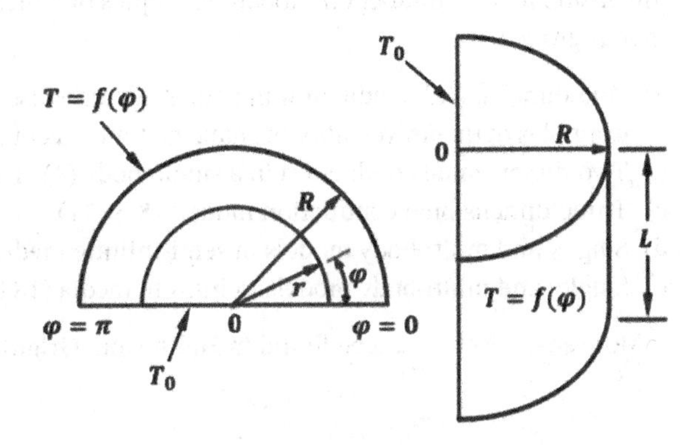

Figure 4.9 Two-dimensional conduction model of half cylinder.

The combined solution of the differential Eqs (4.31) and 4.32, taking into account the new boundary equations, can be written as

$$\phi = (C_1 \cos \lambda\varphi + C_2 \sin \lambda\varphi)(C_3 r^{\lambda} + C_4 r^{-\lambda}) \tag{4.34}$$

The constants in Eq. (4.34) for the special case $f(\varphi) = \text{const} = T_s$, can be evaluated leading to the equation for the temperature distribution within the half cylinder in terms of the variables r, φ (Arpaci 1966)

$$T - T_0 = (T_s - T_0)\frac{2}{\pi}\sum_{n=1}^{\infty}\frac{1-(-1)^n}{n}\left(\frac{r}{R}\right)^n \sin n\varphi \tag{4.35}$$

4.2 Conduction Shape Factor Method

The conduction shape factor method is very useful in solving some two- and three-dimensional problems of two bodies with isothermal (constant temperature) boundaries representing a source and a sink. It provides simple equations for the rate of heat transfer but no means of determining the temperature distribution within the bodies. Examples of the applicability of the shape factor method include heat transfer from buried single pipe or banks of pipes, cables, and spherical containers and heat transfer through rectangular enclosures such as furnaces and kilns.

The rate of heat transfer per unit length in a two-dimensional model is expressed as

$$\dot{Q} = Sk\Delta T \quad W/m \tag{4.36}$$

and in a three-dimensional model as

$$\dot{Q} = Sk\Delta T \quad W \tag{4.37}$$

The shape factor S in Eq. (4.36) is dimensionless, whereas in Eq. (4.37) it has the unit m; ΔT – overall temperature difference between the two isothermal boundaries, °C, or K.

ΔT could be the difference between the temperatures of two bodies of different geometrical shapes, the temperatures of boundaries within a single body, or the temperatures of a geometrical shape and a semi-infinite heat conducting medium.

Conduction shape factors have been obtained analytically for a large number of two- and three-dimensional heat conduction models, examples of which are compiled in Table 4.1 and organised in four groups:

a) One-dimensional conduction in bodies, already discussed in Chapter 2, and presented here as examples of the universality of shape factor concept (1, 2, 3)
b) Two-dimensional conduction in a single-body (4) and two-body (5, 6, 7, 10) models
c) Three-dimensional conduction models (8, 9, 11)
d) Single- and multi-body models in semi-infinite media (12 to 17)
e) Single- and multi-body models in infinite media (18 to 21).

More shape factors can be found in Hahne and Grigull (1975) and Taler and Duda (2006).

Table 4.1 Shape factors for selected heat conduction models: $\dot{Q} = Sk\Delta T = Sk(T_1 - T_2)$.

Shape	Geometry	Shape factor, S
(1) Flat plate or wall	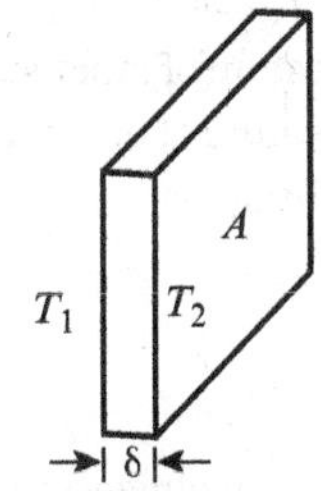	$S = A/\delta$
(2) Long hollow cylinder of length L	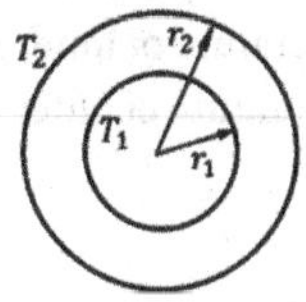	$S = 2\pi L/\ln(r_2/r_1)$ $L >> r$
(3) Hollow sphere	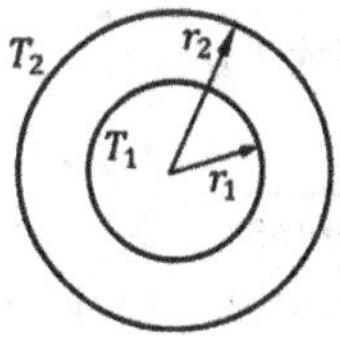	$S = 2\pi r_1 r_2/(r_2 - r_1)$
(4) Square channel of length L	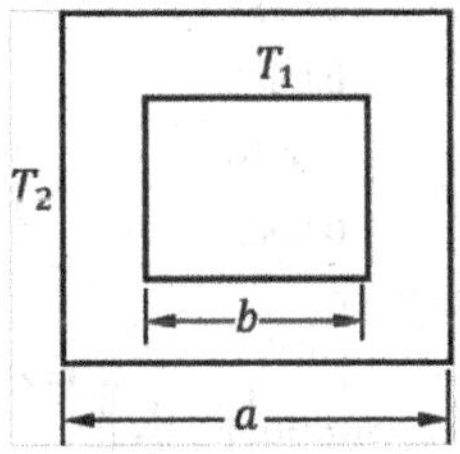	$S = \dfrac{2\pi L}{0.785\ln(a/b)}$ $\quad \dfrac{a}{b} < 1.4$ $S = \dfrac{2\pi L}{0.93\ln(a/b) - 0.05}$ $\quad \dfrac{a}{b} > 1.4$
(5) Long pipe inside rectangular block of equal length L	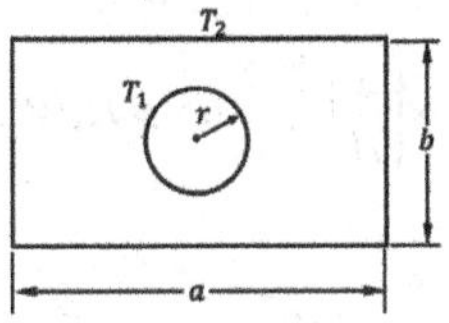	$S = \dfrac{2\pi L}{\ln\left[\dfrac{b}{r}(0.637 - 1.781e^{-2.9a/b})\right]}$ $L >> a, b \quad a \geq b$
(6) Long pipe inside square block of equal length L	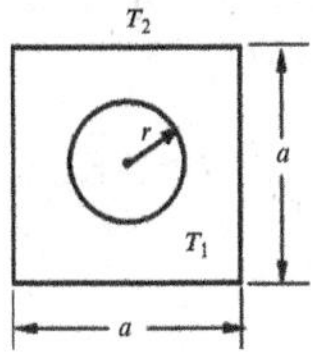	$S = \dfrac{2\pi L}{\ln(0.54\,a/r)}$ $\quad L >> a$
(7) Eccentric cylinder in a larger cylinder of equal length L	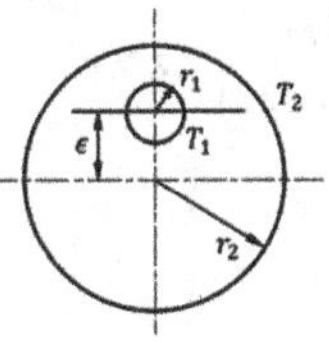	$S = \dfrac{2\pi L}{\cosh^{-1}\left(\dfrac{r_1^2 + r_2^2 - \varepsilon^2}{2r_1 r_2}\right)}$ $\quad L >> r_2$

(Continued)

Table 4.1 (Continued)

Shape	Geometry	Shape factor, S
(8) Corner beam of two adjoining flat walls	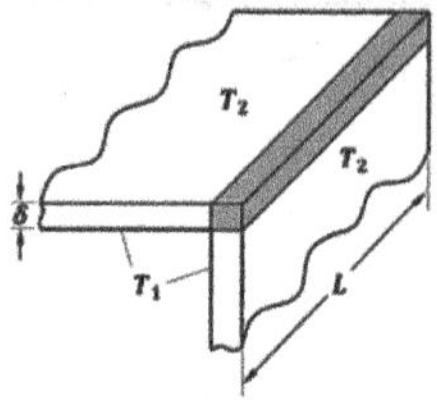	$S = 0.54\,L \quad L > 5\delta$ T_1 – Temperature of inner surfaces T_2 – Temperature of outer surfaces and corner beam
(9) Corner of three adjoining flat walls	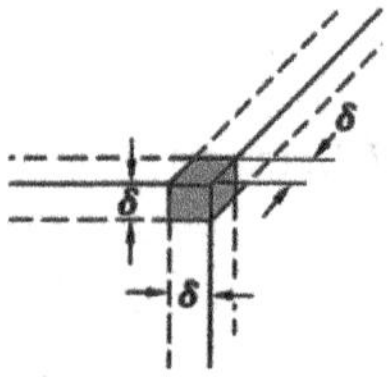	$S = 0.15\delta$ $\delta <<$ inside wall dimensions T_1 – Temperature of inner surfaces T_2 – Temperature of outer surfaces
(10) Rectangular solid with internal wire (rod) of equal length L	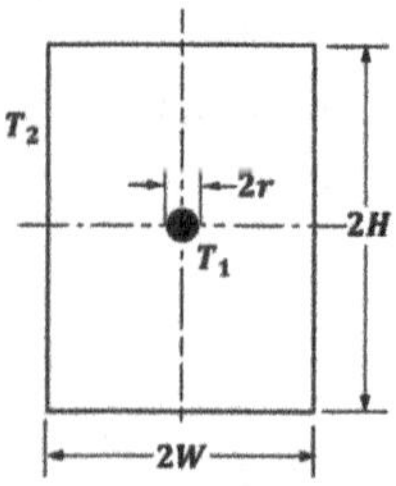	$S = \dfrac{2\pi L}{\ln(4W/\pi r) - C}$ $L >> r$, $W/r \geq 10$ H/W = 1.0, C = 0.1658 H/W = 1.25, C = 0.0793 H/W = 1.5, C = 0.0356 H/W = 2.0, C = 0.0075 H/W = 2.5, C = 0.0016 H/W = 3.0, C = 0.0003 H/W = ∞, C = 0
(11) Rectangular parallelepiped in a semi-infinite medium of conductivity k	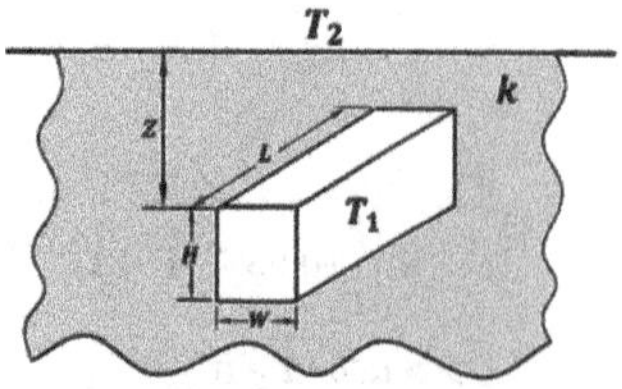	$S = 2.756L\left[\ln\left(1 + \dfrac{z}{W}\right)\right]^{-0.59} \times \left(\dfrac{z}{H}\right)^{-0.078} \quad L >> H, W, z$
(12) Horizontal cylinder of length L in a semi-infinite medium of conductivity k	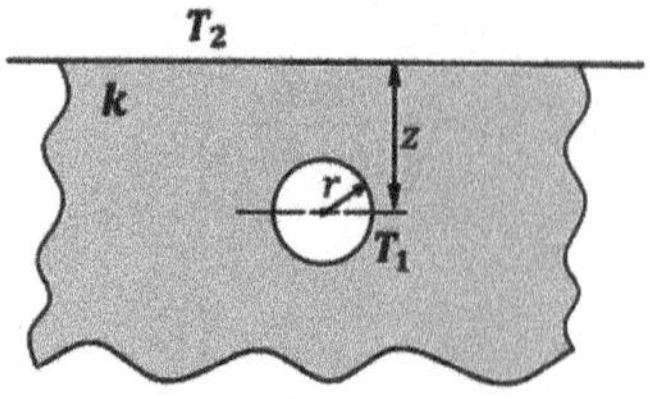	$S = \dfrac{2\pi L}{\cosh^{-1}(z/r)} \quad L >> r$ $S = \dfrac{2\pi L}{\ln(2z/r)} \quad L >> r, z > 3r$
(13) Sphere in a semi-infinite medium of conductivity k	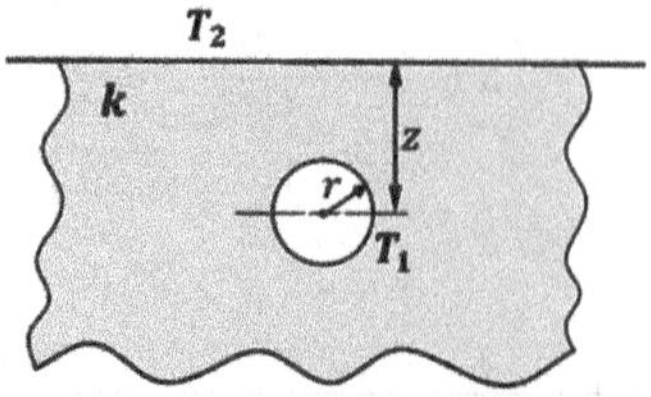	$S = \dfrac{4\pi r}{(1 - r/2z)} \quad z > r$

Table 4.1 (Continued)

Shape	Geometry	Shape factor, S	
(14) Equally spaced identical parallel cylinders of length L in a semi-infinite medium of conductivity k	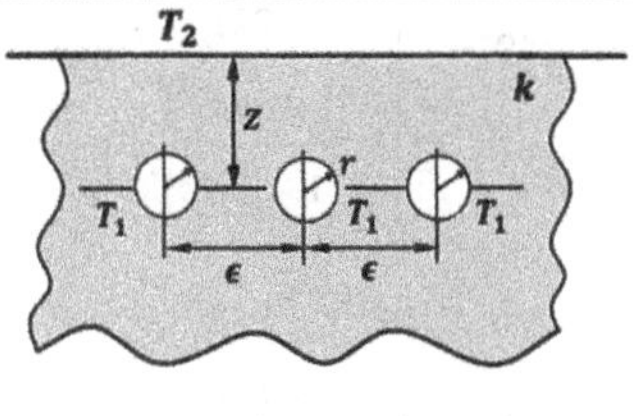	$S=\dfrac{2\pi L}{\ln\left[\dfrac{\varepsilon}{\pi r}\sinh\left(\dfrac{2\pi z}{\varepsilon}\right)\right]}$ $L>>r,\ L>>z,\ \varepsilon\geq 3r$ $\dot{Q}=NSk(T_1-T_2)$ $N-$ number of cylinders	
(15) Thin metal disk in a semi-infinite medium of conductivity k		$S=4r$	$z=0$
		$S=8r$	$\dfrac{z}{r}>>1$
		$S=\dfrac{4\pi r}{\dfrac{\pi}{2}-\tan^{-1}(r/2z)}$	$\dfrac{z}{r}>2$
(16) Vertical cylinder in a semi-infinite medium of conductivity k	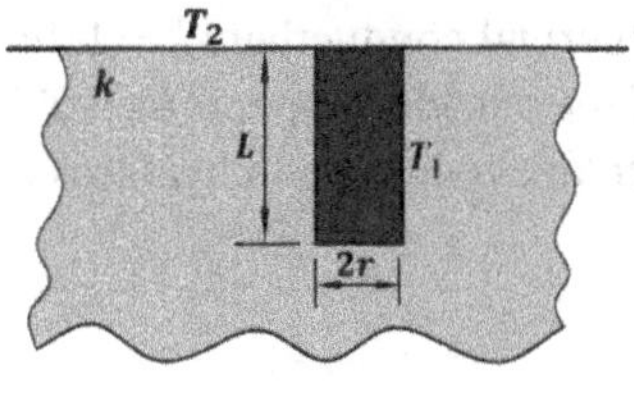	$S=\dfrac{2\pi L}{\ln(2L/r)}$	$L>>r$
(17) Long wires (cables) of length L and radius r in a semi-infinite medium of conductivity k	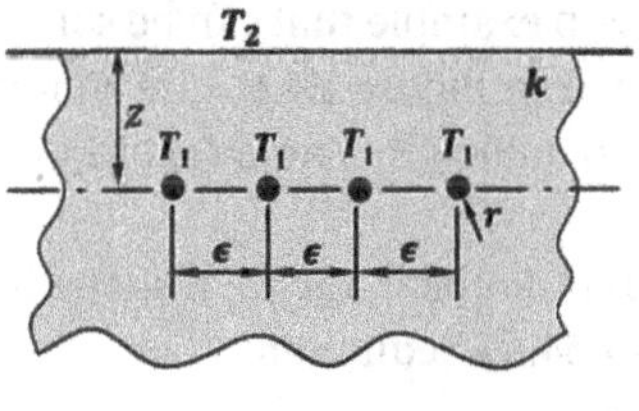	$S=\dfrac{2\pi L}{\dfrac{2\pi z}{\varepsilon}+\ln(\varepsilon/2\pi r)}$ $L>>r, r<<z,\epsilon$ $\dot{Q}=NSk(T_1-T_2)$ $N-$ number of wires	
(18) Two cylinders of length L in an infinite medium of conductivity k	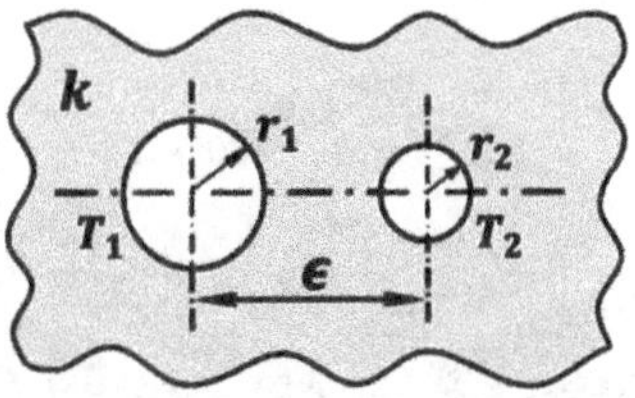	$S=\dfrac{2\pi L}{\cosh^{-1}\left(\dfrac{\varepsilon^2-r_1^2-r_2^2}{2r_1r_2}\right)}$ $L>>r_1,r_2,\ L>>\varepsilon$	
(19) A sphere of temperature T_1 in an infinite medium of conductivity k at temperature T_2	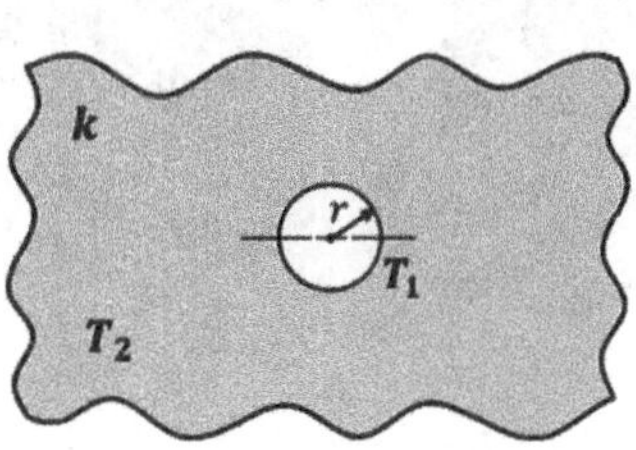	$S=4\pi r$	

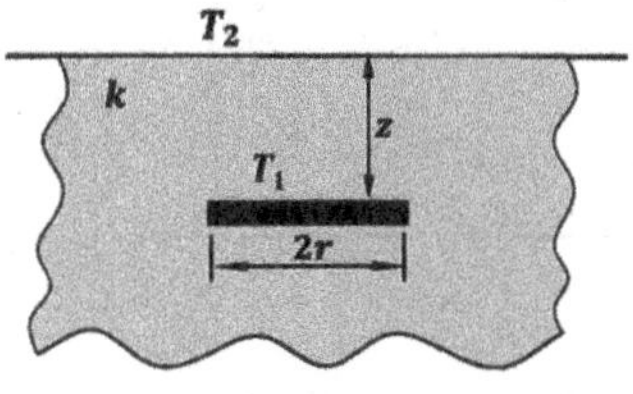

(Continued)

Table 4.1 (Continued)

Shape	Geometry	Shape factor, S
(20) Two spheres of different diameters in an infinite medium of conductivity k		$S=\dfrac{2\pi r_2}{\dfrac{r_2}{r_1}\left[1-\dfrac{(r_1/\varepsilon)^4}{1-(r_2/\varepsilon)^2}\right]-\dfrac{2r_2}{\varepsilon}}$ $r_2 > r_1, \varepsilon > 5r_2$
(21) Two thin parallel disks in an infinite medium of conductivity k		$S=\dfrac{2\pi r}{\pi/2-\tan^{-1}(r/\varepsilon)}$ $\quad \dfrac{\varepsilon}{r}\geq 5$

Example 4.6 An electric furnace has outside dimensions $350 \times 300 \times 200$ mm. The walls are $\delta =$ 8 mm thick and made of insulating material of thermal conductivity $k = 1.2 W/m.K$. The inside wall is maintained at temperature $T_1 = 600°C$ and the outside temperature is $T_2 = 25°C$. Determine the heat lost through the walls. What is the percentage reduction in heat loss if the wall thickness is doubled?

Solution

This is a classical three-dimensional heat conduction example that can be solved by breaking it up into several two-dimensional sub-models, as shown in Figure E4.6. The sub-models can then be solved using shape factors given in items 8 and 9 in Table 4.1. The calculation results are summarised in Table E4.6.

If the thickness of the wall (insulating material) is doubled, the total shape factor will reduce to 21.46, and the heat loss to 14.8 kW, which is about 54.6% reduction.

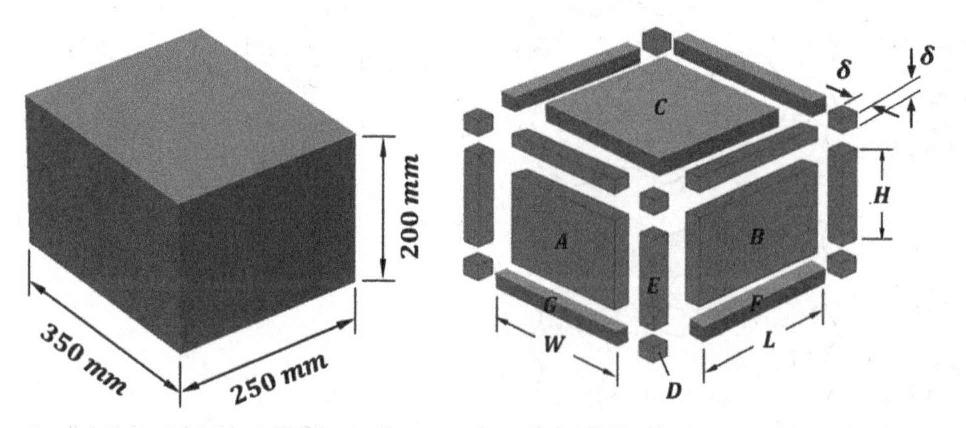

Figure E4.6 Schematic diagram for the furnace in Example 4.6.

Table E4.6 Summary of calculations for Example 4.6.

Sub model	Dimension(s), m^2 or m	Number	Shape factors S_i
Wall A	$W\times H=0.334\times0.184=0.0615$	2	$2\times0.0615/0.008=15.36$
Wall B	$L\times H=0.234\times0.184=0.0431$	2	$2\times0.0431/0.008=10.76$
Wall C	$W\times L=0.334\times0.234=0.0782$	2	$2\times0.0782/0.008=19.55$
Corner beams of two adjoining flat walls G,F,E	$W=(0.35-2\times0.008)=0.334$	4	$4\times0.54\times0.334=0.721$
	$L=(0.25-2\times0.008)=0.234$	4	$4\times0.54\times0.234=0.505$
	$H=(0.2-2\times0.008)=0.184$	4	$4\times0.54\times0.184=0.397$
Corner cubes of three adjoining flat walls D	$\delta=0.008$	8	$8\times0.15\times0.008=0.0096$
Total Shape factor S			47.3
HeatLoss: $\dot{Q}=Sk\Delta T=47.3\times1.2\times(600-25)=32637\,W(32.637\,kW)$			

Example 4.7 Horizontal thin pipes of diameter $d=20mm$ are to be used for residential home floor heating. The pipes are spaced apart at a distance of $\varepsilon=120mm$ and buried at a depth of $z=35mm$ under the floor. The pipes carry hot water at T_1 = 45°C and the exposed floor is to be maintained at T_2 = 20°C. Assuming the combined thermal conductivity of the floor material and the layer of material separating the pipes from the floor is $k=1.15W/m.K$, calculate the rate of heat transfer by conduction from a single pipe per unit length to the floor of the heated room.

Solution

Arrangement 14 in Table 4.1 best approximates the conduction model, with the layer between the floor and pipes taken as a semi-infinite medium with constant thermal conductivity k (Figure E4.7). The rate of conduction heat transfer per pipe per unit length is

$$\dot{Q}=Sk(T_1-T_2)$$

From Table 4.1, for L = 1, S is

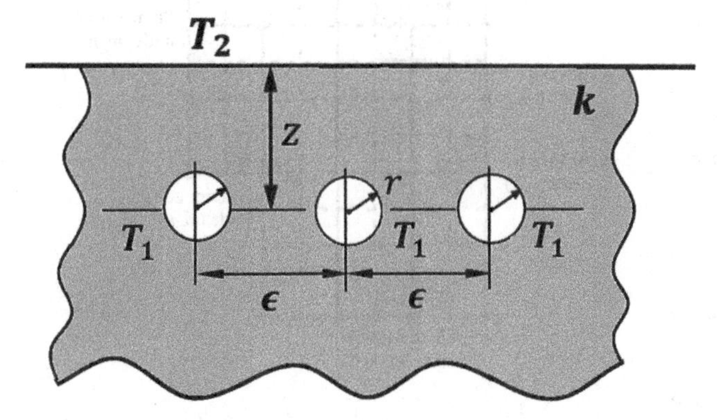

Figure E4.7 Schematic diagram for pipes under the floor.

$$S = \frac{2\pi}{ln\left[\frac{\epsilon}{\pi r} sinh\left(\frac{2\pi z}{\epsilon}\right)\right]} = \frac{2\pi}{ln\left[\frac{120}{10\pi} sinh\left(\frac{2\times\pi\times35}{120}\right)\right]} = 2.56$$

$$\dot{Q} = Sk(T_1 - T_2) = 2.56\times1.15\times(45-20) = 73.62W/m$$

4.3 Numerical Solution of Two-Dimensional Heat Conduction Problems

The solution methods of two-dimensional conduction outlined so far dealt with relatively simple predefined geometries and boundary conditions and cannot be applied to solve problems involving more complex shapes and boundary conditions. The physical shapes of actual engineering systems are usually much more complex with various types of boundary conditions that could be periodic, temperature-dependent, location-dependent, and time-dependent (in transient models). Additionally, thermal conductivity could be temperature-dependent and variable with direction in anisotropic materials. These complex models can only be handled using the finite-difference method, finite element method, or finite volume method. Due to its relative simplicity and wide applicability, only the principles of the finite difference method will be outlined in this section.

Consider the arbitrary two-dimensional plate of thickness δ shown in Figure 4.10. The finite difference method in Cartesian coordinates requires that the physical model be divided into equal increments in the x- and y-directions with nodes (black circles) located at the vertices of the resulting square elements. The nodes could be located inside the model, on the exterior plane surfaces, on

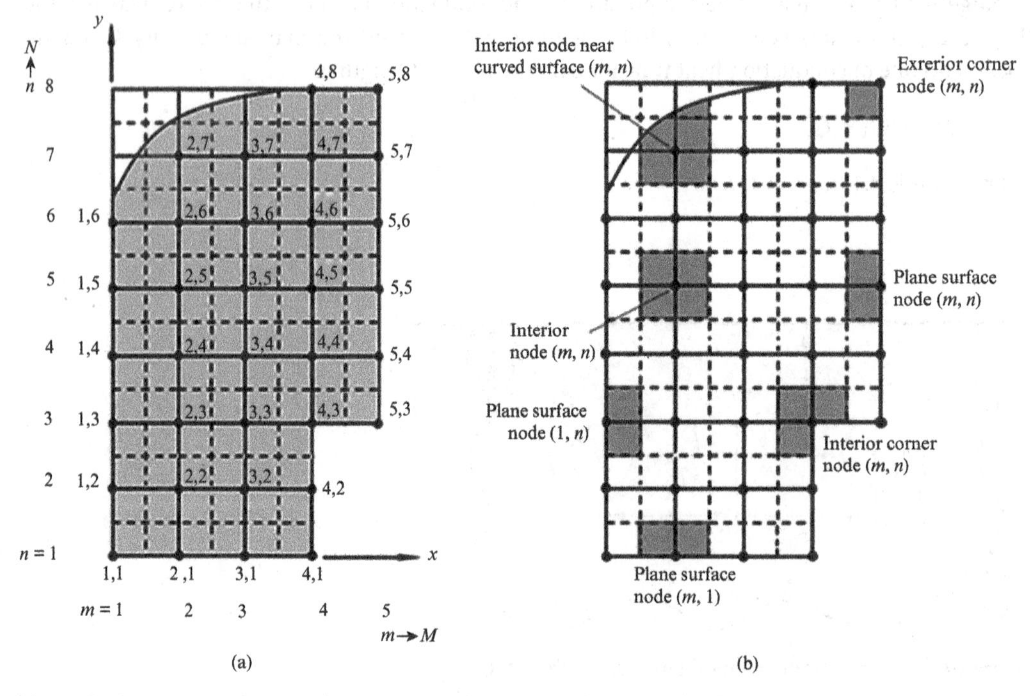

Figure 4.10 Nomenclature used in two-dimensional numerical analysis of heat conduction.

the corner boundaries (interior and exterior), and near the curved exterior boundaries. The x- and y-location of each node is designated by its m-location (changing from 1 to M in the $x-$direction) and n location (changing from 1 to N in the $y-$direction). Determination of the temperature of selected nodes are presented in the sections below.

4.3.1 Interior Node

The principle of conservation of energy (first law of thermodynamics) is applied to the control volume $\Delta v = \Delta x \Delta y \delta$ in Figure 4.11, where δ is the thickness of the highlighted element.

For an element with internal energy generation and no energy storage

$$\dot{Q}_A + \dot{Q}_B - \dot{Q}_C - \dot{Q}_D + \dot{q}\Delta x \Delta y \delta = 0 \tag{4.38}$$

$\dot{q}$ is the heat generated per unit volume (W/m^3)

Assuming the temperature gradient between nodes $m-1,n$ and m,n is linear, the heat conduction across surface A is

$$\dot{Q}_A = -kA_A \frac{\partial T}{\partial x} \approx -k \frac{T_{m,n} - T_{m-1,n}}{\Delta x} \Delta y \delta$$

Similarly, the heat conduction across surface B is

$$\dot{Q}_B = -kA_B \frac{\partial T}{\partial y} \approx -k \frac{T_{m,n} - T_{m,n-1}}{\Delta y} \Delta x \delta$$

The heat conduction across surface C is

$$\dot{Q}_C = -kA_C \frac{\partial T}{\partial x} \approx -k \frac{T_{m+1,n} - T_{m,n}}{\Delta x} \Delta y \delta$$

The heat conduction across surface D is

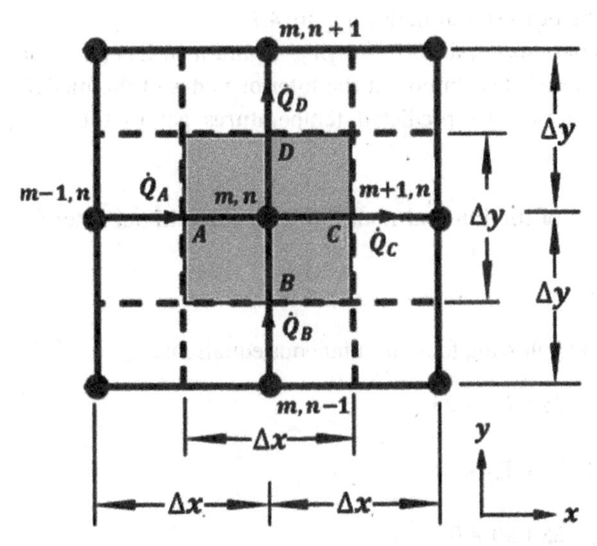

Figure 4.11 Schematic diagram of an interior element and associated node.

$$\dot{Q}_D = -kA_D\frac{\partial T}{\partial y} \approx -k\frac{T_{m,n+1} - T_{m,n}}{\Delta y}\Delta x\delta$$

Substituting the values of $\dot{Q}_A$, $\dot{Q}_B$, $\dot{Q}_C$, and $\dot{Q}_D$ into Eq. (4.38), we obtain

$$-k\frac{T_{m,n} - T_{m-1,n}}{\Delta x}\Delta y\delta - k\frac{T_{m,n} - T_{m,n-1}}{\Delta y}\Delta x\delta + k\frac{T_{m+1,n} - T_{m,n}}{\Delta x}\Delta y\delta + k\frac{T_{m,n+1} - T_{m,n}}{\Delta y}\Delta x\delta + \dot{q}\Delta x\Delta y\delta = 0$$

Dividing by $k\Delta x\Delta y\delta$ and rearranging yields

$$-\frac{T_{m,n} - T_{m-1,n}}{(\Delta x)^2} - \frac{T_{m,n} - T_{m,n-1}}{(\Delta y)^2} + \frac{T_{m+1,n} - T_{m,n}}{(\Delta x)^2} + \frac{T_{m,n+1} - T_{m,n}}{(\Delta y)^2} + \frac{\dot{q}}{k} = 0$$

$$-2\left(\frac{1}{(\Delta x)^2} + \frac{1}{(\Delta y)^2}\right)T_{m,n} + \frac{T_{m-1,n}}{(\Delta x)^2} + \frac{T_{m,n-1}}{(\Delta y)^2} + \frac{T_{m+1,n}}{(\Delta x)^2} + \frac{T_{m,n+1}}{(\Delta y)^2} + \frac{\dot{q}}{k} = 0$$

Multiplying by $(\Delta x)^2$ and rearranging we finally obtain

$$2\left[1 + \left(\frac{\Delta x}{\Delta y}\right)^2\right]T_{m,n} = \left(\frac{\Delta x}{\Delta y}\right)^2(T_{m,n-1} + T_{m,n+1}) + T_{m+1,n} + T_{m-1,n} + \frac{\dot{q}(\Delta x)^2}{k} \tag{4.39}$$

For an element of equal sides $\Delta x = \Delta y$, Eq. (4.39) is reduced to

$$4T_{m,n} = T_{m+1,n} + T_{m-1,n} + T_{m,n-1} + T_{m,n+1} + \frac{\dot{q}(\Delta x)^2}{k} \tag{4.40}$$

If there is no internal heat generation, Eq. (4.40) is further reduced to

$$T_{m,n} = \frac{1}{4}(T_{m+1,n} + T_{m-1,n} + T_{m,n-1} + T_{m,n+1}) \tag{4.41}$$

In other words, Eq. (4.41) states that the temperature at an internal node located at m,n is the arithmetic average of the temperatures at the nearest four nodes (Figure 4.11).

The system of simultaneous algebraic equations obtained by varying m and n in Eqs (4.40) or (4.41) can be solved to estimate the temperature distribution at the interior nodes of the model. The finer the size of the elements, the closer the predicted temperatures get to the true temperatures.

Example 4.8 Determine the temperatures of the internal nodes of the plate without internal head generation shown in Figure E4.8.

Solution

Using Eq. (4.41) for each node, we obtain the following four simultaneous equations:

$$4T_{2,2} = (T_{3,2} + T_{1,2} + T_{2,3} + T_{2,1}) = T_{3,2} + 25 + T_{2,3} + 25$$

$$4T_{3,2} = (T_{4,2} + T_{2,2} + T_{3,3} + T_{3,1}) = 25 + T_{2,2} + T_{3,3} + 25$$

$$4T_{2,3} = (T_{3,3} + T_{1,3} + T_{2,4} + T_{2,2}) = T_{3,3} + 25 + 90 + T_{2,2}$$

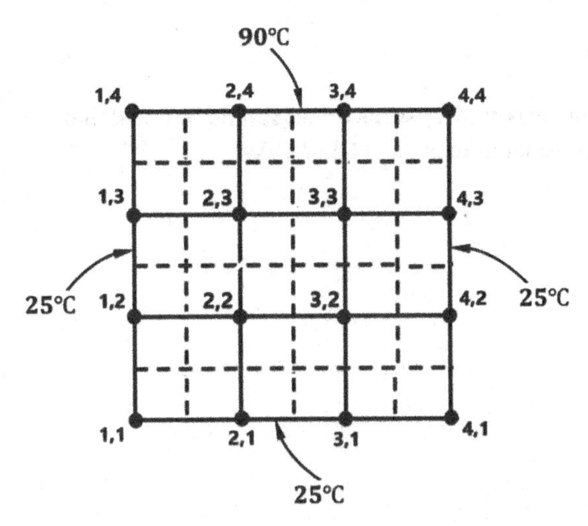

Figure E4.8 Schematic diagram for Example 4.8.

$$4T_{3,3} = (T_{4,3} + T_{2,3} + T_{3,4} + T_{3,2}) = 25 + T_{2,3} + 90 + T_{3,2}$$

Due to symmetry, $T_{2,2} = T_{3,2}$ and $T_{2,3} = T_{3,3}$, the set of four equations can be reduced to two:

$$50 + T_{2,3} - 3T_{2,2} = 0$$

$$115 - 3T_{2,3} + T_{2,2} = 0$$

Solving these equations results in $T_{2,2} = T_{3,2} = 33.1°\text{C}$, $T_{2,3} = T_{3,3} = 49.3°\text{C}$

4.3.2 Plane-Surface Node

Consider node (m,n) located on the right-hand surface of the model in Figure 4.10b and reproduced in more detail in Figure 4.12. The energy equation for the highlighted control volume is

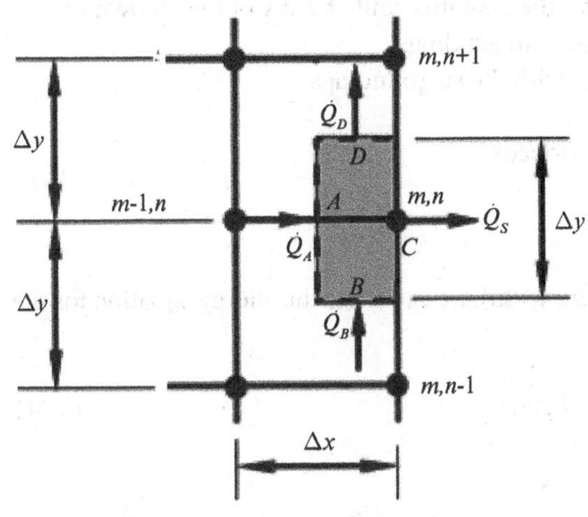

Figure 4.12 Exterior plane-surface node.

$$\dot{Q}_A + \dot{Q}_B - \dot{Q}_D - \dot{Q}_S + \dot{q}(\Delta x/2)\Delta y\delta = 0 \tag{4.42}$$

where $\dot{Q}_S$ represents a prescribed heat crossing the boundary surface C. It should be noted that the control volume for the finite element under consideration is $\Delta v = (\Delta x/2)\Delta y\delta$.

The heat conduction across surface A is

$$\dot{Q}_A = -kA_A\frac{\partial T}{\partial x} \approx -k\frac{T_{M,n} - T_{M-1,n}}{\Delta x}\Delta y\delta$$

The heat conduction across surface B is

$$\dot{Q}_B - kA_B\frac{\partial T}{\partial y} \approx -k\frac{T_{m,n} - T_{m,n-1}}{\Delta y}\frac{\Delta x}{2}\delta$$

The heat conduction across surface D is

$$\dot{Q}_D = -kA_D\frac{\partial T}{dy} \approx -k\frac{T_{m,n+1} - T_{m,n}}{\Delta y}\frac{\Delta x}{2}\delta$$

Substituting into Eq. (4.42) and rearranging as before, we obtain

$$2\left[1+\left(\frac{\Delta x}{\Delta y}\right)^2\right]T_{m,n} - \left(\frac{\Delta x}{\Delta y}\right)^2(T_{m,n-1} + T_{m,n+1}) - 2T_{m-1,n} + \frac{2\Delta x}{k}\left(\frac{\dot{Q}_S}{\Delta y\delta}\right) - \frac{\dot{q}(\Delta x)^2}{k} = 0$$

Heat flux q at C surface is the rate of heat transfer per unit area, i.e.,

$$q = \frac{\dot{Q}_S}{\Delta y\delta}\ W/m^2$$

If we assume that the finite elements are shaped as squares ($\Delta x = \Delta y$), we obtain

$$4T_{m,n} - T_{m,n-1} - T_{m,n+1} - 2T_{m-1,n} + \frac{2(\Delta x)q}{k} - \frac{\dot{q}(\Delta x)^2}{k} = 0 \tag{4.43}$$

Depending on the specified boundary conditions, the heat flux could be any of the following:
$q = h(T_{m,n} - T_\infty)$ – convection heat transfer to the surroundings
$q = \sigma F_{S-\infty}(T_{m,n}^4 - T_\infty^4)$ – radiation heat exchange with the surroundings
q – specified heat flux
$q = 0$ – adiabatic surface conditions (insulated surface).

4.3.3 Interior Node Near Curved Surface

For a node near a curved surface (top left of Figure 4.10b), we can write the energy equation for the highlighted control volume in Figure 4.13 as

$$\dot{Q}_A + \dot{Q}_B - \dot{Q}_C - \dot{Q}_D + \frac{\dot{q}}{4}(\Delta x + a\Delta x)(\Delta y + b\Delta y)\delta = 0 \tag{4.44}$$

The heat conduction across surface A is

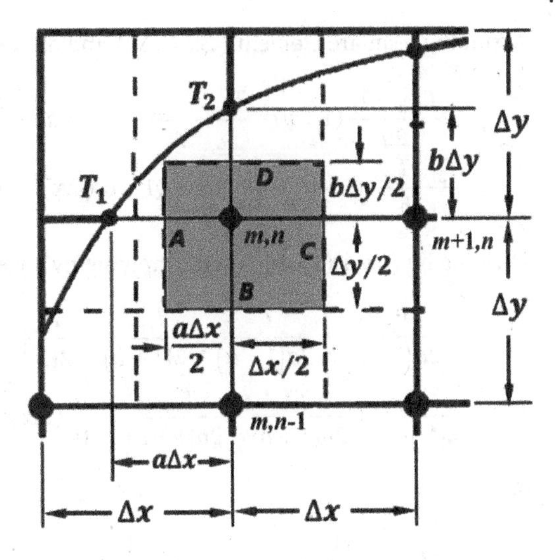

Figure 4.13 Node near curved surface of two-dimensional conduction model.

$$\dot{Q}_A = -kA_A\frac{\partial T}{\partial x} \approx -k\frac{T_{m,n} - T_1}{a\Delta x}\frac{(\Delta y + b\Delta y)}{2}\delta$$

The heat conduction across surface B is

$$\dot{Q}_B = -kA_B\frac{\partial T}{\partial y} \approx -k\frac{T_{m,n} - T_{m,n-1}}{\Delta y}\frac{(\Delta x + a\Delta x)}{2}\delta$$

The heat conduction across surface C is

$$\dot{Q}_C = -kA_C\frac{\partial T}{\partial x} \approx -k\frac{T_{m+1,n} - T_{m,n}}{\Delta x}\frac{(\Delta y + b\Delta y)}{2}\delta$$

The heat conduction across surface D is

$$\dot{Q}_D = -kA_D\frac{\partial T}{\partial y} \approx -k\frac{T_2 - T_{m,n}}{b\Delta y}\frac{(\Delta x + a\Delta x)}{2}\delta$$

Substituting into Eq. (4.44), we obtain

$$-k\frac{T_{m,n} - T_1}{a\Delta x}\frac{(\Delta y + b\Delta y)}{2}\delta - k\frac{T_{m,n} - T_{m,n-1}}{\Delta y}\frac{(\Delta x + a\Delta x)}{2}\delta + k\frac{T_{m+1,n} - T_{m,n}}{\Delta x}\frac{(\Delta y + b\Delta y)}{2}\delta +$$
$$k\frac{T_2 - T_{m,n}}{b\Delta y}\frac{(\Delta x + a\Delta x)}{2}\delta + \frac{\dot{q}}{4}(\Delta x + a\Delta x)(\Delta y + b\Delta y)\delta = 0$$

or

$$-k\frac{T_{m,n} - T_1}{2a}\frac{\Delta y}{\Delta x}(1+b)\delta - k\frac{T_{m,n} - T_{m,n-1}}{2}\frac{\Delta x}{\Delta y}(1+a)\delta + k\frac{T_{m+1,n} - T_{m,n}}{2}\frac{\Delta y}{\Delta x}(1+b)\delta +$$
$$k\frac{T_2 - T_{m,n}}{2b}\frac{\Delta x}{\Delta y}(1+a)\delta + \frac{\dot{q}}{4}(\Delta x)(\Delta y)(1+a)(1+b)\delta = 0$$

Assuming a square element ($\Delta x = \Delta y$) and eliminating δ, we obtain

$$-\frac{T_{m,n} - T_1}{2a}(1+b) - \frac{T_{m,n} - T_{m,n-1}}{2}(1+a) + \frac{T_{m+1,n} - T_{m,n}}{2}(1+b) + \frac{T_2 - T_{m,n}}{2b}(1+a) + \frac{\dot{q}}{4k}(1+a)(1+b)(\Delta x)^2 = 0$$

Dividing by $(1+a)(1+b)$ and multiplying by 4 yields

$$-\frac{T_{m,n}}{2a(1+a)} + \frac{T_1}{2a(1+a)} - \frac{T_{m,n}}{2(1+b)} + \frac{T_{m,n-1}}{2(1+b)} + \frac{T_{m+1,n}}{2(1+a)} - \frac{T_{m,n}}{2(1+a)} + \frac{T_2}{2b(1+b)} - \frac{T_{m,n}}{2b(1+b)} + \frac{\dot{q}}{4k}(\Delta x)^2 = 0$$

or

$$-2T_{m,n}\left[\frac{1}{a(1+a)} + \frac{1}{(1+b)} + \frac{1}{(1+a)} + \frac{1}{b(1+b)}\right] + \frac{2T_1}{a(1+a)} + \frac{2T_{m,n-1}}{(1+b)} + \frac{2T_{m+1,n}}{(1+a)} + \frac{2T_2}{b(1+b)} + \frac{\dot{q}}{k}(\Delta x)^2 = 0$$

Since,

$$\left[\frac{1}{a(1+a)} + \frac{1}{(1+b)} + \frac{1}{(1+a)} + \frac{1}{b(1+b)}\right] = \frac{1}{(1+a)}\left(\frac{1}{a} + 1\right) + \frac{1}{(1+b)}\left(\frac{1}{b} + 1\right) = \left(\frac{1}{a} + \frac{1}{b}\right)$$

we finally get

$$\left(\frac{1}{a} + \frac{1}{b}\right)2T_{m,n} = \frac{2T_{m,n-1}}{(1+b)} + \frac{2T_{m+1,n}}{(1+a)} + \frac{2T_2}{b(1+b)} + \frac{2T_1}{a(1+a)} + \frac{\dot{q}}{k}(\Delta x)^2 \tag{4.45}$$

If there is no internal heat generation, Eq. (4.45) is reduced to

$$\left(\frac{1}{a} + \frac{1}{b}\right)T_{m,n} = \frac{T_{m,n-1}}{(1+b)} + \frac{T_{m+1,n}}{(1+a)} + \frac{T_2}{b(1+b)} + \frac{T_1}{a(1+a)} \tag{4.46}$$

Once the temperatures at the nodes of the model are known, the rates of heat transfer at the edges of the model in the x- and y-directions can be determined from

$$\dot{Q}_x = -\sum_{n=1}^{N} k(\delta\Delta y)\frac{\Delta T}{\Delta x} \tag{4.47}$$

$$\dot{Q}_y = -\sum_{m=1}^{M} k(\delta\Delta x)\frac{\Delta T}{\Delta y} \tag{4.48}$$

Table 4.2 summarises the nodal temperature equations for some of the highlighted elements in Figure 4.10b with various boundary conditions. Heat generation and energy storage are ignored in the analysis and the sides of the elements are equal, $\Delta x = \Delta y$.

Table 4.2 Nodal temperature equations for selected node positions and boundary conditions.

Element model	**Nodal temperature equation**
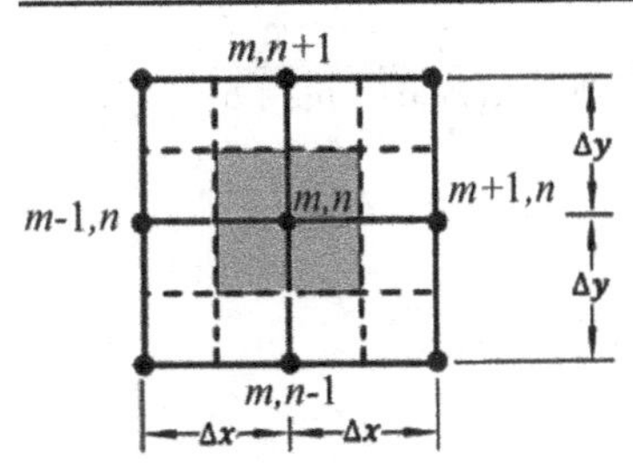 (1) Interior node	$4T_{m,n} = T_{m+1,n} + T_{m-1,n} + T_{m,n+1} + T_{m,n-1}$
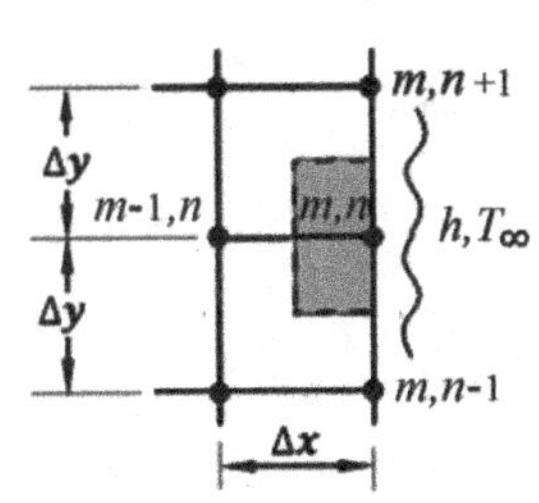 (2) Boundary node on plane surface with convection	$\left(2+\frac{h\Delta x}{k}\right)T_{m,n} = T_{m-1,n} + \frac{1}{2}(T_{m,n-1} + T_{m,n+1}) + \frac{h\Delta x}{k}T_{\infty}$
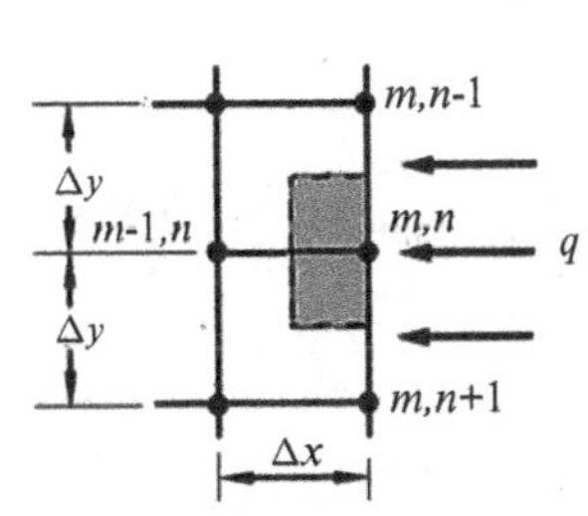 3) Boundary node on plane surface with heat flux	$4T_{m,n} = T_{m,n-1} + T_{m,n+1} + 2T_{m-1,n} + \frac{2\Delta xq}{k}$ q – uniform heat flux, W/m^2
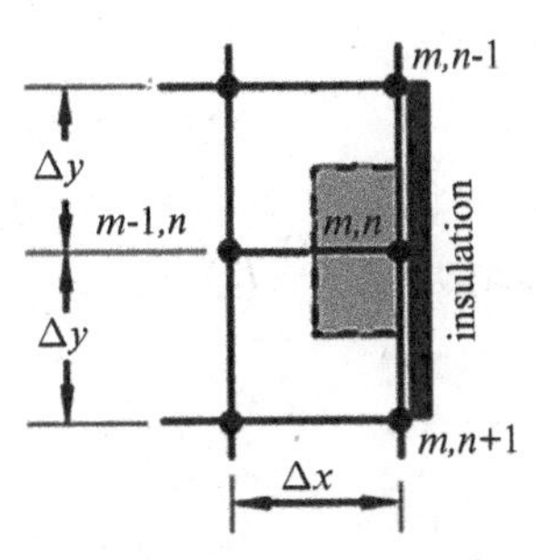 4) Boundary node on insulated plane surface	$4T_{m,n} = T_{m,n+1} + T_{m,n-1} + 2T_{m-1,n}$

(Continued)

Table 4.2 (Continued)

Element model	Nodal temperature equation
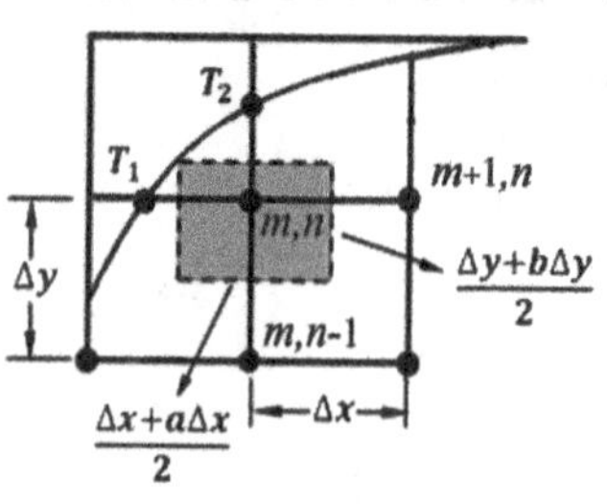 5) Interior node near curved boundary	$\left(\frac{1}{a}+\frac{1}{b}\right)T_{m,n}=\frac{T_{m+1,n}}{(1+a)}+\frac{T_{m,n-1}}{(1+b)}+\frac{T_1}{a(1+a)}+\frac{T_2}{b(1+b)}$
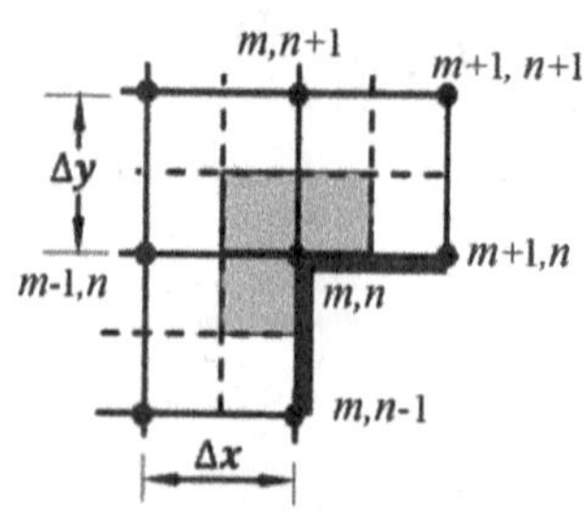 6) Boundary node on insulated interior corner	$6T_{m,n}=T_{m+1,n}+T_{m,n-1}+2T_{m,n+1}+2T_{m-1,n}$
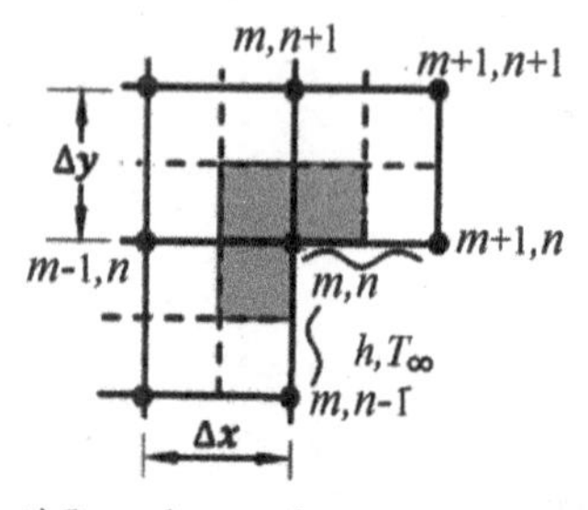 7) Boundary node on interior corner with convection	$\left(3+\frac{h\Delta x}{k}\right)T_{m,n}=T_{m,n+1}+T_{m-1,n}+\frac{1}{2}(T_{m,n-1}+T_{m+1,n})+\frac{h\Delta x}{k}T_\infty$
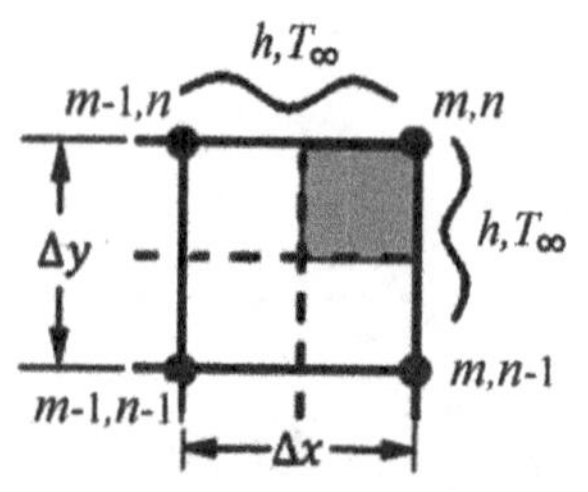 8) Boundary node on exterior corner with convection	$\left(1+\frac{h\Delta x}{k}\right)T_{m,n}=\frac{1}{2}(T_{m,n-1}+T_{m-1,n})+2\frac{h\Delta x}{k}T_\infty$

4.3.4 Finite Difference Formulation in Cylindrical Coordinates

The formulation for the two-dimensional conduction model of one-quarter of a disc of thickness δ or solid cylinder of radius R and length L in the z direction is shown in Figure 4.14. A constant temperature T_s is applied to the circumference and constant temperature T_0 is applied to the base. The unknown temperatures $T(r,\varphi)$ are at the nodes, with coordinates in the ranges $r(0 \rightarrow R)$ and $\varphi(0 \rightarrow \pi)$.

The highlighted area in Figure 4.14 is the area in the $r-\varphi$ plane of the control volume ΔV per unit length in the z-direction

$$\Delta V = \left[\left(r+\frac{\Delta r}{2}\right)^2 - \left(r-\frac{\Delta r}{2}\right)^2\right]\frac{\Delta\varphi}{2}L = r\Delta r\Delta\varphi \tag{4.49}$$

The energy equation for the control volume ΔV with internal energy generation and no energy storage

$$\dot{Q}_A + \dot{Q}_B - \dot{Q}_C - \dot{Q}_D + \dot{q}\Delta V = 0 \tag{4.50}$$

$\dot{q}$ is the heat generated per unit volume (W/m^3).

Assuming the temperature gradients between the central node i,j and adjacent nodes $i-1,j$; $i,j-1$; $i+1,j$; and $i,j+1$, respectively, are linear, we can write Fourier's law for each surface of the highlighted element as follows:

Heat conduction across surface A is

$$\dot{Q}_A = -k\left[\left(r-\frac{\Delta r}{2}\right)\Delta\varphi L\right]\frac{T_{i,j}-T_{i-1,j}}{\Delta r}$$

Heat conduction across surface B is

$$\dot{Q}_B = -k(\Delta r L)\frac{T_{i,j}-T_{i,j-1}}{r\Delta\varphi}$$

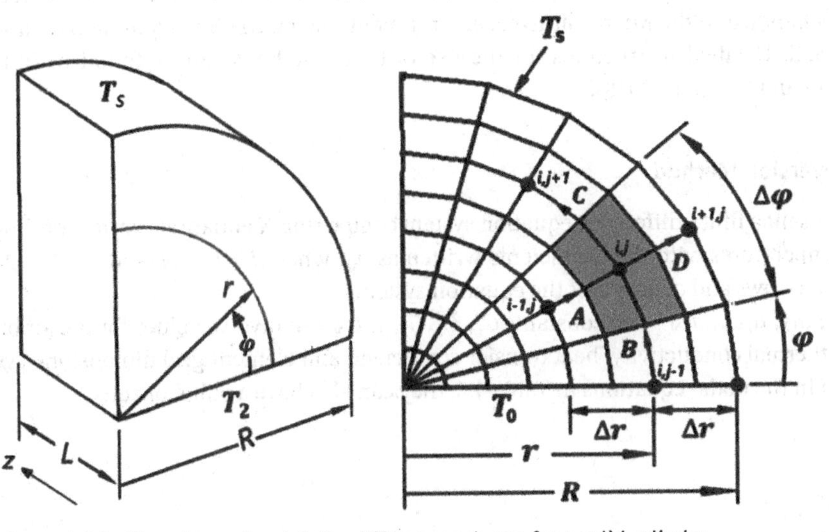

Figure 4.14 Two-dimensional finite-difference scheme for a solid cylinder.

Heat conduction across surface C is

$$\dot{Q}_C = -k(\Delta r L)\frac{T_{i,j+1} - T_{i,j}}{r\Delta\varphi}$$

Heat conduction across surface D is

$$\dot{Q}_D = -k\left[\left(r + \frac{\Delta r}{2}\right)\Delta\varphi L\right]\frac{T_{i+1,j} - T_{i,j}}{\Delta r}$$

Substituting the values of $\dot{Q}_A$, $\dot{Q}_B$, $\dot{Q}_C$, and $\dot{Q}_D$ into Eq. (4.50), ignoring heat generation, and rearranging we obtain

$$-\left[2r(\Delta\varphi)^2 + \frac{2}{r}(\Delta r)^2\right]T_{ij} + \left[\left(r - \frac{\Delta r}{2}\right)(\Delta\varphi)^2\right]T_{i-1,j} + \frac{(\Delta r)^2}{r}T_{i,j-1} +$$
$$\left[\left(r + \frac{\Delta r}{2}\right)(\Delta\varphi)^2\right]T_{i+1,j} + \frac{(\Delta r)^2}{r}T_{i,j+1} = 0 \tag{4.51}$$

For the nodes closest to the outer edges, temperatures $T_{i+1,j}$ and $T_{i,j-1}$ are replaced by T_s and T_o, respectively.

4.4 Solution Methods for Finite-Difference Models

If the number of simultaneous algebraic equations obtained for a two-dimensional conduction problem is small, hand calculations may serve to determine the nodal temperatures in the model. However, as was stated earlier, the finer the size of the elements (the more the number of nodes), the closer the predicted temperatures get to the true temperatures. As a consequence, the number of simultaneous equations may increase and the solution may become cumbersome if not impossible. Fortunately, many techniques have been developed over the years in step with the advances in computer hardware and software developments. In this book, two methods will be considered, as they are readily handled in the Microsoft Excel environment: the matrix inversion method and the iterative method. Detailed instructions for the use of these methods can be found in Billo (2007) and Holman and Holman (2018).

4.4.1 Matrix Inversion Method

Let Eq. (4.52) represent a finite-difference equation system comprising N equations corresponding to N unknown temperatures with the coefficients written as a_{ij}, where $i = 1 \rightarrow N$ and $j = 1 \rightarrow N$ refer, respectively, to rows and columns of the equation system.

Coefficients a_{11}, a_{12}, $a_{13} \ldots, a_{1N}$ and constants C_1, $C_2, C_3, \ldots, C_N$ can involve values for the ambient temperature, thermal conductivity, heat transfer coefficient, and element grid dimensions Δx and Δy, as shown in the nodal equations in Table 4.2; they can also have a value of zero.

$$
\begin{array}{l}
i=1, j=1\rightarrow N \\
j=1, i=1\rightarrow N \downarrow \\
\left.
\begin{array}{l}
a_{11}T_1+a_{12}T_2+a_{13}T_3+\ldots+a_{1j}T_j+\ldots+a_{1N}T_N=C_1 \\
a_{21}T_1+a_{22}T_2+a_{23}T_3+\ldots+a_{2j}T_j+\ldots+a_{2N}T_N=C_2 \\
\cdot \\
\cdot \\
\cdot \\
i=j\rightarrow \quad a_{j1}T_1+a_{j2}T_2+a_{j3}T_3+\ldots+a_{jj}T_j+\ldots+a_{jN}T_N=C_j \\
\cdot \\
\cdot \\
\cdot \\
a_{N1}T_1+a_{N2}T_2+a_{N3}T_3+\ldots+a_{Nj}T_j+\ldots+a_{NN}T_N=C_N
\end{array}
\right\}
\end{array}
\tag{4.52}
$$

In matrix notation

$$[A][T]=[C] \tag{4.53}$$

where

$$A=\begin{bmatrix} a_{11} & a_{12} & \cdot & \cdot & \cdot & a_{1N} \\ a_{21} & a_{22} & \cdot & \cdot & \cdot & a_{2N} \\ \cdot & \cdot & \cdot & \cdot & \cdot & \cdot \\ \cdot & \cdot & \cdot & \cdot & \cdot & \cdot \\ \cdot & \cdot & \cdot & \cdot & \cdot & \cdot \\ a_{N1} & a_{N2} & \cdot & \cdot & \cdot & a_{NN} \end{bmatrix}, \quad T=\begin{bmatrix} T_1 \\ T_2 \\ \cdot \\ \cdot \\ \cdot \\ T_N \end{bmatrix}, \quad C=\begin{bmatrix} C_1 \\ C_2 \\ \cdot \\ \cdot \\ \cdot \\ C_N \end{bmatrix}$$

The solution of Eq. (4.53) is

$$[T]=[A]^{-1}[C] \tag{4.54}$$

where

$[A]^{-1}$ is the inverse matrix of $[A]$ and can be written as

$$[A]^{-1}=\begin{bmatrix} b_{11} & b_{12} & \cdot & \cdot & \cdot & b_{1N} \\ b_{21} & b_{22} & \cdot & \cdot & \cdot & b_{2N} \\ \cdot & \cdot & \cdot & \cdot & \cdot & \cdot \\ \cdot & \cdot & \cdot & \cdot & \cdot & \cdot \\ \cdot & \cdot & \cdot & \cdot & \cdot & \cdot \\ b_{N1} & b_{N2} & \cdot & \cdot & \cdot & b_{NN} \end{bmatrix}$$

Example 4.9 Determine the nodal temperatures of the two-dimensional conduction in the rectangular plate in Figure E4.9.1 using the matrix inversion method. Also, determine the heat rates across the external surfaces per unit depth. Compare the results with the results in Examples 4.1 and 4.2.

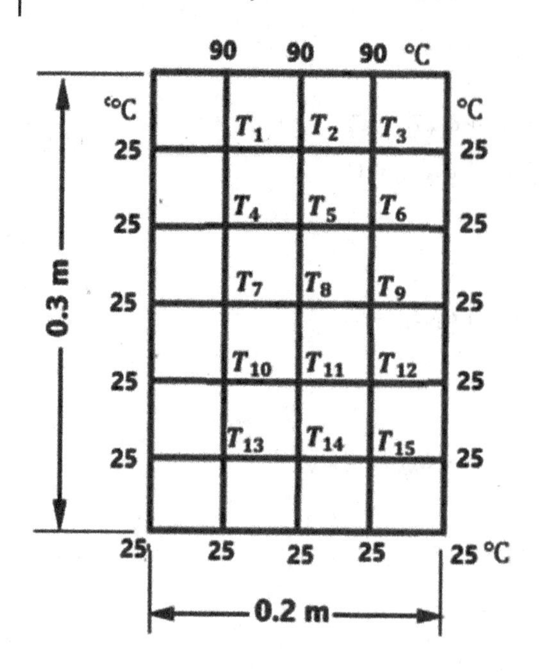

Figure E4.9.1 Finite-difference model for Example 4.9.

Solution

From Eq. (4.41), for the nodal temperature of interior nodes we obtain the following system of simultaneous equations

$$
\begin{aligned}
&-4T_1 + T_2 + T_4 = -115 \\
&T_1 - 4T_2 + T_3 + T_5 = -90 \\
&T_2 - 4T_3 + T_6 = -115 \\
&T_1 - 4T_4 + T_5 + T_7 = -25 \\
&T_2 + T_4 - 4T_5 + T_6 + T_8 = 0 \\
&T_3 + T_5 - 4T_6 + T_9 = -25 \\
&T_4 - 4T_7 + T_8 + T_{10} = -25 \\
&T_5 + T_7 - 4T_8 + T_9 + T_{11} = 0 \\
&T_6 + T_8 - 4T_9 + T_{12} = -25 \\
&T_7 - 4T_{10} + T_{11} + T_{13} = -25 \\
&T_8 + T_{10} - 4T_{11} + T_{12} + T_{14} = 0 \\
&T_9 + T_{11} - 4T_{12} + T_{15} = -25 \\
&T_{10} - 4T_{13} + T_{14} = -50 \\
&T_{11} + T_{13} - 4T_{14} + T_{15} = -25 \\
&T_{12} + T_{14} - 4T_{15} = -50
\end{aligned}
$$

Shown below are, respectively, the input matrices $[A]$ and $[C]$, the output (resultant) matrix$[T]$, and the inverse matrix $[A]^{-1}$.

$$
A=\begin{bmatrix}
-4 & 1 & 0 & 1 & 0 & 0 & 0 & 0 & 0 & 0 & 0 & 0 & 0 & 0 & 0 \\
1 & -4 & 1 & 0 & 1 & 0 & 0 & 0 & 0 & 0 & 0 & 0 & 0 & 0 & 0 \\
0 & 1 & -4 & 0 & 0 & 1 & 0 & 0 & 0 & 0 & 0 & 0 & 0 & 0 & 0 \\
1 & 0 & 0 & -4 & 1 & 0 & 1 & 0 & 0 & 0 & 0 & 0 & 0 & 0 & 0 \\
0 & 1 & 0 & 1 & -4 & 1 & 0 & 1 & 0 & 0 & 0 & 0 & 0 & 0 & 0 \\
0 & 0 & 1 & 0 & 1 & -4 & 0 & 0 & 1 & 0 & 0 & 0 & 0 & 0 & 0 \\
0 & 0 & 0 & 1 & 0 & 0 & -4 & 0 & 0 & 1 & 0 & 0 & 0 & 0 & 0 \\
0 & 0 & 0 & 0 & 1 & 0 & 1 & -4 & 0 & 0 & 1 & 0 & 0 & 0 & 0 \\
0 & 0 & 0 & 0 & 0 & 1 & 0 & 0 & -4 & 0 & 0 & 0 & 0 & 0 & 0 \\
0 & 0 & 0 & 0 & 0 & 0 & 1 & 1 & 0 & -4 & 1 & 0 & 1 & 0 & 0 \\
0 & 0 & 0 & 0 & 0 & 0 & 0 & 0 & 0 & 1 & -4 & 1 & 0 & 1 & 0 \\
0 & 0 & 0 & 0 & 0 & 0 & 0 & 1 & 1 & 0 & 0 & -4 & 0 & 0 & 1 \\
0 & 0 & 0 & 0 & 0 & 0 & 0 & 0 & 0 & 1 & 0 & 0 & -4 & 1 & 0 \\
0 & 0 & 0 & 0 & 0 & 0 & 0 & 0 & 0 & 0 & 1 & 0 & 1 & -4 & 1 \\
0 & 0 & 0 & 0 & 0 & 0 & 0 & 0 & 0 & 0 & 0 & 1 & 0 & 1 & -4
\end{bmatrix}
$$

$$[T]=[A]^{-1}[C]$$

$$
[\mathbf{T}]=[\mathbf{A}^{-1}][C]=[\mathbf{A}^{-1}]\times\begin{bmatrix}-115\\-90\\-115\\-25\\0\\-25\\-25\\0\\-25\\-25\\0\\-25\\-50\\-25\\-50\end{bmatrix}=\begin{bmatrix}T_1\\T_2\\T_3\\T_4\\T_5\\T_6\\T_7\\T_8\\T_9\\T_{10}\\T_{11}\\T_{12}\\T_{13}\\T_{14}\\T_{15}\end{bmatrix}=\begin{bmatrix}53.08\\59.55\\53.08\\37.75\\42.05\\37.75\\30.89\\33.14\\30.89\\27.66\\28.73\\27.66\\26.03\\26.44\\26.03\end{bmatrix}{}^{\circ}\mathrm{C}
$$

$$[A^{-1}]=\begin{bmatrix}
-0.3001 & -0.0996 & -0.0322 & -0.1010 & -0.0661 & -0.0292 & -0.0376 & -0.0346 & -0.0184 & -0.0148 & -0.0164 & -0.0097 & -0.0053 & -0.0064 & -0.0040 \\
-0.0996 & -0.3323 & -0.0996 & -0.0661 & -0.1301 & -0.0661 & -0.0346 & -0.0560 & -0.0346 & -0.0164 & -0.0245 & -0.0164 & -0.0064 & -0.0093 & -0.0064 \\
-0.0322 & -0.0996 & -0.3001 & -0.0292 & -0.0661 & -0.1010 & -0.0184 & -0.0346 & -0.0376 & -0.0097 & -0.0164 & -0.0148 & -0.0040 & -0.0064 & -0.0053 \\
-0.1010 & -0.0661 & -0.0292 & -0.3377 & -0.1342 & -0.0506 & -0.1158 & -0.0825 & -0.0389 & -0.0429 & -0.0410 & -0.0224 & -0.0148 & -0.0164 & -0.0097 \\
-0.0661 & -0.1301 & -0.0661 & -0.1342 & -0.3883 & -0.1342 & -0.0825 & -0.1546 & -0.0825 & -0.0410 & -0.0653 & -0.0410 & -0.0164 & -0.0245 & -0.0164 \\
-0.0292 & -0.0661 & -0.1010 & -0.0506 & -0.1342 & -0.3377 & -0.0389 & -0.0825 & -0.1158 & -0.0824 & -0.0410 & -0.0429 & -0.0097 & -0.0164 & -0.0148 \\
-0.0376 & -0.0346 & -0.0184 & -0.11558 & -0.0825 & -0.0389 & -0.3431 & -0.1406 & -0.0546 & -0.1158 & -0.0825 & -0.0389 & -0.0376 & -0.0346 & -0.0184 \\
-0.0346 & -0.0560 & -0.0346 & -0.0825 & -0.1546 & -0.3425 & -0.1406 & -0.3976 & -0.1406 & -0.0825 & -0.1546 & -0.0825 & -0.0346 & -0.0560 & -0.0346 \\
-0.0184 & -0.0346 & -0.0376 & -0.0389 & -0.0825 & -0.0825 & -0.0546 & -0.1406 & -0.3431 & -0.0389 & -0.0825 & -0.1158 & -0.0184 & -0.0346 & -0.0376 \\
-0.0148 & -0.0164 & -0.0097 & -0.0429 & -0.1546 & -0.0224 & -0.1158 & -0.0825 & -0.0389 & -0.3377 & -0.1342 & -0.0506 & -0.1010 & -0.0661 & -0.0292 \\
-0.0164 & -0.0245 & -0.0164 & -0.0410 & -0.0825 & -0.0410 & -0.0825 & -0.1546 & -0.0825 & -0.1342 & -0.3883 & -0.1342 & -0.0661 & -0.1301 & -0.0661 \\
-0.0097 & -0.0164 & -0.0148 & -0.0224 & -0.0410 & -0.0429 & -0.0389 & -0.0825 & -0.1158 & -0.0506 & -0.1342 & -0.3377 & -0.0292 & -0.0661 & -0.1010 \\
-0.0053 & -0.0064 & -0.0040 & -0.0148 & -0.0164 & -0.0097 & -0.0376 & -0.0346 & -0.0184 & -0.1010 & -0.0661 & -0.0292 & -0.3001 & -0.0996 & -0.0322 \\
-0.0064 & -0.0093 & -0.0064 & -0.0164 & -0.0245 & -0.0164 & -0.0346 & -0.0560 & -0.0346 & -0.0661 & -0.1301 & -0.0661 & -0.0996 & -0.3323 & -0.0996 \\
-0.0040 & -0.0064 & -0.0053 & -0.0097 & -0.0164 & -0.0148 & -0.0184 & -0.0346 & -0.0376 & -0.0292 & -0.0661 & -0.1010 & -0.0322 & -0.0996 & -0.3001
\end{bmatrix}$$

Implementation of the inverse-matrix method for solving the problem in Example 4.10 is outlined in Figure C.1 in Appendix C, with a screen shot of the spreadsheet calculations.

The results of the calculations are presented in Table E4.9.1 together with the results of the analytical solution (Eq. (4.9b)), the Gauss–Seidel method (see section 4.4.2 below), and Excel's inbuilt utility "Solver." Figure E4.9.2 shows the results of computer modelling of the problem for comparison purposes.

For this particular example, at least, all the numerical techniques for solving the finite-element model yield excellent results compared with the results of the exact solution (Eq. 4.9b). The heat rates across surfaces A, B, C, and D are shown in Table E4.9.2 for several elemental sizes obtained by computer modelling.

Table E4.9.1 Solution of Example 4.9 using various methods ($\Delta x = \Delta y = 5mm$).

Nodal temperature °C	Analytical solution	Matrix inversion	Gauss–Seidel	Solver
T_1	53.26385	53.07565	53.07557	53.07602
T_2	60.39180	59.54985	59.54972	59.54751
T_3	53.26385	53.07565	53.07557	53.08009
T_4	37.31405	37.75277	37.75261	37.74734
T_5	41.93190	42.04809	42.04787	42.04713
T_6	37.31405	37.75277	37.75261	37.75993
T_7	30.51373	30.88733	30.88716	30.87491
T_8	32.75086	33.13697	33.13671	33.13537
T_9	30.51373	30.88733	30.88716	30.88892
T_{10}	27.42139	27.65960	27.65944	27.64780
T_{11}	28.41991	28.72511	28.72489	28.72410
T_{12}	27.42139	27.65960	27.65944	27.65865
T_{13}	25.91355	26.02597	26.02588	26.02209
T_{14}	26.29154	26.44426	26.44413	26.44431
T_{15}	25.91355	26.02597	26.02588	26.02560

Table E4.9.2 Effect of grid size on the predicted heat rates.

Δx or Δy (mm)	Surfaces			
	A	B	C	D
	$\dot{Q}(W/m)$			
5	2016	139.86	2016	−4172
2.5	3178	124.6	3178	−6481
2	3550	122.5	3550	−7222
1.25	4330	120.04	4330	−8780
0.625	5478	119.28	5478	−11076
0.5	5847	119.13	5847	−11815

Exact solution result: $\dot{Q}_B = 118.96\,W/m$

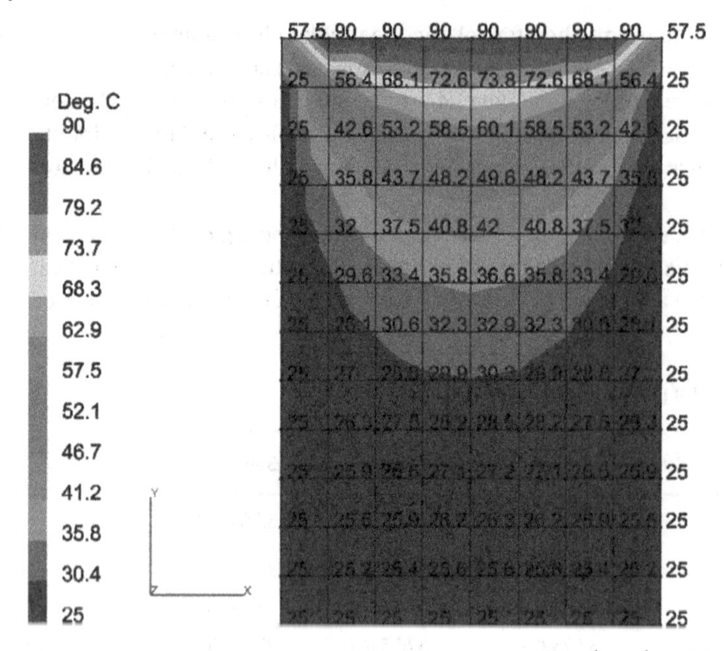

Figure E4.9.2 Computer modelling results of Example 4.9: $\dot{Q}_A = \dot{Q}_C = 3178 W/m$ $\dot{Q}_B = 124.5 W/m$, $\dot{Q}_D = -6488 W/m$

From Tables E4.9.1 and E4.9.2, we can conclude the following:

1) The predicted temperatures by some of the most popular methods of solving simultaneous equations are in excellent agreement with the analytical solution results at the selected element size of 5 mm. Reducing grid size in this case will not have a significant effect on the accuracy of the predicted nodal temperatures.
2) Heat transfer rates across surfaces A, C, and D increase with decreasing element size, which is counter-intuitive as it is well established that the accuracy of the finite difference model should increase with decreasing grid size. This outcome is in agreement with the exact solution's failure to converge when heat rates are calculated at these surfaces. The obtained heat transfer rates are physically meaningless. This is due, as was stated previously, to the stepwise temperature discontinuities at the top corners of the model.
3) Surface B is the exception, in which case the exact solution converges rapidly to a value of $118.96 W/m$. As Table E4.9.2 shows, the initial predicted value of $\dot{Q}_B$ is larger than this value and the grid size must be reduced significantly for the results of the exact solution and the matrix inversion method to converge. At a grid size $\Delta x = 0.5 mm$, the difference is 0.14%. If larger errors can be tolerated, coarser grids can be used for quick and efficient calculations. It should be kept in mind that if a computer program is used for modelling, grid size variation can be handled effectively and efficiently.

4.4.2 Iterative Methods (Gauss–Seidel Method)

Inspection of matrix $[A]$ in Example 4.9 shows a large number of zeros which must be entered carefully, and the larger the number of finite-difference equations the more complex it is to find a quick solution. A more efficient way to solve such equations is to use iterative methods, a good example of which is the popular Gauss–Seidel method. In this method, Eq. (4.52) can be rewritten as

$$T_1 = \frac{1}{a_{11}}\left(C_1 - a_{12}T_2 - a_{13}T_3 - \ldots - a_{1j}T_j - \ldots a_{1N}T_N\right)$$
$$T_2 = \frac{1}{a_{22}}\left(C_2 - a_{21}T_1 - a_{23}T_3 - \ldots - a_{2j}T_j - \ldots a_{2N}T_N\right) \quad (4.55)$$
$$T_j = \frac{1}{a_{jj}}\left(C_j - a_{j1}T_1 - \ldots - a_{j(j-1)}T_{j-1} - a_{j(j+1)}T_{(j+1)} \cdots - a_{jN}T_N\right)$$
$$T_N = \frac{1}{a_{NN}}\left(C_N - a_{N1}T_1 - a_{N2}T_2 - a_{Nj}T_j - \ldots - a_{N(N-1)}T_{(N-1)}\right)$$

The steps involved in the Gauss–Siedel method are:

1) The unknown temperatures are designated as T_j^k, where $j = 1,2,3,\ldots,N$ and k is the order of the iteration (first, second, third,...)
2) An initial value is assumed for each T_j^0 temperature and entered into Eq. (4.55). This value could be a boundary condition for example, but it is usually a zero.
3) Value of T_j^k for $k = 1,2,3,\ldots$ is calculated by solving repeatedly for each unknown value in Eq. (4.55) using the latest resulting values of the variables every time.
4) Convergence is assumed to have occurred when an acceptable accuracy is reached in accordance with the following equation:

$$\frac{T_j^{k+1} - T_j^k}{T_j^k} \le \varepsilon \quad (4.56)$$

ε is a small error in the temperature that is considered to be acceptable.

The Gauss–Siedel method implemented in Excel is demonstrated in the following example.

Example 4.10 Figure E4.10 shows a 0.3-m square bar of unit length having three isothermal surfaces and one surface exposed to convection heat transfer to a fluid at $T_\infty = 30°C$. If the thermal conductivity of the material of the bar is $45\,W/m.K$ and the heat transfer coefficient to the fluid is $h = 50\,W/m^2.K$, determine the temperatures at nodes 1 to 6.

Solution

From Table 4.2 (Item 2), the equation for a surface node can be written as

$$2\left(2 + \frac{h\Delta x}{k}\right)T_{m,n} = 2T_{m-1,n} + T_{m,n-1} + T_{m,n+1} + \frac{2h\Delta x}{k}T_\infty$$

To apply this equation to nodes 3 and 6, the right-hand corner temperatures are assumed to be equal to the average of two adjacent surface temperatures; hence,

$$2 \times \left(2 + \frac{50 \times 0.1}{45}\right)T_3 = 2T_2 + \frac{1}{2}(T_3 + 200) + T_6 + \frac{2 \times 50 \times 0.1}{45} \times 30$$

$$2 \times \left(2 + \frac{50 \times 0.1}{45}\right)T_6 = 2T_5 + \frac{1}{2}(T_6 + 20) + T_3 + \frac{2 \times 50 \times 0.1}{45} \times 30$$

The nodal equations for the interior nodes 1, 2, 4, 5 are found using Eq. (4.41) and the set of simultaneous equations arranged in a more convenient way as follows:

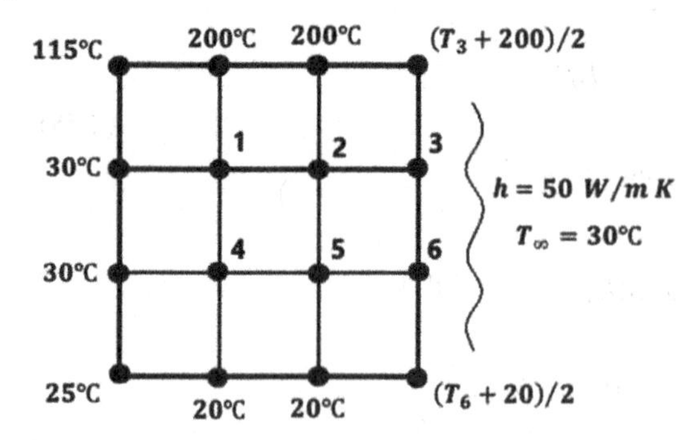

Figure E4.10 Model for Example 4.10: element size $\Delta x = \Delta y = 0.1m$.

Table E4.10 Comparison of the inverse matrix and Gauss–Siedel methods.

Temperature, °C	Gauss–Seidel method	Matrix inversion method
T_1	100.97794	100.97795
T_2	119.67586	119.67587
T_3	111.75977	111.75978
T_4	54.23592	54.23592
T_5	65.96573	65.96574
T_6	69.95117	69.95117

$$4T_1 - T_2 - T_4 = 230$$

$$-T_1 + 4T_2 - T_3 - T_5 = 200$$

$$-2T_2 + 3.722T_3 - T_6 = 106.667$$

$$-T_1 + 4T_4 - T_5 = 50$$

$$-T_2 - T_4 + 4T_5 - T_6 = 20$$

$$-T_3 - 2T_5 + 3.722T_6 = 16.667$$

Screen shots of the implementation of the inverse matrix and Gauss–Siedel methods in a spreadsheet for solving Problem 4.10 are given in Figures C.1 and C.2 in Appendix C. The results are summarised in Table E4.10.

Problems

4.1 A two-dimensional rectangular plate is subjected to the boundary conditions shown in Figure P4.1. Calculate the temperature at the midpoint (1, 0.5).

4.2 Determine the temperatures at node 2 in the rectangular rod of unit depth shown in Figure P4.2 using Eq. (4.9b). The series should converge at $n = 4$. Take $T_1 = 90°C$ and $T = 25°C$.

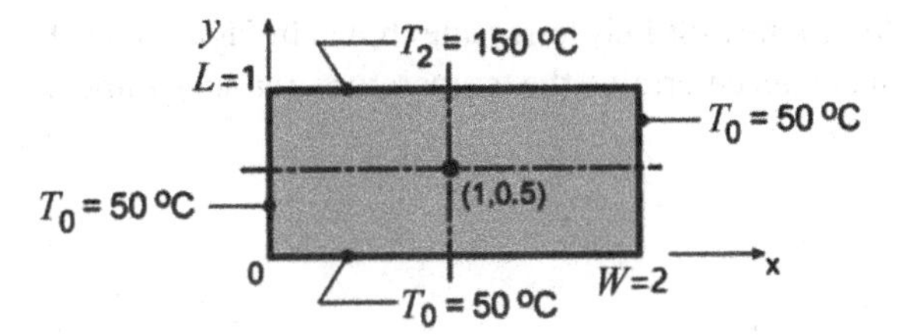

Figure P4.1 Two-dimensional conduction in a rectangular plate.

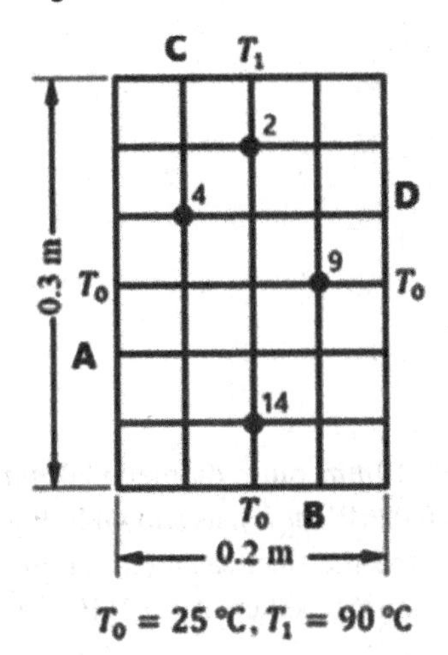

Figure P4.2 Two-dimensional model for Problem 4.2.

4.3 Repeat Problem 4.2 for point 9 in Figure P4.2.

4.4 Repeat Problem 4.2 for point 14 in Figure P4.2.

4.5 Determine the temperatures at nodes 1, 2, and 3 in the rectangular plate shown in Figure P4.5 with two-dimensional conduction. A sinusoidal temperature function $f(x)$ is applied at the top edge of the plate and constant temperature T_0 at the remaining edges.

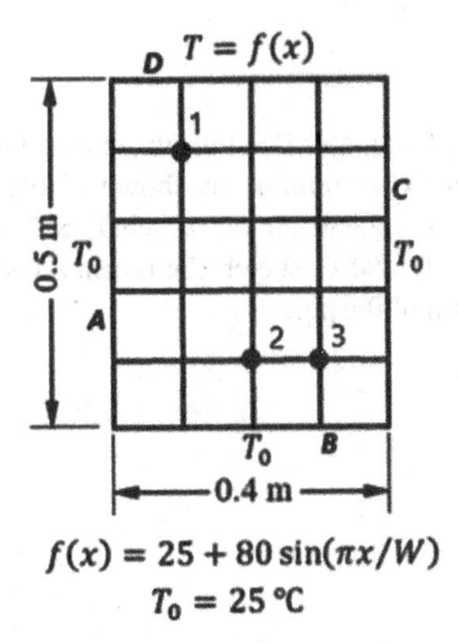

Figure P4.5 Two-dimensional model for Problem 4.5.

4.6 Determine the temperatures at node 1 in the infinitely long plate shown in Figure P4.6. If heat conduction is two-dimensional in the plate, determine the temperature at nodes 1 and 2.

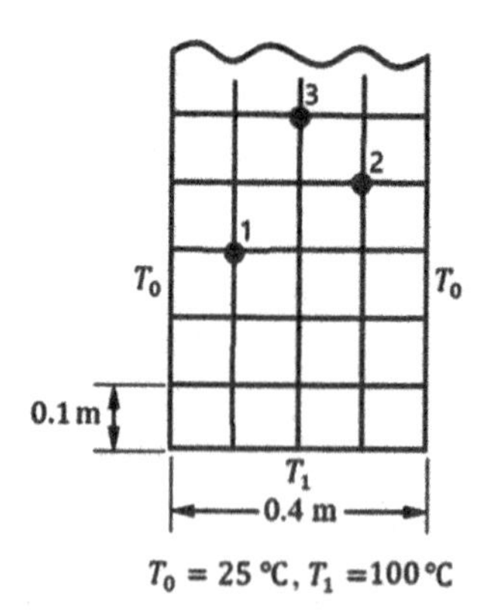

Figure P4.6 Two-dimensional model for Problem 4.6.

4.7 Repeat Problem 4.6 for nodes 2 and 3.

4.8 A steel pipe transporting superheated steam (inner diameter $10mm$, outer diameter $20mm$) is located in a steel case packed with insulation material ($k = 0.026W/m.K$), as shown in Figure P4.8. The temperature of the (superheated) steam is 127°C and the steel casing is exposed to convection at the ambient temperature $T_\infty = 27°C$. with heat transfer coefficient $h = 15W/m^2.K$. Determine the heat loss from the pipe.

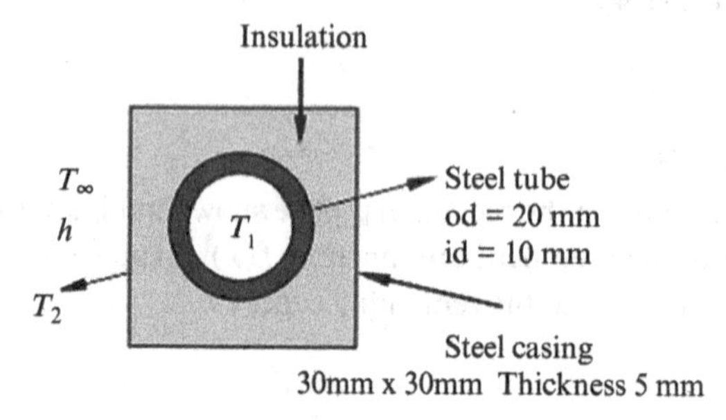

Figure P4.8 Insulated steel pipe.

4.9 Hot water is transported through a thin-wall steel pipe of diameter $D = 150mm$, encased in a concrete block ($k = 1.4W/m.K$) of square cross-section ($H = 300mm$), as shown in Figure P4.9. The outer surfaces of the concrete are exposed to ambient air for which $T_\infty 5°C$ and $h = 30W/m^2.K$. If the inlet temperature of the water is $T_i = 90°C$, sketch the thermal resistance network and determine the heat loss per unit length of the pipe.

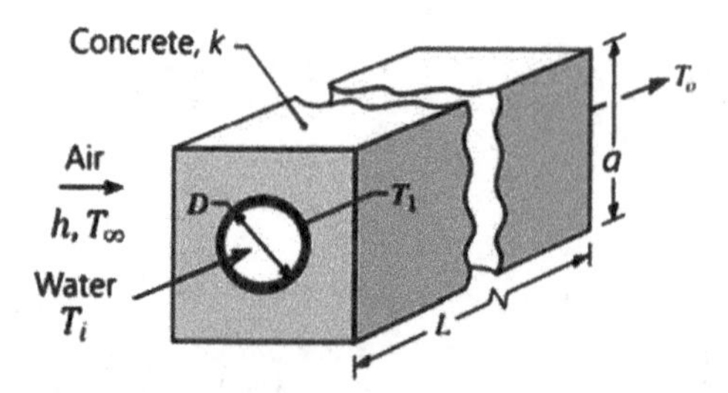

Figure P4.9 Steel pipe encased in concrete block.

4.10 A spherical tank of 1.5 m diameter containing radioactive material is buried in the earth. The distance from the earth surface to the centre of the tank is 2.5 m. The surface of the tank is maintained at $T_2 = 120°C$ as a result of radioactive decay, while the earth surface is at a uniform temperature of $T_1 = 12°C$. If the thermal conductivity of the earth is $0.8 W/m.K$, calculate the rate of heat generated in the tank.

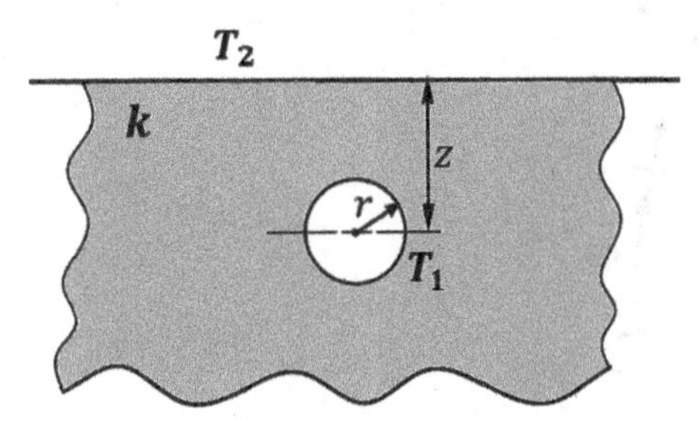

4.11 An oil pipeline has an outside diameter of 30 cm and is buried in damp soil below ground level. The distance from the earth surface to the centre of the pipeline is 1.0 m. The line is 5000 m long, and the oil flows at 2.5 kg s^{-1}. If the inlet temperature of the oil is $T_{in} = 120°C$ and the ground level soil is at $T_s = 23°C$, estimate the oil outlet temperature and the heat loss from the pipe. Take the soil thermal conductivity as $1.5 W/m.K$ and the oil specific heat as $2000 J/kg.K$.

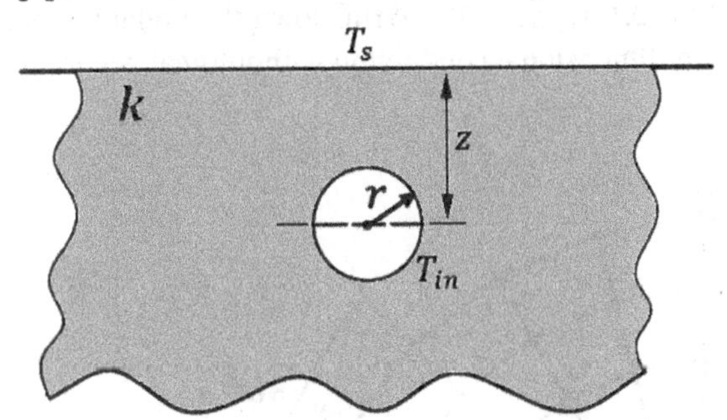

4.12 The nodes of a small thin plate are maintained at constant temperature, as shown in Figure P4.12. Determine the temperature of node 5.

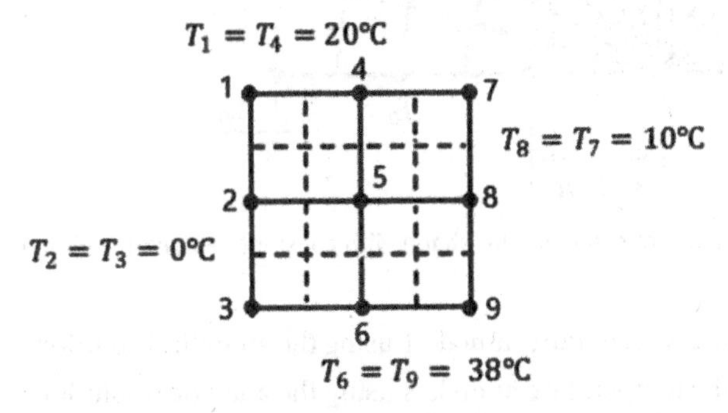

Figure P4.12 Two-dimensional conduction in a small thin plate.

4.13 In a two-dimensional cylindrical configuration shown in Figure P4.13, the radial (Δr) and angular ($\Delta\varphi$) spacings of the nodes are uniform. The boundaries in the radial direction are exposed to convection at h, T_∞ and the boundary in the angular direction is at a constant temperature T_i. Derive the finite-difference equations for surface node $i, j-1$.

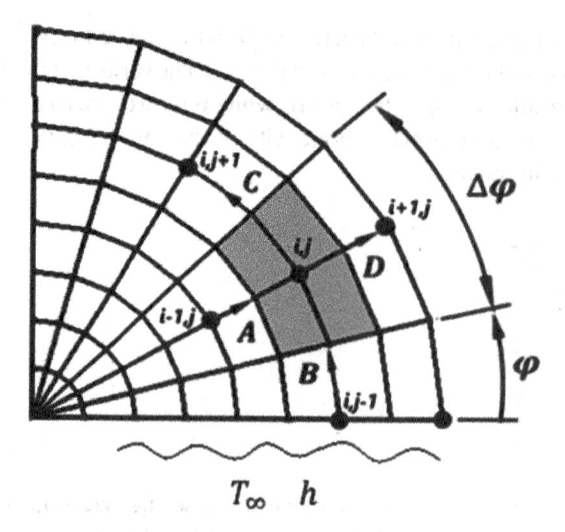

Figure P4.13 Two-dimensional conduction in a plate of cylindrical shape.

4.14 The following data is provided for the semi-circular cylindrical plate shown in Figure P4.14: $r = 0.3m$, $T_0 = 25°C$, $T_s = 100°C$, $\Delta r = 0.1m$, $\Delta\varphi = 30°$. Write down the nodal temperature equations, assuming steady-state two-dimensional conduction without heat generation and storage.

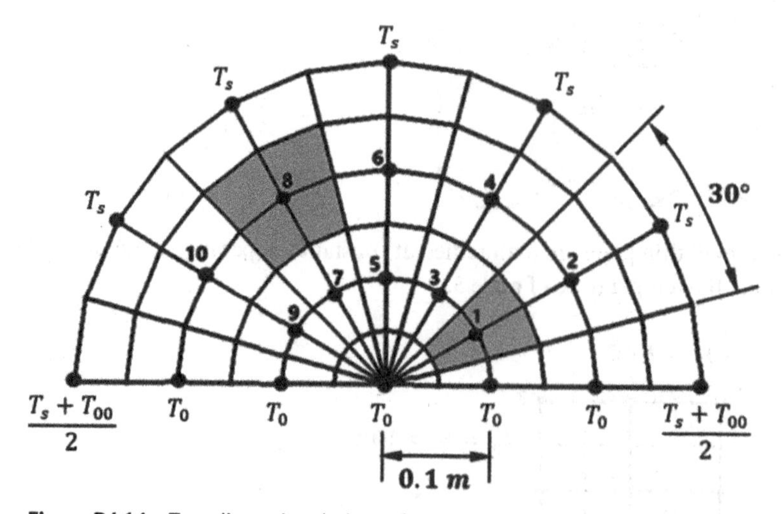

Figure P4.14 Two-dimensional plate of semi-circular shape with constant-temperature boundaries.

4.15 For Problem 4.14, determine the temperature at node 1 using the analytical solution.

4.16 For Problem 4.14, determine the temperature at node 8 using the analytical solution

4.17 For Problem 4.14, determine the temperatures for nodes 1 to 10 using the matrix inversion method.

5

Transient Conduction

When considering steady-state conduction processes in previous chapters, it was assumed that heat was not stored within the body during the process and the boundary conditions were constant and acted instantaneously, causing heat to flow in the body with space-dependent and time-independent temperature distribution. This is obviously far from what occurs in actual heating or cooling processes, in which the boundary conditions change with time, and energy storage or dissipation must be considered when solving for the temperature distribution within a body that will be both time- and space-dependent. Heat conduction under these conditions is known as unsteady, non-stationary, or transient conduction. Examples of such conduction processes include the heating of a body in a furnace, removing a packaged food item from a freezer and thawing it in ambient air, cooling of a heated slab in air, and quench-hardening a steel rod by removing it from a furnace and immersing it into a cold fluid bath. In some cases, the boundary conditions can be periodic in nature, examples of which include periodic heat flow in the cylinder of the internal combustion engine, and heating process of regenerative heat exchangers whose packings are periodically heated and cooled by flue gases and air, respectively.

In any heating or cooling process, heat transfer between a body and a surrounding fluid by convection is influenced by both the internal and surface thermal resistances of the body. The ratio of the internal resistance to the surface resistance is known as the Biot number (Bi)

$$Bi = \frac{R_{internal}}{R_{surface}} = \frac{L_c / k}{1/h} = \frac{hL_c}{k} \tag{5.1}$$

where

h – average heat transfer coefficient

L_c – characteristic length defined generally as the ratio of the volume of the body to the total heat transfer area $(L_c = V / A_s)$

k – thermal conductivity of the solid body.

Transient conduction systems under discussion in this chapter are:

- Distributed systems in which the temperature distribution is both time- and space-dependent. Single and multi-dimensional conduction will be considered
- Lumped-heat capacity systems in which temperature is time-dependent but spatially uniform.

Heat Transfer Basics: A Concise Approach to Problem Solving, First Edition. Jamil Ghojel.

Companion website: www.wiley.com/go/ghojel/heat_transfer

5.1 Analytical Solutions of One-Dimensional Distributed Systems

As with steady-state conduction, exact analytical solutions have been developed for a number of simple geometries, three of which will be discussed briefly in this chapter without full mathematical derivation of the solutions: large plate, long cylinder, and sphere. The characteristic lengths used in the analytical solutions of the transient conduction problems under consideration are:

Long and thin plate of width $2\delta : L_c = \delta$
Long cylinder of radius $R : L_c = R$
Sphere of radius $R : L_c = R$

For the analysis of more geometries and boundary conditions, the reader is referred to the works by Carslaw and Jaeger (1959), VanSant (1983), Latif (2009), and Hahn and Ozisik (2012).

5.1.1 Heating or Cooling of an Infinite Plate

The governing differential equation for a transient heat conduction system without internal heat generation in Cartesian coordinates can be written, based on Eq. (2.14), as

$$\frac{\partial^2 T}{\partial x^2}+\frac{\partial^2 T}{\partial y^2}+\frac{\partial^2 T}{\partial z^2}=\frac{1}{\alpha}\frac{\partial T}{\partial t} \tag{5.2}$$

where α is thermal diffusivity

$$\alpha=\frac{k}{\rho c_p},\ m^2/s$$

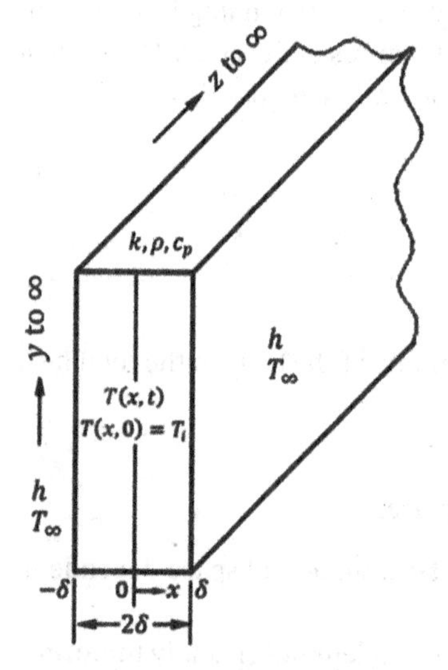

Figure 5.1 One-dimensional transient symmetrical model with convection on two sides.

For the plate, represented schematically in Figure 5.1, it is assumed that the thickness (width) in the $x-$ direction is small compared with the height in the $y-$ direction and length in the $z-$ direction and, consequently, there will be no temperature variations along the y- and z-axes; i.e.,

$$\frac{\partial T}{\partial y}=0 \quad \text{and} \quad \frac{\partial T}{\partial z}=0$$

It is further assumed that the thermophysical properties of the plate k (thermal conductivity), ρ (density), and c_p (specific heat capacity) are constant and the plate is symmetrically heated (or cooled) by convection on two sides with constant heat transfer coefficient h. The surrounding fluid temperature T_∞ is constant and the initial temperature of the plate is T_i.

Equation (5.2) is then reduced to a one-dimensional transient conduction problem

$$\frac{\partial^2 T}{\partial x^2}=\frac{1}{\alpha}\frac{\partial T}{\partial t} \tag{5.3}$$

If the temperature is redefined as $\theta = T - T_\infty$, Eq. (5.3) can be rewritten as

$$\frac{\partial^2 \theta}{\partial x^2} = \frac{1}{\alpha}\frac{\partial \theta}{\partial t} \tag{5.4}$$

The solution of Eq. (5.4) is assumed as the product of two functions: a function of time $\psi(t)$, and a function of distance $X(x)$

$$\theta(x,t) = X(x)\psi(t) \tag{5.5}$$

Substituting Eq. (5.5) into Eq. (5.4) and rearranging we obtain

$$\frac{1}{\alpha\psi(t)}\frac{\partial \psi(t)}{\partial t} = \frac{1}{X(x)}\frac{\partial^2 X(x)}{\partial x^2}$$

Both sides of this expression must be equal to a constant for the equality to hold. If the constant is taken as equal to $-\lambda^2$, we obtain two independent ordinary differential equations

$$\frac{\partial \psi(t)}{\partial t} + \psi(t)\alpha\lambda^2 = 0 \tag{5.6a}$$

$$\frac{\partial^2 X(x)}{\partial x^2} + X(x)\lambda^2 = 0 \tag{5.6b}$$

The general solutions to the differential Eqs (5.6a) and (5.6b) are, respectively,

$$\psi(t) = C_1 e^{-\alpha\lambda^2 t}$$

$$X(x) = C_2 \sin(\lambda x) + C_3 \cos(\lambda x)$$

from which

$$\theta(x,t) = C_1 e^{-\alpha\lambda^2 t} \; C_2 \sin(\lambda x) + C_3 \cos(\lambda x)$$

Mathematically, the problem is equivalent to the problem of a plate of thickness δ, which is perfectly insulated on one side at location $x = 0$ and heated (or cooled) on the outer right-hand side at location $x = \delta$ by convection (Figure 5.1). Taking into account the boundary conditions

At $t = 0$, $\theta = \theta_0 = T_i - T_\infty$
At $x = \delta$, $(d\theta / dx) = -(h/k)\theta_{x=\delta}$
At $x = 0$, $(d\theta / dx)_{x=0} = 0$
the final solution will be

$$\frac{\theta}{\theta_i} = \frac{T - T_\infty}{T_i - T_\infty} = \sum_{n=1}^{\infty} \frac{2\sin\mu_n}{\mu_n + \sin\mu_n \cos\mu_n} \cos\left(\mu_n \frac{x}{\delta}\right) e^{-(\mu_n^2 Fo)} \tag{5.7}$$

or

$$T = T_\infty + (T_i - T_\infty)\sum_{n=1}^{\infty} \frac{2\sin\mu_n}{\mu_n + \sin\mu_n \cos\mu_n} \cos\left(\mu_n \frac{x}{\delta}\right) e^{-(\mu_n^2 Fo)} \tag{5.8}$$

Fo is the Fourier number which is defined as $Fo = \alpha t / \delta^2$, where t is in seconds. The discreet values of μ_n (in radians), known as characteristic values or eigenvalues, are the positive roots of the transcendental equation

$$\mu_n \tan \mu_n = Bi \tag{5.9}$$

where $Bi = h\delta / k$.

The first six roots of Eq. (5.9) in terms of *Bi* are given in Table D.1 in Appendix D.

In addition to the temperature distribution in the plate, it is useful to know the amount of energy stored or rejected resulting from the heat conduction process. If we assume that the initial temperature is higher than the ambient temperature ($T_i > T_\infty$), the total energy removed from the plate Q_e by the time thermal equilibrium is reached at time $t = \infty$ will be

$$Q_e = \rho c_p A\delta (T_i - T_\infty) \tag{5.10}$$

Due to symmetrical temperature distribution in the plate of width 2δ, we consider one half of the plate with width δ and heat transfer area A. The instantaneous energy removed in terms of volume change $(dV = Adx)$ at any moment in time is

$$Q = \rho c_p \int (T_i - T)dV = \rho c_p A \int_0^\delta (T_i - T)dx$$

$$(T_i - T) = (T_i - T_\infty) - (T - T_\infty), \text{ or}$$

$$\frac{(T_i - T)}{(T_i - T_\infty)} = 1 - \frac{(T - T_\infty)}{(T_i - T_\infty)}$$

hence,

$$Q = \rho c_p A (T_i - T_\infty) \int_0^\delta \left(1 - \frac{T - T_\infty}{T_i - T_\infty}\right) dx$$

Substituting for the temperature ratio from Eq. (5.7), integrating over the range $0 - \delta$, and rearranging yields

$$Q = \rho c_p A (T_i - T_\infty) \left[1 - \sum_{n=1}^{\infty} \frac{2(\sin \mu_n)^2}{\mu_n (\mu_n + \sin \mu_n \cos \mu_n)} e^{-(\mu_n^2 Fo)}\right] \tag{5.11}$$

The fraction of energy stored or removed is then

$$\frac{Q}{Q_e} = 1 - \sum_{n=1}^{\infty} \frac{2(\sin \mu_n)^2}{\mu_n (\mu_n + \sin \mu_n \cos \mu_n)} e^{-(\mu_n^2 Fo)} \tag{5.12}$$

For one-term approximation $(Fo > 0.2)$

$$\frac{Q}{Q_e} = 1 - \frac{2(\sin \mu_1)^2}{\mu_1 (\mu_1 + \sin \mu_1 \cos \mu_1)} e^{-(\mu_1^2 Fo)} \tag{5.13}$$

The rate of energy gained or lost from the surface of the entire plate is

$$\dot{Q} = 2Ah(T_s - T_\infty)$$

T_s is the temperature of the plate surface at $x = \pm\delta$.

5.1.2 Analysis of the Plate Solution

Equation (5.8) allows to determine the temperature at any location on the $x-$ axis for the range $-\delta \leq x \leq \delta$ in the heated or cooled infinite plate at any time. Taking a heated plate as an example, several conclusions can be made by examining the plots of temperature versus distance from the centreline at different times and Biot numbers as seen in Figure 5.2:

1) The temperature profile is symmetrical relative to the centreline.
2) For a constant $Bi = 0.5$ and varying time, the temperature profile exhibits a pronounced minimum at the earlier times of the plate's response to the heating process becoming flatter with increasing exposure time until it becomes almost a straight line at $t = 800\,sec$ (Figure 5.2a).
3) For a constant time $t = 50s$ and varying Bi, the temperature profile exhibits very pronounced minimum temperatures at $Bi = 5.0$, becoming almost a straight line at $Bi = 0.01$. As will be discussed later, it is general practice to assume that there is no spatial temperature variation at Biot numbers below 0.2 (dashed line) and the transient conduction processes are only time-dependent.

Figure 5.3 shows a plot of the temperature at the surface of a heated plate at constant Biot number ($Bi = 5.0$) versus the roots (eigenvalues) of the transcendental Eq. (5.9) at different values of Fourier numbers Fo representing the time response of the plate to the heating process. At $Fo > 0.2$, the number of the eigenvalues in the series in Eq. (5.7b) has no effect on the calculated temperature and just the first root of the series will suffice to determine the temperature profile using the so-called one-term approximation method. All roots of the series are required, only if earlier response times are needed.

For $Fo > 0.2$, one-term approximation can be used by writing Eq. (5.7) with the first term μ_1 of the eigenvalue series

$$\frac{T - T_\infty}{T_i - T_\infty} = \frac{2\sin\mu_1}{\mu_1 + \sin\mu_1\cos\mu_1}\cos\left(\mu_1\frac{x}{\delta}\right)e^{-\left(\mu_1^2 Fo\right)} \tag{5.14}$$

The value of μ_1 will depend on the Biot number Bi (see Table D.1 in Appendix D); hence, we can write

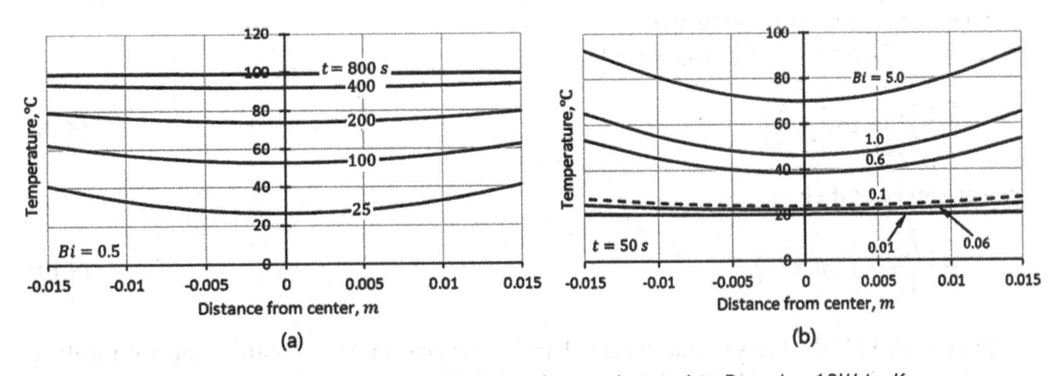

Figure 5.2 Temperature response of heated infinite plate to time and Bi. Data: $k = 12W/m.K$, $c_p = 477\,J/kg.K$, $\rho = 8025kg/m^3$, $2\delta = 0.03m$, $L_c = \delta = 0.015m$, $T_\infty = 100°C$, $T_i = 20°C$.

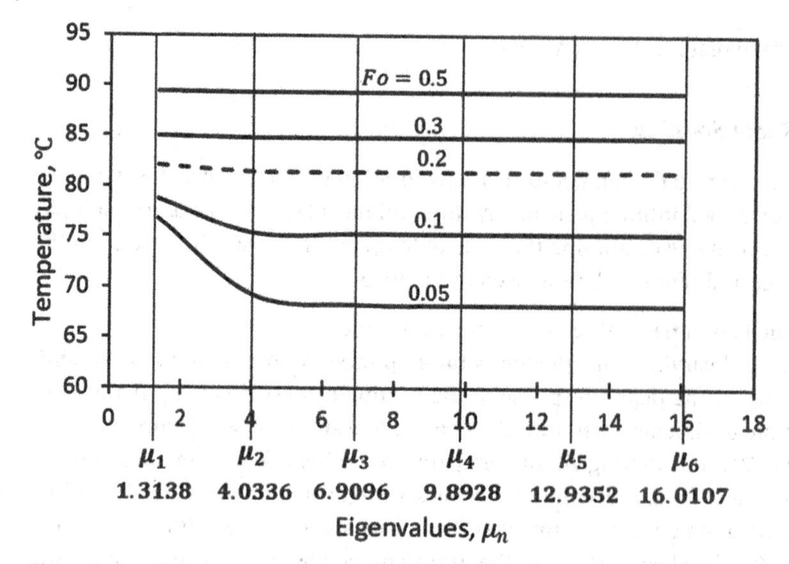

Figure 5.3 Effect of Fourier number and number of roots of the transcendental equation $\mu_n \tan\mu_n = Bi$ on the calculated temperature of an infinite plate: $Bi = 5.0, x = 0.015m$. Other data as in Figure 5.2.

$$\frac{T - T_\infty}{T_i - T_\infty} = f(Bi, x/\delta, Fo) \tag{5.15}$$

From Eq. (5.14), the temperature T_0 at the centreline of the plate $(x = 0)$ is

$$\frac{T_0 - T_\infty}{T_i - T_\infty} = \frac{2\sin\mu_1}{\mu_1 + \sin\mu_1 \cos\mu_1} e^{-(\mu_1^2 Fo)} \tag{5.16}$$

which can be written as

$$\frac{T_0 - T_\infty}{T_i - T_\infty} = f(Bi, Fo) \tag{5.17}$$

Equation (5.17) is the basis of the Heisler chart (Heisler 1947) used to estimate the plate centreline temperature T_0 relative to the initial temperature T_i in terms of Bi and Fo. (Heisler charts can be found in older heat transfer texts.)

Combining Eqs (5.14) and (5.16), we obtain

$$\frac{T - T_\infty}{T_0 - T_\infty} = \cos\left(\mu_1 \frac{x}{\delta}\right) \tag{5.18}$$

which can be written as

$$\frac{T - T_\infty}{T_0 - T_\infty} = f(Bi, x/\delta) \tag{5.19}$$

Equation (5.19) is the basis of the Heisler chart for the determination of the temperature along the x-axis as a function of plate midplane temperature within the range $0 < x < \delta$ in terms of Bi and x/δ.

The surface temperature T_s at $x=\delta$ can be determined from Eq. (5.14)

$$\frac{T_s - T_\infty}{T_i - T_\infty} = \frac{2\sin\mu_1 \cos\mu_1}{\mu_1 + \sin\mu_1 \cos\mu_1} e^{-(\mu_1^2 Fo)} \tag{5.20}$$

Combining Eqs (5.16) and (5.20), the relationship between the centreline and surface temperatures of the infinite plate can be established without the need for the initial temperature T_i

$$\frac{T_s - T_\infty}{T_0 - T_\infty} = \cos\mu_1 = f(Bi) \tag{5.21}$$

By data fitting $\cos(\mu_1) = f(Bi)$, Eq. (5.21) is transformed to

$$\frac{T_s - T_\infty}{T_0 - T_\infty} = \frac{1.0}{1 + 0.6Bi} \tag{5.22}$$

5.1.2.1 Other Boundary Conditions

So far we have considered only the convection heat at the outer surface as a boundary condition. If the heat transfer coefficient tends to infinity, the surface resistance tends to zero and Bi to infinity as can be seen from the definition of the Bi,

$$Bi = \frac{\delta / k}{1/h}$$

As a result, the heating or cooling process will be governed only by the thermophysical properties of the plate and the surface temperature becomes equal to the temperature of the surrounding fluid. The eigenvalues of Eq. (5.9), when $Bi \to \infty$, can be written as (Isachenko 1969)

$$\mu_n = (2n-1)\frac{\pi}{2} \tag{5.23}$$

The first six roots are then

1.570796	4.712389	7.853982	10.99557	14.13717	17.27876

which are the same values given in Table D.1 in Appendix D at $Bi = \infty$.

If the one-term approximation is used, $\mu_n = \pi/2$ and the centreline temperature, given by Eq. (5.16), can be modified to

$$\frac{T_0 - T_\infty}{T_i - T_\infty} = \frac{4}{\pi} e^{-\left(\frac{\pi}{2}\right)^2 Fo} \tag{5.24}$$

Let the temperature ratio at $x=0$ be

$$\frac{T_0 - T_\infty}{T_i - T_\infty} = \Theta_{x=0}$$

hence,

$$Fo = \frac{\alpha t}{\delta^2} = \frac{4}{\pi^2} \ln\left(\frac{4}{\pi} \frac{1}{\Theta_{x=0}}\right) \tag{5.25}$$

The time required for the centreline temperature to reach a prescribed temperature T_o is then

$$t=\frac{1}{\alpha}\left(\frac{2\delta}{\pi}\right)^2 \ln\left(\frac{4}{\pi}\frac{1}{\Theta_{x=0}}\right) sec \tag{5.26}$$

Example 5.1 A large aluminium plate with a thickness of 50 mm is suddenly exposed to convection heat transfer in an environment at 80°C with heat transfer coefficient $h=600\ W/m^2.K$. The initial temperature of the plate is 200°C. The thermophysical properties of aluminium can be taken as $k=204 W/m.K$, $c_p=900 J/kg.K$, $\rho=2700 kg/m^3$. What is the time it takes for the surface temperature to reach 150°C?

Solution

Use Eq. (5.20) to find Fo then t

$$\frac{T_s-T_\infty}{T_i-T_\infty}=\frac{2\sin\mu_1\cos\mu_1}{\mu_1+\sin\mu_1\cos\mu_1}e^{-\left(\mu_1^2 Fo\right)}$$

$$Bi=\frac{\delta/k}{1/h}=\frac{0.025\times 600}{204}=0.0735$$

For $Bi=0.0735$, the eigenvalue $\mu_1=0.2672$ (Table D.1 in Appendix D)

$$\frac{150-80}{200-80}=\frac{2\times\sin 0.2672\times\cos 0.2672}{0.2672+\sin 0.2672\cos 0.2672}e^{-\left(\mu_1^2 Fo\right)}$$

$$0.5833=0.9759e^{-(0.0714 Fo)}$$

$$Fo=\frac{\alpha t}{\delta^2}=7.2088$$

$$\alpha=\frac{k}{\rho c_p}=\frac{204}{2700\times 900}=8.395\times 10^{-5}$$

$$\therefore t=\frac{7.2088\times 0.025^2}{8.395\times 10^{-5}}=53.7s$$

5.1.3 Heating or Cooling of an Infinite Solid Cylinder

A heated or cooled solid cylinder of radius R and length L can be considered an infinite cylinder with no temperature variation along the length in the z-direction $\left(\partial T/\partial z=0\right)$ if $L>>R$ (Figure 5.4). If the cylinder thermophysical properties k, ρ, c_p are constant and internal heat generation is ignored, the general differential equation for two-dimensional conduction in cylindrical coordinates given by Eq. (2.15) can be simplified to

$$\frac{1}{r}\frac{\partial}{\partial r}\left(r\frac{\partial T}{\partial r}\right)+\frac{1}{r^2}\frac{\partial^2 T}{\partial \varphi^2}=\frac{1}{\alpha}\frac{\partial T}{\partial t} \tag{5.27}$$

If it is further assumed that there is no temperature variation in the $\varphi-$ direction $\left(\partial T/\partial\varphi=0\right)$, we obtain the differential equation for one-dimensional transient conduction in an infinite cylinder

$$\frac{\partial^2 T}{\partial r^2}+\frac{1}{r}\frac{\partial T}{\partial r}=\frac{1}{\alpha}\frac{\partial T}{\partial t} \tag{5.28}$$

where $\alpha = k/\rho c_p$

The entire cylinder is at an initial temperature T_i and the outer surface is heated (or cooled) by convection with a heat transfer coefficient h and the surrounding fluid is at a constant temperature T_∞.

If the temperature is redefined as $\theta = T(r,t)$, Eq. (5.28) can be rewritten as

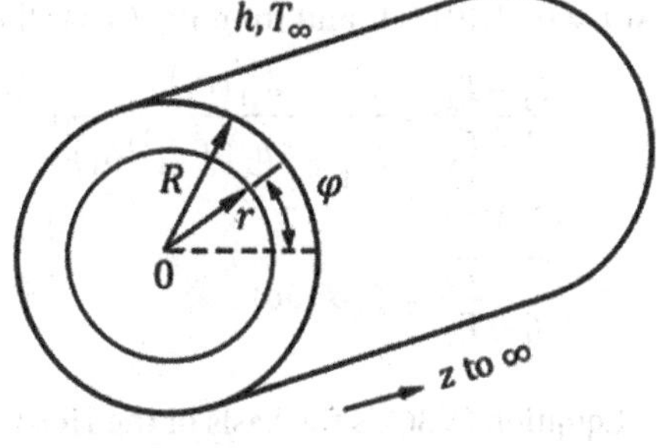

Figure 5.4 Transient conduction model for a long cylinder.

$$\frac{\partial^2 \theta}{\partial r^2}+\frac{1}{r}\frac{\partial \theta}{\partial r}=\frac{1}{\alpha}\frac{\partial \theta}{\partial t} \tag{5.29}$$

The boundary and initial conditions are

At $t=0$ and $0 \le r \le R$, $\theta = \theta_i = T_i - T_\infty$

At $r=0$ and $t>0$, $(\partial\theta/\partial r)_{r=0} = 0$ (the temperature profile is symmetrical)

At $r=R$ and $t>0$, $(\partial\theta/\partial r)_{r=R} = -(h/k)\theta_{r=R}$

The final equation for the temperature distribution on the basis of the method of separation of variables is

$$\frac{T-T_\infty}{T_i-T_\infty}=\sum_{n=1}^{\infty}\frac{2J_1(\mu_n)}{\mu_n\left[J_0^2(\mu_n)+J_1^2(\mu_n)\right]}J_0\left(\mu_n\frac{r}{R}\right)e^{-(\mu_n^2 Fo)} \tag{5.30}$$

or

$$T=T_\infty+(T_i-T_\infty)\sum_{n=1}^{\infty}\frac{2J_1(\mu_n)}{\mu_n\left[J_0^2(\mu_n)+J_1^2(\mu_n)\right]}J_0\left(\mu_n\frac{r}{R}\right)e^{-(\mu_n^2 Fo)} \tag{5.31}$$

Solution of Eq. (5.30) requires data of Bessel functions of the first kind J_0 and J_1, which can be found in Table B.1 in Appendix B. Parameter μ_n $(n=1,2,3,\ldots)$ represent the positive roots of the transcendental equation

$$\mu_n J_1(\mu_n) - Bi J_0(\mu_n) = 0 \tag{5.32}$$

The Biot number here is defined as $Bi = hR/k$ and the first six roots of Eq. (5.32) are given in Table D.2 in Appendix D.

Substituting for $J_1(\mu_n)$ from Eq. (5.32) into Eq. (5.31) yields a relatively simpler expression for the temperature distribution in a long cylinder

$$\frac{T-T_\infty}{T_i-T_\infty}=\sum_{n=1}^{\infty}\frac{J_0\left(\mu_n\frac{r}{R}\right)}{J_0(\mu_n)}\frac{2Bi}{\left(\mu_n^2+Bi^2\right)}e^{-(\mu_n^2 Fo)} \tag{5.33}$$

The Fourier number for a solid cylinder $Fo = \alpha t/R^2$. If the one-term approximation is used $(Fo > 0.2)$, Eq. (5.31) is reduced to

$$\frac{T-T_\infty}{T_i-T_\infty}=\frac{2J_1(\mu_1)}{\mu_1\left[J_0^2(\mu_1)+J_1^2(\mu_1)\right]}J_0\left(\mu_1\frac{r}{R}\right)e^{-(\mu_1^2 Fo)} \tag{5.34}$$

At $r=0$, $J_0(0)=1$, and from Eq. (5.34) the centre temperature $T=T_0$ is given by

$$\frac{T_0-T_\infty}{T_i-T_\infty}=\frac{2J_1(\mu_1)}{\mu_1\left[J_0^{\,2}(\mu_1)+J_1^{\,2}(\mu_1)\right]}e^{-(\mu_1^2 Fo)} \tag{5.35}$$

$$\frac{T_0-T_\infty}{T_i-T_\infty}=f(Bi,Fo) \tag{5.36}$$

Equation (5.36) is the basis of the Heisler chart used for the determination of the temperature at the centre of T_0 relative to the initial temperature T_i in infinite cylinders in terms of Bi and Fo.

Combining Eqs (5.34) and (5.35), we obtain

$$\frac{T-T_\infty}{T_0-T_\infty}=J_0\left(\mu_1\frac{r}{R}\right) \tag{5.37}$$

which can be written as

$$\frac{T-T_\infty}{T_0-T_\infty}=f(Bi,r/R) \tag{5.38}$$

Equation (5.38) is the basis of the Heisler chart for the determination of the temperature along the r-axis as a function of cylinder centreline temperature within the range $0<r<R$ in terms of Bi and x/δ.

The surface temperature T_s at $r=R$ can be obtained from Equation (5.34)

$$\frac{T_s-T_\infty}{T_i-T_\infty}=\frac{2J_1(\mu_1)J_0(\mu_1)}{\mu_1\left[J_0^{\,2}(\mu_1)+J_1^{\,2}(\mu_1)\right]}e^{-(\mu_1^2 Fo)} \tag{5.39}$$

Combining Eqs (5.35) and (5.39), the relationship between the centreline and surface temperatures of the cylinder can be established without the need for the initial temperature T_i from

$$\frac{T_s-T_\infty}{T_0-T_\infty}=J_0(\mu_1) \tag{5.40a}$$

where

$$J_0(\mu_1)\cong\frac{1}{\left(1+0.65Bi+0.0039Bi^2\right)} \tag{5.40b}$$

The cylinder reaches thermal equilibrium with the surroundings when its temperature becomes equal to temperature of the surroundings T_∞ and the energy lost or gained is

$$Q_e=\pi R^2 L\rho c_p(T_i-T_\infty) \tag{5.41}$$

For a cylinder of length L, the energy stored in or lost from the elemental volume $dV=(2\pi r dr)L$ at any moment in time is

$$Q=\rho c_p\int(T_i-T)dV=2\pi L\rho c_p\int_0^R(T_i-T)rdr$$

$$(T_i - T) = (T_i - T_\infty) - (T - T_\infty),\ \text{or}$$

$$\frac{(T_i - T)}{(T_i - T_\infty)} = 1 - \frac{(T - T_\infty)}{(T_i - T_\infty)}$$

hence,

$$Q = 2\pi L\rho c_p (T_i - T_\infty)\int_0^R \left(1 - \frac{T - T_\infty}{T_i - T_\infty}\right) r dr$$

Using Eq. (5.31), we can write

$$Q = 2\pi L\rho c_p (T_i - T_\infty)\left[\int_0^R r dr - \sum_{n=1}^{\infty} \frac{2J_1(\mu_n)e^{-(\mu_n^2 Fo)}}{\mu_n\left[J_0^{\,2}(\mu_n) + J_1^{\,2}(\mu_n)\right]}\int_0^R rJ_0\left(\mu_n \frac{r}{R}\right)dr\right]$$

Since,

$$\int_0^R r dr = \frac{R^2}{2} \quad \text{and} \quad \int_0^R rJ_0\left(\mu_n \frac{r}{R}\right)dr = \left[\frac{r}{\mu_n / R} J_1\left(\mu_n \frac{r}{R}\right)\right]_0^R = \frac{R^2}{\mu_n} J_1(\mu_n)$$

we finally obtain

$$Q = \rho c_p L\pi R^2 (T_i - T_\infty)\left[1 - 2\sum_{n=1}^{\infty} \frac{2\left[J_1(\mu_n)\right]^2}{\mu_1^2\left[J_0^{\,2}(\mu_n) + J_1^{\,2}(\mu_n)\right]} e^{-(\mu_n^2 Fo)}\right] \tag{5.42}$$

The fractional energy stored or released is

$$\frac{Q}{Q_e} = 1 - 2\sum_{n=1}^{\infty} \frac{2\left[J_1(\mu_n)\right]^2}{\mu_1^2\left[J_0^{\,2}(\mu_n) + J_1^{\,2}(\mu_n)\right]} e^{-(\mu_n^2 Fo)} \tag{5.43}$$

For one-term approximation $(Fo > 0.2)$

$$\frac{Q}{Q_e} = 1 - \frac{4\left[J_1(\mu_1)\right]^2}{\mu_1^2\left[J_0^{\,2}(\mu_1) + J_1^{\,2}(\mu_1)\right]} e^{-(\mu_1^2 Fo)} \tag{5.44}$$

The rate of heat loss (or gain) at the surface of the cylinder is

$$\dot{Q}_s = 2\pi RLh(T_s - T_\infty)\ W \tag{5.45}$$

where T_s is the temperature of the cylinder surface at $r = R$.

Example 5.2 A long solid cylinder of radius $R = 0.03\ m$ at initial $T_i = 27°C$ is heated in a furnace at constant temperature $T_\infty = 500°C$. The convection coefficient of heat transfer from the furnace to the cylinder is $h = 1000\ W/m^2.K$. Determine the time it takes for the surface of the cylinder to reach 300°C and the corresponding temperature at the centreline axis of the cylinder. The thermo-physical properties of the cylinder are: $\rho = 8000\ kg/m^3$, $c_p = 500\ J/kg.K$, $k = 40\ W/m.K$.

Solution

Equation (5.39) can be used to determine Fo, from which time t for the surface temperature to reach 573 K can be calculated

$$\frac{T_s - T_\infty}{T_i - T_\infty} = \frac{2J_1(\mu_1)J_0(\mu_1)}{\mu_1\left[J_0^2(\mu_1) + J_1^2(\mu_1)\right]}e^{-(\mu_1^2 Fo)}$$

The order of calculations are as follows:

$$Bi = \frac{hR}{k} = \frac{1000 \times 0.03}{40} = 0.75$$

From Table D.1 at $Bi = 0.75$, $\mu_1 = 1.11815$
From Table B.1, $J_0(\mu_1) = 0.717$, $J_1(\mu_1) = 0.471$

$$\frac{573 - 773}{300 - 773} = \frac{2 \times 0.471 \times 0.717}{1.11815\left[0.717^2 + 0.471^2\right]}e^{-(\mu_1^2 Fo)}$$

$$0.515 = e^{-(1.25 Fo)}$$

$$Fo = \frac{\alpha t}{R^2} = \frac{1}{1.25}\ln\left(\frac{1}{0.515}\right) = 0.5309$$

$$\alpha = \frac{k}{\rho c_p} = \frac{40}{8000 \times 500} = 1.0 \times 10^{-5}$$

$$\therefore t = \frac{0.5309 \times 0.03^2}{1.0 \times 10^{-5}} = 47.8 s$$

From Eq. (5.40a)

$$\frac{T_s - T_\infty}{T_0 - T_\infty} = J_0(\mu_1)$$

$$T_0 = T_\infty + \frac{T_s - T_\infty}{J_0(\mu_1)} = 773 + \frac{1}{0.717}(573 - 773) = 494 K$$

A screen shot of the Excel spreadsheet of the solution using Eq. (5.31) is given in Figure D.1 in Appendix D.

5.1.4 Heating or Cooling of a Sphere

Consider the heated or cooled sphere of radius R shown in Figure 5.5 with temperature variation in the radial direction only ($\partial T / \partial \theta = 0, \partial T / \partial \varphi = 0$). If the thermophysical properties k, ρ, c_p are constant and internal heat generation is ignored, the general differential equation for three-dimensional conduction in spherical coordinates given by Eq. (2.16) can be simplified to

$$\frac{1}{r^2}\frac{\partial}{\partial r}\left(r^2\frac{\partial T}{\partial r}\right) = \frac{1}{\alpha}\frac{\partial T}{\partial t}$$

or

$$\frac{\partial^2 T}{\partial r^2}+\frac{2}{r}\frac{\partial T}{\partial r}=\frac{1}{\alpha}\frac{\partial T}{\partial t} \tag{5.46}$$

If the temperature is redefined as $\theta = T(r,t) - T_\infty$, Eq. (5.46) can be rewritten as

$$\frac{\partial^2 \theta}{\partial r^2}+\frac{2}{r}\frac{\partial \theta}{\partial r}=\frac{1}{\alpha}\frac{\partial \theta}{\partial t} \tag{5.47}$$

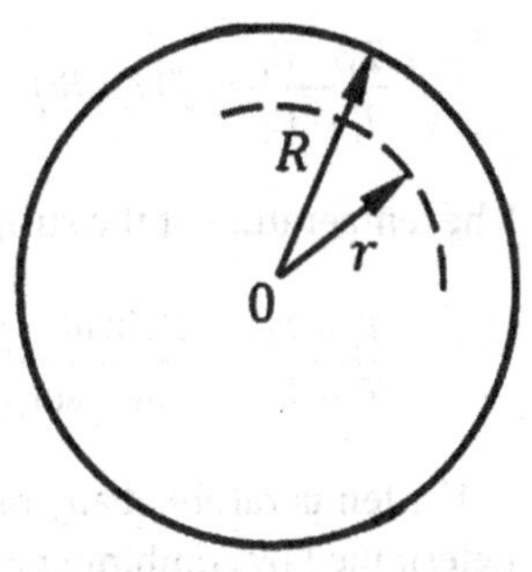

Figure 5.5 One-dimensional transient model for a sphere.

The boundary and initial conditions are

At $t=0$ and $0 \le r \le R$, $\theta = \theta_i = T_i - T_\infty$

At $r=0$ and $t>0$, $(\partial\theta/\partial r)_{r=0}=0$ (the temperature profile is symmetrical)

At $r=R$ and $t>0$, $(\partial\theta/\partial r)_{r=R} = -(h/k)\theta_{r=R} = -(h/k)(T_{r=R}-T_\infty)$

The final equation for the temperature distribution on the basis of the method of separation of variables is

$$\frac{T-T_\infty}{T_i-T_\infty}=\sum_{n=1}^{\infty}\frac{2(\sin\mu_n-\mu_n\cos\mu_n)}{\mu_n-\sin\mu_n\cos\mu_n}\frac{\sin\left(\mu_n\frac{r}{R}\right)}{\left(\mu_n\frac{r}{R}\right)}e^{-(\mu_n^2 Fo)} \tag{5.48}$$

or

$$T=T_\infty+(T_i-T_\infty)\sum_{n=1}^{\infty}\frac{2(\sin\mu_n-\mu_n\cos\mu_n)}{\mu_n-\sin\mu_n\cos\mu_n}\frac{\sin\left(\mu_n\frac{r}{R}\right)}{\left(\mu_n\frac{r}{R}\right)}e^{-(\mu_n^2 Fo)} \tag{5.49}$$

The Fourier number $Fo=\alpha t/R^2$ and the eigenvalues μ_n $(n=1,2,3,\ldots 0)$ are the positive roots of the transcendental equation

$$Bi = 1-\mu_n\cot\mu_n \tag{5.50}$$

The first six positive roots μ_n of the transcendental Eq. (5.42) are given in Table D.3 in Appendix D.

The series in Eq. (5.48) usually converges quickly and for $Fo>0.2$ only the first term of the eigenvalues is required to obtain a result. Equation (5.48) can then be written as

$$\frac{T-T_\infty}{T_i-T_\infty}=\frac{2(\sin\mu_1-\mu_1\cos\mu_1)}{\mu_1-\sin\mu_1\cos\mu_1}\frac{\sin\left(\mu_1\frac{r}{R}\right)}{\left(\mu_1\frac{r}{R}\right)}e^{-(\mu_1^2 Fo)} \tag{5.51}$$

As r tends to 0, $(\mu_1 r/R)$ becomes very small and $\sin(\mu_1 r/R)\approx(\mu_1 r/R)$; hence, $\sin(\mu_1 r/R)/(\mu_1 r/R)$ tends to 1. The temperature at the centre of the sphere T_0 is then

$$\frac{T_0-T_\infty}{T_i-T_\infty}=\frac{2(\sin\mu_1-\mu_1\cos\mu_1)}{\mu_1-\sin\mu_1\cos\mu_1}e^{-(\mu_1^2 Fo)} \tag{5.52}$$

$$\frac{T_0 - T_\infty}{T_i - T_\infty} = f(Fo, Bi) \tag{5.53}$$

The temperature at the surface of the sphere T_s (at $r = R$) is

$$\frac{T_s - T_\infty}{T_i - T_\infty} = \frac{2(\sin\mu_1 - \mu_1 \cos\mu_1)}{\mu_1 - \sin\mu_1 \cos\mu_1} \frac{\sin(\mu_1)}{(\mu_1)} e^{-(\mu_1^2 Fo)} \tag{5.54}$$

The temperature at any radial distance from the centre relative to the centre temperature can be determined by combining Eqs (5.51) and (5.52)

$$\frac{T - T_\infty}{T_0 - T_\infty} = \frac{\sin\left(\mu_1 \dfrac{r}{R}\right)}{\left(\mu_1 \dfrac{r}{R}\right)} \tag{5.55}$$

or

$$\frac{T - T_\infty}{T_0 - T_\infty} = f(Bi, r/R) \tag{5.56}$$

The Heisler charts for one-dimensional transient conduction in spheres are based on Eqs (5.53) and (5.56).

Combining Eqs (5.52) and (5.54), the relationship between the centre and surface temperatures of the sphere can be established without the need for the initial temperature T_i

$$\frac{T_s - T_\infty}{T_0 - T_\infty} = \frac{\sin(\mu_1)}{(\mu_1)} \tag{5.57}$$

For estimation of the surface temperature in the absence of data for the eigenvalue $\mu_1(Bi)$, Eq. (5.58) can be used.

$$\frac{\sin(\mu_1)}{(\mu_1)} \cong \frac{1.003}{\left(1 + 0.611278Bi + 0.0221Bi^2\right)} \tag{5.58}$$

The sphere reaches thermal equilibrium with the surroundings when its temperature becomes equal to temperature of the surroundings T_∞ and the energy accumulated is

$$Q_e = \rho c_p \left(\frac{4}{3}\pi R^3\right)(T_\infty - T_i) \tag{5.59}$$

The cumulative energy entering or leaving the model at any moment in time is

$$Q = \rho c_p \int (T - T_i) dV = \int_0^R (T - T_i) r^2 dr$$

Substituting for T from Eq. (5.49) and integrating yields

$$Q = \rho c_p \left(\frac{4}{3}\pi R^3\right)(T_i - T_\infty)\left[1 - 3\sum_{n=1}^{\infty} \frac{2(\sin\mu_n - \mu_n \cos\mu_n)^2}{\mu_n^3(\mu_n - \sin\mu_n \cos\mu_n)} e^{-(\mu_n^2 Fo)}\right] \tag{5.60}$$

The fractional energy stored or released at any moment in time is then

$$\frac{Q}{Q_e}=1-3\sum_{n=1}^{\infty}\frac{2\left(\sin\mu_n-\mu_n\cos\mu_n\right)^2}{\mu_n^3\left(\mu_n-\sin\mu_n\cos\mu_n\right)}e^{-\left(\mu_n^2 Fo\right)} \tag{5.61}$$

For the approximate solution ($Fo > 0.2$)

$$\frac{Q}{Q_e}=1-\frac{6\left(\sin\mu_1-\mu_1\cos\mu_1\right)^2}{\mu_1^3\left(\mu_1-\sin\mu_1\cos\mu_1\right)}e^{-\left(\mu_1^2 Fo\right)} \tag{5.62}$$

The rate of heat loss (or gain) at the surface of the sphere is

$$\dot{Q}_s=4\pi R^2 h\left(T_s-T_\infty\right)W \tag{5.63}$$

where T_s is the temperature of the surface of the sphere at $r=R$.

5.1.5 Heisler Charts

Several references were made to the Heisler charts (Heisler 1947) in the previous discussion. These charts, which feature in many heat-transfer references, are useful tools for quick graphical determination of temperatures in transient conduction in infinite plates, long cylinders, and in spheres.

However, Heisler charts are difficult to read, particularly at low Fourier and high Biot numbers and the results overall are inaccurate. The analytical solutions presented in this text provide accurate results and are relatively easily handled in spreadsheets. All necessary data required for this purpose, such as Bessel functions and eigenvalues of different transcendental functions with screen shots of their implementation, are provided in Appendices B and D.

Example 5.3 A standard 58 g egg has a volume of about 50 cm^3. If it is approximated as a sphere, what are the surface and core temperatures 10 minutes after dropping in boiling water at 100°C. Take the initial temperature $T_i=10°C$ and the heat transfer coefficient for boiling water $h=1000\,W/m^2.K$. The thermal properties of the egg are: thermal conductivity $k=0.6\,W/m.K$ and specific heat $c_p=4400\,J/kg.K$.

Solution

We need to find three unknowns to solve the problem: Bi, Fo, μ_n.

$$R=\left(\frac{3V}{4\pi}\right)^{1/3}=\left(\frac{3\times 50}{4\pi}\right)^{1/3}=2.285\,cm$$

$$\rho_e=\frac{58\times 10^{-3}}{50\times 10^{-6}}=1160\,kg/m^3$$

$$Bi=\frac{hR}{k}=\frac{1000\times 0.02285}{0.6}=38.08$$

$$\alpha=\frac{k}{\rho c_p}=\frac{0.6}{1160\times 4200}=1.23\times 10^{-7}$$

$$Fo = \frac{\alpha t}{R^2} = \frac{1.23 \times 10^{-7} \times 600}{0.02285^2} = 0.2413$$

From Table D.3 (Appendix D), the first root of function $Bi = 1 - \mu_n \cot \mu_n$ at $Bi = 38.08$ is $\mu_1 = 3.0584$.

From Eq. (5.51)

$$\frac{T - T_\infty}{T_i - T_\infty} = \frac{2(\sin \mu_1 - \mu_1 \cos \mu_1)}{\mu_1 - \sin \mu_1 \cos \mu_1} \frac{\sin\left(\mu_1 \frac{r}{R}\right)}{\left(\mu_1 \frac{r}{R}\right)} e^{-(\mu_1^2 Fo)}$$

at $r = 0.00001$, $T_0 = 52.25°C$

at $r = 0.02285$, $T_s = 98.7°C$

When calculating the core temperature T_0, the value of r in the equation should be a non-zero number (e.g. 0.00000001). Using two roots or more of function $Bi = 1 - \mu_n \cot \mu_n$ results in $T_0 = 53.14°C$, and $T_s = 98.68°C$.

5.2 Time-Dependent and Spatially Uniform Temperature Distribution

It is generally accepted that if the temperature distribution in a transient conduction system for a small or thin body is spatially independent, a simple analysis technique known as *lumped heat capacity* or *lumped capacitance method* can be used to predict the average temperature-time profile of the system, and the condition for applying this method is $Bi < 0.1$. To check the validity of this statement, Eq. (5.8) is used to solve the one-dimensional transient heat conduction in the infinite plate shown in Figure 5.1 at $x = 0$ and $x = \delta$ using the following data: Plate width $2\delta = 0.03m$, height H and length L of the plate are large compared to the width. The plate is initially at $T_i = 20°C$ and is exposed to the ambient temperature $T_\infty = 100°C$ with convection heat transfer coefficient h between the ambient and two sides of the plate. The plate material has thermal conductivity $k = 12 W/m.K$, density $\rho = 8025\ kg/m^3$, and specific heat capacity $c_p = 477 J/kg.K$. To change the Biot number Bi the heat transfer coefficient h is varied, as shown in Table 5.1.

It is evident from Table 5.1 that the larger the Biot number, the greater the temperature variation across the plate. The least variation occurs at $Bi = 0.01$ and the most at $Bi = 5$. At $Bi = 5$ and beyond, the rate of temperature variation starts decreasing slowly. The temperature variation in the plate is around 10%, even at $Bi = 0.06$, with the difference decreasing with time. For the specific conduction model under consideration, the Biot number at which the conduction process could be considered spatially uniform is much smaller than the generally accepted threshold of $Bi = 0.1$ if an earlier time

Table 5.1 Variation of Biot number with heat transfer coefficient in an infinite plate at $t = 50$ *sec.*

$h,\ W/m^2.K$	8	48	80	480	800	4,000
Bi	0.01	0.06	0.1	0.6	1.0	5
Centre to surface temperature change, °C	0.4	2.27	3.65	14.58	18.58	22.23
% Temperature change	1.96	10.11	15.2	37.57	39.92	31.68

Table 5.2 Spatial temperature variation with material type at Biot number $Bi = 0.06$ and $t = 50$ *sec.*

	Aluminium alloy 6063-T6	Alloy 42	Carbon steel 0.3–0.5% C	Inconel X-750	Titanium	Glass
h $W/m^2.K$	804	50.24	186.8	46.8	87.6	3.12
k $W/m.K$	201	12.56	46.7	11.7	21.9	0.78
ρ kg/m^3	2692	8099	7831	8510	4500	2700
c_p $J/kg.K$	879	507	419	436	522	840
Centre to surface temperature change °C	0.78	2.27	1.96	2.27	2.09	1.46
% Temperature change	1.06	10.14	5.96	10.08	7.35	7.30

of the model's response is important. Generally, for a fixed Bi, the spatial temperature variation also depends on the thermophysical properties of the plate material, as shown in Table 5.2.

From the data in Tables 5.1 and 5.2, it can be concluded that the condition for assuming spatially uniform temperature distribution in a transient conduction system at earlier response time may be better stated as $Bi << 0.1$.

5.2.1 Lumped Capacitance Method

This method will be developed on the basis of the same physical model shown in Figure 5.1 for a heated plate with finite dimensions in the y- and z-directions (H and L, respectively). Considering the whole plate as a system, the first law of thermodynamics for the rate of heat input $\dot{Q}$ through the side surfaces $(A_s = 2HL)$ by convection can be written in terms of the change in internal energy with time (heat accumulation or heat storage)

$$\dot{Q} = \frac{dU}{dT}$$

$$A_s h(T_\infty - T) = \rho V c_p \frac{dT}{dt}$$

or

$$-A_s h(T - T_\infty) = \rho V c_p \frac{dT}{dt} \tag{5.64}$$

where
A_s – total heat transfer area, m^2
h – average heat transfer coefficient, $W/m^2.K$
T_∞ – ambient temperature (temperature of the heat source heating the plate), °C
T_i – initial uniform temperature of the plate, °C
ρ – density of the plate material, kg/m^3
c_p – specific heat of the plate material, $J/kg.K$
V – volume of the plate, m^3

Equation (5.64) is a first-order differential equation which can be rewritten, after separating the variable, as

$$-\frac{A_s h}{\rho V c_p} dt = \frac{dT}{T - T_\infty}$$

Integrating both sides

$$-\frac{A_s h}{\rho V c_p} \int_0^t dt = \int_{T_i}^{T} \frac{dT}{T - T_\infty}$$

and rearranging yields

$$\ln \frac{T - T_\infty}{T_i - T_\infty} = -\frac{A_s h}{\rho V c_p} t$$

$$\frac{T - T_\infty}{T_i - T_\infty} = \exp\left(-\frac{A_s h}{\rho V c_p} t\right) \tag{5.65}$$

$$T = T_\infty + (T_i - T_\infty) \exp\left(-\frac{A_s h}{\rho V c_p} t\right) \tag{5.66}$$

Using the characteristic length definition for the plate $L_c = V / A_s$, the exponential term on the right side of Eq. (5.66) can be rearranged in terms of non-dimensional parameters as follows

$$\frac{A_s h}{\rho V c_p} t = \left(\frac{h}{\rho L_c c_p} t\right) = \left(\frac{h}{\rho L_c c_p} t\right) \frac{k}{k} \frac{L_c}{L_c} = \left(\frac{h L_c}{k}\right) \frac{k}{L_c 2} \frac{t}{\rho c_p} = \left(\frac{h L_c}{k}\right) \left(\frac{k}{\rho c_p}\right) \frac{t}{L_c 2} = \left(\frac{h L_c}{k}\right) \left(\frac{\alpha t}{L_c 2}\right)$$

where

$Bi = \frac{h L_c}{k}$ is the Biot number

$\alpha = \frac{k}{\rho c_p}$ is the thermal diffusivity, m^2/s

$Fo = \frac{\alpha t}{L_c 2}$ is Fourier number

Equations (5.65) and (5.66) can now be written as

$$\frac{T - T_\infty}{T_i - T_\infty} = \exp(-Bi Fo) \tag{5.67}$$

$$T = T_\infty + (T_i - T_\infty) \exp(-Bi Fo) \tag{5.68}$$

Figure 5.6 compares the temperature-time profiles of the heated infinite plate predicted by Eq. (5.8) (exact solution) and Eq. (5.68) (approximate solution) for different Biot numbers. The temperatures predicted by the lump capacitance method at $Bi = 0.01$ exhibit very good agreement with the analytical results from Eq. (5.8) and the agreement improves with increasing time.

The instantaneous heat transfer rate on the basis of Eq. (5.66) is

$$\dot{Q}_i = \rho V c_p \frac{dT}{dt} = \rho V c_p \frac{d}{dt}\left[T_\infty + (T_i - T_\infty) \exp\left(-\frac{A_s h}{\rho V c_p} t\right)\right]$$

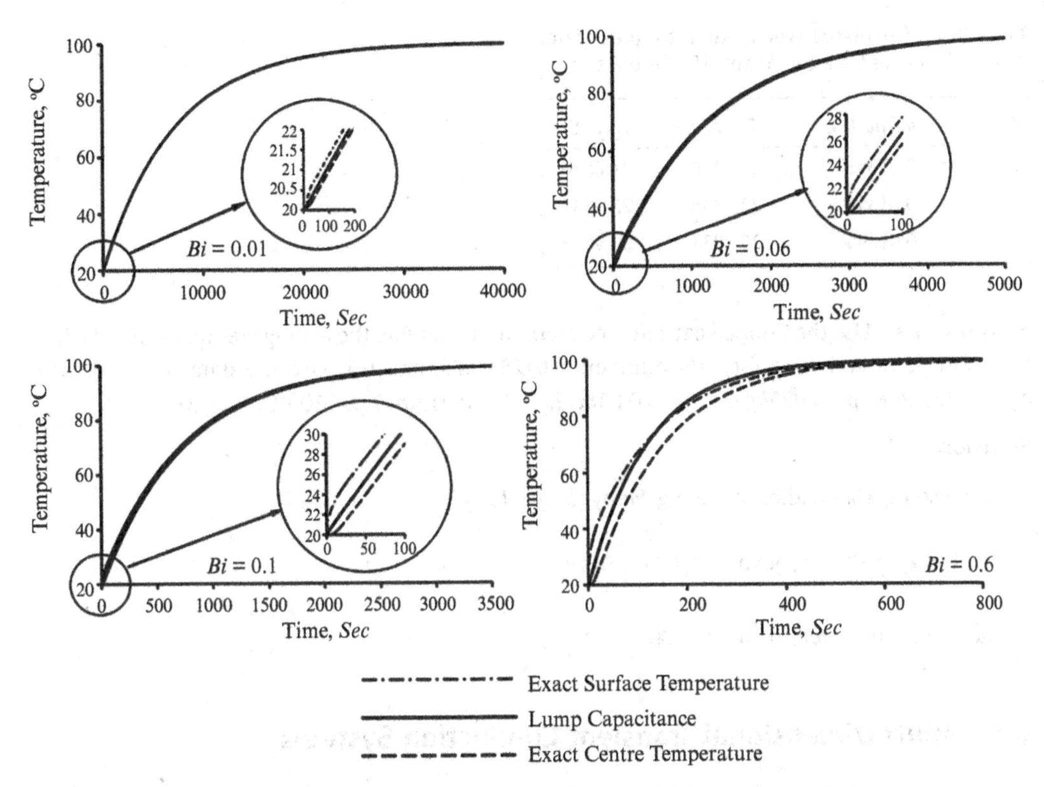

Figure 5.6 Effect of Biot number on the time response of a heated infinite plate under transient conduction conditions.

$$\dot{Q}_i = -A_s h (T_i - T_\infty) \exp\left(-\frac{A_s h}{\rho V c_p} t\right) \tag{5.69}$$

or

$$\dot{Q}_i = -A_s h (T_i - T_\infty) \exp(-BiFo) \tag{5.70}$$

The cumulative (total) heat Q stored in the plate from $t = 0$ to any instant of time t is then

$$Q = \int_0^t \dot{Q}_i dt = -A_s h (T_i - T_\infty) \int_0^t \exp\left(-\frac{A_s h}{\rho V c_p} t\right) dt$$

$$Q = \rho V c_p (T_i - T_\infty) \left[\exp\left(-\frac{A_s h}{\rho V c_p} t\right)\right]_0^t$$

$$Q = \rho V c_p (T_i - T_\infty) \left[\exp\left(-\frac{A_s h}{\rho V c_p} t\right) - 1\right] \tag{5.71}$$

or

$$Q = \rho V c_p (T_i - T_\infty) [\exp(-BiFo) - 1] \tag{5.72}$$

Table E5.4 Calculated average temperatures of the plate in Example 5.4 after 50 and 100 seconds.

Bi	$k / \rho c_p L_c 2$	T_{50s} °C	T_{100s} °C
0.015	0.01393	20.8316	21.6546
0.05	0.01393	22.7386	25.3834
0.1	0.01393	25.3834	30.4045

Example 5.4 Use the lumped capacitance method to calculate the average temperature of a large plate after 50s and 100s for Biot numbers 0.0125, 0.05, and 0.1. Use the data: $k = 12W/m.K$, $c_p = 477J/kg.K$, $\rho = 8025kg/m^3$, $2\delta = 0.03m$, $L_c = \delta = 0.015m$, $T_\infty = 100°C$, $T_i = 20°C$.

Solution

We can use Eq. (5.68) after replacing Fo by $\left(k / \rho c_p L_c 2\right)t$

$$T = T_\infty + (T_i - T_\infty)\exp\left[-Bi\left(k / \rho c_p L_c 2\right)t\right]$$

The results are given in Table E5.4.

5.3 Multi-Dimensional Transient Conduction Systems

One-dimensional transient conduction problems considered in previous sections dealt with unrealistic physical shapes such as infinite plates and very long cylinders in which heat flows axially along the x-axis in plates or radially along radius r in cylinders. In real physical systems, heat can additionally flow in the other direction of the applicable coordinate system, which renders mathematical analysis extremely complex. Simple two- and three-dimensional transient conduction problems can be solved by a technique known as the *product solution method* on the basis of superposition (or combination) of the solutions of one-dimensional conduction problems of different geometries assuming:

1) Heat transfer to or from the body is by convection, with convective coefficient h and the surrounding fluid is at constant temperature T_∞
2) Constant initial temperature of the body T_i
3) Constant thermophysical properties.

The product solution method is only applicable to simple geometries under special boundary conditions. The overwhelming majority of multi-dimensional problems are nowadays solved by means of computational techniques using specialized computer software.

5.3.1 Long Rectangular Bar

A long rectangular bar is formed by the intersection of two infinitely long plates (slabs) of thicknesses $2\delta_1$ and $2\delta_2$ at right-angles, as shown by the highlighted shape in Figure 5.7. The rectangular bar is two-dimensional and the heat conduction governing equation without heat generation from Eq. (2.14) is

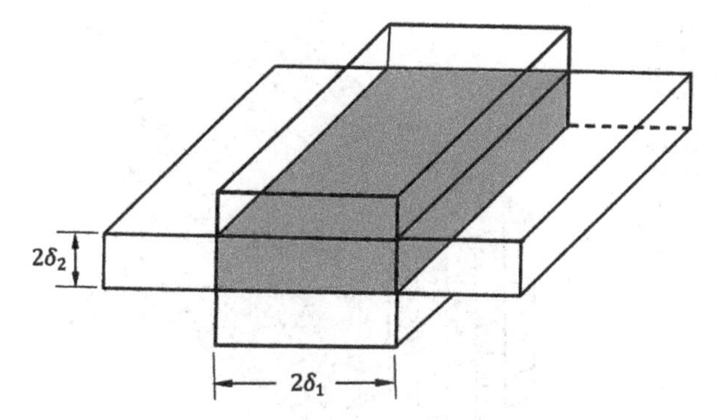

Figure 5.7 Formation of long rectangular bar from two infinite plates.

$$\frac{\partial^2 T}{\partial x^2}+\frac{\partial^2 T}{\partial y^2}=\frac{1}{\alpha}\frac{\partial T}{\partial t} \tag{5.73}$$

This equation is solved by the method of separation of variables and it is found that the product of the solutions of the two intersecting infinite plates satisfies Eq. (5.73)

$$\Theta_{LB}(x,y,t)=\Theta_{IP1}(x,t)\Theta_{IP2}(y,t) \tag{5.74}$$

where *LB*, *IP*1, *IP*2 denote long bar, infinite plate 1, and infinite plate 2, respectively.

The terms in Eq. (5.74) can be written as

$$\Theta_{IP1}(x,t)=\frac{T(x,t)-T_\infty}{T_i-T_\infty} \tag{5.75}$$

$$\Theta_{IP2}(y,t)=\frac{T(y,t)-T_\infty}{T_i-T_\infty} \tag{5.76}$$

$$\Theta_{LB}(x,y,t)=\frac{T(x,y,t)-T_\infty}{T_i-T_\infty} \tag{5.77}$$

or

$$T(x,y,t)=T_\infty+(T_i-T_\infty)\Theta_{LB}(x,y,t)$$

The dimensionless temperature distributions in Eqs (5.75) and (5.76) can be determined, as already discussed in the previous section, using Eq. (5.7) or, if available, the Heisler charts.

The fractional energy loss or gain in the two-dimensional bar, as shown by Langston (1982), can be written as

$$\left(\frac{Q}{Q_e}\right)_{LB}=\left(\frac{Q}{Q_e}\right)_{IP1}+\left(\frac{Q}{Q_e}\right)_{IP2}\left[1-\left(\frac{Q}{Q_e}\right)_{IP1}\right] \tag{5.78}$$

The fractional energy released or gained in the infinite plates *IP*1 and *IP*2 can be calculated from Eq. (5.12) or Eq. (5.13).

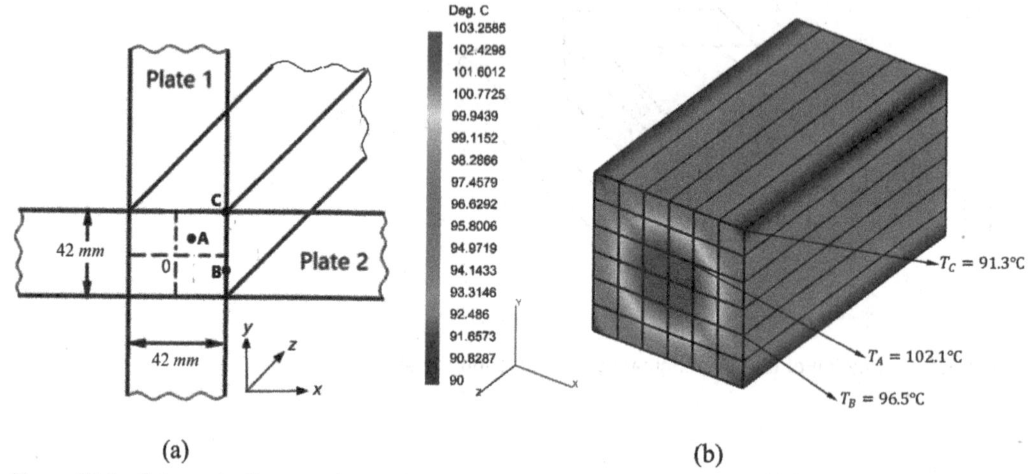

Figure E5.5 Schematic diagram of two-dimensional conduction in a long bar (a), and corresponding finite difference computer modelling results (b).

Example 5.5 A very long square metal bar of dimensions $40\,mm \times 40\,mm$ is heated to 120°C, then cooled by immersion in a coolant at 18°C with heat transfer coefficient $h = 600\,W/m^2.K$. The thermophysical properties of the bar can be taken as $\rho = 7870\,kg/m^3$, $c_p = 450\,J/kg.K$, and $k = 80\,W/m.K$. Determine the temperature at point A in Figure E5.5a after 15 seconds of cooling. The x, y coordinates of A relative to the centre 0 in metres are (0.007, 0.007).

Solution

The square bar is formed from the intersection of two plates of thickness 0.042 m at right-angles, as shown in Figure E5.5a. Heat conduction is two-dimensional in the $x - y$ =plane and invariable in the z -direction.

$$\Theta_{LB}(x,y,t) = \Theta_{IP1}(x,t)\Theta_{IP2}(y,t)$$

Solving Eq. (5.7) for the eigen values

μ_{1-6} : 0.381078 3.19081 6.308128 9.441438 12.5789 15.71798

yields
Plate 1: $x = 0.007m$, $\Theta_{IP1}(0,15) = 0.9086$
Plate 2: $y = 0.007$, $\Theta_{IP2}(0,15) = 0.9086$
For the long bar

$$\Theta_{LB}(0,0,15) = \frac{T(0.007,0.007,15) - 18}{120 - 18} = 0.9086 \times 0.9086 = 0.82557$$

$$\therefore T(0,0,15) = T_A = 102.2°C$$

5.3.2 Short Cylinder

A short finite cylinder is formed when an infinite cylinder intersects with an infinite plate at right-angles, as shown by the highlighted shape in Figure 5.8.

The governing equation of heat conduction without heat generation for this two-dimensional shape from Eq. (2.15) is

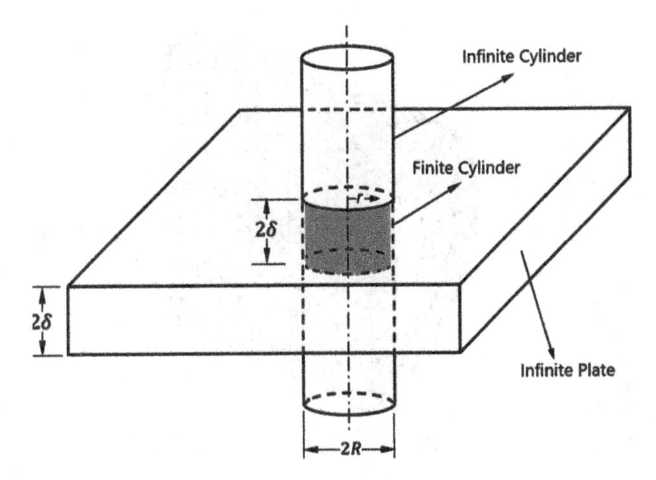

Figure 5.8 Formation of solid cylinder of radius R and height 2δ from infinite cylinder and plate.

$$\frac{1}{r}\frac{\partial}{\partial r}\left(r\frac{\partial T}{\partial r}\right)+\frac{\partial^2 T}{\partial z^2}=\frac{1}{\alpha}\frac{\partial T}{\partial t} \tag{5.79}$$

The solution to Eq. (5.79) can be written as

$$\Theta_{sc}(r,z,t)=\Theta_{IP}(z,t)\Theta_{IC}(r,t) \tag{5.80}$$

where SC, IP, IC denote short cylinder, infinite plate, and infinite cylinder, respectively, and

$$\Theta_{sc}(r,z,t)=\frac{T(r,z,t)-T_\infty}{T_i-T_\infty} \tag{5.81}$$

$$\Theta_{IP}(z,t)=\frac{T(z,t)-T_\infty}{T_i-T_\infty} \tag{5.82}$$

$$\Theta_{IC}(r,t)=\frac{T(r,t)-T_\infty}{T_i-T_\infty} \tag{5.83}$$

The dimensionless temperature ratios in Eqs (5.82) and (5.83) can be determined using Eqs (5.7) and (5.30) or, if available, the Heisler charts.

The fractional energy loss or gain in the two-dimensional cylinder can be written as

$$\left(\frac{Q}{Q_e}\right)_{SC}=\left(\frac{Q}{Q_e}\right)_{IP}+\left(\frac{Q}{Q_e}\right)_{IC}\left[1-\left(\frac{Q}{Q_e}\right)_{IP}\right] \tag{5.84}$$

The fractional energy released or gained in the infinite plates IP can be calculated from Eq. (5.12) or Eq. (5.13). The fractional energy for an infinite cylinder IC can be calculated from Eq. (5.43) or Eq. (5.44).

Example 5.6 A heated stainless-steel cylinder of radius $R = 4\,cm$ and length $2\delta = 6\,cm$ is dropped in a fluid bath at temperature $T_\infty = 25°\text{C}$. The cylinder is initially at temperature $T_i = 300°\text{C}$ and the heat transfer coefficient $h = 400\,W/m^2.K$. The thermophysical properties of the cylinder are

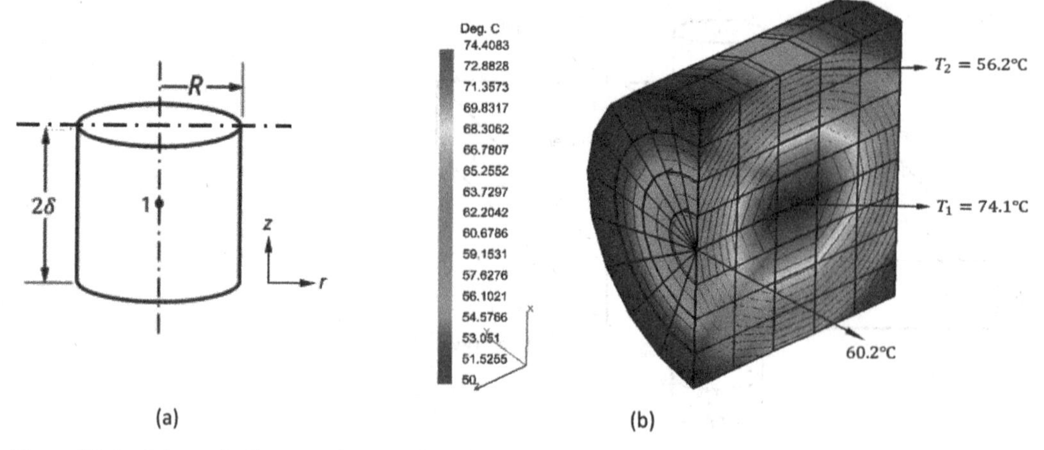

Figure E5.6 Schematic diagram of a two-dimensional conduction in a cylinder (a), and corresponding finite difference computer modelling results (b).

$\rho = 7800\,kg/m^3$, $c_p = 500\,J/Kg.K$, $k = 15.5\,W/m.K$. Determine the temperatures at point 1 in Figure E5.6a five minutes after the start of the cooling process.

Solution

$$\Theta_{sc}(r,z,t) = \Theta_{IP}(z,t)\Theta_{IC}(r,t)$$

Centre of the cylinder at point 1

Infinite plate:

Plate thickness is equal to the cylinder length $2\delta = 0.06m$
At point 1, $z = 0$, $Bi = 0.774, Fo = 1.3248$

μ_{1-6} : 0.780574 3.36751 6.403503 9.506016 12.62759 15.75707

From Eq. (5.7), $\Theta_{IC}(z,t) = \Theta_{IP}(0,\ 300) = 0.4903$

Cylinder:

At point 1, $r = 0, Bi = 1.0322, Fo = 0.74519$

μ_{1-6} : 1.268774 4.086642 7.160155 10.18018 13.40078 16.46862

From Eq. (5.30), $\Theta_{IP}(r,t) = \Theta_{IC}(0,300) = 0.36417$

$$\Theta_{sc}(r,z,t) = T_1 = \frac{T(0,0,300) - T_\infty}{T_i - T_\infty} = 0.4903 \times 0.36417 = 0.17855$$

$$\therefore T(0,0,300) = 25 + 0.17855 \times (300 - 25) = 74.1°C$$

5.3.3 Rectangular Parallelepiped

A parallelepiped, also known as solid block or rectangular solid, is a three-dimensional shape that is formed by the intersection of three infinite plates at right-angles to each other. The result of intersection of two infinite plates of thicknesses $2\delta_1$ and $2\delta_2$ (Figure 5.7) is a very long (infinite)

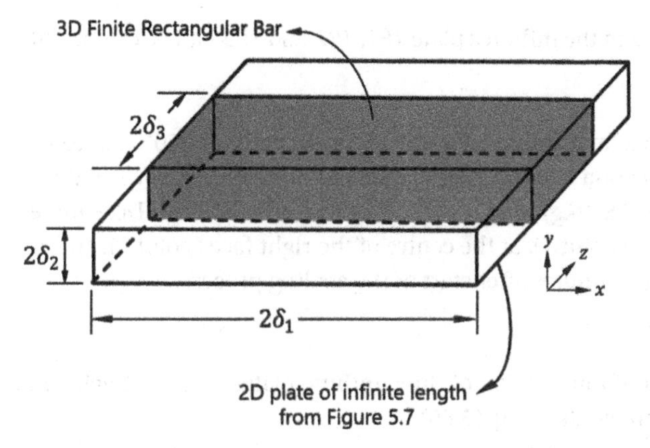

Figure 5.9 The formation of a parallelepiped from infinite plates.

rectangular bar. If a third plate of thickness $2\delta_3$ intersects with this rectangular bar at right-angles, a solid block of dimensions $2\delta_1 \times 2\delta_2 \times 2\delta_3$ is formed, as illustrated by the highlighted shape in Figure 5.9.

Ignoring the heat generation term in Eq. (5.14), the governing heat conduction equation for this shape is reduced to

$$\frac{\partial^2 T}{\partial x^2}+\frac{\partial^2 T}{\partial y^2}+\frac{\partial^2 T}{\partial z^2}=\frac{1}{\alpha}\frac{\partial T}{\partial t} \tag{5.85}$$

The solution to Eq. (5.85) can be written as

$$\Theta_{PP}(x,y,z,t)=\Theta_{IP1}(x,t)\Theta_{IP2}(y,t)\Theta_{IP3}(z,t) \tag{5.86}$$

where $PP, IP1, IP2, IP3$ denote parallelepiped, infinite plate 1, infinite plate 2, and infinite plate 3, respectively and

$$\Theta_{PP}(x,y,z,t)=\frac{T(x,y,z,t)-T_\infty}{T_i-T_\infty} \tag{5.87}$$

$$\Theta_{IP1}(x,t)=\frac{T(x,t)-T_\infty}{T_i-T_\infty} \tag{5.88}$$

$$\Theta_{IP2}(y,t)=\frac{T(y,t)-T_\infty}{T_i-T_\infty} \tag{5.89}$$

$$\Theta_{IP3}(z,t)=\frac{T(z,t)-T_\infty}{T_i-T_\infty} \tag{5.90}$$

The dimensionless temperature distributions in Eqs (5.88) to (5.90) can be determined using Eq. (5.7) or the Heisler charts.

The fractional energy loss or gain in the three-dimensional parallelepiped can be written as

$$\left(\frac{Q}{Q_e}\right)_{PP}=\left(\frac{Q}{Q_e}\right)_{IP1}+\left(\frac{Q}{Q_e}\right)_{IP2}\left[1-\left(\frac{Q}{Q_e}\right)_{IP1}\right]+\left(\frac{Q}{Q_e}\right)_{IP3}\left[1-\left(\frac{Q}{Q_e}\right)_{IP1}\right] \tag{5.91}$$

The fractional energy released or gained in the infinites plate $IP1$, $IP2$ and $IP3$ can be calculated from Eq. (5.12) or Eq. (5.13).

Example 5.7 The steel cube shown schematically in Figure E5.7a, initially at 450°C, is to be cooled in an oil bath at 120°C with convection heat transfer coefficient of 220 $W/m^2.K$. The thermophysical properties of the cube are: $\rho = 7850 kg/m^3, c_p = 0.5 kJ/kg.K, k = 44 W/m.K.$ Determine the temperatures at the centre of the cube (point 1), at the centre of the right face (point 2), and at the top-right corner of the cube (point 3) 3 minutes after start of the cooling process.

Solution

The cube can be considered as the result of intersection of three infinite plates, each of thickness 0.01 m. We use Eq. (5.7) to solve the components in Eq. (5.86).

$$\Theta_{PP}(x,y,z,t) = \Theta_{IP1}(x,t)\Theta_{IP2}(y,t)\Theta_{IP3}(z,t)$$

Point 1: Coordinates (0,0,0); $Bi = 0.25, Fo = 0.8071$

$$\mu_{1-6}: 0.4773 \quad 3.219 \quad 6.32265 \quad 9.4512 \quad 12.58625 \quad 15.72385$$

$$\Theta_{IP1}(0,180) = \Theta_{IP2}(0,180) = \Theta_{IP3}(0,180) = 0.8634$$

$$\Theta_{PP}(0.05,0,0,180) = \frac{T(0.05,0,0,180) - 120}{450 - 120} = 0.8634^3 = 0.6436$$

$$\therefore T(0,0,0,180) = T_1 = 120 + 0.6436 \times (450 - 120) = 332.3°\text{C}$$

Point 2: Coordinates (0.05,0,0)

$$\Theta_{IP1}(0.05,180) = 0.7669$$

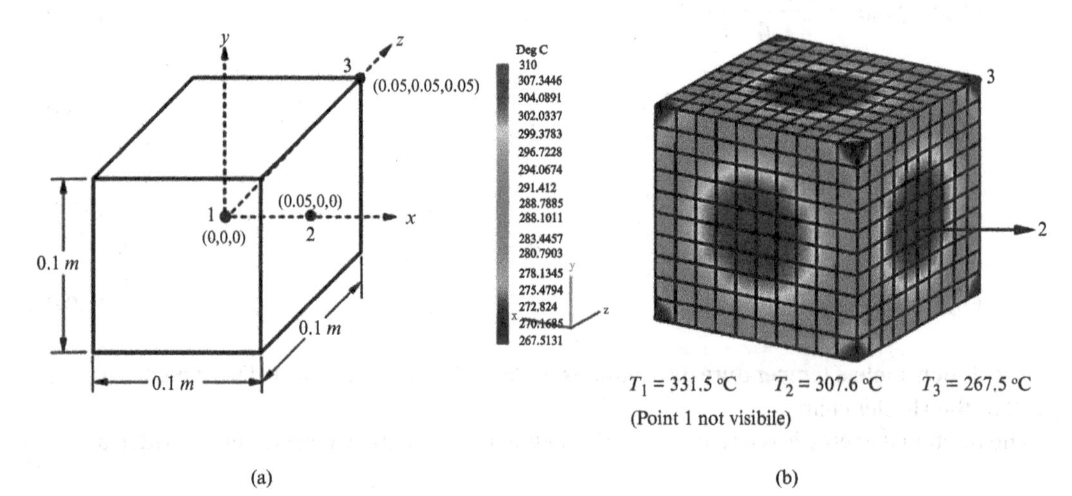

Figure E5.7 Schematic diagram of three-dimensional conduction in a cube (a) and corresponding finite difference computer modelling results (b).

$$\Theta_{IP2}(0,180) = \Theta_{IP3}(0,180) = 0.8634$$

$$\Theta_{pp}(0.05,0,0,180) = 0.7669 \times 0.8634^2 = 0.5717$$

$$\therefore T(0.05,0,0,180) = T_2 = 120 + 0.5717 \times (450 - 120) = 308.7°\text{C}$$

Point 3: coordinates (0.05,0.05,0.05)

$$\Theta_{IP1}(0.05,180) = \Theta_{IP2}(0.05,180) = \Theta_{IP3}(0.05,180) = 0.7669$$

$$\Theta_{PP}(0.05,0.05,0.05,180) = 0.7669^3 = 0.451$$

$$\therefore T(0.05,0.05,0.05,180) = T_3 = 120 + 0.4511 \times (450 - 120) = 268.8°\text{C}$$

The predicted temperatures by the computer model are shown in Figure E5.7b. The results are very close, confirming the accuracy of the product solution method.

5.4 Finite-Difference Method for Solving Transient Conduction Problems

As was the case with steady-state conduction, mathematical solutions of transient conduction problems are limited to very simple geometries, constant thermal properties, and linear boundary conditions. More complex problems are best solved using numerical methods, particularly the well-established finite-difference method. This method combined with digital computers provide a powerful and versatile tool for solving geometrically complex systems with variable or time- and/or location-dependent thermophysical properties, and nonlinear boundary conditions that could also be time- or location-dependent. The exact solutions to transient conduction in Examples 5.5–5.7 were duplicated by finite-difference computer modelling solutions as an illustration. In this section, two variations of the method will be presented in a concise form, namely, the explicit and the implicit methods.

5.4.1 Explicit Finite-Difference Method

The finite-difference method introduced previously for steady-state conduction can be applied to transient conduction by adding the element of time-dependence of temperature to the mathematical model and calculations are performed at time steps $\tau, \tau+1, \tau+2, \ldots$ for each node in the physical model until the temperature distribution is obtained at the required final state. In a two-dimensional conduction model, for example, at time interval $\tau+1$, the temperature $T_{m,n}^{\tau+1}$ of a node is determined in terms of the temperatures at the previous time step $T_{m,n}^{\tau}$. At the start of the process, using the prescribed initial temperature $T_{m,n}^{0}$ at time $t=0$, the new calculated temperature will be $T_{m,n}^{1}$. The process is repeated for successive time increments until a final solution is obtained. No iteration is required with the explicit method and the solution progresses with the time steps; hence, the origin of the term "forward-difference" method. The subscripts in the temperature terms denote the location of the nodes in the directions of the coordinates x and y.

5.4.1.1 One-Dimensional Transient Conduction

The governing equation of one-dimensional transient conduction without heat generation and with constant thermophysical properties can be deduced from Eq. (2.14) and written as

$$k\frac{\partial^2 T}{\partial x^2} = \rho c_p \frac{\partial T}{\partial t} \tag{5.92}$$

To solve this equation numerically by the finite-difference method, we consider the conduction scheme shown in Figure 5.10. The physical body is divided into nodes and elements, with each element having width Δx in the x-direction, height of $1m$ in the y-direction, and length $1m$ in the z-direction. Each node is surrounded by a control volume $\Delta V = dx.1.1$ (highlighted) with heat inflow rate $\dot{Q}_A$ and outflow rate of $\dot{Q}_B$ across surfaces A, B, respectively.

The energy conservation equation for the control volume is

$$\dot{Q}_A - \dot{Q}_B = \frac{dU}{dt}$$

where dU / dt is the rate of increase of the internal energy of the control volume.

For the time interval Δt from time step τ to time step $\tau + 1$, the energy equation becomes

$$\dot{Q}_A - \dot{Q}_B = \rho c_p \Delta V \frac{\left(T_m^{\tau+1} - T_m^{\tau}\right)}{\Delta t}$$

or, taking into account that $\Delta V = \Delta x \times 1 \times 1$

$$\dot{Q}_A - \dot{Q}_B = \rho c_p \Delta x \frac{\left(T_m^{\tau+1} - T_m^{\tau}\right)}{\Delta t} \tag{5.93}$$

Substituting for $\dot{Q}_A$ and $\dot{Q}_B$ at time step τ, we obtain

$$-kA_A \frac{T_m^{\tau} - T_{m-1}^{\tau}}{\Delta x} + kA_B \frac{T_{m+1}^{\tau} - T_m^{\tau}}{\Delta x} = \rho c_p \Delta x \frac{\left(T_m^{\tau+1} - T_m^{\tau}\right)}{\Delta t} \tag{5.94}$$

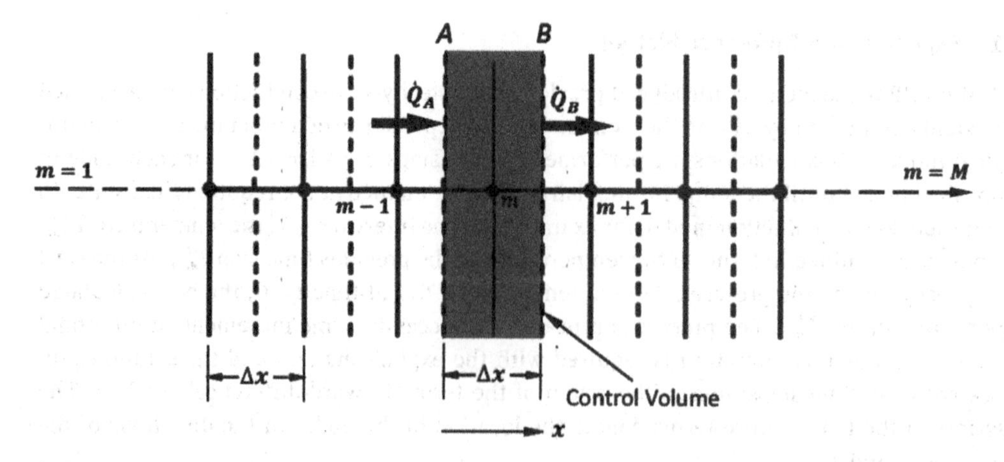

Figure 5.10 One-dimensional transient conduction scheme showing interior control volume.

$$A_A = A_B = 1 \times 1 = 1$$

Multiplying by Δx, dividing by k, and rearranging results in

$$T_{m-1}^{\tau} - T_m^{\tau} + T_{m+1}^{\tau} - T_m^{\tau} = \frac{\rho c_p}{k} \frac{\Delta x^2}{\Delta t}\left(T_m^{\tau+1} - T_m^{\tau}\right)$$

Using the definitions of thermal diffusivity $\alpha = k / \rho c_p$ and element Fourier number $Fo = \alpha \Delta t / \Delta x^2$, we obtain

$$T_{m-1}^{\tau} - T_m^{\tau} + T_{m+1}^{\tau} - T_m^{\tau} = \frac{1}{Fo}\left(T_m^{\tau+1} - T_m^{\tau}\right) \tag{5.95}$$

Rearranging, we finally obtain the equation for the interior node m in Figure 5.10

$$T_m^{\tau+1} = (1 - 2Fo)T_m^{\tau} + Fo\left(T_{m-1}^{\tau} + T_{m+1}^{\tau}\right) \tag{5.96}$$

For stable solution for the interior node, the coefficient of T_m^{τ} must be positive, i.e., $(1 - 2Fo) \geq 0$; hence, the criterion for stable solution is

$$Fo \leq \frac{1}{2} \tag{5.97}$$

Equation (5.97) sets the limit to the time step to $\Delta t \leq \Delta x^2 / 2\alpha$. The accuracy of the solution can be increased with the decrease of both time increment and element size. High accuracy will require more elements (nodes) and longer run time.

Equation (5.96) is called the explicit difference equation and can be used to calculate the temperatures at successive time steps starting from the specified initial temperature at $t = 0$.

We now consider the surface node m and the corresponding highlighted control volume $(\Delta V = \Delta x / 2)$, as shown in Figure 5.11.

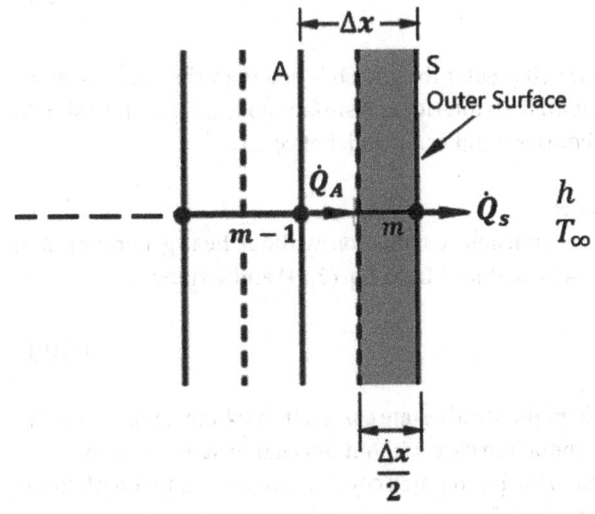

Figure 5.11 Surface control volume for the scheme in Figure 5.10.

The energy conservation equation for this element is

$$\dot{Q}_A - \dot{Q}_S = \rho c_p \frac{\Delta x}{2} \frac{\left(T_m^{\tau+1} - T_m^{\tau}\right)}{\Delta t}$$

$$-kA_A \frac{T_m^{\tau} - T_{m-1}^{\tau}}{\Delta x} - hA_S\left(T_m^{\tau} - T_{\infty}^{\tau}\right) = \rho c_p \frac{\Delta x}{2} \frac{\left(T_m^{\tau+1} - T_m^{\tau}\right)}{\Delta t} \tag{5.98}$$

$$A_A = A_S = 1$$

Multiplying by Δx, dividing by k, and rearranging results in

$$\left(T_{m-1}^{\tau} - T_m^{\tau}\right) + \frac{h\Delta x}{k}\left(T_{\infty}^{\tau} - T_m^{\tau}\right) = \frac{1}{2} \frac{\rho c_p}{k} \frac{\Delta x^2}{\Delta t}\left(T_m^{\tau+1} - T_m^{\tau}\right)$$

Using the definitions $Bi = h\Delta x / k$, $\alpha = k / \rho c_p$, and $Fo = \alpha \Delta t / \Delta x^2$, we obtain

$$\left(T_{m-1}^{\tau} - T_m^{\tau}\right) + Bi\left(T_{\infty}^{\tau} - T_m^{\tau}\right) = \frac{1}{2Fo}\left(T_m^{\tau+1} - T_m^{\tau}\right)$$

Bi and *Fo* are respectively the mesh Biot and mesh Fourier numbers in which the characteristic length is taken as Δx.

Rearranging, we finally obtain equation for the surface node m in Figure 5.11

$$T_m^{\tau+1} = T_m^{\tau}(1 - 2Fo - 2BiFo) + 2Fo\left(T_{m-1}^{\tau} + BiT_{\infty}^{\tau}\right) \tag{5.99}$$

For a stable solution for the surface node, the coefficient of T_m^{τ} must not be negative, i.e.,

$$1 - 2Fo - 2BiFo \geq 0$$

or

$$Fo \leq \frac{1}{2(1 + Bi)} \tag{5.100}$$

Since *Bi* is always positive and could be much greater than 1, *Fo* is less than the value given by Eq. (5.97). In a physical conduction model with both interior and surface nodes, Eq. (5.100) should be used to select the maximum allowable Fourier number *Fo* and, hence Δt.

5.4.1.2 Two-Dimensional Transient Conduction

The governing equation of two-dimensional transient conduction without heat generation and with constant thermophysical properties can be deduced from Eq. (2.14) and written as

$$k\frac{\partial^2 T}{\partial x^2} + k\frac{\partial^2 T}{\partial y^2} = \rho c_p \frac{\partial T}{\partial t} \tag{5.101}$$

The schematic diagram for the interior node in the steady-state conduction scheme, reproduced in Figure 5.12, can be used for the transient conduction case without internal heat generation.

The energy conservation equation applied to the highlighted control volume of unit length in the z direction $(\Delta V = \Delta x \Delta y)$ for the time interval Δt from time step τ to time step $\tau + 1$ is

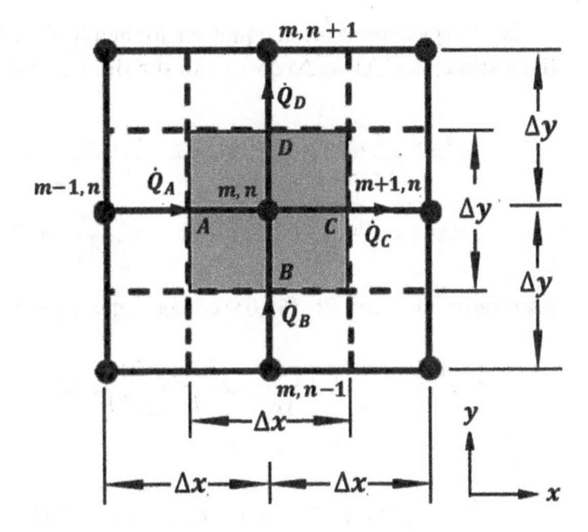

Figure 5.12 Schematic diagram of an interior element and associated node in transient conduction mode.

$$-k\frac{T_{m,n}^{\tau}-T_{m-1,n}^{\tau}}{\Delta x}\Delta y-k\frac{T_{m,n}^{\tau}-T_{m,n-1}^{\tau}}{\Delta y}\Delta x+k\frac{T_{m+1,n}^{\tau}-T_{m,n}^{\tau}}{\Delta x}\Delta y+k\frac{T_{m,n+1}^{\tau}-T_{m,n}^{\tau}}{\Delta y}\Delta x=\rho c_p\Delta x\Delta y\frac{\left(T_{m,n}^{\tau+1}-T_{m,n}^{\tau}\right)}{\Delta t}$$

Assuming the element is of equal sides $(\Delta x=\Delta y)$ and dividing by $\Delta x^2 k\Delta t$, we obtain

$$-\frac{T_{m,n}^{\tau}-T_{m-1,n}^{\tau}}{\Delta x^2}-\frac{T_{m,n}^{\tau}-T_{m,n-1}^{\tau}}{\Delta x^2}+\frac{T_{m+1,n}^{\tau}-T_{m,n}^{\tau}}{\Delta x^2}+\frac{T_{m,n+1}^{\tau}-T_{m,n}^{\tau}}{\Delta x^2}=\frac{\rho c_p}{k\Delta t}\left(T_{m,n}^{\tau+1}-T_{m,n}^{\tau}\right) \tag{5.102}$$

Rearranging finally yields

$$T_{m,n}^{l+1}=\frac{\alpha\Delta t}{\Delta x^2}\left(T_{m+1,n}^{\tau}+T_{m-1,n}^{\tau}+T_{m,n+1}^{\tau}+T_{m,n-1}^{\tau}\right)+\left(1-\frac{4\alpha\Delta t}{\Delta x^2}\right)T_{m,n}^{\tau} \quad \text{or}$$

$$T_{m,n}^{\tau+1}=Fo\left(T_{m+1,n}^{\tau}+T_{m-1,n}^{\tau}+T_{m,n+1}^{\tau}+T_{m,n-1}^{\tau}\right)+(1-4Fo)T_{m,n}^{\tau} \tag{5.103}$$

where $\alpha=k/\rho c_p, Fo=\alpha\Delta t/\Delta x^2$

The criterion for stable solution for the interior node in Figure 5.12 is

$$(1-4Fo)\geq 0,$$

or

$$Fo\leq\frac{1}{4} \tag{5.104}$$

Consider now a node on a plane surface with convection heat transfer to a fluid at temperature T_∞, as shown in Figure 5.13

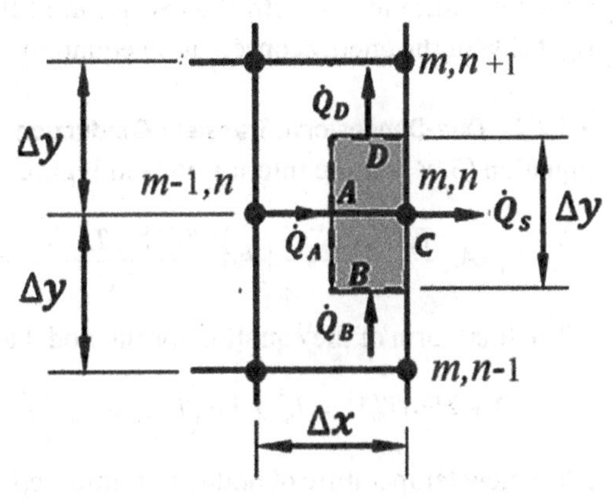

Figure 5.13 Exterior plane-surface node.

The energy conservation equation applied to the highlighted control volume with unit length in the z-direction $(\Delta V = \Delta x \Delta y / 2)$ for the time interval Δt from time step τ to time step $\tau + 1$ is

$$-k\frac{T_{m,n}^{\tau} - T_{m-1,n}^{\tau}}{\Delta x}\Delta y \Delta t - k\frac{T_{m,n}^{\tau} - T_{m,n-1}^{\tau}}{\Delta y}\frac{\Delta x}{2}\Delta t - h\left(T_{m,n}^{\tau} - T_{\infty}^{\tau}\right)$$
$$\Delta y \Delta t + k\frac{T_{m,n+1}^{\tau} - T_{m,n}^{\tau}}{\Delta y}\frac{\Delta x}{2}\Delta t = \rho c_p \frac{\Delta x}{2}\Delta y\left(T_{m,n}^{\tau+1} - T_{m,n}^{\tau}\right) \tag{5.105}$$

Assuming $\Delta x = \Delta y$, Eq. (5.105) can be rearranged yielding

$$T_{m,n}^{\tau+1} = T_{m,n}^{\tau}\left(1 - 4\frac{\alpha \Delta t}{\Delta x^2} - 2\frac{h\Delta x}{k}\frac{\alpha \Delta t}{\Delta x^2}\right) + \left(2T_{m-1,n}^{\tau} + T_{m,n-1}^{\tau} + T_{m,n+1}^{\tau} + 2\frac{h\Delta x}{k}T_{\infty}^{\tau}\right)\frac{\alpha \Delta t}{\Delta x^2}$$

or

$$T_{m,n}^{\tau+1} = \left(2T_{m-1,n}^{\tau} + T_{m,n-1}^{\tau} + T_{m,n+1}^{\tau} + 2BiT_{\infty}^{\tau}\right)Fo + T_{m,n}^{\tau}\left(1 - 4Fo - 2BiFo\right) \tag{5.106}$$

or

$$T_{m,n}^{\tau+1} = T_{m,n}^{\tau} + \left(2T_{m-1,n}^{\tau} + T_{m,n-1}^{\tau} + T_{m,n+1}^{\tau} - 2BiT_{m,n}^{\tau} + 2BiT_{\infty}^{\tau} - 4T_{m,n}^{\tau}\right)Fo \tag{5.107}$$

The criterion for stable solution for the surface node in Figure 5.13 is

$$1 - 4Fo - 2BiFo \geq 0$$

$$Fo \leq \frac{1}{4 + 2Bi} \tag{5.108}$$

5.4.2 Implicit Finite-Difference Method

The explicit finite-difference method is easy to implement and the solution of transient conduction problems are obtained in a straightforward fashion without the need for iteration. However, the need to meet certain stability criteria in selecting the time interval, and by association, element size, has led to the development of the implicit or backward-difference method for use in the solution of transient problems requiring fine spatial mesh size without the need to simultaneously decrease the time interval. The method is effected by replacing the time step τ with step $\tau + 1$ in all temperature terms, with the exception of the term in the internal energy component on the right side of the energy conservation equation.

5.4.2.1 One-Dimensional Transient Conduction

Equation (5.94) for the interior node in Figure 5.10 is rewritten as

$$-kA_A\frac{T_m^{\tau+1} - T_{m-1}^{\tau+1}}{\Delta x} + kA_B\frac{T_{m+1}^{\tau+1} - T_m^{\tau+1}}{\Delta x} = \rho c_p \frac{\Delta x}{\Delta t}\left(T_m^{\tau+1} - T_m^{\tau}\right) \tag{5.109}$$

The final form of the equation for the nodal temperature is

$$(1 + 2Fo)T_m^{\tau+1} = T_m^{\tau} + Fo\left[\left(T_{m+1}^{\tau+1} + T_{m-1}^{\tau+1}\right)\right] \tag{5.110}$$

The new temperature of node m at time step $\tau + 1$ now depends on the new unknown temperatures of the adjacent nodes, and the nodal equations must be solved simultaneously. This can be done, as was the case with the steady-state conduction, using matrix inversion method.

The equation for temperature of a plane-surface node can be obtained by rewriting Eq. (5.98) as

$$-kA_A\frac{T_m^{\tau+1}-T_{m-1}^{\tau+1}}{\Delta x}-hA_S\left(T_m^{\tau+1}-T_\infty^{\tau+1}\right)=\rho c_p\frac{\Delta x}{2\Delta t}\left(T_m^{\tau+1}-T_m^{\tau}\right)$$

Rearranging

$$(1+2Fo+2BiFo)T_m^{\tau+1}=2Fo\left(T_{m-1}^{\tau+1}+BiT_\infty^{\tau+1}+\right)+T_m^{\tau} \tag{5.111}$$

5.4.2.2 Two-Dimensional Transient Conduction

For an interior node, Eq. (5.102) is rewritten as

$$-\frac{T_{m,n}^{\tau+1}-T_{m-1,n}^{\tau+1}}{\Delta x^2}-\frac{T_{m,n}^{\tau+1}-T_{m,n-1}^{\tau+1}}{\Delta x^2}+\frac{T_{m+1,n}^{\tau+1}-T_{m,n}^{\tau+1}}{\Delta x^2}+\frac{T_{m,n+1}^{\tau+1}-T_{m,n}^{\tau+1}}{\Delta x^2} \tag{5.112}$$
$$=\frac{\rho c_p}{k\Delta t}\left(T_{m,n}^{\tau+1}-T_{m,n}^{\tau}\right)$$

Rearranging

$$(1+4Fo)T_{m,n}^{\tau+1}=Fo\left(T_{m+1,n}^{\tau+1}+T_{m-1,n}^{\tau+1}+T_{m,n+1}^{\tau+1}+T_{m,n-1}^{\tau+1}\right)+T_{m,n}^{\tau}=0 \tag{5.113}$$

For a plane-surface node and square meshes (Figure 5.13), we can rewrite Eq. (5.105) after rearranging as

$$T_{m-1,n}^{\tau+1}-T_{m,n}^{\tau+1}-0.5T_{m,n}^{\tau+1}+0.5T_{m,n-1}^{\tau+1}-Bi\left(T_{m,n}^{\tau+1}-T_\infty^{\tau+1}\right)+0.5T_{m,n+1}^{\tau+1}-0.5T_{m,n}^{\tau+1}=\frac{1}{2Fo}\left(T_{m,n}^{\tau+1}-T_{m,n}^{\tau}\right)$$

or, finally,

$$(1+4Fo+2BiFo)T_{m,n}^{\tau+1}-Fo\left(2T_{m-1,n}^{\tau+1}+T_{m,n-1}^{\tau+1}+T_{m,n+1}^{\tau+1}+2BiT_\infty^{\tau+1}\right)-T_{m,n}^{\tau}=0 \tag{5.114}$$

Equations (5.111), (5.113), and (5.114) show that the coefficient of $T_{m,n}^{i}$ for each of the transient conduction cases discussed in this section can never be negative and, therefore, solution instability is not an issue with the implicit formulation. This means that the time interval and mesh size can be selected independently from each other.

5.4.3 Finite Difference Formulation in Cylindrical Coordinates

The formulation for the two-dimensional conduction model of one-quarter of a disc of thickness δ or solid cylinder of radius R and length L is shown in Figure 5.14. The unknown temperatures $T(r,\varphi)$ are at the nodes with coordinates $r(0\rightarrow R)$ and $\varphi(0\rightarrow\pi/2)$. The locations of the nodes are determined by the index i in the radial direction and index j in the angular direction.

The energy equation for the highlighted control volume ΔV with energy storage and no internal energy generation for the time step from τ to $\tau+1$ is

$$\dot{Q}_A+\dot{Q}_B-\dot{Q}_C-\dot{Q}_D=\rho c_p\Delta V\frac{\left(T_{i,j}^{\tau+1}-T_{i,j}^{\tau}\right)}{\Delta t} \tag{5.115}$$

The volume ΔV of the highlighted element of the cylinder per unit length is

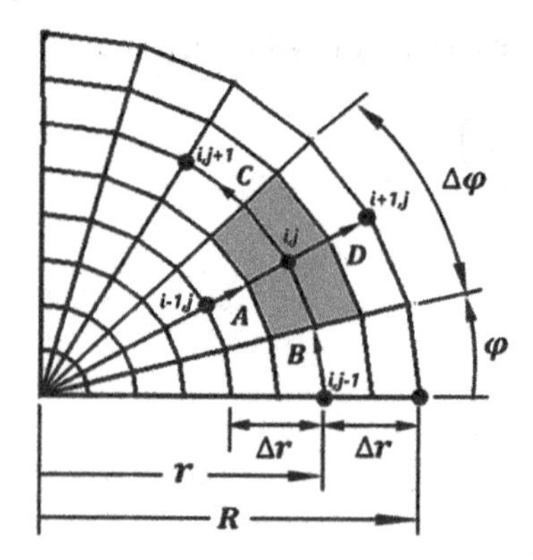

Figure 5.14 Two-dimensional finite-difference scheme for a solid cylinder.

$$\Delta V=\left[\left(r+\frac{\Delta r}{2}\right)^2-\left(r-\frac{\Delta r}{2}\right)^2\right]\frac{\Delta\varphi}{2}=r\Delta\varphi\Delta r \tag{5.116}$$

Assuming the temperature gradients between the central node i,j and adjacent nodes $i-1,j$, $i,j-1$, $i+,j$, and $i,j+1$, respectively, are linear, we can write Fourier's law for each surface of the highlighted element as follows:

The heat conduction across surface A is

$$\dot{Q}_A=-k\left[\left(r-\frac{\Delta r}{2}\right)\Delta\varphi\right]\frac{T_{i,j}-T_{i-1,j}}{\Delta r}$$

The heat conduction across surface B is

$$\dot{Q}_B=-k(\Delta r)\frac{T_{i,j}-T_{i,j-1}}{r\Delta\varphi}$$

The heat conduction across surface C is

$$\dot{Q}_C=-k(\Delta r)\frac{T_{i,j+1}-T_{i,j}}{r\Delta\varphi}$$

The heat conduction across surface D is

$$\dot{Q}_D=-k\left[\left(r+\frac{\Delta r}{2}\right)\Delta\varphi\right]\frac{T_{i+1,j}-T_{i,j}}{\Delta r}$$

Substituting the values of $\dot{Q}_A$, $\dot{Q}_B$, $\dot{Q}_C$, and $\dot{Q}_D$ into Eq. (5.115) and rearranging we obtain the nodal temperature equation in the explicit formulation

$$\begin{aligned}&-\left[2r(\Delta\varphi)^2+\frac{2}{r}(\Delta r)^2\right]T_{i,j}^{\tau}+\left[\left(r-\frac{\Delta r}{2}\right)(\Delta\varphi)^2\right]T_{i-1,j}^{\tau}+\frac{(\Delta r)^2}{r}T_{i,j-1}^{\tau}\\&+\left[\left(r+\frac{\Delta r}{2}\right)(\Delta\varphi)^2\right]T_{i+1,j}^{\tau}+\frac{(\Delta r)^2}{r}T_{i,j+1}^{\tau}\\&=\frac{1}{\alpha}\frac{r\Delta\varphi\Delta r}{\Delta t}\left(T_{i,j}^{\tau+1}-T_{i,j}^{\tau}\right)\end{aligned} \tag{5.117}$$

To obtain the nodal temperature equation in the implicit formulation, all superscripts on the left side of Eq. (5.117) should be replaced with the time step $\tau+1$.

The nodal equations for transient two-dimensional conduction systems in explicit and implicit formulations are listed in Tables 5.3 and 5.4. To make the equations more generalized, internal heat generation and the following boundary conditions are applied to the external nodes:

- Convection
- Constant heat flux
- Insulation
- Prescribed temperature.

Table 5.3 Explicit nodal equations for transient one- and two-dimensional conduction for selected node positions and boundary conditions $\Delta x = \Delta y$.

One-dimensional transient conduction	
Element model	**Nodal temperature equation**
(1) Interior node	$T_m^{\tau+1} = Fo\left(T_{m-1}^{\tau} + T_{m-1}^{\tau} + \frac{\Delta x^2}{k}\dot{q}\right) + (1-2Fo)T_m^{\tau}$ $Fo = \alpha\Delta t / \Delta x^2$ $\dot{q}$ – rate of energy generation, W/m^3 Stability criterion: $Fo \leq 1/2$
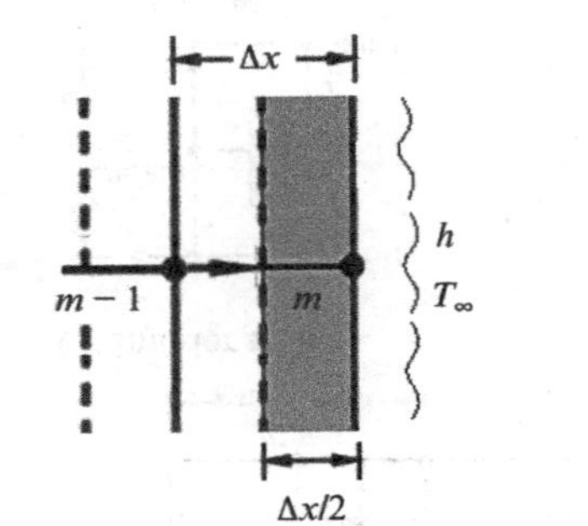 (2) Boundary node on plane surface with convection	$T_m^{\tau+1} = Fo\left(2T_{m-1}^{\tau} + 2BiT_{\infty}^{\tau} + \frac{\Delta x^2}{k}\dot{q}\right)(1-2BiFo-2Fo)T_m^{\tau}$ $Bi = h\Delta x / k$ Stability criterion: $Fo \leq 1/2(1+Bi)$

(Continued)

Table 5.3 (Continued)

Two-dimensional transient conduction	
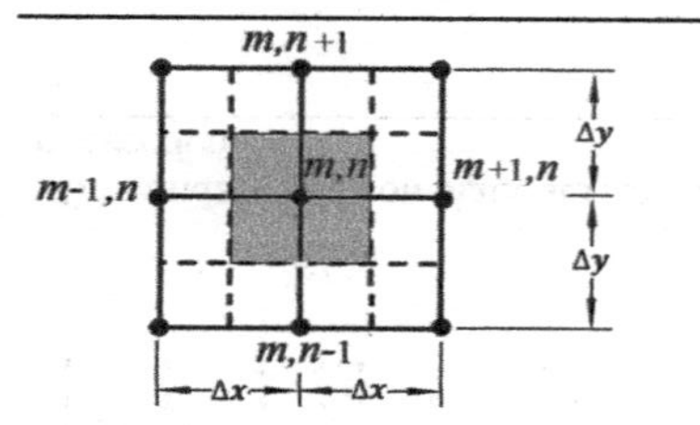 (3) Interior node	$T^{i+1}_{m,n} = Fo\left(T^{\tau}_{m+1,n} + T^{\tau}_{m-1,n} + T^{\tau}_{m,n+1} + T^{\tau}_{m,n-1} + \frac{\Delta x^2}{k}\dot{q}\right) + (1-4Fo)T^{\tau}_{m,n}$ Stability criterion: $Fo \le 1/4$
(4) Boundary node on plane surface with convection	$T^{\tau+1}_{m,n} = Fo\left(2T^{\tau}_{m-1,n} + T^{\tau}_{m,n-1} + T^{\tau}_{m,n+1} + 2BiT^{\tau}_{\infty} + \frac{\Delta x^2}{k}\dot{q}\right) + (1-4Fo-2BiFo)T^{\tau}_{m,n}$ Stability criterion: $Fo \le 1/2(2+Bi)$
(5) Boundary node on plane surface with heat flux	$T^{\tau+1}_{m,n} = Fo\left(2T^{\tau}_{m-1,n} + T^{\tau}_{m,n-1} + T^{\tau}_{m,n+1} + \frac{\Delta x^2}{k}\dot{q} - \frac{2\Delta x}{k}q\right) + (1-4Fo)T^{\tau}_{m,n}$ Stability criterion: $Fo \le 1/4$ q – Specified heat flux

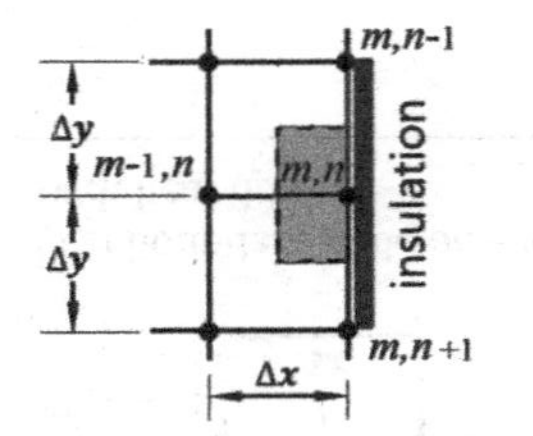

(6) Boundary node on insulated plane surface

$$T_{m,n}^{\tau+1} = Fo\left(2T_{m-1,n}^{\tau} + T_{m,n-1}^{\tau} + T_{m,n+1}^{\tau} + \frac{\Delta x^2}{k}\dot{q}\right) + (1-4Fo)T_{m,n}^{\tau}$$

Stability criterion: $Fo \leq 1/4$

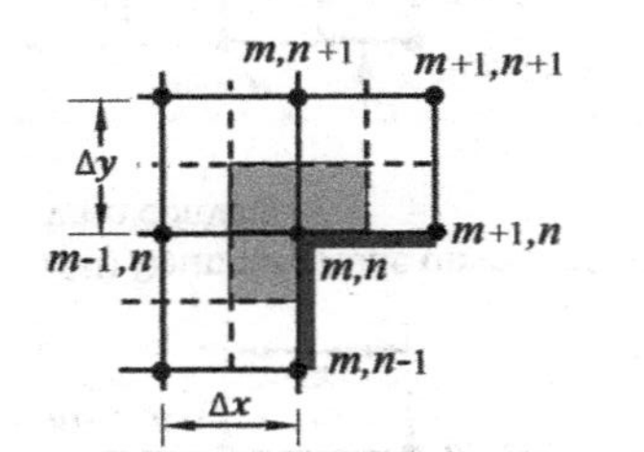

(7) Boundary node on insulated interior corner

$$T_{m,n}^{\tau+1} = Fo\left[\frac{2}{3}\left(2T_{m-1,n}^{\tau} + T_{m+1,n}^{\tau} + T_{m,n-1}^{\tau} + T_{m,n+1}^{\tau}\right) + \frac{\Delta x^2}{k}\dot{q}\right] + (1-4Fo)T_{m,n}^{\tau}$$

Stability criterion: $Fo \leq 1/4$

(8) Boundary node on interior corner with heat flux

$$T_{m,n}^{\tau+1} = Fo\left[\frac{2}{3}\left(2T_{m-1,n}^{\tau} + T_{m+1,n}^{\tau} + T_{m,n-1}^{\tau} + T_{m,n+1}^{\tau}\right) + \frac{\Delta x^2}{k}\dot{q} - \frac{4}{3}\frac{\Delta x}{k}q\right] + (1-4Fo)T_{m,n}^{\tau}$$

Stability criterion: $Fo \leq 1/4$

(Continued)

Table 5.3 (Continued)

<table>
<tr>
<td>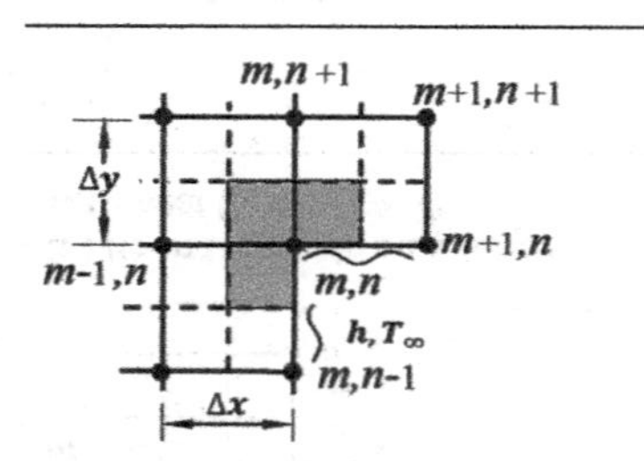

(9) Boundary node on interior corner with convection</td>
<td>$$T_{m,n}^{\tau+1} = Fo\left[\frac{2}{3}\left(2T_{m-1,n}^{\tau} + T_{m+1,n}^{\tau} + T_{m,n-1}^{\tau} + 2T_{m,n+1}^{\tau} + 2BiT_{\infty}^{\tau}\right) + \frac{\Delta x^2}{k}\dot{q}\right] + T_{m,n}^{\tau}\left(1 - 4Fo - \frac{4}{3}BiFo\right)$$
Stability criterion: $Fo \leq 3/4(3 + Bi)$</td>
</tr>
<tr>
<td>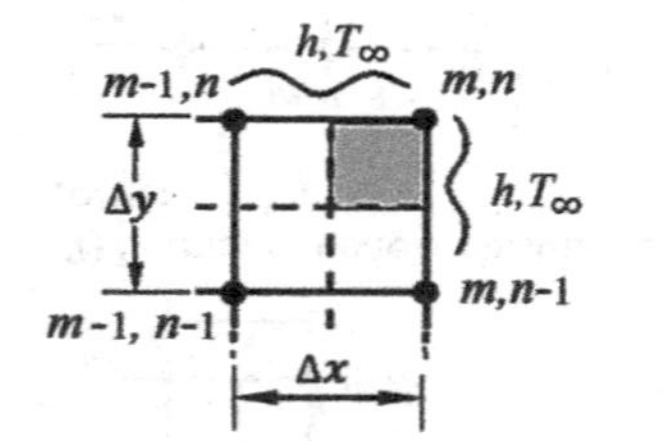

(10) Boundary node on exterior corner with convection</td>
<td>$$T_{m,n}^{\tau+1} = Fo\left(2T_{m-1,n}^{\tau} + 2T_{m,n-1}^{\tau} + 4BiT_{\infty}^{\tau} + \frac{\Delta x^2}{k}\dot{q}\right) + (1 - 4Fo - 4BiFo)T_{m,n}^{\tau}$$
Stability criterion: $Fo \leq 1/4(1 + Bi)$</td>
</tr>
<tr>
<td>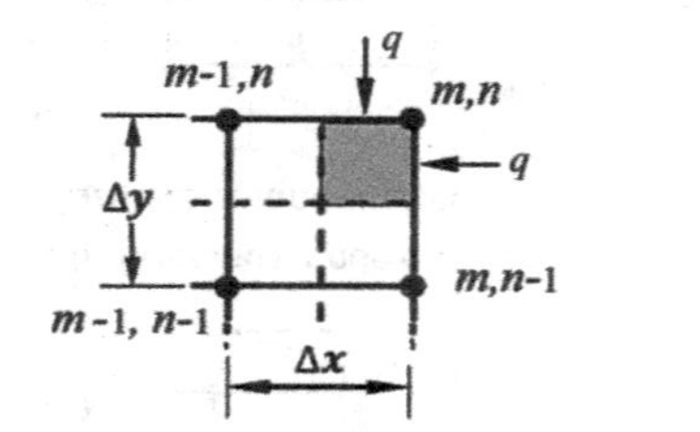

(11) Boundary node on exterior corner with heat flux</td>
<td>$$T_{m,n}^{\tau+1} = T_{m,n}^{\tau}(1 - 4Fo) + Fo\left(2T_{m-1,n}^{\tau} + 2T_{m,n-1}^{\tau} + \frac{\Delta x^2}{k}\dot{q} - \frac{4\Delta x}{k}q\right)$$
Stability criterion: $Fo \leq 1/4$</td>
</tr>
</table>

Table 5.4 Implicit nodal equations for transient one- and two-dimensional conduction for selected node positions and boundary conditions ($\Delta x = \Delta y$).

One-dimensional transient conduction	
Element model	**Nodal temperature equation**
$m-1$, m, $m+1$, Δx (1) Interior node	$(1+2Fo)T_m^{\tau+1} = T_m^{\tau} + Fo\left(T_{m-1}^{\tau+1} + T_{m+1}^{\tau+1} + \frac{\Delta x^2}{k}\dot{q}\right)$
Δx, $m-1$, m, h, T_∞, $\Delta x/2$ (2) Boundary node on plane surface with convection	$(1+2Fo+2BiFo)T_m^{\tau+1} = 2Fo\left(T_{m-1}^{\tau+1} + BiT_\infty^{\tau+1} + \frac{\Delta x^2}{k}\dot{q}\right) + T_m^{\tau}$

(Continued)

Table 5.4 (Continued)

Two-dimensional transient conduction	
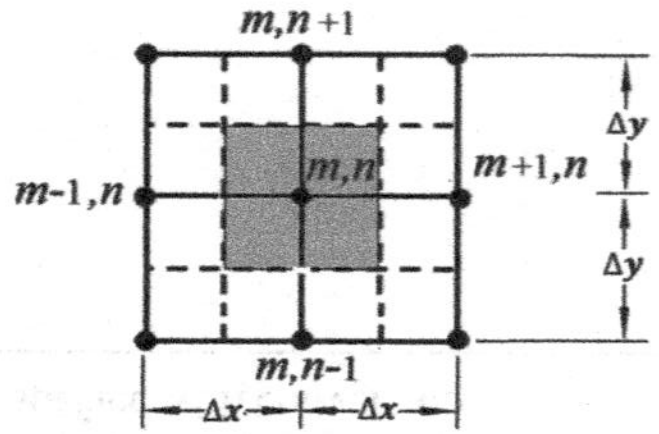 $m,n+1$; $m-1,n$; m,n; $m+1,n$; $m,n-1$; Δy; Δy; Δx; Δx (3) Interior node	$\left(1+4Fo\right)T_{m,n}^{\tau+1}=Fo\left(T_{m-1,n}^{\tau+1}+T_{m,n-1}^{\tau+1}+T_{m+1,n}^{\tau+1}+T_{m,n+1}^{\tau+1}+\frac{\Delta x^2}{k}\dot{q}\right)+T_{m,n}^{\tau}$
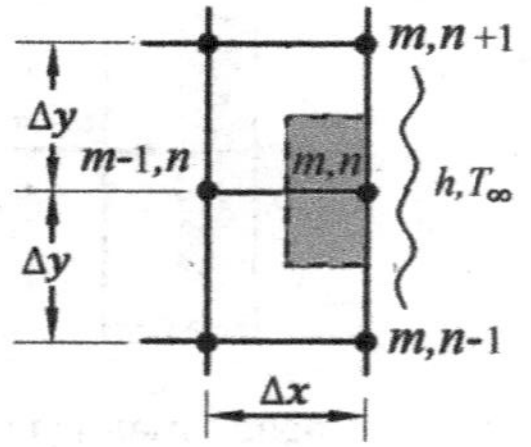 $m,n+1$; Δy; $m-1,n$; m,n; h,T_{∞}; Δy; $m,n-1$; Δx (4) Boundary node on plane surface with convection	$\left(1+4Fo+2BiFo\right)T_{m,n}^{\tau+1}=Fo\left(2T_{m-1,n}^{\tau+1}+T_{m,n-1}^{\tau+1}+T_{m,n+1}^{\tau+1}+2BiT_{\infty}^{\tau+1}+\frac{\Delta x^2}{k}\dot{q}\right)+T_{m,n}^{\tau}$
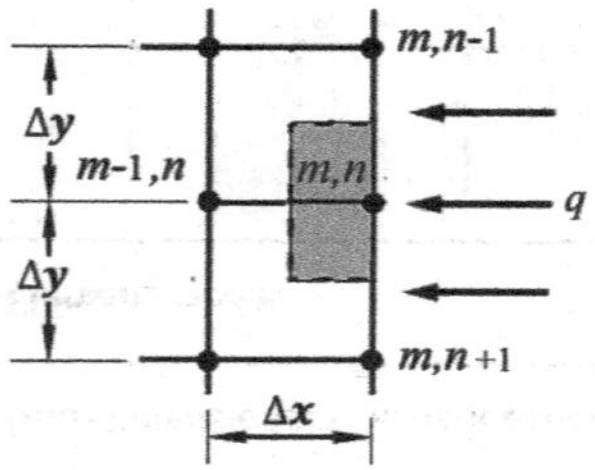 $m,n-1$; Δy; $m-1,n$; m,n; q; Δy; $m,n+1$; Δx (5) Boundary node on plane surface with heat flux	$\left(1+4Fo\right)T_{m,n}^{\tau+1}=Fo\left(2T_{m-1,n}^{\tau+1}+T_{m,n-1}^{\tau+1}+T_{m,n+1}^{\tau+1}+\frac{\Delta x^2}{k}\dot{q}-\frac{2\Delta x}{k}q\right)+T_{m,n}^{\tau}$

(6) Boundary node on insulated plane surface

$$(1+4Fo)T_{m,n}^{\tau+1} = Fo\left(2T_{m-1,n}^{\tau+1} + T_{m,n-1}^{\tau+1} + T_{m,n+1}^{\tau+1} + \frac{\Delta x^2}{k}\dot{q}\right) + T_{m,n}^{\tau}$$

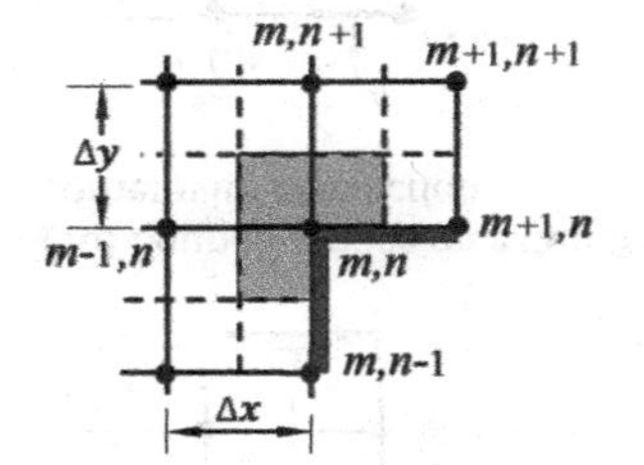

(7) Boundary node on insulated interior corner

$$(1+4Fo)T_{m,n}^{\tau+1} = Fo\left[\frac{2}{3}\left(2T_{m-1,n}^{\tau+1} + T_{m+1,n}^{\tau+1} + T_{m,n-1}^{\tau+1} + 2T_{m,n+1}^{\tau+1}\right) + \frac{\Delta x^2}{k}\dot{q}\right] + T_{m,n}^{\tau}$$

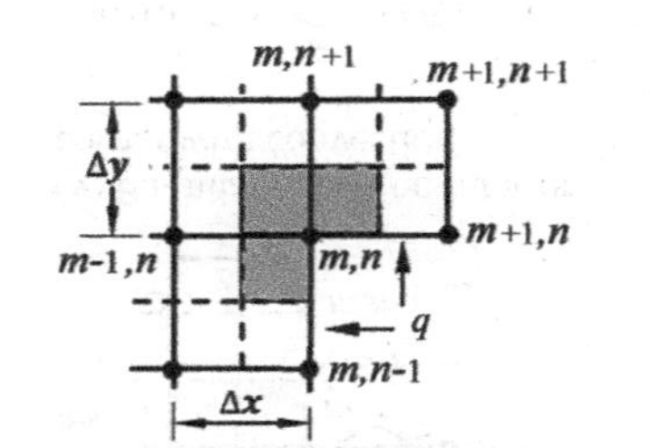

(8) Boundary node on interior corner with heat flux

$$(1+4Fo)T_{m,n}^{\tau+1} = Fo\left[\frac{2}{3}\left(2T_{m-1,n}^{\tau+1} + T_{m+1,n}^{\tau+1} + T_{m,n-1}^{\tau+1} + 2T_{m,n+1}^{\tau+1}\right) + \frac{\Delta x^2}{k}\dot{q} - \frac{4}{3}\frac{\Delta x}{k}q\right] + T_{m,n}^{\tau}$$

(Continued)

Table 5.4 (Continued)

Node	Equation
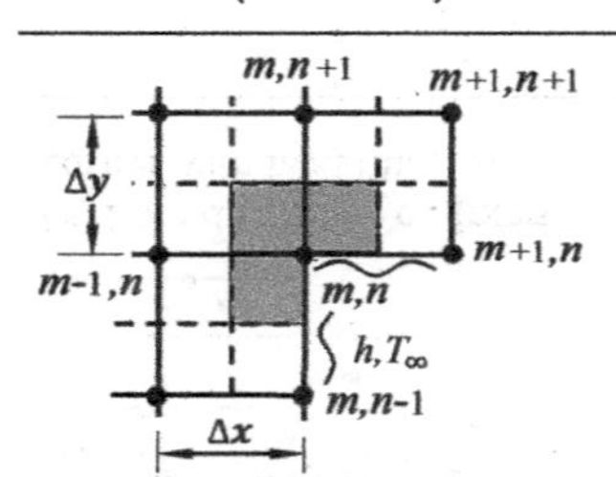 (9) Boundary node on interior corner with convection	$\left[1+4Fo\left(1+\frac{Bi}{3}\right)\right]T_{m,n}^{\tau+1}=Fo\left[\frac{2}{3}\left(2T_{m-1,n}^{\tau+1}+T_{m+1,n}^{\tau+1}+T_{m,n-1}^{\tau+1}+2T_{m,n+1}^{\tau+1}+2BiT_{\infty}^{\tau+1}\right)+\frac{\Delta x^2}{k}\dot{q}\right]+T_{m,n}^{\tau}$
(10) Boundary node on exterior corner with convection	$\left[1+4Fo(1+Bi)\right]T_{m,n}^{\tau+1}=2Fo\left(T_{m-1,n}^{\tau+1}+T_{m,n-1}^{\tau+1}+2BiT_{\infty}^{\tau+1}+\frac{\Delta x^2}{k}\dot{q}\right)+T_{m,n}^{\tau}$
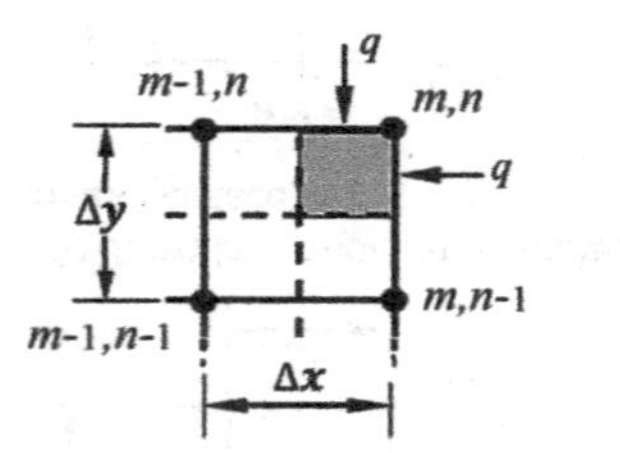 (11) Boundary node on exterior corner with heat flux	$\left[1+4Fo\right]T_{m,n}^{\tau+1}=2Fo\left(T_{m-1,n}^{\tau+1}+T_{m,n-1}^{\tau+1}+\frac{\Delta x^2}{k}\dot{q}-\frac{4\Delta x}{k}q\right)+T_{m,n}^{\tau}$

Example 5.8 A very long square metal bar of dimensions $42mm \times 42mm$ is heated to 120°C, then cooled by immersing in a coolant at 18°C with heat transfer coefficient $h = 600\,W/m^2.K$. The thermophysical properties of the bar can be taken as $\rho = 7870\,kg/m^3, c_p = 450J/kg.K$, and $k = 80\,W/m.K$. Determine the temperatures at points A, B, and C after 15 seconds from start of cooling using the forward-difference (explicit) method.

Solution

Figure E5.8 shows the meshing in the $x-y$ plane $(\Delta x = \Delta y = 14mm)$ for the square bar. The mesh size is selected so that points A, B, and C coincide with the node locations $T_{3,3}$, $T_{4,2}$, and $T_{4,4}$, respectively.

$$Bi = \frac{h\Delta x}{k} = \frac{600 \times 0.014}{80} = 0.105$$

If we were to write the full suite of nodal temperatures for the entire mesh, it will look as follows:

Interior nodes: Nodal equations from item (3) in Table 5.3 without internal heat generation

$$T_{2,3}^{\tau+1} = Fo\left(T_{3,3}^{\tau} + T_{1,3}^{\tau} + T_{2,4}^{\tau} + T_{2,2}^{\tau}\right) + (1-4Fo)T_{2,3}^{\tau}$$

$$T_{3,3}^{\tau+1} = Fo\left(T_{4,3}^{\tau} + T_{2,3}^{\tau} + T_{3,4}^{\tau} + T_{3,2}^{\tau}\right) + (1-4Fo)T_{3,3}^{\tau}$$

$$T_{2,2}^{\tau+1} = Fo\left(T_{3,2}^{\tau} + T_{1,2}^{\tau} + T_{2,3}^{\tau} + T_{2,1}^{\tau}\right) + (1-4Fo)T_{2,2}^{\tau}$$

$$T_{3,2}^{\tau+1} = Fo\left(T_{4,2}^{\tau} + T_{2,2}^{\tau} + T_{3,3}^{\tau} + T_{3,1}^{\tau}\right) + (1-4Fo)T_{3,2}^{\tau}$$

Stability criterion: $Bi \le 0.25$

Exterior surface nodes with convection: Nodal equations from item (4) in Table 5.3 without heat generation

$$T_{1,3}^{\tau+1} = Fo\left(T_{1,4}^{\tau} + T_{1,2}^{\tau} + 2T_{2,3}^{\tau} + 2BiT_{\infty}^{\tau}\right) + (1-4Fo-2BiFo)T_{1,3}^{\tau}$$

$$T_{1,2}^{\tau+1} = Fo\left(T_{1,3}^{\tau} + T_{1,1}^{\tau} + 2T_{2,2}^{\tau} + 2BiT_{\infty}^{\tau}\right) + (1-4Fo-2BiFo)T_{1,2}^{\tau}$$

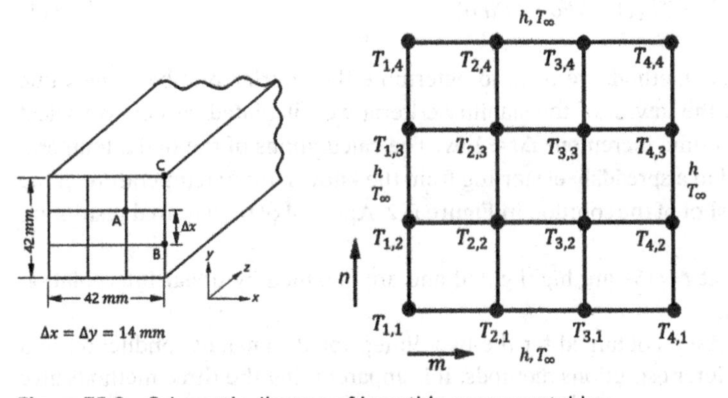

Figure E5.8 Schematic diagram of long thin square metal bar.

$$T_{2,1}^{\tau+1} = Fo\left(T_{1,1}^{\tau} + T_{3,1}^{\tau} + 2T_{2,2}^{\tau} + 2BiT_{\infty}^{\tau}\right) + (1 - 4Fo - 2BiFo)T_{2,1}^{\tau}$$

$$T_{3,1}^{\tau+1} = Fo\left(T_{2,1}^{\tau} + T_{4,1}^{\tau} + 2T_{3,2}^{\tau} + 2BiT_{\infty}^{\tau}\right) + (1 - 4Fo - 2BiFo)T_{3,1}^{\tau}$$

$$T_{4,2}^{\tau+1} = Fo\left(T_{4,1}^{\tau} + T_{4,3}^{\tau} + 2T_{3,2}^{\tau} + 2BiT_{\infty}^{\tau}\right) + (1 - 4Fo - 2BiFo)T_{4,2}^{\tau}$$

$$T_{4,3}^{\tau+1} = Fo\left(T_{4,2}^{\tau} + T_{4,4}^{\tau} + 2T_{3,3}^{\tau} + 2BiT_{\infty}^{\tau}\right) + (1 - 4Fo - 2BiFo)T_{4,3}^{\tau}$$

$$T_{3,4}^{\tau+1} = Fo\left(T_{2,4}^{\tau} + T_{4,4}^{\tau} + 2T_{3,3}^{\tau} + 2BiT_{\infty}^{\tau}\right) + (1 - 4Fo - 2BiFo)T_{3,4}^{\tau}$$

$$T_{2,4}^{\tau+1} = Fo\left(T_{1,4}^{\tau} + T_{3,4}^{\tau} + 2T_{2,3}^{\tau} + 2BiT_{\infty}^{\tau}\right) + (1 - 4Fo - 2BiFo)T_{2,4}^{\tau}$$

Stability criterion: $Fo \leq 1/2(2+Bi)$, $Fo \leq 0.237$

Exterior corner nodes with convection: Nodal equations from item (10) in Table 5.3 without heat generation

$$T_{1,4}^{\tau+1} = Fo\left(2T_{1,3}^{\tau} + 2T_{2,4}^{\tau} + 4BiT_{\infty}^{\tau}\right) + (1 - 4Fo - 4BiFo)T_{1,4}^{\tau}$$

$$T_{1,1}^{\tau+1} = Fo\left(2T_{2,1}^{\tau} + 2T_{1,2}^{\tau} + 4BiT_{\infty}^{\tau}\right) + (1 - 4Fo - 4BiFo)T_{1,1}^{\tau}$$

$$T_{4,1}^{\tau+1} = Fo\left(2T_{4,2}^{\tau} + 2T_{3,1}^{\tau} + 4BiT_{\infty}^{\tau}\right) + (1 - 4Fo - 4BiFo)T_{4,1}^{\tau}$$

$$T_{4,4}^{\tau+1} = Fo\left(4T_{4,3}^{\tau} + 4BiT_{\infty}^{\tau}\right) + (1 - 4Fo - 4BiFo)T_{4,4}^{\tau}$$

Stability criterion: $Fo \leq 1/4(1+Bi)$, $Fo \leq 0.226$

However, due to the symmetry resulting from the applied boundary conditions, the three unknown temperatures can be calculated using just three equations:

$$T_{3,3}^{\tau+1} = Fo\left(2T_{4,3}^{\tau} + 2T_{3,3}^{\tau}\right) + (1 - 4Fo)T_{3,3}^{\tau} \tag{i}$$

$$T_{4,3}^{\tau+1} = Fo\left(T_{4,4}^{\tau} + T_{4,3}^{\tau} + 2T_{3,3}^{\tau} + 2BiT_{\infty}^{\tau}\right) + (1 - 4Fo - 2BiFo)T_{4,3}^{\tau} \tag{ii}$$

$$T_{4,4}^{\tau+1} = Fo\left(4T_{4,3}^{\tau} + 4BiT_{\infty}^{\tau}\right) + T_{4,4}^{\tau}(1 - 4Fo - 4BiFo) \tag{iii}$$

To apply the finite-difference method, we need to determine the Fourier number. The value should be slightly lower than the lowest of the stability criteria we calculated earlier. We select $Fo = 0.22$, which determines a time increment $\Delta t \approx 1.9s$. The calculations of the nodal temperatures can readily be conducted in a spreadsheet starting from the known initial temperature of the bar $T_i = 120°C$ (see the screen shot of the solution in Figure D.2, Appendix D). Required results are shown in Table E5.8.1.

The required temperatures at $t = 15s$ are highlighted and are obtained by linear interpolation between $t = 13.3$ and $t = 15.2s$,

Table E5.8.2 compares the results obtained for the two-dimensional transient conduction in a long square bar from three different solutions methods. It is apparent that the three methods give very close results.

Table E5.8.1 Selected results of the explicit solution.

t, s	$T_{3,3}^{i+1}$	$T_{4,2}^{T+1}$	$T_{4,4}^{T+1}$
0	120	120	120
1.9	120.00	115.29	110.58
3.8	117.93	111.83	106.17
5.7	115.24	108.93	103.00
7.6	112.47	106.20	100.37
9.5	109.71	103.60	97.89
11.4	107.02	101.08	95.53
13.3	104.41	98.64	93.25
15	102.13	96.51	91.27
15.2	101.87	96.26	91.04

Table E5.8.2 Comparison of results from three different methods.

Location	Product solution Example 5.5	Explicit finite difference Example 5.8	Computer modelling (Solaria)
A	102.2	102.1	101.8
B	96.8	96.5	96.3
C	91.7	91.3	91.2

Problems

5.1 A large aluminium plate of 50mm thickness is suddenly exposed to convection heat transfer in an environment at 80°C with heat transfer coefficient $h = 600 W/m^2.K$. The initial temperature of the plate is 200°C. The thermophysical properties of aluminium can be taken as $k = 204 W/m.K, c_p = 900 J/kg.K$, and $\rho = 2700 kg/m^3$. How long would it take for the surface temperature to cool to 180°C?

5.2 A large aluminium plate of 40mm thickness is suddenly exposed to convection heat transfer in an environment at 100°C with heat transfer coefficient $h = 700 W/m^2.K$. The initial temperature of the plate is 200°C. The thermophysical properties of aluminium can be taken as $k = 204 W/m.K$, $c_p = 900 J/kg.K$, and $\rho = 2700 kg/m^3$. What is the time it takes for the surface temperature to reach 150°C?

5.3 A large plate of 3cm thickness ($k = 12 W/m.K$) at an initial temperature of 100°C is cooled by convection heat transfer in an environment at 30°C with heat transfer coefficient $h = 480 W / m^2.K$. Take the density of the plate material $\rho = 8025 kg/m^3$ and specific heat $= 477 J/kg.K$. Calculate the temperature at the centre of the plate 50 seconds after start of cooling.

5.4 Determine the temperature at the surface of the plate for the data in Problem 5.3.

5.5 A 6-mm metal plate made from a new alloy is to be used in the nose section of a spacecraft that could be subjected upon re-entering the earth's atmosphere to an environment at $T_\infty = 2150°C$ and

$h=3395\,W/m^2.K$. The thermophysical properties of the alloy are: $\rho=788\,kg/m^3, c_p=460\,J/kg.K$, $k=55\,W/m.K$. How long will it take for the outer surface of the plate to reach temperature $T_s=1100°C$?

5.6 A long metal rod of $60\,mm$ diameter, density $\rho=8000\,kg/m^3$, specific heat $c_p=500\,J/kg.K$, and thermal conductivity $k=50\,W/m.K$ is initially at an unknown uniform temperature and is heated in a furnace maintained at 750 K. The convection coefficient is estimated to be $h=1000\,W/m^2.K$. Determine the centreline temperature of the rod if the surface temperature is 550 K.

5.7 For the data in Problem 5.5, how long will it take to raise the centreline temperature of the rod from the initial temperature of $T_i=300K$ to $500K$?

5.8 A long cylindrical bar of 80 mm diameter at initial temperature $T_i=830°C$ is cooled by quenching in a large bath at $T_\infty=40°C$ with heat transfer coefficient $h=180W/m^2.K$. Take the thermal conductivity of the bar as $k=17.5W/m.K$ and thermal diffusivity as $\alpha=5.2\times10^{-6}\ m^2/s$. Determine the time it takes to cool the bar to $T_0=120°C$ and the temperature of the surface of the bar at that moment.

5.9 A small chicken weighing 1.5 kg is initially at a temperature of 15°C. The chicken is to be frozen in a refrigerator freezer in an environment of $T_\infty=-10°C$ and $h=400\,W/m^2.K$. The following thermophysical properties can be assumed for the chicken: $\rho-950\,kg/m^3$, $k=0.45\,W/m.K$, and $c_p=3640\,J/kg.K$. Assuming the chicken is of spherical shape, estimate the temperature of the centre and the surface of the chicken.

5.10 Use the lumped capacitance method to calculate the average temperature of a large plate after 50 s and 100 s for Biot numbers 0.0125, 0.05, and 0.1. Use the data: $k=12\,W/m.K$, $c_p=477\,J/kg.K, \rho=8025\,kg/m^3, 2\delta=0.03m, L_c=\delta=0.015m, T_\infty=100°C, T_i=20°C$.

5.11 A standard 58g egg has a volume of about $50cm^3$. The egg, which is approximated as a sphere is placed in boiling water at 100°C. Take the initial temperature of the egg $T_i=18°C$, the heat transfer coefficient $h=1000\,W/m^2.K$, specific heat capacity $c_p=4200\,J/kg.K$, and the thermal conductivity within the egg $k=0.8\,W/m.K$. Estimate the temperatures at the centre and surface of the egg if it is to be served either soft-boiled (3 minutes) or hard-boiled (10 minutes).

5.12 Determine the cooking time required to raise the temperature of the centre of a spherical dumpling $80mm$ in diameter from initial uniform temperature of 20 to 70°C. The dumpling is heated with saturated steam at 95°C. The heat capacity, density, and thermal conductivity are estimated to be $c_p=3500\,J/kg.K, \rho=1000\,kg/m^3$, and $k=0.5\,W/m.K$. The heat transfer coefficient of condensing steam is $1000\,W/m^2.K$.

5.13 A cylindrical steel ingot 100 mm diameter and 300 mm long is heat treated in a 6-m long furnace. The initial temperature of the ingot is 90°C and the final temperature when it comes out of the furnace is 800°C. The furnace is at 1250°C and the combined convection and radiation heat transfer coefficient is $h_t=500\,W/m^2$. The thermophysical properties of steel can be taken as: $k=41\,W/m.K, \rho=8130\,kg/m^3, c_p=434\,J/kg.K$. Determine:

(a) time required for the ingot to reach 800°C

(b) maximum speed at which the ingot should travel in the furnace to attain the required temperature.

5.14 A very long square metal bar of dimensions $40mm\times40mm$ is heated to 120°C, then cooled by immersing in a coolant at 18°C with heat transfer coefficient $h=600\,W/m^2.K$. The thermophysical properties of the bar can be taken as $\rho=7870\,kg/m^3$, $c_p=450\,J/kg.K$, and $k=80\,W/m.K$. Determine the temperature at point B shown in Figure P5.14 after 15 seconds of cooling. The $x-y$ coordinates of point B relative to the centre 0 in metres is $B(0.021,-0.007)$.

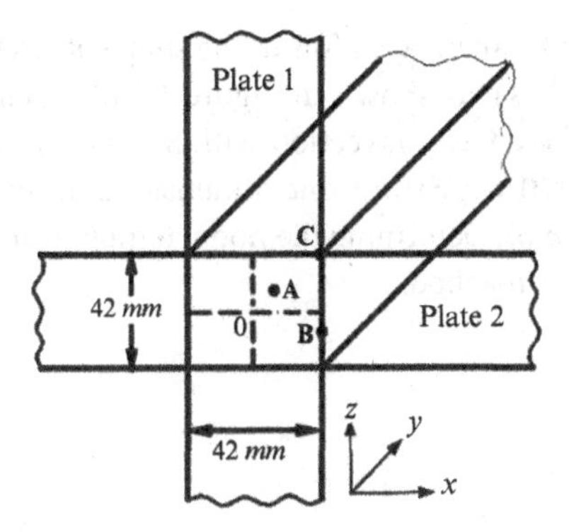

Figure P5.14 Schematic diagram of 2D conduction in a long bar.

5.15 Repeat Problem 5.14 for point C in Figure P5.14.

5.16 A heated stainless-steel cylinder of radius $R = 4\,cm$ and length $2\delta = 6\,cm$ is placed in a fluid bath at temperature $T_\infty = 25°C$. The cylinder is initially at temperature $T_i = 300°C$ and the heat transfer coefficient $h = 400\,W/m^2.K$. The thermophysical properties of the cylinder are $\rho = 7800\,kg/m^3$, $c_p = 500\,J/Kg.K$, $k = 15.5\,W/m.K$. Determine the surface temperature at point 2 in Figure P5.16 five minutes after the start of the cooling process.

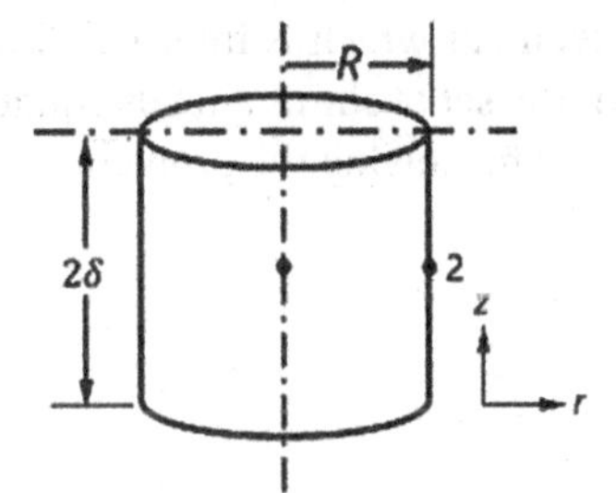

Figure P5.16 Heated short cylinder.

5.17 In the two-dimensional cylindrical configuration for transient conduction shown in Figure P5.17, the radial (Δr) and angular $(\Delta\varphi)$ increments of the nodes are uniform. The boundaries in the radial direction are exposed to convection at h, T_∞ and the boundary in the angular direction is at a constant temperature T_{s2}. Derive the finite-difference equation for surface node i, j.

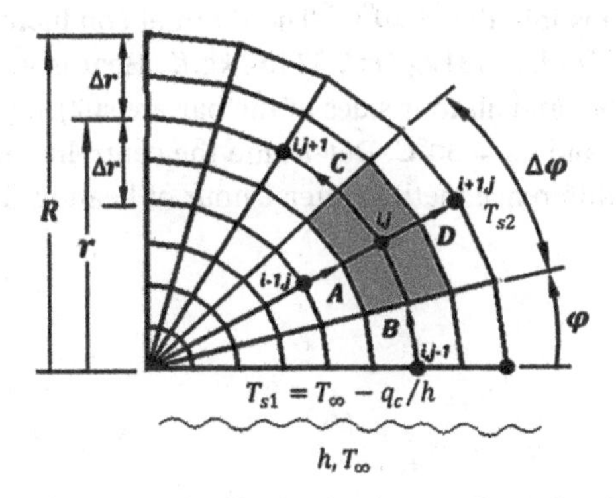

Figure P5.17 Conduction in two-dimensional semi-circular plate.

5.18 A surface at temperature $T_0 = 120°C$ is to be cooled by an array of fins in the shape of circular aluminium rod ($k = 237 W/m.K$ $\alpha = 9.71 \times 10^5\ m^2/s$), as shown in Figure P5.18. Each fin is $80 mm$ long and $8 mm$ in diameter. The fin is cooled by convection with $h = 35 W/m^2.K$ and $T_\infty = 15°C$. The temperature at the base $T_0 = 120°C$. Assume one-dimensional transient conduction and take the nodal increment $\Delta x = 20 mm$. Determine the nodal temperatures 1, 2, 3, and 4 at the moment steady-state conduction is reached.

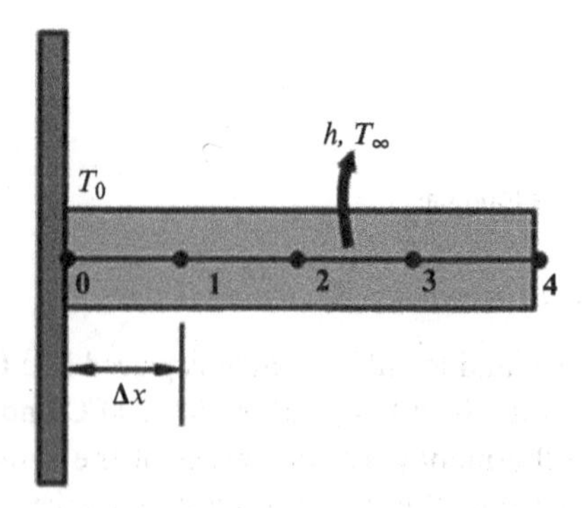

Figure P5.18 Fin of circular cross-section.

5.19 Figure P5.19 shows the square cross-section of a long metal bar which is internally heated and subjected to convection on all four sides. Write down the set of simultaneous equations required if the explicit finite difference method is to be used for solving the problem.

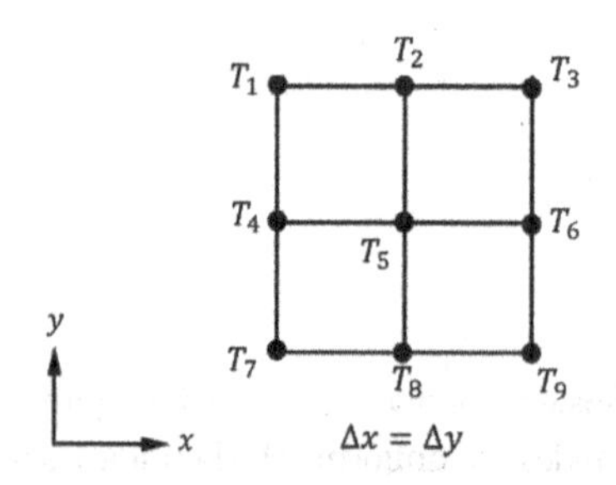

Figure P5.19 Cross-section of square metal bar.

5.20 The bar in Problem 5.19 ($200 mm \times 200 mm$) is initially at 20°C.. The thermal conductivity, density, and specific are, respectively, $41 W/m.K$, $8131 kg/m^3$, $434 J/kg.K$. Heat is generated internally at a rate of $\dot{q} = 9.0 \times 10^5 W/m^3$ and all four sides of the bar are subjected to convection heat transfer at $h = 60 W/m^2.K$ and $T_\infty = 30°C$. Determine the centreline temperature of the bar using the explicit finite difference method after 1 hour of heating. Take $\Delta x = \Delta y = 100 mm$.

6

Fundamentals of Convection Heat Transfer

It was stated in Chapter 1 that the process of heat transfer is energy in motion in the presence of a temperature gradient. In conduction, the agitated layer of molecules of the transmitting body interacts with the cooler neighbouring layer of molecules causing heat energy to "flow." In convection, heat energy is predominantly transmitted in gas or liquid medium from or to a surface of a solid body. Heat transmission within the fluid will be mainly by conduction in the presence of a temperature gradient. Convection is always accompanied by fluid motion, and depending on the intensity of motion, the convection heat transfer process could by natural or forced. In natural convection, the microscopic motion of the fluid molecules is caused by the presence of a temperature gradient that causes a density differential setting the fluid in motion. In forced convection, the flow of the fluid is more definite and usually caused by wind, a compressor, pump, or a fan. As we have seen in the previous chapters dealing with conduction, convection was repeatedly referred to as a boundary condition acting on the external surfaces of known geometries of steady-state and transient conduction systems. Hence, convection is mainly concerned with heat transfer to or from a surface of a solid body. In the presence of a solid surface, the motion of the fluid will be affected significantly by the interaction of the fluid and this surface, and it is essential to look into this interaction to understand convection heat transfer. The science of convection is a convergence of the principles of heat conduction and fluid mechanics and the mathematical analysis of convection is based on the concepts of fluid mechanics that are discussed briefly in the following sections.

6.1 Convection Governing Equation

The equation used to determine the rate of heat transfer by convection between a fluid at temperature T_∞ and a solid surface at temperature T_s is

$$\dot{Q} = Ah(T_s - T_\infty)W \tag{6.1}$$

where
A – heat transfer area
h – heat transfer coefficient (convection coefficient)

In heat transfer practice, this equation is generally known as Newton's law of cooling, with some prominent dissenting voices (Bejan 2013) crediting Fourier with the formulation of the equation. In this book, the equation will be referred to as Newton's law of cooling.

Heat Transfer Basics: A Concise Approach to Problem Solving, First Edition. Jamil Ghojel.

Companion website: www.wiley.com/go/ghojel/heat_transfer

Equation (6.1) can be written in terms of the heat flux as

$$q=\frac{\dot{Q}}{A}=h\left(T_s-T_\infty\right) \qquad W/m^2 \tag{6.2}$$

6.2 Viscosity

When a fluid flows over the surface of a solid, the velocity may vary across that surface and the adjacent fluid layer at distance *dy*, as shown in Figure 6.1. The fluid at the surface is at rest and its velocity increases continuously as the distance increases away from the surface. Viscosity is a property of the fluid that causes resistance to the velocity variations by creating shear stresses between the solid surface (stationary) and moving fluid layer and between adjacent layers of fluid in relative motion farther away.

The shear stress, determined experimentally, is directly proportional to the velocity gradient normal to the flow dU/dy and the constant of proportionality μ is the coefficient of viscosity

$$\tau=\mu\frac{dU}{dy} \tag{6.3}$$

The coefficient μ is generally known as the dynamic viscosity with units $N.s/m^2$ or $kg/m.s$ to distinguish it from the kinematic viscosity defined as the ratio of dynamic viscosity to fluid density $\nu=\mu/\rho$ with units $N.m/kg$ or m^2/s. Viscosity is temperature-dependent decreasing sharply with temperature in some fluids (e.g., glycerine, motor oils) and moderately in others (e.g., kerosene, mercury, water). Viscosity of air, helium, and carbon dioxide increases somewhat with temperature.

The shear stress is also proportional to the coefficient of friction f expressed as

$$\tau=f\frac{\rho U^2}{2} \tag{6.4}$$

6.3 Types of Flow

It was stated earlier that convection is always accompanied by fluid motion of varying degrees of intensity. At low flow velocities, the fluid travels along continuous parallel lines known as streamlines and the flow is said to be laminar. If the flow velocity is gradually increased, there comes a

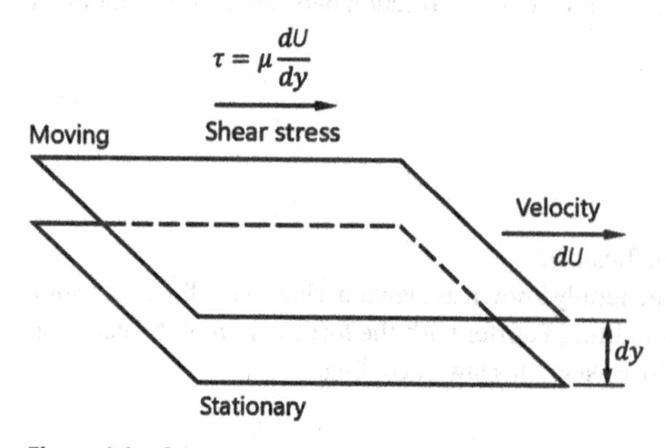

Figure 6.1 Schematic diagram of viscous flow.

moment when the flow is disrupted and the parallel streamlines are replaced by irregular motion in the form of eddies and the flow becomes turbulent. The transition from laminar flow to turbulent flow always occurs with increasing flow velocity. However, the transition point is also dependent on the physical shape and dimensions of the flow environment and physical properties of the fluid. The general criterion that determines the transition point from laminar flow to turbulent flow is the Reynolds number, defined as the ratio of inertia force to viscous force and expressed mathematically as

$$Re = \frac{U \rho L_c}{\mu} \tag{6.5}$$

U – mean flow velocity
ρ – fluid density
L_c – characteristic or reference length of the flow
μ – dynamic viscosity of the fluid.

For a flow along a flat plate of length L, $L_c = L$; for flow in the passage between two parallel horizontal plates of width h, $L_c = 2h$; and for flow in a circular passage (tube) or over a cylinder or sphere of diameter D, $L_c = D$. For flows inside irregular passages, the characteristic length is defined as four times the ratio of the flow cross-section to the wetted perimeter ($4A / P$).

At $Re \geq 2300$, a fully developed flow in a tube becomes turbulent. At $Re \geq 5 \times 10^5$, flow over a flat plate becomes turbulent.

6.4 The Hydrodynamic (Velocity) Boundary Layer

6.4.1 Flow Over a Flat Plate

Consider the long plate shown in Figure 6.2 with a continuous stream of fluid flowing parallel to the surface at free stream velocity U_∞. The velocity of the fluid in contact with the plate surface will be zero, increasing progressively in the y-direction until it become constant and equal to the free stream velocity at a distance of $y = \delta(x)$. The velocity profile of the fluid flow changes with x, as shown in Figure 6.2, as well as the distance $\delta(x)$, which increases continuously forming the curve drawn as a dashed line. The space enclosed by this line and the plate surface is known as the boundary layer, which can theoretically be of indefinite extent as it approaches the free stream velocity asymptotically. The boundary layer thickness limit is usually taken as the value of $\delta(x)$ when the following equality is achieved

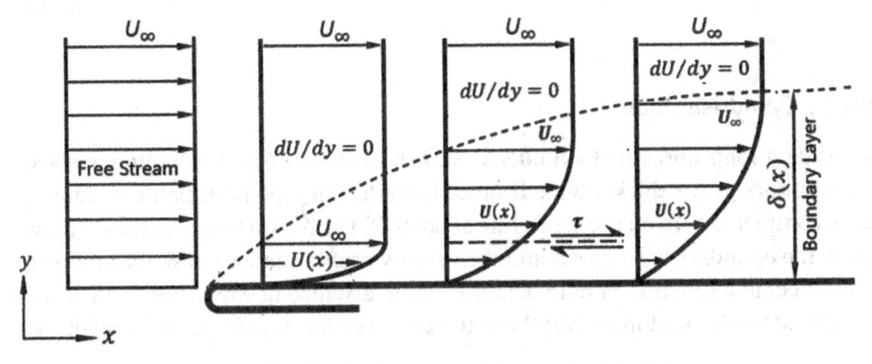

Figure 6.2 Flow velocity profiles in the boundary layer of a flat plate.

$$U(x) = 0.99U_\infty \tag{6.6}$$

The rate of flow velocity within the boundary layer is $dU/dy \neq 0$, and without is $dU/dy = 0$. The thickness of the laminar boundary layer on a flat plate as a function of distance x from the leading edge of the plate and Reynolds number is

$$\delta = \frac{4.64x}{Re_x^{1/2}} \tag{6.7}$$

The Reynolds number in Eq. (6.7) is based on the linear distance x, i.e., $L_c = x$,

$$Re_x = \frac{\rho U_\infty x}{\mu}$$

The velocity profile in the laminar boundary layer at any distance x from the leading edge of the plate can be approximated by a cubic parabola (Rogers and Mayhew 1992)

$$\frac{U}{U_\infty} = \frac{3}{2}\left(\frac{y}{\delta}\right) - \frac{1}{2}\left(\frac{y}{\delta}\right)^3 \tag{6.8}$$

The local coefficient of friction at distance x and the average value over the length x are, respectively

$$f_x = \frac{0.664}{Re_x^{1/2}} \tag{6.9}$$

$$f = \frac{1.328}{Re_x^{1/2}} \tag{6.10}$$

When the flow becomes turbulent, the local coefficient of friction, boundary layer thickness, and flow velocity are, respectively

$$f_x = \frac{0.0592}{Re_x^{1/5}} \tag{6.11}$$

$$\delta = \frac{0.38x}{Re_x^{1/5}} \tag{6.12}$$

$$\frac{U}{U_\infty} = \left(\frac{y}{\delta}\right)^{1/7} \tag{6.13}$$

6.4.2 Flow Inside a Cylindrical Tube

The flow profile inside a long horizontal cylinder is shown in Figure 6.3. The boundary layer formed on the inner surface grows thicker with distance from the entry plane at point A, causing the free flow area to contract to zero on the centreline at point B. During this stage of flow, known as entry flow region, the boundary layer grows and the velocity profile changes in shape and magnitude with the latter being a function of both x and r. Some distance downstream from B, the flow becomes dominated by viscous forces over the entire cross-section at point C and the velocity

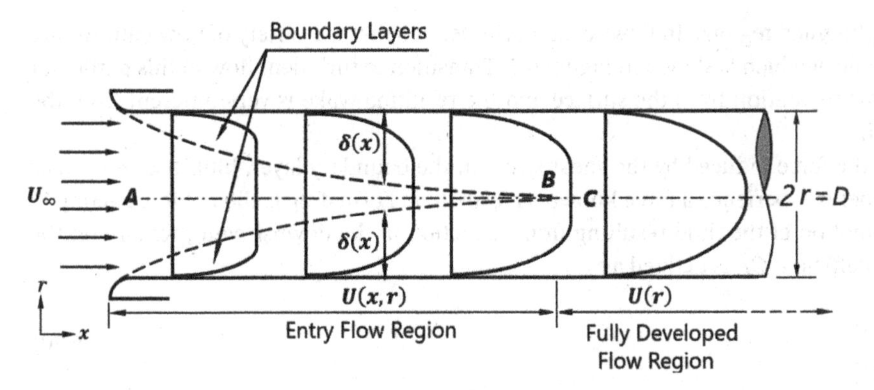

Figure 6.3 Flow velocity profiles and boundary layers in a tube (hollow cylinder).

profile changes to parabolic for laminar flow and varies only with r. The flow region downstream from point C is known as the fully developed region.

The length of the entry flow region is approximated as

$$\frac{x_e}{D} = 0.288 Re_D \tag{6.14}$$

$$Re_d = \frac{\rho U_\infty D}{\mu} \tag{6.15}$$

For fully developed laminar flow, the flow velocity relative to the velocity at the axis of the tube is given by

$$\frac{U}{U_r} = 2\left(\frac{y}{r}\right) - \left(\frac{y}{r}\right)^2 \tag{6.16}$$

and the coefficient of friction is

$$f = \frac{16}{Re_D} \tag{6.17}$$

For fully developed turbulent flow, the flow velocity and coefficient of friction are, respectively

$$\frac{U}{U_r} = \left(\frac{y}{r}\right)^{1/7} \tag{6.18}$$

$$f = \frac{0.0791}{Re_D^{0.25}} \tag{6.19}$$

6.4.3 Flow Over Tube or Sphere

If the flow over a tube or sphere is considered, the flow pattern shown in Figure 6.4 is obtained for a laminar flow. The boundary layer divides at the front of the tube symmetrically as it grows up to the point where the flow stream separates from the surface of the tube then reverses direction forming bubbles comprising circulating flow streams (vortices) in the wake behind the tube, as shown by the highlighted regions. In flows over a cylinder or sphere, a variety of flow patterns are formed, an example of which is shown in Figure 6.5. Transition to turbulent flow in this particular case is before flow separation from the surface and the resulting wake is very different from the one in Figure 6.4.

In addition to the force induced by the shear stress in the boundary layer, bluff bodies such as cylinders and spheres experience a force known as form drag. Form drag is the net force acting in the direction of motion of the fluid resulting from reduction in the downstream pressure on the body. The drag coefficient C_D is defined as

$$C_D = \frac{F/A}{\frac{1}{2}\rho U_\infty^2} \tag{6.20}$$

where F is the exerted drag force and A is the frontal area.

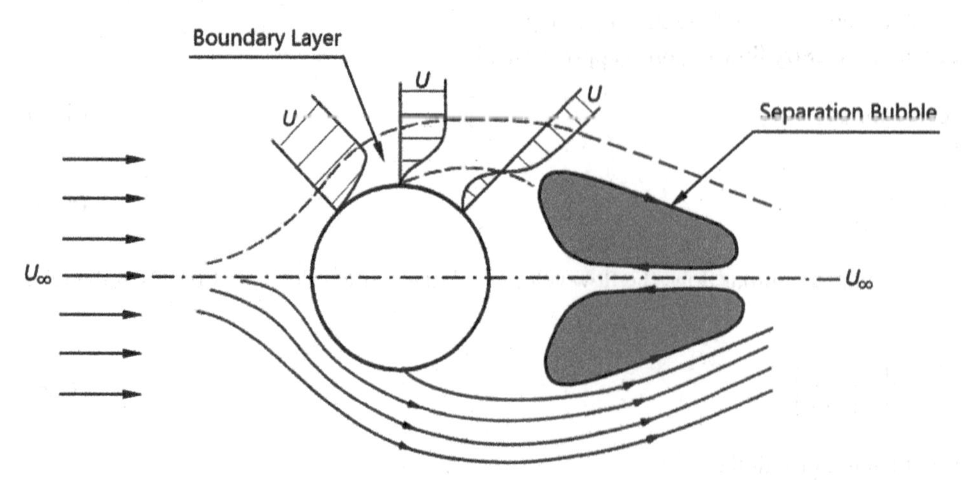

Figure 6.4 Cross laminar flow over a tube (solid circular cylinder) or a sphere with separation ($5 < Re < 40$).

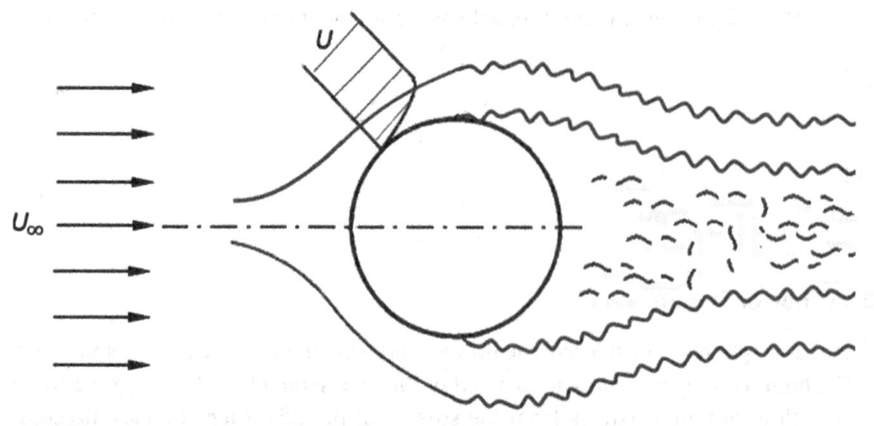

Figure 6.5 Cross flow over a tube (solid circular cylinder) or a sphere with early transition to turbulent flow ($3\times10^5 < Re < 3.5\times10^6$).

6.5 The Thermal Boundary Layer

In addition to the velocity distribution in the vicinity of a solid surface, temperature distribution needs to be considered within the context of convection heat transfer. Hence the concept of thermal boundary layer within which the temperature varies between the wall and the fluid at a distance of $\delta_T(x)$ in the y-direction (Figure 6.6). The thickness δ_T increases with distance x and the shape of the temperature profile within the layer depends on the relative values of the free stream temperature and the surface temperature. The temperature of the fluid in contact with the surface is equal to the temperature of the surface and temperature gradient develops in the boundary layer. If the surface temperature T_s is lower than the free stream temperature T_∞, the fluid temperature increases until it becomes equal to T_∞ at the edge of the thermal boundary layer (Figure 6.6b). Conversely, if the surface temperature T_s is higher than the free stream temperature T_∞, the fluid temperature decreases until it becomes equal to T_∞ at the edge of the thermal boundary layer (Figure 6.6c).

The thickness of the thermal boundary layer could be less, equal, or greater than the thickness of the hydrodynamic boundary layer, albeit similar in shape. If heating of the plate starts at a distance of x_0 from the leading edge of the plate, as shown in Figure 6.7, the thermal boundary layers will be a function of x_0, x, and Prandtl number Pr (Holman 2010)

$$\delta_T = \frac{\delta}{1.026} Pr^{-1/3} \left[1 - \left(\frac{x_0}{x} \right)^{3/4} \right]^{1/3} \tag{6.21}$$

The Prandtl number which figures extensively in fluid mechanics and heat transfer is defined as the ratio of rate of diffusion of viscous effects to the rate of diffusion of heat

$$Pr = \frac{c_p \mu}{k} = \frac{\nu}{\alpha} \tag{6.22}$$

The fluid at the surface ($y = 0$) is at rest and heat transfer to the layer adjacent to the surface is by conduction only

$$q = -k_f \left(\frac{dT}{dy} \right)_{y=0} \tag{6.23}$$

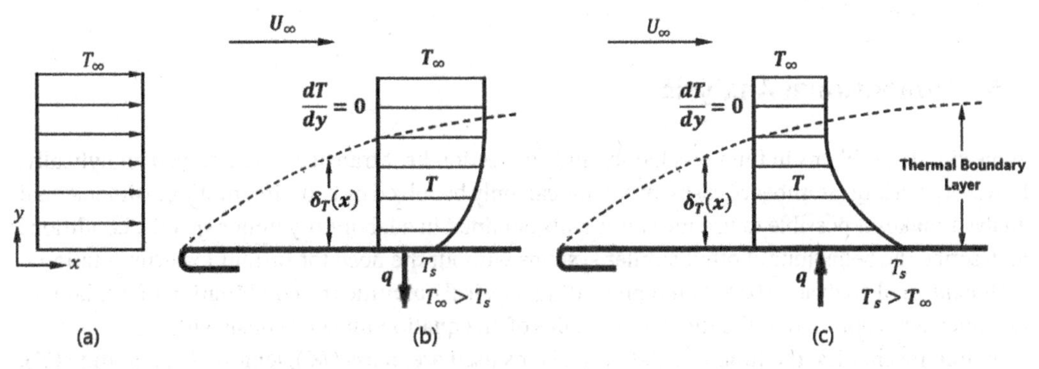

Figure 6.6 Temperature profiles of the free stream and in the thermal boundary layer: (a) free stream; (b) hot stream and cold surface; (c) cold stream and hot surface.

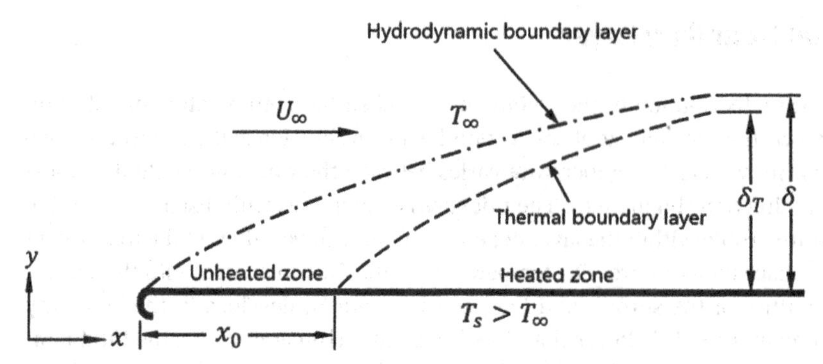

Figure 6.7 Relative positions of the hydrodynamic and thermal boundary layers with delayed start of heating from the leading edge of a heated flat plate.

Combining Eq. (6.23) with Eq. (6.2) we obtain

$$h = \frac{-k_f \left(\frac{dT}{dy}\right)_{y=0}}{T_s - T_\infty} \tag{6.24}$$

If heat flux is taken as positive when $T_s > T_\infty$ and as negative when $T_s < T_\infty$, Eq. (6.24) can be rewritten for the heat transfer coefficient at distance x from the leading edge of the plate in terms of the thermal boundary layer thickness δ_T at x as

$$h(x) \sim \frac{k_f \frac{\Delta T}{\delta_T(x)}}{\Delta T} \sim \frac{k_f}{\delta_T(x)} \tag{6.25}$$

$\Delta T = (T_s - T_\infty)$ is the temperature change within the thermal boundary layer.

Equation (6.25) shows that for a constant thermal conductivity of the fluid, the heat transfer coefficient $h(x)$ decreases with increasing $\delta_T(x)$ as x increases. As a consequence, the convection heat flux from Eq. (6.2) also decreases with increasing x. The use of a variable heat transfer coefficient is cumbersome and it is usually replaced by an average value over the length of the plate

$$h = \frac{1}{L}\int_0^L h(x)dx \tag{6.26}$$

6.6 Dimensional Analysis

Some simple problems in fluid mechanics and convection heat transfer can be solved analytically; however, in a large number of cases, problems can only be solved by experimentation. Dimensional analysis makes it possible to use measurements obtained in a laboratory under specific conditions to describe the behaviour of other similar systems without the need for further experimentation.

Dimensional analysis is based on representing physical quantities by combination of fundamental dimensions noting that the units of two sides of an equation must be consistent.

In fluid mechanics, the fundamental dimensions used are: mass (M), length (L), and time (T). In convection heat transfer, one more fundamental dimension, namely temperature (θ) is required.

Generally, any physical quantity can be represented by these four fundamental dimensions including complex thermodynamic quantities. For example, heat ($Joule \rightarrow N.m \rightarrow kg.m^2 / s^2$) can be represented dimensionally as ML^2T^{-2} and thermal conductivity $\left(W / m.K\right)$ as $MLT^{-3}\theta^{-1}$. To simplify the dimensional representation of the quantities involving heat, a fundamental dimension H for heat can be added. This is justifiable since, unlike thermodynamic processes in which energy can undergo conversion from one form to another, in convective processes, energy transfer is the dominant phenomenon, particularly for slow flows. Then heat will simply have a dimension H and thermal conductivity will be transformed to $HM^{-1}\theta^{-1}$. These fundamental dimensions can now be used to provide qualitative descriptions of various physical quantities, including thermodynamic quantities that are relevant to convection heat transfer. Table 6.1 lists the symbols, units, and dimensions of common physical quantities used in fluid mechanics and heat transfer. For effective application of dimensional analysis, it is essential to state which independent variables are relevant to the problem.

Table 6.1 Symbols, units, and dimensions of common physical quantities based on the (M, L, T, H, θ) and (M, L, T, θ) fundamental dimensions systems.

Quantity	Symbol	Units (m, s, kg, K)	Dimensions (M, L, T, H, θ)	Dimensions (M, L, T, θ)
Length	l	m	L	L
Time	t	s	T	T
Mass	m	kg	M	M
Temperature	T	K	θ	θ
Velocity	C or U	m / s	LT^{-1}	LT^{-1}
Volume	V	m^3	L^3	L^3
Acceleration	a	m / s^2	LT^{-2}	LT^{-2}
Angular velocity	ω	rad	T^{-1}	T^{-1}
Area	A	m^2	L^2	L^2
Angular acceleration	$\dot{\omega}$	rad / s^2	T^{-2}	T^{-2}
Volume flow rate	$\dot{V}$	m^3 / s	L^3T^{-1}	L^3T^{-1}
Mass flow rate	$\dot{m}$	kg / s	MT^{-1}	MT^{-1}
Pressure	p	N / m^2	$ML^{-1}T^{-2}$	$ML^{-1}T^{-2}$
Density	ρ	kg / m^3	ML^{-3}	ML^{-3}
Force	F	$N\left(kg.m / m^2\right)$	MLT^{-2}	MLT^{-2}
Specific weight	γ	N / m^3	$ML^{-2}T^{-2}$	$ML^{-2}T^{-2}$
Dynamic viscosity	μ	$N.s / m^2$	$ML^{-1}T^{-1}$	$ML^{-1}T^{-1}$
Kinematic viscosity	ν	m^2 / s	L^2T^{-1}	L^2T^{-1}
Surface tension	σ	N / m	MT^{-2}	MT^{-2}
Torque, moment of force	T, M	$N.m$	ML^2T^{-2}	ML^2T^{-2}
Shear stress	τ	N / m^2	$ML^{-1}T^{-2}$	$ML^{-1}T^{-2}$
Heat	Q	$J\left(N.m\right)$	H	ML^2T^{-2}
Heat rate	$\dot{Q}$	$W\left(J / s\right)$	HT^{-1}	ML^2T^{-3}
Thermal resistance	R	K / W	$T\theta H^{-1}$	$L^{-2}M^{-1}T^3\theta$
Thermal conductivity	k	$W / m.K$	$HL^{-1}T^{-1}\theta^{-1}$	$MLT^{-3}\theta^{-1}$

(Continued)

Table 6.1 (Continued)

Quantity	Symbol	Units (m,s,kg,K)	Dimensions (M,L,T,H,θ)	Dimensions (M,L,T,θ)
Specific heat	c_p	$J/m.K$	$HM^{-1}\theta^{-1}$	$L^2T^{-2}\theta^{-1}$
Heat transfer coefficient	h	$W/m^2.K$	$HL^{-2}T^{-1}\theta^{-1}$	$MT^{-3}\theta^{-1}$
Thermal diffusivity	α	m^2/s	L^2T^{-1}	L^2T^{-1}
Coefficient of expansion	β	$1/K$	θ^{-1}	θ^{-1}
Gas constant	R	$J/m.K$	$HM^{-1}\theta^{-1}$	$L^2T^{-2}\theta^{-1}$

In cases where temperature is a basic physical quantity and it is preferable to avoid using an extra fundamental dimension such as θ, the gas constant is usually lumped together with the temperature and the combined variable $RT = p/\rho$ (from the equation of state) will have the dimensions

$$\frac{ML^{-1}T^{-2}}{ML^{-3}} = L^2T^{-2}$$

6.6.1 The Rayleigh Method

Consider the functional relationship for the dependent variable u expressed in terms of already known independent variables $u_1,\ u_2,\ u_3, u_4, u_5$ as

$$u = f(u_1,\ u_2,\ u_3, u_4, u_5) \tag{6.27}$$

The Rayleigh method, also known as the method of indices, states that this functional relationship can be expressed as the sum of a series of terms each raised to a power that may have any value

$$u = A\left(u_1^{a_1} u_2^{a_2} u_3^{a_3} u_4^{a_4} u_5^{a_5}\right) + B\left(u_1^{b_1} u_2^{b_2} u_3^{b_3} u_4^{b_4} u_5^{b_5}\right) + \ldots \tag{6.28}$$

where A and B are non-dimensional constants.

In practice, only the first term in the series in Eq. (6.28) is required to establish the functional relationships of convection problems.

According to the principle of dimensional consistency, each term in each of the series on the right side of Eq. (6.28) must have the same dimensions as the dependent variable u.

Consider the heat transfer coefficient for forced flow in a circular tube of diameter D. The functional relationship of the following form can be assumed

$$h = f\left(D, U, \rho, \mu, k, c_p, \Delta\theta\right) \tag{6.29}$$

where U is the fluid velocity; ρ, μ, c_p are the fluid density, dynamic viscosity, and specific heat respectively; k is the thermal conductivity; and $\Delta\theta$ is the temperature relative to the wall surface in contact with the fluid.

The first Rayleigh series for the functional relationship of Eq. (6.29) is

$$h = A\left(D^{a_1}U^{a_2}\rho^{a_3}\mu^{a_4}k^{a_5}c_p^{a_6}\Delta\theta^{a_7}\right) \tag{6.30}$$

A – non-dimensional constant

Expressing all quantities in terms of fundamental dimensions system M, L, T, θ, H (see Table 6.1), we can write

$$\left[HL^{-2}T^{-1}\theta^{-1}\right] = A\left[(L)^{a_1}\left(LT^{-1}\right)^{a_2}\left(ML^{-3}\right)^{a_3}\left(ML^{-1}T^{-1}\right)^{a_4}\left(HL^{-1}T^{-1}\theta^{-1}\right)^{a_5}\left(HM^{-1}\theta^{-1}\right)^{a_6}(\theta)^{a_7}\right]$$

Observing homogeneous dimensionality, we obtain five simultaneous equations

M: $0 = a_3 + a_4 - a_6$
L: $-2 = a_1 + a_2 - 3a_3 - a_4 - a_5$
T: $-1 = -a_2 - a_4 - a_5$
θ: $-1 = -a_5 - a_6 + a_7$
H: $1 = a_5 + a_6$

There are seven unknowns in these equations and the solution can be expressed in terms of two arbitrarily selected indices. We select for the current case a_3 and a_6. The remaining indices can be found in terms of these two indices as follows:

$$a_5 = 1 - a_6$$
$$a_7 = 0$$
$$a_4 = a_6 - a_3$$
$$a_2 = a_3$$
$$a_1 = -1 + a_3$$

Equation (6.30) can now be rewritten as

$$h = A\left(D^{-1+a_3}U^{a_3}\rho^{a_3}\mu^{a_6-a_3}k^{1-a_6}c_p^{a_6}\theta^0\right) = A\left(\frac{D^{a_3}}{D}U^{a_3}\rho^{a_3}\frac{\mu^{a_6}}{\mu^{a_3}}\frac{k}{k^{a_6}}c_p^{a_6}\right)$$

Multiplying both sides by D/k and rearranging results in three-dimensional groups

$$\frac{hD}{k} = A\left(\frac{\rho UD}{\mu}\right)^{a_3}\left(\frac{c_p\mu}{k}\right)^{a_6} \tag{6.31}$$

or

$$Nu = A\left(Re^{a_3}Pr^{a_6}\right)$$

or

$$Nu = f(Re, Pr) \tag{6.32}$$

where

$$Nu = \frac{hD}{k} \text{ – Nusselt number}$$

$$Re = \frac{\rho UD}{\mu} \text{ – Reynolds number}$$

$$Pr = \frac{c_p\mu}{k} \text{ – Prandtl number}$$

The analysis has allowed us to reduce the unknown dependent variable and seven independent variables in Eq. (6.29) to one non-dimensional constant and three non-dimensional variable groups (Eq. 6.31). Constant A and indices a_3 and a_6 can be determined experimentally by correlating data in terms of Reynolds, Nusselt, and Prandtl numbers. The definitions and physical meanings of these numbers (dimensionless groups) and others that are used in heat transfer are shown in Table 6.2.

Table 6.2 Dimensionless numbers.

Group	Definition	Meaning
Biot number, Bi	$\frac{hL_c}{k} = \frac{L_c/k}{1/h}$	Conduction resistance/convection resistance
Brinkman number, Br	$Br = \frac{U^2\mu}{k(T_s - T_\infty)}$	Viscous dissipation/conductive heat flux ($Br = EcPr$)
Eckert number, Ec	$\frac{U^2}{c_p(T_s - T_\infty)}$	Kinetic energy/enthalpy change
Euler number	$Eu = \frac{\Delta p}{\rho U^2}$	Pressure force/inertia force
Fourier number, Fo	$\frac{\alpha t}{L_c^2}$	Rate of heat conduction/rate of thermal energy storage in a solid
Froud number, Fr	$\frac{U^2}{gL_c}$	Inertial force/gravitational force
Grashof number, Gr	$Gr = \frac{\beta g(T_s - T_\infty)L_c^3}{\nu^2}$	Buoyancy force/viscous force
Jakob number	$Ja = \frac{c_p(T_s - T_{sat})}{h_{fg}}$	Sensible energy/latent energy absorbed during liquid–vapour phase change
Nusselt number, Nu	$Nu = \frac{hL_c}{k} = \frac{h(T_s - T_\infty)}{k(T_s - T_\infty)/L_c}$	Convection heat flux/conduction heat flux
Peclet number, Pe	$Pe = \frac{UL_c}{\alpha}$	$RePr$
Prandtl Number, Pr	$\frac{c_p\mu}{k} = \frac{\nu}{\alpha}$	Rate of diffusion of viscous effects/rate of diffusion of heat
Rayleigh number, Ra	$Ra = \frac{\beta g(T_s - T_\infty)L_c^3}{\nu\alpha}$	$GrPr$

Table 6.2 (Continued)

Group	Definition	Meaning
Reynolds number, Re	$\frac{U\rho L_c}{\mu}=\frac{UL_c}{\nu}$	Inertia force/viscous force
Schmidt number	$Sc=\frac{\nu}{D_{AB}}$	Momentum/mass diffusivity
Sherwood number	$Sh=\frac{h_m L}{D_{AB}}$	Represents effectiveness of mass transfer at the surface
Stanton number, St	$St=\frac{h}{\rho c_p U}$	$Nu/RePr$

T_s – surface or wall temperature (T_w), T_∞ – fluid temperature

6.6.2 Buckingham Pi (Π or π) Theorem

The Buckingham Pi theorem states that if there are m dependent and independent variables in a dimensionally homogeneous functional relationship that describes a process and if these variables can be expressed by n fundamental dimensions (M,L,T,θ or M,L,T,θ,H), then the process can be described by a new functional relationship comprising $m-n$ dimensionless groups called Π-terms or π-terms.

If we have the functional relationship

$$u_1=f(u_2,\ u_3,\ u_4,u_5\ldots u_m),\ \text{or}$$

$$f(\ u_1,u_2,\ u_3,\ u_4,u_5\ldots u_m)=0 \tag{6.33}$$

the Π-theorem stipulates that it can be written in terms of dimensionless groups as

$$\Pi_1=f(\Pi_2,\ \Pi_3,\ \Pi_4,\ldots\Pi_{m-n}),\ \text{or}$$

$$f(\Pi_1,\Pi_2,\ \Pi_3,\ \Pi_4,\ldots\Pi_{m-n})=0 \tag{6.34}$$

Each of the Π-terms in Eq. (6.34) will have the form

$$\Pi_i=\left[u_p u_1^{a_1} u_2^{a_2}\ u_3^{a_3}\ u_4^{a_4} u_5^{a_5}\ldots u_j^{a_i}\right] \tag{6.35}$$

$$i=1,2,3,\ldots,m-n,\ j=1,2,3,\ldots,n$$

u_p is one of $m-n$ primary variables selected so that each dimensionless product (Π-term) characterizes some distinct feature of the flow under consideration. For example, if the problem involves forced convection, flow velocity U could be taken as a primary variable. Since fluid properties affect heat transfer in the fluid, we may consider specific heat as a primary variable. Another choice may be the dependent variable we want to determine such as heat flux or heat transfer coefficient. In natural convection flow on a vertical plate, gravity can be selected as a primary variable. Any primary variable must appear only in one Π-term. The remaining n variables are repeated in all Π-terms.

The Π-Theorem can manage larger number of dimensional variables compared with the Rayleigh method and the resulting functional relationship is easier to determine experimentally.

As an illustration of the application of this method, we can rework the problem of heat transfer coefficient for forced flow in a circular tube examined above as follows:

The functional relationship is written as

$$f\left(h,U,c_p,D,\rho,\mu,k,\Delta\theta\right)=0$$

and the fundamental dimensions M,L,T,θ,H will be used in the analysis.

The number of dimensional variables including the dependent variable h is $m=8$, and the number of fundamental dimensions is $n=5$.. Hence, the number of Π-terms is $m-n=3$.

We select $h,U,\ C_p$ as the primary variables.

$$\Pi_1=\left[hD^{a_1}\rho^{a_2}\mu^{a_3}k^{a_4}\left(\Delta\theta\right)^{a_5}\right]$$

Expressing all quantities in terms of the fundamental dimensions system M,L,T,θ,H, we can write

$$M^0L^0T^0\theta^0H^0=\left(HL^{-2}T^{-1}\theta^{-1}\right)(L)^{a_1}\left(ML^{-3}\right)^{a_2}\left(ML^{-1}T^{-1}\right)^{a_3}\left(HL^{-1}T^{-1}\theta^{-1}\right)^{a_4}(\theta)^{a_5}$$

Observing homogeneous dimensionality, we obtain five simultaneous equations

$M:0=a_2+a_3$
$L:\ 0=-2+a_1-3a_2-a_3-a_4$
$T:\ 0=-1-a_3-a_4$
$\theta:\ 0=-1-a_4+a_5$
$H:0=1+a_4$

Solving the five simultaneous equations yields

$$a_1=1\ a_2=0\ a_3=0\ a_4=-1\ a_5=0$$

$$\Pi_1=\left[\frac{hD}{k}\right]$$

The process for the second Π term

$$\Pi_2=\left[UD^{b_1}\rho^{b_2}\mu^{b_3}k^{b_4}\left(\Delta\theta\right)^{b_5}\right]$$

$$M^0L^0T^0\theta^0H^0=\left(LT^{-1}\right)(L)^{b_1}\left(ML^{-3}\right)^{b_2}\left(ML^{-1}T^{-1}\right)^{b_3}\left(HL^{-1}T^{-1}\theta^{-1}\right)^{b_4}(\theta)^{b_5}$$

$M:0=b_2+b_3$
$L:\ 0=1+b_1-3b_2-b_3-b_4$
$T:\ 0=-1-b_3-b_4$
$\theta:\ 0=-b_4+b_5$
$H:0=b_4$

$$b_1=1\ b_2=1\ b_3=-1\ b_4=0\ b_5=0$$

$$\Pi_2 = \left[\frac{UD\rho}{\mu}\right]$$

For the third Π term

$$\Pi_3 = \left[c_p D^{c_1} \rho^{c_2} \mu^{c_3} k^{c_4} (\Delta\theta)^{c_5}\right]$$

$$M^0 L^0 T^0 \theta^0 H^0 = \left(HM^{-1}\theta^{-1}\right)(L)^{c_1}\left(ML^{-3}\right)^{c_2}\left(ML^{-1}T^{-1}\right)^{c_3}\left(HL^{-1}T^{-1}\theta^{-1}\right)^{c_4}(\theta)^{c_5}$$

$M: 0 = -1 + c_2 + c_3$
$L: \; 0 = c_1 - 3c_2 - c_3 - c_4$
$T: \; 0 = -c_3 - c_4$
$\theta: \; 0 = -1 - c_4 + c_5$
$H: 0 = 1 + c_4$

$$c_1 = 0 \; c_2 = 0 \; c_3 = 1 \; c_4 = -1 \; c_5 = 0$$

$$\Pi_3 = \left[\frac{c_p \mu}{k}\right]$$

hence, the equation for the heat transfer coefficient can be written as

$$\Pi_1 = f(\Pi_2, \Pi_3)$$

$$\frac{hD}{k} = f\left(\frac{UD\rho}{\mu}, \frac{c_p \mu}{k}\right)$$

or

$$Nu = f(Re, Pr)$$

This is the same result obtained with the method of indices. The correlations obtained using both the indices method and Buckingham's π-method show that the heat transfer coefficient is independent of the temperature relative to the wall surface $(\Delta\theta)$. This is to be expected, since thermophysical properties of the fluid were assumed constant. If these properties are temperature dependent, the correlations may not hold anymore. If $\Delta\theta$ is ignored, the functional relationship for h becomes

$$h = f\left(D, U, \rho, \mu, k, c_p\right) \tag{6.36}$$

and the analysis above can be repeated with the four fundamental dimensions M, L, T, θ yielding the same correlation $Nu = f(Re, Pr)$.

If the flow is categorized as high-speed, in which the kinetic energy of the fluid can be converted to heat, dimension H can no longer be used as a fundamental dimension and we are left with dimensions M, L, T, θ. So, if $m = 8$ and $n = 4$, the functional relationship will include the dimensional groups (Π-groups) Nu, Re, Pr, and a fourth one known as the Eckert number $E = U^2 \mu / \Delta\theta$. The latter is determined using the method of indices and rearranged to account for the dissipation of kinetic energy by viscous friction in the boundary layer resulting in the following correlation (Mills 1999)

$$Nu = f(Re, Pr, EcPr) \tag{6.37}$$

The product $EcPr$ is the Brinkman number Br

$$Br = \frac{U^2 \mu}{k(T_s - T_\infty)} \tag{6.38}$$

6.7 Geometric Similarity and Other Considerations

1) The last two examples illustrate the effectiveness of dimensional-analysis techniques in reducing the total number of dependent and independent variables (eight in this case) to just three, and written as

$$Nu = CRe^a Pr^b \tag{6.39}$$

C, a, and b are constants that can be determined from experiments for a given geometry (tube in this case) and fluid by varying, for example, the flow velocity and specific heat, and plotting the results. Once Eq. (6.39) is established for flows inside a circular tube, it can be used for any tube size as long as the length is much greater than the diameter. It cannot be used for short tubes and other configurations, such as flow across a circular tube or along flat or inclined surfaces.

2) The functional relationship for the heat flux q in natural convection flow on a vertical plane can be expressed as $q = f(L_c, \beta, g, \rho, \mu, k, c_p, \Delta\theta)$ and the non-dimensional function comprising four Π-terms is

$$Nu = f\left(Pr, \beta\Delta\theta, \frac{gL^3}{\nu^2}\right) \tag{6.40}$$

where β is the volumetric coefficient of thermal expansion (θ^{-1}), g is the gravitational constant (LT^{-2}), and ν is the kinematic viscosity. The Nusselt number in this case can be rewritten as

$$Nu = \frac{q\Delta\theta L_c}{k}$$

3) Thermophysical properties of the fluid in the dimensionless groups such as density ρ, dynamic viscosity μ, thermal conductivity k, and specific heat c_p, are temperature dependent to varying degrees. Density for liquids remains almost unchanged with temperature, whereas for gases it decreases with increasing temperature. The specific heats for air and mercury increase significantly with temperature but only marginally for glycerine. Specific heat for water reaches a peak at around $400K$, dropping continuously afterwards. Increasing the temperature causes the specific heat for engine oil to decrease. Viscosity decreases sharply with temperature in some fluids (e.g., glycerine, motor oils) and moderately in others (e.g., kerosene, mercury, water). Viscosity of air, helium, and carbon dioxide increases somewhat with temperature. Accounting for all variations of the properties with temperature is not practicable and the properties in engineering calculations are usually evaluated at a reference temperature T_r also known as the mean film temperature T_f. Often, this temperature is taken as equal to the arithmetic mean of the surface and free stream temperatures.

$$T_f = \frac{T_\infty + T_s}{2} \tag{6.41}$$

In practice, the temperature at which properties are evaluated for a specific convection heat transfer problem is usually stated alongside the correlation or group of correlations.

4) Dimensional analysis can provide satisfactory solutions to convective problems that are too complex to solve analytically but cannot provide any insight into the physics of the problems.
5) Dimensional analysis based on the Rayleigh method and Buckingham Π-theorem can yield useful correlations in relatively simple convection heat transfer problems; however, difficulties arise when analysing more complex problems involving large number of variables and different physical phenomena. In the latter case, good physical insight, experience, and probably some experimentation will be required for the selection of appropriate non-dimensional parameters. Correlation (6.37) obtained for high-speed flow in a tube is an example of the problem stated here.

Problems

6.1 Write down the Reynolds number for flow over flat plate, flow across circular cylinder, flow inside a pipe, and flow inside a channel.

6.2 Air at 300 K and 1.0atm flows over a flat plate 1.2m long. The hydrodynamic boundary layer thickness at the trailing edge of the plate is 15mm. If the flow is laminar, what is the flow velocity at that point?

6.3 A plate (1.0m × 0.3m) is heated electrically to a constant temperature of 230°C. Air at 25°C and 1 bar flows at 60m/s over the plate. Determine whether the flow is laminar or turbulent at the tailing edge of the plate.

6.4 Air at 1.0atm and 350 K flows over the top surface of a flat plate of length 200mm and width 1.0m. Determine the total drag force acting over the plate surface when the flow speed is 30 m/s.

6.5 Air at 1.0atm and 300 K flows over the top surface of a flat plate of 0.6 m length with flow speed of is 8 m/s. The plate is maintained at 400 K. Determine:

(a) the boundary layer thickness at a distance of 0.6m from the leading edge

(b) the position at the point of transition to turbulent flow

(c) total drag force on the plate per unit width at the moment of transition.

6.6 The local heat transfer coefficient h_x for turbulent flow over the flat plate in Figure P6.6 is given as $h(x) = Ax^{-0.2}$, where x is the distance measured from the leading edge of the plate and A is constant of proportionality. Derive an expression for the average heat transfer coefficient for a plate of length L and width w (6.1).

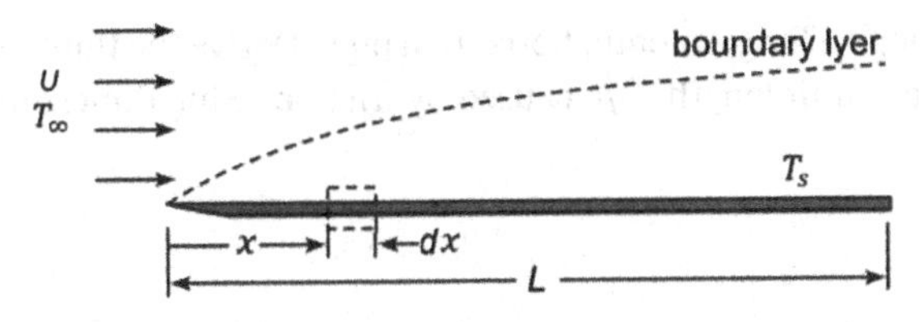

Figure P6.6 Schematic diagram of flow over flat plate.

6.7 Air at 1 atm and $300K$ is flowing over a flat plate. If the flow speed is $30m/s$, determine the distance from the leading edge of the plate up to which the flow remains laminar. Take the properties of air at 1 atm and $300K$ as $\rho = 1.161 kg/m^3$, $\mu = 1.846 \times 10^{-5} kg/m.s$.

6.8 Experimental results for heat transfer over a flat plate with an extremely rough surface were found to be correlated by an expression of the form $Nu_x = 0.04Re^{0.9}Pr^{1/3}$, where Nu_x is the local value of Nusselt number at a position x measured from the leading edge of the plate. Derive an expression for ratio of average heat transfer coefficient $\bar{h}$ over length L to local heat transfer coefficient h_x.

6.9 If the correlation in Problem 6.8 is replaced by $Nu_x = 0.332Re_x^{1/2}Pr^{1/3}$, what is the ratio of the average heat transfer coefficient $\bar{h}$ to the local heat transfer coefficient h_x?

6.10 Air at $1.0atm$ and $300K$ flows over the top surface of a flat plate of length with flow speed of $8\ m/s$. The plate is maintained at $400K$. Determine the local heat transfer coefficient at a distance of 0.6 m from the plate leading edge. The correlation for forced laminar flow of air over a horizontal plate is $Nu_x = 0.332Re_x^{1/2}Pr^{1/3}$

6.11 A thermocouple bead is used to measure the temperature of a gas flowing through a hot channel, as shown in Figure P6.11. The heat transfer coefficient between the gas and thermocouple is given by the function $h = aU^{0.8}$, where U is the gas velocity and heat transfer rate by radiation from the walls to the thermocouple is proportional to temperature difference $\dot{Q}_r = b(T_w - T)$. When the gas is flowing at $5\ m/s$, the thermocouple reads $323\ K$ and when it is flowing at $10\ m/s$, it reads $313\ K$. Calculate:

(a) the wall temperature at a gas temperature of $298\ K$

(b) the temperature the thermocouple indicates when the gas velocity is $20\ m/s$.

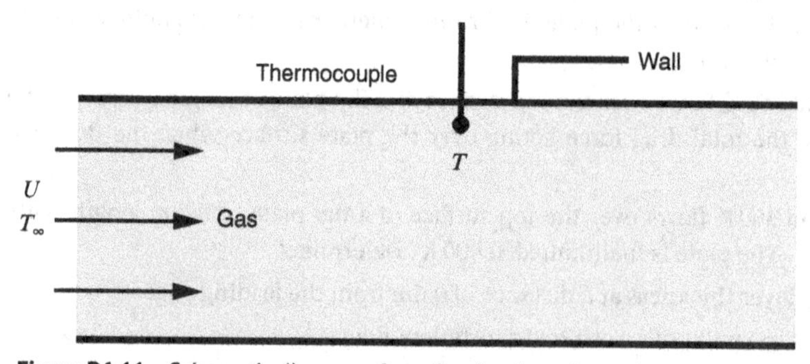

Figure P6.11 Schematic diagram of gas flowing in a channel.

6.13 Heat is generated in a 2-m long conductor and 3-mm diameter as a result of electrical resistance heating at the rate of $90W$. If the temperature of the surface of the conductor is to be maintained at 150°C within an environment at 20°C, what is the coefficient of convection heat transfer between the conductor and the surroundings. Ignore the effect of radiation on the heat transfer process.

6.14 An incompressible fluid flows through a long, smooth, horizontal pipe. Derive the functional relationship for the pressure loss per unit length Δp caused by friction using dimensional analysis.

6.15 Derive the actual correlation for the pressure loss of the flow in Problem 5.14 for a 1500-mm long pipe of 12.5-mm diameter using the following experimental data of total pressure drop versus flow speed at 20°C:

$U, m/s$	0.375	0.594	0.887	1.780	3.392	5.157	7.114	8.757
$\Delta p_t, Pa$	300	747	1480	5075	15753	32606	57456	82832

The properties of water at 20°C are: $\rho = 998.2 kg/m^3$, $\mu = 1.005 \times 10^{-3} N.s/m^2$

7

Forced Convection – External Flows

In a forced external flow over a solid object, a boundary layer forms on the surface of the object without being influenced by other surfaces. The boundary layer formed could be laminar or turbulent, depending on the flow velocity. The flow outside the boundary layer is characterized by constant velocity (flow stream velocity) and temperature. The fluid motion is normally driven by external mechanical means such as compressor, pump, or blower, but could also be caused by wind and sea waves. Three geometries will be considered: flat (horizontal) plate, circular cylinder, and sphere.

7.1 Flow Over a Flat Plate

7.1.1 Laminar Flow Over a Flat Plate

One of the few problems in convection heat transfer that can be solved analytically is the two-dimensional parallel flow over a flat plate. To evaluate the convection heat transfer coefficient for such a flow, the equations of conservation of mass, momentum, and energy in the boundary layer are solved for the temperature gradient at the interface between the fluid and the solid surface. The three differential equations, collectively known as Prandtl's boundary-layer approximation, are:

Conservation of mass (continuity equation)

$$\frac{\partial u}{\partial x}+\frac{\partial u}{\partial y}=0 \tag{7.1}$$

Conservation of momentum

$$u\frac{\partial u}{\partial x}+v\frac{\partial u}{\partial y}=\left(\frac{\mu}{\rho}\right)\frac{\partial^2 u}{dy^2} \tag{7.2}$$

Conservation of energy

$$u\frac{\partial T}{\partial x}+v\frac{\partial T}{\partial y}=\left(\frac{k}{\rho c_p}\right)\frac{\partial^2 T}{dy^2} \tag{7.3}$$

Equations (7.2) and (7.3) are very similar and become identical for $\mu / \rho = k / \rho c_p$, or $c_p\mu / k = Pr = 1$. This implies that at $Pr = 1$ the velocity and temperature distribution are identical and the

Heat Transfer Basics: A Concise Approach to Problem Solving, First Edition. Jamil Ghojel.

Companion website: www.wiley.com/go/ghojel/heat_transfer

hydrodynamic and thermal boundary layers are of the same thickness. The ratio of the two boundary layers at other Prandtl number values is equal to $Pr^{-1/3}$ (Ede 1967).

The simultaneous solution of the governing equations yields the thickness of the hydrodynamic boundary layer δ, the velocity gradient in the fluid at the surface, the shear stress per unit area at the solid surface, and the Nusselt number (heat transfer coefficient). The results of the analysis are presented without proof and the reader can find detailed solutions in advanced references.

Combining Eq. (6.7) for the hydrodynamic boundary layer thickness with Eq. (6.8) for the velocity profile in the x-direction in the boundary layer (see Chapter 6), we obtain

$$\frac{U}{U_\infty} \approx \frac{1}{3.09}\left[\left(\frac{y}{x}\right)Re_x^{1/2}\right] - \frac{1}{200}\left[\left(\frac{y}{x}\right)Re_x^{1/2}\right]^3 \tag{7.4}$$

The plot for U/U_∞ versus $(y/x)Re_x^{1/2}$ is shown in Figure 7.1 as a solid line. The plot based on the Blasius simultaneous solution of Eqs (7.1) and (7.2) from Forsberg (2021) is shown as a broken line. The two plots correlate well up to $(y/x)Re_x^{1/2} = 4.64$.

The shear stress per unit area of the wall surface is

$$\tau = 0.332\frac{\mu U_\infty}{x}Re_x^{1/2} \tag{7.5}$$

The local coefficient of friction at distance x and the average value over the length x were given in Chapter 6 by Eqs (6.9) and (6.10), respectively.

The temperature distribution in the thermal boundary layer can also by written as a cubic parabola featuring the thermal boundary layer δ_T (Rogers and Mayhew 1992; Thomas 1999)

$$\frac{T - T_s}{T_\infty - T_s} = \frac{3}{2}\left(\frac{y}{\delta_T}\right) - \frac{1}{2}\left(\frac{y}{\delta_T}\right)^3 \tag{7.6}$$

If we assume that the plate is heated over its entire length, $x_0 = 0$ (see Figure 6.7) and Eq. (6.20) can be rewritten as

$$\delta_T = \frac{\delta}{1.026}Pr^{-1/3} \tag{7.7a}$$

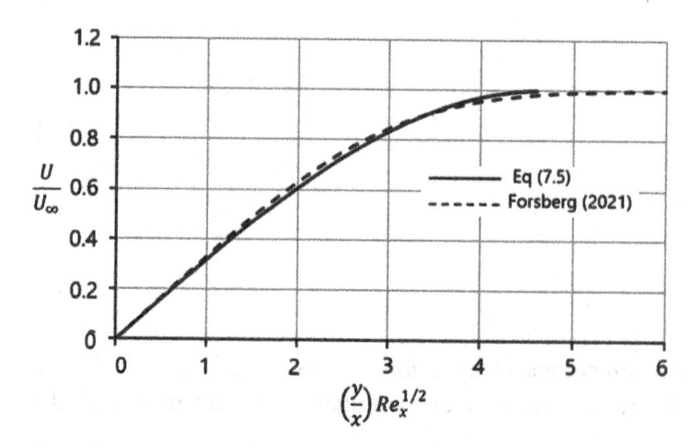

Figure 7.1 Velocity profile in the boundary layer of flow over a flat plate.

$$\frac{\delta_T}{\delta} \sim Pr^{-1/3} \tag{7.7b}$$

Equation (7.7b) shows that the ratio of thermal to hydrodynamic boundary layers is inversely proportional to the Prandtl number $(Pr = c_p\mu / k = \nu / \alpha)$ of the fluid.

Combining Eqs. (6.7), (7.6), and (7.7a) we obtain

$$\frac{T - T_s}{T_\infty - T_s} = 0.332Pr^{1/3}\left[\left(\frac{y}{x}\right)Re_x^{1/2}\right] - 5.414\times10^{-3}Pr\left[\left(\frac{y}{x}\right)Re_x^{1/2}\right]^3 \tag{7.8}$$

Figure 7.2 shows the temperature distribution for different Prandtl numbers calculated using Eq. (7.8).

As stated above, at $Pr = 1$ the velocity and temperature distribution are identical (velocity profile is designated by the black circular markers in Figure (7.2) and the hydrodynamic and thermal boundary layers are of the same thickness).

Local conduction and convection heat fluxes at a heated wall are equal

$$q_x = h_x\left(T_s - T_\infty\right) = -k\left(\frac{dT}{dy}\right)_{y=0}$$

hence,

$$h_x = \frac{-k\left(\frac{dT}{dy}\right)_{y=0}}{\left(T_s - T_\infty\right)} \tag{7.9}$$

Differentiating Eq. (7.6) and putting $y = 0$ yields

$$\left(\frac{dT}{dy}\right)_{y=0} = \frac{3}{2}\frac{1}{\delta_T}\left(T_\infty - T_s\right) \tag{7.10}$$

Combining Eqs (7.10), (6.21), and (6.7) we obtain

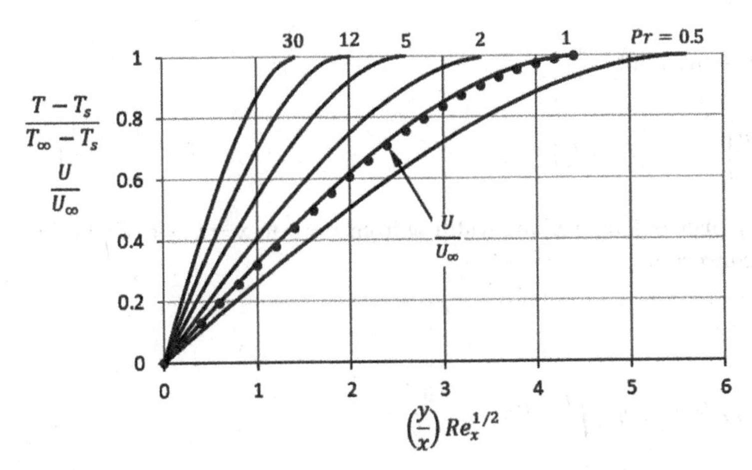

Figure 7.2 Temperature distribution in laminar boundary layer.

$$\left(\frac{dT}{dy}\right)_{y=0} = \frac{3}{2}\frac{1.026Re_x^{1/2}}{4.64x}Pr^{1/3}\left[1-\left(\frac{x_0}{x}\right)^{3/4}\right]^{-1/3}\left(T_\infty - T_s\right)$$

$$\left(\frac{dT}{dy}\right)_{y=0} = 0.332\,x^{-1/2}\left(\frac{U_\infty \rho}{\mu}\right)^{1/2} Pr^{1/3}\left[1-\left(\frac{x_0}{x}\right)^{3/4}\right]^{-1/3}\left(T_\infty - T_s\right) \quad (7.11)$$

The local coefficient of convection heat transfer at the wall can now be found from Eqs (7.9) and (7.11)

$$h_x = 0.332kx^{-1/2}\left(\frac{U_\infty \rho}{\mu}\right)^{1/2} Pr^{1/3}\left[1-\left(\frac{x_0}{x}\right)^{3/4}\right]^{-1/3} \quad (7.12)$$

If we multiply both sides of Eq. (7.12) by x / k, we obtain its non-dimensional form

$$Nu_x = 0.332Re_x^{1/2}Pr^{1/3}\left[1-\left(\frac{x_0}{x}\right)^{3/4}\right]^{-1/3} \quad (7.13)$$

where $Nu_x = h_x x / k$

For a flat plate heated over its entire length $x_0 = 0$, the local Nusselt number is

$$Nu_x = 0.332Re_x^{1/2}Pr^{1/3} \quad \text{for } 0.6 \le Pr \le 10,\ Re_x < 5\times10^5 \quad (7.14a)$$

and the average Nusselt number is

$$\overline{Nu_L} = \frac{\bar{h}L}{k} = 0.664Re_L^{1/2}Pr^{1/3} \quad \text{for } 0.6 < Pr < 10,\ Re_x < 5\times10^5 \quad (7.14b)$$

The local heat transfer coefficient can be written as

$$h_x = 0.332kx^{-1/2}\left(\frac{U_\infty \rho}{\mu}\right)^{1/2} Pr^{1/3} \quad (7.15)$$

The local heat flux is equal to $\left[-\left(dT / dy\right)_{y=0}\right]$

$$q_x = 0.332kx^{-1/2}\left(\frac{U_\infty \rho}{\mu}\right)^{1/2} Pr^{1/3}\left(T_\infty - T_s\right) \quad (7.16)$$

Integrating Eq. (7.16) over a plate of length L and width w from $x = 0$ to $x = L$, we obtain the total rate of heat transfer by convection

$$\dot{Q} = w\int_0^L qxdx$$

$$\dot{Q} = 0.322kw\left(\frac{U_\infty \rho}{\mu}\right)^{1/2} Pr^{1/3}\left(T_\infty - T_s\right)\int_0^L x^{-1/2}dx$$

$$\dot{Q}=0.322kw\left(\frac{U_{\infty}\rho}{\mu}\right)^{1/2}Pr^{1/3}\left(T_s-T_{\infty}\right)\left(2L^{1/2}\right)$$

$$\dot{Q}=0.644kw\left(\frac{U_{\infty}\rho L}{\mu}\right)^{1/2}Pr^{1/3}\left(T_s-T_{\infty}\right)$$

$$\dot{Q}=0.644kwRe_L^{1/2}Pr^{1/3}\left(T_s-T_{\infty}\right) \tag{7.17}$$

It can be shown, on the basis of the equations deduced so far, that fluid friction and heat transfer are linked as per the following relationship:

$$\frac{f}{2}=St_xPr^{2/3} \tag{7.18}$$

St_x is the Stanton number, defined as

$$St_x=\frac{h_x}{\rho c_p U_{\infty}}=\frac{Nu_x}{Re_x Pr} \tag{7.19}$$

Example 7.1 A solar collector 0.4 m wide and 2.5 m long is mounted horizontally on a house roof. Wind at average temperature T_{∞} = 15°C and velocity $U_{\infty}=4m/s$ is blowing over the collector. Assume that the sun heats the entire collector to a temperature of $T_s=65°C$. Determine the heat transfer coefficient h_x at $x=1.25m$ and the heat loss rate from the surface up to that point. The thermophysical properties of air: $\rho=1.225kg/m^3$, $c_p=1.007kJ/kg.K$, $\mu=1.802\times10^{-5}\,kg/m.s$, $k=0.02476W/m.K.$

Solution

From Eq. (7.15)

$$h_x=0.332kx^{-1/2}\left(\frac{U_{\infty}\rho}{\mu}\right)^{1/2}Pr^{1/3}$$

$$Pr=\frac{c_p\mu}{k}=\frac{1007\times1.802\times10^{-5}}{0.02476}=0.732$$

$$h_{1.25m}=0.332\times0.02476\times1.25^{-1/2}\left(\frac{4\times1.225}{1.802\times10^{-5}}\right)^{1/2}0.732^{1/3}=3.455W/m^2.K$$

For the plate up to $x=1.25m$,

$$Re_{1.25}=\frac{U_{\infty}\rho L}{\mu}=\frac{4\times1.225\times1.25}{1.802\times10^{-5}}=3.4\times10^5$$

Since $Re_L<5\times10^5$, the flow is laminar and we use Eq. (7.17) for the heat loss rate

$$\dot{Q}=0.644kwRe_L^{1/2}Pr^{1/3}\left(T_s-T_{\infty}\right)$$

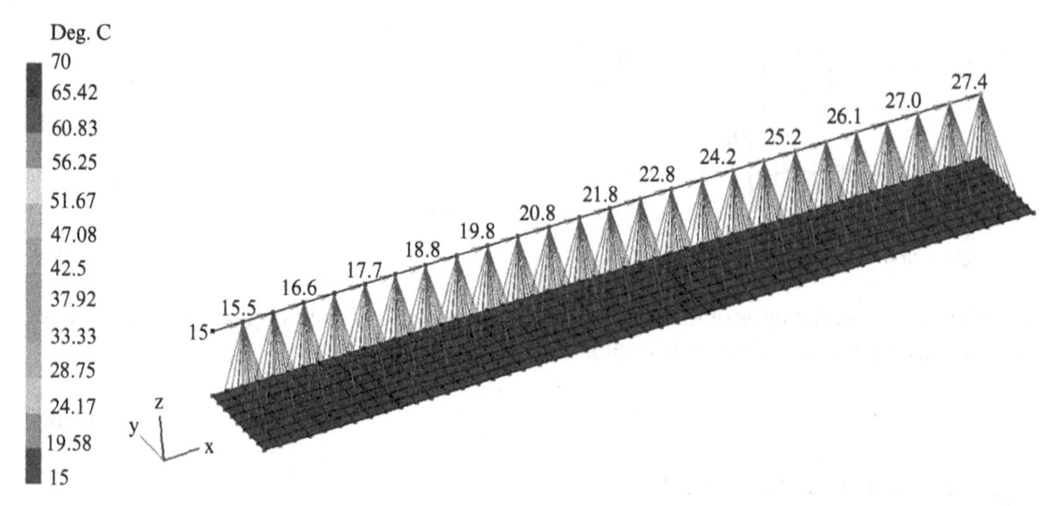

Figure E7.1 Temperature variation in a heated solar collector.

$$\therefore\ \dot{Q} = 0.644 \times 0.02476 \times 0.4 \times \left(3.4 \times 10^5\right)^{1/2} \times 0.732^{1/3} \times (70 - 15) = 184.3W$$

Figure E7.1 shows Solaria modelling of this problem with linear temperature variation along the collector length in degrees C. The dark thin lines symbolize convection heat transfer from the collector to the flowing air.

7.1.2 Turbulent Flow Over a Flat Plate

The flow in the boundary layer over a flat plate in most engineering applications is turbulent. The development of the turbulent flow over a flat plate is illustrated in Figure 7.3 and shows that a flow is initially laminar over a critical distance x_{cr} defining the laminar region followed by a transition region before it develops to a full turbulent flow. Transition between laminar and turbulent flow occurs at Reynolds number $Re = 5 \times 10^5$, which corresponds to a critical distance $x_{cr} = 5 \times 10^5 \mu / \rho U_\infty$.

The flow velocity or stream velocity U outside the boundary layer is constant in a laminar flow but acquires a fluctuating character as a function of time in a turbulent flow, as shown in Figure 7.4.

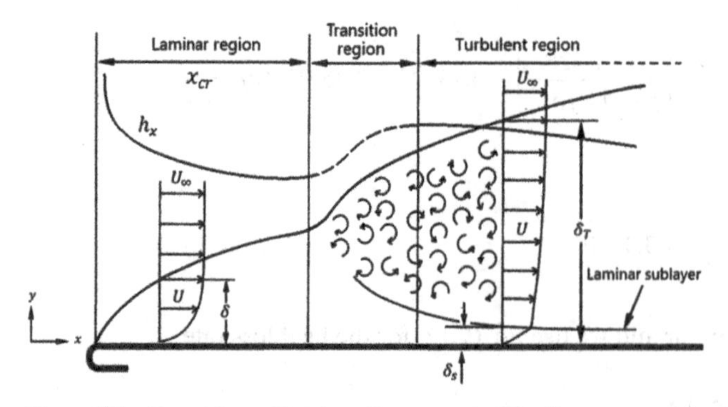

Figure 7.3 Transition of laminar flow to turbulent flow over a flat plate and its effect on the velocity profile and heat transfer coefficient.

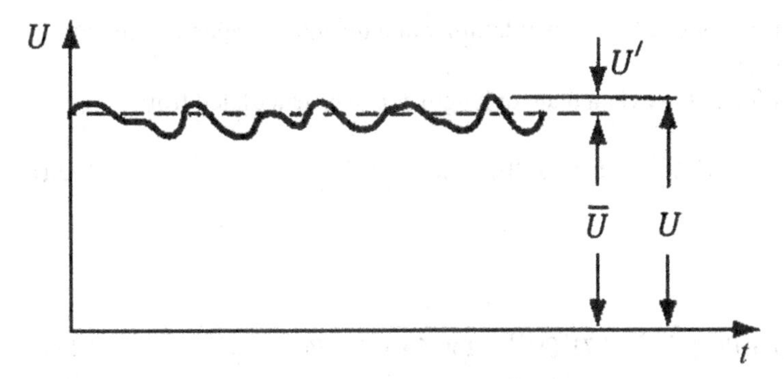

Figure 7.4 Velocity fluctuations in a turbulent flow.

Therefore, only a mean value of velocity $\bar{U}$ can be specified which then determines the instantaneous velocity as

$$U = \bar{U} + U' \tag{7.20}$$

The mixing action of turbulence causes heat and momentum transfer to change significantly and the degree of turbulence to vary with location such that turbulence near the solid surface may be laminar, even in a fully developed turbulent flow. The laminar boundary layer near the wall is known as the laminar sublayer (Figure 7.3). All these complexities render exact analytical analysis based on solving boundary layer differential equations of even the simplest flows almost impossible. Approximate solution based on the integral equation method and empirical data based on experiments provide the equations below for turbulent flow analysis.

Equation (6.13) presented in Chapter 6 as the velocity profile in turbulent flows should be applicable only outside the laminar sublayer

$$\frac{U}{U_\infty} = \left(\frac{y}{\delta}\right)^{1/7} \tag{7.21}$$

The thickness of the turbulent boundary layer is

$$\delta_T = 0.38x\,Re_x^{-1/5} \tag{7.22}$$

The friction factor is

$$f = 0.0592\,Re_x^{-1/5} \qquad \text{for } 5\times10^5 < Re_x < 10^7 \tag{7.23a}$$

$$f = 0.37\left(\log Re_x\right)^{-2.584} \qquad \text{for } 10^7 < Re_x < 10^9 \tag{7.23b}$$

The local heat transfer coefficient can be calculated from

$$Nu_x = \frac{hx}{k} = 0.0296\,Re_x^{4/5}Pr^{1/3} \qquad \text{for } 5\times10^5 < Re_x < 10^7,\quad 0.6 < Pr < 60 \tag{7.24a}$$

$$Nu_x = \frac{hx}{k} = 0.285\,Re_L\left(\log Re_L\right)^{-2.584}Pr^{1/3} \qquad \text{for } 10^7 < Re_x < 10^9,\quad 0.6 < Pr < 60 \tag{7.24b}$$

The variation of the local heat transfer coefficient over laminar and turbulent regions of the flow is shown schematically in Figure 7.3.

The average heat transfer coefficient $\bar{h}$ over a plate of length L can be calculated from

$$\overline{Nu_L} = \frac{\bar{h}L}{k} = \left(0.037Re_L^{4/5} - 871\right)Pr^{1/3} \quad \text{for} \quad 5\times10^5 < Re_L < 10^7 \tag{7.25a}$$

where $Re_L = U\rho L/\mu$

$$\overline{Nu_L} = \frac{\bar{h}L}{k} = \left[0.228Re_L\left(\log Re_L\right)^{-2.584} - 871\right]Pr^{1/3} \quad \text{for} \quad 5\times10^7 < Re_L \le 10^9 \tag{7.25b}$$

Example 7.2 Calculate the average heat loss rate from the solar collector in Example 7.1 for wind velocities $25m/s$ and $65m/s$.

Solution

$W = 0.4m$, $L = 2.5m$, $T_\infty = 15°C$, $T_s = 65°C$, $\rho = 1.25kg/m^3$, $c_p = 1.005kJ/kg.K$, $\mu = 1.9\times10^{-5}\,kg/m.s$, $k = 0.0265\,W/m.K$, $Pr = 0.72$

At $U_\infty = 25m/s$

$$Re_L = \frac{U\rho L}{\mu} = \frac{25\times1.25\times2.5}{1.9\times10^{-5}} = 4.11\times10^6 \text{ (flow is turbulent)}$$

We use Eq. (7.25a) to calculate the average heat transfer coefficient $\bar{h}$

$$\overline{Nu_L} = \frac{\bar{h}L}{k} = \left(0.037Re_L^{4/5} - 871\right)Pr^{1/3} = \left[0.037\times\left(4.11\times10^6\right)^{0.8} - 871\right]\times0.72^{1/3}$$

$$\overline{Nu_L} = 5701$$

$$\bar{h} = \frac{\overline{Nu_L}k}{L} = \frac{5701\times0.0265}{2.5} = 60.4W/m^2.K$$

$$\dot{Q} = \bar{h}A\left(T_s - T_\infty\right) = 60.4\times0.4\times2.5\times50 = 3020W$$

At $U_\infty = 65m/s$

$$Re_L = \frac{U\rho L}{\mu} = \frac{65\times1.25\times2.5}{1.9\times10^{-5}} = 1.07\times10^7 \text{ (flow is turbulent)}$$

We use Eq. (7.25b) to calculate the average heat transfer coefficient $\bar{h}$

$$\overline{Nu_L} = \frac{\bar{h}L}{k} = \left[0.228Re_L\left(\log Re_L\right)^{-2.584} - 871\right]Pr^{1/3}$$

$$\overline{Nu_L} = \left[0.228\times1.15\times10^7\times\left(\log\left(1.07\times10^7\right)\right)^{-2.584} - 871\right]\times0.72^{1/3} = 13388$$

$$\bar{h} = \frac{\overline{Nu_L}k}{L} = \frac{13388\times0.0265}{2.5} = 141.9W/m^2.K$$

$$\dot{Q} = \bar{h}A(T_s - T_\infty) = 141.9 \times 0.4 \times 2.5 \times 50 = 7095\,W$$

7.2 Flow Over a Cylindrical Tube

Schematic representations of the fluid flow field around a cylinder or tube in cross flow are shown in Figure 7.5. The flow splits into two streams that flow around the cylinder forming vortices and a wake at the rear of the cylinder (Figure 7.5a). A mixed laminar and turbulent boundary layer develops around the cylinder that separates at a point downstream determined by the Reynolds number Re_D. Velocity in the boundary layer decreases from the free stream velocity U_x at the edge of the boundary layer to zero at the surface of the cylinder (Figure 7.5b). Following separation, the velocity in the boundary layer near the surface reverses and a wake is formed downstream of the flow. This wake is characterized by the formation of vortices and highly irregular motion and is the cause of the pressure drop in the flow at the rear of the cylinder.

In addition to the force induced by the shear stress in the boundary layer, bluff bodies such as cylinders and spheres experience a force known as form drag. Form drag is the net force acting in the direction of motion due to a pressure gradient in the flow between the stagnation point and separation point.

A dimensionless drag coefficient C_D is defined as

$$C_D = \frac{F_D / A_f}{\frac{1}{2}\rho U_\infty^2} \tag{7.26}$$

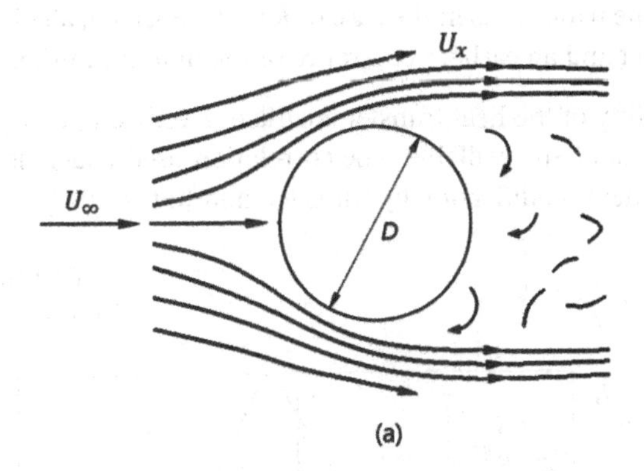

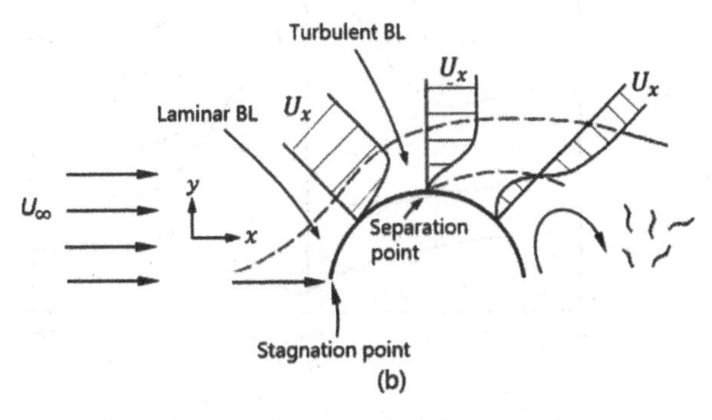

Figure 7.5 Horizontal cylinder (tube) in cross flow.

where F_D is the drag force and A_f is the frontal area of the bluff body (for a cylinder diameter D and length L, $A_f = LD$). For smooth surfaces, C_D varies within 1.1 to 1.2 for a wide range of Reynolds numbers.

As a result of the complexity of the flow field of the circular cylinder, an analytical solution based on differential equations is not feasible. However, experiments show that the heat transfer coefficient can be represented by empirical equations of the form

$$Nu = f\left(Re_D, Pr\right)$$

where

$$Nu = \frac{hD}{k}, Re_D = \frac{U_\infty \rho D}{\mu}, Pr = \frac{c_p \mu}{k}$$

A representation of the effect of Reynolds number Re_D on the local heat transfer from a heated cylinder to water is shown in Figure 7.6. Each of the three curves represents the change of the local heat transfer coefficient at constant Re_D as the flow progresses over the cylinder.

- At the lower Reynolds number Re_{D1}, the heat transfer coefficient decreases from the stagnation point reaching a minimum at the separation point towards the rear of the cylinder ($\theta > 90$), then starts increasing steadily in the vortex region
- At the intermediate Reynolds number Re_{D2}, two minima are observed, with the first occurring at the earlier separation point followed by a sharp increase in the heat transfer coefficient in the turbulent boundary layer reaching a maximum followed by a drop as the turbulent boundary layer separates in the wake
- At the higher Reynolds number Re_{D3}, the same patterns as in the case of Re_{D2} is repeated, albeit at higher values of the heat transfer coefficient and an earlier occurrence of the first minimum.

Due to the complexities of the flow and variability of the heat transfer coefficient, only empirical correlations for the average heat transfer coefficient are available. The correlation used widely is based on work by McAdams (1954) and subsequent modification by Knusden and Katz (1958)

$$\overline{Nu}_L = C Re_D^m Pr^n \tag{7.27}$$

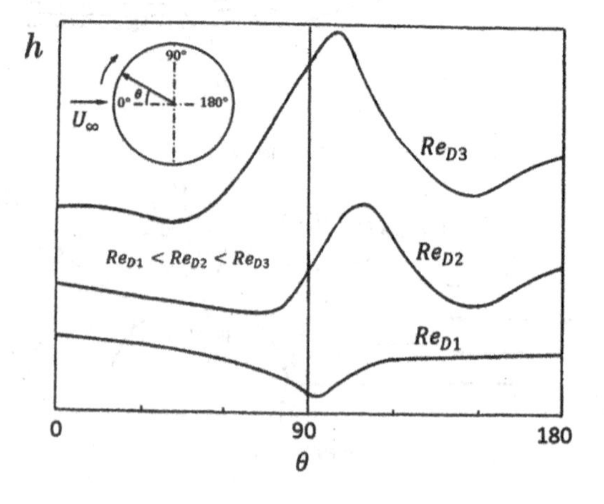

Figure 7.6 Variation of local heat transfer coefficient with location on cylinder surface and Reynolds number.

Table 7.1 Constant C and exponent m for Eq. (7.27).

Re_D	C	m
$0.4-4$	0.989	0.33
$4-40$	0.911	0.385
$40-4000$	0.683	0.466
$4000-40{,}000$	0.193	0.618
$40{,}000-400{,}000$	0.027	0.805

Adapted from Hilpert (1933) and Knusden and Katz (1958).

Exponent n is usually assumed constant and equal to one-third. The values of C and exponent m vary with the Reynolds number, as specified in Table 7.1. The properties for use with Eq. (7.27) are evaluated at the film (mean) temperature $T_f = (T_s + T_\infty)/2$.

The following empirical correlations, based on work by Zukauskas et al. (1968), Zukauskas and Ulinskas (1988), can also be used for cylindrical tubes in cross flow.

$$\overline{Nu}_L = 0.56\, Re_D^{0.5} Pr^{0.38} \left(\frac{Pr}{Pr_s}\right)^{0.25} \quad \text{for} \quad 5 < Re_D < 10^3 \tag{7.28}$$

$$\overline{Nu}_L = 0.28\, Re_D^{0.6} Pr^{0.38} \left(\frac{Pr}{Pr_s}\right)^{0.25} \quad \text{for} \quad 10^3 < Re_D < 2\times10^5 \tag{7.29}$$

For use in flow of air across cylindrical tubes, Eqs (7.28) and (7.29) are simplified to

$$\overline{Nu}_L = 0.49\, Re_D^{0.5} \quad \text{for} \quad Re_D < 10^3 \tag{7.30}$$

$$\overline{Nu}_L = 0.24\, Re_D^{0.6} \quad \text{for} \quad Re_D > 10^3 \tag{7.31}$$

All properties in Eqs (7.28) to (7.31) are evaluated at the flow stream temperature T_∞ with the exception of Pr_s, which is evaluated at the wall temperature T_s.

A more complicated correlation that can be used for a wide range of Re_D and Pr is proposed by Churchill and Bernstein (1977)

$$\overline{Nu}_D = 0.3 + \frac{0.62 Re_D^{1/2} Pr^{1/3}}{\left[1+(0.4/Pr)^{2/3}\right]^{1/4}} \left[1+\left(\frac{Re_D}{282{,}000}\right)^{5/8}\right]^{4/5} \tag{7.32}$$

All properties are evaluated at the film temperature $T_f = (T_s + T_\infty)/2$.

7.3 Tube Banks in Crossflow

Problems of heat transfer to or from single tubes are not common in engineering practice, as it is more likely to have systems in which a number of tubes are arranged together forming an array of tubes known as "tube bank" or "bank of tubes." A typical example would be a heat exchanger with one fluid flowing inside the tubes and another fluid flowing over the tubes. The tubes in the tube banks are arranged horizontally in line or a staggered order in the direction of flow, as shown in Figure 7.7. The tubes form squares in the inline configuration (Figure 7.7a) and triangles in the staggered configuration (Figures 7.7b and 7.7c).

The configurations are characterized by the tube diameter D and three pitches that are measured between the tube centres: longitudinal pitch S_L, transverse pitch S_T, and diagonal pitch S_D. When evaluating the empirical correlations for tube banks, the free flow velocity U_∞ can no longer be used and instead a maximum mean velocity U_{max} is used, which is defined below and summarized in Table 7.2.

For inline arrangement of tubes of length L, $A_1 = (S_T - D)L$, the Continuity equation is

$$U_{max}\rho A_1 = U_\infty \rho S_T L$$

hence,

$$U_{max} = \frac{U_\infty S_T}{S_T - D}. \tag{7.33}$$

For a staggered arrangement of tubes of length L, $A_2 = (S_D - D)L$, where S_D is the diagonal pitch

$$S_D = \sqrt{\left[(S_T/2)^2 + S_L^2\right]}$$

The Continuity equation is

$$2U_{max}\rho A_2 = U_\infty \rho S_T L$$

hence,

$$U_{max} = \frac{U_\infty S_T/2}{\sqrt{\left[(S_T/2)^2 + S_L^2\right]} - D} \tag{7.34}$$

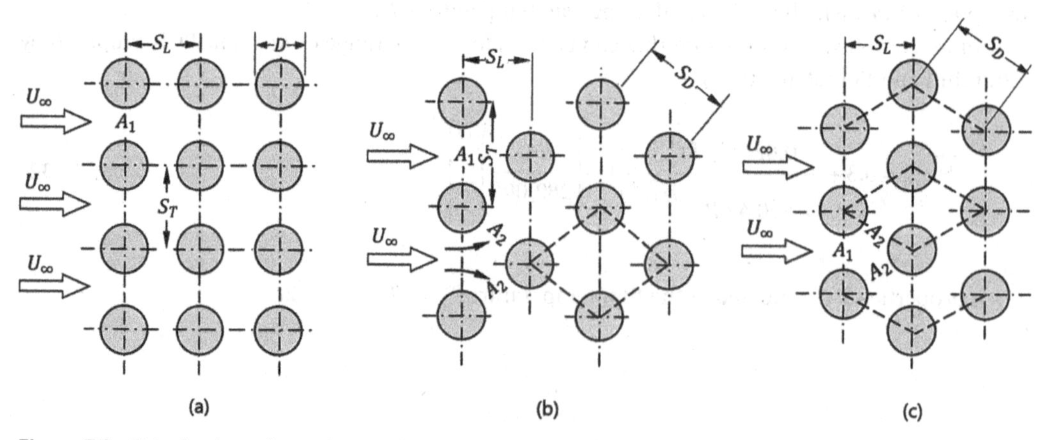

Figure 7.7 Tube bank configurations (a) inline tube rows; (b and c) staggered tube rows.

Table 7.2 Determination of U_{max} in tube banks.

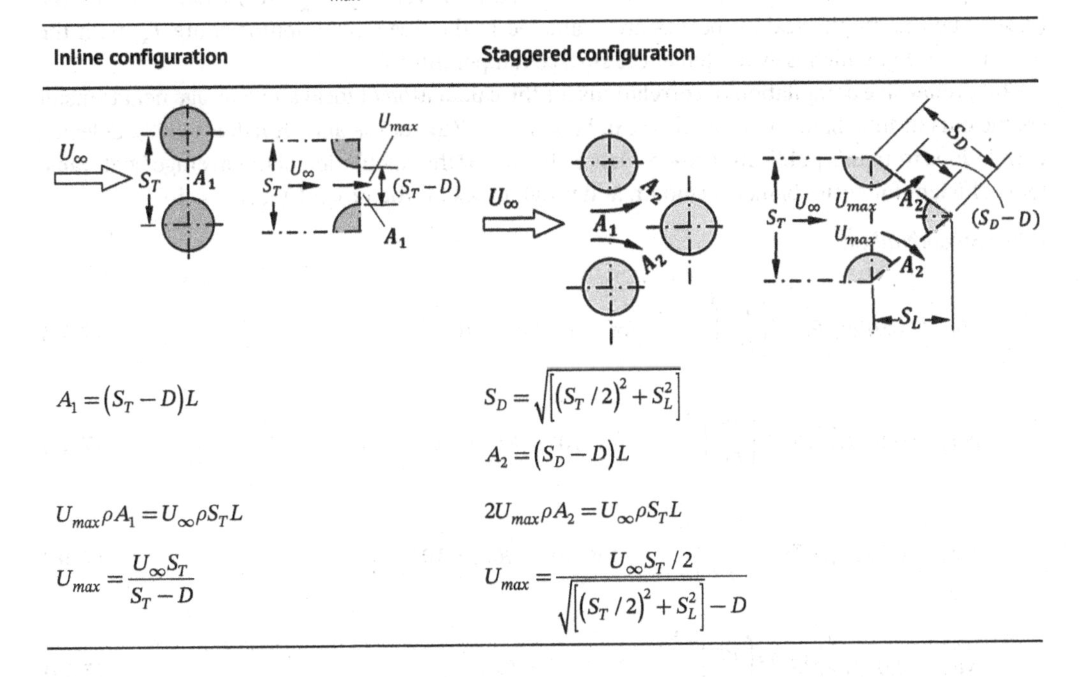

Inline configuration	Staggered configuration
$A_1 = (S_T - D)L$	$S_D = \sqrt{\left[(S_T/2)^2 + S_L^2\right]}$
	$A_2 = (S_D - D)L$
$U_{max}\rho A_1 = U_\infty \rho S_T L$	$2U_{max}\rho A_2 = U_\infty \rho S_T L$
$U_{max} = \dfrac{U_\infty S_T}{S_T - D}$	$U_{max} = \dfrac{U_\infty S_T/2}{\sqrt{\left[(S_T/2)^2 + S_L^2\right]} - D}$

Equation (7.34) is used for the staggered tube bank if

$$2A_2 < A_1, \text{ or } 2\times\left\{\sqrt{\left[(S_T/2)^2 + S_L^2\right]} - D\right\} < (S_T - D);$$

If this condition is not met, the maximum velocity is calculated from Eq. (7.33).
Depending on the transverse pitch, the flow fields across the front tube row of an inline tube bank and a single tube are generally similar and the heat transfer coefficients approximately equal. At high Reynolds numbers, the tube rows downstream of the flow are in the wakes of the upstream rows and in contact with a flow of a high degree of turbulence, as a result of which the heat transfer coefficient increases first then becomes constant somewhere between the third and fifth rows. A decrease of heat transfer coefficient could occur at low Reynolds numbers in inline configurations. The staggered configuration provides better mixing for the tubes downstream of the flow, resulting in a more effective heat transfer process in comparison with the inline configuration. As a general rule, heat transfer intensifies in the inner rows of tube banks with decreasing longitudinal pitch, with the exception of inline banks in which intensity could decrease with decreasing pitch.

7.3.1 Banks of Smooth Tubes

The functional relationship for heat transfer in banks of smooth tubes in dimensionless form encompasses the Nusselt number, Reynolds number, Prandtl number, thermophysical properties of the fluid of the free stream and in the vicinity of the wall surface, and the geometric parameters in terms of longitudinal S_L/D and traverse S_T/D pitches

$$Nu_D = f\left(Re_D, \frac{Pr}{Pr_s}, \frac{\mu_\infty}{\mu_s}, \frac{k_\infty}{k_s}, \frac{c_{p\infty}}{c_{ps}}, \frac{\rho_\infty}{\rho_s}, \frac{S_L}{D}, \frac{S_T}{D}\right) \tag{7.35}$$

The Reynolds number Re_D is evaluated at the maximum velocity U_{max} from Eq. (7.33) or Eq. (7.34). The thermophysical properties are evaluated at the free stream temperature T_∞ with the exception of Pr_s, which is evaluated at the surface temperature T_s.

Comprehensive compilation of correlations for the calculation of local and average heat transfer coefficients in tube banks in cross flow can be found in Zukauskas and Ulinskas (1988). Selected correlations from this publication are presented below for the calculation of the average heat transfer coefficient for entire banks of smooth and rough tubes for the range $10 < Re_D < 10^6$.

Inline Smooth tubes:

$$\overline{Nu}_L = 0.9 Re_D^{0.4} Pr^{0.36} \left(\frac{Pr}{Pr_s}\right)^{0.25} \quad \text{for } 10 < Re_D < 10^2 \tag{7.36}$$

$$\overline{Nu}_L = 0.52 Re_D^{0.5} Pr^{0.36} \left(\frac{Pr}{Pr_s}\right)^{0.25} \quad \text{for } 10^2 < Re_D < 10^3 \tag{7.37}$$

$$\overline{Nu}_L = 0.27 Re_D^{0.65} Pr^{0.36} \left(\frac{Pr}{Pr_s}\right)^{0.25} \quad \text{for } 10^3 < Re_D < 10^5 \tag{7.38}$$

$$\overline{Nu}_L = 0.033 Re_D^{0.8} Pr^{0.4} \left(\frac{Pr}{Pr_s}\right)^{0.25} \quad \text{for } 10^5 < Re_D < 10^6 \tag{7.39}$$

Staggered Smooth tubes:

$$\overline{Nu}_L = 1.04 Re_D^{0.4} Pr^{0.36} \left(\frac{Pr}{Pr_s}\right)^{0.25} \quad \text{for } 10 < Re_D < 1.6 \times 10^2 \tag{7.40}$$

$$\overline{Nu}_L = 0.71 Re_D^{0.5} Pr^{0.36} \left(\frac{Pr}{Pr_s}\right)^{0.25} \quad \text{for } 1.6 \times 10^2 < Re_D < 10^3 \tag{7.41}$$

$$\overline{Nu}_L = 0.35 Re_D^{0.6} Pr^{0.36} \left(\frac{S_T}{S_L}\right)^{0.2} \left(\frac{Pr}{Pr_s}\right)^{0.25} \quad \text{for } \frac{S_T}{S_L} < 2,\ 10^3 < Re_D < 2 \times 10^5 \tag{7.42}$$

$$\overline{Nu}_L = 0.4 Re_D^{0.65} Pr^{0.36} \left(\frac{Pr}{Pr_s}\right)^{0.25} \quad \text{for } \frac{S_T}{S_L} > 2, 10^3 < Re_D < 2 \times 10^5 \tag{7.43}$$

$$\overline{Nu}_L = 0.031 Re_D^{0.8} Pr^{0.4} \left(\frac{S_T}{S_L}\right)^{0.2} \left(\frac{Pr}{Pr_s}\right)^{0.25} \quad \text{for } 2 \times 10^5 < Re_D < 2 \times 10^6 \tag{7.44}$$

The relatively low local heat transfer coefficient in the first three rows of a tube bank can have appreciable effect on the mean heat transfer coefficient of the tube bank. Hence, it is recommended to multiply correlations in Eqs (7.36) to (7.44) by a correction factor C_N if the number of longitudinal tube rows (in the direction of flow) is less than 16. The values of C_N are dependent on the type of tube bank (inline or staggered) and on the Reynolds number, as shown in Table 7.3. The table is based on graphical data in Zukauskas and Ulinskas (1988).

Table 7.3 Correction factor for the mean heat transfer coefficient for tube banks with N rows in the direction of flow less than 16.

	C_N			
	Inline bank		Staggered bank	
N	$10^2 < Re_D < 10^3$	$Re_D > 10^3$	$10^2 < Re_D < 10^3$	$Re_D > 10^3$
1	1	0.698	0.834	0.618
2	1	0.809	0.878	0.766
3	1	0.868	0.914	0.847
4	1	0.904	0.937	0.895
5	1	0.929	0.953	0.925
6	1	0.947	0.965	0.945
7	1	0.959	0.973	0.959
8	1	0.969	0.979	0.969
9	1	0.977	0.985	0.976
10	1	0.983	0.988	0.982
11	1	0.988	0.991	0.987
12	1	0.992	0.994	0.990
13	1	0.995	0.996	0.994
14	1	0.997	0.998	0.997
15	1	0.999	0.999	0.999
16	1	1	1	1

7.3.2 Banks of Rough Staggered Tubes

The functional relationship for heat transfer in banks of rough tubes in a staggered tube bank in dimensionless form encompasses the Nusselt number, Reynolds number, Prandtl number, relative roughness of the wall surface ε / D. (ε is the average roughness height), and the geometric parameters in terms of longitudinal S_L / D and traverse S_T / D pitches (Zukauskas and Ulinkas 1988)

$$Nu_D = f\left(Re_D, Pr, \frac{Pr}{Pr_s}, \frac{\varepsilon}{D}, \frac{S_L}{D}, \frac{S_T}{D}\right) \tag{7.45}$$

The Reynolds number Re_D is evaluated at the maximum velocity U_{max} from Eq. (7.33) or Eq. (7.34). The thermophysical properties are evaluated at the free stream temperature T_∞ with the exception of Pr_s, which is calculated at the surface temperature T_s.

For

$$1.25 \le S_T / D \le 1.5$$

$$0.935 \le S_L / D \le 2.0$$

$$6.67 \times 10^{-3} \le \varepsilon / D \le 40 \times 10^{-3}$$

$$\overline{Nu}_L = 0.5\left(\frac{S_T}{S_L}\right)^{0.2}\left(\frac{\varepsilon}{D}\right)^{0.1} Re_D^{0.65} Pr^{0.36}\left(\frac{Pr}{Pr_s}\right)^{0.25} \quad \text{for} \quad 10^3 < Re_D < 10^5 \tag{7.46}$$

$$\overline{Nu}_L = 0.1\left(\frac{S_T}{S_L}\right)^{0.2}\left(\frac{\varepsilon}{D}\right)^{0.15} Re_D^{0.8} Pr^{0.4}\left(\frac{Pr}{Pr_s}\right)^{0.25} \quad \text{for} \quad 10^5 < Re_D < 2\times10^6 \tag{7.47}$$

For the range $10^3 < Re_D < 10^5$, increasing surface roughness (ε / D) of a tube bank can increase the heat transfer coefficient up to 50% compared with smooth tubes. Similar improvement can be achieved with lower surface roughness in the range $10^5 < Re_D < 1.2\times10^6$.

Example 7.3 An air-to-water heat exchanger is arranged as a staggered tube bank of six longitudinal rows of tubes with outside diameter $D = 16mm$. The transverse pitch is $S_T = 30mm$ and longitudinal pitch is $S_L = 34mm$. The free stream temperature of air flowing across the tube bank is $T_\infty = 15°C$, the outside surface temperature of the tubes is $T_s = 65°C$, and the stream flow velocity is $U_\infty = 7m/s$. Calculate the average heat transfer coefficient for the air side of the tube bank.

Solution

Thermophysical properties of air

at $T_\infty = 15°C$

$\rho = 1.2096\ kg/m^3$, $c_p = 1004\ J/kg.K$, $\mu = 1.789\times10^{-5}\ kg/s.m$, $k = 0.02602\,W/m.K$

at $T_s = 65°C$ for $Pr_s = c_p\mu/k$

$\rho = 1.0368\ kg/m^3$, $c_p = 1007.45\,J/kg.K$, $\mu = 2.043\times10^{-5}\,kg/s.m$, $k = 0.02949\ W/m.K$

$$2\times\left\{\sqrt{\left[(S_T/2)^2 + S_L^2\right]} - D\right\} = 2\times\left\{\sqrt{\left[(0.03/2)^2 + 0.034^2\right]} - 0.016\right\} = 0.0423$$

$$(S_T - D) = 0.03 - 0.016 = 0.014$$

Since $2\times\left\{\sqrt{\left[(S_T/2)^2 + S_L^2\right]} - D\right\} > (S_T - D)$, we use Eq. (7.33) for U_{max}

$$U_{max} = \frac{U_\infty S_T}{S_T - D} = \frac{7\times0.03}{0.03 - 0.016} = 15m/s$$

$$Re_D = \frac{U_{max} D\rho}{\mu} = \frac{15\times0.016\times1.2096}{1.789\times10^{-5}} = 16227$$

$$Pr = \frac{c_p\mu}{k} = \frac{1004\times1.789\times10^{-5}}{0.02602} = 0.69$$

$$Pr_s = \frac{c_p\mu}{k} = \frac{1007.45\times2.043\times10^{-5}}{0.02949} = 0.698$$

The correction factor for six rows from Table 7.2 at $Re_D > 10^3$ $C_N = 0.945$

Since $S_T/S_L < 2$ and $Re_D = 16227$, we use Eq. (7.42)

$$\overline{Nu_L} = C_N\left[0.35Re_D^{0.6}Pr^{0.36}\left(\frac{S_T}{S_L}\right)^{0.2}\left(\frac{Pr}{Pr_s}\right)^{0.25}\right] = 0.945\times16227^{0.6}\times0.69^{0.36}\left(\frac{0.03}{0.034}\right)^{0.2}\times\left(\frac{0.69}{0.698}\right)^{0.25} = 94.5$$

$$h = \frac{\overline{Nu_L}k}{D} = \frac{94.5\times0.02602}{0.016} = 153.6W/m^2.K$$

7.4 Flow Over Non-Circular Tubes

Tubes of non-circular shapes are frequently encountered in engineering practice, and knowledge of the effect of these shapes on forced convection heat transfer is essential. Collated results of experimental results from two sources by Jakob (1962) for tubes with various cross-sectional shapes in cross flow of air can be estimated by the simple correlation

$$\overline{Nu}_L = CRe_D^n \tag{7.48}$$

with the constant C and exponent n as given for selected shapes in Table 7.4. All properties are evaluated at the film temperature $T_f = (T_s + T_\infty)/2$ and the characteristic dimension D for use in the Reynolds number is shown for each shape in the table. Compared with the results for circular tubes in cross flow of air, some of the non-circular profiles provide higher heat transfer coefficients.

7.5 Flow Over Spheres

The flow characteristics of flow over a sphere are similar to the flow across a cylindrical tube. For a sphere immersed in infinite medium, the functional relationship for steady heat transfer by forced convection, as stated by Whitaker (1977), is

Table 7.4 Constant n and exponent n in the correlation $\overline{Nu}_L = CRe_D^n$ for non-circular tubes in gas cross flow.

Shape	Re_D	n	C
U_∞ → D	5000 – 100,000	0.675	0.102
U_∞ → D	5000 – 100,000	0.588	0.246
U_∞ → D	2500 – 15,000	0.612	0.224
U_∞ → D	3000 – 15,000	0.804	0.085
U_∞ → D	5000 – 100,000	0.638	0.138
U_∞ → D	5000 – 19,500	0.638	0.144
	19,500 – 100,000	0.782	0.035
U_∞ → D	4000 – 15,000	0.731	0.205

$$\left(\overline{Nu}_L - 2\right) = f\left(Re_D, Pr,\ \mu_\infty / \mu_s\right)$$

which acquires the following form

$$\overline{Nu}_L = 2 + \left(0.4Re_D^{1/2} + 0.06Re_D^{2/3}\right)Pr^{0.4}\left(\frac{\mu_\infty}{\mu_s}\right)^{1/4} \tag{7.49}$$

This correlation is valid for the ranges:

$$3.5 < Re_D < 7.6 \times 10^4$$

$$30.71 < Pr < 380$$

$$1.0 < \mu_\infty / \mu_s < 3.2$$

The equation was derived on the basis that a steady conduction from a spherical surface of a stationary fluid ($Re_D = 0$) exists for which $\overline{Nu_D} = 2$.

The reason for including the Reynolds number twice with different exponents is due to the assumption that the heat transfer from the sphere is considered as two parallel processes: one in the laminar boundary layer region, and the other in the wake region.

All properties are evaluated at T_∞, with the exception of μ_s, which is evaluated at the temperature of the surface of the sphere.

A special case of forced convection heat transfer from spheres relates to liquid droplets moving in a gaseous medium, examples of which include atomized hydrocarbon fuel droplets from the injected sprays into the combustion chamber of internal combustion engines and gas turbines, water droplets falling in the cooling tower of a steam power plant, and liquid droplets in spray drying chambers. Convection heat transfer to droplets moving at very high speeds under high-pressure in high-temperature compressed gaseous media (ICE and gas turbines) are too complicated to estimate by a single empirical correlation. However, it is possible to estimate the convection heat transfer coefficient of falling liquid droplets using the correlation by Ranz and Marshall (1952)

$$\overline{Nu}_L = 2 + 0.6Re_D^{1/2}Pr_D^{1/3} \tag{7.50}$$

Problems

7.1 Hot air at average temperature $T_\infty = 70°C$ and velocity $U_\infty = 4m/s$ is blowing over a plate (0.4 m wide and 2.5 m long) mounted horizontally. Assume that the plate surface is at uniform temperature $T_s = 15°C$. Determine the heat transfer coefficient h_x at $x = 1.25m$ and the total heat loss rate from the surface. The thermophysical properties of air: $\rho = 1.028 kg/m^3$, $c_p = 1.007 kJ/kg.K$, $\mu = 2.052 \times 10^{-5} kg/m.s$, $k = 0.02881 W/m.K$.

7.2 Air at 1 atm and 20°C is flowing over a horizontal flat plate at a speed of $3m/s$. The plate is $0.28m$ wide and $1.0\,m$ long and is maintained at 56°C. At a distance $x = 0.28m$ from the leading edge of the plate, determine:

(a) the local heat transfer coefficients
(b) the average heat transfer coefficients
(c) the rate of heat loss from the plate by convection.

Take the properties of air as: $\rho = 1.137 kg/m^3$, $c_p = 1.007 kJ/kg.K$, $\mu = 1.906 \times 10^{-5} kg/m.s$, $k = 0.0266 W/m.K$, $Pr = 0.722$.

7.3 The underside of the crankcase of a passenger vehicle can is modelled as a flat plate of 1.0m long and 0.25 m wide. The surface temperature of the crankcase is $T_s = 75°C$, the ambient temperature $T_\infty = 25°C$, and the vehicle is travelling at a speed of 72km/hr. Assuming the air flow is turbulent and ignoring radiation, estimate the rate of heat loss.

7.4 Calculate the Nusselt number for the following cases:

(a) A horizontal electronic component with a surface temperature of 35°C, 5 mm wide and 10 mm long, dissipating 0.1 W of heat by free convection from its one side into air at 20°C. Take for air $k = 0.026 W/m.K$

(b) A 1 kW central heating radiator 1.5m long and 0.6 m high with a surface temperature of 80°C, dissipating heat by radiation and convection into a room at 20°C ($k = 0.026 W/mK$), assume blackbody radiation and $\sigma = 5.67 \times 10^{-8} W/m^2.K^4$

(c) Air at 6°C ($k_a = 0.026 W/m.K$) adjacent to a wall 3 m high and thickness $\delta = 0.15$ m made of brick with ($k_b = 0.3 W/m.K$), the inside temperature of the wall is 18°C and the outside wall temperature is 12°C.

7.5 Engine oil at 100°C flows over the surface of a flat plate 1 m long at velocity 0.1m/s. The plate is maintained at $T_s = 20°C$. Determine:

(a) the velocity and thermal boundary layer at the tailing edge of the plate

(b) the local heat flux and surface shear stress at the tailing edge

(c) the heat transfer rate per unit width of the plate.

Take the properties of the oil at 60°C as: $\rho = 864 kg/m^3$, $\mu = 0.074 kg/m.s$, $k = 0.1404 W/m.K$, $Pr = 1080$.

7.6 Air at 1 atm and 300 K flows over a flat plate of length 1 m. If the boundary layer thickness at the tailing edge of the plate is 12 mm, determine the flow velocity, assuming the flow to be laminar.

7.7 Air at 1 atm and 20°C flows with velocity 35 m/s over the surface of a flat plate that is maintained at 300°C. Treating the flow as laminar, calculate the rate at which the heat is transferred per metre width from both sides of the plate over a distance of 0.5 m from the leading edge.

7.8 A long 8 m pipe of 80 mm diameter transporting steam has an external temperature of 90°C. Wind (air temperature $T_\infty = 10°C$) is blowing across the pipe at a speed of 14 m/s causing the pipe surface to cool. Determine the rate of heat loss from the pipe using the correlation by Knusden and Katz given by Eq. (7.27).

7.9 Repeat Problem 7.8 using Churchill and Bernstein correlation given by Eq. (7.32).

7.10 A circular pipe of 25-mm outside diameter is placed in an airstream at $T_\infty = 20°C$ and 1-atm pressure. The air moves in cross flow over the pipe at 15 m/s, while the outer surface of the pipe is maintained at $T_s = 100°C$. Determine:

(a) the drag force exerted on the pipe per unit length

(b) the rate of heat transfer from the pipe per unit length.

7.11 Air at 1 atm and 30°C flows across a 5.0-cm-diameter cylinder at a velocity of 50 m/s. The cylinder surface is maintained at a temperature of 150°C. Calculate the heat loss per unit length of the cylinder.

7.12 Figure P7.12 shows the cross-sections of three long beams of equal length subjected to air in a cross flow at $U_\infty = 10 m/s$ and $T_\infty = 130°C$. The cylindrical beam has a diameter of $D = 15 mm$, and the cross-sectional area is the same for all three configurations. Determine which beam has the largest heat transfer rate per unit length. The initial temperature of the surface of all beams is $T_\infty = 30°C$.

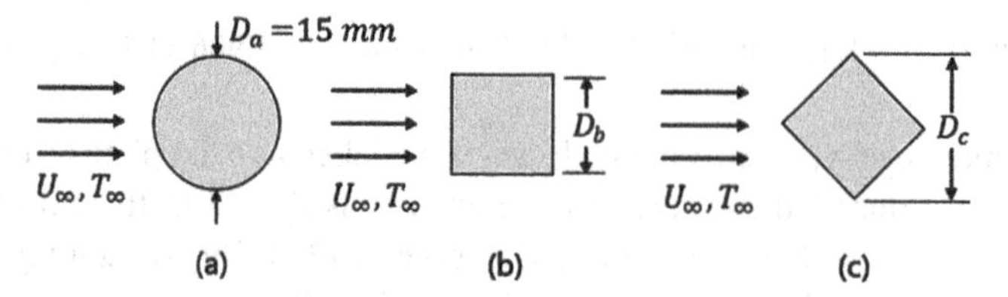

Figure P7.12 Air flow across beams of different cross-sections.

7.13 A copper sphere of 10 mm diameter is heated to 75°C and then cooled in an air stream at 25°C, average unknown convection heat transfer coefficient $\bar{h}$, and a speed of $U_\infty = 10m/s$. How long will it take for the sphere to cool down to 30°C?

Take the properties of copper as: $\rho = 8933kg/m^3$, $c_p = 387J/kg.K$, $k = 399W/m.K$, $Pr = 0.7154$.

7.14 Air flows across a long square beam ($200mm \times 200mm$) at a velocity of 12 *m/s*. The surface temperature is maintained at 75°C. Free-stream air conditions are 25°C and 0.6 atm. Calculate the heat loss from the cylinder per unit length.

7.15 Air at 1.0 atm and temperature of 30°C flows at a speed of 4 m/s across two long beams, as shown in Figure P7.15. The two beams have an identical cross-sectional area of 0.03142 m^2. The surface temperature of each beam is maintained at 70°C. Calculate the heat loss from the beams per unit length.

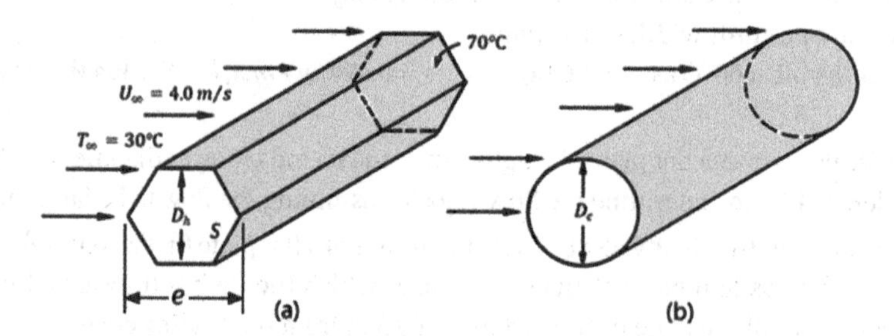

Figure P7.15 Air flow across two beams of different cross-sections.

7.16 Determine the rate of heat transfer to the hexagonal beam in Problem 5.15 if it is turned 90^0 about its axis, as shown in Figure P7.16

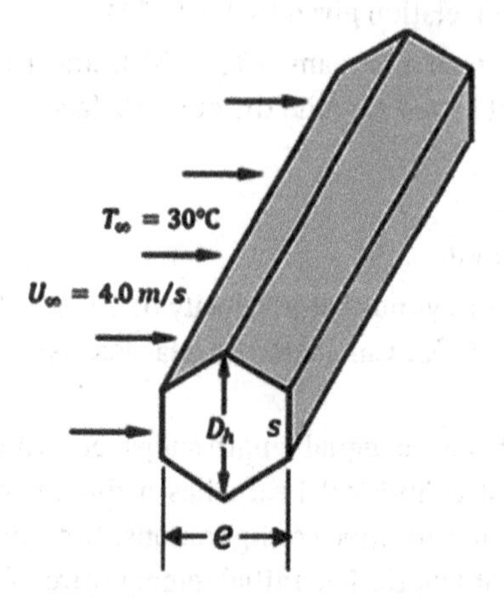

Figure P7.16 Air flow across hexagonal beam.

7.17 Air flows over a spherical tank of 1.8 m diameter containing iced water. Air at 20°C is flowing at a speed of $6m/s$. Determine the heat transfer coefficient, rate of heat transfer to the tank, and rate of melting of the ice.

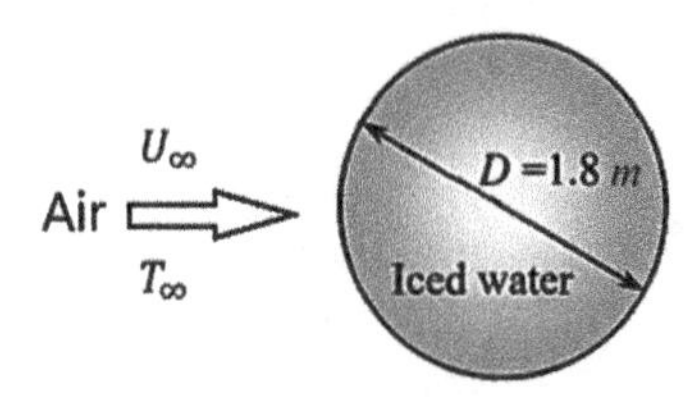

Figure P7.17 Air flows over a spherical tank.

7.18 In a staggered tube bundle of a heat exchanger, water is passed through the tubes and air is passed in cross flow over the tubes. The tube outside diameter is 9.5 mm and the longitudinal and transverse pitches are $S_L = 11.9mm$ and $S_T = 14.3mm$. There are 10 rows of tubes in the airflow direction and 8 tubes per row. Under typical operating conditions, the cylinder surface temperature is at 75°C, while the air (at atmospheric pressure) upstream temperature and velocity are 15°C and $7m/s$, respectively. Determine the air-side convection coefficient and the rate of heat transfer for the tube bundle.

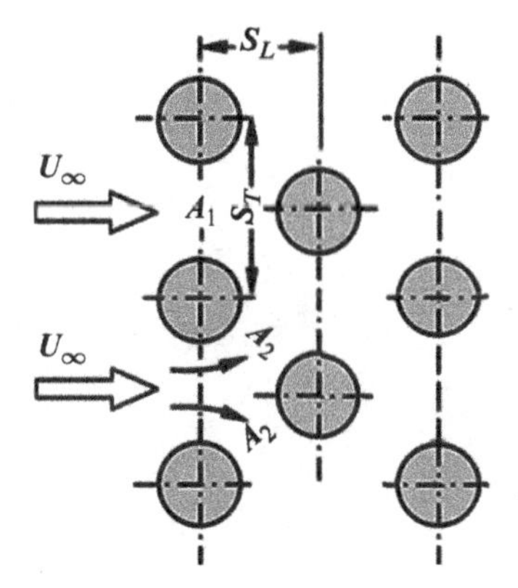

7.19 Air at 1 atm flows normal to a square in-line bank of 400 tubes having diameters of 6 mm and lengths of $0.5\ m$. $S_T = S_L = 9mm$. The air enters the tube bank at 300 K and at a velocity such that the Reynolds number based on inlet properties and the maximum velocity at inlet is 50,000. If the outside wall temperature of the tubes is 400 K, calculate the air temperature rise as it flows through the tube bank.

7.20 A tube bank consists of a square array of 144 tubes arranged in an in-line position. The tubes have a diameter of 1.5 cm and length of 1.0 m; the centre-to-centre tube spacing is 2.0 cm. If the surface temperature of the tubes is maintained at 350 K and air enters the tube bank at 1 atm, 300 K, and $U_\infty = 6\ m/s$, calculate the total heat lost by the tubes.

8

Forced Convection – Internal Flows

Any heating or cooling process in a flowing fluid inside conduits of various shapes and sizes is known as internal forced convection heat transfer or, in short, internal forced convection. Examples of this heat transfer mode can be seen in Figure 8.1, which is an expanded reproduction of Figure 1.1 in Chapter 1. Heat is transferred from the combustion products in the cylinder as the piston moves downwards during the expansion (power) stroke, albeit with complex motion of the gases whose pressure and temperature vary within wide ranges. The cooling water in the water jacket surrounding the cylinder is flowing at reasonably high velocity in the annular conduit, picking up most of the heat from the hot cylinder wall and transferring it to the cooling water that is circulated with the aid of a pump (not shown) through a pipe to a heat exchanger, popularly known as a radiator, and back to the engine at a lower temperature. The reduction of the temperature in the radiator is the result of the combined effect of internal forced convection in the tube bank and external forced convection to the surroundings.

8.1 Forced Convection Inside Tubes

Analysis of the process of heat transfer by forced convection in tubes is a prerequisite to the design of equipment where forced internal flows are involved. In addition to the cases of forced convection illustrated in Figure 8.1, there are many more applications such as air-conditioning, heating, and refrigeration equipment, and power plants. The effectiveness of the forced convective process is evaluated in terms of the Nusselt number, which depends primarily on the Reynolds number. Figure 8.2 is a qualitative representation of fluid flow in a tube in the form of Nu_D versus (Re_D) on log scale showing three regions with their approximate ranges of the Reynolds numbers: laminar flow region $(10 < Re_D < 2.3\times10^3)$, transitional flow region $(2.3\times10^3 < Re_D < 10^4)$, and turbulent flow region $(10^4 < Re_D < 10^7)$. Correlations for the flow regions will be presented in the following sections.

Heat Transfer Basics: A Concise Approach to Problem Solving, First Edition. Jamil Ghojel.

Companion website: www.wiley.com/go/ghojel/heat_transfer

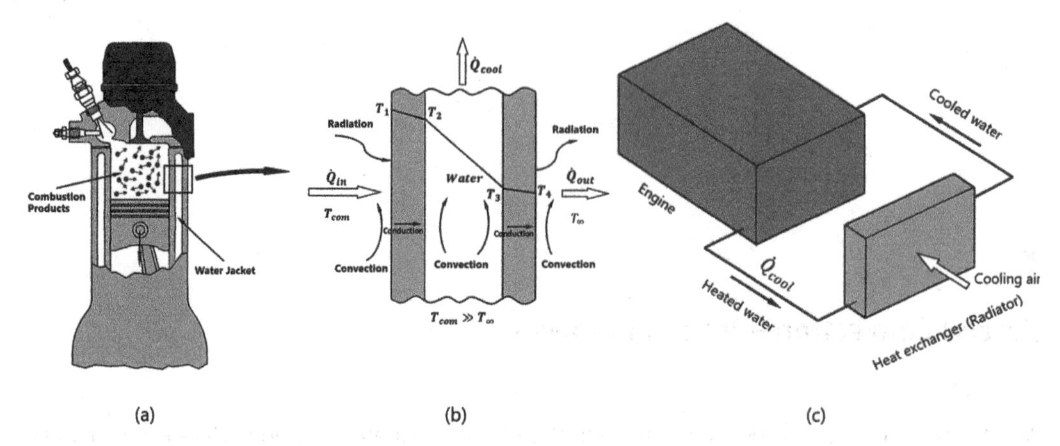

Figure 8.1 Schematic diagram of heat transfer processes in internal combustion engine: (a) Engine cylinder; (b) Control volume of water jacket; (c) Cooling water recirculation.

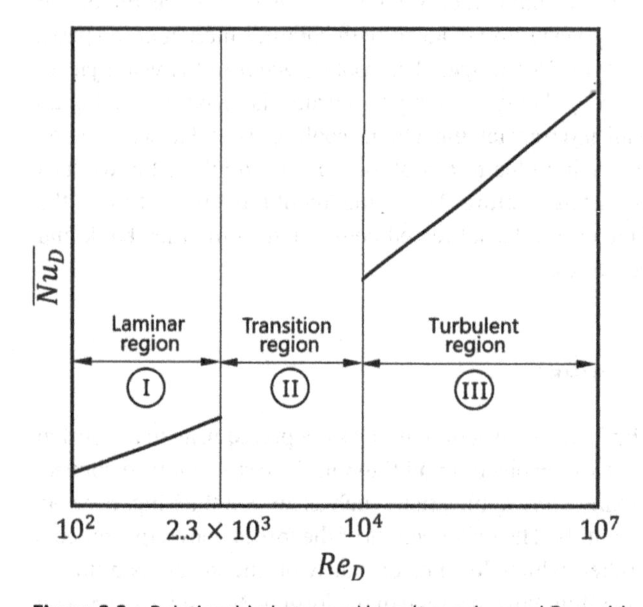

Figure 8.2 Relationship between Nusselt number and Reynolds number for typical flow in a conduit.

8.2 Laminar Forced Convection (Region I)

The laminar flow and temperature profiles inside a heated long horizontal cylinder are shown in Figure 8.3. Both hydrodynamic and thermal boundary layers formed on the inner surface grow thicker with distance from the entry plane at point A, contracting to zero on the centreline at point B. During this stage of flow, known as entry flow region, the velocity and temperature profiles change in shape and magnitude with both x and r. Some distance downstream from B, the flow becomes dominated by viscous forces over the entire cross-section at point C and the profiles change to parabolic varying only with r. The flow region downstream from point C is known as the fully-developed region. The length of the entry flow region (tube length required for the velocity distribution to become fully develop) is approximately equal to $0.05Re_D$.

8.2.1 Fully Developed Flow

Analytical solutions for the problem of forced convection in circular tubes are generally limited to fully-developed laminar flows only. Based on steady-state mass and momentum conservation equations, these solutions yield the value of the Nusselt number from which the heat transfer coefficient can be determined and rate of heat transfer calculated using Newton's law of cooling. The Nusselt number thus obtained will depend on the boundary conditions on the outer surface of the tube. Unlike flow over a flat plate, there is no free stream temperature in the fully-developed region (see the temperature profile at C in Figure 8.3b); therefore, a temperature known as the *bulk temperature* is used in the analysis of flow in ducts. This temperature can be conceptualized as follows (Ede 1967). If the fluid emerging at a cross-section of the tube is collected in a cup and thoroughly mixed, its final equilibrium temperature will be the bulk or mixed-cup temperature of the fluid at that position (T_b). This is the temperature often measured in experimental investigations. Mathematically, the bulk temperature is determined by integration over the cross-sectional area A of the duct if the velocity and temperature distributions are known.

$$T_b = \frac{\rho}{\dot{m}} \int UT dA \tag{8.1}$$

For flow of fluid with constant properties in a circular tube

$$T_b = \frac{2\pi\rho}{\dot{m}} \int_0^R UTr dr \tag{8.2}$$

where ρ is fluid density, $\dot{m}$ is the mass flow rate, and R is the radius of the tube.

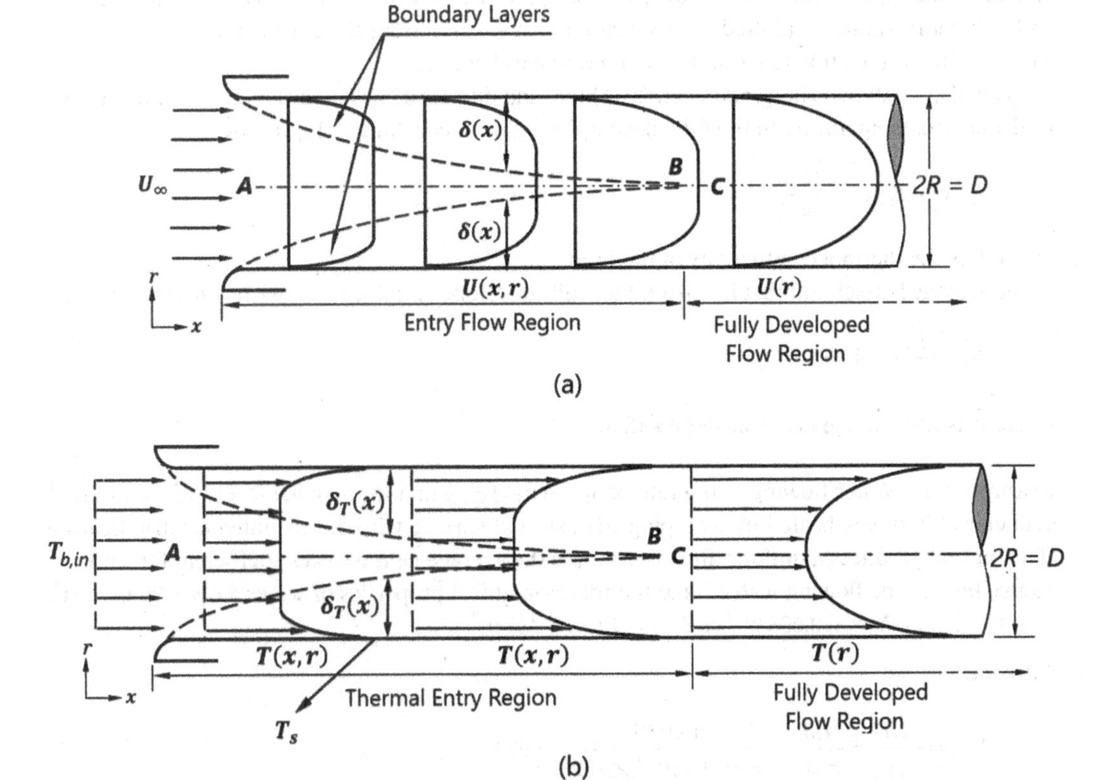

Figure 8.3 Flow in a heated tube: (a) velocity profiles and hydrodynamic boundary layer; (b) temperature profiles and thermal boundary layer.

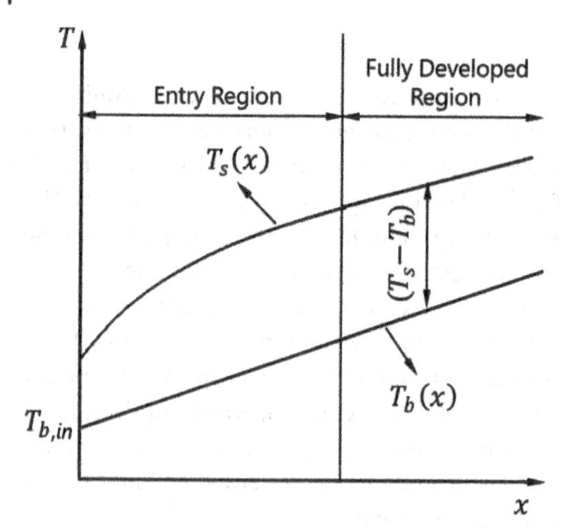

Figure 8.4 Effect of application of constant heat flux on the surface and fluid bulk temperatures.

For uniform heat flux q applied on the tube surface and constant specific heat (c_p) of the fluid, the variation of the bulk temperature with location x measured from the entry to the tube is

$$T_b = T_{b,in} + \frac{2\pi qR}{\dot{m}c_p}x \tag{8.3}$$

Figure 8.4 shows the temperature variation with distance from the inlet x for uniform heat flux applied to the outer surface of the tube. The temperatures shown are the surface temperature T_s and bulk temperature T_b plotted versus location x measured from the entrance to the tube. For a constant flux q, the bulk temperature changes linearly with x.

It can also be shown (Bergman et al. 2011) that the difference between the temperatures of the wall and bulk temperature in the fully developed region is constant and given by

$$T_s(x) - T_b(x) = \frac{22}{48}\frac{qR}{k} \tag{8.4}$$

where k is the thermal conductivity of the fluid.

The average Nusselt number is constant for fully developed laminar flow with constant heat flux

$$\overline{Nu} = \frac{\bar{h}D}{k} = 4.36 \tag{8.5}$$

where $\bar{h}$ is the average heat transfer coefficient.

Example 8.1 Water flowing at the rate of $\dot{m} = 0.012 kg/s$ in a thin-walled steel tube of internal diameter of $25 mm$ is heated by wrapping the external surface with an insulated electric-heating element that produces a uniform flux. Determine the average heat transfer coefficient for the convective heat to the flowing water. Take the thermophysical properties of water as: $\rho = 998 kg/m^3$, $c_p = 4.17 kJ/kg.K$, $k = 0.608 W/m.K$, $\mu = 9.1\times10^{-4} s/m^2$.

Solution

$$Re_D = \frac{\rho UD}{\mu} = \frac{4\dot{m}}{\pi\mu D} = \frac{4\times0.012}{\pi\times9.1\times10^{-4}\times0.025} = 672$$

The flow is laminar and Eq. (8.5) can be used to calculate $\bar{h}$

$$\frac{\bar{h}D}{k} = 4.36$$

$$\bar{h} = \frac{4.36 \times 0.608}{0.025} = 106\,W/m^2.K$$

If the boundary condition is a constant surface temperature, only the bulk temperature will change with location x, as shown in Figure 8.5. The ratio of temperature differences at the inlet and outlet of the tube may be expressed as (Bergman et al. 2011)

$$\ln\left(\frac{\Delta T_{out}}{\Delta T_{in}}\right) = -\frac{PL}{\dot{m}c_p}\bar{h} \tag{8.6}$$

where P and L are the perimeter and length of the tube respectively, and $\dot{m}$ and c_p are the mass flow rate and specific heat of the fluid respectively.

The heat transfer by convection to the fluid is equal to the net heat flow rate $(\dot{Q}_{in} - \dot{Q}_{out})$

$$\dot{Q} = (\dot{Q}_{in} - \dot{Q}_{out})$$

$$\dot{Q} = \dot{m}c_p\left[(T_s - T_{in}) - (T_s - T_{out})\right]$$

$$\dot{Q} = \dot{m}c_p(\Delta T_{out} - \Delta T_{in}) \tag{8.7}$$

Combining Eqs (8.6) and (8.7) results in

$$\dot{Q} = \bar{h}PL\left[\frac{\Delta T_{out} - \Delta T_{in}}{ln\left(\frac{\Delta T_{out}}{\Delta T_{in}}\right)}\right] = \bar{h}PL(LMTD) \tag{8.8}$$

LMTD (log mean temperature difference) is equal to the expression in the square bracket.

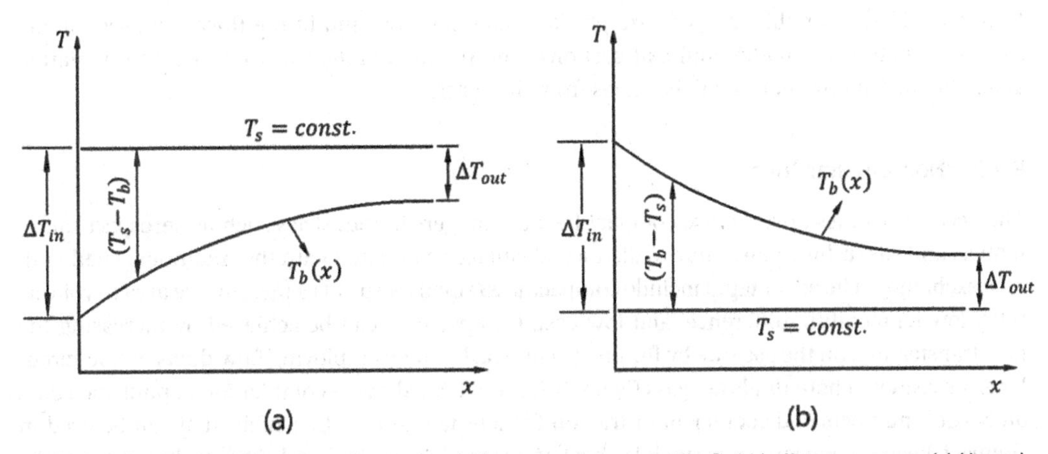

Figure 8.5 Effect of application of constant surface temperature on the fluid bulk temperature: (a) Heated fluid; (b) Cooled fluid.

The average Nusselt number is constant for the fully-developed laminar flow with constant surface temperatures

$$\overline{Nu} = \frac{\bar{h}D}{k} = 3.66 \tag{8.9}$$

where $\bar{h}$ is the average heat transfer coefficient.

Example 8.2 Water flows in a thin-walled copper tube of 50 mm internal diameter and 1.2 m length at a rate of $0.025 kg/s$. The bulk temperatures of the water at the inlet and outlet are, respectively, $300K$ and $328K$. The temperature of the outer surface of the tube is maintained at $370K$ by condensing steam. Determine the coefficient of heat transfer and total heat transfer rate to the flowing water. Take the thermophysical properties of water as: $\rho = 998\ kg/m^3$, $c_p = 4.179\ kJ/kg.K$, $k = 0.634\ W/m.K$, $\mu = 6.31 \times 10^{-4} s/m^2$.

Solution

$$Re_D = \frac{\rho UD}{\mu} = \frac{4\dot{m}}{\pi \mu D} = \frac{4 \times 0.025}{\pi \times 6.31 \times 10^{-4} \times 0.05} = 1009$$

The flow is laminar and Eq. (8.9) can be used

$$\bar{h} = \frac{3.66 \times 0.634}{0.05} = 46.4\ W/m^2.K$$

$$LMTD = \frac{\Delta T_{out} - \Delta T_{in}}{\ln\left(\frac{\Delta T_{out}}{\Delta T_{in}}\right)} = \frac{(370-328)-(370-300)}{\ln\left(\frac{370-328}{370-300}\right)} = \frac{-28}{\ln\left(\frac{42}{70}\right)} = 54.8°\text{C}$$

From Eq. (8.8)

$$\dot{Q} = \bar{h}PL(LMTD) = 46.4 \times \pi \times 0.05 \times 1.2 \times 54.8 = 479\ W$$

Figure 8.6 is an illustration of computer modelling of laminar flow in the circular tube in Example 8.2. The variable temperature profiles along the flow and in the flow cross-section are shown for entrance, middle, and exit sections. The plotted graphs show temperature variation along the tube at different radial distances from the centre.

8.2.2 Non-Circular Tubes

The need sometimes arises to use compact heat exchangers in industries such as aerospace, automotive, and gas turbine power due to distinct advantages compared with the traditional shell-and-tube exchanger. The advantages include compactness (small footprint), large surface area to volume ratio, low-temperature difference, and low cost. Compactness can be achieved by increasing the heat transfer area on the gas side by finning (Figure 8.7b), or by employing flow ducts of non-circular cross-section if both fluids are gas (Figure 8.7c). Table 8.1 shows a compilation of published data on Nusselt numbers and coefficient of friction for selected non-circular tubes that can be used in compact heat exchangers. In principle, heat transfer area in shell-and-tube exchangers can be increased by increasing the length of the circular tubes arranged, as shown in Figure 8.7a, or reducing their diameter. However, both measures lead to increase in fabrication difficulties and cost and

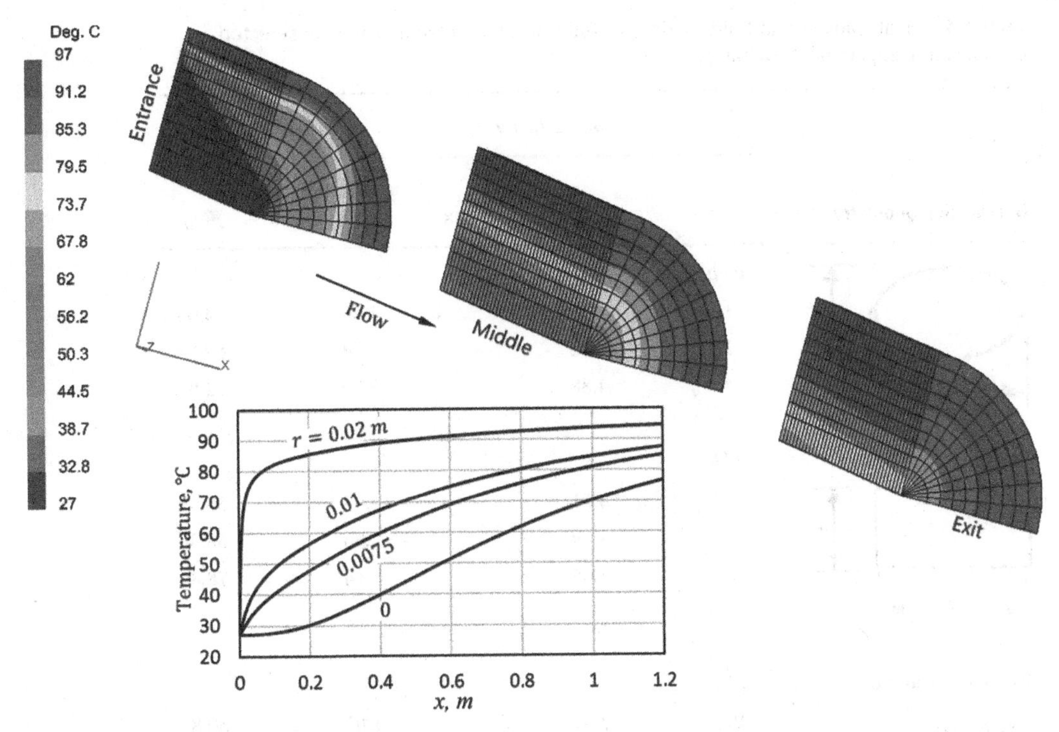

Figure 8.6 Computer modelling of water flow in a straight circular tube.

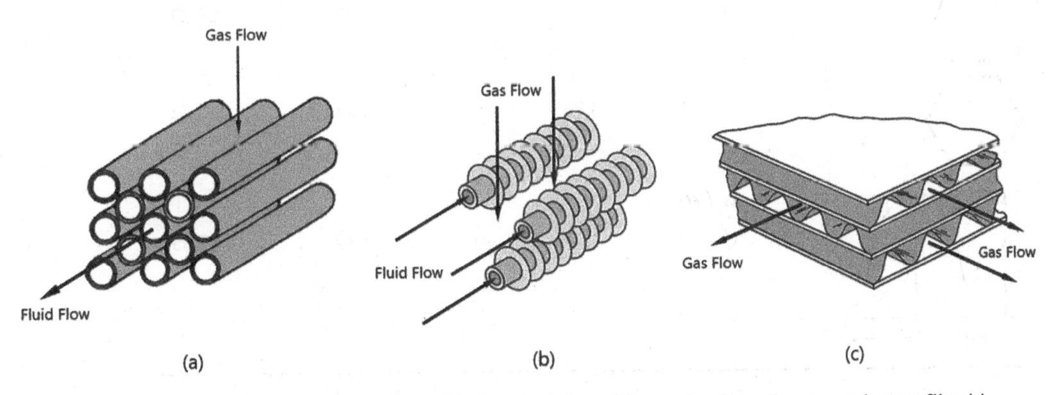

Figure 8.7 Heat exchanger configurations: (a) circular tubes; (b) annular fins of rectangular profile; (c) triangular tubes.

increase in friction power (increase in running cost). A balance between heat rate and friction power can be achieved in compact exchangers by proper selection of flow channels, examples of which are given in Table 8.1. Comprehensive coverage of compact heat exchangers can be found in the original and revised editions of the work of Kays and London (1964, 1984, 2018).

8.2.3 Laminar Forced Convection Correlations

Equations (8.5) and (8.9) can be used only if the flow at the entrance is already fully developed and heat transfer is independent of the velocity of the fluid (constant Reynolds number). In reality, the

Table 8.1 Heat transfer and fluid friction for fully-developed laminar flow in selected channels in compact heat exchangers.

Duct section geometry		$Nu_D = hL/k$		
		Constant heat flux	**Constant wall temperature**	fRe_D
(Ellipse, axes a and b)	a/b			
	1	4.36	3.66	64.00
	2	4.56	3.74	67.28
	4	4.88	3.79	72.96
Rectangle	a/b			
	1	3.61	2.98	56.92
	3	4.79	3.96	68.36
	6	6.05	5.14	78.80
Isosceles Triangle	$\varphi(°)$			
	10	2.45	1.61	50.8
	30	2.91	2.26	52.28
	60	3.11	2.47	53.32
Hexagon		3.86	3.34	60.22
(Sine-shaped duct, $2a$ wide, $2b$ high)	$2a/2b$			
	$\frac{\sqrt{3}}{2}$	3.014	2.39	12.63

Nusselt number falls from a very high value at the entrance to the constant values of 4.36 or 3.66 when the flow becomes fully developed, which can occur at a significant distance from the entrance. It is found that a more realistic solution which takes into account entrance effects is of the general form

$$\overline{Nu}_D = f\left(Re_D, Pr, \frac{D}{L}\right) \tag{8.10}$$

The term D/L in this equation is important for short tubes. A relatively simple correlation for heat transfer of liquids in tubes with constant surface temperature proposed by Sieder and Tate (1936) has the form

$$\overline{Nu}_D = 1.86\left(Re_D Pr \frac{D}{L}\right)^{1/3}\left(\frac{\mu_b}{\mu_s}\right)^{0.14} \tag{8.11}$$

Equation (8.11) is valid for $Re_D \leq 2300$, $0.48 < Pr < 16700$, and $0.0044 < \mu_b / \mu_s < 9.75$. The physical properties are evaluated at the arithmetic mean of the bulk fluid temperature $T_m = (T_{b,in} + T_{b,out})/2$, except for the viscosity term μ_s, which is evaluated at the mean surface (wall) temperature.

Another correlation suggested in Mikheyev and Mikheyeva (1977) for circular tubes is of the form

$$Nu_{D,f} = 1.4\left(Re_D \frac{D}{L}\right)^{0.4} Pr^{0.33}\left(\frac{Pr}{Pr_s}\right)^{0.25} \tag{8.12}$$

for $10 < Re_D < 2\times10^4$ and $L/D > 10$. Thermophysical properties are evaluated at the average bulk temperature $T_m = (T_{b,in} + T_{b,out})/2$, except for Pr_s, which is evaluated at the surface temperature T_s. The range of Reynolds numbers includes the transition region.

Gnielinski (2010) proposed the following two correlations for fully developed laminar flows in circular tubes of all lengths:

Constant wall temperature:

$$\overline{Nu}_{D,T} = \left[\overline{Nu}_{T,1}^{\,3} + 0.7^3 + \left(\overline{Nu}_{T,2} - 0.7\right)^3 + \overline{Nu}_{T,3}^{\,3}\right]^{1/3} \tag{8.13}$$

where

$$\overline{Nu}_{T,1} = 3.66$$

$$\overline{Nu}_{T,2} = 1.615\left(\frac{Re_D Pr D}{L}\right)^{1/3}$$

$$\overline{Nu}_{T,3} = \left(\frac{2}{1+22Pr}\right)^{1/6}\left(\frac{Re_D Pr D}{L}\right)^{1/2}$$

Constant heat flux:

$$\overline{Nu}_{D,q} = \left[\overline{Nu}_{q,1}^{\,3} + 0.6^3 + \left(\overline{Nu}_{q,2} - 0.6\right)^3 + \overline{Nu}_{q,3}^{\,3}\right]^{1/3} \tag{8.14}$$

where

$$\overline{Nu}_{q,1} = 4.364$$

$$\overline{Nu}_{q,2} = 1.953\left(\frac{Re_D Pr D}{L}\right)^{1/3}$$

$$\overline{Nu}_{q,3} = 0.924(Pr)^{1/3}\left(\frac{Re_D D}{L}\right)^{1/2}$$

8.3 Turbulent Forced Convection (Region III)

When the Reynolds number exceeds 2000, the flow in conduits becomes turbulent. Figure 8.8 shows a simplified structure of turbulent flow in a circular tube, which shows that the flow is fully turbulent with the eddies in close proximity to the viscous sublayer indicated by the dashed lines.

The length of the entry flow region is independent of the Reynolds number and approximately equal to 50 D. Heat transfer across the viscous sublayer is by conduction and beyond this layer by intensive mixing. The temperature in the cross-section of the tube is essentially constant and temperature gradient is across the viscous sublayer. The velocity varies insignificantly in the core region of the tube, with steep reduction near the wall region in the viscous sublayer.

Unlike laminar flow inside circular and non-circular tubes discussed in Section 8.1, there are no analytical and theoretical solutions for the problem of internal turbulent forced convection. The approach to solving the problem is to evaluate the Nusselt number from empirical equations based on experimental results.

The simplest and most widely used correlation is the Dittus–Boetler equation for the average Nusselt number of a heated or cooled fluid

$$\overline{Nu}_D = \frac{\bar{h}D}{k} = 0.023Re_D^{0.8}Pr^n \tag{8.15}$$

All properties are evaluated at the bulk temperature T_b, $n = 0.4$ for heating $(T_s > T_b)$, and $n = 0.3$ for cooling $(T_s < T_b)$. This equation is valid for the ranges:

$$6000 < Re_D < 10^7$$

$$0.5 < Pr < 120$$

$$L/D > 60$$

If thermophysical fluid properties are likely to vary significantly because of large temperature differences, the correlation proposed by Sieder and Tate (1936) is recommended

$$\overline{Nu}_D = 0.027Re_D^{0.8}Pr^{1/3}\left(\frac{\mu_b}{\mu_s}\right)^{0.14} \tag{8.16}$$

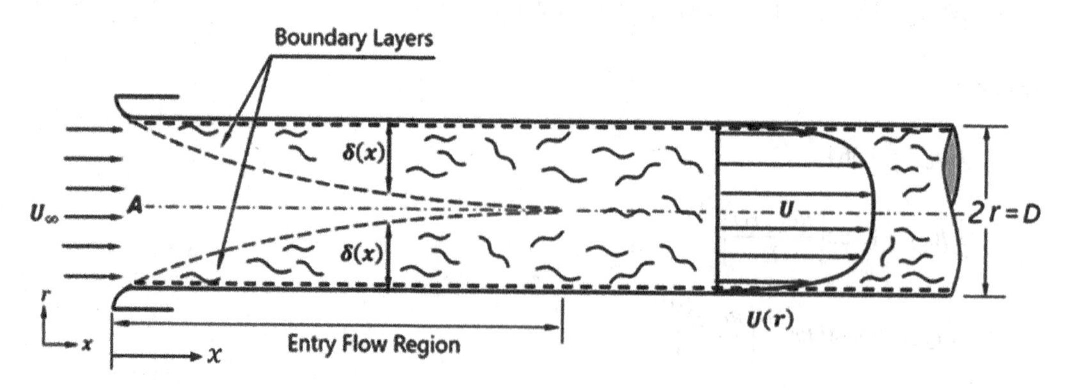

Figure 8.8 Flow structure and fully-developed velocity profile of turbulent flow in circular tube.

All properties in Eq. (8.16) are evaluated at the mean bulk temperature, except for the dynamic viscosity term μ_s, which is evaluated at the wall temperature. The ranges of applicability of the equation are

$$6000 < Re_D < 10^7$$

$$0.7 < Pr < 10000$$

$$L/D > 60$$

A correlation for circular and non-circular conduits of any length is proposed in Mikheyev and Mikheyeva (1977)

$$\overline{Nu_D} = 0.021 Re_D^{0.8} Pr^{0.43} \left(\frac{Pr_b}{Pr_s}\right)^{0.25} \in_L \tag{8.17}$$

The parameter $\in_L$ accounts for entrance length effects and varies depending on the length-to-diameter ratio of the tube L/D. For $1 \leq L/D \leq 50$, $\in_L$ varies with both L/D and Re_D, as shown in Figure 8.9. For $L/D \geq 50$, $\in_L = 1$. The characteristic length used in this equation is calculated from $D = 4A/P$, where A is the cross-sectional area and P is the perimeter of the conduit. Eq. (8.17) is applicable for the ranges

$$10^4 < Re_D < 5 \times 10^6$$

$$0.6 < Pr < 2500$$

A correlation with two independent variables of the following form fits the data in Figure 8.9 rather well and can be used for any value within the ranges $1 \leq L/D \leq 50$ and $10^4 \leq Re_D \leq 5 \times 10^6$

$$\in_L = a + b\ln(L/D) + c/Re_D + d[\ln(L/D)]^2 + e/Re_D^2 + f\ln(L/D)/Re_D \tag{8.18}$$

where $a = 1.205358$, $b = -7.32 \times 10^{-2}$, $c = 6.21 \times 10^3$, $d = 4.19 \times 10^{-3}$, $e = -1.8 \times 10^7$, and $f = -1.13 \times 10^3$.

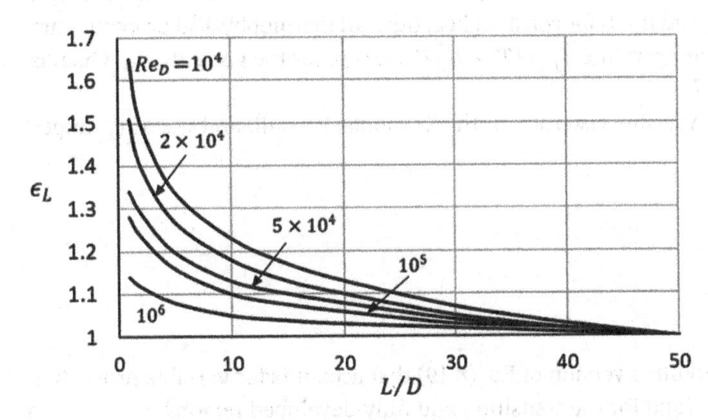

Figure 8.9 Correction factor $\in_L$. for the Nusselt number in Eq. (8.17).

Petukhov (1970) proposed two rather complicated correlations for fully developed turbulent flow in tubes that are applicable within large ranges of Prandtl and Reynolds numbers.

Constant thermophysical properties

$$\overline{Nu_D} = \frac{(f/8)Re_D Pr}{K_1 + K_2 (f/8)^{1/2}\left(Pr^{2/3} - 1\right)} \tag{8.19}$$

or a simplified version thereof

$$\overline{Nu_D} = \frac{(f/8)Re_D Pr}{1.07 + 12.7 (f/8)^{1/2}\left(Pr^{2/3} - 1\right)} \tag{8.20}$$

Equations (8.19) to (8.20) are applicable for constant thermophysical properties for the Reynolds and Prandtl ranges

$$10^4 < Re_D < 5 \times 10^6$$

$$0.5 < Pr < 2000$$

Variable viscosity

$$\overline{Nu_D} = \frac{(f/8)Re_D Pr}{K_1 + K_2 (f/8)^{1/2}\left(Pr^{2/3} - 1\right)}\left(\frac{\mu_b}{\mu_s}\right)^n \tag{8.21}$$

The friction coefficient f for smooth tubes and constants K_1 and K_2 in Eqs (8.19) to (8.21) are determined from

$$f = \frac{1}{(1.82 \log_{10} Re_D - 1.64)^2}$$

$$K_1 = 1.0 + 3.4f$$

$$K_2 = 11.7 + \frac{1.8}{Pr^{1/3}}$$

The exponent of the ratio of dynamic viscosities in Eq. (8.21) is $n = 0.11$ for heated fluid $(T_s > T_b)$, $n = 0.25$ for cooled fluid $(T_s < T_b)$, and $n = 0$ for constant heat flux. All thermophysical properties are evaluated at the mean arithmetic temperature $T_f = (T_s + T_b)/2$, except for the viscosity μ_s, which is evaluated at the wall temperature T_s.

Equation (8.21) can be used for variable viscosity for the Reynolds, Prandtl, and viscosity ranges

$$10^4 < Re_D < 1.25 \times 10^6$$

$$2 < Pr < 140$$

$$0.025 < \mu_b / \mu_s < 12.5$$

Gnielinski (1976) proposed a modified version of Eq. (8.19) that accounts for variable properties and entrance length effects and is valid for the transition and fully-developed regions

$$\overline{Nu_D} = \frac{(f/8)(Re_D - 1000)Pr}{1 + 12.7 (f/8)^{1/2}\left(Pr^{2/3} - 1\right)}\left[1 + \left(\frac{D}{L}\right)^{2/3}\right]K \tag{8.22}$$

where

$K = (Pr_b / Pr_s)^{0.11}$ for liquids, $K = (T_b / T_s)^{0.45}$ for gases, D is the tube diameter, and L is the tube length.

$$f = \frac{1}{(1.80\log_{10} Re_D - 1.5)^2}$$

Equation (8.22) is valid for the ranges

$$2300 < Re_D < 10^6$$

$$0.5 < Pr < 200$$

$$0 < D/L < 1$$

The application of the various correlations to the flow of water in a 2-m long tube of 12-mm diameter is shown in Figure 8.10. The correlations used are:

Laminar region: Eqs (8.11) to (8.14)
Turbulent region: Eqs (8.16), (8.18), and (8.21)
$D = 12\,mm$, $L = 2\,m$ $T_b = 50°C$, $T_s = 70°C$, properties of water:
$\mu_b = 5.44\times10^{-4} N.m/s$, $\mu_s = 4.0\times10^{-4} N.m/s$, $Pr_b = 3.53$, $Pr_s = 2.53$

The correlations used in the calculations give significantly divergent results over much of the Reynolds number range in the laminar flow region. Gnielinski correlations (Eqs 8.13 and 8.14) overestimate the predicted Nusselt numbers for both constant temperature and constant heat flux boundary conditions. In the turbulent region, the results for the Nusselt number for all three correlations show similar results. It should be noted that the calculation results from correlation (Eq. 8.21) by Gnielinski and correlation (Eq. 8.12) by Mikheyev and Mikheyeva in the transition region do not correspond and a method of estimating the Nusselt number in this region is described in the following section.

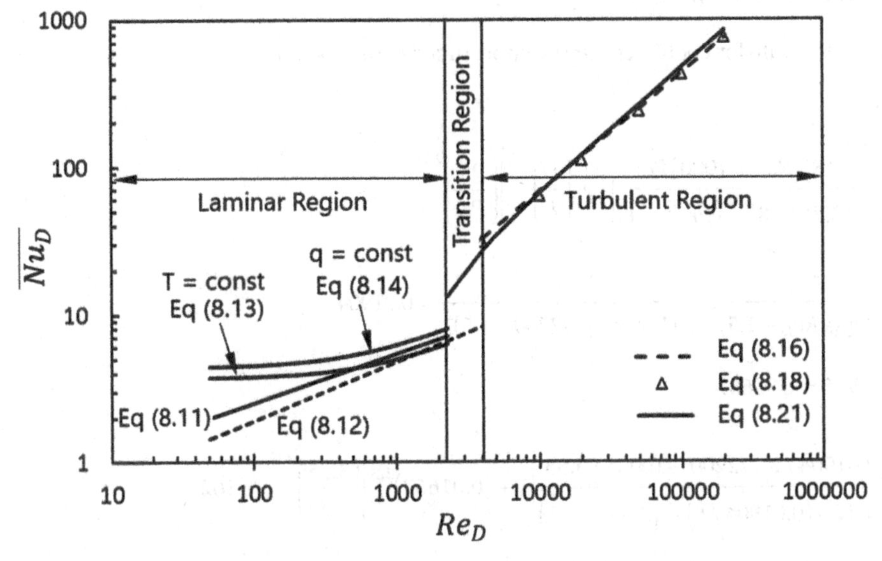

Figure 8.10 Laminar and turbulent flow of heated water in circular tube.

Example 8.3 Water flows in a thin tube of diameter $D = 50\,mm$ at a uniform speed of $U = 0.8\,m/s$. The average bulk temperature of the water is $T_b = 50°C$ and the temperature of the wall is $T_s = 70°C$. Assuming fully-developed flow, calculate the mean heat transfer coefficient $\bar{h}$ for two tube lengths: $L = 3\,m$ and $L = 1.2$. The properties of water: $\rho = 988.1\,kg/m^3$, $k = 0.644\,W/m.K$, $\mu_b = 5.47\times10^{-4}\,N.m/s$, $\mu_s = 4.04\times10^{-4}\,N.m/s$, $Pr_b = 3.55$, $Pr_s = 2.55$.

Solution

The Reynolds number for the flow is

$$Re_D = \frac{U\rho D}{\mu} = \frac{0.8\times988.1\times0.05}{5.47\times10^{-4}} = 7.22\times10^4 \text{ (turbulent flow)}$$

The effect of the tube length is accounted for in Eqs (8.17) and (8.22). From Eq. (8.17)

$$\overline{Nu}_D = 0.021 Re_D^{0.8} Pr^{0.43}\left(\frac{Pr_b}{Pr_s}\right)^{0.25} \in_L$$

For $L = 3\,m$, $L/D = 60$
hence, $\in_L = 1$

$$\overline{Nu}_D = 0.021\times72200^{0.8}\times3.55^{0.43}\left(\frac{3.55}{2.55}\right)^{0.25} = 303$$

$$\bar{h} = \frac{\overline{Nu}_D\times k}{D} = \frac{303\times0.644}{0.05} = 3902\,W/m.K$$

For $L = 1.2\,m$, $L/D = 24$, Eq. (8.18) gives $\in_L = 1.0478$

$$\overline{Nu}_D = 303\times1.0478 = 317$$

$$\bar{h} = 3902\times1.0478 = 4088\,W/m.K$$

The increase in heat transfer coefficient in the short tube is $\Delta\bar{h} = 4.75\%$

From Eq. (8.22)

$$\overline{Nu}_D = \frac{(f/8)(Re_D - 1000)Pr}{1 + 12.7(f/8)^{1/2}\left(Pr^{2/3} - 1\right)}\left[1 + \left(\frac{D}{L}\right)^{2/3}\right]\left(\frac{Pr_b}{Pr_s}\right)^{0.11}$$

$$f = \frac{1}{(1.80\log_{10} Re_D - 1.5)^2} = \frac{1}{(1.80\log_{10} 72200 - 1.5)^2} = 0.01904$$

$$D/L = 0.05/3 = 0.01667$$

$$\overline{Nu}_D = \frac{(0.01904/8)(72200 - 1000)\times3.55}{1 + 12.7(0.01904/8)^{1/2}\left(3.55^{2/3} - 1\right)}\left[1 + (0.01667)^{2/3}\right]\left(\frac{3.55}{2.55}\right)^{0.11} = 362$$

$$\bar{h}=\frac{\overline{Nu_D}\times k}{D}=\frac{362\times0.644}{0.05}=4663W/m.K$$

$$D/L=0.05/1.2=0.0417$$

$$\overline{Nu_D}=381$$

$$\bar{h}=4903$$

The increase in heat transfer coefficient in the short tube is $\Delta\bar{h}=5.15\%$

8.3.1 Forced Convection for Flow in the Transition Region (Region II)

Heat exchangers often operate in the transition region between laminar and turbulent flows. The difficulty in evaluating the flow in this region is due to the complexity and instability of the flow and possible influence of factors other than the Reynolds number. One approach to solving this problem, as proposed by Gnielinski (2013), is to use linear interpolation between one correlation in the laminar region and another in the turbulent region. The relationship obtained for the Nusselt number in the transient region is then

$$Nu_{tran}=(1-\gamma)Nu_{lam,2300}+\gamma Nu_{turb,4000} \tag{8.23}$$

where

$$\gamma=\frac{Re_{tran}-2300}{4000-2300}\qquad 0\leq\gamma\leq1.0 \tag{8.24}$$

$Nu_{lam,2300}$ is determined from Eq. (8.13) or Eq. (8.14) and $Nu_{turb,4000}$ from Eq. (8.21). The variation of Nu_{tran} in the range $2300\leq Re_D\leq4000$ in accordance with Eq. (8.23) is shown as broken line in Figure 8.11.

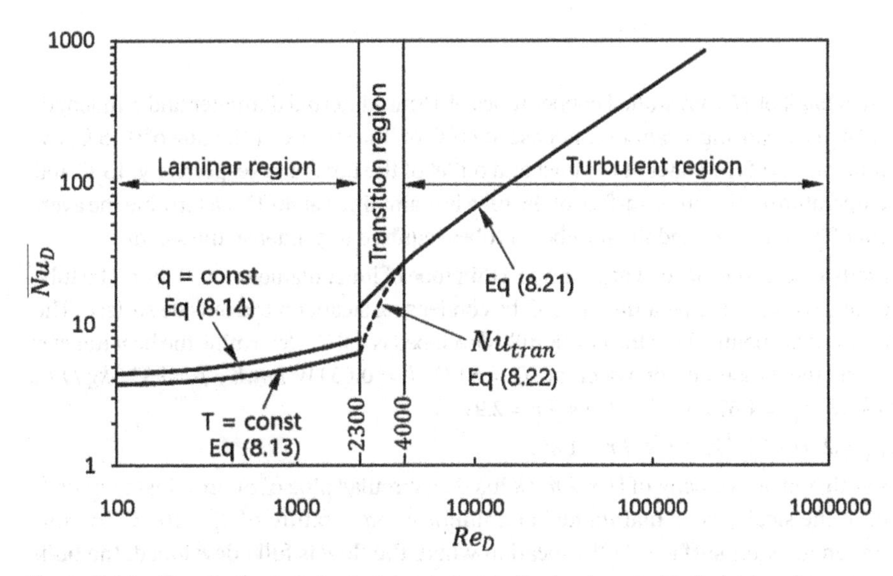

Figure 8.11 Flow in the transient region between the laminar and turbulent regions.

Example 8.4 A constant heat flux is imposed at the surface of a tube 1 mm in diameter, in which water at 18°C flows at a uniform speed of $0.2\ m/s$. Assuming the flow is fully developed, determine the distance at which the tube surface temperature reaches 74°C. The properties of water at $T_m = (18+72)/2 = 45°C$: $k = 0.637\ W/m.K$, $\rho = 990\ kg/m^3$, $c_p = 4180\ J/kg.K$, $\mu = 5.96\times10^{-4} kg/m.s$

Solution

$$Re = \frac{\rho UD}{\mu} = \frac{990\times0.2\times0.001}{5.96\times10^{-4}} = 332$$

The flow is laminar and Eqs (8.3) and (8.4) can be applied.

Let the temperature of the surface reach 74°C at $x = L$. The bulk temperature at that point from Eq. (8.3) is

$$T_b(x) = T_{b,in} + \frac{2\pi qR}{\dot{m}c_p}x = 18 + \frac{2\pi qR}{\pi R^2 \rho U c_p}L$$

$$T_b(x) = 18 + \frac{2q}{R\rho U c_p}L = 18 + \frac{2\times6000}{0.0005\times990\times0.2\times4180}L = 18 + 29L$$

The temperature of the surface at $x = L$ from Eq. (8.4)

$$T_s(x) = T_b(x) + \frac{22}{48}\frac{qR}{k}$$

$$72 = (18 + 29L) + \frac{22\times6000\times0.0005}{48\times0.637} = 18 + 29\,L + 2.158$$

$\therefore L = 1.788\ m$ measured from the entrance

Problems

8.1 Water flows in a bank of N thin-walled copper tubes of 15 mm internal diameter and 5 m length at a velocity of $4\ m/s$, causing steam to condense at 40°C on the surface at the rate of $0.16\ kg/s$. The bulk temperatures of the water at the inlet and outlet of the tubes are, respectively, 15°C and 35°C. The temperature of the outer surface of the tube is maintained at 40°C. Determine the average coefficient of heat transfer and the number of tubes required to condense the steam.

8.2 Water flows with a mean velocity of $2\ m/s$ inside a thin tube of inner diameter of 50 mm. The tube is maintained at a constant temperature of 95°C by condensing steam on the outside surface. The bulk temperature at the point where the flow is fully developed is 60°C. Determine the heat transfer coefficient. Take the properties of water at $T_b = 60°C$: $k = 0.654\,W/m.K$, $\rho = 983.3\,kg/m^3$, $c_p = 4185\,J/kg.K$, $\mu_b = 4.67\times10^{-4} kg/m.s$, $Pr = 2.99$

At 95°C: $\mu_s = 2.97\times10^{-4} kg/m.s$, $Pr = 1.85$.

8.3 Water flows with a mean velocity of $U = 2\ m/s$ inside a circular pipe of 50 mm inside diameter. The wall of the steel pipe is maintained at a uniform temperature of $T_s = 100°C$ by condensing steam on its outer surface. At the location where the flow is fully developed, the bulk temperature of water is $T_b = 60°C$. Determine the heat transfer coefficient using the Petukhov Eq. (8.21)

8.4 Atmospheric air at 0.012 kg *m/s* flows in a smooth tube of 25 mm diameter. Use an appropriate correlation to calculate the average heat-transfer coefficient for properties evaluated at 25, 125, 230, and 530°C.

8.5 Water entering at 10°C is heated to 40°C in the tube of 0.02 m internal diameter at a mass flow rate of 0.01 kg/s. The outside of the tube is covered with an insulated electrical heating element that produces a uniform heat flux of 15,000 W/m^2 over the surface. Determine

(a) the heat transfer coefficient

(b) the length of pipe needed for a 30°C increase in average temperature

(c) The bulk temperature at the outlet.

8.6 Water flows at the rate of $2cm^3$ in a copper tube 25 mm in diameter and 2 m long. The inlet and outlet bulk temperatures of the water are 20°C and 60°C, respectively. The water is heated by adding heat at a constant rate and the flow is fully developed hydrodynamically and thermally at the tube inlet. Determine the rate of heat addition and the maximum wall temperature.

8.7 Consider a 10 m long smooth rectangular tube $(25mm \times 25mm)$ that is maintained at a constant surface temperature T_s (Figure P8.7). Liquid water enters the tube at 20°C with a mass flow rate of $0.01 kg/s$. Determine the tube surface temperature necessary to heat the water to the desired outlet temperature of 80°C.

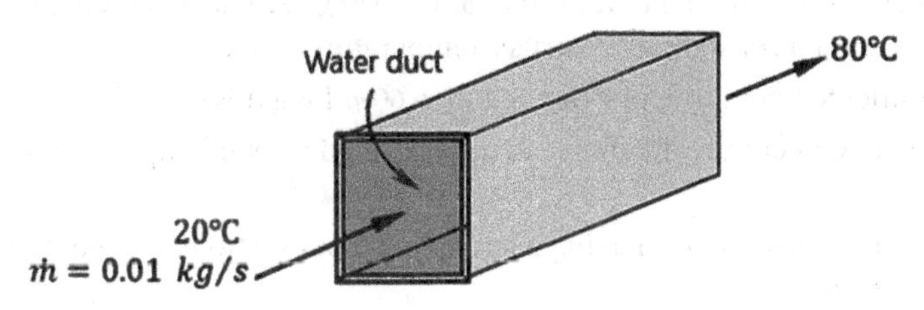

Figure P8.7 Flow in a long rectangular tube.

8.8 Hot air at 55°C enters a 12 m long thin metal sheet duct of rectangular cross-section $(0.2m \times 0.2m)$ at an average velocity of 4 m/s and leaves at a temperature of 45°C, as shown in Figure P8.8. Determine:

(a) the duct surface temperature

(b) the rate of heat loss from the air to the surroundings.

Use the Dittus–Boetler correlation (Eq. 8.15) to estimate the heat transfer coefficient if the flow is turbulent.

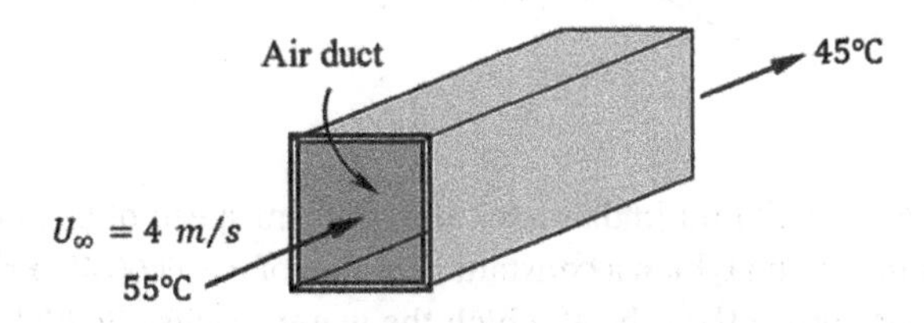

Figure P8.8 Flow of air in a long rectangular tube.

8.9 Repeat Problem 8.8 using correlation (Eq. 8.17).

8.10 Air with a mass flow rate of 0.075 kg m/s flows through a tube of diameter $0.225m$. The air enters at 100°C and, after a distance of $L = 6m$, cools to 60°C. Determine the heat transfer coefficient of the air.

8.11 Engine oil enters an $8m$ long tube at 20°C. The tube diameter is $20\,mm$, and the flow rate is $0.4\,kg/s$. Calculate the outlet temperature of the oil if the tube surface temperature is maintained at 80°C. Take the properties of oil as: $k = 0.145 W/m.K$, $\rho = 888 kg/m^3$, $c_p = 1880 J/kg.K$, $\mu = 0.7992 kg/m.s$, $Pr = 10{,}400$.

8.12 Repeat Problem 8.11 for ethylene glycol having the following properties at 20°C: $k = 0.249 W/m.K$, $\rho = 1116.6 kg/m^3$, $c_p = 2382 J/kg.K$, $\mu = 0.02142 kg/m.s$, $Pr = 204$.

8.13 Liquid ammonia flows through a $25mm$-diameter smooth tube $2.5m$ long at a rate of 0.4 kg/s. The ammonia enters at 10°C and leaves at 40°C, and a constant heat flux is imposed on the tube wall. Calculate the average wall temperature necessary to heat the ammonia.

8.14 Repeat Problem 8.13 with engine oil as the fluid.

8.15 A short tube is $6.4mm$ in diameter and $15cm$ long. Water enters the tube at $1.5m/s$ and 40°C, and a constant-heat-flux condition is maintained such that the tube wall temperature remains at 25°C above the water bulk temperature. Calculate the heat transfer rate and exit water temperature.

8.16 Engine oil flows through a $25mm$ diameter tube at a rate of $0.5kg/s$. The oil enters the tube at a temperature of 25°C. For a constant tube surface temperature of 100°C:

(a) determine the oil outlet temperature for a $6m$ and for a $60m$ long tube

(b) compare the log mean temperature difference to the arithmetic mean temperature difference for each case.

8.17 Repeat part (a) of Problem 8.16 for a tube having a hexagonal cross-section area that is equal to the area of the circular tube.

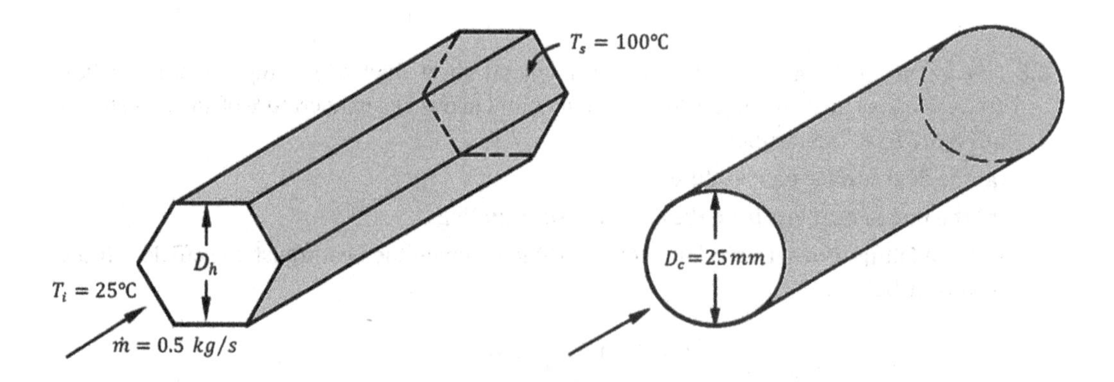

8.18 Water at 20°C flows through a small tube, 1 mm in diameter at a uniform speed of $0.2m/s$. The flow is fully developed at a point beyond which a constant heat flux of $q = 6000 W/m^2$ is imposed. Determine the point farther down the tube at which the water reaches its highest temperature of 74°C.

8.19 Steam is generated on the surface of heat-exchanger tubes, each carrying pressurized water Each tube is 2.5m long and 25mm diameter. The water flows in the tubes at 3m / s, and the tube surfaces are maintained at 250°C. Assume the bulk temperature of the pressurized water over the length of the tube is 280°C. Estimate the heat transfer coefficient using the Dittus–Boetler correlation and Sieder and Tate correlation.

8.20 Repeat Problem 8.19 using the correlations by Mikheyev and Mikheyeva, Gnielinski, and Petukhov.

9

Natural (Free) Convection

Natural convection is a mode of heat transfer that does not require an external driver such as a pump, blower, or compressor to drive the heat transfer process in the presence of a temperature gradient. Examples of the occurrence of natural convection include power generation (cooling transmission lines, transformers, heat losses from pipes carrying steam or other heated fluids), electronic devices (power transistors in electronic devices, TVs), building heating systems (steam radiators, walls of the building, and bodies of human occupants). Natural convection is also present in commonplace phenomena such as cooling of a cup of tea or heating of a cool drink on a kitchen table by the surrounding air. Natural convection is the main driver in oceanic and atmospheric motions and the related heat and mass transfer processes forming weather patterns and other natural phenomena.

Rapid rotation is another field of force that can be utilized to produce natural convection as the case with the cooling of gas turbine rotors. In this scheme, rotational forces generate large centrifugal forces that are proportional to the fluid density and are associated with strong convection currents for effective cooling.

9.1 Boundary Layer in Free Convection

Natural convection arises when a surface at a certain temperature is maintained in a still fluid at a temperature higher or lower than the surface temperature. A density gradient is created in the layer of fluid adjacent to the surface as a result of the fluid being heated up or cooled down. The density difference introduces a buoyant force which, combined with gravity force, causes the fluid near the surface to flow. The flow velocities for this mode of heat transfer are lower in comparison with flow velocities in external forced convection leading to much lower heat transfer coefficients. If high rates of heat transfer are required, larger surface areas need to be provided to compensate for the low heat transfer coefficients.

Figure 9.1 shows the development of the hydrodynamic boundary layer on an isothermal (constant temperature) vertical plate in still fluid at temperature T_∞ and density ρ_∞. Despite the fact that the natural convective currents generated by gravity are low, the flow characteristics are similar to those in forced convection, albeit with different velocity profile, which exhibits zero velocity at the wall and at the edge of the boundary layer. If the plate temperature is higher than the fluid temperature, the convective flow under the influence of buoyancy will move vertically upwards. If the plate temperature is lower than the fluid temperature, the flow will be downwards.

Heat Transfer Basics: A Concise Approach to Problem Solving, First Edition. Jamil Ghojel.

Companion website: www.wiley.com/go/ghojel/heat_transfer

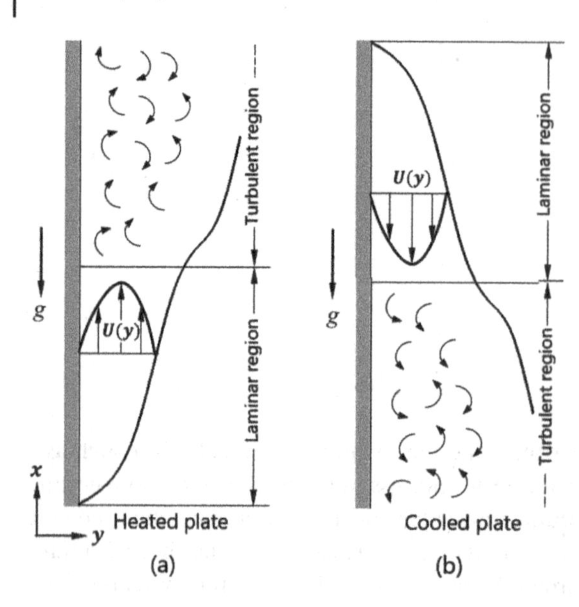

Figure 9.1 Development of boundary layer on vertical heated (a) and cooled (b) plates.

9.2 Governing Equation for Laminar Boundary Layer

Consider the schematic diagram shown in Figure 9.2 of the processes in the laminar hydrodynamic boundary layer of a long isothermal vertical plate at temperature T_s in still fluid at temperature T_∞. The plate in the z direction is very wide so that the flow can be assumed to be two-dimensional. The length of the plate, normally designated L, is measured in the x-direction.

The analysis of the flow is based on the following assumptions:

1) The fluid is incompressible
2) Inertia forces are negligible in comparison with the gravity and viscous forces
3) Heat conduction and convection in the x-direction is negligible
4) Pressure gradients across the boundary layer are small
5) The shear stresses on element area $dydz$ are ignored
6) Thermophysical properties of the fluid are constant, except the density which varies linearly with temperature.

The governing differential equations for the conservation of mass, momentum, and energy are respectively

$$\frac{\partial u}{\partial x}+\frac{\partial v}{\partial y}=0 \tag{9.1}$$

$$\rho u\frac{\partial u}{\partial x}+\rho v\frac{\partial u}{\partial y}=\rho g\beta(T-T_\infty)+\mu\frac{\partial^2 u}{\partial y^2} \tag{9.2}$$

$$\rho u\frac{\partial T}{\partial x}+\rho v\frac{\partial T}{\partial y}=\frac{k}{c_p}\frac{\partial^2 T}{\partial T^2} \tag{9.3}$$

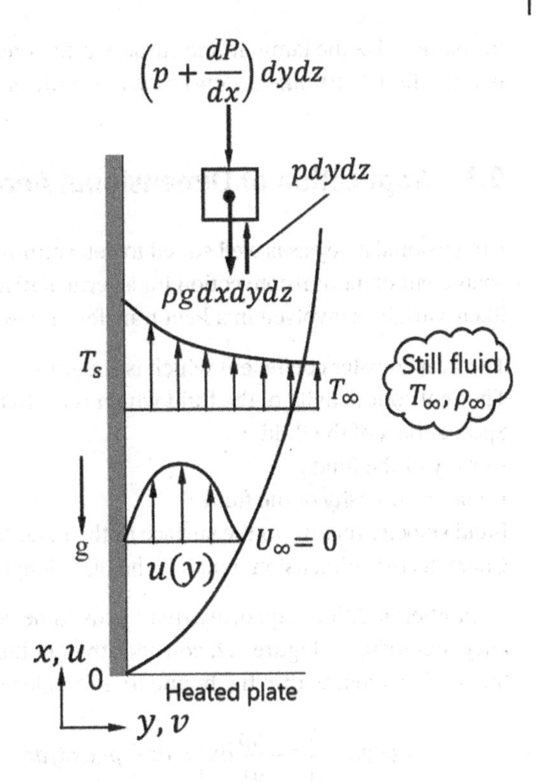

Figure 9.2 Force balance on a fluid element in natural convection flow.

The solution of Eqs (9.1) to (9.3) yields the local convection coefficient h at location x from origin 0

$$Nu_x = \frac{hx}{k} = \left(\frac{Gr_x}{4}\right)^{1/4} F(Pr) \tag{9.4}$$

In this equation, Gr_x is the local Grashof number and $F(Pr)$ is some function of the Prandtl number.

The average convection coefficient $\bar{h}$ for a plate of length L is

$$\overline{Nu_L} = \frac{\bar{h}L}{k} = Ra_L^{1/4}\left(\frac{Pr}{2.5 + 5\sqrt{Pr} + 5Pr}\right)^{1/4} \tag{9.5}$$

The Rayleigh number Ra_L is the product of the Prandtl and Grashof numbers

$$Ra_L = Pr\ Gr_L \tag{9.6}$$

Gr_L is the Grashof number defined as

$$Gr_L = \frac{g\beta\rho^2 (T_s - T_\infty)L^3}{\mu^2} = \frac{\beta g (T_s - T_\infty)L^3}{\vartheta^2} \tag{9.7}$$

β is the coefficient of bulk expansion and ϑ is the kinematic viscosity ($= \mu / \rho$).

The dimensionless Grashof number represents the ratio of the buoyancy forces to the viscous forces. Equation (9.5) is applicable to both heated and cooled vertical plates.

As was the case with forced convection, fluid flow in natural convection could be laminar, transitional, or turbulent. However, instead of the Reynolds number, the Rayleigh number Ra_L is used to characterize the laminar and turbulent flow regions. The critical point (point of transition from laminar flow to turbulent flow) for gases with $Pr = 1$ occurs at $Ra_L \approx 10^9$.

9.3 Application of Dimensional Analysis to Natural Convection

Dimensional analysis is well suited to determining the functional relationship for the heat transfer coefficient of natural convection for a vertical plate on the basis of previous knowledge of the most likely variables involved in a heat transfer process. These are:

The heat transfer coefficient which is sought h
Thermal conductivity of the fluid which is exchanging heat with the plate k
Specific heat of the fluid c_p
Density of the fluid ρ
Dynamic viscosity of the fluid μ
Fluid velocity relative to the surface of the plate U
Characteristic dimension which is the plate length in the x direction, L.

Another additional quantity that needs to be considered in this case is some measure of buoyancy. Referring to Figure 9.2, consider the net body force Φ due to pressure and gravity acting on the fluid element within the boundary layer close to the surface

$$\Phi = pdydz - \left(p + \frac{dp}{dx}dx\right)dydz - \rho gdxdydz \tag{9.8}$$

$$\Phi = -\frac{dp}{dx}dxdydz - \rho gdxdydz \tag{9.9}$$

The rate of pressure variation in the x-direction is

$$\frac{dp}{dx} = -\rho_\infty g$$

hence,

$$\Phi = g(\rho_\infty - \rho)dxdydz$$

and the body force per unit volume ($\varphi = \Phi / dxdydz$) within the boundary layer in term of densities (or specific volumes) is

$$\varphi = g(\rho_\infty - \rho) = g\frac{v - v_\infty}{vv_\infty}$$

$$\text{If } v \approx \frac{v_s + v_\infty}{2}$$

then

$$\varphi \approx g\frac{v_s - v_\infty}{2vv_\infty} \tag{9.10}$$

Since the fluid is incompressible and density is assumed to vary only with temperature, it is convenient to express the buoyancy force in terms of temperature difference by introducing the coefficient of bulk expansion β defined in terms of the specific volume v as

$$\beta = \frac{1}{v}\left(\frac{dv}{dT}\right)_{p=const}$$

For an ideal gas

$$\beta = \frac{1}{T}$$

The average value of β for the temperatures range between the wall surface (T_s) and ambient (T_∞)

$$\beta_{av} = \frac{v_s - v_\infty}{v_\infty (T_s - T_\infty)} \tag{9.11}$$

Combining Eqs (9.10) and (9.11) results in

$$\varphi \approx \frac{1}{2}\frac{\beta_{av} g (T_s - T_\infty)}{v} \approx \frac{1}{2}\beta_{av} g \rho (T_s - T_\infty) \tag{9.12}$$

or, more generally,

$$\varphi \propto \beta_{av} g \rho (T_s - T_\infty) \tag{9.13}$$

Equation (9.13) shows that for the free convection model for a flat vertical plate, the buoyancy force is proportional to the expansion coefficient, fluid density, and temperature difference between the plate surface and the fluid. The subscript to the coefficient of expansion will be dropped henceforth giving

$$\varphi \propto \beta g \rho (T_s - T_\infty) \tag{9.14}$$

Since ρ is usually taken as an independent variable in heat transfer analysis, we can reduce the term on the right side of Eq. (9.14) to $g\beta(T_s - T_\infty)$ and take it as the measure of the buoyancy force in the functional relationship for a natural convection for a vertical plate, written as follows:

$$h = f(k, c_p, \rho, \mu, U, l, \beta g \Delta T)$$

or

$$f(h, k, c_p, \rho, \mu, U, l, \beta g \Delta T) = 0 \tag{9.15}$$

The 8 quantities $h, k, c_p, \rho, \mu, U, l, g\beta\Delta T$ can be defined in terms of the 4 fundamental units M, L, T, θ. According to the Buckingham theorem, the minimum number of independent dimensionless products is 4, hence Eq. (9.15) is rewritten as

$$f(\Pi_1, \Pi_2, \Pi_3, \Pi_4) = 0 \tag{9.16}$$

With the variables h, U, $g\beta\Delta T, c_p$ taken as the prime variables, the independent Π product can be written as

$$\Pi_1 = h k^{a_1} \mu^{a_2} \rho^{a_3} l^{a_4}$$

$$\Pi_2 = U k^{b_1} \mu^{b_2} \rho^{b_3} l^{b_4}$$

$$\Pi_3 = (\beta g \Delta T) k^{c_1} \mu^{c_2} \rho^{c_3} l^{c_4}$$

$$\Pi_4 = c_p k^{d_1} \mu^{d_2} \rho^{d_3} l^{d_4}$$

For the independent Π products to be dimensionless, the sum of the exponents of each primary dimension must sum up to zero. Thus, for Π_1

$$\left(MT^{-3}\theta^{-1}\right)\left(MLT^{-3}\theta^{-1}\right)^{a_1}\left(ML^{-1}T^{-1}\right)^{a_2}\left(ML^{-3}\right)^{a_3}(L)^{a_4} = M^0L^0T^0\theta^0 \tag{9.17}$$

Equating the exponents on both sides of Eq. (9.17), we obtain

For $M: 1+a_1+a_2+a_3=0$
For $L:\ a_1-a_2-3a_3+a_4=0$
For $T:\ -3-3a_1-a_2=0$
For $\theta:\ -1-a_1=0$

Solving the above four simultaneous equations yields

$$a_1=-1,\ a_2=0,\ a_3=0, a_4=1$$

Therefore,

$$\Pi_1 = hk^{-1}\mu^0\rho^0 l^1$$

or

$\Pi_1 = (hl/k) = Nu$ (Nusselt number)

Similar solutions for the remaining Π products result in

$\Pi_2 = (Ul\rho/\mu) = Re$ (Reynolds number)

$\Pi_3 = (\beta g\Delta T\rho^2 l^3/\mu^2) = \beta g\Delta T l^3/\vartheta^2 = Gr$ (Grashof number), $\Delta T = T_s - T_\infty$

$\Pi_4 = (c_p\mu/k) = Pr$ (Prandtl number)

Therefore, the final functional relationship is of the form

$f(Nu, Re, Gr, \mathrm{Pr}) = 0$, or

$$Nu = f(Re, Gr, Pr) \tag{9.18}$$

Since the velocity U in natural convection is small, the Reynolds number can be dropped from Eq. (9.18), which is then simplified to

$$Nu = f(Gr, Pr) \tag{9.19}$$

The properties in Eq. (9.19) are evaluated at the mean film temperature $T_f = (T_s + T_\infty)/2$.

9.4 Empirical Correlations for Natural Convection

Empirical equations obtained by correlating experimental data for different flow scenarios in terms of the dimensionless variables in Eq. (9.19) are useful for estimating the convection heat transfer coefficient in engineering applications and are generally of the form

$$\overline{Nu}_L = C(Gr_L Pr)^n = C(Ra_L)^n \tag{9.20}$$

where Ra_L is the Rayleigh number ($Ra_L = Gr_L Pr$) and C and n are constants that vary with the geometry of the body, nature of the convection process, and range of Ra_L.

9.4.1 Vertical Plates

The correlations for vertical plates that are to follow are generally applicable to vertical plates at constant surface temperature ($T_s = \text{const}$), but can also be applied to vertical plates at constant surface heat flux ($q_s = \text{const.}$). However, since the temperature in the latter case increases with height, temperature T_s should be replaced by the temperature at the midpoint of the plate $T_{L/2}$. Temperature $T_{L/2}$ can be determined iteratively (Çengel and Ghajar 2011) using Newton's law of cooling and a correlation as per Eq. (9.20).

$$q_s = \bar{h}\left(T_{L/2} - T_\infty\right)$$

For a known constant heat flux q_s,

$$\overline{Nu}_L = \frac{\bar{h}L}{k} = \frac{q_s L}{k\left(T_{L/2} - T_\infty\right)} \tag{9.21}$$

$T_{L/2}$ is varied until a value is reached at which the average Nusselt numbers from Eqs (9.20) and (9.21) match to the required degree of accuracy.

Reasonable agreement with experiments can be obtained for various fluids and Prandtl numbers in the range from 0.01 to 1000 using the following earlier correlations for vertical plates:

$$\overline{Nu}_L = 0.62\left(Gr_L Pr\right)^{1/4} \text{ for laminar region } 10^5 < Gr_L Pr < 10^9$$
$$\overline{Nu}_L = 0.12\left(Gr_L Pr\right)^{1/3} \text{ for turbulent region } 10^9 < Gr_L Pr < 10^{13}$$

The thermophysical properties are evaluated at the film temperature $T_f = \left(T_s + T_\infty\right)/2$.

More accurate, albeit more complex, correlations were proposed by Churchill and Chu (1975a) of the forms

$$\overline{Nu}_L = 0.68 + \frac{0.67 Ra_L^{1/4}}{\left[1 + \left(0.492/Pr\right)^{9/16}\right]^{4/9}} \tag{9.22}$$

for the laminar region ($Ra \leq 10^9$)

$$\overline{Nu}_L = \left\{0.825 + \frac{0.387 Ra_L^{1/6}}{\left[1 + \left(0.492/Pr\right)^{9/16}\right]^{8/27}}\right\}^2 \tag{9.23}$$

for both the laminar and turbulent regions ($10^4 < Ra < 10^{13}$).

The thermophysical properties in Eqs (9.22) and (9.23) are evaluated at the film temperature $T_f = \left(T_s + T_\infty\right)/2$. If the properties vary significantly with temperature, they are evaluated at the mean of the surface temperature and the bulk temperature.

Correlations accounting for the effect of temperature variation through the ratio of the Prandtl numbers for vertical plates and cylinders can be found in Mikheyev and Mikheyeva (1977)

$$\overline{Nu}_L = 0.76\left(Gr_L Pr\right)^{0.25}\left(\frac{Pr}{Pr_s}\right)^{0.25} \quad \text{for } 10^3 < Ra_L < 10^9 \tag{9.24}$$

$$\overline{Nu}_L = 0.15\left(Gr_L Pr\right)^{0.33}\left(\frac{Pr}{Pr_s}\right)^{0.25} \quad \text{for } Ra_D > 10^9 \tag{9.25}$$

The thermophysical properties in Eqs (9.24) and (9.25) are evaluated at the temperature of the surroundings T_∞. Equation (9.25) gives results close to Eq. (9.23) if the exponent 0.33 is replaced by 0.32.

A vertical cylinder can be treated as a vertical plate if the following condition is satisfied:

$$D \geq \frac{35L}{Gr_L^{0.25}}$$

The characteristic length L in the equations for plates is replaced by the diameter of the tube D when applied to vertical tubes.

Example 9.1 A vertical plate 4 m high and 8 m wide is maintained at a constant temperature of 70°C. If the surrounding atmosphere is at 10°C, what is the rate of heat loss from the plate?

Solution

The properties of air at $T_f = (70+10)/2 = 40°C$: $\rho = 1.127\, kg/m^3, c_p = 1007\, J/kg.K, k = 0.02662\, W/m.K,$ $\mu = 1.918 \times 10^{-5}\ kg/m.s\ \left(\nu = 1.708 \times 10^{-5}\right), Pr = 0.7255, \beta = 1/T_f = 3.195 \times 10^{-3}\ K^{-1}$

$$Ra = Gr_D Pr = \left(\frac{L_c^3 (g\beta\Delta T)}{\nu^2}\right)\left(\frac{\mu c_p}{k}\right)$$

$$Ra = \left(\frac{4^3\left(9.81 \times 3.195 \times 10^{-3} \times 60\right)}{\left(1.708 \times 10^{-5}\right)^2}\right)(0.7255) = 3.01 \times 10^{11}$$

Since $Ra > 9 \times 10^9$, correlation (9.23) by Churchill and Chu (1975a) for $10^4 < Ra < 10^{13}$ can be used

$$\overline{Nu}_L = \left\{0.825 + \frac{0.387 Ra_L^{1/6}}{\left[1 + (0.492/Pr)^{9/16}\right]^{8/27}}\right\}^2$$

$$\overline{Nu}_L = \left\{0.825 + \frac{0.387 \times \left(3.01 \times 10^{11}\right)^{1/6}}{\left[1 + (0.492/0.7255)^{9/16}\right]^{8/27}}\right\}^2 = 752.5$$

$$\bar{h} = \frac{\overline{Nu}_L k}{L} = \frac{752.5 \times 0.02662}{4} = 5W/m^2K$$

$$\dot{Q} = \bar{h}A(T_s - T_\infty) = 5 \times 4 \times 8 \times 60 = 9600W$$

9.4.2 Horizontal Plates

Unlike flow over a vertical plate, the gravity and buoyant vectors in the case of a horizontal plate are perpendicular to the plate surface which affects the flow patterns significantly. Unimpeded flow resulting from immersing a heated horizontal plate at temperature T_s into a fluid at temperature T_∞, such that $T_s > T_\infty$ is an upwards motion perpendicular to the surface with a plume forming as shown in Figure 9.3. Natural convection flows over large surfaces are more complex with combined upwards and downwards motions of fluid parcels under different density gradients, resulting in the flow patterns shown in Figure 9.4a for a plate with the hot surface up, and in Figure 9.4d for a plate with the cold surface down.

For a finite plate with the hot surface down (Figure 9.4b), the upwards motion of the fluid is impeded by the plate and the flow occurs within a thin horizontal layer near the hot surface before it rises at the edges. For a finite plate with the cold surface up (Figure 9.4c), dense fluid accumulates near the cold surface in the form of a thin layer which then flows sideways and spills over at the edges.

The following correlations are recommended for horizontal plates under different conditions:

Figure 9.3 Convection flow over small plate with heated surface facing up.

Hot surface up and cold surface down (a and d):

$$\overline{Nu}_L = 0.54 Ra_L^{1/4} \quad \text{for } 10^4 \le Ra_L \le 10^7, Pr \ge 0.7 \tag{9.26}$$

$$\overline{Nu}_L = 0.154 Ra_L^{1/3} \quad \text{for } 10^7 \le Ra_L \le 10^{11}, \text{ all Pr} \tag{9.27}$$

Hot surface down and cold surface up (b and c):

$$\overline{Nu}_L = 0.52 Ra_L^{1/5} \quad \text{for } Ra_L \le 10^5, Pr \ge 0.7 \tag{9.28}$$

$$\overline{Nu}_L = 0.27 Ra_L^{1/4} \quad \text{for } 10^5 \le Ra_L \le 10^{10}, Pr \ge 0.7 \tag{9.29}$$

The characteristic length L depends on the shape of the horizontal plate and is given by $L = A/P$ where A is the area of the plate and P is the perimeter. The thermophysical properties of the fluid are evaluated at the film temperature $T_f = (T_s + T_\infty)/2$.

Equations (9.26) to (9.29) can be used for horizontal surfaces with constant temperature or constant flux. In the latter case, the temperature averaged over the plate is determined as explained in Section 9.4.1.

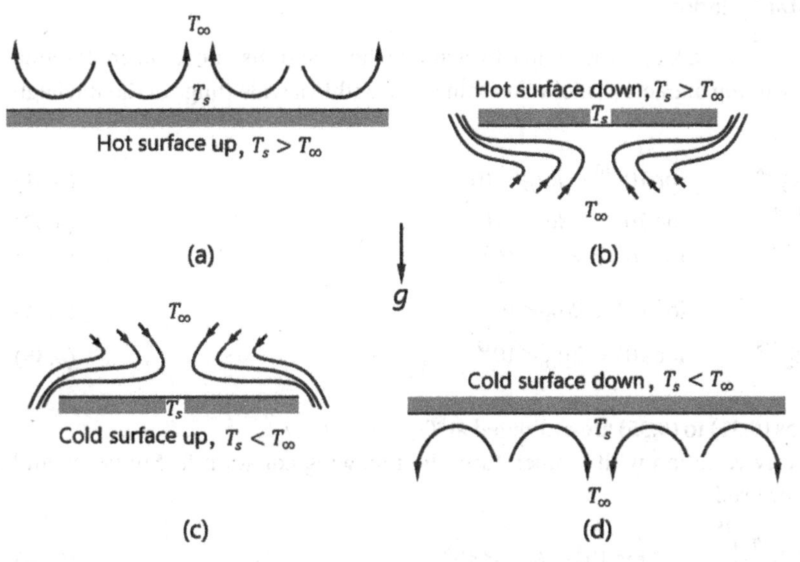

Figure 9.4 Convection flow patterns over heated and cooled horizontal surfaces facing up and down.

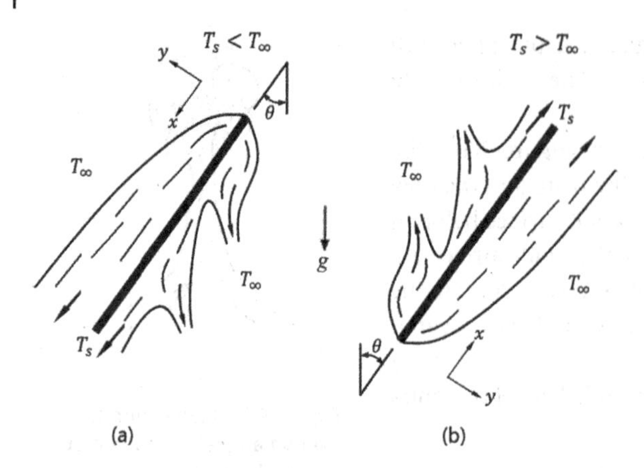

Figure 9.5 Convection flow patterns over heated and cooled inclined flats surfaces: (a) cold plate, $T_s < T_\infty$; (b) hot plate, $T_s > T_\infty$.

9.4.3 Inclined Plates

The plates shown in Figure 9.5 are inclined to the gravity vector at angle θ. As a result, the buoyancy force will have two components: one along the plate, and the other perpendicular to the plate. The flow patterns arising from the direction of the force perpendicular to the plate cause the heat transfer coefficient to increase on the lower surface of the cold plate and on the upper surface of the hot plate.

For flows over the upper surface of an inclined cold plate (Figure 9.5a) and the lower surface of an inclined hot plate (Figure 9.5b), Eqs (9.22) and (9.23) by Churchill and Chu (1975a) can be used after replacing g with $g\cos\theta$ in the Rayleigh number and provided θ does not exceed 88°. The Rayleigh number in these equations can be written as

$$Ra_L = \frac{\beta(T_s - T_\infty) g \cos\theta L^3}{\alpha\vartheta} \tag{9.30}$$

For the lower side of the cold inclined plate and the upper side of the hot inclined plate, the same equations can be used, provided θ does not exceed 60° (Mills 1999).

9.4.4 Long Horizontal Cylinder

Correlations of the form $\overline{Nu_D} = CRa_D^n$ are available to estimate the heat transfer coefficient for long horizontal cylinders of varying diameters, including thin wires within a wide range of the Rayleigh number

$$\overline{Nu}_D = 0.675Ra_D^{0.058} \quad \text{for } 10^{-10} < Ra_D < 10^{-2} \tag{9.31}$$

$$\overline{Nu}_D = 1.02Ra_D^{0.148} \quad \text{for } 10^{-2} < Ra_D < 10^{2} \tag{9.32}$$

$$\overline{Nu}_D = 0.85Ra_D^{0.188} \quad \text{for } 10^{2} < Ra_D < 10^{4} \tag{9.33}$$

$$\overline{Nu}_D = 0.48Ra_D^{0.25} \quad \text{for } 10^{4} < Ra_D < 10^{7} \tag{9.34}$$

$$\overline{Nu}_D = 0.125Ra_D^{0.333} \quad \text{for } 10^{7} < Ra_D < 10^{12} \tag{9.35}$$

All the properties in Eqs (9.31) to (9.35) are evaluated at $T_f = (T_s + T_\infty)/2$.

To account for property variation with temperature, the following equation in Mikheyev and Mikheyeva (1977) can be used

$$\overline{Nu}_D = 0.5Ra_D^{0.25}\left(\frac{Pr}{Pr_s}\right)^{.25} \quad \text{for } 10^{3} < Ra_D < 10^{8} \tag{9.36}$$

The thermophysical properties in Eq. (9.36) are evaluated at the temperature of the surroundings T_∞.

For thin horizontal wires, the following correlation can be used for low Rayleigh numbers (Isachenko et al. 1969).

$$\overline{Nu}_D = 1.18 Ra_D^{1/8} \qquad \text{for } Ra_D < 5\times10^2 \tag{9.37}$$

A rather complicated correlation that is applicable for a wide range of Rayleigh numbers is recommended by Churchill and Chu (1975b)

$$\overline{Nu}_D^{\,1/2} = 0.6 + 0.387\left\{\frac{Ra_D}{\left[1+\left(0.559/Pr\right)^{9/16}\right]^{16/9}}\right\}^{1/6} \qquad \text{for} \quad Ra_D \le 10^{12} \tag{9.38}$$

The thermophysical properties in Eqs (9.36) to (9.38) are evaluated at the film temperature $T_f = (T_s + T_\infty)/2$. If the properties vary significantly with temperature, they are evaluated as the mean of the surface temperature and the bulk temperature.

9.4.5 Spheres

For a sphere immersed in a fluid, Churchill (1983) proposed the following correlation

$$\overline{Nu}_D = 2 + \frac{0.589 Ra_D^{1/4}}{\left[1+\left(0.469/Pr\right)^{9/16}\right]^{4/9}} \qquad \text{for } Ra_D \le 10^{11}, Pr \ge 0.7 \tag{9.39}$$

The properties are evaluated as for cylinders.

9.4.6 Flow in Channels

An example of this flow is the flow between two cooled or heated parallel plates of height H and width δ, which is open to the ambient temperature T_∞ at the top and bottom of the channel (Figure 9.6). Flow boundary layers forming on both channel walls at the inlet grow in thickness with increasing distance from the inlet until they eventually merge. The most frequently quoted correlation in heat transfer references for this flow is a correlation proposed by Elenbaas (1942) for air

$$\overline{Nu}_\delta = \frac{1}{4} Ra_\delta \left(\frac{\delta}{H}\right)\left\{1-\exp\left[-\frac{35}{Ra_\delta(\delta/H)}\right]\right\}^{3/4} \tag{9.40}$$

Equation (9.40) is valid for symmetrically heated constant-temperature plates ($T_1 = T_2 = T_s$) within the range $10^{-1} \le (\delta/H) Ra_\delta \le 10^5$. The Nusselt number and Rayleigh number in Eq. (9.40) are defined, respectively, as

$$\overline{Nu}_\delta = \left(\frac{\dot{Q}/A}{T_s - T_\infty}\right)\frac{\delta}{k}$$

$$Ra_\delta = \frac{g\beta(T_s - T_\infty)\delta^3}{\alpha\vartheta}$$

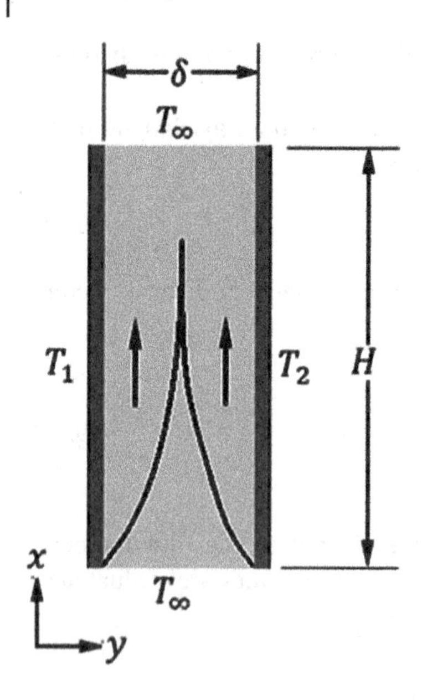

Figure 9.6 Natural convection flow between two heated parallel plates.

9.4.7 Flow in Closed Spaces

9.4.7.1 Vertical Rectangular Cavity

Figure 9.7 shows a vertical rectangular cavity with heated or cooled side walls and adiabatic top and bottom surfaces. This arrangement is widely used in building walls in which air gaps in the form of a thin layer are used as thermal insulators.

El Sherbiny et al. (1982) proposed the following set of equations to estimate the convective heat transfer coefficient for fluid layers in vertical enclosures (Figure 9.7) for $10^3 < Ra_\delta < 10^7$.

$$\overline{Nu}_{\delta 1} = 0.0605 Ra_\delta^{1/3} \tag{9.41}$$

$$\overline{Nu}_{\delta 2} = \left\{ 1 + \left[\frac{0.104 Ra_\delta^{0.293}}{1 + \left(\frac{6310}{Ra_\delta} \right)^{1.36}} \right]^3 \right\}^{1/3} \tag{9.42}$$

$$\overline{Nu}_{\delta 3} = 0.242 \left[\frac{Ra_\delta}{(H/\delta)} \right]^{0.272} \tag{9.43}$$

$$\overline{Nu}_{\delta} = \max\left\{ \overline{Nu_{\delta 1}}, \overline{Nu_{\delta 1}}, \overline{Nu_{\delta 1}} \right\} \tag{9.44}$$

Equation (9.44) states that the required Nusselt number is the maximum of the three values obtained from Eqs (9.41) to (9.43). For $Ra_\delta \le 10^3$, $\overline{Nu_\delta} \cong 1.0$.

Jackob (1962) recommended the following correlations based on the conductive layer model:

$$\varepsilon_{eff} = 0.18 Gr_\delta^{1/4} \left(\frac{H}{\delta} \right)^{-1/9} \quad H/\delta \ge 3,\ 2\times10^4 \le Gr_\delta \le 2\times10^5 \tag{9.45}$$

$$\varepsilon_{eff} = 0.065 Gr_\delta^{1/3} \left(\frac{H}{\delta} \right)^{-1/9} \quad H/\delta \ge 3,\ 2\times10^5 \le Gr_\delta \le 10^7 \tag{9.46}$$

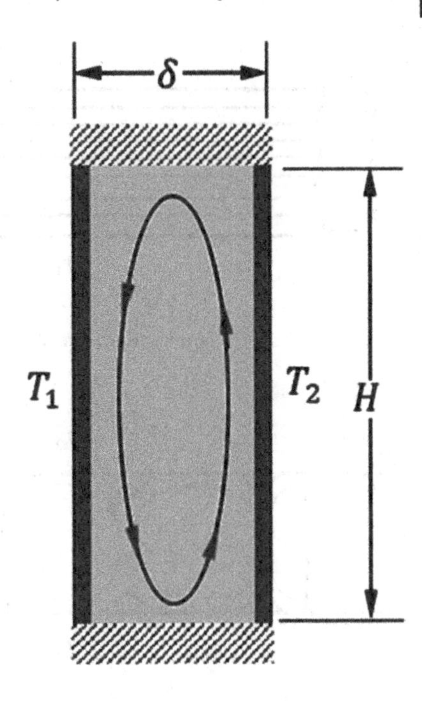

Figure 9.7 Natural convection flow in a vertical cavity.

Parameter ε_{eff} is nondimensional quantity defined as

$$\varepsilon_{eff} = \frac{k_{eff}}{k} \tag{9.47}$$

where k is the thermal conductivity of the fluid and k_{eff} is the effective conductivity of a fictitious quiescent fluid that transfers the same amount of heat as the actual moving fluid and is defined as

$$k_{eff} = \frac{\dot{Q}}{\Delta T}\frac{\delta}{A} \tag{9.48}$$

9.4.7.2 Horizontal Fluid Layer

Figure 9.8 shows two configurations of horizontal fluid layers with small aspect ratios.

When the fluid is heated from the top (Figure 9.8a), circulation is absent and the fluid is almost quiescent with the formation of thermally stratified field. Heat transfer through the fluid layer will be by conduction only and $\overline{Nu}_\delta = 1$. When the fluid is heated from below, counter-rotating recirculation cells form (Figure 9.8b) and the average heat transfer coefficient can be estimated from the following correlation (Globe and Dropkin 1959)

$$\overline{Nu}_\delta = 0.069 Ra_\delta^{1/3} Pr^{0.074} \quad \text{for} \, 3\times10^5 \le Ra_\delta \le 7\times10^9 \tag{9.49}$$

Mikheyev and Mikheyeva (1977) proposed the following correlations for natural convection heat transfer in vertical and horizontal cavities, annular spaces between long concentric cylinders, and spaces between concentric spheres based on the conductive layer model.

$$\varepsilon_{eff} = 0.105 Ra_\delta^{0.3} \qquad \text{for } 10^3 \le Ra_\delta \le 10^6 \tag{9.50}$$

$$\varepsilon_{eff} = 0.4 Ra_\delta^{0.2} \qquad \text{for } 10^6 \le Ra_\delta \le 10^{10} \tag{9.51}$$

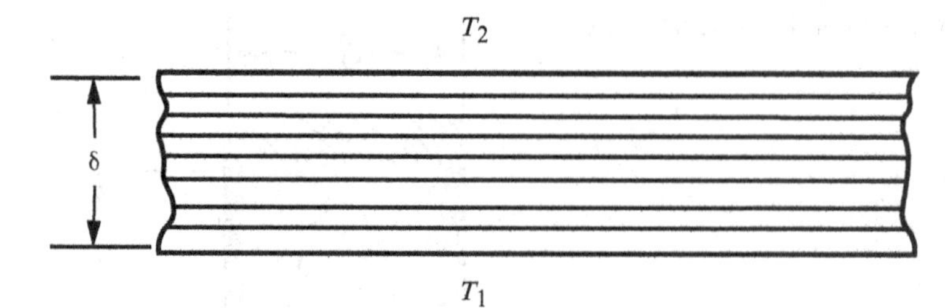

(a) Heating from above $T_2 > T_1$

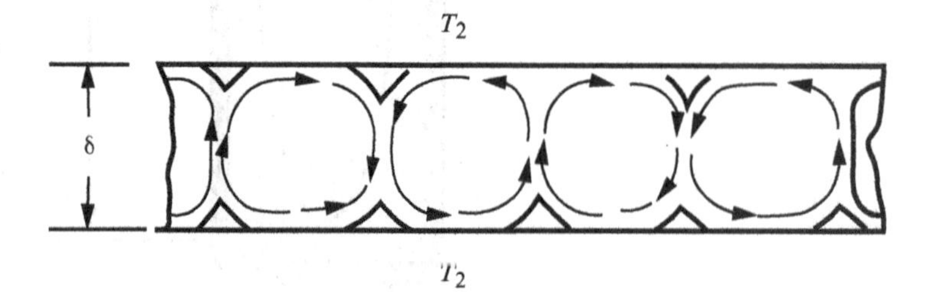

(b) Heating from below $T_1 > T_2$

Figure 9.8 Horizontal fluid layers with small aspect ratios: (a) heated from above; (b) heated from below.

The thermophysical properties in Eqs (9.50) and (9.51) are evaluated at the mean temperature $T_m = (T_1 + T_2)/2$

Example 9.2 A double vertical wall of height $H = 2.7m$ has a gap $\delta = 0.15m$ and the walls are at temperatures $T_1 = 20°C$ and $T_2 = 40°C$, as shown in Figure E9.2. Estimate the rate of heat transfer through the structure. Take the thermophysical properties as: $\rho = 1.1497\,kg/m^3$, $k = 0.027075\,W/m.K$, $\mu = 1.86 \times 10^{-5}\,kg/m.s$, $c_p = 1004.89\,J/kg.K$.

Solution

$$\beta = \frac{1}{303} = 0.0033$$

$$\alpha = \frac{k}{\rho c_p} = \frac{0.02707}{1.1497 \times 1004.89} = 2.343 \times 10^{-5}$$

$$\vartheta = \frac{\mu}{\rho} = \frac{1.86 \times 10^{-5}}{1.1497} = 1.6178 \times 10^{-5}$$

$$Pr = \frac{\mu}{\rho\alpha} = \frac{1.86 \times 10^{-5}}{1.1497 \times 2.343 \times 10^{-5}} = 0.69$$

$$Ra_\delta = \frac{g\beta(T_1 - T_2)\delta^3}{\alpha\vartheta} = \frac{9.81 \times 0.0033 \times 20 \times 0.15^3}{2.343 \times 10^{-5} \times 1.6178 \times 10^{-5}} = 5.765 \times 10^6$$

The results of calculations from two methods are shown in Table E9.2.

The maximum values from Eqs (9.41) to (9.44) are rewritten in the fourth row of Eq. (9.44).

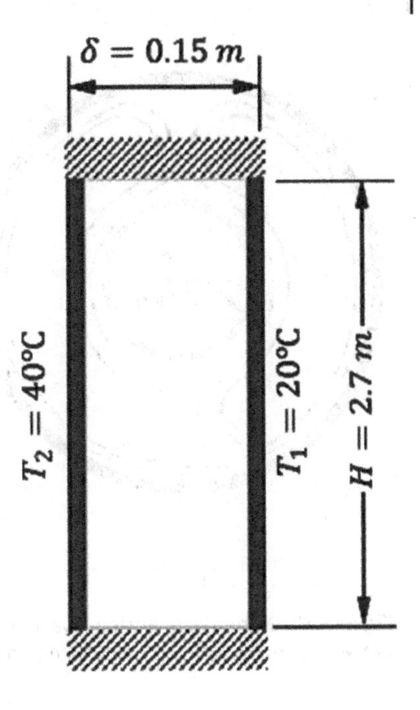

Figure E9.2 Double vertical wall of finite height.

Table E9.2 Calculation results using the equations from El Sherbiny et al. (1982) and Mikheyev and Mikheyeva (1977).

	$\overline{Nu_\delta}$	$\bar{h}$, $W/m.K$	$\dot{Q}/A$, W/m^2
Eq. (9.41)	10.85	1.96	39.16
Eq. (9.42)	9.96	1.8	25.94
Eq. (9.43)	7.61	1.37	27.47
Eq. (9.44)	**10.85**	**1.96**	**39.16**
	ε_{eff}/k	ε_{eff}	$\dot{Q}/A, W/m^2$
Eq. (9.51)	9.0	0.24	32.49

9.4.7.3 Concentric Cylinders

In concentric spheres and concentric horizontal cylinders, the flow is driven by the temperature gradient between the hot and cold surfaces, as shown in Figure 9.9. In the laminar flow regime, two counter-rotating cells are formed symmetrically about the vertical plane, with the direction of flow being dependent on whether the outer plate is heated or cooled.

The heat transfer rate between two cylinders per unit length can be estimated in terms of effective conductivity k_{eff}, defined in Eq. (9.47), as the independent variable by the correlation suggested by Raithby and Holland (1975) with the inner cylinder heated (Figure 9.9a).

$$\varepsilon_{eff} = \frac{k_{eff}}{k} = 0.386 Ra_c^{1/4} \left(\frac{Pr}{0.861 + Pr} \right)^{1/4} \quad (9.52)$$

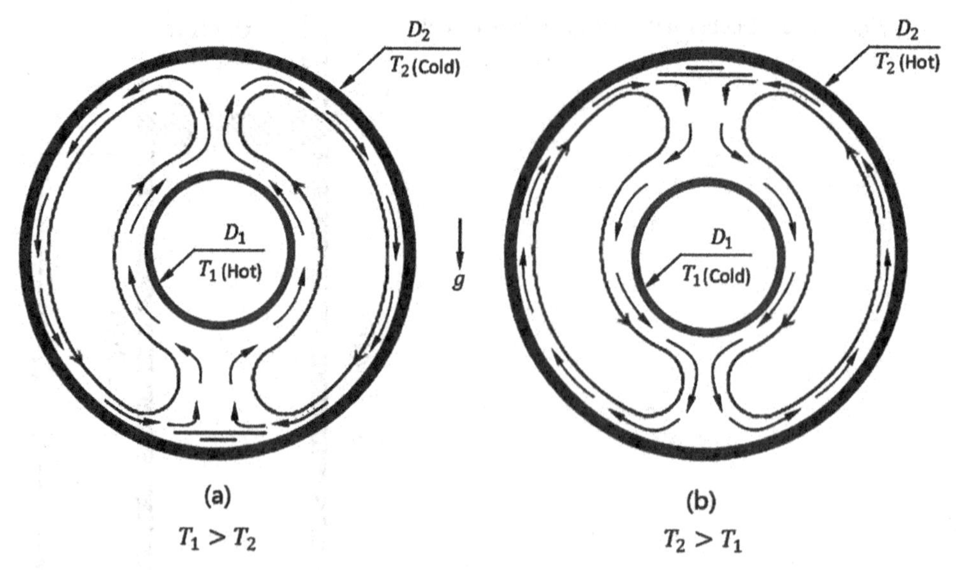

Figure 9.9 Natural convection flow in the annular space between two long concentric horizontal cylinders or two concentric spheres: (a) heating on the inside; (b) heating on the outside.

In this equation

$$Ra_c = \frac{\left[\ln(D_2/D_1)\right]^4}{\left(\frac{1}{D_1^{3/5}} + \frac{1}{D_2^{3/5}}\right)^5} \frac{Ra_\delta}{\delta^3} \tag{9.53}$$

where $\delta = 0.5(D_o - D_i)$, and

$$Ra_\delta = \frac{g\beta(T_1 - T_2)\delta^3}{\alpha\vartheta} \tag{9.54}$$

The rate of heat transfer per unit length of a cylinder of length L is

$$\frac{\dot{Q}}{L} = \frac{2\pi k_{eff}(T_1 - T_2)}{ln(D_2/D_1)} \tag{9.55}$$

Equation (9.52) is valid over the ranges

$0.7 \le Pr \le 6000$

$10 \le Ra_c \le 10^7$

9.4.7.4 Concentric Spheres

For concentric spheres, the effective conductivity (Raithby and Holland 1975) is

$$\varepsilon_{eff} = \frac{k_{eff}}{k} = 0.74 Ra_s^{1/4} \left(\frac{Pr}{0.861 + Pr}\right)^{1/4} \tag{9.56}$$

In this equation

$$Ra_s = \frac{\frac{1}{(D_1D_2)^4}}{\left(\frac{1}{D_1^{7/5}} + \frac{1}{D_2^{7/5}}\right)^5} \delta Ra_\delta \quad (9.57)$$

where $\delta = 0.5(D_o - D_i)$, and

$$Ra_\delta = \frac{g\beta(T_1 - T_2)\delta^3}{\alpha\vartheta} \quad (9.58)$$

The total rate of heat transfer is then

$$\dot{Q} = \frac{2\pi k_{eff}(T_1 - T_2)}{\left(\frac{1}{D_1} - \frac{1}{D_2}\right)} \quad (9.59)$$

Equation (9.56) is valid over the ranges

$$0.7 \leq Pr \leq 4200$$
$$10 \leq Ra_s \leq 10^7$$

Example 9.3 A long thin-walled cylindrical tube of diameter $D_1 = 0.1m$ is maintained at $T_1 = 120°C$ by passing superheated steam through it (Figure E9.3). Insulation is to be provided by a layer of air between the tube and another concentric tube of diameter $D_o = 0.12m$ at temperature $T_2 = 35°C$. Determine the rate of heat transfer by natural convection from the tube per unit length.

Solution

Thermophysical properties at $T_m = (T_1 + T_2)/2 = 350K$: $\rho = 0.995 kg/m^3$, $k = 0.03 W/m.K$, $\vartheta = 2.092\times10^{-5}\ m^2/s$, $\alpha = 2.99\times10^{-5}\ m^2/s$, $\beta = 0.00285\ K^{-1}$.

The heat rate from Eq. (9.55) is

$$\frac{\dot{Q}}{L} = \frac{2\pi k_{eff}(T_1 - T_2)}{\ln(D_2/D_1)}$$

The effective conductivity can be determined from Eq. (9.52) if the Rayleigh number Ra_c is known. We calculate Ra_δ from Eq. (9.54), then Ra_c from Eq. (9.53)

$$Ra_\delta = \frac{g\beta(T_1 - T_2)\delta^3}{\alpha\vartheta} = \frac{9.81\times0.00285\times(120-35)\times0.01^3}{29.9\times10^{-6}\times20.92\times10^{-6}} = 5.38\times10^3$$

$$Ra_c = \frac{[\ln(D_2/D_1)]^4}{\left(\frac{1}{D_1^{3/5}} + \frac{1}{D_2^{3/5}}\right)^5}\frac{Ra_\delta}{\delta^3} = \frac{[\ln(0.12/0.1)]^4}{\left(\frac{1}{0.01^{3/5}} + \frac{1}{0.012^{3/5}}\right)^5}\frac{5.38\times10^3}{0.01^3} = 242$$

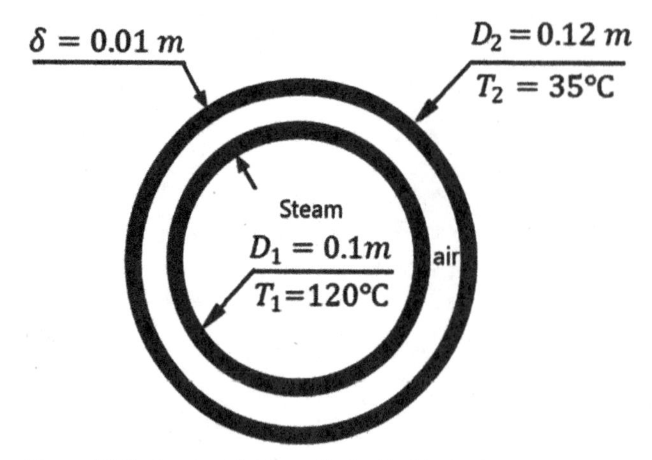

Figure E9.3 Insulated thin-walled cylindrical tube.

From Eq. (9.52)

$$\frac{k_{eff}}{k} = 0.386 Ra_c^{1/4}\left(\frac{Pr}{0.861+Pr}\right)^{1/4} = 0.366\times 214^{0.25}\times\left(\frac{0.7}{0.861+0.7}\right)^{0.25} - 1.25$$

$$k_{eff} = 1.25\times 0.03 = 0.0375 W/m.K$$

Hence,

$$\frac{\dot{Q}}{L} = \frac{2\pi\times 0.0375\times(120-35)}{\ln(0.12/0.1)} = 109.8 W/m$$

Example 9.4 Determine the rate of heat transfer by natural convection in an enclosure formed by two concentric spheres with the same diameters, boundary conditions, and air properties, as in Example 9.3.

Solution

Ra_δ is unchanged $Ra_\delta = 5380$

$$Ra_s = \frac{\frac{1}{(D_1 D_2)^4}}{\left(\frac{1}{D_1^{7/5}}+\frac{1}{D_2^{7/5}}\right)^5}\delta Ra_\delta = 14.72$$

From Eq. (9.56)

$$\varepsilon_{eff} = \frac{k_{eff}}{k} = 0.74 Ra_s^{1/4}\left(\frac{Pr}{0.861+Pr}\right)^{1/4} = 1.186$$

$$k_{eff} = 0.0356 W/m.K$$

$$\dot{Q} = \frac{2\pi k_{eff}(T_1 - T_2)}{\left(\frac{1}{D_1}-\frac{1}{D_2}\right)} = 11.4 W$$

9.5 Mixed Free and Forced Convection

In the models of convection heat transfer considered so far, the fluid near a hot surface was assumed to have an externally driven motion (forced convection), or motionless in a quiescent reservoir (natural convection). However, in engineering applications, depending on the intensity of the forced motion, the heat transfer from the wall to the surroundings may be natural convection, forced convection, or a combination of both. As a first approximation, the overall heat transfer coefficient for mixed flows can be determined from (Bergman et al. 2011)

$$\left(\overline{Nu_M}\right)^n = \left(\overline{Nu_F}\right)^n \mp \left(\overline{Nu_N}\right)^n \tag{9.60}$$

Subscripts M, F, and N stand for mixed, forced, and natural flows, respectively. The index $n = 3$ can be used for all flow schemes. The plus sign in Eq. (9.60) applies to assisting and transverse flows (Figures 9.10a and 9.10c) and the minus sign applies to opposing flow (Figure 9.10b).

The contribution of natural convection to the overall heat transfer coefficient of mixed flows diminishes as the flow becomes turbulent.

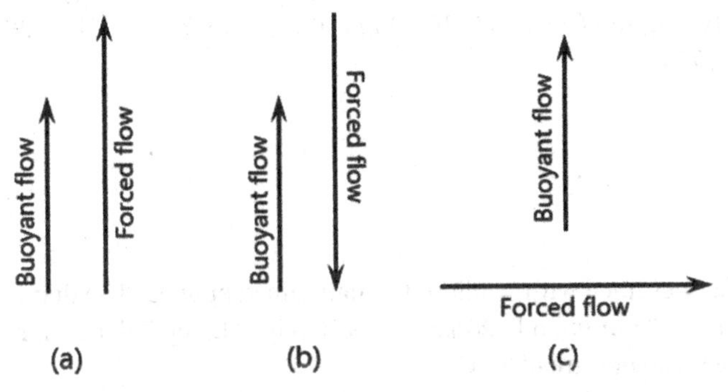

Figure 9.10 Relative directions of buoyancy-induced and forced flows: (a) assisting flow; (b) opposing flow; (c) transverse flow.

Problems

9.1 The second Π product in the functional relationship for free convection on a vertical plate of height l is $\Pi_2 = Uk^{b_1}\mu^{b_2}\rho^{b_3}l^{b_4}$. Using the dimensional analysis method, show that $\Pi_2 = Re$.

9.2 The third Π product in the functional relationship for free convection on a vertical plate of height l is $\Pi_3 = (\beta g \Delta T)k^{c_1}\mu^{c_2}\rho^{c_3}l^{c_4}$. Show that $\Pi_3 = Gr$.

9.3 The fourth Π product in the functional relationship for free convection on a vertical plate of height l is $\Pi_4 = c_p k^{d_1}\mu^{d_2}\rho^{d_3}l^{d_4}$. Show that $\Pi_4 = Pr$.

9.4 A vertical plate 4 m high and 8 m wide is maintained at a constant temperature of 90°C. If the surrounding atmosphere is at 30°C, what is the rate of heat loss from the plate?

9.5 A vertical tube 100 *mm* diameter and 4 m long is carrying superheated steam. The surface of the tube is maintained at a constant temperature of 150°C and the ambient air temperature is 30°C. Estimate the rate of heat loss from the tube.

9.6 A vertical plate of length $0.9m$ and width $0.5m$, maintained at a uniform temperature of $500K$, is cooled in still air at $300K$ and 1 atm. Determine the average heat transfer coefficient using the appropriate correlation by Churchill and Chu (1975b).

9.7 A vertical tube 100 mm diameter and 4 m long carrying hot gas is losing heat to quiescent air at 30°C by free convection. If the surface of the tube is to be maintained at a constant temperature of 170°C, what is the rate of heat transfer from the tube?

9.8 A horizontal rod of length $1.3m$ and diameter $25mm$, maintained at a constant surface temperature of 60°C, is submerged in still water at 20°C. Determine the heat loss from the rod.

9.9 The vertical door of a hot oven is 0.5 m high and is maintained at 200°C and is exposed to quiescent atmospheric air at 20°C. Find:

(a) local heat transfer coefficient half way up the door

(b) average heat transfer coefficient for the entire door.

9.10 A horizontal copper tube of $10mm$ outer diameter carries liquid freon at – 30°C. If a $2m$ length of this tube must pass uninsulated through the still air at 40°C, determine the total heat loss when outside tube surface emissivity is 0.8. Use the following properties and correlations for determination of convection coefficient:

The properties of air at $T_f = (40-30)/2 = 5°C$: $\rho = 1.2538 kg/m^3$, $c_p = 1003 J/kg.K$, $k = 0.02422\ W/m.K$, $\mu = 1.74\times10^{-5}\ kg/m.s$ ($\nu = 1.387\times10^{-5}$), $Pr = 0.72$, $\beta = 1/T_f = 1/278 = 3.97\times 10^{-3} K^{-1}$. At $T_s = -30°C$, $Pr_s = 0.738$

$$\overline{Nu}_D = 0.85 Ra_D^{0.188}$$

$$\overline{Nu}_D = 0.5 Ra_D^{0.25}\left(\frac{Pr}{Pr_s}\right)^{0.25}$$

9.11 The surface of a cylindrical container filled with hot liquid is maintained at 98°C. The diameter and height of the cylinder are $250mm$ and $120mm$, respectively. The cylinder is surrounded by still air at an average temperature of 25°C.

(a) Determine the rate of heat loss from the cylinder by convection

(b) Determine the rate of heat loss by radiation if the emissivity of the outside surface of the cylinder is 0.8.

9.12 A plate 0.5 m high and 0.6 m wide is heated to a constant temperature $Pr_s = 0.738$ on one side and is insulated on the other side. The plate is in a room at $T_\infty = 20°C$. Determine the rate of heat loss from the exposed surface by convection for the following positions of the plate: upright, horizontal with hot surface up, and horizontal with the hot surface down.

9.13 Cooling processes are sometimes conducted in an inert medium. In such a process involving a vertical heated plate ($1.0\ m\times1.0m$), a choice must be made between helium and nitrogen. Which cooling medium should be selected if the plate is to be maintained at constant temperature $T_s = 900K$ and the fluid is at temperature T_∞? Take the film temperature as $T_f = (T_s + T_\infty)/2 = 700K$.

9.14 The simple correlation $\overline{Nu}_L = 0.6 Ra^{1/4}$ can be used to estimate the average heat transfer coefficient of free convection for a cube immersed in quiescent medium.

(a) Estimate the average heat transfer coefficient $\bar{h}$ for a cube ($1.0m\times1.0m\times1.0m$) maintained at a temperature of $T_s = 100°C$ and surrounded by still air at $T_\infty = 40°C$

(b) Estimate $\bar{h}$ from the values of heat transfer coefficient obtained for each surface separately using the appropriate correlations and compare the outcomes of (a) and (b)

9.15 Preheated small ceramic spheres, each of $45 mm$ diameter, are maintained at a temperature of 150°C while subjected to hot still air at 230°C. Determine the convection heat transfer coefficient and the rate of heat transferred to a sphere.

9.16 A double vertical wall of height $H = 3.0\ m$ has a gap $\delta = 200 mm$ and the walls are at temperatures $T_1 = 20°C$ and $T_2 = 40°C$. as shown in Figure P9.16. Estimate the rate of heat transfer through the gap per unit width (normal to the plane of the paper). Use Eq. (9.44) if applicable.

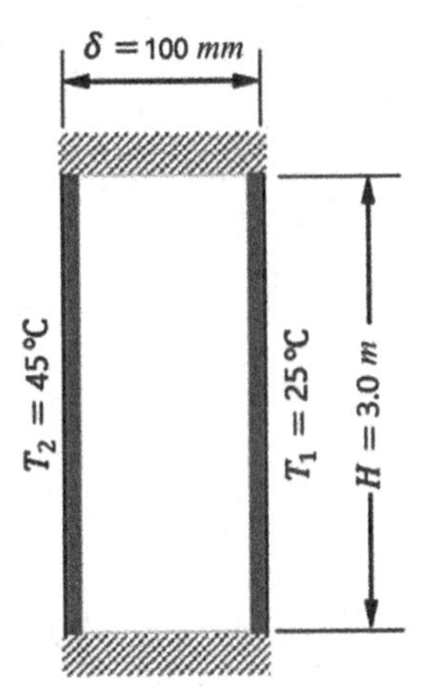

Figure P9.16 Double vertical wall of finite height *H*.

9.17 Repeat Problem 9.16 using Jacob's applicable correlation and note the difference from the previous result.

9.18 Repeat Problem 9.16 for horizontally oriented geometry.

9.19 Figure P9.19 shows a long thin-walled cylindrical tube of diameter $D_1 = 0.15 m$ which is maintained at $T_1 = 120°C$ by passing superheated steam through it. Insulation is to be provided by a layer of air between the tube and outer concentric tube of diameter $D_o = 0.18 m$ at temperature $T_2 = 35°C$. Determine the rate of heat transfer by natural convection from the tube per unit length.

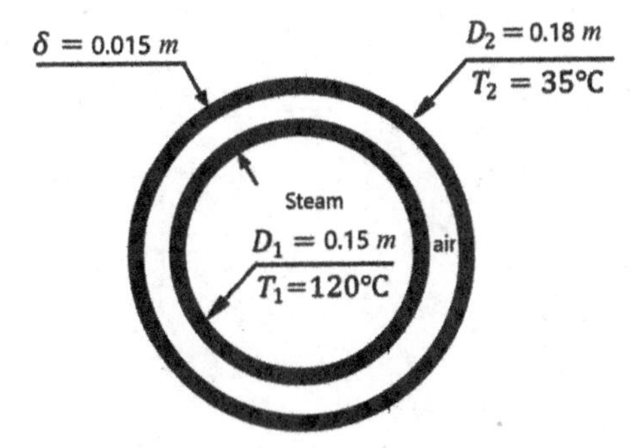

Figure P9.19 Steam flow in insulated thin-walled cylindrical tube.

9.20 Repeat Problem 9.19 using Eq. (9.50) and compare the results in the two problems.

10

Thermal Radiation

Properties and Shape Factors

Heat is energy in transition under the motive force of temperature difference. In conduction heat transfer, energy transfer is on the molecular scale with no movement of macroscopic portions of the matter. There is direct exchange of kinetic energy between particles of matter and the rate of heat transfer depends on the temperature difference to the first power. In convection heat transfer, redistribution of energy is due partly to conduction and partly to transport of enthalpy by the motion of the fluid. As for conduction, the rate of heat transfer by convection depends on the temperature difference to the first power.

Thermal radiation is energy emission from any heated matter that is sustained by thermal energy at a temperature above absolute zero, and the higher the temperature the greater the amount of energy radiated. Radiation can be viewed as electromagnetic waves emitted by the matter which then propagate at the speed of light ($2.998 \times 10^8 m/s$). Radiation heat transfer is the exchange of thermal radiation between two bodies or more at different temperatures. During this exchange process, each body converts random molecular energy of the matter into outgoing electromagnetic waves, and absorbs incoming electromagnetic waves which are then converted into random molecular energy causing a temperature increase.

Electromagnetic wave propagation does not require a medium and can take place in a vacuum and the rate at which heat is radiated is proportional to the absolute temperature of the body raised to the fourth power. Thermal radiation is strongly linked to:

- combustion processes in heat engines (internal combustion engines, gas turbines, rockets, etc.)
- furnace technology
- fires and fire engineering
- thermal control in space technology (space probes, satellites)
- cryogenic insulation
- studies of the energy balance of earth and global warming.

Heat Transfer Basics: A Concise Approach to Problem Solving, First Edition. Jamil Ghojel.

Companion website: www.wiley.com/go/ghojel/heat_transfer

10.1 The Electromagnetic Spectrum

Radiation is commonly viewed as propagation of electromagnetic waves at the speed of light c ($2.998 \times 10^8 m/s$), which can be written as

$$c = \lambda v$$

In this equation λ is wavelength (m) and v is frequency (s^{-1}). The micron or micrometre (μm), which is equal to 10^{-6} m, is widely used as the unit of wavelength in engineering.

Thermal radiation is part of the electromagnetic spectrum within the wavelength range of 0.1 to 1000 μm, as shown by the highlighted portion in Figure 10.1. The range of wavelengths visible to the human eye ($0.4-0.7\mu m$) lies within the thermal radiation part as does the infrared range and parts of the ultraviolet and microwave ranges.

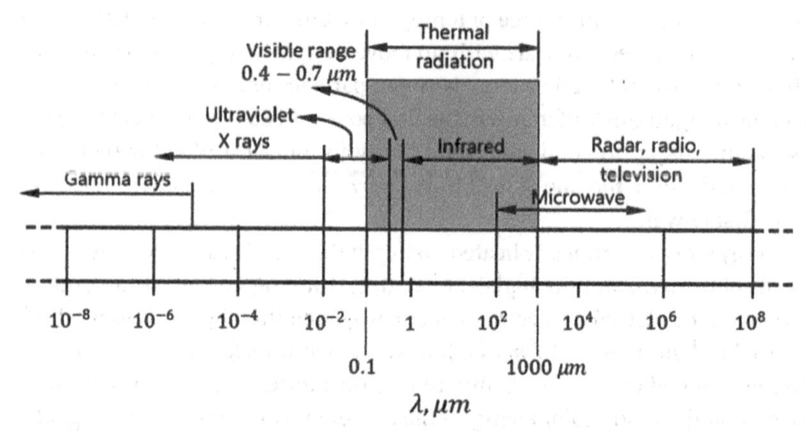

Figure 10.1 Electromagnetic wavelength spectrum with the thermal radiation part highlighted.

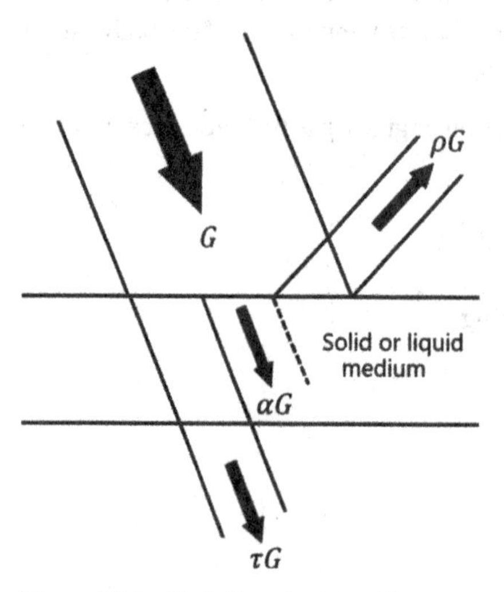

Figure 10.2 Radiation on the surface of a semi-transparent medium.

10.2 Definitions and Radiation Properties

When radiation falls on the surface of a solid or liquid medium, a part could be absorbed, part reflected, and part transmitted (Figure 10.2)

$$\rho G + \alpha G + \tau G = G \tag{10.1}$$

In this equation, G is irradiation (incident thermal radiation energy), ρG is the reflected radiation, αG is the absorbed radiation, and τG is the transmitted radiation. Dividing both sides of Eq. (10.1) by G we obtain

$$\alpha + \rho + \tau = 1 \tag{10.2}$$

where α is absorptivity, ρ is reflectivity, and τ is transmissivity of the medium.

A medium is said to be opaque if it reflects or absorbs incident radiation but does not transmit radiation ($\alpha = 1-\rho, \tau = 0$). Most solids and liquids are opaque objects and are sometimes referred to as optically thick participating media. A transparent medium normally reflects or transmits incident radiation but does not absorb radiation ($\tau = 1-\rho, \alpha = 0$). Glass and gases are transparent to thermal radiation but most gases are not reflective. A semi-transparent medium reflects, absorbs, and transmits incident thermal radiation ($\alpha + \rho + \tau = 1$).

An idealized medium that absorbs all the incident radiation energy (no reflection nor transmission) is known as a black body or black surface. All the radiation energy leaving this surface is emitted by the surface and no surface can emit more radiation than a black body. A black body is a *diffuse surface*, i.e., the intensity of the reflected thermal radiation is constant for all angles of irradiation and reflection as depicted by the parallel lines in Figure 10.2 (most real unpolished surfaces are often considered diffuse also). A black body emits radiant energy over the entire thermal radiation wavelength spectrum. The emitted power per unit surface area per unit wavelength $E_{b\lambda}$ is given by Planck's law

$$E_{b\lambda} = \frac{C_1 \lambda^{-5}}{\exp(C_2 / \lambda T) - 1}. \tag{10.3}$$

in consistent SI units

$$\lambda - m$$

$$C_1 = 3.742 \times 10^{-16}\ W.m^4 / m^2 \text{ or} (W.m^2)$$

$$C_2 = 1.4389 \times 10^{-2}\ m.K$$

$$E_{b\lambda} - W / (m^2.m) \text{ or } W / m^3$$

If the wavelength is in micrometres (microns)

$$\lambda - \mu m$$

$$C_1 = 3.742 \times 10^{8}\ W.\mu m^4 / m^2$$

$$C_2 = 1.4389 \times 10^{4}\ \mu m.K$$

$$E_{b\lambda} - W / m^2.\mu m$$

Equation (10.3) yields the radiant power emitted at any specified wavelength, which is known as the monochromatic emissive power of a black body. The logarithmic plots of $E_{b\lambda}$ versus λ at various temperatures are shown in Figures 10.3 and 10.4 using both sets of units and constants.

The highlighted area represents the visible wavelength range and associated temperatures. The plots show that the emissive power increases with temperature with the peak values shifting to lower wavelength region. The relationship between the wavelength at maximum emissive power (λ_{max}) and the temperature is given by Wien's law

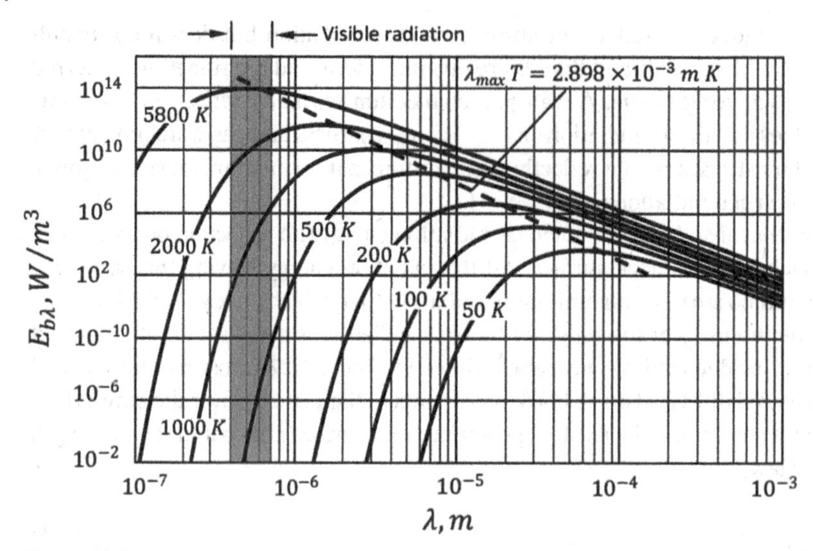

Figure 10.3 Monochromatic emissive power versus wavelength at various temperatures ($E_{b\lambda}$ in W/m^3, λ in μm).

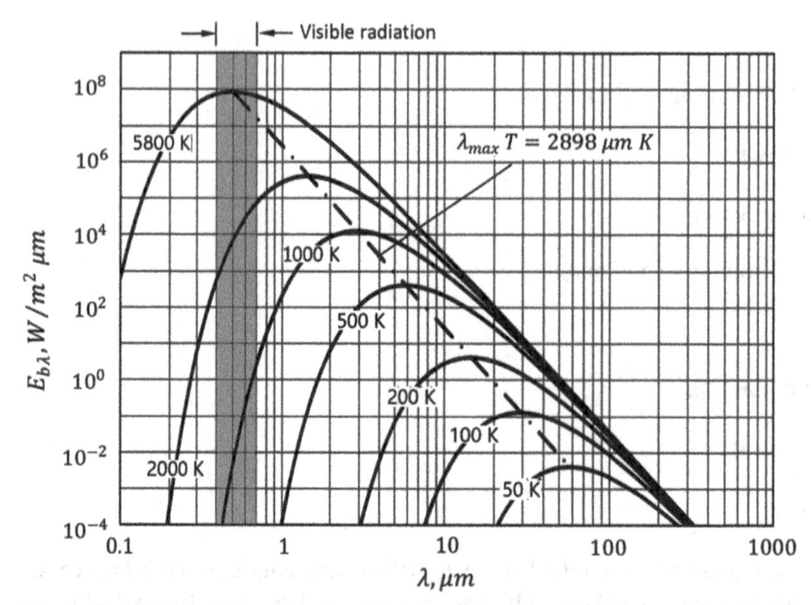

Figure 10.4 Monochromatic emissive power versus wavelength at various temperatures ($E_{b\lambda}$ in $W/m^2\ \mu m$, λ in μm).

$$\lambda_{max}T = 2.898 \times 10^{-3}\ m.K \tag{10.4}$$

or

$$\lambda_{max} = \frac{2.898 \times 10^{-3}}{T}\ m$$

$$\lambda_{max}T = 2898\ \mu m.K \tag{10.5}$$

Example 10.1 Determine the blackbody emissive power E_b in SI units and the corresponding peak wavelength λ_{max} for surfaces at 200 K, 800 K, and 2000 K.

Solution

$$E_b = \sigma T^4 = 5.67\times10^{-8}\,T^4.W/m^2$$

$$\lambda_{max} = \frac{2.898\times10^{-3}}{T}\,m$$

At $T = 200\,K,\ E_b = 90.7\,W/m^2;\ \lambda_{max} = 1.45\times10^{-5}\,m$
At $T = 800\,K,\ E_b = 2.32\times10^4\,W/m^2;\ \lambda_{max} = 3.62\times10^{-6}\,m$
At $T = 2000\,K,\ E_b = 9.07\times10^5\,W/m^2;\ \lambda_{max} = 1.45\times10^{-6}\,m$

The overall emissive power of a black body can be determined by integrating Eq. (10.3) over the entire wavelength spectrum

$$E_b = \int_0^{\infty} E_{b\lambda}d\lambda = \sigma T^4 \quad W/m^2 \tag{10.6}$$

Equation (10.6) is the Stefan–Boltzmann law and $\sigma(=5.67\times10^{-8}\,W/m^2.K^4)$ is the Stefan–Boltzmann constant.

The emissive power of a black body per unit area over a wavelength band $0-\lambda$ is

$$E_{b,0-\lambda}(T) = \int_0^{\lambda} E_{b\lambda}(\lambda,T)d\lambda \quad W/m^2 \tag{10.7}$$

An alternative way of representing the emissive power over a wavelength band $0-\lambda$ is the fractional blackbody function f_λ, defined as $E_{b,0-\lambda}(T)$ divided by the total radiation E_b emitted from a black body over the entire wavelength spectrum

$$f_\lambda = \frac{\int_0^{\lambda} E_{b\lambda}(\lambda,T)d\lambda}{\sigma T^4} = \int_0^{\lambda T} \frac{E_{b\lambda}(\lambda,T)d(\lambda T)}{\sigma T^5} \tag{10.8}$$

The integral of Eq. (10.8) may be evaluated to obtain $f_{0-\lambda}$ as a function of λT and the results presented in tabular form, as in Table 10.1. The data in the table can be used to obtain the fraction of radiation between any two wavelengths λ_1 and λ_2

$$f_{\Delta\lambda} = f_{0-\lambda_2} - f_{0-\lambda_1} \tag{10.9}$$

Table 10.1 Blackbody radiation function $f_\lambda = f(\lambda T)$.

λT $\mu m.K$	f_λ	λT $\mu m.K$	f_λ
200	0.000000	6200	0.754140
400	0.000000	6400	0.769234
600	0.000000	6600	0.783199
800	0.000016	6800	0.796129
1000	0.000321	7000	0.808109

(Continued)

Table 10.1 (Continued)

λT $\mu m.K$	f_λ	λT $\mu m.K$	f_λ
1200	0.002134	7200	0.819217
1400	0.007790	7400	0.829527
1600	0.019718	7600	0.839102
1800	0.039341	7800	0.848005
2000	0.066728	8000	0.856288
2200	0.100888	8050	0.874608
2400	0.140256	9000	0.890029
2600	0.183120	9050	0.903085
2800	0.227897	10,000	0.914199
3000	0.273232	10,500	0.923710
3200	0.318102	11,000	0.931890
3400	0.361735	11,500	0.939959
3600	0.403607	12,000	0.945098
3800	0.443382	13,000	0.955139
4000	0.480877	14,000	0.962898
4200	0.516014	15,000	0.969981
4400	0.548796	16,000	0.973814
4600	0.579280	18,000	0.980860
4800	0.607559	20,000	0.985602
5000	0.633747	25,000	0.992215
5200	0.658970	30,000	0.995340
5400	0.680360	40,000	0.997967
5600	0.701046	50,000	0.998953
5800	0.720158	75,000	0.999713
6000	0.737818	100,000	0.999905

Example 10.2 The temperature of the filament of an incandescent bulb is $T_s = 2600K$. Determine the fraction of radiant energy emitted by the filament within the visible range of the electromagnetic wave spectrum ($0.4 - 0.76\ \mu m$). Assume the filament is a black body.

Solution

The visible range of the electromagnetic wave spectrum is from $\lambda_1 = 0.4\ \mu m$ to $\lambda_1 = 0.76\ \mu m$.

From Table 10.1, the blackbody radiations functions corresponding to $\lambda_1 T$ and $\lambda_2 T$ at $T_s = 2600K$ are

$\lambda_1 T = 0.4 \times 2600 = 1040,\ f_{\lambda_1} = 0.0006836$ (interpolated)

$\lambda_2 T = 0.76 \times 2600 = 1976,\ f_{\lambda_2} = 0.063442$ (interpolated)

$$\therefore\ \Delta\lambda_{1-2} = 0.063442 - 0.0006836 = 0.06276\ (6.276\%)$$

The light bulb has conversion efficiency of 6.276%.

Consider a non-black body of surface area A and absorptivity α inside a blackbody enclosure with the two bodies at equilibrium at temperature T (Figure 10.5). The body will continuously radiate to the enclosure and receive energy from the enclosure at an equal rate. The energy balance for this case is

$$EA = \alpha GA \tag{10.10}$$

If the inner body is a black body, $\alpha = 1$ and

$$E_b A = GA \tag{10.11}$$

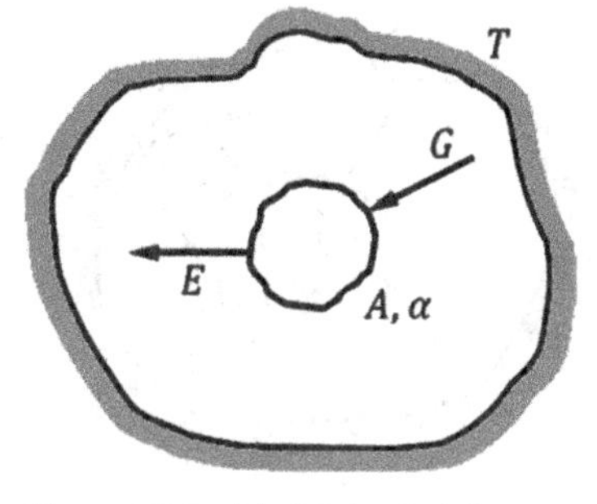

Figure 10.5 Model for deriving Kirchoff's law.

Dividing Eq. (10.10) by Eq. (10.11) gives

$$\frac{E}{E_b} = \alpha \tag{10.12}$$

The ratio E / E_b is the emissivity of the body ε, hence

$$\varepsilon = \alpha \tag{10.13}$$

Equation (10.13) is called Kirchoff's identity.

For a non-black body $\alpha < 1$; hence, $\varepsilon < 1$.

A real body can be approximated by a non-black body, known as a grey body, whose monochromatic emissivity ε_λ can change with temperature but independent of the wavelength and defined as

$$\varepsilon_\lambda = \frac{E_\lambda}{E_{b\lambda}}$$

Monochromatic (spectral) refers to radiation at one wavelength.

10.3 Shape Factors

Figure 10.6a shows a convex black object 1 of area A_1 at temperature T_1 in black enclosure at temperature T_2. A convex object is an object that cannot "see" itself and can radiate in all directions but not to itself. A concave object (the enclosure in Figure 10.6a) can see itself and can radiate to other objects it can see as well as to itself. All the energy emitted by object 1 is absorbed by enclosure 2 and vice-versa. Objects 1 and 2 could also be two long concentric cylinders or two concentric spheres. The rate of energy exchange between objects 1 and 2 is

$$\dot{Q}_{1-2} = A_1\left(\sigma T_1^4 - \sigma T_2^4\right) \tag{10.14}$$

If the inner object 1 is grey, its rate of energy emission will be $\varepsilon\sigma T_1^4$ and will continue to receive energy from the enclosure at the rate of σT_2^4, which will be absorbed only partly at the rate $\alpha\sigma T_2^4 = \varepsilon\sigma T_2^4$ (absorptivity equals emissivity according to Kirchoff's law); hence,

$$\dot{Q}_{1-2} = A_1\varepsilon\sigma\left(T_1^4 - T_2^4\right).$$

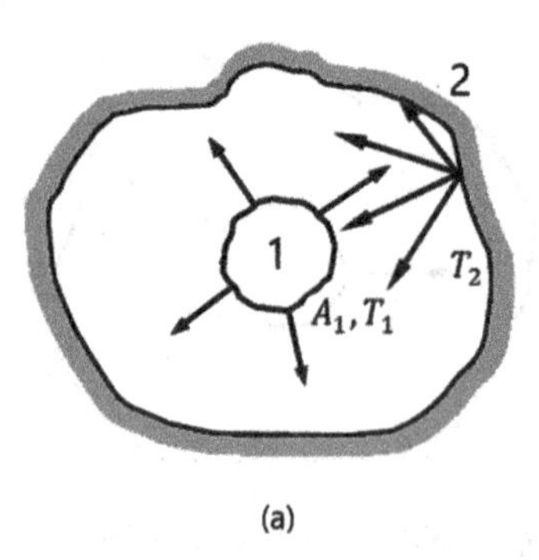

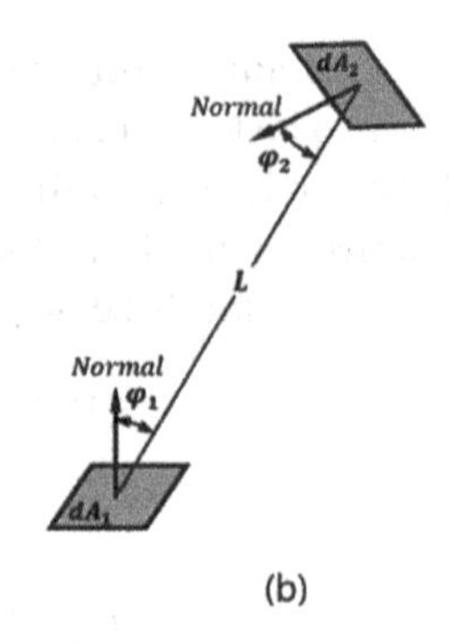

Figure 10.6 Thermal radiation exchange between: (a) convex object in a black enclosure; and (b) two finite elemental objects.

Now consider the two finite surfaces of elemental areas dA_1 and dA_2 separated by distance L, as shown in Figure 10.6b. The amount of energy leaving surface dA_1 and directly intercepted by surface dA_2 is known as the shape factor, view factor, or configuration factor. In the current text, the term "shape factor" will be used throughout.

The shape factor for the two elemental areas in Figure 10.6b is expressed as $dF_{dA_1-dA_2}$. For two finite areas, A_1 and A_2, it is expressed as $F_{A_1-A_2}$ or, as in the current text, as F_{1-2}. The magnitude of the shape factor between two surfaces depends on surface sizes, distance separating them, and orientation relative to each other. Generally, the calculation of a shape factor between any two finite surfaces requires the solution to a double area integral, which presents difficulties if the integrals are complicated. Many view factors for relatively complex configurations can be determined without reverting to complicated integration by using certain simple rules.

10.3.1 Reciprocity Rule

First consider two black finite surfaces of different areas and temperature, as shown in Figure 10.7. The rate of radiation energy leaving surface 1 and falling on surface 2 is

$$A_1 E_{b1} F_{1-2}$$

and the rate of radiation energy leaving surface 2 and falling on surface 1 is

$$A_2 E_{b2} F_{2-1}.$$

Then the net energy exchange between surfaces A_1 and A_2 is

$$\dot{Q}_{1-2} = A_1 E_{b1} F_{1-2} - A_2 E_{b2} F_{2-1} \tag{10.15}$$

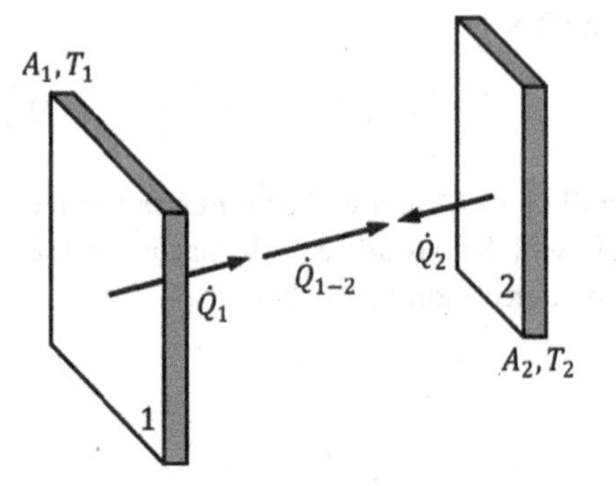

Figure 10.7 Radiation energy exchange between two black finite surfaces.

If both surfaces are at the same temperature, $E_{b1} = E_{b2}$ and $\dot{Q}_{1-2} = 0$; hence,

$$A_1 F_{1-2} = A_2 F_{2-1} \tag{10.16}$$

Equation (10.16) is known as the *reciprocity rule* for shape factors and is valid even when the surfaces are at different temperatures. Combining Eqs (10.15) and (10.16)

$$\dot{Q}_{1-2} = A_1 F_{1-2}\left(E_{b1} - E_{b2}\right) \tag{10.17}$$

Since $E_{b1} = \sigma T_1^4$ and $E_{b2} = \sigma T_2^4$, Eq. (10.17) acquires the form

$$\dot{Q}_{1-2} = A_1 F_{1-2}\left(\sigma T_1^4 - \sigma T_2^4\right) \tag{10.18}$$

10.3.2 Summation Rule

Consider an enclosure consisting of n surfaces of areas $A_1, A_2, A_3, \ldots, A_i \ldots A_n$ with surface i emitting radiation to the other surfaces, as shown in Figure 10.8. The net radiation energy exchange for this scheme is

$$\dot{Q}_i = \dot{Q}_{i-i} + \dot{Q}_{i-1} + \dot{Q}_{i-2} + \dot{Q}_{i-3} + \ldots + \dot{Q}_{i-j} + \ldots \dot{Q}_{i-n}$$

which can be rewritten as

$$\dot{Q}_i = A_i E_{bi} F_{i-i} + A_i E_{bi} F_{i-1} + A_i E_{bi} F_{i-2} + A_i E_{bi} F_{i-3} + \ldots + A_i E_{bi} F_{i-j} + \ldots + A_i E_{bi} F_{i-n}$$

or

$$E_{bi} A_i = E_{bi} A_i \left(F_{i-i} + F_{i-1} + F_{i-2} + F_{i-3} + \ldots + F_{i-j} + \ldots + F_{i-n}\right)$$

from which

$$\sum_{j=1}^{n} F_{i-j} = 1 \tag{10.19}$$

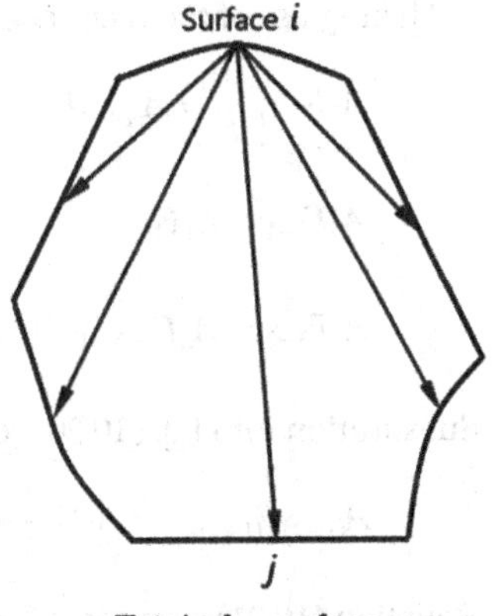

Figure 10.8 Enclosure with n surfaces with surface i radiating to the other surfaces.

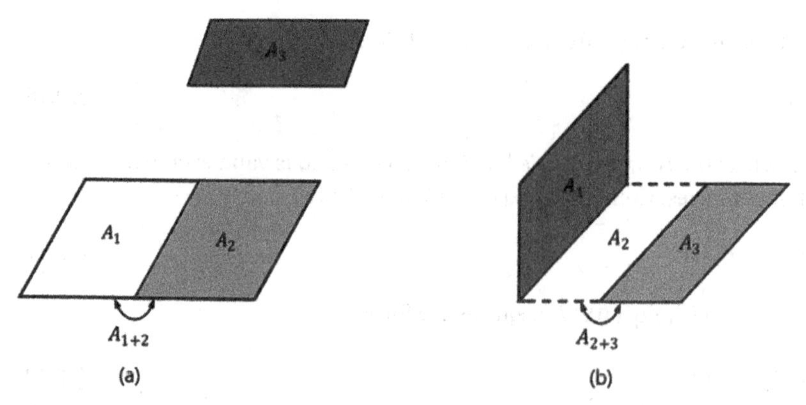

Figure 10.9 Radiation exchange between a surface and a composite surface.

The terms $\dot{Q}_{i-i}$ and F_{i-i} are included in the equations because surface i is depicted as concave and can see itself. If surface i is a plane or convex surface and cannot see itself, these terms will be omitted. Equation (10.19) is known as the *summation rule* for shape factors.

10.3.3 Superposition Rule

If the shape factor of a certain geometry is not available in analytical or graphical form, it can be expressed in terms of known shape factors of constituent parts of the geometrical configuration using the *superposition rule* which states that "the total shape factor is the sum of its parts." As an example, consider Figure 10.9a comprising three surfaces: A_1, A_2, and A_3. Applying this rule gives

$$F_{3-(1+2)} = F_{3-1} + F_{3-2}$$

Multiplying both sides by A_3, we obtain

$$A_3F_{3-(1+2)} = A_3F_{3-1} + A_3F_{3-2} \tag{10.20}$$

In other words, Eq. (10.20) states that "the radiation leaving surface A_3 and intercepted by the combined surfaces A_1 and A_2 is equal to the sum of the radiation from surface A_3 to the surfaces A_1 and A_2 taken separately."

Making use of the reciprocal rule, we can write

$$A_3F_{3-(1+2)} = A_{(1+2)}F_{(1+2)-3}$$

$$A_3F_{3-1} = A_1F_{1-3}$$

$$A_3F_{3-2} = A_2F_{2-3}$$

Substitution into Eq. (10.20) results in

$$A_{(1+2)}F_{(1+2)-3} = A_1F_{1-3} + A_2F_{2-3} \tag{10.21}$$

Equation (10.21) states that "the total radiation leaving combined surfaces A_1 and A_2 ($A_{(1+2)}$) and intercepted by surface A_3 is equal to the sum of the radiations from surfaces A_1 and A_2 taken separately."

For the configuration in Figure 10.9b, the fraction of radiation leaving surface A_1 and falling onto surface A_{2+3} in accordance with Eq. (10.20) is

$$F_{1-(2+3)} = F_{1-2} + F_{1-3} \tag{10.22}$$

The only unknown factor in this equation is F_{1-3}

$$F_{1-3} = F_{1-(2+3)} - F_{1-2} \tag{10.23}$$

$F_{1-(2+3)}$ and F_{1-2} can be determined from the equation in item 9 (Table 10.2) or the graph presented in Figure 10.21.

10.3.4 Symmetry Rule

This rule is the reciprocity relationship that can be derived from the symmetry of opposing surfaces of a geometry. In Figure 10.10a

$A_1 = A_3$, $A_2 = A_4$ and

$$A_1F_{1-4} = A_3F_{3-2}$$

$$A_2F_{2-3} = A_4F_{4-1}$$

In Figure 10.10b

$A_1 = A_5$, $A_2 = A_6$, $A_3 = A_7$, $A_4 = A_8$, and

$$A_1F_{1-8} = A_5F_{5-4}$$

$$A_2F_{2-7} = A_6F_{6-3}$$

$$A_3F_{3-6} = A_2F_{2-7}$$

$$A_4F_{4-7} = A_8F_{8-3}$$

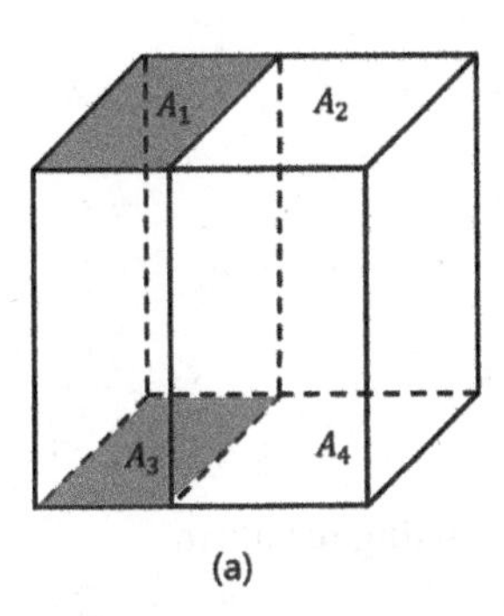

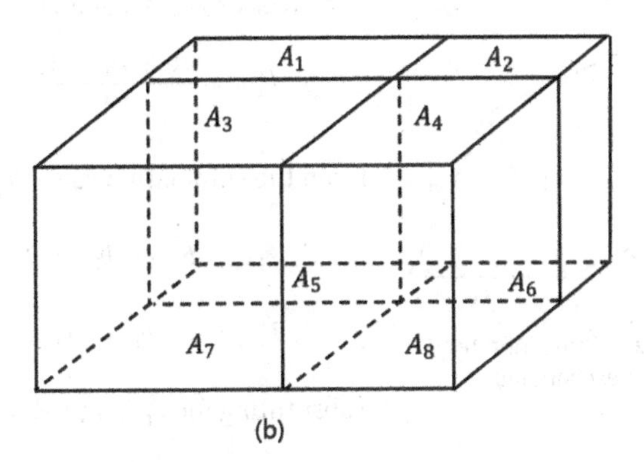

Figure 10.10 Representation of symmetrical surfaces for the symmetry rule.

10.3.5 String Rule

Another rule can be defined for two-dimensional geometries from the consideration of the three infinitely long surfaces forming a prism, as shown in Figure 10.11. For the radiation exchange between surface A_1 and the other two surfaces

$$A_1E_{b1} = A_1E_{b1}F_{1-2} + A_1E_{b1}F_{1-3}$$

$$A_1 = A_1F_{1-2} + A_1F_{1-3} \tag{10.24}$$

Similarly, application of the reciprocity rule from Eq. (10.16) results in the following for surfaces A_2 and A_3

$$A_2 = A_2F_{2-1} + A_2F_{2-3} = A_1F_{1-2} + A_2F_{2-3} \tag{10.25}$$

$$A_3 = A_3F_{3-1} + A_1F_{3-2} = A_1F_{1-3} + A_2F_{2-3} \tag{10.26}$$

The simultaneous solution of Eqs (10.24), (10.25), and (10.26) yields

$$F_{1-2} = \frac{A_1 + A_2 - A_3}{2A_1}$$

$$F_{1-3} = \frac{A_1 + A_3 - A_2}{2A_1}$$

$$F_{2-3} = \frac{A_2 + A_3 - A_1}{2A_2}$$

Consider now the two infinitely long surfaces A_1 and A_2 exchanging radiation, as shown in Figure 10.12. The dashed lines represent fictitious infinitely long surfaces forming shapes with triangular cross-sections.

For surfaces A_1, A_3, and A_5

$$F_{1-3} = \frac{A_1 + A_3 - A_5}{2A_1}$$

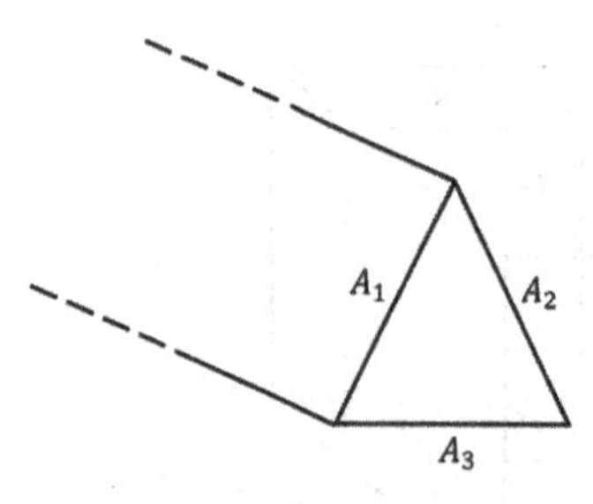

Figure 10.11 Three infinitely long surfaces exchanging radiation.

For surfaces A_1, A_4, and A_6

$$F_{1-6} = \frac{A_1 + A_6 - A_4}{2A_1}$$

From the summation rule (Eq. 10.19)

$$F_{1-2} + F_{1-3} + F_{1-6} = 1$$

$$F_{1-2} = 1 - F_{1-3} - F_{1-6}$$

Substituting for F_{1-3} and F_{1-6} and rearranging results in

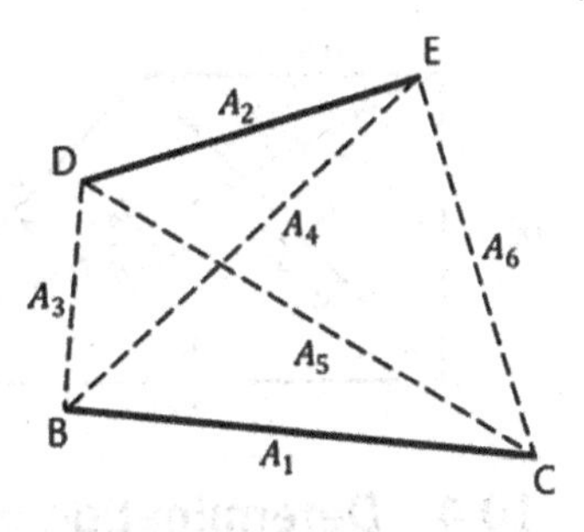

Figure 10.12 Radiation exchange between two infinitely long surfaces.

$$F_{1-2} = \frac{(A_4 + A_5) - (A_3 + A_6)}{2A_1} \text{ which can be rewritten as}$$

$$F_{1-2} = \frac{1}{2\overline{BC}}\left[\left(\overline{BE} + \overline{CD}\right) - \left(\overline{BD} + \overline{CE}\right)\right]$$

Let $\overline{BC} = L_1$, hence

$$F_{1-2} = \frac{1}{2L_1}\left[\sum (\text{diagonals}) - \sum (\text{sides})\right] \qquad (10.27)$$

Equation (10.27) is known as the *string rule* or *crossed-strings rule.*

Example 10.3 A long heating duct has a cross-section, as shown in Figures (E10.3a) and (E10.3b). What fraction of the radiation leaving A_1 falls on A_2, A_3, and A_4? What fraction of radiation leaving A_3 is directed towards A_1?

Solution

From the triangle formed by surfaces A_1, A_2, A_5

$$F_{1-2} = \frac{A_1 + A_2 - A_5}{2A_1} = \frac{0.7071 + 0.5 - 1.118}{2 \times 0.7071} = 0.063$$

For F_{1-3} the string rule, Eq. (10.20) yields

$$F_{1-3} = \frac{A_5 + A_6 - A_2 - A_4}{2A_1} = \frac{1.118 + 0.7071 - 0.5 - 1}{2 \times 0.7071} = 0.2299$$

From the triangle formed by surfaces A_1, A_4, A_6

$$F_{1-4} = \frac{A_1 + A_4 - A_6}{2A_1} = \frac{0.7071 + 1 - 0.7071}{2 \times 0.7071} = 0.7071$$

The fraction of radiation leaving A_3 and intercepted by surface A_1 can now be determined from the reciprocity rule using Eq. (10.16)

$$F_{3-1} = \frac{A_1}{A_3} F_{1-3} = \frac{0.7071 \times 0.2299}{0.5} = 0.3251$$

Figure E10.3a Long heating duct.

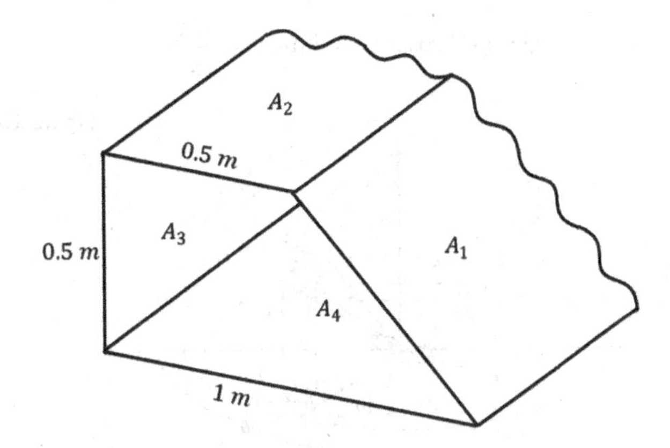

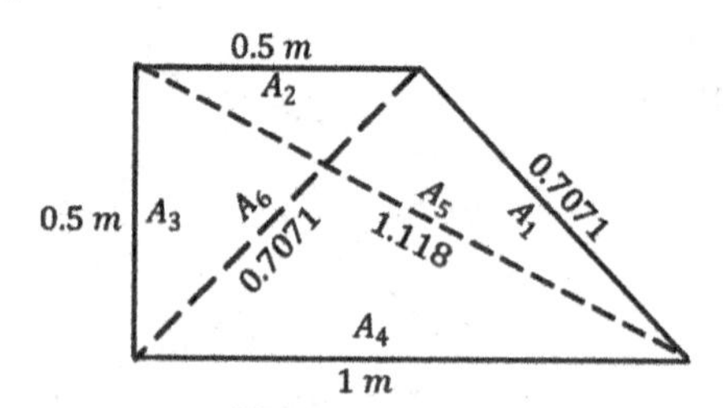

Figure E10.3b Cross-section of the long heating duct.

10.4 Determination of Shape Factors for Finite Surfaces

Figure 10.13 shows radiation exchange between two blackbody differential elements in spherical coordinates. Element dA is fixed and element dA_n is on the surface of a hemisphere of radius R. Element dA_n is defined by radius R and angles θ and φ and is equal to

$$dA_n = (Rd\varphi)(R\sin\varphi d\theta) = R^2\ (\sin\varphi d\varphi d\theta) \tag{10.28}$$

By analogy with plane geometry with differential angle $d\varphi$ in Figure (10.14a), differential conical or prismatic solid angle $d\omega$ in "steradian" (sr) is defined as

$$d\omega = \frac{dA_n}{R^2}$$

Substituting for dA_n from Eq. (10.28) into this definition we obtain

$$d\omega = \sin\varphi d\varphi d\theta\ sr \tag{10.29}$$

Illustrations of the solid angle $d\omega$ for both conical and prismatic differential elements are shown in Figures 10.14b and 10.14c.

The intensity of radiation i_b is defined as the radiation rate emitted per unit area per unit solid angle seen from element dA_n. For incremental projected area $dA\cos\varphi$ and incremental solid angle $d\omega$ the intensity is then

$$i_b = \frac{d\dot{Q}}{dA\cos\varphi d\omega}\quad W/m^2.sr \tag{10.30}$$

Substituting $d\omega$ from Eq. (10.29) into Eq. (10.30) and rearranging yields

$$d\dot{Q} = i_b dA\cos\varphi d\omega = i_b dA\cos\varphi\frac{dA_n}{R^2} \tag{10.31}$$

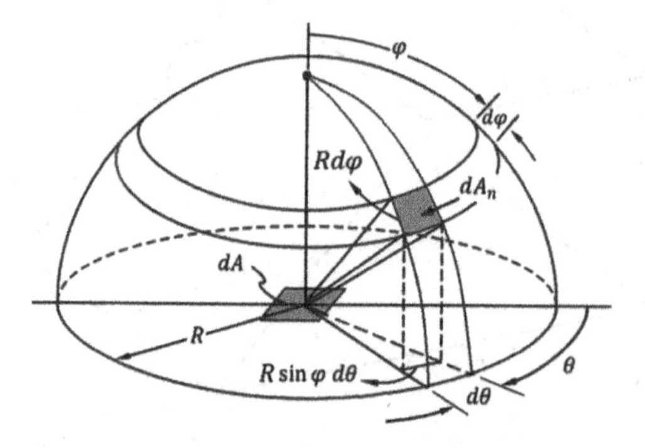

Figure 10.13 Definition of intensity of radiation.

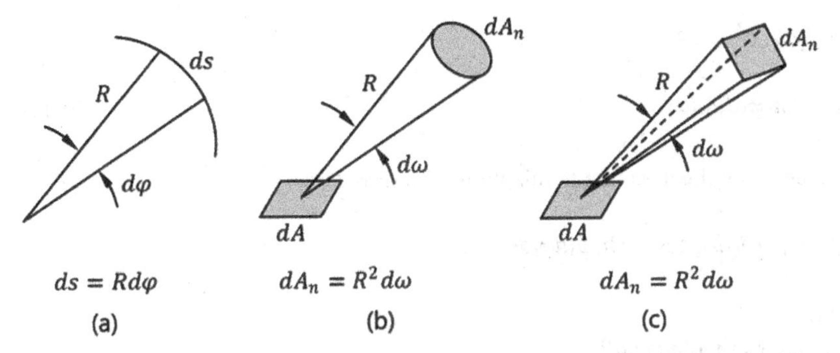

Figure 10.14 Plane and solid angles: (a) plane angle; (b) conical solid angle; (c) prismatic solid angle.

The definition of i_b can be seen in elevation view in Figure 10.15 with dA_1 replacing dA.

Now consider the two arbitrary finite surfaces exchanging radiation, as shown in Figure 10.16. The two elements dA_1 and dA_2 are located at a distance L from each other and angles φ_1 and φ_2 are the angles between L and the normal to the surfaces. The net radiation energy exchange between the two surfaces, applying the reciprocity rule, is

$$\dot{Q}_{1-2} = A_1 F_{1-2}(E_{b1} - E_{b2}) = A_2 F_{2-1}(E_{b1} - E_{b2}) \tag{10.32}$$

From Eq. (10.31), the energy emitted from surface dA_1 to surface dA_n (Figure 10.16) will be

$$d\dot{Q}_1 = i_b dA_1 \cos\varphi_1 \frac{dA_n}{R^2} \tag{10.33}$$

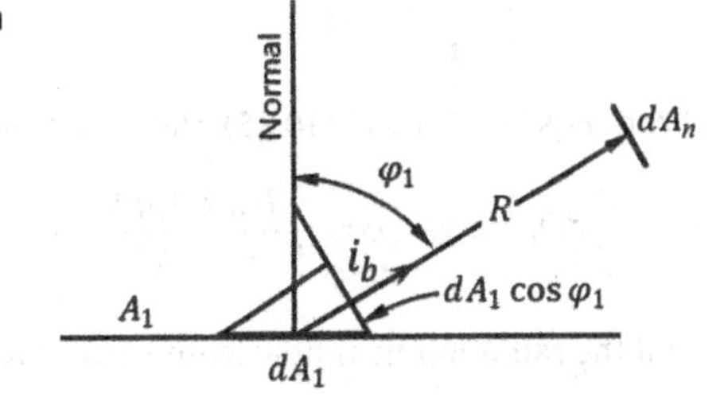

Figure 10.15 Schematic diagram of radiation intensity seen from differential area dA_1.

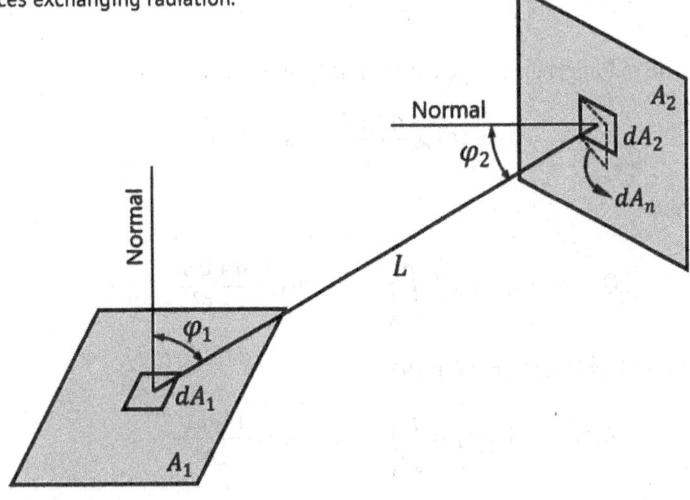

Figure 10.16 Two arbitrary surfaces exchanging radiation.

Since $dA_n = R^2 \sin\varphi_1 d\varphi_1 d\theta$,

$$d\dot{Q}_1 = i_b dA_1 \cos\varphi_1 \sin\varphi_1 d\varphi_1 d\theta \qquad (10.34)$$

The emissive power received by the area of a hemisphere of radius R is

$$E_b dA_1 = \int d\dot{Q}_1 = \int_0^{2\pi}\int_0^{\pi/2} i_b dA_1 \cos\varphi_1 \sin\varphi_1 d\varphi_1 d\theta$$

$$E_b dA_1 = i_b dA_1 \int_0^{2\pi}\int_0^{\pi/2} \cos\varphi_1 \sin\varphi_1 d\varphi_1 d\theta$$

Finally,

$$E_b = \pi i_b \qquad (10.35)$$

Combining Eqs (10.6) and (10.35) yields the radiation intensity in terms of the temperature of surface A_1

$$i_b = \frac{\sigma T^4}{\pi} \qquad (10.36)$$

From Figure (10.16), the relationship between dA_2 and dA_n (elemental area of the surface of the hemisphere) for $L = R$ is

$$dA_n = dA_2 \cos\varphi_2$$

From Eqs (10.34) and (10.35), the radiation heat flow from surface dA_1 to surface dA_2

$$d\dot{Q}_1 = \cos\varphi_1 \cos\varphi_2 \frac{E_{b1}}{\pi}\frac{dA_1 dA_2}{R^2} \qquad (10.37)$$

and the radiation heat flow from surface dA_2 to surface dA_1

$$d\dot{Q}_2 = \cos\varphi_2 \cos\varphi_1 \frac{E_{b2}}{\pi}\frac{dA_2 dA_1}{R^2} \qquad (10.38)$$

The radiation exchange between dA_1 and dA_2 is then

$$d\dot{Q}_{1-2} = \cos\varphi_1 \cos\varphi_2 \frac{dA_1 dA_2}{\pi R^2}(E_{b1} - E_{b2})$$

and

$$\dot{Q}_{1-2} = (E_{b1} - E_{b2}) \int_{A_1}\int_{A_2} \cos\varphi_1 \cos\varphi_2 \frac{dA_1 dA_2}{\pi R^2} \qquad (10.39)$$

From Eqs (10.32) and (10.39)

$$A_1 F_{1-2} = A_2 F_{2-1} = \int_{A_1}\int_{A_2} \cos\varphi_1 \cos\varphi_2 \frac{dA_1 dA_2}{\pi R^2} \qquad (10.40)$$

Equation (10.40) can be used for the determination of shape factors for two- and three-dimensional geometrical configurations involving two surfaces exchanging radiation. The use of this equation could be quite challenging unless the geometries are simple enough to easily evaluate the double integral. For more complex shapes, tables, graphs, and computer codes can be consulted.

Example 10.4 Determine the shape factor $F_{dA_1-A_2}$ between a differential area dA_1 centred parallel to a finite disk of radius R_2 and area A_2 as shown in Figure E10.4.

Solution

$$\varphi_1 = \varphi_2 = \varphi$$

$$dA_2 = r_2 d\theta dr_2$$

$$R^2 = H^2 + r_2^2$$

$$\cos\varphi = \frac{H}{\left(H^2 + r_2^2\right)^{1/2}}$$

From Eq. (10.34)

$$dA_1 F_{dA_1-A_2} = \int_{A_1}\int_{A2} \cos\varphi_1 \cos\varphi_2 \frac{dA_1 dA_2}{\pi R^2}$$

Since dA_1 is a differential element

$$dA_1 F_{dA_1-A_2} = \frac{dA_1}{\pi}\int_0^{R_2}\int_0^{2\pi} \cos^2\varphi \frac{dA_2}{R^2}$$

Substituting for dA_2 and R^2 and rearranging yields

$$F_{dA_1-A_2} = \frac{1}{\pi}\int_0^{R_2}\int_0^{2\pi} H^2 d\theta \frac{r_2 dr_2}{\left(H^2 + r_2^2\right)^2} = \frac{1}{\pi}\int_0^{2\pi} H^2 d\theta \int_0^{R_2} \frac{r_2 dr_2}{\left(H^2 + r_2^2\right)^2} = 2H^2 \int_0^{R_2} \frac{r_2 dr_2}{\left(H^2 + r_2^2\right)^2}$$

The last integral can be solved by setting $H^2 + r_2^2 = \lambda$

$$F_{dA_1-A_2} = H^2 \int_{H^2}^{H^2+R_2^2} \frac{d\lambda}{\lambda^2} = \left[-\frac{H^2}{\lambda}\right]_{H^2}^{H^2+R_2^2} = H^2\left(\frac{1}{H^2} - \frac{1}{H^2 + R_2^2}\right)$$

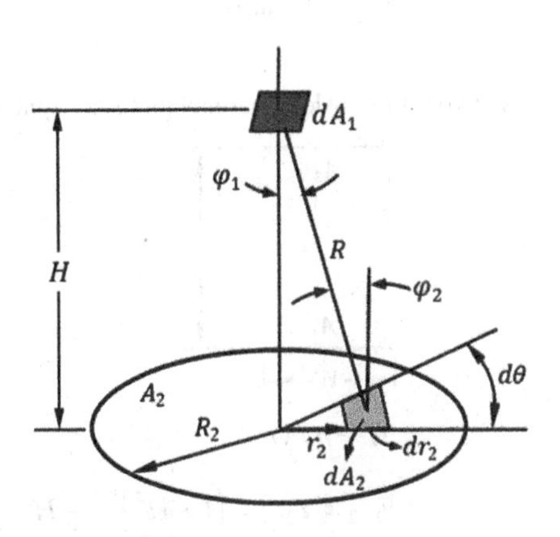

Figure E10.4 Schematic diagram for the determination of shape factor $F_{dA_1-A_2}$.

Finally,

$$F_{dA_1-A_2} = \frac{H^2}{H^2 + R_2^2}$$

10.5 Shape Factor Equations

Derived equations of shape factors for selected geometries are provided in Table 10.2 with the more complex equations presented separately in graphical form in Figures 10.18–10.24 and magnified version thereof in Appendix E. More comprehensive compilations of shape factor equations can be found in Wong (1977), Howell (1982), Siegel and Howell (1992), Rohsenow, et al (1998), and Modest (2013).

Table 10.2 Shape factor equations for selected geometries.

1) Infinitely long three-sided enclosure

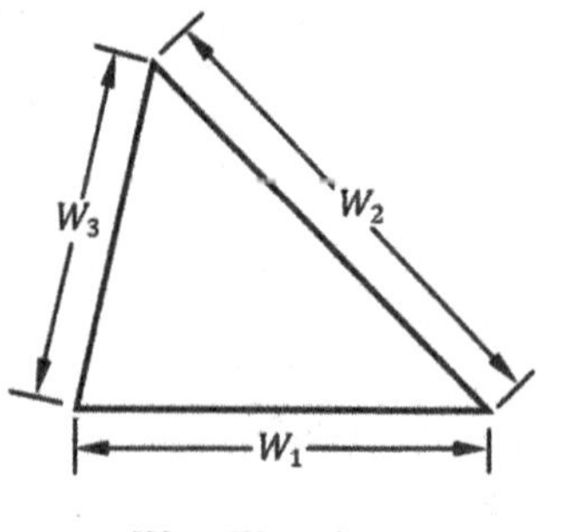

$$F_{1-2} = \frac{W_1 + W_2 - W_3}{2W_1}, F_{1-3} = \frac{W_1 + W_3 - W_2}{2W_3}, F_{2-3} = \frac{W_2 + W_3 - W_1}{2W_2}$$

2) Two infinitely long plates of equal width with common edge

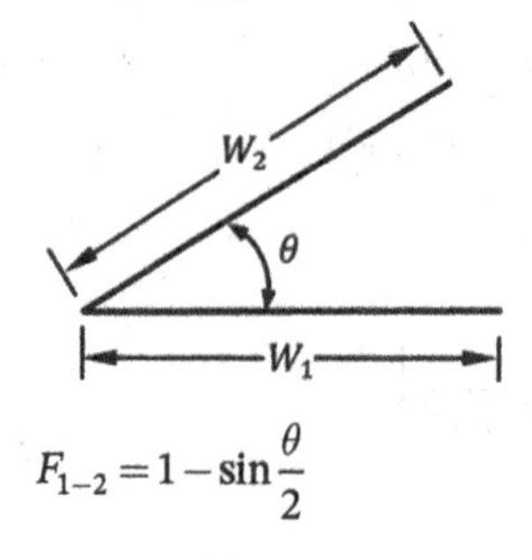

$$F_{1-2} = 1 - \sin\frac{\theta}{2}$$

3) Two infinitely long parallel plates of equal width

A_2

h

A_1

W

$H = h/W$

$$F_{1-2} = F_{2-1} = \left(1 + H^2\right)^{1/2} - H$$

Table 10.2 (Continued)

4) Infinite parallel plates of different widths connected at midpoints

$$X = W_1 / h$$

$$Y = W_2 / h$$

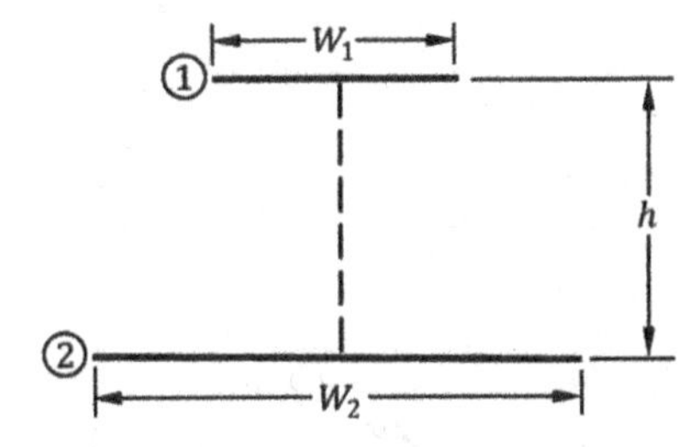

$$F_{1-2} = \frac{1}{2X}\left\{\left[(X+Y)^2+4\right]^{1/2} - \left[(Y-X)^2+4\right]^{1/2}\right\}$$

5) Surface of upright cylinder to annular disk at lower end of cylinder

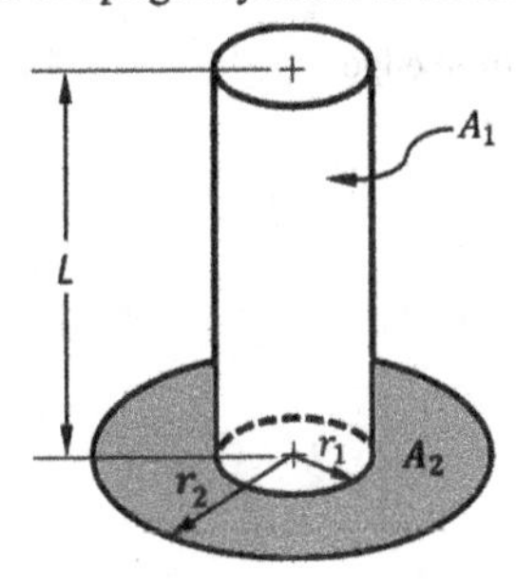

$$X = r_1 / r_2 \; Y = L / r_2$$

$$A = Y^2 + X^2 - 1$$

$$B = Y^2 - X^2 + 1$$

$$F_{1-2} = \frac{B}{8XY} + \frac{1}{2\pi}\left\{\cos^{-1}\frac{A}{B} - \frac{1}{2Y}\left[\frac{(A+2)^2}{X^2} - 4\right]^{1/2}\cos^{-1}\frac{AX}{B} - \frac{A}{2XY}\sin^{-1}X\right\}$$

Plot of $F_{1-2} = f(r_1 / r_2, L / r_2)$ in Figure 10.18

(Continued)

Table 10.2 (Continued)

6) Parallel strip element and rectangular plate of equal lengths

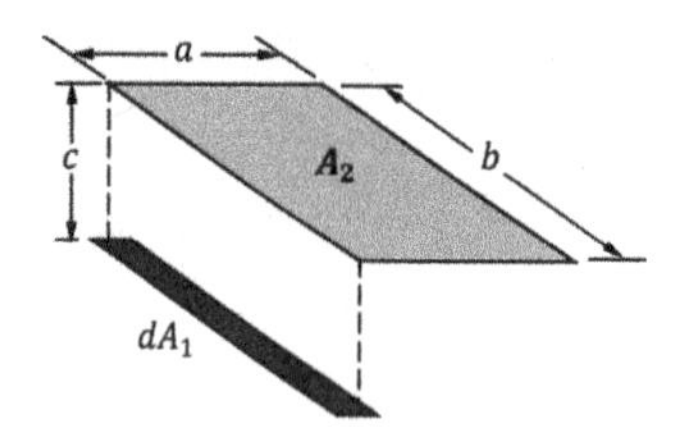

$X = a/c$

$Y = b/c$

$$F_{d1-2} = \frac{1}{\pi Y}\left[\left(1+Y^2\right)^{1/2}\tan^{-1}\frac{X}{\left(1+Y^2\right)^{1/2}} - \tan^{-1}X + \frac{XY}{\left(1+X^2\right)^{1/2}}\tan^{-1}\frac{Y}{\left(1+X^2\right)^{1/2}}\right]$$

Plot of $F_{1-2} = f\left(a/c, b/c\right)$ in Figure 10.19

7) Two infinitely long perpendicular plates of unequal width and common edge

h

90°

W

$$H = \frac{h}{W}$$

$$F_{1-2} = \frac{1}{2}\left[1 + H - \left(1+H^2\right)^{1/2}\right]$$

8) Two parallel and equal finite rectangles

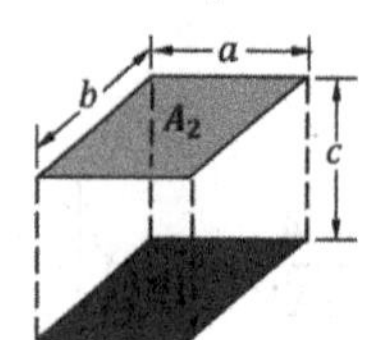

$X = a/c$

$Y = b/c$

$$F_{1-2} = \frac{2}{\pi XY}\left\{\begin{array}{l} ln\left[\dfrac{\left(1+X^2\right)\left(1+Y^2\right)}{1+X^2+Y^2}\right]^{0.5} + X\left(1+Y^2\right)^{1/2}\tan^{-1}\dfrac{X}{\left(1+Y^2\right)^{1/2}} \\ +Y\left(1+X^2\right)^{1/2}\tan^{-1}\dfrac{Y}{\left(1+X^2\right)^{1/2}} - X\tan^{-1}X - Y\tan^{-1}Y \end{array}\right\}$$

Plot of $F_{1-2} = f\left(a/c, b/c\right)$ in Figure 10.20

Table 10.2 (Continued)

9) Two perpendicular finite rectangles of same length with one common edge

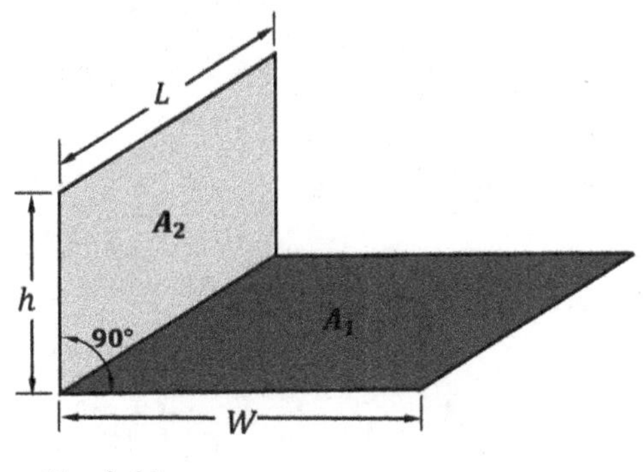

$X = h/L$
$Y = W/L$

$$F_{1-2} = \frac{1}{\pi Y}\left(\begin{array}{l} Y\tan^{-1}\frac{1}{Y} + X\tan^{-1}\frac{1}{X} - \left(X^2+Y^2\right)^{1/2}\tan^{-1}\frac{1}{\left(X^2+Y^2\right)^{1/2}} \\ +\frac{1}{4}\ln\left\{\left[\frac{\left(1+Y^2\right)\left(1+X^2\right)}{1+X^2+Y^2}\right]\left[\frac{Y^2\left(1+X^2+Y^2\right)}{\left(1+Y^2\right)\left(Y^2+X^2\right)}\right]^{Y^2}\left[\frac{X^2\left(1+X^2+Y^2\right)}{\left(1+X^2\right)\left(Y^2+X^2\right)}\right]^{X^2}\right\} \end{array} \right)$$

Plot of $F_{1-2} = f(h/L, W/L)$ in Figure 10.21

10) Parallel circular disk and plane element with its normal passing through disk centre

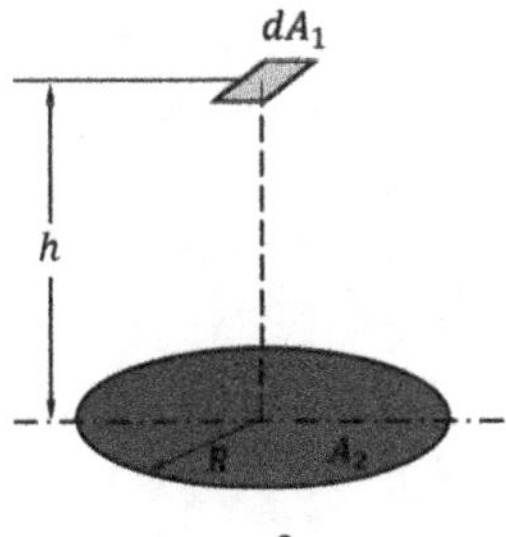

$$F_{d1-2} = \frac{R^2}{R^2+h^2}$$

(Continued)

Table 10.2 (Continued)

11) Parallel elliptical plate and plane element with its normal passing through plate centre

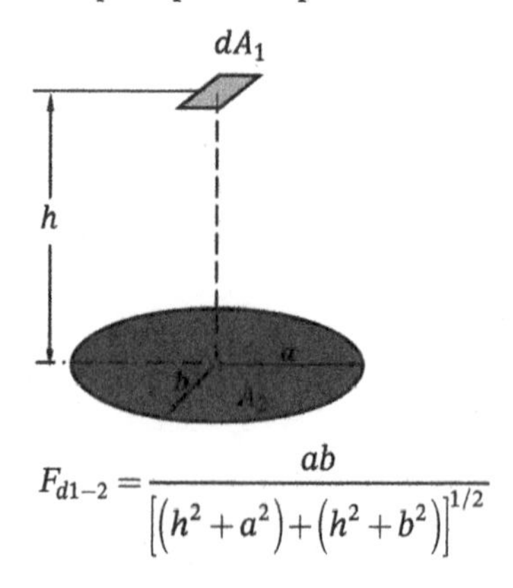

$$F_{d1-2} = \frac{ab}{\left[\left(h^2+a^2\right)+\left(h^2+b^2\right)\right]^{1/2}}$$

12) Coaxial parallel disks of different diameters

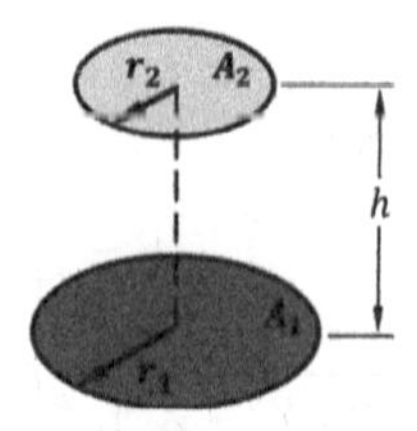

$$R_1 = r_1 / h \; R_2 = r_2 / h$$

$$X = 1 + \frac{1+R_2^2}{R_1^2}$$

$$F_{1-2} = \frac{1}{2}\left\{X - \left[X^2 - 4\left(\frac{R_2}{R_1}\right)^2\right]^{1/2}\right\}$$

Plot of $F_{1-2} = f\left(h/r_1, r_2/h\right)$ in Figure 10.22

13) Infinitely long parallel cylinders of equal diameters

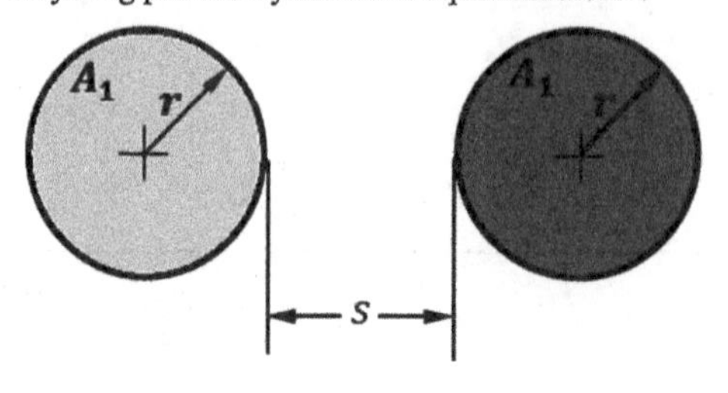

$$X = 1 + \frac{s}{2r}$$

$$F_{1-2} = F_{2-1} = \frac{1}{\pi}\left[\left(X^2-1\right)^{1/2} + \sin^{-1}\frac{1}{X} - X\right]$$

Table 10.2 (Continued)

14) Infinitely long parallel cylinders of unequal diameters

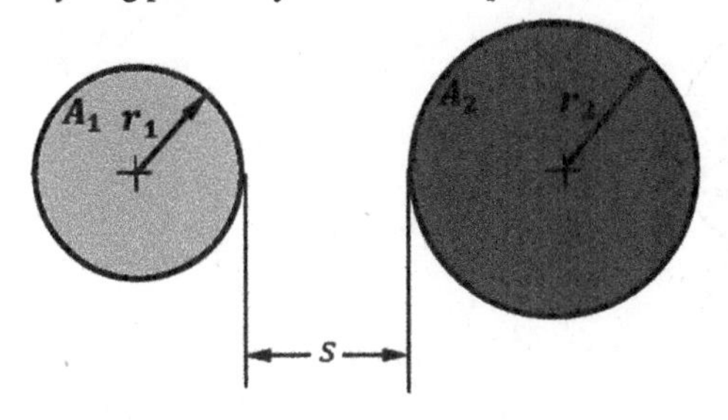

$$X = r_2 / r_1 \quad Y = s / r_1$$
$$C = 1 + X + Y$$

$$F_{1-2} = \frac{1}{2\pi}\left\{\pi + \left[C^2 - (X+1)^2\right]^{1/2} + (X-1)\cos^{-1}\left(\frac{X}{C} - \frac{1}{C}\right) - (X+1))\cos^{-1}\left(\frac{X}{C} + \frac{1}{C}\right)\right\}$$

15) Concentric cylinders of infinite length

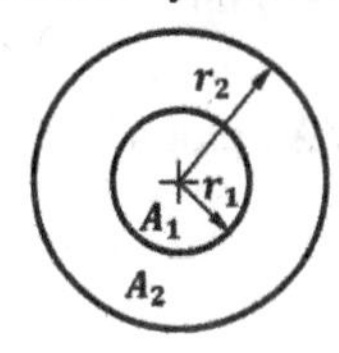

$F_{1-2} = 1 \qquad F_{2-1} = r_1 / r_2 \qquad F_{2-2} = 1 - r_1 / r_2$

16) Concentric Spheres

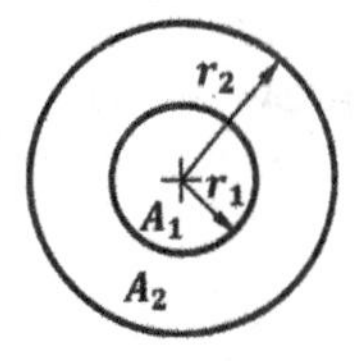

$F_{1-2} = 1 \qquad F_{2-1} = (r_1 / r_2)^2 \qquad F_{2-2} = 1 - (r_1 / r_2)^3$

(Continued)

Table 10.2 (Continued)

17) Two concentric cylinders of equal finite lengths

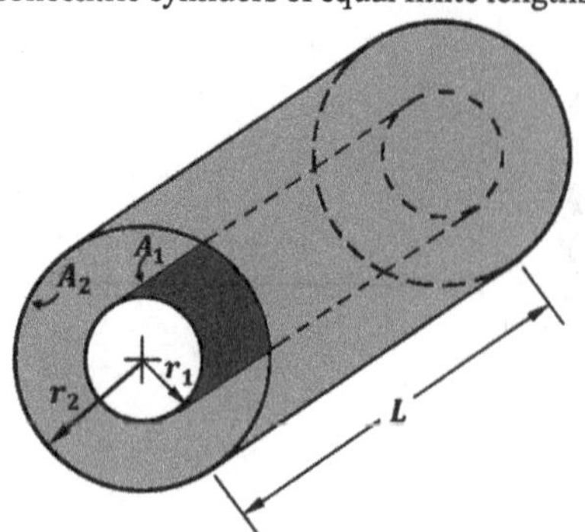

$$X = r_2/r_1 \quad Y = L/r_1$$

$$A = Y^2 + X^2 - 1$$

$$B = Y^2 - X^2 + 1$$

$$F_{2-1} = \frac{1}{X} - \frac{1}{\pi X}\left\{\cos^{-1}\frac{B}{A} - \frac{1}{2Y}\left[\left(A^2 + 4A - 4X^2 + 4\right)^{1/2}\cos^{-1}\frac{B}{XA} + B\sin^{-1}\frac{1}{X} - \frac{\pi A}{2}\right]\right\}$$

Plot of $F_{2-1} = f\left(r_1/r_2, L/r_2\right)$ or, since $r_1/r_2 = 1/X$ and $L/r_2 = Y/X$, plot of $F_{2-1} = f\left(1/X, Y/X\right)$ in Figure 10.23

$$F_{2-2} = 1 - \frac{1}{X} + \frac{2}{\pi X}\tan^{-1}\frac{2\sqrt{X^2-1}}{Y} - \frac{Y}{2\pi X}$$

$$\left[\frac{\sqrt{4X^2+Y^2}}{Y}\sin^{-1}\frac{4\left(X^2-1\right)+\left(Y/X\right)^2\left(X^2-2\right)}{Y^2+4\left(X^2-1\right)} - \sin^{-1}\frac{X^2-2}{X^2} + \frac{\pi}{2}\left(\frac{\sqrt{4X^2+Y^2}}{Y} - 1\right)\right]$$

Plot of $F_{2-2} = f\left(r_1/r_2, L/r_2\right)$ or $F_{2-1} = f\left(1/X, Y/X\right)$ in Figure 10.24

18) Infinite plate and row of parallel tubes

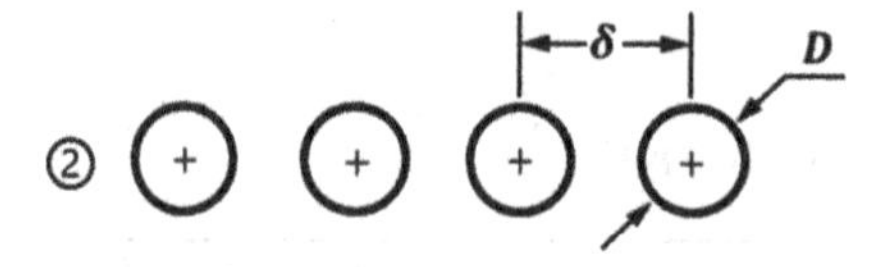

$$F_{1-2} = 1 - \left[1 - \left(\frac{D}{\delta}\right)^2\right]^{1/2} + \frac{D}{\delta}\tan^{-1}\left[\left(\frac{\delta}{D}\right)^2 - 1\right]^{1/2}, \text{ or}$$

$$F_{1-2} = \frac{D}{\delta}\cos^{-1}\frac{D}{\delta} + 1 - \left[1 - \left(\frac{D}{\delta}\right)^2\right]^{1/2}$$

$$F_{2-1} = \frac{1}{\pi}\left\{\frac{\delta}{D} - \left[\left(\frac{\delta}{D}\right)^2 - 1\right]^{1/2} + \tan^{-1}\left[\left(\frac{\delta}{D}\right)^2 - 1\right]^{1/2}\right\} = \frac{\delta}{\pi D}F_{1-2}$$

The shape factor F_{1-2}^{n} for a bank of tubes of n rows (Figure 10.17), from Yurenev and Lebedev (1976), is

$$F_{1-2}^{n} = 1 - (1 - F_{1-2})^{n} \tag{10.41}$$

where F_{1-2} is the shape factor for a single row given in Table 10.2 (18).

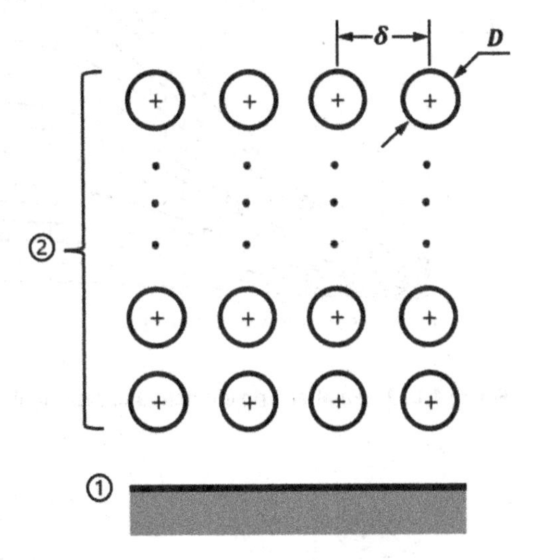

Figure 10.17 Infinite plate with multi-row tube bank of parallel tubes.

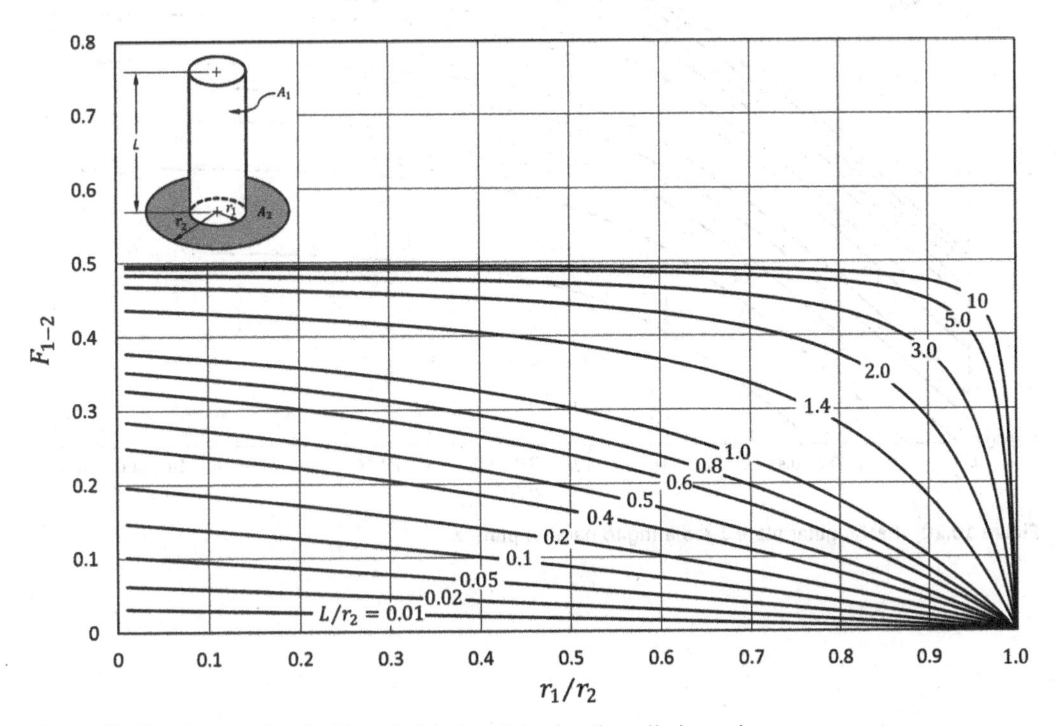

Figure 10.18 Concentric cylinder and disk plate joined at the cylinder end.

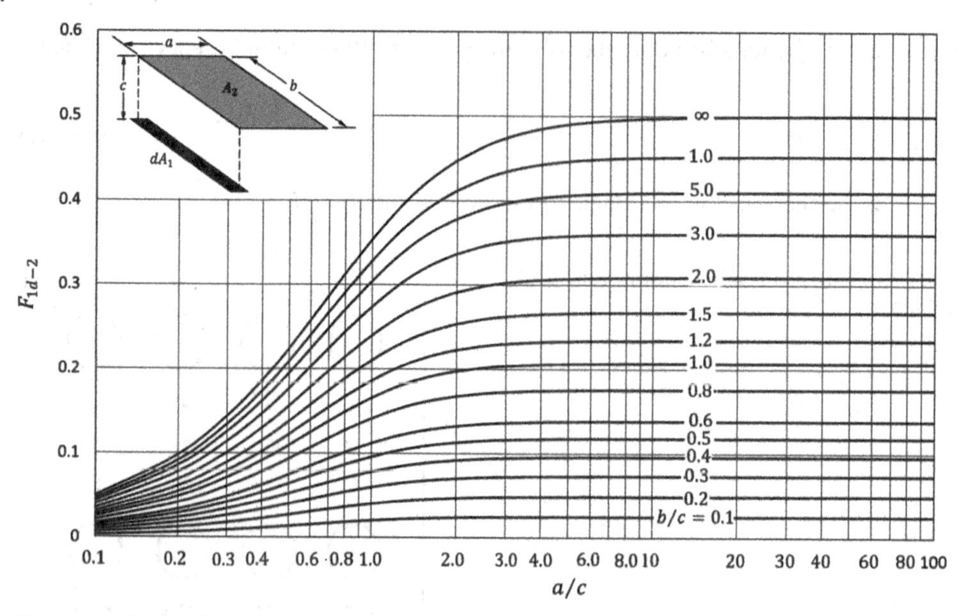

Figure 10.19 Parallel strip element and rectangular plate of equal lengths.

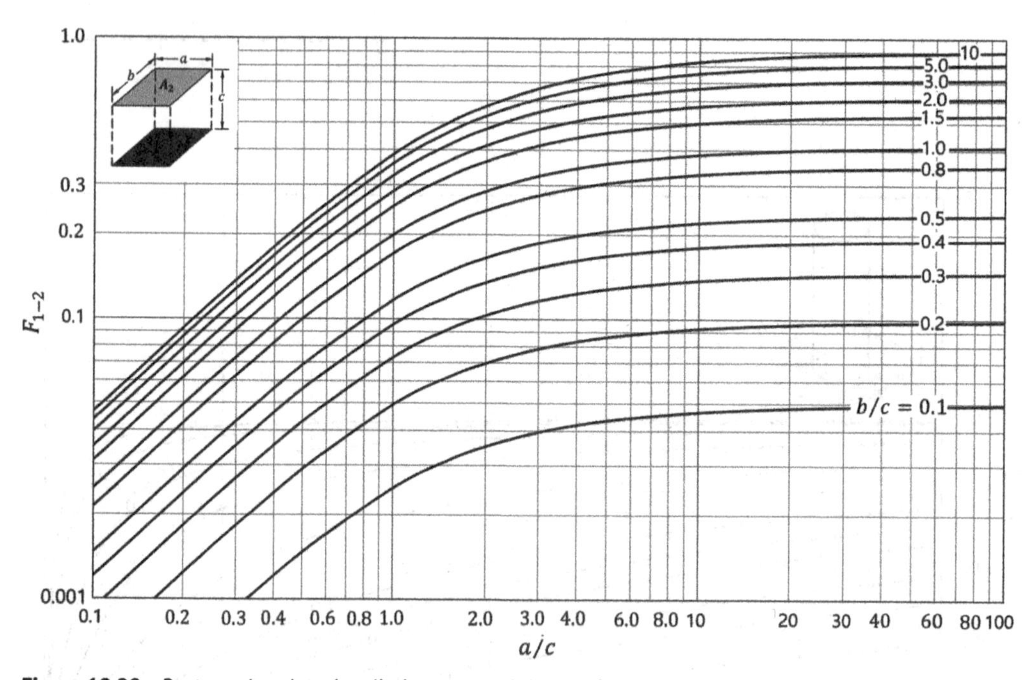

Figure 10.20 Rectangular plate 1 radiating to parallel plate 2.

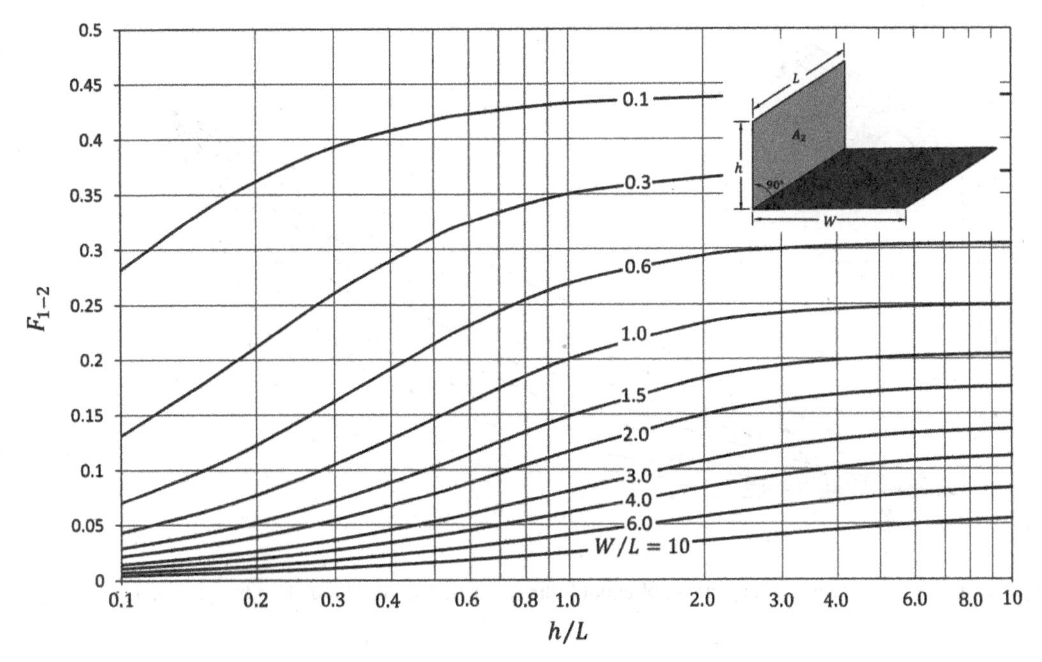

Figure 10.21 Rectangle 1 radiating to rectangle 2 with which it has a common edge.

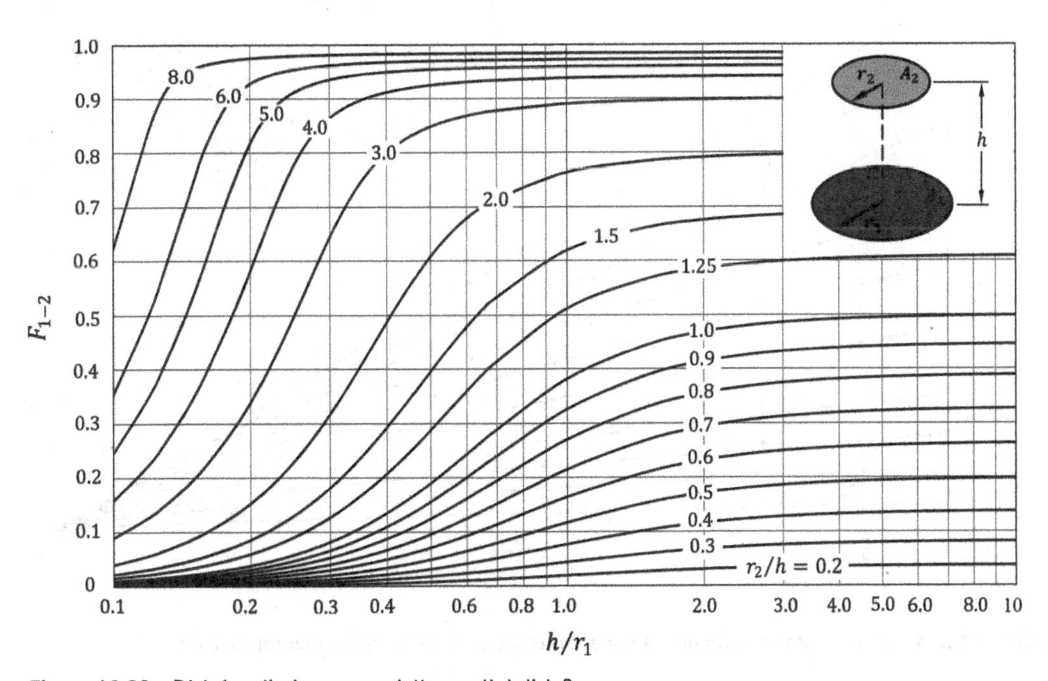

Figure 10.22 Disk 1 radiating to coaxially parallel disk 2.

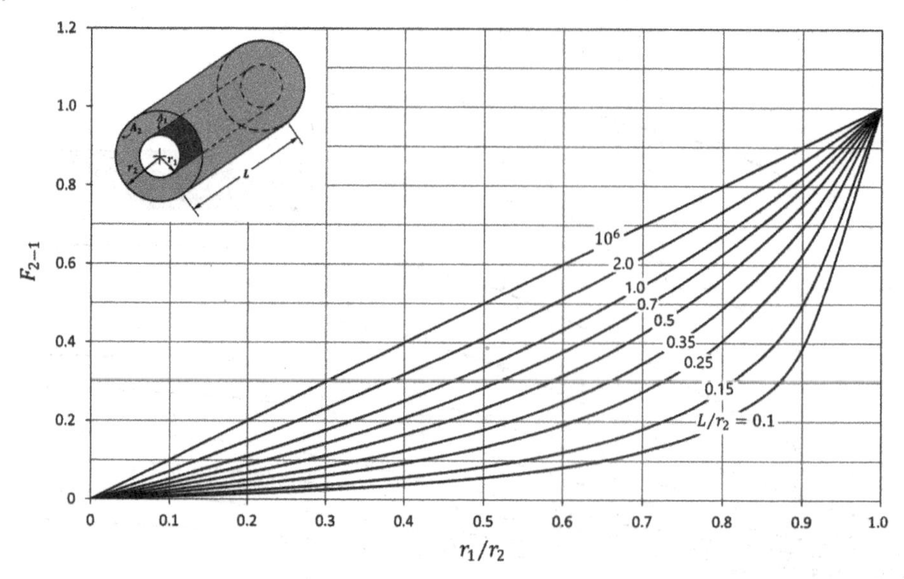

Figure 10.23 Two concentric cylinders of equal finite lengths: cylinder 2 radiating to cylinder 1.

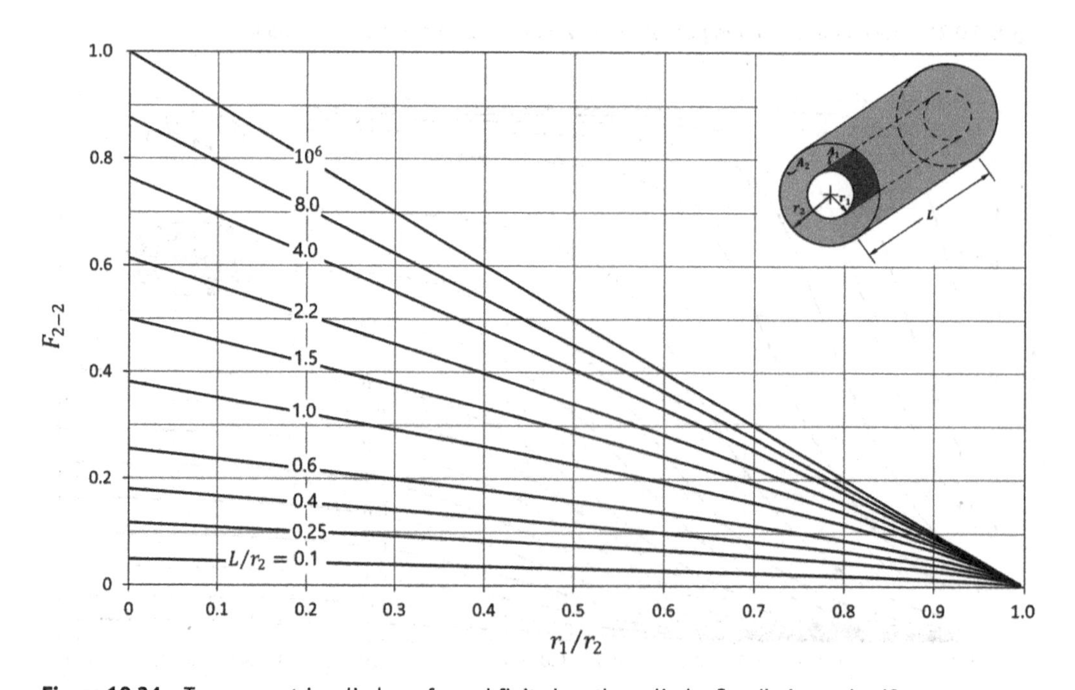

Figure 10.24 Two concentric cylinders of equal finite lengths: cylinder 2 radiating to itself.

Problems

10.1 A thermal radiation source provides 1.2 kW/m^2 to a plate insulated on one side and having absorptivity $\alpha = 0.8$ and emissivity $\varepsilon = 0.5$. It is then placed in a large enclosure at $T_\infty = 20°C$. Determine the temperature at the surface of the plate.

10.2 The enclosure in Problem 10.1 is filled with air at $T_\infty = 20°C$ and the convective heat transfer coefficient between the plate and ambient air is 10 $W/m^2.K$. What is the percentage drop in plate temperature as a result of the combined radiation and convection heat transfer?

10.3 A coated plate ($350\,mm \times 350\,mm$) is placed in a large room and heated to 525°C. If the surface of the plate is assumed to be a blackbody, determine:

(a) the total blackbody emissive power
(b) the total of thermal radiation energy from the plate in 10 minutes
(c) the emissive power emitted at a wavelength of 5 μm.

10.4 Solar flux at the outer edge of the earth's atmosphere is approximately 1350 W/m^2. Considering the sun–earth geometry in Figure P10.4, determine,

(a) the sun's emissive power
(b) temperature and wavelength of maximum emission from the sun
(c) the temperature of the earth's atmosphere.

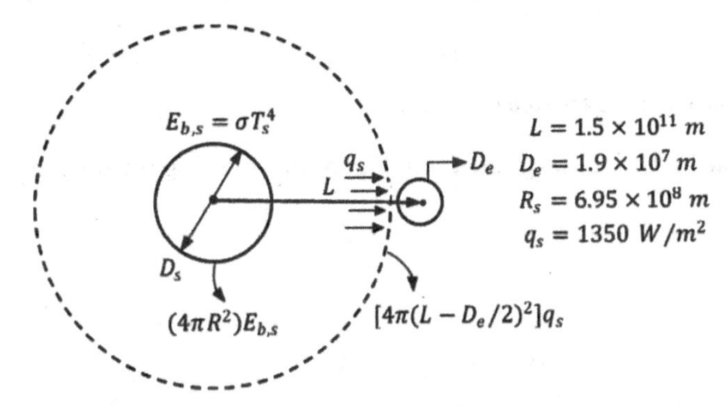

Figure P10.4 Schematic diagram of the Sun–Earth geometry.

10.5 An incandescent light bulb emits 10% of its energy at wavelengths shorter than 0.7 μm. If it is assumed that the filament is a blackbody emitter, what is the temperature of the filament?

10.6 Three different surfaces are heated to a temperature $T_s = 1200\,K$. The total radiosity (radiation heat flux leaving these surfaces) from the surfaces are $E_1 = 6760\,W/m^2$, $E_2 = 21{,}800\,W/m^2$ and $E_3 = 48{,}850\,W/m^2$. Assuming the surfaces are grey and no radiation is reflected, determine the total emissivities for the three surfaces.

10.7 The temperature of the tungsten filament of an incandescent bulb is $T_s = 2500K$. Determine the fraction of radiant energy emitted by the filament within the visible range of the electromagnetic wave spectrum. Assume the filament is a black body.

10.8 To make the bulb in Problem 10.7 more efficient, the tungsten filament is heated to $T_s = 3250K$. Determine the gain in the efficiency of converting heat to visible light. Assume the filament is a black body.

10.9 For practical purposes, the sun can be considered to be a blackbody radiator with an effective temperature of 5800 K. Determine the fraction of energy emitted by the sun that falls in the visible region of the electromagnetic spectrum.

10.10 The temperature of a black surface of area 0.3 m^2 is 560°C. Calculate:

(a) the rate of energy emission

(b) the wavelength of maximum monochromatic emission power.

10.11 An opaque grey surface has a reflectivity of 0.2. What is the absorptivity and the total emissive power at 2000 K?

10.12 Use the string rule to derive the expressions for the shape factor F_{1-2} for the long surfaces 1 and 2 in the three configurations in Figure P10.12.

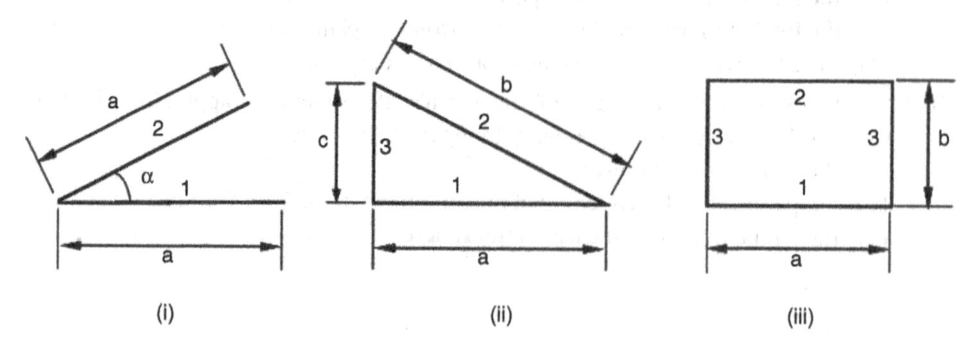

Figure P10.12 Three configurations for Problem 10.12.

10.13 Determine the shape factor F_{1-4} in the geometry in Figure P10.13 using the superposition rule.

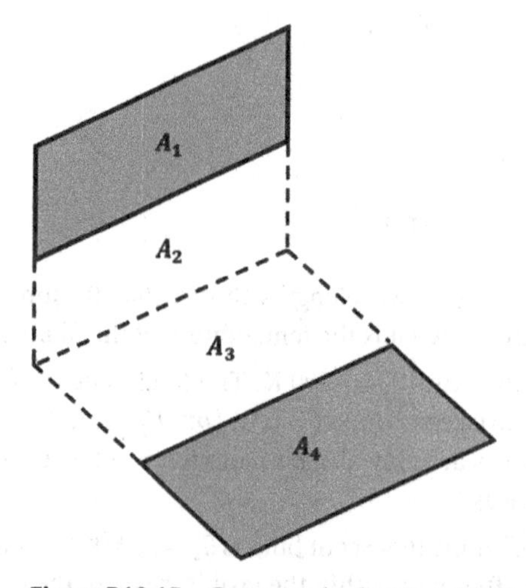

Figure P10.13 Geometry for Problem 10.13.

10.14 Find the shape factor for the radiation exchange between the elemental area dA_1 and finite rectangular plate (surface 2), as shown in Figure P10.14.

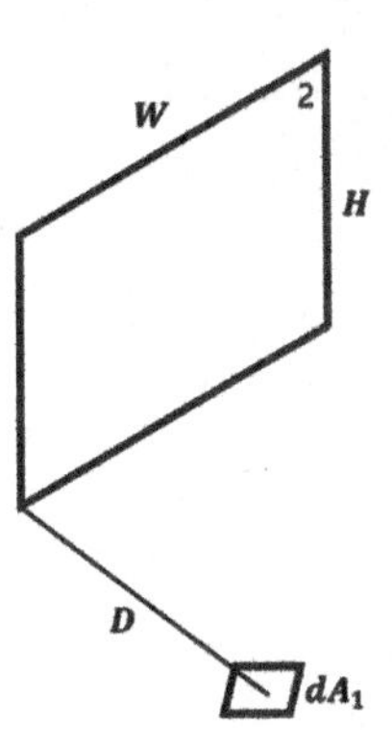

Figure P10.14 Radiation exchange between elemental area dA_1 and rectangular plate.

10.15 Work out from first principles the shape factors F_{1-2} and F_{2-1} for two infinitely long parallel cylinders of equal diameter separated centre-to-centre by distance S, as shown in Figure P10.15.

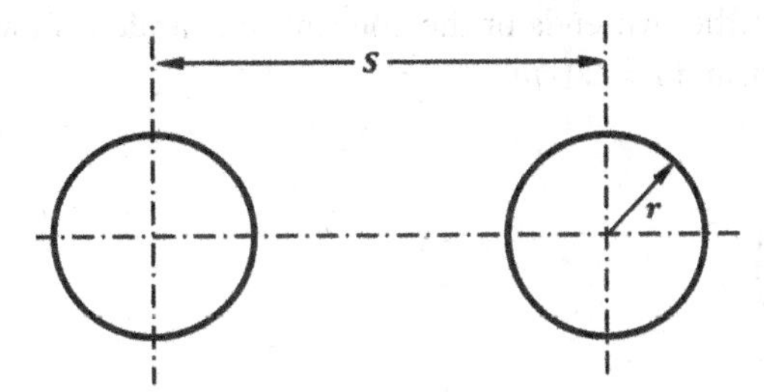

Figure P10.15 Two long parallel cylinders of equal diameter separated centre-to-centre by distance S.

10.16 Determine the F_{1-2} for the two cylinders if S is the distance separating the surfaces of the cylinder, as shown in Figure P10.16.

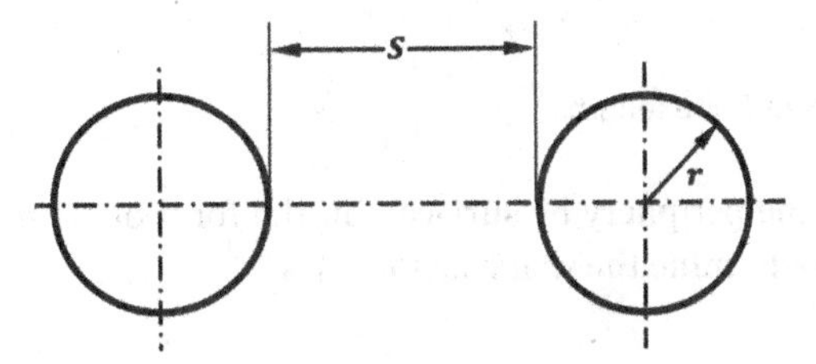

Figure P10.16 Two long parallel cylinders of equal diameter with distance S separating inner surfaces.

10.17 Determine the fraction of radiation leaving the surface of the disc 1 at the bottom of the cylinder (Figure P10.17) that escapes through the coaxial ring opening at the top of the cylinder. $L = 10\ cm$, $r_1 = 10\ cm$, $r_2 = 5\ cm$, $r_3 = 8\ cm$.

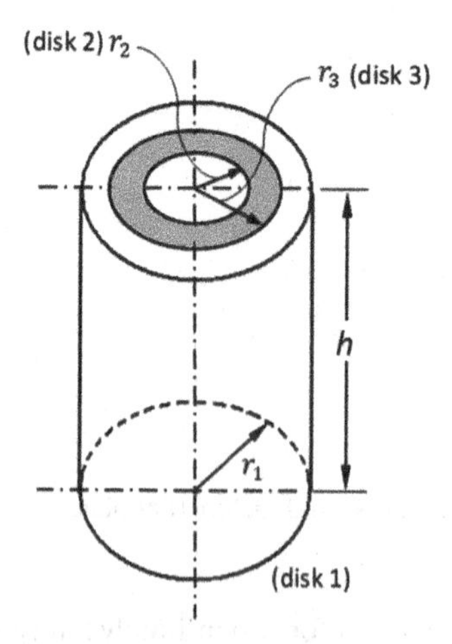

Figure P10.17 Geometrical representation of Problem 10.17.

10.18 Determine the shape factor between the two ends of the concentric cylinders shown in Figure P10.18 if $r_1 = 10\ cm$, $r_2 = 20\ cm$, and $L = 20\ cm$.

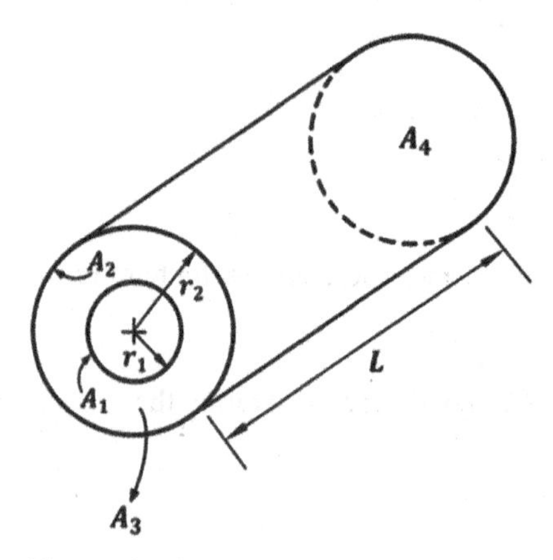

Figure P10.18 Two concentric cylinders of finite length.

10.19 Metal discs 1 and 2 are bounded at the periphery by surface 3 in the form of 0.1 m wide shroud, as shown in Figure P10.19. Determine the shape factors F_{1-2}.

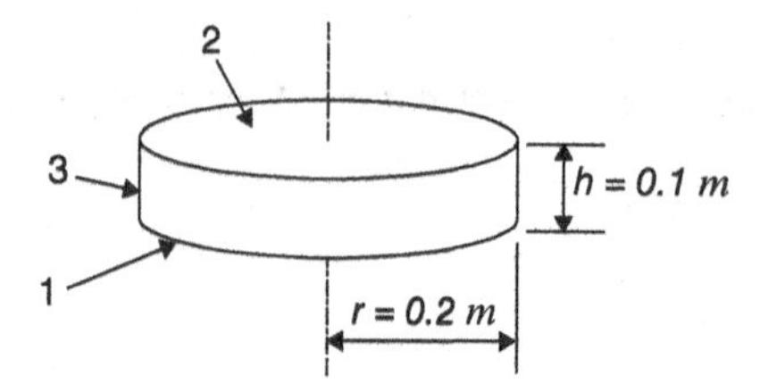

Figure P10.19 Two coaxial discs separated by distance h.

10.20 Determine the remaining shape factors in the configuration in Problem 10.19.

10.21 Two parallel rectangular surfaces 1 and 2 have the dimensions $1\,m \times 2\,m$ and are located opposite each other at a distance of $L = 4\,m$ (Figure P10.21). The surfaces are at temperatures 100°C and 200°C, respectively. Assuming the surfaces behave as a black body, what is the rate of radiative heat exchange between the two surfaces?

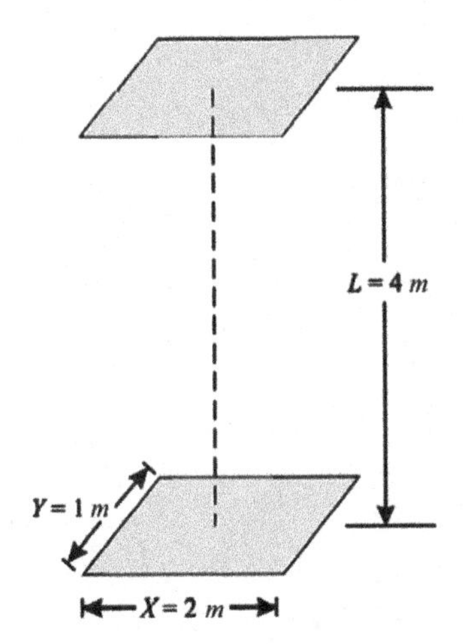

Figure P10.21 Parallel rectangles separated by distance L.

11

Thermal Radiation

Radiation Exchange Between Non-Black Bodies

In the previous Chapter 10, blackbody-blackbody and blackbody-grey body radiation exchanges were considered, accounting for the appropriate shape factors. But in practical applications, we are concerned with radiation exchange between nonblack (grey or real) surfaces. In such cases, radiation leaving a surface is due to both reflection and emission, and the receiving surface(s) (if opaque, $\tau = 0$) experience partial reflection and absorption. In analysing such problems, it is usually assumed that all surfaces involved are grey, isothermal (uniform constant temperature), and reflectivity and emissivity are constant over all the surfaces. Two new concepts are to be introduced in the analyses in the following sections, namely, radiosity and irradiation that are assumed uniform over all surfaces. Grey surfaces are also known as diffuse surfaces, which indicates that the intensity of emitted radiation as well as the directional emissivity and absorptivity are independent of direction.

11.1 Radiation Exchange Between Two Grey Surfaces

The relationship $\dot{Q}_{1-2} = A_1 \varepsilon \sigma (T_1^4 - T_2^4)$ for radiation exchange between a grey body and blackbody enclosure does not hold if the enclosure is also non-black (grey), as some of the radiation it emits will be reflected from the enclosure back to it. Consider the schematic diagram of the grey surface shown in Figure 11.1, depicting a real plate surface i and two imaginary surfaces at close proximity on both sides of the plate, u and s. Two heat fluxes can be defined:

Irradiation:

G – total radiation energy incident upon surface s per unit time and per unit area (W/m^2) from one surface or multiple surfaces.

Radiosity:

J – total energy leaving surface s per unit time and per unit area (W/m^2) that can be intercepted by one or multiple surfaces and is the sum of emitted radiation εE_b from surface u and reflected radiation ρG from surface i.

Net energy balances at surfaces s and u respectively

$$q = J - G \tag{11.1}$$

Heat Transfer Basics: A Concise Approach to Problem Solving, First Edition. Jamil Ghojel.

Companion website: www.wiley.com/go/ghojel/heat_transfer

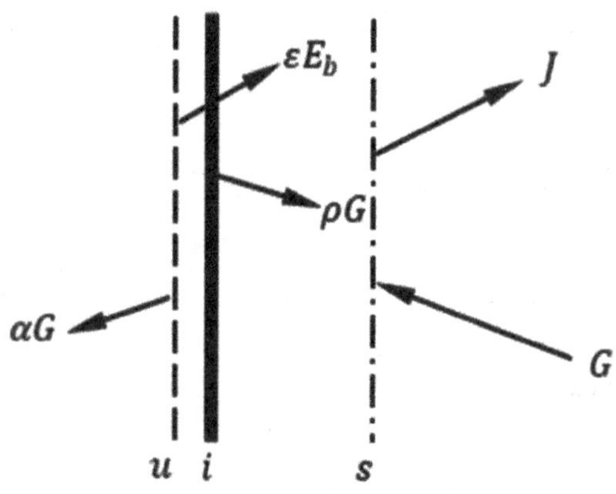

Figure 11.1 Schematic diagram of radiant heat flux for a grey surface.

$$q = \varepsilon E_b - \alpha G \tag{11.2}$$

Eliminating G and solving for q and recalling that $\alpha = \varepsilon$, we obtain the net heat flux emitted by the plate

$$q = \frac{\varepsilon}{1-\varepsilon}\left(E_b - J\right) \quad W/m^2 \tag{11.3}$$

If the surface area of the plate is A, the emitted heat rate is

$$\dot{Q} = \frac{\varepsilon A}{1-\varepsilon}\left(E_b - J\right) \quad W \tag{11.4}$$

$$\dot{Q} = \frac{\left(E_b - J\right)}{\dfrac{1-\varepsilon}{\varepsilon A}} = \frac{\left(\sigma T^4 - J\right)}{\dfrac{1-\varepsilon}{\varepsilon A}} \quad W \tag{11.5}$$

A similar result can be obtained using the definition of J,

$$J = \varepsilon E_b + \rho G$$

where $\rho = 1 - \alpha = 1 - \varepsilon$.

Using electrical analogy, Eq. (11.5) can be represented by the radiation resistance circuit in Figure 11.2, in which $\sigma T^4 - J$ is the radiation driving potential, and $1-\varepsilon/\varepsilon A$ is the "surface resistance."

Consider surface i and j having radiosities J_i and J_j, as shown in Figure 11.3.

The net radiation exchange rate between the two surfaces is

$$\dot{Q}_{i-j} = J_i A_i F_{i-j} - J_j$$

Since $A_i F_{i-j} = A_j F_{j-i}$

$$\dot{Q}_{i-j} = A_i F_{i-j}\left(J_i - J_j\right) = A_j F_{j-i}\left(J_i - J_j\right)$$

$$\dot{Q}_{i-j} = \frac{J_i - J_j}{\dfrac{1}{A_i F_{i-j}}} = \frac{J_i - J_j}{\dfrac{1}{A_j F_{j-i}}} \tag{11.6}$$

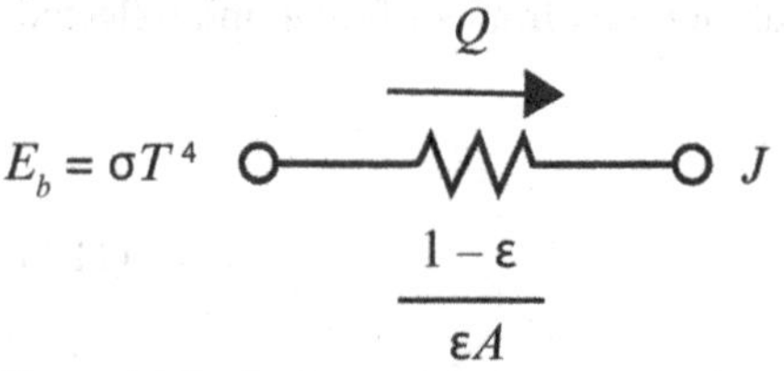

$\dot{Q}$

$E_b = \sigma T^4$ — J

$\dfrac{1-\varepsilon}{\varepsilon A}$

Figure 11.2 Surface resistance representing net energy radiation from a plate.

The electrical circuit for Eq. (11.6) is shown in Figure 11.4. The radiation driving potential in this case is $J_i - J_j$ and the resistance is $1/A_iF_{i-j}$ or $1/A_jF_{j-i}$, which is known as "space resistance" or "shape factor resistance."

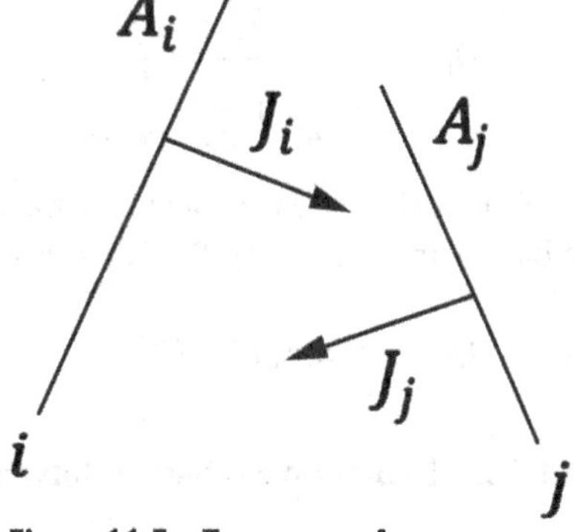

Figure 11.3 Two grey surfaces radiating to each other.

11.2 Thermal Radiation Networks

Radiation networks are useful tools for solving radiation heat transfer problems and can be built from combinations of surface and space resistances for two or more grey surfaces:

1) A surface resistance is assigned to each surface by connecting one end to a node representing the emissive power $E_b = \sigma T^4$ and the other to a node representing radiosity J
2) The radiosity node is then connected to each of the radiosity nodes of other surfaces it can see by space resistances.

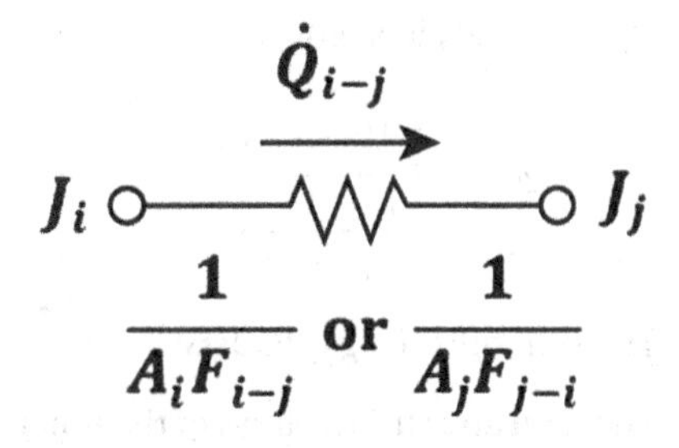

Figure 11.4 Space resistance representing radiation exchange between two grey surfaces.

The following are some examples of constructing radiation networks.

11.2.1 Grey Object in Grey Enclosure

The two surfaces see each other and nothing else, as shown in Figure 11.5.

The radiation network for this arrangement is shown in Figure 11.6.

The radiative heat rate exchange is

$$\dot{Q}_{i-j} = \frac{J_i - J_j}{\dfrac{1}{A_iF_{i-j}}} = \frac{J_i - J_j}{\dfrac{1}{A_jF_{j-i}}}$$

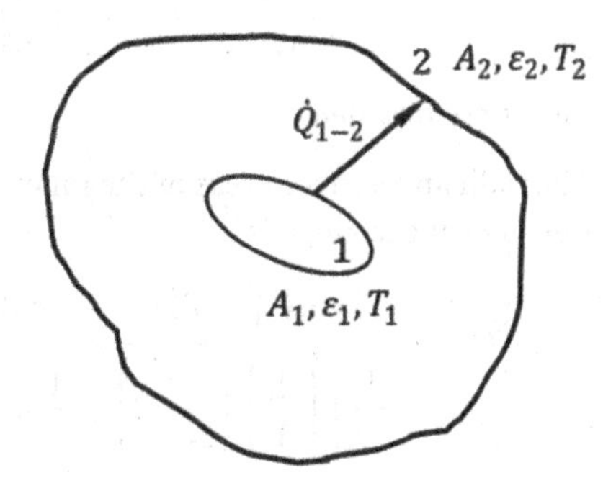

Figure 11.5 Grey object in grey enclosure.

From the summation rule

$$F_{1-1} + F_{1-2} = 1$$

If the inner object 1 is convex, $F_{1-1} = 0$; hence, $F_{1-2} = 1$ and the heat rate can be rewritten as

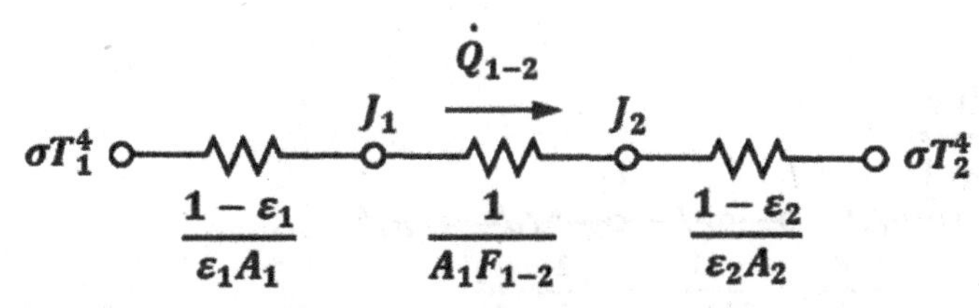

Figure 11.6 Radiation network for grey object in grey enclosure.

$$\dot{Q}_{1-2} = \frac{\sigma\left(T_1^4 - T_2^4\right)}{\frac{1-\varepsilon_1}{\varepsilon_1 A_1} + \frac{1}{A_1} + \frac{1-\varepsilon_2}{\varepsilon_2 A_2}} \tag{11.7}$$

If the surface area A_2 of enclosure 2 is much larger than the surface area A_1 of object 1, or A_2 is a black surface $(\varepsilon_2 = 1)$, Eq. (11.7) is reduced to

$$\dot{Q}_{1-2} = \varepsilon_1 A_1 \sigma\left(T_1^4 - T_2^4\right) \tag{11.8}$$

11.2.2 Radiation Exchange Between Two Grey Surfaces

Two very large parallel planes:

The radiative heat rate between the two surfaces is the same as in Eq. (11.7); however, since $A_1 = A_2 = A$, it is reduced to

$$\dot{Q}_{1-2} = \frac{\sigma A\left(T_1^4 - T_2^4\right)}{\frac{1}{\varepsilon_1} + \frac{1}{\varepsilon_2} - 1} \tag{11.9a}$$

Two concentric long cylinders:

The radii and surface areas of the inner and outer cylinders are, respectively, r_1, r_2, A_1, and A_2. The rate of heat exchange is

$$\dot{Q}_{1-2} = \frac{\sigma A_1\left(T_1^4 - T_2^4\right)}{\frac{1}{\varepsilon_1} + \frac{A_1}{A_2}\left(\frac{1}{\varepsilon_2} - 1\right)} = \frac{\sigma A_1\left(T_1^4 - T_2^4\right)}{\frac{1}{\varepsilon_1} + \frac{r_1}{r_2}\left(\frac{1}{\varepsilon_2} - 1\right)} \tag{11.9b}$$

Two concentric spheres:

The radii and surface areas of the inner and outer spheres are, respectively, r_1, r_2, A_1, and A_2. The rate of heat exchange is

$$\dot{Q}_{1-2} = \frac{\sigma A_1\left(T_1^4 - T_2^4\right)}{\frac{1}{\varepsilon_1} + \left(\frac{A_1}{A_2}\right)^2\left(\frac{1}{\varepsilon_2} - 1\right)} = \frac{\sigma A_1\left(T_1^4 - T_2^4\right)}{\frac{1}{\varepsilon_1} + \left(\frac{r_1}{r_2}\right)^2\left(\frac{1}{\varepsilon_2} - 1\right)} \tag{11.9c}$$

11.2.3 Three Infinitely Long Parallel Planes

The arrangement is shown in Figure 11.7, with plate 3 representing a radiation shield with high reflectivity (low emissivity). The plates are of equal size $A_1 = A_2 = A_3 = A$; hence, $F_{1-3} = F_{3-2} = 1$.

If the emissivities of the two sides of the shield are the same, the radiation network and the heat rate are respectively

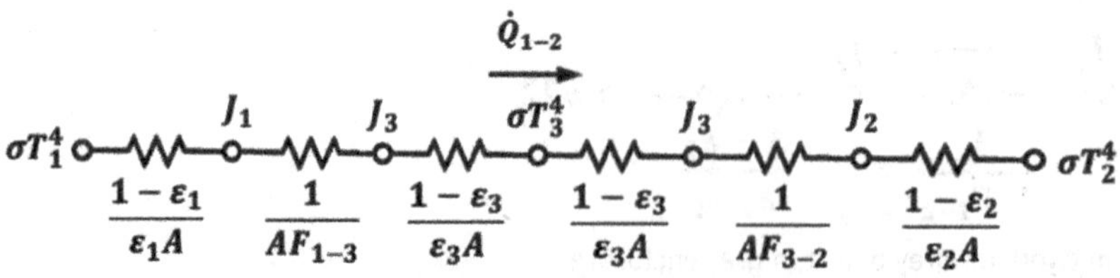

$\dot{Q}_{1-2}$

σT_1^4 — $\frac{1-\varepsilon_1}{\varepsilon_1 A}$ — J_1 — $\frac{1}{AF_{1-3}}$ — J_3 — $\frac{1-\varepsilon_3}{\varepsilon_3 A}$ — σT_3^4 — $\frac{1-\varepsilon_3}{\varepsilon_3 A}$ — J_3 — $\frac{1}{AF_{3-2}}$ — J_2 — $\frac{1-\varepsilon_2}{\varepsilon_2 A}$ — σT_2^4

$$\dot{Q}_{1-2} = \frac{\sigma A\left(T_1^4 - T_2^4\right)}{\frac{1}{\varepsilon_1} + \frac{2}{\varepsilon_3} + \frac{1}{\varepsilon_2} - 2} \tag{11.10}$$

If, on the other hand, the emissivities of the two sides of the shield are different, the radiation network and the heat rate are respectively

$\dot{Q}_{1-2}$ →

σT_1^4 — $\frac{1-\varepsilon_1}{\varepsilon_1 A}$ — J_1 — $\frac{1}{AF_{1-3}}$ — J_3 — $\frac{1-\varepsilon_3}{\varepsilon_{3,1} A}$ — σT_3^4 — $\frac{1-\varepsilon_3}{\varepsilon_{3,2} A}$ — J_3 — $\frac{1}{AF_{3-2}}$ — J_2 — $\frac{1-\varepsilon_2}{\varepsilon_2 A}$ — σT_2^4

$$\dot{Q}_{1-2} = \frac{\sigma A\left(T_1^4 - T_2^4\right)}{\frac{1}{\varepsilon_1} + \frac{1-\varepsilon_{3,1}}{\varepsilon_{3,1}} + \frac{1-\varepsilon_{3,2}}{\varepsilon_{3,2}} + \frac{1}{\varepsilon_2}} \tag{11.11a}$$

where $\varepsilon_{3,1}$ is the emissivity of the shield facing plane 1 and $\varepsilon_{3,2}$ is that facing plane 2.

If there are N shields having the same emissivity ε_s, the radiative heat rate is

$$\dot{Q}_{1-2} = \frac{A\sigma\left(T_1^4 - T_2^4\right)}{\frac{1}{\varepsilon_1} + \frac{1}{\varepsilon_2} - 1 + N\left(\frac{2}{\varepsilon_s} - 1\right)} \tag{11.11b}$$

If the wall and shield emissivities are the same and equal to ε

$$\dot{Q}_{1-2} = \frac{A\sigma\left(T_1^4 - T_2^4\right)}{(N+1)\left(\frac{2}{\varepsilon} - 1\right)} = \frac{\dot{Q}_{1-2(no\ shield)}}{N+1} \tag{11.11c}$$

Example 11.1 Two very large parallel walls at temperatures 40°C and 20°C with emissivities of 0.4 and 0.7, respectively, are exchanging radiation. Determine:

a) the net radiation exchange rate between the two walls
b) the percentage reduction in heat transfer if an aluminium radiation shield with uniform emissivity $\varepsilon_3 = 0.035$ is inserted between the walls, as shown in Figure 11.7
c) the effect on heat transfer of different emissivities of the two sides of the shield $\varepsilon_{3,1} = 0.035$ and $\varepsilon_{3,2} = 0.07$.

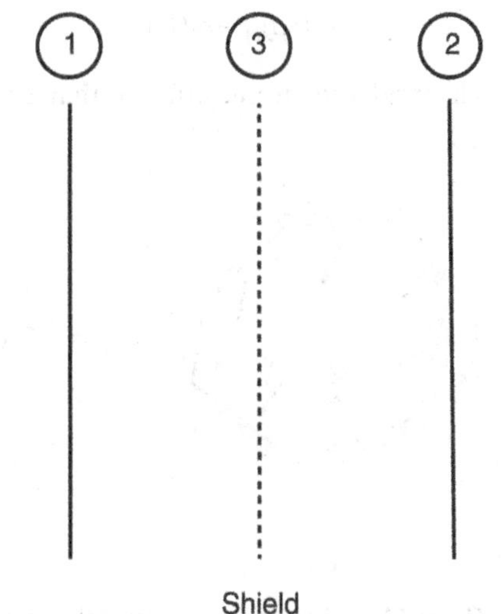

Figure 11.7 Two parallel plates with radiation shield in-between.

Solution

a) Without radiation shield

From Eq. (11.7) for $A_1 = A_2 = A$ and $F_{1-2} = 1$.

$$\dot{Q}_{1-2} = \frac{A\sigma\left(T_1^4 - T_2^4\right)}{\frac{1}{\varepsilon_1} + \frac{1}{\varepsilon_2} - 1}$$

$$\frac{\dot{Q}_{1-2}}{A}=\frac{5.67\times10^{-8}\left(313^4-293^4\right)}{\frac{1}{0.4}+\frac{1}{0.7}+1}=\frac{126.3}{4.93}=25.6\,W/m^2$$

b) With radiation shield of uniform emissivity
From Eq. (11.10) and Figure 11.7

$$\frac{\dot{Q}_{1-2}}{A}=\frac{\sigma\left(T_1^4-T_2^4\right)}{\frac{1}{\varepsilon_1}+\frac{2}{\varepsilon_3}+\frac{1}{\varepsilon_2}-2}=\frac{5.67\times10^{-8}\left(313^4-293^4\right)}{\frac{1}{0.4}+\frac{2}{0.035}+\frac{1}{0.7}-2}=\frac{126.3}{59}=2.14\,W/m^2$$

Reduction in heat transfer is 91.6%.

c) With radiation shield of non-uniform emissivity
From Eq. (11.11)

$$\dot{Q}_{1-2}=\frac{\sigma A\left(T_1^4-T_2^4\right)}{\frac{1}{\varepsilon_1}+\frac{1-\varepsilon_{3,1}}{\varepsilon_{3,1}}+\frac{1-\varepsilon_{3,2}}{\varepsilon_{3,2}}+\frac{1}{\varepsilon_2}}$$

$$\frac{\dot{Q}_{1-2}}{A}=\frac{5.67\times10^{-8}\left(313^4-293^4\right)}{\frac{1}{0.4}+\frac{1-0.035}{0.035}+\frac{1-0.07}{0.07}+\frac{1}{0.7}}=\frac{126.3}{44.78}=2.82\,W/m^2$$

Reduction in heat transfer is 89.0%. This is an example of shield surface facing wall 2 being less polished, causing smaller reduction in heat transfer rate than in (b).

11.2.4 Radiation Exchange Between Several Grey Surfaces

Consider the enclosure with n grey surfaces and the corresponding radiation network (Figures 11.8a and 11.8b). From Eq. (11.2), the net radiation energy exchange between surfaces i and k can be written as

$$\dot{Q}_i=A_i\left(\varepsilon_i E_{bi}-\alpha_i G_k\right) \tag{11.12}$$

The irradiation from surface k that is intercepted by surface i can be expressed as

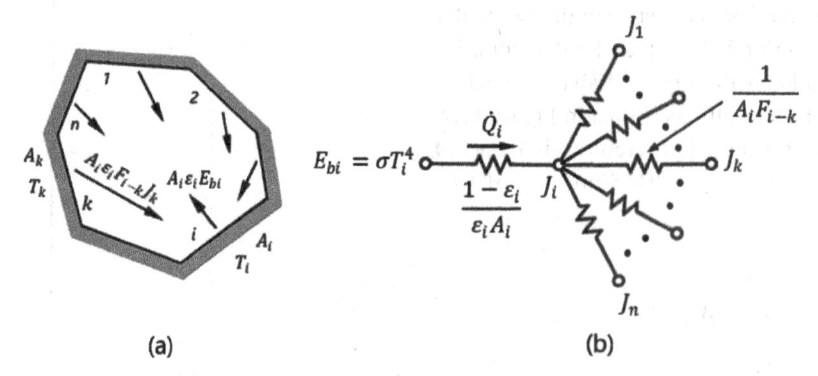

Figure 11.8 Enclosure with several grey surfaces (a) and the corresponding radiation network (b).

$$A_iG_k = A_kF_{k-i}J_k$$

and since $\alpha_i = \varepsilon_i$ and $A_kF_{k-i} = A_iF_{i-k}$, $G_k = F_{i-k}J_k$
hence, Eq. (11.12) becomes

$$\dot{Q}_i = \varepsilon_i A_i \left(E_{bi} - F_{i-k}J_k\right) \tag{11.13}$$

Considering the total irradiation received by surface i from the other k surfaces $(k = 1, 2, 3, \ldots, n)$, the energy balance from Eq. (11.13) can be written as

$$\dot{Q}_i = \varepsilon_i A_i \left(E_{bi} - \sum_{k=1}^{n} F_{i-k}J_k\right) \tag{11.14}$$

If a surface is blackbody surface $(\varepsilon_i = 1)$, the heat rate becomes

$$\dot{Q}_i = A_i \left(E_{bi} - \sum_{k=1}^{n} F_{i-k}J_k\right) \tag{11.15}$$

Equating Eqs (11.1) and (11.2), we can write the radiosity for surface i as follows:

$$J_i = \varepsilon_i E_{bi} + (1 - \varepsilon_i)G_k$$

From Figure 11.8b, the heat rate can be written as

$$\dot{Q}_i = \frac{E_{bi} - J_i}{\dfrac{1 - \varepsilon_i}{\varepsilon_i A_i}} \tag{11.16}$$

Equating Eq. (11.14) and Eq. (11.16) and rearranging we obtain

$$J_i = \varepsilon_i E_{bi} + (1 - \varepsilon_i)\sum_{k=1}^{n} J_k F_{i-k}$$

or

$$J_i - (1 - \varepsilon_i)\sum_{k=1}^{n} J_k F_{i-k} = \varepsilon_i E_{bi} \tag{11.17}$$

where $i = 1, 2, 3, \ldots, n$ $k = 1, 2, 3, \ldots, n$

Referring to the radiation network in Figure 11.8b, the heat rate can also be written in terms of the radiosities and space resistances as

$$\dot{Q}_i = \left(\frac{E_{bi} - J_i}{\dfrac{1 - \varepsilon_i}{\varepsilon_i A_i}}\right) = \sum_{k=1}^{n} \left(\frac{J_i - J_k}{\dfrac{1}{F_{i-k}A_i}}\right) \tag{11.18}$$

This equation states that the radiation heat rate to i through its surface resistance is equal to the net heat rate from surface i to all other surfaces in the enclosure through corresponding space resistances.

If the temperatures of all surfaces are known, then

$$E_{bi} = \sigma T_i^4$$

Equation (11.17) can be used for any number of surfaces in an enclosure by writing a system of linear simultaneous equations that can be solved for n unknown radiosities J_i. The radiation heat rates can then be determined from Eq. (11.4), which can be rewritten for each unknown radiosity node as

$$\dot{Q}_i = \frac{\varepsilon_i A_i}{1-\varepsilon_i}\left(E_{bi} - J_i\right) \ W \tag{11.19}$$

If a surface is insulated, there will be no heat transfer $\left(\dot{Q}_i = 0\right)$ and

$$E_{bi} = J_i \tag{11.20}$$

The general form of a set of linear simultaneous equations for n grey surfaces in terms of the radiosities is given below

$$\left.\begin{array}{l} a_{11}J_1 + a_{12}J_2 + \ldots + a_{1k}J_k + \ldots + a_{1n}J_n = C_1 \\ a_{21}J_1 + a_{22}J_2 + \ldots + a_{2k}J_k + \ldots + a_{2n}J_n = C_2 \\ \cdot \\ \cdot \\ \cdot \\ a_{k1}J_1 + a_{k2}J_2 + \ldots + a_{kk}J_i + \ldots + a_{1kn}J_n = C_k \\ \cdot \\ \cdot \\ \cdot \\ a_{n1}J_1 + a_{n2}J_2 + \ldots + a_{nk}J_k + \ldots + a_{nn}J_n = C_n \end{array}\right\} \tag{11.21}$$

The equations can be solved by the Gauss–Siedel method or matrix inversion method discussed in Chapter 4.

11.2.5 Enclosure With Four Long Grey Surfaces That See Each Other

This enclosure (a long furnace) and the radiation network are shown in Figure 11.9. Surface 1 is at temperature T_1, surface 2 at temperature T_2, surface 3 at temperature T_3, and surface 4 at temperature

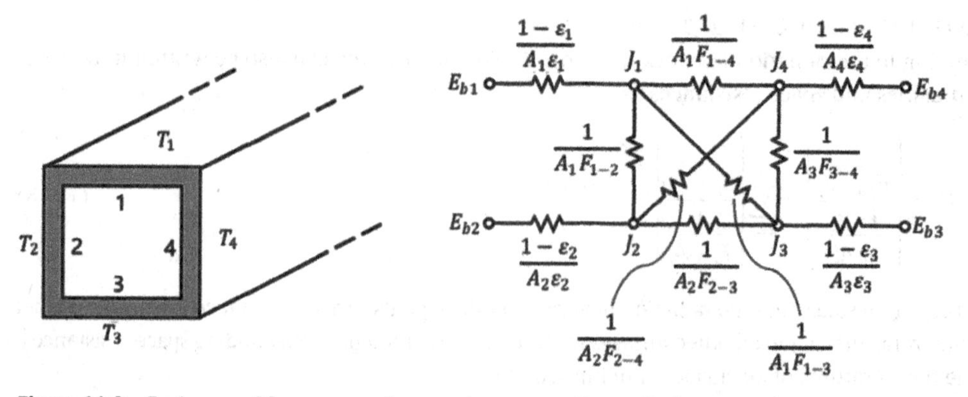

Figure 11.9 Enclosure of four grey surfaces and corresponding radiation network.

T_4. The system of linear simultaneous equations with four unknown radiosities J_1, J_2, J_3, and J_3 can be written based on Eq. (11.17) as follows:

$$\left.\begin{aligned} J_1-(1-\varepsilon_1)(j_1F_{1-1}+j_2F_{1-2}+j_3F_{1-3}+j_4F_{1-4})=\varepsilon_1E_{b1}\\ J_2-(1-\varepsilon_2)(j_1F_{2-1}+j_2F_{2-2}+j_3F_{2-3}+j_4F_{2-4})=\varepsilon_2E_{b2}\\ J_3-(1-\varepsilon_3)(j_1F_{3-1}+j_2F_{3-2}+j_3F_{3-3}+j_4F_{3-4})=\varepsilon_3E_{b3}\\ J_4-(1-\varepsilon_4)(j_1F_{4-1}+j_2F_{4-2}+j_3F_{4-3}+j_4F_{4-4})=\varepsilon_4E_{b3} \end{aligned}\right] \quad (11.22)$$

11.2.6 Enclosure With Three Long Grey Surfaces That See Each Other

This enclosure and the radiation network are shown in Figure 11.10. Surface 1 is at temperature T_1, surface 2 at temperature T_2, and surface 3 at temperature T_3. To solve this problem, we use Eq. (11.17) to set up a system of linear simultaneous equations with three unknown radiosities J_1, J_2, and J_3

$$\begin{aligned} &J_1-(1-\varepsilon_1)(j_1F_{1-1}+j_2F_{1-2}+j_3F_{1-3})=\varepsilon_1E_{b1}\\ &J_2-(1-\varepsilon_2)(j_1F_{2-1}+j_2F_{2-2}+j_3F_{2-3})=\varepsilon_2E_{b2}\\ &J_3-(1-\varepsilon_3)(j_1F_{3-1}+j_2F_{3-2}+j_3F_{3-3})=\varepsilon_3E_{b3} \end{aligned} \quad (11.23)$$

For known surface temperatures $E_{bi}=\sigma T_i^4$, the radiation heat rate can be determined from Eq. (11.19).

Example 11.2 The following data is provided for the enclosure in Figure 11.10. The triangle has equal sides of 1.0 m, $T_1=1400°C$, $T_2=320°C$, $T_3=1120°C$, and all three surfaces have an emissivity of 0.3. Determine the radiant heat transfer to surface 2.

Solution

$$E_{b1}=\sigma T_1^4=5.67\times10^{-8}\times1673^4=444.2kW/m^2$$

$$E_{b2}=\sigma T_2^4=5.67\times10^{-8}\times593^4=7.0kW/m^2$$

$$E_{b3}=\sigma T_3^4=5.67\times10^{-8}\times1393^4=213.5kW/m^2$$

For unit length of enclosure

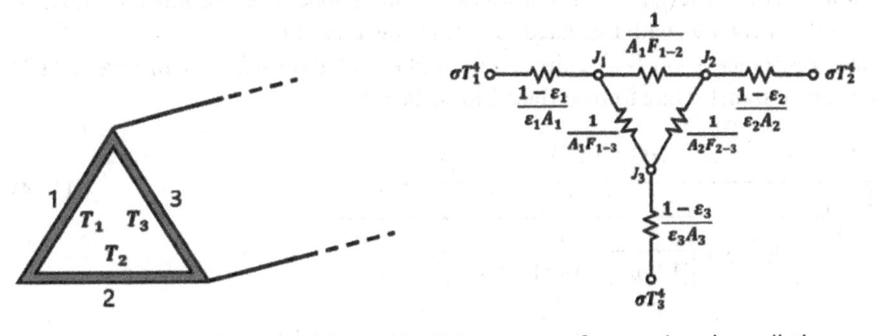

Figure 11.10 Long three-sided duct with all three grey surfaces exchanging radiation.

$$F_{1-2} = \frac{A_1 + A_2 - A_3}{2A_1} = \frac{1+1-1}{2} = 0.5$$

$$F_{2-1} = 0.5$$

$$F_{1-3} = 1 - F_{1-2} = 0.5,\ F_{3-1} = 0.5$$

Substituting in the three simultaneous radiosity Eqs (11.23) and rearranging yields

$$J_1 - 0.35J_2 - 0.35J_3 = 133.26$$

$$-0.35J_1 + J_2 - 0.35J_3 = 2.1$$

$$-0.35J_1 - 0.35J_2 + J_3 = 64.05$$

Solving,

$J_1 = 271.03 W/m^2$ $J_2 = 173.88 W/m^2$ $J_3 = 219.77 W/m^2$
Using Eq. (11.19),

$$\frac{\dot{Q}_2}{A_2} = \frac{\varepsilon_2}{1-\varepsilon_2}(E_{b2} - J_2) = \frac{0.3}{1-0.3}(7 - 173.88) = -71.52 kW/m^2$$

$$\frac{\dot{Q}_1}{A_1} = \frac{\varepsilon_1}{1-\varepsilon_1}(E_{b1} - J_1) = \frac{0.3}{1-0.3}(444.2 - 271.03) = 74.21 kW/m^2$$

$$\frac{\dot{Q}_3}{A_3} = \frac{\varepsilon_3}{1-\varepsilon_3}(E_{b3} - J_3) = \frac{0.3}{1-0.3}(213.5 - 219.77) = -2.687 kW/m^2$$

$$\sum_{i=1}^{3} \frac{\dot{Q}_i}{A_i} = 74.21 - 71.52 - 2.687 = 0.003 \approx 0$$

11.2.7 Three Surfaces With One of Them Insulated

Consider the furnace in Figure 11.11, which can be regarded as a three-sided enclosure if the two side surfaces are taken as a single surface 2. If surface A_3 is perfectly insulated, it will absorb and reradiate all the incident energy to the other two surfaces and will have zero heat flow rate, $\dot{Q}_3 = 0$ (adiabatic surface). The surface resistance attached to node 3 is not equal to zero and node J_3, which does not draw any current ($\dot{Q}_3 = 0$), is known as a floating node. The resistance $1-\varepsilon_3/\varepsilon_3 A_3$ will not be displayed in networks with insulated surfaces henceforward.

Ignoring surface resistance $1-\varepsilon_3/\varepsilon_3 A_3$, the series-parallel radiation network in Figure 11.11 yields the radiative heat transfer rate from surface 1 to surface 2

$$\dot{Q}_{1-2} = \frac{\sigma\left(T_1^4 - T_2^4\right)}{\dfrac{1-\varepsilon_1}{\varepsilon_1 A_1} + \dfrac{1}{A_1F_{1-2} + \dfrac{1}{\left[(1/A_1F_{1-3}) + (1/A_2F_{2-3})\right]}} + \dfrac{1-\varepsilon_2}{\varepsilon_2 A_2}} \tag{11.24}$$

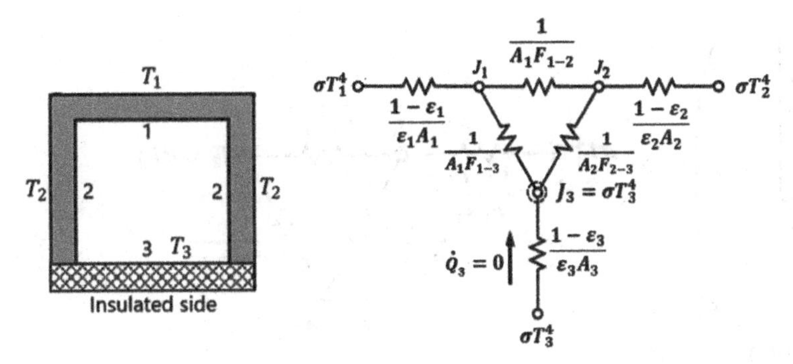

Figure 11.11 Three surfaces exchanging radiation with one insulated surface.

11.2.8 Two Parallel Flat Plates of Equal Finite Size in Very Large Room

Two equal and parallel flat plates (1 and 2) exchange radiation with each other and with a very large room (3), as shown in Figure 11.12.

Since A_3 is very large, the surface resistance attached to node 3 in the radiation network becomes almost negligible and behaves like a black body with $\varepsilon_3 = 1$

$1 - \varepsilon_3 / \varepsilon_3 A_3 \approx 0$, and $J_3 = \sigma T_3^4$.

The radiative heat transfer rate from surface 1 to surface 2 can be worked out from the radiation series-parallel network in Figure 11.12.

$$\dot{Q}_{1-2} = \frac{\sigma A_1\left(T_1^4 - T_2^4\right)}{\dfrac{A_1 + A_2 - 2A_1F_{1-2}}{A_2 - A_1\left(F_{1-2}\right)^2} + \left(\dfrac{1}{\varepsilon_1} - 1\right) + \dfrac{A_1}{A_2}\left(\dfrac{1}{\varepsilon_2} - 1\right)} \tag{11.25}$$

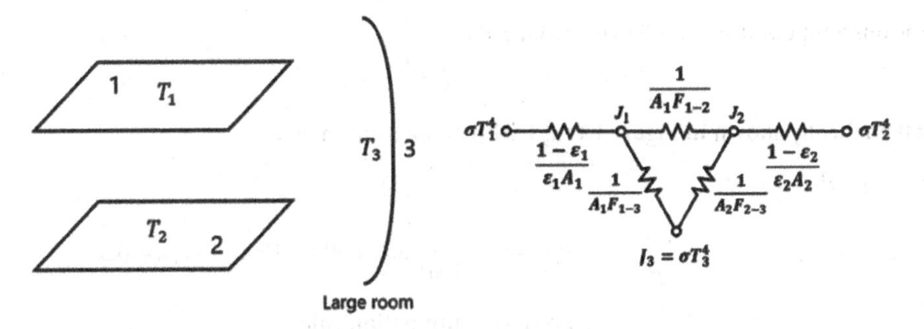

Figure 11.12 Two parallel plates in a large room.

11.2.9 Two Surfaces With One of Them Insulated in Large Room

This arrangement in the form of two perpendicular plates 1 and 2, with common edge in a very large room representing surface 3, is shown in Figure 11.13.

Node 2: Surface resistance is $1 - \varepsilon_2 / \varepsilon_2 A_2$, $J_2 = \sigma T_2^4$, and $\dot{Q}_2 = 0$
Node 3: A_3 is very large, surface resistance $1 - \varepsilon_3 / \varepsilon_3 A_3 \approx 0$, and $J_3 = \sigma T_3^4$.

The radiative heat transfer rate from surface 1 to surface 2 can be worked out from the radiation series-parallel network in Figure 11.13.

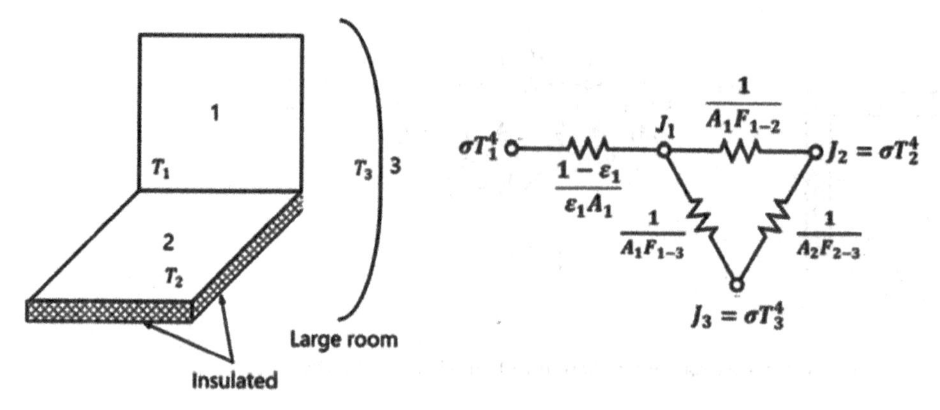

Figure 11.13 Two perpendicular plates with common edge in a very large room.

$$\dot{Q}_{1-2} = \frac{\sigma\left(T_1^4 - T_2^4\right)}{\frac{1-\varepsilon_1}{\varepsilon_1 A_1} + \frac{1}{A_1F_{1-2} + \frac{1}{\left[\left(1/A_1F_{1-3}\right)+\left(1/A_2F_{2-3}\right)\right]}}} \tag{11.26}$$

Example 11.3 A very long drying duct, 400 mm wide and 300 mm deep, has a heating element embedded in the top surface that dissipates 1000 W in the form of thermal radiation per metre length of duct. The bottom surface of the duct is well insulated and the two sides are conducting and maintained at $T_3 = 350\,K$. If the emissivity of the top surface is 0.9 and that of the sides is 0.6, determine:

a) the shape factors F_{1-2}, F_{1-3}, F_{3-1}, and F_{2-3}
b) the radiation energy leaving the top surface that falls directly on the bottom insulated surface $\dot{Q}_{1-2}$
c) the equilibrium temperature of the lower surface T_2.

Solution

The duct is schematically shown in Figure E11.3.1 with all measurements.

a) From the string rule

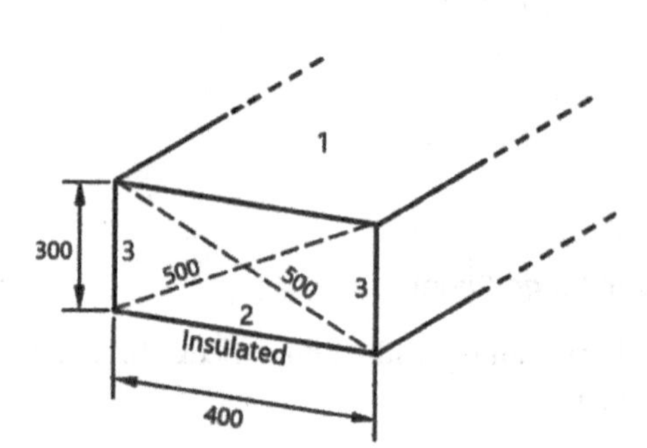

Figure E11.3.1 Cross section of the drying duct.

$$F_{1-2} = \frac{1}{2\times 400}\left[(500+500)-(300+300)\right] = 0.5$$

From the summation rule

$F_{1-2} + F_{1-3} = 1$; hence, $F_{1-3} = 0.5$

$$A_1F_{1-3} = A_2F_{3-1}$$

$$F_{3-1} = \frac{A_1}{A_3}F_{1-3} = \frac{0.5\times 400}{2\times 300} = 0.333$$

Due to symmetry $F_{2-3} = F_{1-3} = 0.5$

b) The radiation series-parallel network and its equivalent series network are shown in Figure E11.3.2.
The equivalent space resistance between J_1 and J_3 is

$$\frac{1}{R_{eq}}=\frac{1}{\frac{1}{A_1F_{1-3}}}+\frac{1}{\frac{1}{A_1F_{1-2}}+\frac{1}{A_2F_{2-3}}}$$

For unit length of the duct

$$\frac{1}{R_{eq}}=0.4\times0.5+\frac{1}{\frac{1}{0.4\times0.5}+\frac{1}{0.4\times0.5}}=0.3$$

and the equivalent resistance $R_{eq}=3.333$
Since $\dot{Q}_1=\dot{Q}_3$

$$\dot{Q}_1=\frac{\sigma\left(T_1^4-T_3^4\right)}{\frac{1-\varepsilon_1}{\varepsilon_1A_1}+R_{eq}+\frac{1-\varepsilon_3}{\varepsilon_3A_3}}$$

$$1000=\frac{5.67\times10^{-8}\left(T_1^4-350^4\right)}{\frac{1-0.9}{0.9\times0.4}+3.333+\frac{1-0.6}{0.6\times0.6}}$$

Hence, $T_1=560K$
From Eq. (11.5), for node J_1

$$\dot{Q}_1=\frac{\left(\sigma T_1^4-J_1\right)}{\frac{1-\varepsilon_1}{\varepsilon_1A_1}}$$

from which

$$J_1=\sigma T_1^4-\dot{Q}_1\left(\frac{1-\varepsilon_1}{\varepsilon_1A_1}\right)=5.67\times10^{-8}\times560^4-\frac{1000\times(1-0.9)}{0.9\times0.4}=5298W/m^2$$

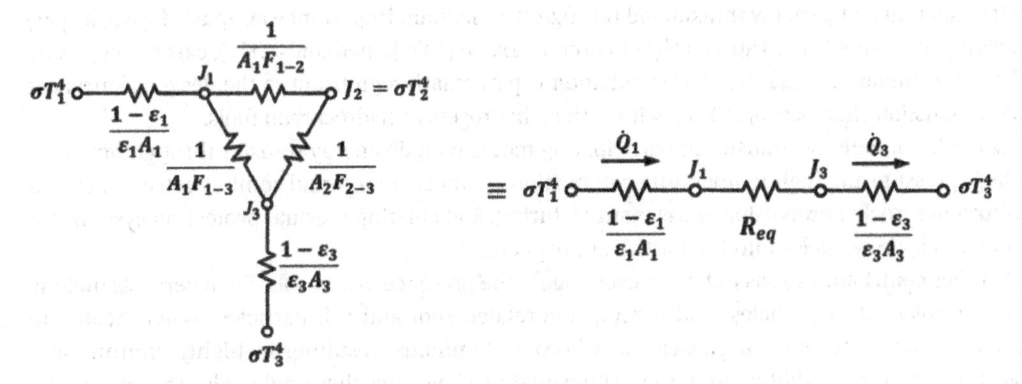

Figure E11.3.2 Radiation network representations with one insulated side.

Similarly, for node J_3

$$\dot{Q}_3 = \dot{Q}_1 = \frac{-\left(\sigma T_3^4 - J_3\right)}{\dfrac{1-\varepsilon_3}{\varepsilon_3 A_3}}$$

from which

$$J_3 = \sigma T_3^4 + \dot{Q}_1\left(\frac{1-\varepsilon_3}{\varepsilon_3 A_3}\right) = 5.67\times10^{-8}\times350^4 + \frac{1000\times(1-0.6)}{0.6\times0.6} = 1962 W/m^2$$

From Eq. (11.6)

$$\dot{Q}_{1-3} = \frac{J_1 - J_3}{\dfrac{1}{A_1 F_{1-3}}} = \frac{5298-1962}{\dfrac{1}{0.5\times0.4}} = 667 W/m$$

$\dot{Q}_1$ is the sum of $\dot{Q}_{1-2}$ and $\dot{Q}_{1-3}$

$\therefore \dot{Q}_{1-2} = \dot{Q}_1 - \dot{Q}_{1-3} = 1000 - 667 = 333\ W/m$, and

$$J_2 = J_1 - \dot{Q}_{1-2}\left(\frac{1}{A_1 F_{1-2}}\right) = 5298 - \frac{333}{0.5\times0.4} = 3633 W/m^2$$

Since $\dot{Q}_2 = 0$, $J_2 = \sigma T_2^4$ and

$$T_2 = \left(\frac{J_2}{\sigma}\right)^{1/4} = \left(\frac{3633}{5.67\times10^{-8}}\right)^{1/4} = 503\ K$$

11.3 Radiation Exchange With Participating Medium

In this chapter, we have so far considered radiation exchange between surfaces separated by vacuum, which is a perfectly transparent or nonparticipating medium. Air at normal temperatures and pressures is considered to be a transparent medium. In practice, spaces between surfaces exchanging radiation are more likely than not to be filled with substances, particularly gaseous substances, that are neither opaque nor transparent to thermal radiation, as the incident radiation is partially absorbed by the medium and partially transmitted through the medium. Important examples of gaseous participating media are water vapour (H_2O), carbon dioxide (CO_2), methane (CH_4), carbon monoxide (CO), and sulphur dioxide (SO_2). Gas radiation is particularly significant to the design of furnaces and combustion chambers operating with carbon, hydrogen, or hydrocarbon fuels.

Examples of radiative transfer in participating media include energy transfer through hot gases in high-pressure and high-temperature combustion chambers in internal combustion engines and gas turbines, rocket propulsion, glass manufacturing, and ablating thermal protection systems for re-entry of space vehicles into the Earth's atmosphere.

Another contributor to gas radiation exchange is the presence of aerosols. Such aerosols include liquid droplets, dust particles, and combustion-related soot and ash particles. When heated to incandescence, soot and ash particles can become luminous, resulting in highly emitting and absorbing centres. In addition to the absorbing quality of aerosols, they tend to also enhance radiation scatter.

11.3.1 Absorption of Radiation

Consider a layer of gas subjected to irradiation of intensity I_0, as shown in Figure 11.14. On passing through an absorbent gas layer, the incident radiation intensity is attenuated (weakened) due to absorption. If the intensity is I at a position x in the medium, then at distance $x+dx$ from the origin $(x=0)$ it will be $I+dI$. Since the amount absorbed is proportional to the thickness of the gas layer and to the incident intensity, we can write

$$dI=-KIdx \tag{11.27}$$

where the constant of proportionality K is called the coefficient of absorption.

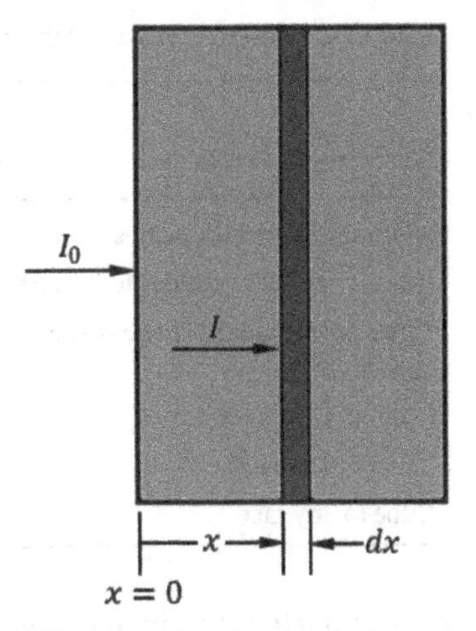

Figure 11.14 Absorption of radiation in a gas layer.

The cumulative intensity at position x can be found by separating the variables, rearranging, and integrating

$$\int_{I_0}^{I_x}\frac{dI}{I}=-\int_0^x Kdx$$

$$[\ln I]_{I_0}^{I_x}=\ln I_x-\ln I_0=\ln\left(\frac{I_x}{I_0}\right)=-Kx$$

$$\frac{I_x}{I_0}=\tau=e^{-Kx} \tag{11.28}$$

where τ is the transmissivity (portion of radiation that is transmitted through a gas layer). If the gas does not reflect radiation, the portion of radiation absorbed by the gas will be $\alpha=1-\tau=1-e^{-Kx}$ and since $\alpha=\varepsilon$, we finally get

$$\varepsilon=1-e^{-Kx} \tag{11.29}$$

11.3.2 Gaseous Emission

A gas which absorbs radiation will also emit radiation and the amount absorbed or emitted by a layer of gas depends on layer thickness, the radiation wavelength band, gas pressure, and gas temperature. The amount of radiation absorbed by a gas is particularly dependent on the path (or beam) length traversed by radiation. From Eq. (11.29), it is apparent that when x reaches infinity, the fraction absorbed tends to unity, which means that a thick gas layer behaves like a black body.

Table 11.1 shows selected computed values of the mean beam length of some gas-bounding geometries.

More data on effective path length for various geometries can be found in specialized references such as Brewster (1992), Siegel and Howell (1992), and Modest (2013). For geometries for which exact L values have not been calculated, the following approximation can be used (Hottel and Sarofim, 1967):

$$L\cong 0.9\left(\frac{4V_g}{A_g}\right) \tag{11.30}$$

where V_g and A_g are the volume and surface area of the gas mass, respectively.

Table 11.1 Selected values of the mean beam length of some gas-bounding geometries.

Geometry of gas volume	Characteristic dimension	Average mean beam length
Infinite slab to both planes	Spacing between slabs δ	1.76δ
Infinite circular cylinder to its surface	Diameter D	$0.95D$
Circular cylinder to entire surface	Diameter D	$0.6D$
Length / Diameter = 1		$0.80D$
Length / Diameter = 2		
Sphere to its surface	Diameter D	$0.65D$
Cube to any face	Side L	$0.6L$

Computing the effective emissivity for a finite body of gas is very complex due to the many possible paths that radiation will take. However, data for total emissivity of a limited number of gases are available in the literature for engineering applications. The data most frequently used in heat transfer references are the Hottel charts published as a chapter in a book edited by McAdams (1954). These charts are for the average emissivities for water vapour and carbon dioxide in mixtures with non-radiating gases (i.e., hydrogen, oxygen, or nitrogen) at 1 atm total pressure. The charts are presented as $\varepsilon = f(T, pL)$, where p is the partial pressure of the gas and L is the mean beam length (L here should be distinguished from the geometrical symbol of length).

The mean beam length L, according to Hottel (McAdams, 1954) is an approximation in which radiation exchange between an actual gas volume and its surface is replaced by radiation exchange between hemispherical gas volume of radius L and a black element dA located at the base of the hemisphere, as depicted in Figure 11.15. The gas is at temperature T_g and the elemental surface is at temperature T_s.

Hottel charts are also supplemented by graphs to account for:

- correction of emissivities at total pressures other than $P_1 = 1$ atm
- correction of emissivity of mixtures of water vapour and carbon dioxide at selected temperatures.

Work by other researchers (Leckner 1972; Docherty 1982; Farag 1982), based on later spectral data, reported significant deviations from the Hottel data at low and high temperatures. More recent work by Alberti et al. (2015, 2016, 2018, 2020) lead to updated charts on the basis of comprehensive and more accurate spectroscopic data for carbon dioxide, water vapour, and carbon monoxide molecules. Abridged versions of the Alberti charts for CO_2, H_2O, and CO, in mixtures with non-radiating gases such as nitrogen, at a total equivalent pressure of 1 bar, are shown in Figures 11.16–11.18. The charts, extended to higher pressures and temperatures, are standard emissivity ε^0 plotted as a function of the gas temperature at constant values of the product of the partial pressure and mean beam length $p_g.L$. More detailed and enlarged charts can be found in Appendices F, G, and H (can be accessed online at "www.wiley.com/go/Ghojel/heat transfer"). Note that the "effective pressures," defined by Eqs (11.32 to 11.34), are used in the charts in lieu of the total pressures used in the Hottel charts.

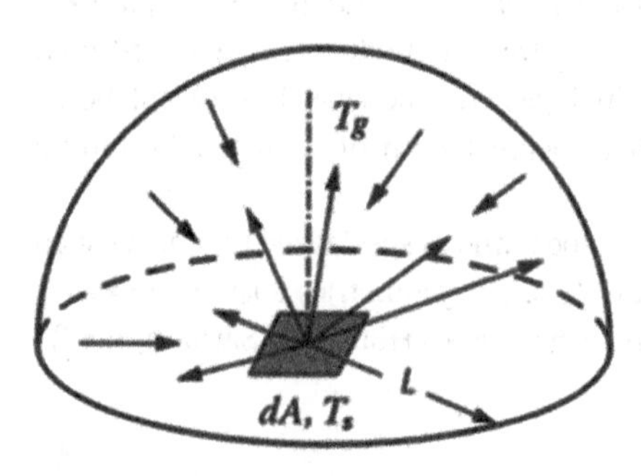

Figure 11.15 Radiation from hemispherical gas mass.

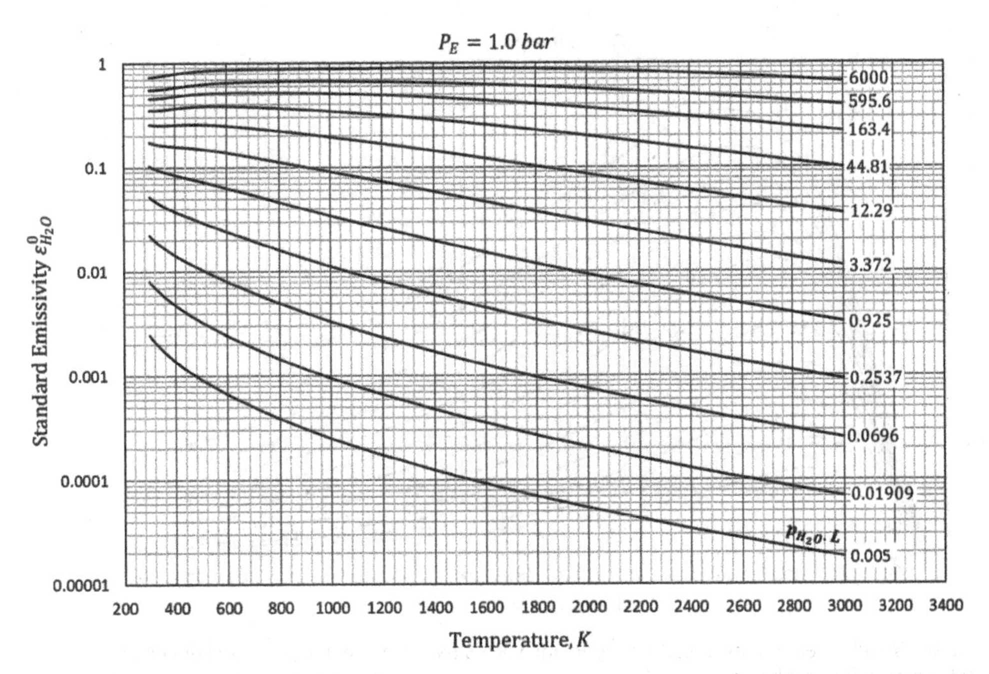

Figure 11.16 Standard emissivity of water vapour versus temperature and $p_{H_2O}.L$ (in *bar.cm*) at equivalent pressure $P_E = 1.0\,bar$.

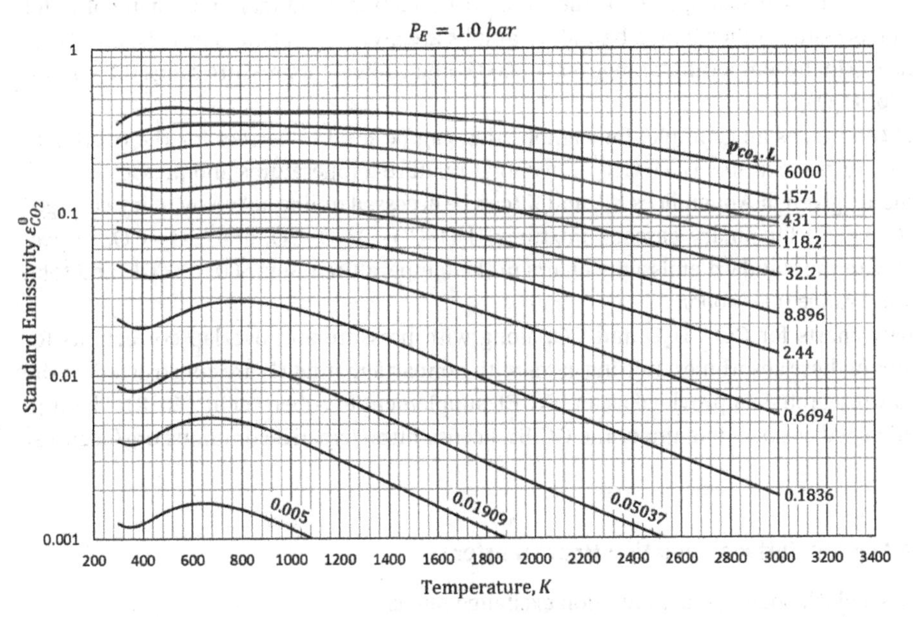

Figure 11.17 Standard emissivity of carbon dioxide versus temperature and $p_{CO_2}.L$ (in *bar.cm*) at equivalent pressure $P_E = 1.0\,bar$.

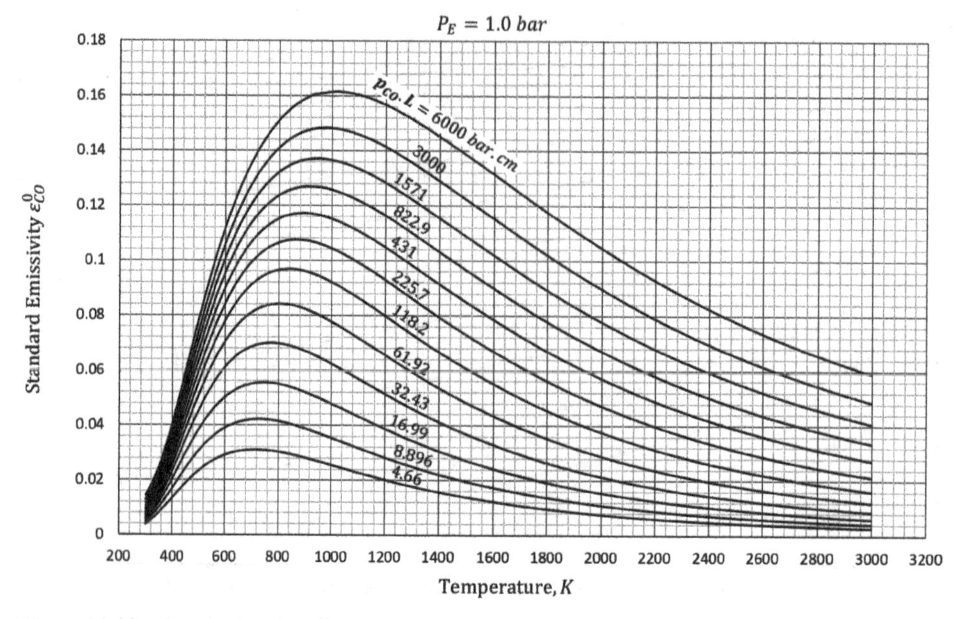

Figure 11.18 Standard emissivity of carbon monoxide versus temperature and $p_{CO}.L$ (in *bar.cm*) at equivalent pressure $P_E = 1.0\,bar$.

Figure 11.19 shows, as an example, two graphs for the correction of standard emissivity of water vapour for total pressures other than 1 bar. More graphs are given in Appendices F, G, and H for H_2O, CO_2, and CO in the temperature range $500-3000K$ and equivalent pressure range $0.1-50bar$ ($0.1-150bar$ for H_2O).

Figure 11.20 shows, as an example, four graphs that can be used to correct for the overlap of gases in the binary gas mixtures H_2O–CO_2 at two total-mixture temperatures and pressures.

More graphs are given in Appendices I, J, and K (can be accessed online at "www.wiley.com/go/ghojel/heat_transfer") for binary gaseous mixtures H_2O–CO_2, H_2O–CO, and CO_2–CO, respectively, for use with the standard emissivity charts in the temperature range $500-3000K$ and total mixture pressure range $0.1-40bar$.

All emissivity charts for CO_2, H_2O and CO, along with pressure and overlap corrections for binary mixtures of these absorbing/emitting gases with nonparticipating N_2 are constructed on the basis of the raw research data provided as an EXCEL supplement to Alberti et al. (2018). The supplement is published as open data under the CC BY license http://creativecommons.org/licenses/by/4.0.

11.3.3 Gas-Mass to Surface Radiation Heat Transfer

If the surface is a black body, the net radiation exchange rate is

$$\dot{Q}_{net} = \varepsilon_g A_s \sigma T_g^4 - \alpha_g A_s \sigma T_s^4 \tag{11.31}$$

where ε_g is the total emissivity of the gas mixture, and α_g is the total absorptivity of the same gas mixture at temperature T_g for radiation emitted from the blackbody surface A_s at T_s.

To determine the total emissivity ε_g and absorptivity α_g using the aforementioned charts, the equivalent pressure of each gas in the mixture is determined as follows (Alberti et al. 2018):

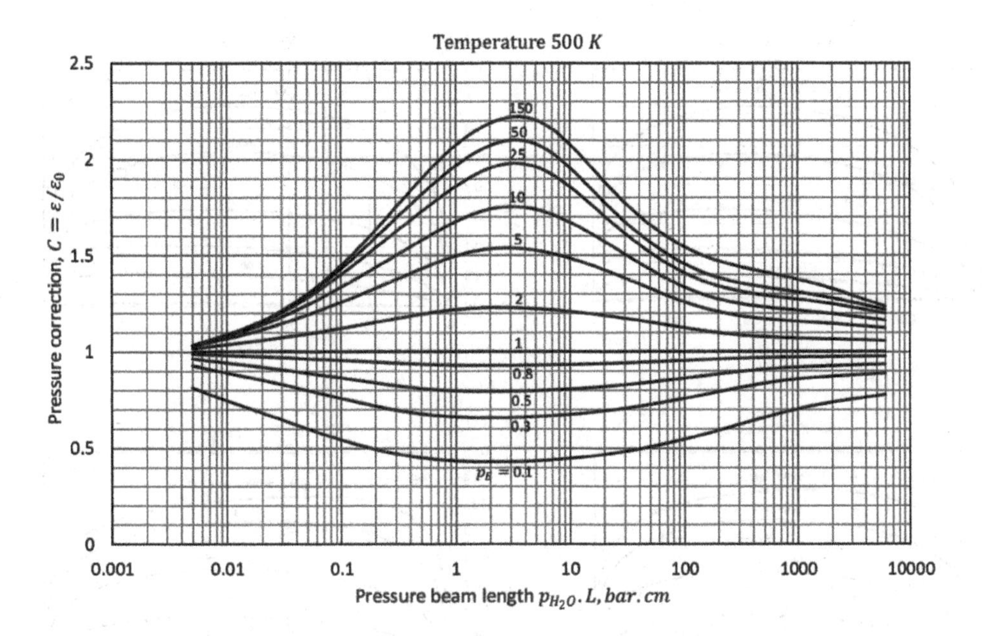

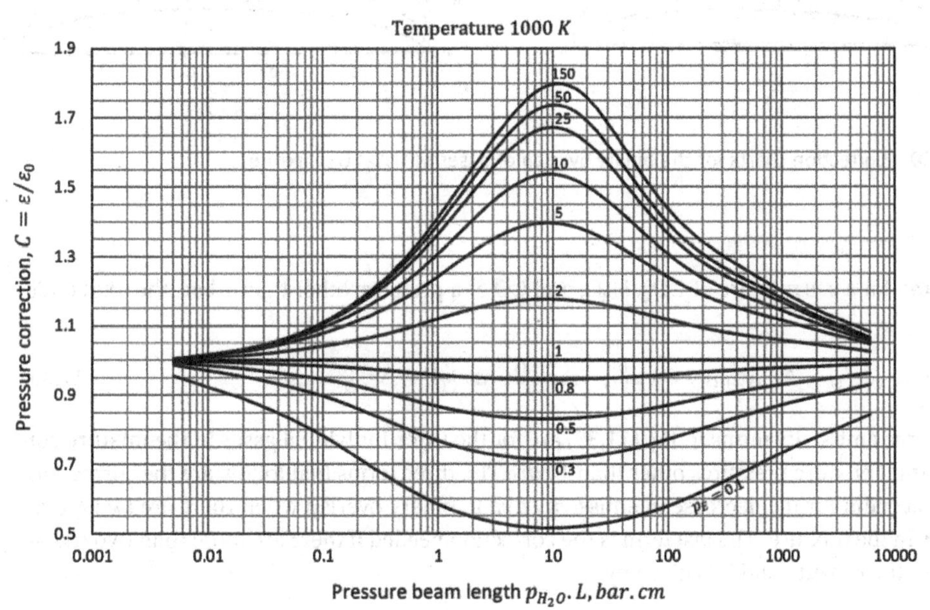

Figure 11.19 Pressure correction charts for emissivity of water vapour at two temperatures.

$$P_E^{CO_2} = P_t + 0.28p_{CO_2} = P_t(1 + 0.28x_{CO_2}) \tag{11.32}$$

$$P_E^{H_2O} = P_t + 5p_{H_2O} = P_t(1 + 5x_{H_2O}) \tag{11.33}$$

$$P_E^{CO} = P_t \tag{11.34}$$

where p_{CO_2}, p_{H_2O}, and p_{CO} are partial pressures and x_{CO_2} and x_{H_2O} are volume concentrations.

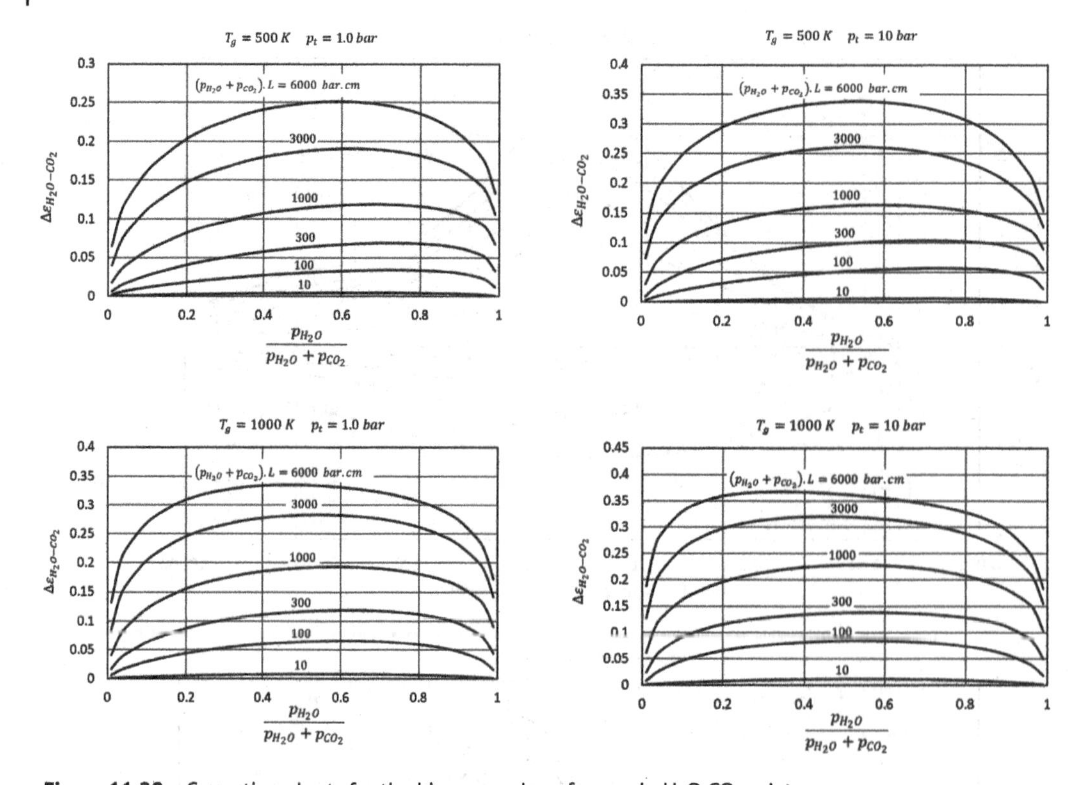

Figure 11.20 Correction charts for the binary overlap of gases in H_2O CO_2 mixtures.

In general, for a gas mixture of CO_2, H_2O, and CO at a pressure other than 1 bar, the total emissivity is

$$\varepsilon_g = C_{CO_2}\varepsilon^0_{CO_2} + C_{H_2O}\varepsilon^0_{H_2O} + C_{CO}\varepsilon^0_{CO} - \Delta\varepsilon^{H_2O}_{CO_2} - \Delta\varepsilon^{H_2O}_{CO} - \Delta\varepsilon^{CO_2}_{CO} + \Delta\varepsilon \tag{11.35}$$

The first three emissivity terms in Eq. (11.35) are for the three individual gases in the mixture corrected for total mixture pressures other than 1 *bar*. The three terms that follow are the corrections needed to account for the decrease in emissivity due to binary overlap when there are two radiating species in the mixture. The last term is the correction needed if there are more than two radiating gases in the mixture and it is given by

$$\Delta\varepsilon = max\left[\Delta\varepsilon^{H_2O}_{CO} + \Delta\varepsilon^{CO_2}_{CO} - \varepsilon^{CO} : \varepsilon^{H_2O}\varepsilon^{CO_2}\varepsilon^{CO}\right] \tag{11.36}$$

The absorptivity of a radiating component α_{gi} in a gaseous mixture can be evaluated from the emissivity as

$$\alpha^i_g\left(T_s, T_g, p_i.L, P_t\right) = C_{gi}.\varepsilon^0_i\left[T_s, p_i.L.\left(\frac{T_s}{T_g}\right)^s, P_t\right].\left(\frac{T_g}{T_s}\right)^n \tag{11.37}$$

In other words, the absorptivity of radiation absorbing gas at temperature T_g and beam length $p_i.L$ for black-body radiation emitted from a source at the temperature T_s (enclosure temperature) is equal to $(T_g/T_s)^n$ multiplied by the emissivity of the gas at temperature T_s for an effective beam length $p_l.L.(T_s/T_g)^s$.

In Eq. (11.37), p_l is the partial pressure of the gaseous component, L is the mean beam length, P_t is the total pressure of the mixture, and C_{gl} is the pressure correction factor for the standard emissivity. Hottel (1954) suggests a value of $s=1$, $n=0.45$ for water vapour, and $n=0.65$ for carbon dioxide. Penner (1959) suggests a value of $s=1$ and $n=0.5$ for both gases. Alberti et al. (2018) suggest a value of $s=0.5$ and a complex relationship for s of the form $s=f(p_l.L,.T_g/T_s, P_t)$ that varies for CO_2, H_2O, and CO. For engineering calculations, the value of s and n suggested by Hottel (1954) for CO_2 and H_2O are still widely used.

The total absorptivity of a three-gas mixture is then

$$\alpha_g = \alpha_g^{CO_2} + \alpha_g^{H_2O} + \alpha_g^{CO} - \Delta\alpha_{CO_2}^{H_2O} - \Delta\alpha_{CO}^{H_2O} - \Delta\alpha_{CO}^{CO_2} + \Delta\varepsilon$$

$$\alpha_g^{CO_2} = C_{CO_2}\varepsilon_{CO_2}^0\left[T_s, p_l.L.\left(\frac{T_s}{T_g}\right)^s, P_t\right].\left(\frac{T_g}{T_s}\right)^{0.5} \tag{11.38}$$

$$\alpha_g^{H_2O} = C_{H_2O}\varepsilon_{H_2O}^0\left[T_s, p_l.L.\left(\frac{T_s}{T_g}\right)^s, P_t\right].\left(\frac{T_g}{T_s}\right)^{0.5} \tag{11.39}$$

$$\alpha_g^{CO} = C_{CO}\varepsilon_{CO}^0\left[T_s, p_l.L.\left(\frac{T_s}{T_g}\right)^s, P_t\right].\left(\frac{T_g}{T_s}\right)^{0.5} \tag{11.40}$$

Since $\alpha_l = \varepsilon_l$,

$$\alpha_g = \alpha_g^{CO_2} + \alpha_g^{H_2O} + \alpha_g^{CO} - \Delta\varepsilon_{CO_2}^{H_2O} - \Delta\varepsilon_{CO}^{H_2O} - \Delta\varepsilon_{CO}^{CO_2} + \Delta\varepsilon \tag{11.41}$$

The procedure of determining the emissivity and absorptivity of a gas mixture consisting of CO_2, H_2O, CO, and N_2, using Alberti charts is as follows:

1) Calculate the partial pressures $p_{CO_2} = x_{CO_2}P_t$, $p_{H_2O} = x_{H_2O}P_t$, and $p_{CO} = x_{CO}P_t$
2) Calculate the pressure beam length for each component $p_{CO_2}.L$, $p_{H_2O}.L$, and $p_{CO}.L$
3) Calculate the equivalent pressures $p_E^{CO_2} = p_t(1+0.28x_{CO_2})$, $p_E^{H_2O} = p_t(1+5x_{H_2O})$, and $p_E^{CO} = p_t$
4) Calculate the mixing ratios MR:
   - H_2O in H_2O+CO_2: $MR = p_{H_2O}/(p_{H_2O}+p_{CO_2}) = x_{H_2O}/(x_{H_2O}+x_{CO_2})$
   - H_2O in H_2O+CO: $MR = p_{H_2O}/(p_{H_2O}+p_{CO}) = x_{H_2O}/(x_{H_2O}+x_{CO})$
   - CO in CO_2+CO: $MR = p_{CO}/(p_{CO_2}+p_{CO}) = x_{CO}/(x_{CO_2}+x_{CO})$.
5) Look up the individual standard emissivities $\varepsilon_{CO_2}^0$, $\varepsilon_{H_2O}^0$, and ε_{CO}^0 at the given temperature from the charts in Appendices F, G, and H (interpolation may be required)
6) Look up the pressure correction factors C_{CO_2}, C_{H_2O}, C_{CO} from the relevant graphs in Appendices F, G, and H (interpolation may be required)
7) Determine the corrected emissivities $\varepsilon_{CO_2} = C_{CO_2}\varepsilon_{CO_2}^0$, $\varepsilon_{H_2O} = C_{H_2O}\varepsilon_{H_2O}^0$, and $\varepsilon_{CO} = C_{CO}\varepsilon_{CO}^0$
8) Determine the binary overlap corrections $\Delta\varepsilon_{CO_2}^{H_2O}$, $\Delta\varepsilon_{CO}^{H_2O}$, and $\Delta\varepsilon_{CO}^{CO_2}$ from graphs in Appendices I, J, and K (three-way interpolation may be required: logarithmic interpolation between two $p_E.L$ values followed by logarithmic interpolation between two successive P_t values and, finally, linear interpolation between two successive temperatures)

9) Calculate the correction needed for the presence of more than two radiating gases in the mixture $\Delta\varepsilon$ using Eq. (11.36)
10) Calculate the overall emissivity of the gas mixture using Eq. (11.35)
11) The procedure from 1 to 9 is applied to determine the absorptivity of the gas mixture using Eqs (11.38) to (11.41).

The procedure above requires linear and/or and logarithmic interpolation between two values. The linear interpolation between two points, (X_1, Y_1) and (X_2, Y_2) for a given point X is calculated from

$$Y = Y_1 + \frac{Y_2 - Y_1}{X_2 - X_1}(X - X_1) \tag{11.42}$$

and the logarithmic interpolation from

$$Y = Y_1\left(\frac{X}{X_1}\right)^b \tag{11.43}$$

where

$$b = \frac{\log\left(\frac{Y_2}{Y_1}\right)}{\log\left(\frac{X_2}{X_1}\right)} \tag{11.44}$$

In Eqs (11.42) to (11.44), Y denotes emissivity ε and X denotes $p_{gas}.L$.

Logarithmic interpolation is required in the following cases:

a) Two $(p_{H_2O}.L)$ successive values for the given temperature
b) Two p_E successive values when performing emissivity pressure correction for each species
c) Two p_T successive values when performing binary overlap correction.

Example 11.4 Determine the emissivity ε_{H_2O} for water vapour at $p_t = 1.0\ bar$ and $T_g = 1600\ K$ and mean beam length $L = 0.42\ cm$ from the chart for water vapour in Figure F.1 in Appendix F.

Solution

$$p_{H_2O} = x_{H_2O} p_t = 1.0 \times 1.0 = 1.0\ bar$$

$$p_{H_2O}.L = 1.0 \times 0.42 = 0.42\ bar.cm$$

$$p_E = p_t\left(1 + 5x_{H_2O}\right) = 1\left(1 + 5 \times 1.0\right) = 6\ bar$$

For

$0.3506 < p_{H_2O}.L = 0.42\ bar.cm < 0.4845$ at $T_g = 1600\ K$ by logarithmic interpolation

$$\varepsilon_{H_2O} = C_{H_2O}\varepsilon^0_{H_2O} = 1.05 \times 0.007 = 0.00735$$

Example 11.5 The products of combustion of hydrogen contain 10% water vapour by volume and 90% nitrogen. The mixture is at a pressure of $P_t = 2\ bar$ and $T_g = 1000\ K$. Assuming a mean beam length of 50 cm, determine the total emissivity of the mixture.

Solution

$$p_{H_2O} = x_{H_2O} P_t = 0.1 \times 2 = 0.2\ bar$$

$$p_{H_2O}.L = 0.2 \times 50 = 10\ bar.cm$$

$$p_E = P_t\left(1 + 5x_{H_2O}\right) = 2\left(1 + 5 \times 0.1\right) = 3\ bar$$

$12.29 > \left(p_{H_2O}.L = 10\ bar.cm\right) > 8.896$; and at $T_g = 1000\ K$ from Figure F.1 by logarithmic interpolation

$$\varepsilon^0_{H_2O} \approx 0.139$$

The pressure correction factor can be determined for $P_E = 3$ at $T_g = 1000K$ at $p_{H_2O}.L = 10$ from Figure F.3 in Appendix F by interpolating for P_E between 2 bar and 5 bar: $C_{H_2O} = 1.3$; hence,

$$\varepsilon_{H_2O} = C_{H_2O}\varepsilon^0_{H_2O} = 0.139 \times 1.3 \approx 0.18$$

Example 11.6 The composition by volume of the combustion products in a cylindrical chamber is 8% carbon dioxide and 6% water vapour. The pressure and temperature of the products are 1 bar and 1200 K, respectively. If the mean path length is 73 cm, determine

a) The emissivity of the gas mixture
b) The heat radiated from the mass of combustion products.

Solution

(a)

$$p_{CO_2} = x_{CO_2} P_t = 0.08 \times 1 = 0.08\ bar$$

$$p_{CO_2}.L = 0.08 \times 73 = 5.84\ bar.cm$$

$$P_E^{CO_2} = P_t\left(1 + 0.28x_{CO_2}\right) = 1 \times \left(1 + 0.28 \times 0.08\right) = 1.0224\ bar$$

From the charts in Appendix G, $\varepsilon_{CO_2} = C_{CO_2}\varepsilon^0_{CO_2} = 0.086 \times 1.0 \cong 0.086$

$$p_{H_2O} = x_{H_2O} P_t = 0.06 \times 1 = 0.06\ bar$$

$$p_{H_2O}.L = 0.06 \times 73 = 4.38\ bar.cm$$

$$p_E = P_t\left(1 + 5x_{H_2O}\right) = 1 \times \left(1 + 5 \times 0.06\right) = 1.3\ bar$$

From the charts in Appendix F, $\varepsilon_{H_2O} = C_{H_2O}\varepsilon^0_{H_2O} = 0.072 \times 1.05 \cong 0.076$.

Total emissivity for the mixture accounting for the effect of overlap at 1200 K can be determined from Figures I.2 and I.3 in Appendix I and Eq. (11.35)

$$\varepsilon_g \approx (0.086 + 0.076) - 0.007 \cong 0.155$$

(b) The rate of radiant heat per unit area of the cylindrical gas mass is

$$\frac{\dot{Q}_g}{A_g} = \varepsilon_g E_b = 0.155 \times 5.67 \times 10^{-8} \times 1200^4 = 18224\ W/m^2 \left(18.224\ kW/m^2\right)$$

11.4 Combined Radiation and Convection

In practice, heat is very rarely transferred solely by conduction, convection, or radiation and most likely is transferred by combination of two or three modes simultaneously. A general case will involve conduction through a wall, convection over inner and outer surfaces, and radiation exchange between the walls and the surrounding environment(s). Convection and radiation take place simultaneously and in parallel, while conduction occurs in series prior to or following the convection/radiation process. More specific examples include:

- Flow of high-temperature combustion products in the combustion chambers of reciprocating internal combustion engines and gas turbines
- Hot steam transported through an insulated pipe indoors or outdoors
- Heat transfer through a wall with radiation shield in the wall cavity
- Heat transfer through a double-glazed window
- Industrial furnaces.

The total rate of heat transferred by convection and radiation is

$$\dot{Q}_t = \dot{Q}_c + \dot{Q}_r$$

$$\dot{Q}_t = h_c A (T_1 - T_2) + h_r A (T_1 - T_2)$$

$$\dot{Q}_t = A (h_c + h_r)(T_1 - T_2) \tag{11.45}$$

$\dot{Q}_r$ is the rate of radiation heat transfer from surface A at temperature T_1, having emissivity ε_1 and the surroundings at temperature 2 (assumed black body), and h_c and h_r are the average heat transfer coefficients of convection and radiation respectively.

$$\dot{Q}_r = A\varepsilon_1\sigma\left(T_1^4 - T_2^4\right) = h_r A (T_1 - T_2)$$

$$h_r = \frac{\varepsilon_1\sigma\left(T_1^4 - T_2^4\right)}{(T_1 - T_2)} = \frac{\varepsilon_1\sigma\left(T_1^2 - T_2^2\right)\left(T_1^2 + T_2^2\right)}{(T_1 - T_2)}$$

or

$$h_r = \varepsilon_1\sigma(T_1 + T_2)\left(T_1^2 + T_2^2\right) \tag{11.46}$$

For radiation heat exchange between two very large plates at temperatures T_1 and T_2, having emissivities ε_1 and ε_2 $(F_{1-2} = 1)$ from Eq. (11.9a)

$$\frac{\sigma\left(T_1^4-T_2^4\right)}{\frac{1}{\varepsilon_1}+\frac{1}{\varepsilon_2}-1}=h_r\left(T_1-T_2\right)$$

hence,

$$h_r=\frac{\sigma\left(T_1+T_2\right)\left(T_1^2+T_2^2\right)}{\frac{1}{\varepsilon_1}+\frac{1}{\varepsilon_2}-1} \tag{11.47}$$

For radiation heat exchange between two long concentric cylinders of surface areas A_1 and A_2, temperatures T_1 and T_2, and emissivities ε_1 and ε_2 $(F_{1-2}=1)$ from Eq. (11.9b)

$$\frac{\sigma\left(T_1^4-T_2^4\right)}{\frac{1}{\varepsilon_1}+\frac{A_1}{A_2}\left(\frac{1}{\varepsilon_2}-1\right)}=h_r\left(T_1-T_2\right)$$

hence,

$$h_r=\frac{\sigma\left(T_1+T_2\right)\left(T_1^2+T_2^2\right)}{\frac{1}{\varepsilon_1}+\frac{A_1}{A_2}\left(\frac{1}{\varepsilon_2}-1\right)} \tag{11.48}$$

Problems

11.1 The long duct shown in Figure P11.1 has a heating source embedded in surface A_4, which dissipates radiation energy at the rate of 1200 W/m. The lower surface A_3 is conducting and maintained at $350K$, and the two sides A_1 and A_2 are insulated and are combined together to form a single surface A_{1+2}. The emissivity of surface A_4 is 0.9 and that of surface A_3 is 0.6.
(a) Determine the shape factors F_{3-1}, F_{1-3}, $F_{4-(1+2)}$, $F_{3-(1+2)}$, and F_{4-3}
(b) Sketch the radiation network for the configuration.

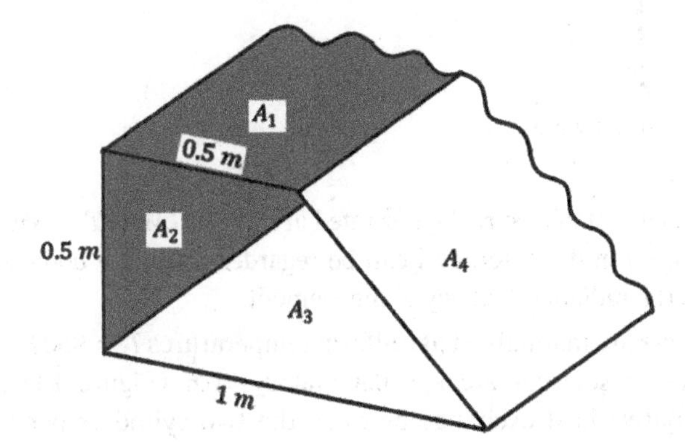

Figure P11.1 Long duct with a heating source embedded in surface A_4.

11.2 Use the shape factor data and the radiation network in Problem 11.1 to determine:

(a) the radiative energy leaving A_4 and directly intercepted by the combined surfaces A_{1+2}

(b) the equilibrium temperature $T_{(1+2)}$ of the combined surface surfaces $A_{(1+2)}$.

11.3 Two very large parallel walls at temperatures 50°C and 25°C with emissivities 0.42 and 0.65, respectively, are exchanging radiation. Determine the net radiation flux between the two walls.

11.4 Determine the percentage reduction in heat flux if an aluminium radiation shield with uniform emissivity $\varepsilon_3 = 0.035$ is inserted between the walls in Problem 11.3.

11.5 The radiation shield in Problem 11.4 is replaced with one having non-uniform emissivity: emissivity of the surface facing wall 1 is $\varepsilon_{3,1} = 0.03$, and emissivity of the surface facing wall 2 is $\varepsilon_{3,2} = 0.075$. What effect would this shield have on the radiation heat flux between the two walls?

11.6 Two large parallel walls at temperatures 300°C and 30°C with emissivities 0.45 and 0.85, respectively are exchanging radiation. Determine:

(a) the net radiation heat flux between the two walls

(b) the percentage reduction in heat transfer if three aluminium radiation shields each having uniform emissivity $\varepsilon_3 = 0.5$ are inserted between the walls.

11.7 What is the percentage reduction in heat transfer with the three radiation shields in place if the emissivities of the walls and the shields are equal to $\varepsilon_s = 0.5$ and the temperatures remain the same?

11.8 Using the data in Problem 11.7, calculate the steady-state temperatures of the three shields. Use the configuration in Figure P11.8.

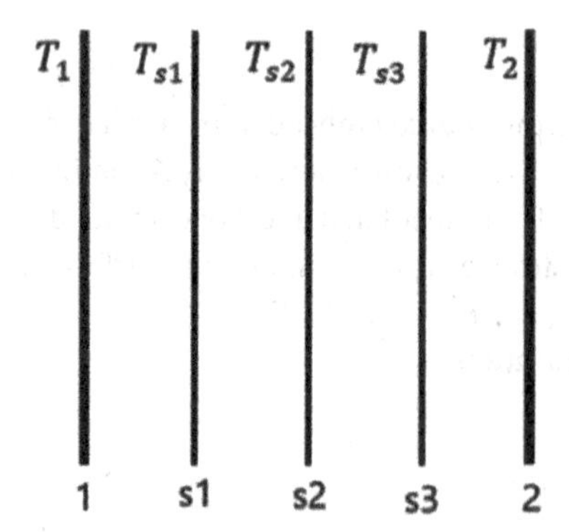

Figure P11.8 Three shields between two walls.

11.9 The temperature of the heating element of electric heater rated at $\dot{Q}_e = 1.0 kW$ is $T_s = 850°C$. The element is $300 mm$ long, $10 mm$ in diameter, and can be regarded as a grey body with emissivity of $\varepsilon = 0.92$. Calculate the radiant efficiency of the element.

11.10 Two very long concentric cylinders are maintained at uniform temperatures $T_1 = 950 K$ and $T_2 = 500 K$ and the emissivities of surfaces are $\varepsilon_1 = 0.8$ and $\varepsilon_2 = 0.6$ (Figure P11.10). Determine the steady-state radiative heat exchange between the two cylinders per unit length.

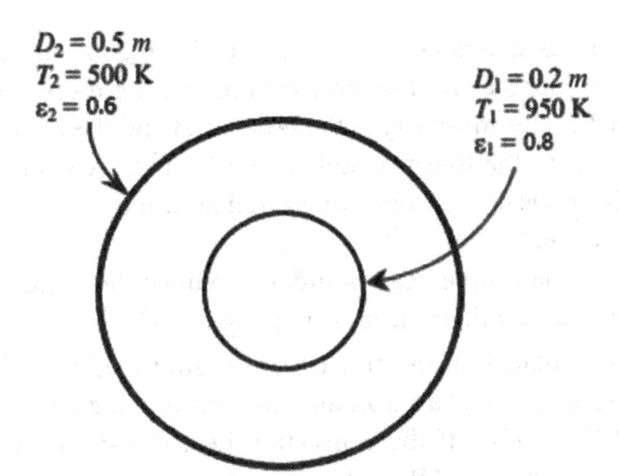

Figure P11.10 Two isothermal concentric cylinders.

11.11 Two concentric spheres are maintained at uniform temperatures $T_1 = 950K$ and $T_2 = 500K$ and the emissivities of surfaces are $\varepsilon_1 = 0.8$ and $\varepsilon_2 = 0.6$. Determine the steady-state radiative heat exchange between the two spheres.

11.12 Figure P11.12 shows two concentric cylinders located in a large room (surface A_3) at temperature $T_3 = 300K$. The cylinders have the dimensions $L = 200mm$, $r_1 = 50mm$, and $r_2 = 100mm$. Determine the following shape factors: F_{2-1}, F_{2-2}, F_{1-2}, and F_{2-3}.

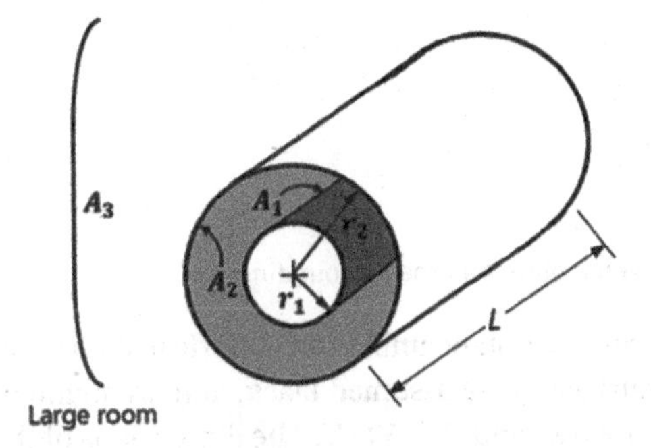

Figure P11.12 Schematic representation of two concentric cylinders in a large room.

11.13 For the configuration in Problem 11.12, the temperature and emissivity of the inner cylinders are $T_1 = 900\,K$ and $\varepsilon_1 = 0.8$, and the emissivity of the outer cylinder is $\varepsilon_2 = 0.2$. Sketch the radiation network for the configuration and determine:
(a) the temperature of the outer cylinder
(b) the total heat lost by the inner cylinder.

11.14 A thermocouple is inserted in heated gas flowing through a duct having a wall temperature of $t_w =$ 140°C (see Figure P11.14). Under steady-state flow conditions, the thermocouple indicates a temperature of $t_c =$ 280°C. If the emissivity of the bead is 0.4 and convection heat transfer coefficient from the gas to the thermocouple is $150\,W/m^2.K$, find the true gas temperature. Assume the duct wall to be a black body and that the gas is non-participating gas.

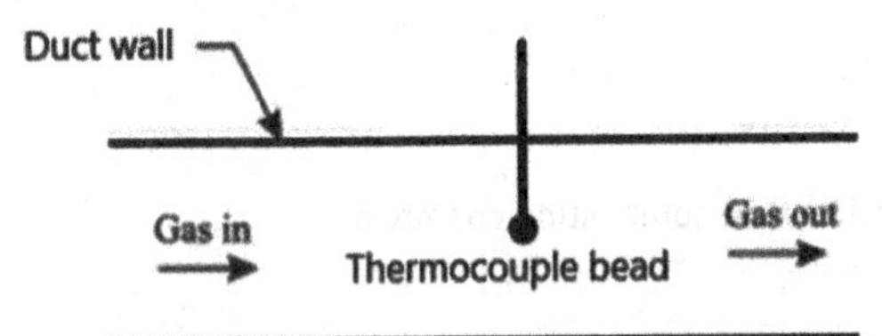

Figure P11.14 Thermocouple in heated gas flow.

11.15 A thermocouple is inserted in heated gas at temperature $T_g = 820°C$ flowing through a duct having a wall temperature of T_w. Under steady-state flow conditions, the thermocouple indicates a temperature of $T_c = 800°C$. The emissivity of the bead is 0.3 and the convection heat transfer coefficient from the gas to the thermocouple is $90 W/m^2.K$. Assuming the duct wall to be a black body and the gas is non-participating gas, determine:
(a) the temperature of the wall of the duct
(b) the required emissivity of the thermocouple bead in order to reduce the temperature difference between the gas temperature and thermocouple reading to 5°C.

11.16 Metallic plate 1 (Figure P11.16) is insulated on its left side and is subjected to heat flux $q_t = 1500 W/m^2$. The surface emissivity of the plate is 0.7 and the surroundings (designated as surface 2) at a temperature of $T_\infty = 30°C$. If the convection heat transfer coefficient between the plate and the surroundings is $h_c = 15 W/m^2.K$, what is the temperature of the plate wall? Assume steady-state conditions and ignore the effect of ambient air.

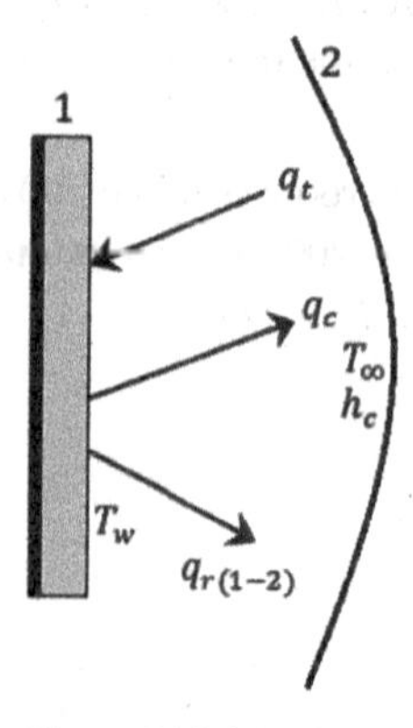

Figure P11.16 Radiation exchange between a plate and the surroundings.

11.17 Figure P11.17 represents a simplified combustion chamber of cylindrical shape. The base (surface 1), top (surface 2), and side (surface 3) are assumed black, and are maintained at uniform temperatures $T_1 = 700 K$, $T_1 = 1400 K$, and $T_1 = 500 K$. The dimensions of the combustion chamber are: $D = 4m$, $H = 2m$. Ignoring convection heat transfer and assuming steady-state conditions, determine the net rate of radiation heat transfer from surface 1.

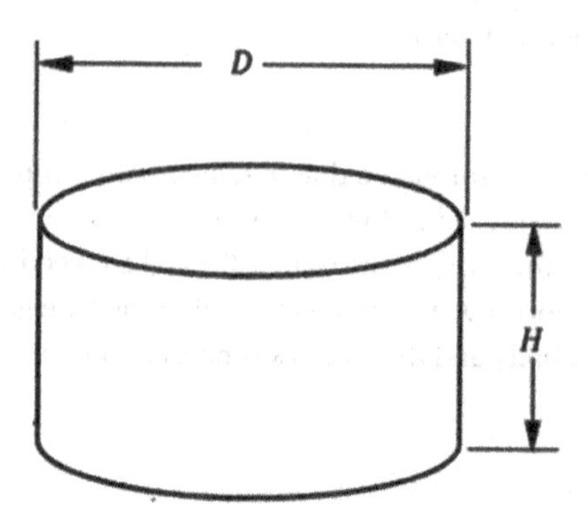

Figure P11.17 Schematic diagram of cylindrical combustion chamber.

11.18 Two parallel discs, 1 and 2, $D = 500\,mm$ in diameter, are spaced $H = 400\,mm$ apart with one disc located directly above the other disc, as shown in Figure P11.18. One disc is maintained at 500°C and the other at 227°C. The emissivities of the discs are 0.2 and 0.4, respectively. The curved cylindrical surface approximates a black body and is maintained at a temperature of 67°C. Determine the rate of heat loss by radiation from the inside surfaces of each disc

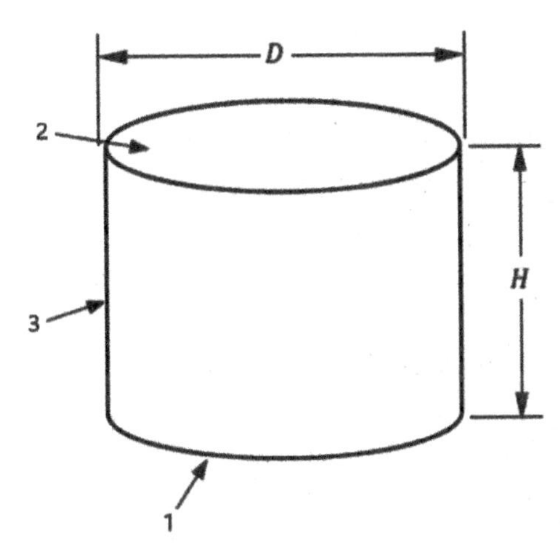

Figure P11.18 Radiation exchange between two coaxial parallel discs.

11.19 Use the emissivity charts to determine the standard emissivities at a temperature of $1000\,K$ and total effective pressure of 1 bar for the following gases:
- Water vapour
- Carbon dioxide
- Carbon monoxide.

Take the mean path length as $0.3\,cm$.

11.20 The products of combustion of hydrogen contain 15% water vapour by volume and 85% nitrogen. The mixture is at a pressure of $P_t = 2.86\,bar$ and $T_g = 2000\,K$. Assuming a mean beam length of 60 cm, determine the total emissivity of the mixture.

11.21 Calculate the heat radiated from a mass of combustion products (carbon dioxide, water vapour, and nitrogen) characterized by the following data:
- Gas temperature $T_g = 1000\,K$
- Total gas pressure $p_T = 1\,bar$
- Partial pressure of carbon dioxide in the products $p_{CO_2} = 0.1\,bar$
- Partial pressure of water in the products $p_{H_2O} = 0.08\,bar$
- Mean beam length $L = 396\,cm$.

11.22 The cylindrical furnace shown schematically in Figure P11.22 contains combustion gases at $1600\,K$ and a total pressure of $2\,bar$. The composition of the combustion gases by volume is as follows: $10\%H_2O$, $6\%CO_2$, $80\%N_2$, $4\%O_2$. The furnace wall temperature is $800\,K$. Determine:

(a) the emissivity of the combustion gases
(b) the absorptivity of the combustion gases
(c) the rate of radiation heat transfer from the combustion gases to the furnace walls.

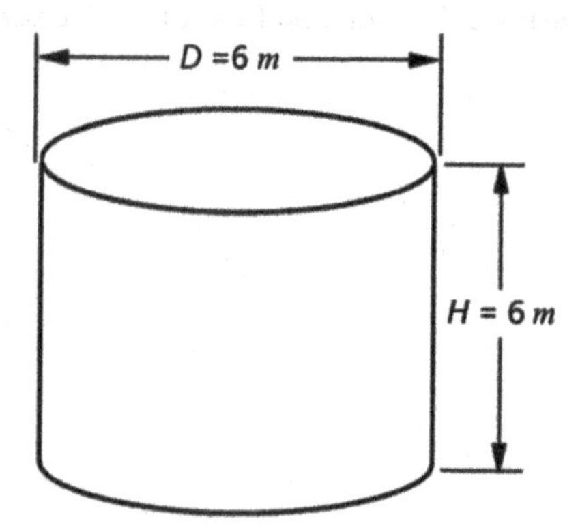

Figure P11.22 Schematic diagram of cylindrical furnace.

12

Heat Exchangers

Heat exchangers are devices that are used to transfer heat from one fluid to another: liquid to liquid, liquid to gas, gas to gas. The design and/or performance analysis of heat exchangers is based on the principles of conservation of energy and heat transfer. For the latter, mainly conductive and convective heat transfer principles are required and radiation effects are usually negligible.

Heat exchangers are widely used in engineering practice in the areas of refrigeration, heating and air conditioning, food processing, paper manufacturing, power generation, chemical processing, oil refining, and mobility.

Three basic types of heat exchangers can be distinguished:

1) The two fluids flow in different channels and are spatially separated by a common wall in such a way that good indirect thermal contact is facilitated. These exchangers are often known as recuperators
2) The fluids flow alternatively through the same channel with temporal (time) separation. These exchangers are known as regenerators. The heat exchange process is essentially absorption of heat from one fluid and subsequent release of heat to the other fluid
3) The fluids are allowed to mix in a common space, facilitating direct thermal contact.

Only the following types of recuperators will be considered in this chapter: double-pipe, shell-and-tube, and cross-flow heat exchangers. Thermal analysis of these exchanger types will be based on the following simplifying assumptions:

- The heat exchanger operates under steady-state conditions
- Heat losses to the surroundings are negligible
- There are no thermal energy sources and sinks in the exchanger walls or fluids
- In counter-flow and parallel-flow exchangers, the temperature of each fluid is uniform over all flow cross-sections
- Thermophysical properties of the fluids are constant throughout the exchanger
- The individual and overall heat transfer coefficients are independent of temperature, time, and location throughout the exchanger.

12.1 Overall Heat Transfer Coefficient

The simplest heat exchanger is the double-pipe, which is also known as the concentric-tube exchanger. The exchanger is comprised of two concentric cylindrical tubes with the fluids flowing in

Heat Transfer Basics: A Concise Approach to Problem Solving, First Edition. Jamil Ghojel.

Companion website: www.wiley.com/go/ghojel/heat_transfer

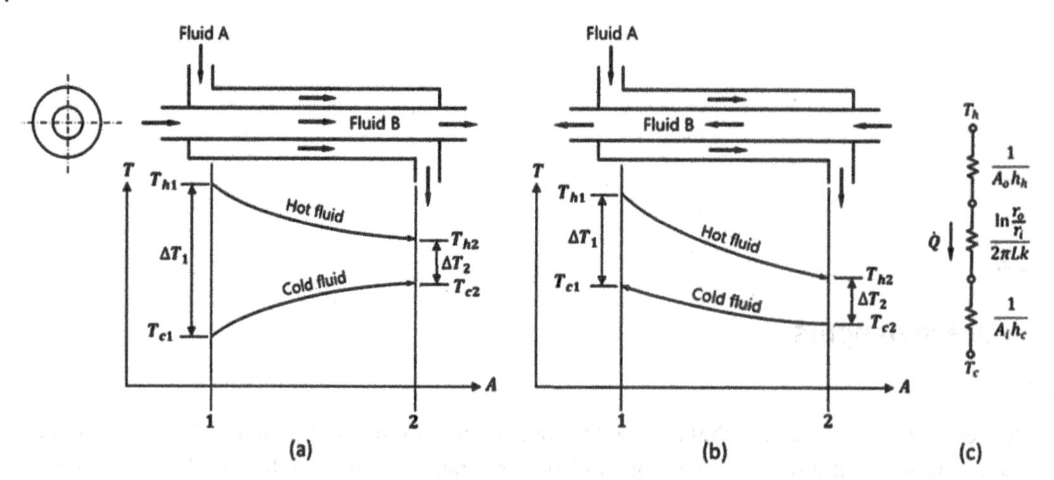

Figure 12.1 (a) Parallel-flow heat exchanger; (b) counter-flow heat exchanger; (c) thermal resistance network.

parallel in the inner and outer tubes. If the two flows are in the same direction, the exchanger is called a co-current or parallel-flow exchanger, as shown in Figure 12.1a. If the two flows are in opposite directions, the exchanger is called a counter-flow heat exchanger, as shown in Figure 12.1b.

In the analysis of the exchangers in Figure 12.1, fluid A is assumed to be the hot fluid and B the cold fluid. As heat is transferred from the hot fluid to the cold fluid, heat is lost by the former and its temperature decreases and heat is gained by the latter and its temperature increases steadily. In a parallel-flow heat exchanger (Figure 12.1a), the exit temperature T_{c2} of the cold fluid cannot exceed the exit temperature of the hot fluid T_{h2}, but it is theoretically possible for T_{c2} to become equal to T_{h2} if the exchanger is infinitely large. In the counter-flow heat exchanger depicted in Figure 12.1b, the exit temperature of the cold fluid T_{c1} can exceed the exit temperature of the hot fluid T_{h2} if the exchanger is sufficiently large. Therefore, if a double-pipe exchanger is to be used in an application, it is preferable to select a counter-flow heat exchanger.

The rate of heat transfer across surface A separating the two tubes can be written in terms of the overall heat transfer coefficient U as

$$\dot{Q} = UA\left(T_h - T_c\right) = \frac{\left(T_h - T_c\right)}{1/UA} \tag{12.1}$$

For length L of the heat transfer tube, the rate of heat transfer can be written in terms of conduction and convection thermal resistances in terms of the inner and outer radii of the inner tube

$$\dot{Q} = \frac{T_h - T_c}{\dfrac{1}{A_o h_o} + \dfrac{\ln \frac{r_o}{r_i}}{2\pi L k} + \dfrac{1}{A_i h_i}} \tag{12.2}$$

The network of the thermal resistance R_t for heat transfer through the cylindrical wall separating the hot and cold fluids is shown in Figure 12.1c.

Equating Eqs (12.1) and (12.2)

$$\frac{1}{UA} = R_t = \frac{1}{A_o h_o} + \frac{\ln\frac{r_o}{r_i}}{2\pi Lk} + \frac{1}{A_i h_i} \tag{12.3}$$

where r_i and r_o are the inner and outer radii of the inner tube, and h_i and h_o are the convection heat transfer coefficients at the inner and outer surfaces of the inner tube. Area A in Eq. (12.3) could be either A_o or A_i.

For the outer wall area A_o of the inner tube

$$\frac{1}{U_o} = \frac{A_o}{A_i h_i} + \frac{A_o \ln\frac{r_o}{r_i}}{2\pi Lk} + \frac{1}{h_o} \tag{12.4}$$

or

$$U_o = \frac{1}{\frac{A_o}{A_i h_i} + \frac{A_o \ln\frac{r_o}{r_i}}{2\pi Lk} + \frac{1}{h_o}} \tag{12.5}$$

For the inner wall area A_i of the inner tube

$$\frac{1}{U_i} = \frac{1}{h_i} + \frac{A_i \ln\frac{r_o}{r_i}}{2\pi Lk} + \frac{A_i}{A_o h_o} \tag{12.6}$$

or

$$U_i = \frac{1}{\frac{1}{h_i} + \frac{A_i \ln\frac{r_o}{r_i}}{2\pi Lk} + \frac{A_i}{A_o h_o}} \tag{12.7}$$

The overall heat transfer coefficients U_i and U_o, defined by Eqs (12.5) and (12.7), are valid for clean heat transfer surfaces in the exchanger. However, during normal operation, fouling deposits in the form of scale, corrosion, bacteria, etc., are accumulated on the surfaces, causing additional thermal resistance on both sides of the tube. As a result, the overall heat transfer coefficient increases leading to reduction in the performance of the heat exchanger. In the presence of fouling, Eqs (12.3), (12.5), and (12.7) are rewritten as

$$\frac{1}{UA} = R_{f,o} + \frac{1}{A_o h_o} + \frac{\ln\frac{r_o}{r_i}}{2\pi Lk} + \frac{1}{A_i h_i} + R_{f,i} \tag{12.8}$$

$$U_o = \frac{1}{A_o R_{f,o} + \frac{1}{h_o} + \frac{A_o \ln\frac{r_o}{r_i}}{2\pi Lk} + \frac{A_o}{A_i h_i} + A_o R_{f,i}} \tag{12.9}$$

Table 12.1 Fouling factors for selected gases and liquids.

Type of fluid	F_f, $m^2.K/W$
Sea water:	
below 50°C	8.84×10^{-5}
above 50°C	1.76×10^{-4}
Treated boiler feedwater above 50°C	8.84×10^{-4}
Quenching oil	7.07×10^{-4}
Steam, non-oil-bearing	8.84×10^{-5}
Industrial air	3.53×10^{-4}
Refrigerating liquid	1.77×10^{-4}
Hydraulic fluid	1.77×10^{-4}
Engine exhaust gas	1.77×10^{-3}
Transformer oil	1.77×10^{-4}
Engine lubricating oil	1.77×10^{-4}

$$U_i = \frac{1}{A_i R_{f,i} + \dfrac{1}{h_i} + \dfrac{A_i \ln \dfrac{r_o}{r_i}}{2\pi Lk} + \dfrac{A_i}{A_o h_o} + A_i R_{f,o}} \tag{12.10}$$

$R_{f,i}$ and $R_{f,o}$ are the fouling thermal resistances on the inner and outer surfaces of the tube respectively. The fouling thermal resistance multiplied by the surface area is called the fouling factor or fouling resistance F_f, selected values for which are listed in Table 12.1 for various gases and fluids. F_f is calculated from the following equation

$$F_f = \frac{1}{U_{\text{fouled}}} - \frac{1}{U_{\text{clean}}} \tag{12.11}$$

where U_{clean} and U_{fouled} are, respectively, the experimentally determined overall heat transfer coefficients of the clean and fouled heat transfer surfaces.

12.2 The LMTD Method of Heat Exchanger Analysis

12.2.1 Double-Pipe Heat Exchangers

Consider the temperature profiles in the parallel-flow exchanger shown in Figure 12.2. The temperature gradient ($\Delta T = T_h - T_c$) is not constant, as the temperatures of both fluids vary continuously throughout.

The heat balance equation for the entire exchanger is

$$\dot{Q} = \dot{m}_h c_{ph}\left(T_{h1} - T_{h2}\right) = \dot{m}_c c_{pc}\left(T_{c2} - T_{c1}\right)$$

and for the highlighted elemental area dA, taking into consideration that dT_h is negative,

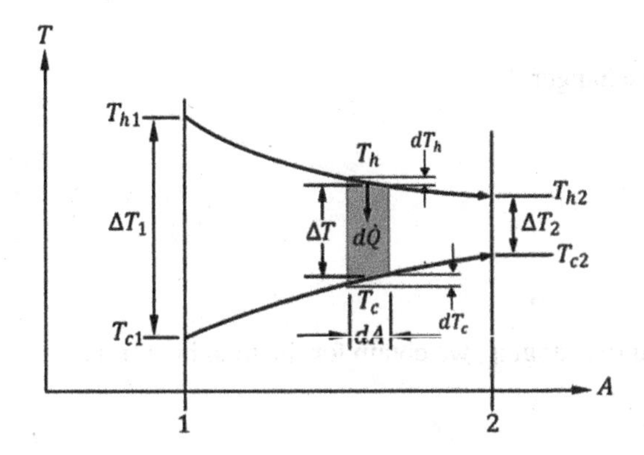

Figure 12.2 Temperature profiles in parallel-flow heat exchanger.

$$d\dot{Q} = -\dot{m}_h c_{ph} dT_h = \dot{m}_c c_{pc} dT_c \tag{12.12}$$

from which

$$dT_h = -\frac{d\dot{Q}}{\dot{m}_h c_{ph}}$$

$$dT_c = \frac{d\dot{Q}}{\dot{m}_c c_{pc}}$$

$$dT_h - dT_c = d(\Delta T) = -d\dot{Q}\left(\frac{1}{\dot{m}_h c_{ph}} + \frac{1}{\dot{m}_c c_{pc}}\right) \tag{12.13}$$

where $\dot{m}_h$, c_{ph} and $\dot{m}_c$, c_{pc} are the mass flow rate and specific heat at constant pressure of the hot and cold fluids respectively and $\Delta T = T_h - T_c$, which varies along the abscissa in Figure 12.2.

The heat transferred can also be written as

$$d\dot{Q} = U\Delta T dA \tag{12.14}$$

Combining Eqs (12.13) and (12.14) and rearranging yields

$$\frac{d(\Delta T)}{\Delta T} = -U\left(\frac{1}{\dot{m}_h c_{ph}} + \frac{1}{\dot{m}_c c_{pc}}\right)dA \tag{12.15}$$

From Figure 12.2, $\Delta T_1 = (T_{h1} - T_{c1})$ at point 1, and $\Delta T_2 = (T_{h2} - T_{c2})$ at point 2. Integrating Eq. (12.15) between points 1 and 2 results in

$$\int_{\Delta T_1}^{\Delta T_2} \frac{d(\Delta T)}{\Delta T} = \int^{A} -U\left(\frac{1}{\dot{m}_h c_{ph}} + \frac{1}{\dot{m}_c c_{pc}}\right)dA$$

$$\ln\left(\frac{T_{h2} - T_{c2}}{T_{h1} - T_{c1}}\right) = -UA\left(\frac{1}{\dot{m}_h c_{ph}} + \frac{1}{\dot{m}_c c_{pc}}\right) \tag{12.16}$$

In this equation, $A = A_i$ or $A = A_o$.

From Eq. (12.12) for the entire length of the exchanger

$$\dot{m}_h c_{ph} = \frac{\dot{Q}}{T_{h1} - T_{h2}}$$

$$\dot{m}_c c_{pc} = \frac{\dot{Q}}{T_{c2} - T_{c1}}$$

Combining these equations with Eq. (12.16) and rearranging, we obtain for the total heat transfer rate

$$\dot{Q} = UA \frac{(T_{h2} - T_{c2}) - (T_{h1} - T_{c1})}{\ln\left[\frac{(T_{h2} - T_{c2})}{(T_{h1} - T_{c1})}\right]} \tag{12.17}$$

Based on Eq. (12.1), Eq. (12.17) can be rewritten as

$$\dot{Q} = UA\Delta T_m \tag{12.18}$$

where ΔT_m is an average temperature difference for the heat exchanger. Equating Eqs (12.17) and (12.18), we obtain

$$\Delta T_m = \frac{(T_{h2} - T_{c2}) - (T_{h1} - T_{c1})}{\ln\left[\frac{(T_{h2} - T_{c2})}{(T_{h1} - T_{c1})}\right]}$$

or

$$\Delta T_m = \frac{(T_{h1} - T_{c1}) - (T_{h2} - T_{c2})}{\ln\left[\frac{(T_{h1} - T_{c1})}{(T_{h2} - T_{c2})}\right]} \tag{12.19a}$$

or

$$\Delta T_m = \frac{\Delta T_1 - \Delta T_2}{\ln\left(\frac{\Delta T_1}{\Delta T_2}\right)} = \frac{\Delta T_2 - \Delta T_1}{\ln\left(\frac{\Delta T_2}{\Delta T_1}\right)} \tag{12.19b}$$

where ΔT_1 and ΔT_2 are the temperature differences at points 1 and 2 in Figure 12.2, respectively, which can be written as

$$\Delta T_1 = T_{h1} - T_{c1} = T_{h,i} - T_{c,i}$$

$$\Delta T_2 = T_{h2} - T_{c2} = T_{h,o} - T_{c,o}$$

Using the same analysis for a counter-flow heat exchanger, with temperature profiles as in Figure 12.1b, it can be shown that Eq. (12.9b) also applies in this case with the proviso that the temperature differences are written as

$$\Delta T_1 = T_{h1} - T_{c1} = T_{h,i} - T_{c,o}$$

$$\Delta T_2 = T_{h2} - T_{c2} = T_{h,o} - T_{c,l}$$

The subscripts i and o are for "inlet" and "outlet," respectively.

The temperature difference ΔT_m, known as the *log mean temperature difference* (LMTD), is useful for heat-exchanger analysis when the inlet and outlet temperatures are known or are easily determined (e.g., new exchanger design). Knowing ΔT_m, it is then easy to calculate the rate of heat flow, surface area, or overall heat-transfer coefficient, depending on the available initial data.

Example 12.1 The operating conditions of a counter-flow heat exchanger, having an effective area of 1.0 m^2, are as follows:

Fluid A: $T_{h1} = 60°C$, $T_{h2} = 35°C$, mass flow rate $\dot{m}_h = 1.2\, kg/s$, specific heat

$$c_{ph} = 1200\, J/kg.K.$$

Fluid B: $T_{c2} = 18°C$, mass flow rate $\dot{m}_c = 2.0\, kg/s$, specific heat $c_{pc} = 1600\, J/kg.K$

Determine the overall heat transfer coefficient.

Solution

The outlet temperature of fluid B T_{c1} can be calculated using the heat balance equation

$$\dot{Q} = \dot{m}_c c_{pc}\left(T_{c1} - T_{c2}\right) = \dot{m}_h c_{ph}\left(T_{h1} - T_{h2}\right)$$

$$T_{c1} = T_{c2} + \frac{\dot{m}_h c_{ph}}{\dot{m}_c c_{pc}}\left(T_{h1} - T_{h2}\right) = 18 + \frac{1.2 \times 1200}{2 \times 1600}(60 - 35) = 29.3°C$$

$$\Delta T_m = \frac{\left(T_{h1} - T_{c1}\right) - \left(T_{h2} - T_{c2}\right)}{\ln\left[\dfrac{\left(T_{h1} - T_{c2}\right)}{\left(T_{h2} - T_{c1}\right)}\right]} = \frac{(60 - 29.3) - (35 - 18)}{\ln\left[\dfrac{(60 - 29.2)}{(35 - 18)}\right]} = 23.2°C$$

The overall heat transfer coefficient U can now be determined from Eq. (12.18).

$$= \frac{\dot{Q}}{A\Delta T_m} = \frac{\dot{m}_h c_{ph}\left(T_{h1} - T_{h2}\right)}{A\Delta T_m} = \frac{1.2 \times 1200(60 - 35)}{1 \times 23.2} = 1551.7\, W/m^2.K$$

Example 12.2 The exchanger in Example 12.1 is thoroughly cleaned after being in service for a long time, increasing the overall heat transfer coefficient by 15%. All temperatures remain the same as before, with the exception of the outlet temperature T_{h2} of fluid A. Determine the fouling factor and the new outlet temperature of fluid A. Assume further that the thermophysical properties of the fluids remain unchanged.

Solution

From Eq. (12.11)

$$F_f = \frac{1}{U_{fouled}} - \frac{1}{U_{clean}} = \frac{1}{1551.7} - \frac{1}{1551.7 \times 1.15} = 8.4 \times 10^{-5}$$

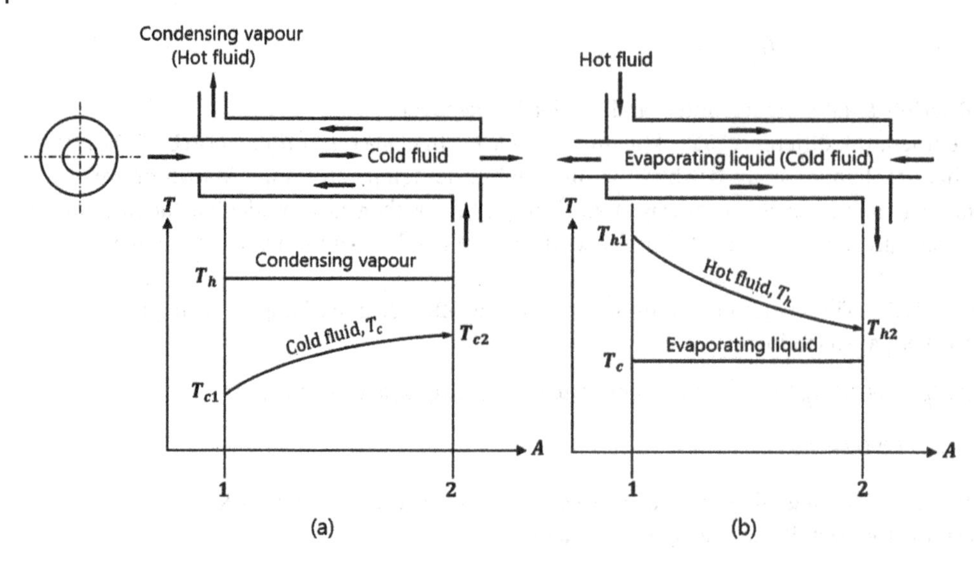

Figure 12.3 Temperature profiles for condensing vapour (a) and evaporating liquid (b) in a counter-flow double-pipe heat exchanger.

the ratios of the rate of heat transfer after and before cleaning the heat exchanger is

$$\frac{U_{clean}A\Delta T_{m(clean)}}{U_{fouled}A\Delta T_{m(fouled)}} = \frac{\dot{m}_h c_{ph}\left(T_{h1}-T_{h2}\right)_{clean}}{\dot{m}_h c_{ph}\left(T_{h1}-T_{h2}\right)_{fouled}} = \frac{\dot{m}_c c_{pc}\left(T_{c2}-T_{c1}\right)_{clean}}{\dot{m}_c c_{pc}\left(T_{c2}-T_{c1}\right)_{fouled}}$$

$$\frac{U_{clean}\Delta T_{m(clean)}}{U_{fouled}\Delta T_{m(fouled)}} = \frac{\left(T_{h1}-T_{h2}\right)_{clean}}{\left(T_{h1}-T_{h2}\right)}$$

$$1.15\times\frac{\Delta T_{m(clean)}}{23.2} = \frac{\left(60-T_{h2}\right)}{25}$$

$$\Delta T_{m(clean)} = 0.8\times(60-T_{h2})$$

$$\frac{\left(60-29.3\right)-\left(T_{h2}-18\right)}{\ln\left[\frac{\left(60-29.3\right)}{\left(T_{h2}-18\right)}\right]} = 0.8\times(60-T_{h2})$$

Solving this equation by trial and error, we obtain $T_{h2} = 32.8°C$.

Heat exchangers are sometimes used as evaporators and condensers in power, refrigeration, and air-conditioning systems, as well as in process industries. The temperature of only one stream changes in the exchanger, while the temperature of the evaporating or condensing fluid remains constant (Figure 12.3). The direction in which the fluids flow in this case is immaterial.

12.2.2 Shell-and-Tube Heat Exchangers

Double-pipe heat exchangers are most suitable for low-capacity heat exchangers with small heat transfer areas. If a large surface area is required, the shell-and-tube heat exchanger, shown in its simplest configuration in Figure 12.4, can be used with either a parallel-flow or a counter-flow arrangement. The exchanger has one shell-side pass and one tube-side pass and is designated as a "1:1 exchanger." A more effective modification of this exchanger is the one shell-side pass and two tube-side passes (1:2 exchanger) shown in Figure 12.5. Better results can be obtained by increasing the number of shell and/or tubes with configurations such as, for example, 1:4, 1:6, 2:8, 3:6, and 3:12. It is also normal practice to install baffles to increase heat transfer process in the heat exchanger by diverting the flow on the shell side across the tube bundles back and forth between baffle pairs, as shown in Figure 12.6.

In multi-pass shell-and-tube heat exchangers, the driving force (over-all temperature gradient) is always less than the LMTD, derived earlier for counter-flow and parallel-flow heat exchangers. Therefore, the rate of heat transfer given by Eq. (12.8) is rewritten as

$$\dot{Q} = UAF\Delta T_m \tag{12.20}$$

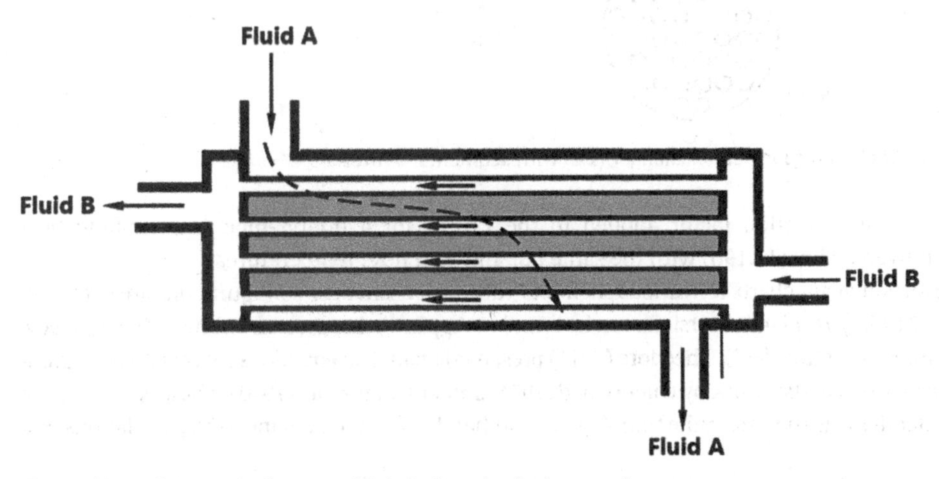

Figure 12.4 Schematic diagram of 1:1 shell-and-tube heat exchanger.

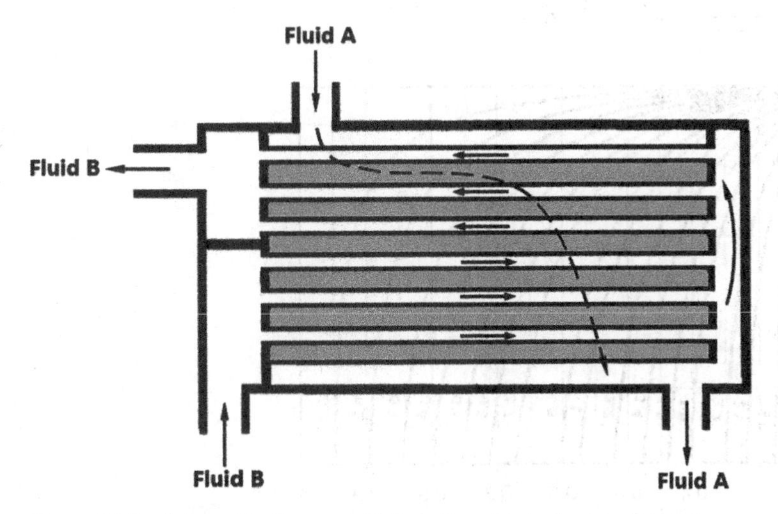

Figure 12.5 Schematic diagram of 1:2 shell-and-tube heat exchanger.

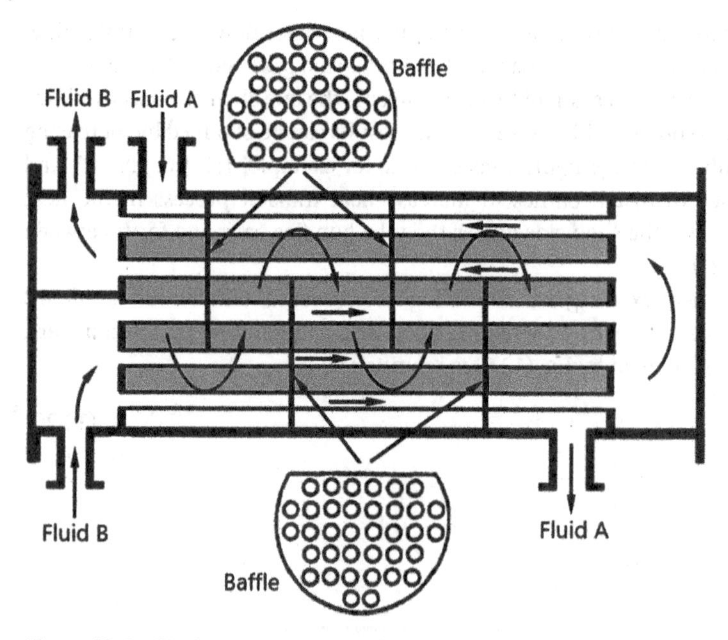

Figure 12.6 Shell-and-tube heat exchanger (1:2) with segmental baffles.

where F is a correction factor applied to the LMTD for a double-pipe counter-flow heat exchanger given by Eq. (12.19b) with the same hot and cold fluid temperatures.

The correction factor charts for various shell-and-tube heat exchanger configurations are shown in Figures 12.7 to 12.10 (enlarged versions can be found in Appendix L, accessed online at "www.wiley.com/go/Ghojel/heat transfer"). Theodore (2011) presented analytical equations for several configurations on the basis of earlier work by Underwood (1934) and Bowman et al. (1940). These equations are not presented here as they are rather cumbersome to handle for quick engineering calculations and

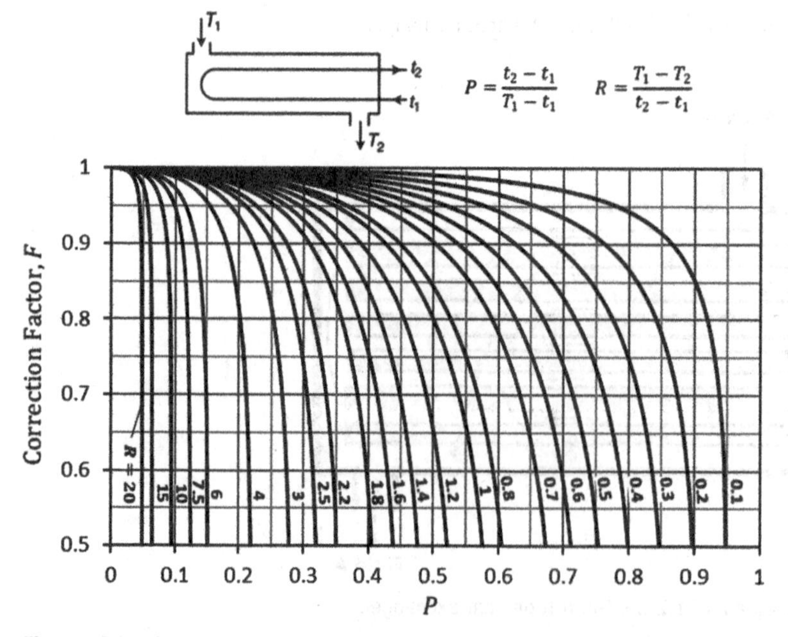

Figure 12.7 Correction factor for one-shell pass; 2 or more tube passes.

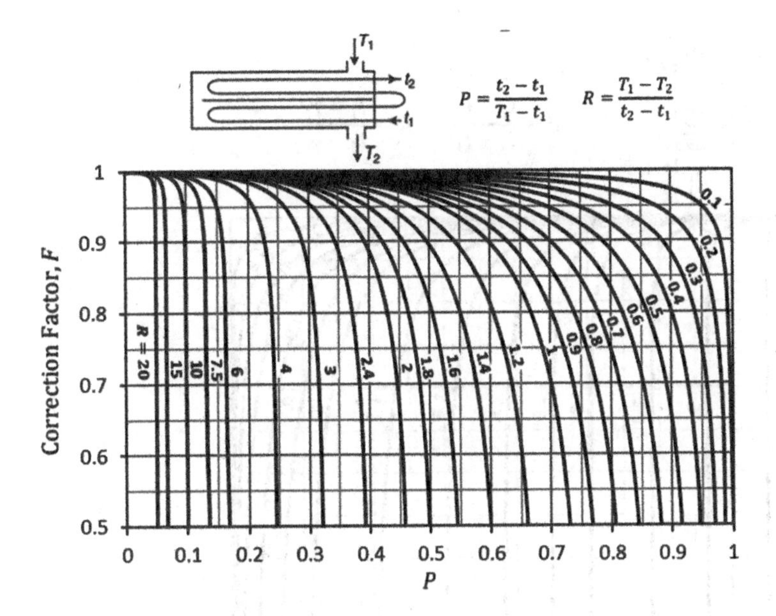

Figure 12.8 Correction factor for two shell passes; four or more tube passes.

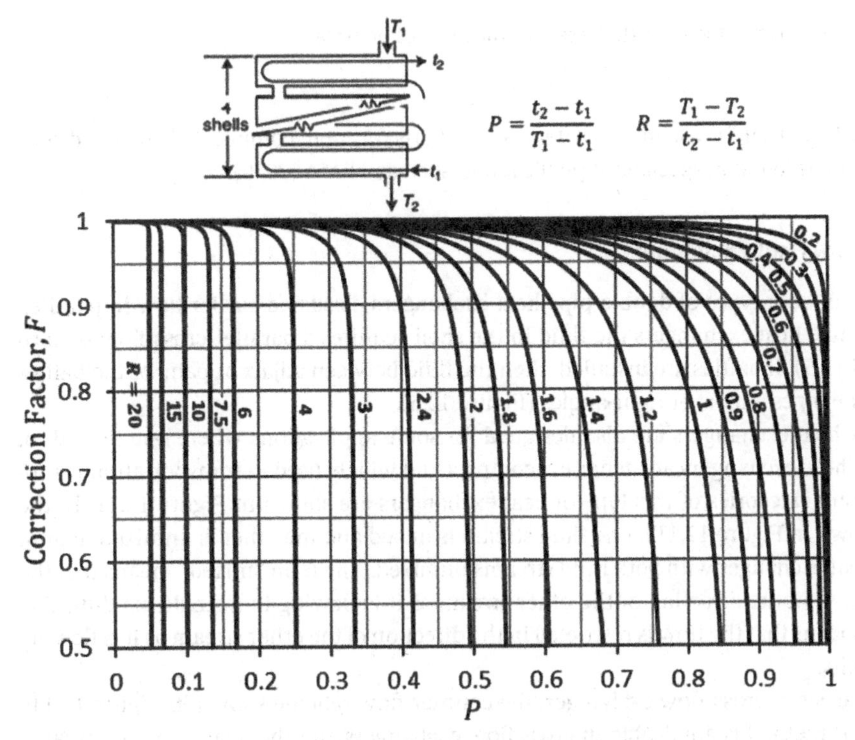

Figure 12.9 Correction factor for four shell passes; eight tube passes.

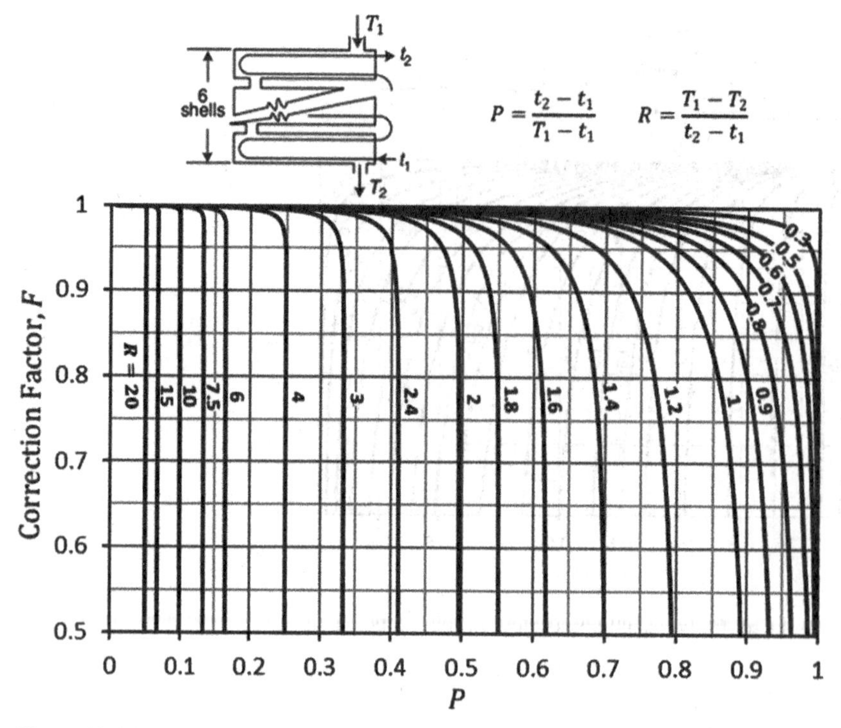

Figure 12.10 Correction factor for six shell passes; 12 tube passes or more.

they may be more beneficial for use in computer codes. Charts for more complicated shell-and-tube heat exchangers can be found in specialized publications, such as TEMA (2019).

12.2.3 Cross-Flow Heat Exchangers

The fluid streams in both types of double-pipe heat exchangers discussed earlier flow in parallel, and in shell-and-tube heat exchangers the fluid in the shell acquires a parallel/cross-flow pattern (Figures 12.4 and 12.5). If baffles are installed, then the fluid between adjacent pairs of the baffles will flow across the pipes almost at right-angles (Figure 12.6).

Pure cross-flow heat exchangers are also designed for some applications where one or both of the fluids is gas. These exchangers are generally compact and widely used in transportation equipment. The two main categories of this type of heat exchangers are shown in Figure 12.11. In the configuration shown in Figure 12.11a, one fluid stream is mixed and the other is unmixed. Figure 12.11b shows a heat exchanger with both fluid streams unmixed. The term "mixed" means that the flow is not restricted in the direction of the other stream as it is flowing in a single conduit. The term "unmixed" means that the flow is restricted in the direction of the other stream as it is flowing in multiple conduits.

For thermal analysis of cross-flow exchanger, the counter-flow scheme shown in Figure 12.11c is used. The LMTD method is applicable in cross-flow exchangers and the heat rate is calculated using Eq. (12.20) with the correction factor determined from the curves in Figures 12.12 to 12.14.

The correction factor chart for a cross-flow heat exchanger with both fluids unmixed is shown in Figure 12.12. The curves are plotted using the web application CheGuide (https://CheGuide.com/lmtd.html)

Figure 12.11 Cross-flow heat exchangers: (a) fluid A is mixed and B unmixed; (b) both fluids A and B are unmixed.

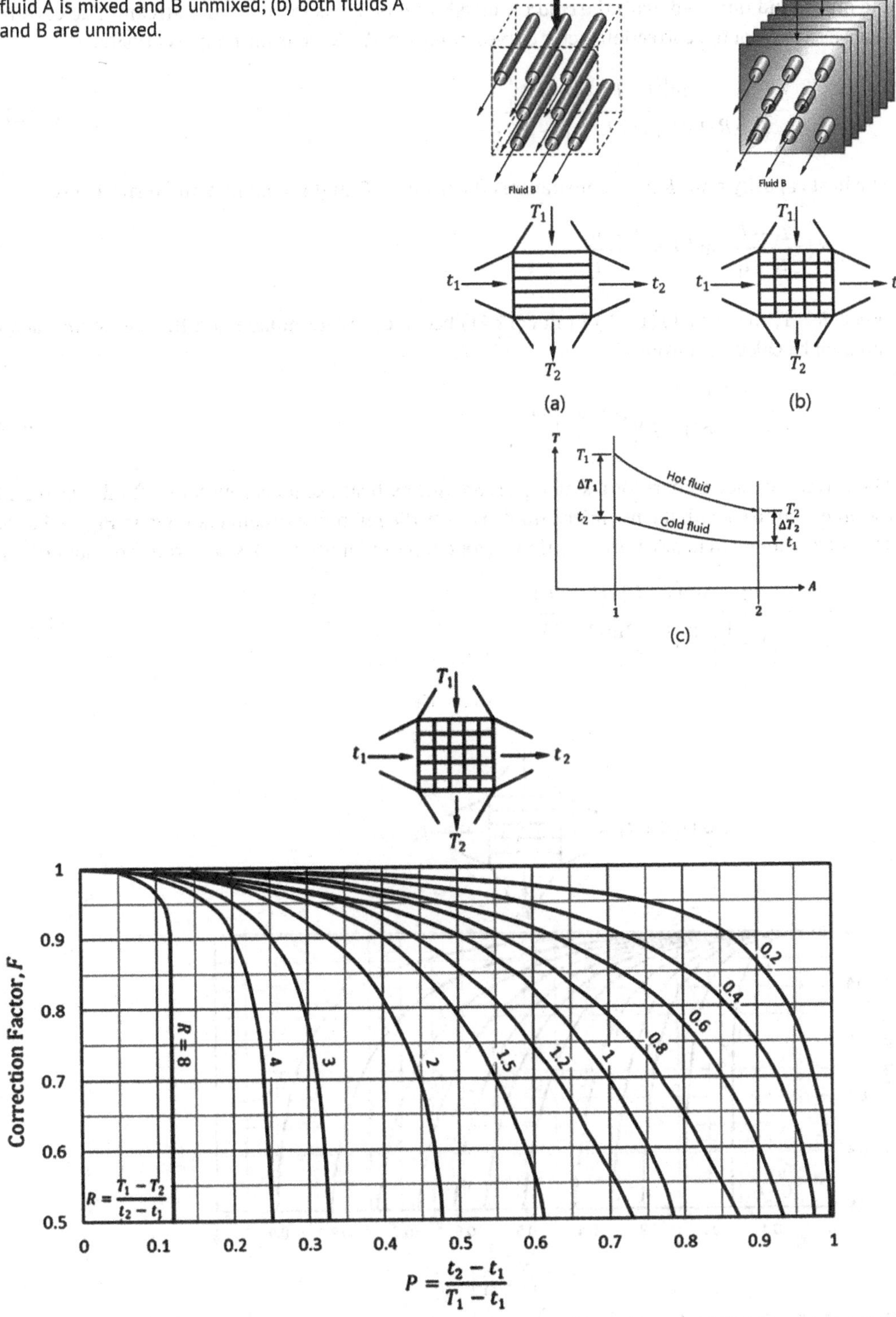

Figure 12.12 Correction factor for a single-pass cross-flow heat exchanger with both fluids unmixed.

Figure 12.13 shows the correction factor chart for cross-flow exchanger with one fluid mixed and the other fluid unmixed. with the cold unmixed fluid taken as the reference stream. The correlations used to plot the correction factor curves in Figure 12.13 are from Thomas (1999).

$$F = -\frac{1}{(1-R)} \frac{\ln[(1-PR)/(1-P)]}{\ln[1+(1/R)\ln(1-PR)]} \tag{12.21}$$

The heat capacity ratio R and temperature effectiveness P in this equation are, respectively,

$$R = \frac{T_1 - T_2}{t_2 - t_1} \text{ and } P = \frac{t_2 - t_1}{T_1 - t_1}$$

When $R = 1$, the term $1/(1-R)$ in Eq. (12.21) becomes indeterminate and the correction factor can then be calculated from

$$F = -\frac{P/(1-P)}{\ln[1+\ln(1-P)]} \text{ (for } R = 1) \tag{12.22}$$

The correction factor chart for a single pass cross-flow heat exchanger with one fluid mixed and the other unmixed with the hot mixed fluid taken as the reference stream is shown in Figure 12.14. The correlation used to plot the correction factor curves in Figure 12.14 is also from Thomas (1999).

$$F = -\frac{1}{\left(1-\frac{1}{R}\right)} \frac{\ln[(1-P)/(1-PR)]}{\ln[1+R\ln(1-P)]} \tag{12.23}$$

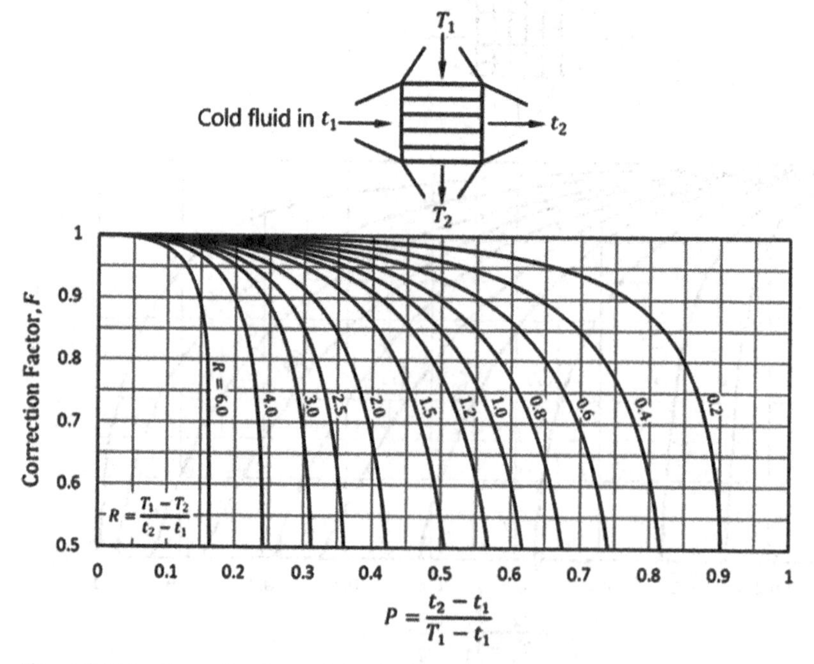

Figure 12.13 Correction factor for a single-pass cross-flow heat exchanger with one fluid mixed and the other unmixed; cold unmixed fluid is the reference stream.

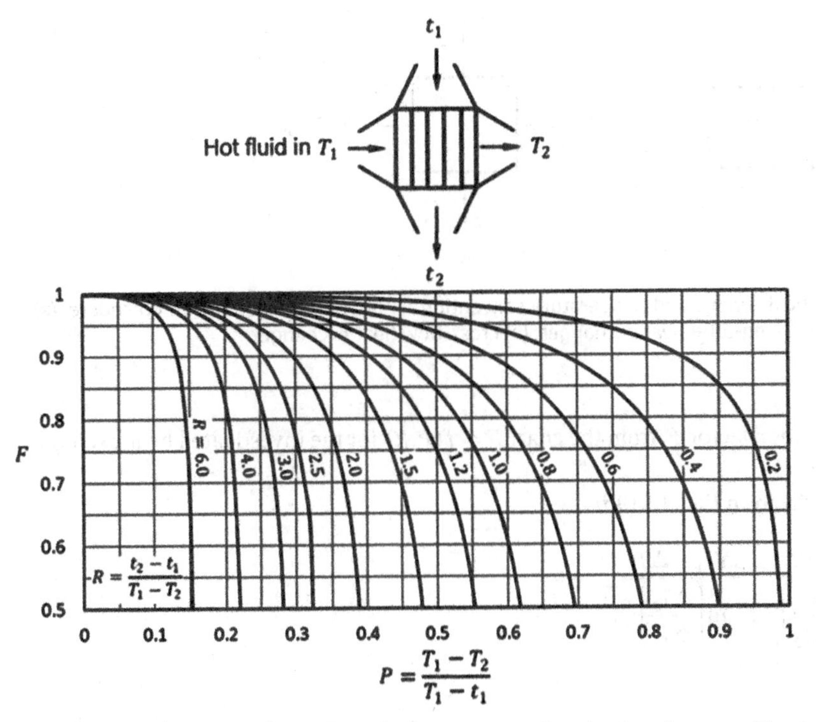

Figure 12.14 Correction factor for a single-pass cross-flow heat exchanger with one fluid mixed and the other unmixed; hot mixed fluid is the reference stream.

where

$$R = \frac{t_2 - t_1}{T_1 - T_2} \text{ and} P = \frac{T_1 - T_2}{T_1 - t_1}$$

For $R = 1$, the correction factor is calculated from Eq. (12.22) after replacing P with PR.

Enlarged versions of the correction factor charts for cross-flow heat exchangers are given in Appendix L, which can be accessed online at "www.wiley.com/go/Ghojel/heat transfer."

12.2.4 LMTD Thermal Design Procedure

The procedures outlined below are for all shell-and-tube heat exchangers with overall counter-flow arrangement and single-pass cross-flow heat exchangers, as shown schematically in Figure 12.15. Following the temperature convention used in the $F = f(P,R)$ charts, the inlet and outlet temperatures for the hot fluid are given as uppercase letters and for the cold fluid as lowercase letters. For double-pipe heat exchangers, the first two steps below are skipped.

If it is required to determine the exchanger heat transfer area A for a newly designed exchanger for the given data $U, T_1, t_1, T_2, t_2, C_h = \dot{m}_h c_{ph}, C_c = \dot{m}_c c_{pc}$, the following direct solution procedure is used:

1) Calculate R and P from the given temperatures from

$$R = \frac{T_1 - T_2}{t_2 - t_1}, P = \frac{t_2 - t_1}{T_1 - t_1}$$

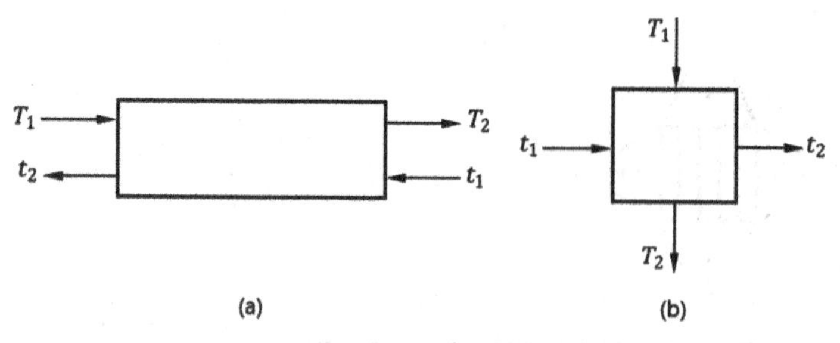

Figure 12.15 Schematic diagrams and temperature convention for heat exchanger design procedure: (a) overall counter-flow shell-and-tube heat exchanger; (b) cross-flow heat exchanger.

2) Estimate the correction factor F from the chart $F = f(P, R)$ for the investigated heat-exchanger flow arrangement
3) Calculate the LMTD from Eq. (12.19b)

$$\Delta T_m = \frac{\Delta T_1 - \Delta T_2}{\ln\left(\frac{\Delta T_1}{\Delta T_2}\right)} = \frac{\Delta T_2 - \Delta T_1}{\ln\left(\frac{\Delta T_2}{\Delta T_1}\right)}$$

where

$$\Delta T_1 = T_1 - t_2 = T_{h,i} - t_{c,o}$$

$$\Delta T_2 = T_2 - t_1 = T_{h,o} - t_{c,i}$$

4) Calculate the heat rate $\dot{Q}$ from the terminal temperatures and capacity rate for either the hot or cold fluid using the energy equation

$$\dot{Q} = \dot{m}_c c_{pc}\left(t_2 - t_1\right) = \dot{m}_h c_{ph}\left(T_1 - T_2\right)$$

5) Calculate the exchanger heat transfer area A from Eq. (12.20)

$$A = \frac{\dot{Q}}{UF\Delta T_m}.$$

Example 12.3 A shell-and-tube heat exchanger with one shell pass and four tube passes is to be designed to heat water from 18°C to 80°C by hot oil entering the exchanger shell at 160°C and leaving at 95°C. The water flows through 10 metal tubes with a total flow rate of 3 kg/s. The overall heat transfer coefficient for the tube wall is 350 $W/m^2.K$. Calculate the rate of heat transfer to the water and the total heat transfer area of the tubes. Take the specific heat of water as c_{pw}= 4180 $J/kg.K$.

Solution

$$\dot{Q} = \dot{m}_c c_{pw}\left(t_2 - t_1\right) = 3 \times 4180 \times \left(80 - 18\right) = 777{,}480\,W\ \left(777.48\,kW\right)$$
$$\Delta T_1 = T_1 - t_2 = 160 - 80 = 80°\text{C}$$

$$\Delta T_2 = T_2 - t_1 = 95 - 18 = 77$$

$$\Delta T_m = \frac{\Delta T_1 - \Delta T_2}{\ln\left(\frac{\Delta T_1}{\Delta T_2}\right)} = \frac{80 - 77}{\ln\left(\frac{80}{77}\right)} = 78.5°C$$

$$R = \frac{T_1 - T_2}{t_2 - t_1} = \frac{160 - 95}{80 - 18} = 1.048 \; P = \frac{t_2 - t_1}{T_1 - t_1} = \frac{80 - 18}{160 - 18} = 0.437$$

From the chart in Figure 12.7, $F \cong 0.87$

$$\therefore A = \frac{\dot{Q}}{UF\Delta T_m} = \frac{777.48 \times 10^3}{350 \times 0.87 \times 78.5} = 32.5\, m^2$$

Example 12.4 Superheated steam is used to heat water in a single-pass cross-flow heat exchanger with one fluid mixed and the other unmixed. Steam at a flow rate of $0.7\,kg/s$ enters the exchanger at temperature 240°C and leaves at 120°C. Water flowing at a flow rate of $0.6\,kg/s$ enters the exchanger at temperature 17°C. If the overall heat transfer coefficient is 600 $W/m^2.K$, determine the heat transfer area of the exchanger.

Solution

Water exit temperature can be determined from the energy equation

$$\dot{m}_w c_{pw}(t_2 - t_1) = \dot{m}_s c_{ps}(T_1 - T_2)$$

From thermophysical property tables, at $t_w = 17°C$ $c_{pw} = 4178\,J/kg.K$; at $t_s = 180°C$

$$c_{ps} = 1980\,J/kg.K$$

$$\therefore t_2 = 17 + \frac{0.7 \times 1980 \times (240 - 120)}{0.6 \times 4178} \approx 83°C$$

$$R = \frac{T_1 - T_2}{t_2 - t_1} = \frac{240 - 120}{83 - 17} = 1.818 \; P = \frac{t_2 - t_1}{T_1 - t_1} = \frac{83 - 17}{240 - 17} = 0.296$$

From the chart in Figure 12.13, $F = 0.93$

$$\Delta T_1 = T_1 - t_2 = 240 - 83 = 157°C$$

$$\Delta T_2 = T_2 - t_1 = 120 - 17 = 103°C$$

$$\Delta T_m = \frac{\Delta T_1 - \Delta T_2}{\ln\left(\frac{\Delta T_1}{\Delta T_2}\right)} = \frac{157 - 103}{\ln\left(\frac{157}{103}\right)} = 128°C$$

$$\dot{Q} = \dot{m}_w c_{pw}(t_2 - t_1) = 0.6 \times 4{,}178 \times (83 - 17) = 165.4\,kW$$

The heat transfer area can now be calculated from

$$A = \frac{\dot{Q}}{UF\Delta T_m} = \frac{165.4 \times 10^3}{600 \times 0.93 \times 128} = 2.32\, m^2$$

If it is required to determine the outlet temperatures of an existing exchanger T_2, t_2 for the given data $A, U, T_1, t_1, C_h = \dot{m}_h c_{ph}, C_c = \dot{m}_c c_{pc}$, the following direct solution procedure can be used:

1) Calculate R from the energy equation $\dot{Q} = C_c(t_2 - t_1) = C_h(T_1 - T_2)$

$$R = \frac{T_1 - T_2}{t_2 - t_1} = \frac{C_c}{C_h}$$

2) Assume a value of the unknown temperature t_2 and calculate P from

$$P = \frac{t_2 - t_1}{T_1 - t_1}$$

3) Estimate the correction factor from the chart $F = f(P,R)$ for the investigated heat-exchanger flow arrangement
4) Calculate the LMTD from Eq. (12.19b)

$$\Delta T_m = \frac{\Delta T_1 - \Delta T_2}{\ln\left(\frac{\Delta T_1}{\Delta T_2}\right)} = \frac{\Delta T_2 - \Delta T_1}{\ln\left(\frac{\Delta T_2}{\Delta T_1}\right)}$$

   where $\Delta T_1 = T_1 - t_2, \Delta T_2 = T_2 - t_1$
5) Calculate the heat rate from Eq. (12.20) $\dot{Q} = UAF\Delta T_m$
6) Calculate t_2 from $t_2 = t_1 + \dot{Q}/C_c$ and compare with the assumed value in step 2
7) Repeat steps 2 to 6 until satisfactory agreement between the assumed value of t_2 in step 2 and the calculated value in step 6 is obtained.

12.3 The Effectiveness-NTU Method of Heat-Exchanger Analysis

The LMTD approach to heat-exchanger analysis is useful when the inlet and outlet temperatures are known or are easily determined. The LMTD can then be calculated and corrected using charts similar to the ones in Figures 12.7 to 12.10 and 12.12 to 12.14. Based on the final value of the LMTD, the heat flow rate, surface area, or overall heat-transfer coefficient may be determined. However, when the inlet or exit temperatures are to be evaluated for a given heat exchanger, the LMTD/Correction Factor analysis becomes more complicated involving an iterative procedure. In such a situation, heat-exchanger analysis can be performed using a method based on the effectiveness of the heat exchanger in transferring a given amount of heat. This method is known as the Effectiveness-NTU Method of Heat Exchanger Analysis, where NTU stand for "number of transfer units."

The effectiveness of a heat-exchanger ε, which is a measure of its performance, is defined as

$$\varepsilon = \frac{\dot{Q}}{\dot{Q}_{max}} \tag{12.24}$$

$\dot{Q}$ in this equation is the actual heat transfer rate and $\dot{Q}_{max}$ is the maximum possible heat transfer rate. Q may be calculated as energy lost by the hot fluid or energy gained by the cold fluid. $\dot{Q}_{max}$ is the maximum energy that could be obtained when the fluid with the minimum value of the product $\dot{m}c_p$ (minimum fluid) undergoes the maximum temperature change, which is between the initial temperatures of the hot and cold fluids.

$$\dot{Q}_{max} = \left(\dot{m}c_p\right)_{min}\left(T_{h,in} - T_{c,in}\right)$$

Let $C_h = \dot{m}_h c_{ph}$, $C_c = \dot{m}_c c_{pc}$, where C_h and C_c are the hot and cold fluid heat capacity rates, respectively, then

$$\dot{Q}_{max} = C_{min}\left(T_{h,in} - T_{c,in}\right) \tag{12.25}$$

$C_{min} = \left(\dot{m}c_p\right)_{min}$ in Eq. (12.25) is the smaller of the two values of the hot and cold fluid heat capacity rates C_h and C_c, respectively.

The number of transfer units is defined as

$$NTU = \frac{UA}{C_{min}} \tag{12.26}$$

The equations for the heat rates and effectiveness for double-pipe heat exchangers are summarised in Figure 12.16.

12.3.1 Effectiveness-NTU Relation for Parallel-Flow Exchanger

Let the hot fluid be the fluid with minimum heat capacity rate $C_h = C_{min}$, the heat capacity ratio $C_r = C_{min} / C_{max}$, and the *number of transfer units* $NTU = UA / C_{min}$.

Equation (12.16) can be rewritten as

$$\ln\left(\frac{T_{h2} - T_{c2}}{T_{h1} - T_{c1}}\right) = -\frac{UA}{C_{min}}\left(1 + \frac{C_{min}}{C_{max}}\right)$$

Defining the heat capacity ratio $C_r = C_{min} / C_{max}$, we obtain

$$\ln\left(\frac{T_{h2} - T_{c2}}{T_{h1} - T_{c1}}\right) = -NTU(1 + C_r)$$

or

$$\frac{T_{h2} - T_{c2}}{T_{h1} - T_{c1}} = \exp\left[-NTU(1 + C_r)\right] \tag{12.27}$$

From the energy balance

$$\dot{Q} = C_{min}(T_{h1} - T_{h2}) = C_{max}(T_{c2} - T_{c1})$$

$$T_{c2} = T_{c1} + C_r(T_{h1} - T_{h2})$$

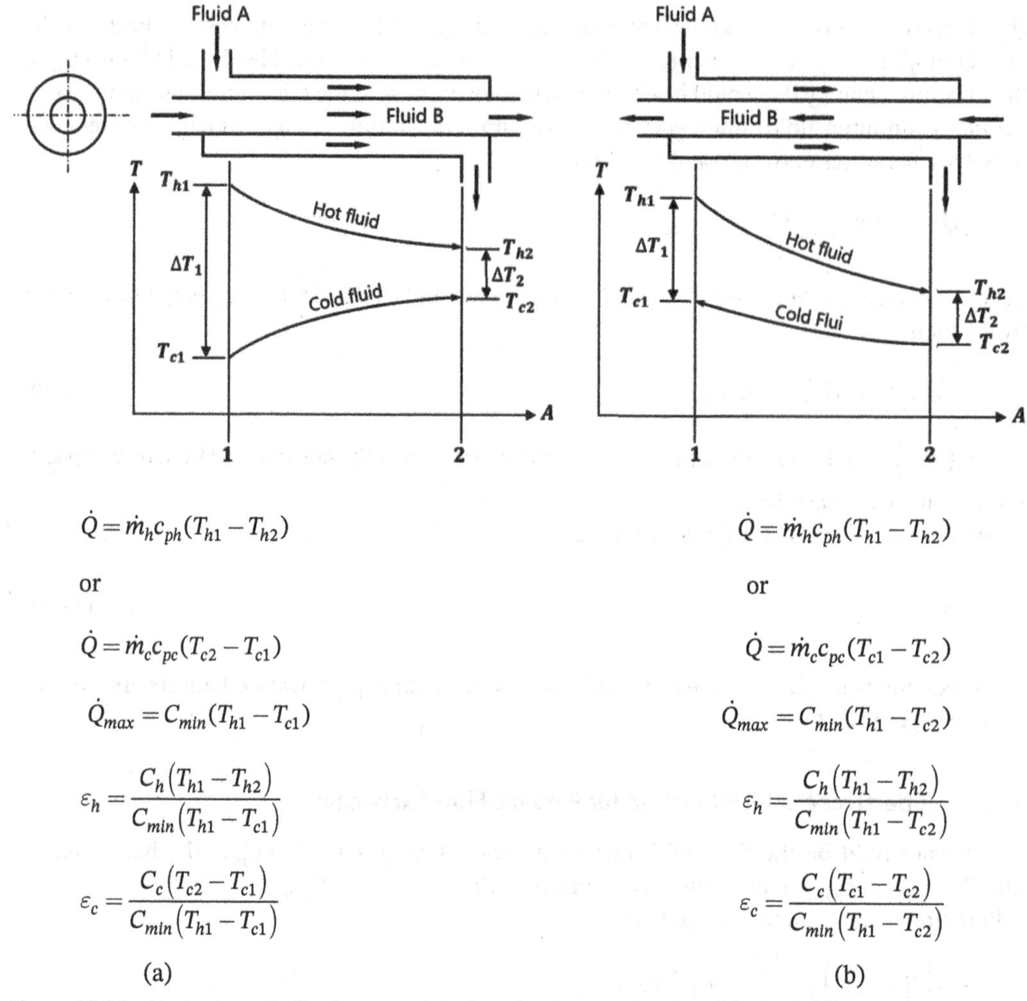

Figure 12.16 Heat rate and effectiveness of double-pipe heat exchangers: (a) parallel-flow; (b) counter-flow.

Substituting for T_{c2} into Eq. (12.27), we obtain

$$\frac{T_{h2} - T_{c1} - C_r(T_{h1} - T_{h2})}{T_{h1} - T_{c1}} = \exp[-NTU(1 + C_r)]$$

Adding and subtracting T_{h1} and re-arranging the left side of this equation, we obtain

$$\frac{-(T_{h1} - T_{h2}) + (T_{h1} - T_{c1}) - C_r(T_{h1} - T_{h2})}{T_{h1} - T_{c1}} = \exp[-NTU(1 + C_r)] \tag{12.28}$$

From the definition of $\varepsilon = \varepsilon_h$ from Figure 12.16a and assuming the hot fluid is the minimum fluid

$$\varepsilon = \frac{C_{min}\left(T_{h1} - T_{h2}\right)}{C_{min}\left(T_{h1} - T_{c1}\right)} = \frac{\left(T_{h1} - T_{h2}\right)}{\left(T_{h1} - T_{c1}\right)}$$

Equation (2.28) can now be rewritten as

$$-\varepsilon + 1 - \varepsilon C_r = \exp\left[-NTU(1 + C_r)\right]$$

Solving for ε finally yields

$$\varepsilon = \frac{1 - \exp\left[-NTU\left(1 + C_r\right)\right]}{1 + C_r} \tag{12.29}$$

It should be noted that the same result would be obtained if the cold fluid is taken as the minimum fluid; i.e., $C_{min} = C_c$.

12.3.2 Effectiveness-NTU Relation for Counter-Flow Exchanger

The temperature profiles of the cold and hot fluids in the heat exchanger are shown in Figure 12.17.

From the energy balance in the highlighted elemental area noting the negative temperature gradients dT_h and dT_c

$$d\dot{Q} = -\dot{m}_h c_{ph} dT_h = -\dot{m}_c c_{pc} dT_c \tag{12.30}$$

from which

$$dT_h = -\frac{d\dot{Q}}{\dot{m}_h c_{ph}}$$

$$dT_c = -\frac{d\dot{Q}}{\dot{m}_c c_{pc}}$$

$$dT_h - dT_c = d\left(\Delta T\right) = d\dot{Q}\left(-\frac{1}{\dot{m}_h c_{ph}} + \frac{1}{\dot{m}_c c_{pc}}\right) \tag{12.31}$$

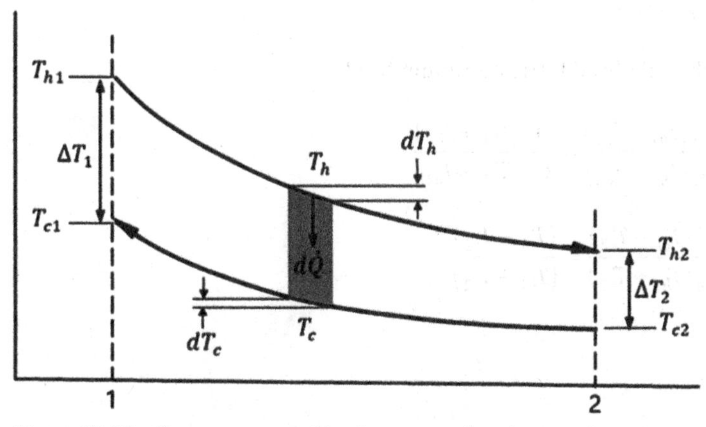

Figure 12.17 Temperature profiles in counter-flow heat exchanger.

Since,

$$d\dot{Q} = U\Delta T dA \tag{12.32}$$

combining Eqs (12.31) and (12.32) we obtain

$$\frac{d(\Delta T)}{\Delta T} = U\left(-\frac{1}{\dot{m}_h c_{ph}} + \frac{1}{\dot{m}_c c_{pc}}\right)dA \tag{12.33}$$

$\Delta T_1 = (T_{h1} - T_{c1})$ at point 1, and $\Delta T_2 = (T_{h2} - T_{c2})$ at point 2. Integrating Eq. (12.33) between points 1 and 2 results in

$$\int_{\Delta T_2}^{\Delta T_1} \frac{d(\Delta T)}{\Delta T} = \int^{A} U\left(-\frac{1}{\dot{m}_h c_{ph}} + \frac{1}{\dot{m}_c c_{pc}}\right)dA$$

$$\ln\left(\frac{T_{h2} - T_{c2}}{T_{h1} - T_{c1}}\right) = UA\left(-\frac{1}{\dot{m}_h c_{ph}} + \frac{1}{\dot{m}_c c_{pc}}\right) = UA\left(-\frac{1}{C_h} + \frac{1}{C_c}\right) \tag{12.34}$$

Let $C_c = C_{min}$, $C_h = C_{max}$, and $C_r = C_{min} / C_{max}$; hence,

$$\ln\left(\frac{T_{h2} - T_{c2}}{T_{h1} - T_{c1}}\right) = \frac{UA}{C_{min}}\left(1 - \frac{C_{min}}{C_{max}}\right) = NTU(1 - C_r)$$

or

$$\ln\left(\frac{T_{h1} - T_{c1}}{T_{h2} - T_{c2}}\right) = -NTU(1 - C_r)$$

or

$$\frac{T_{h1} - T_{c1}}{T_{h2} - T_{c2}} = \exp\left[-NTU(1 - C_r)\right] \tag{12.35}$$

Rewrite the left-hand side of Eq. (12.35) as follows:

$$\frac{T_{h1} - T_{c1}}{T_{h2} - T_{c2}} = \frac{(T_{h1} - T_{c1})/(T_{h1} - T_{c2})}{(T_{h2} - T_{c2})/(T_{h1} - T_{c2})}. \tag{12.36}$$

From Figure 12.16b, assuming the cold fluid is the minimum fluid

$$\varepsilon = \frac{C_h(T_{h1} - T_{h2})}{C_{min}(T_{h1} - T_{c2})} = \frac{C_{max}(T_{h1} - T_{h2})}{C_{min}(T_{h1} - T_{c2})} = \frac{1}{C_r}\frac{(T_{h1} - T_{h2})}{(T_{h1} - T_{c2})}$$

$$\varepsilon = \frac{C_c(T_{c1} - T_{c2})}{C_{min}(T_{h1} - T_{c2})} = \frac{C_{min}(T_{c1} - T_{c2})}{C_{min}(T_{h1} - T_{c2})} = \frac{(T_{c1} - T_{c2})}{(T_{h1} - T_{c2})}$$

$$T_{h2} = T_{h1} - C_r\varepsilon(T_{h1} - T_{c2})$$

$$T_{c1} = T_{c2} + \varepsilon(T_{h1} - T_{c2})$$

Substituting for T_{h2} and T_{c1} into Eq. (12.36) and rearranging yields

$$\frac{T_{h1} - T_{c1}}{T_{h2} - T_{c2}} = \frac{1-\varepsilon}{1-C_r\varepsilon} \tag{12.37}$$

Equating Eqs (12.35) and (12.37) yields

$$\frac{1-\varepsilon}{1-\varepsilon C_r} = \exp\left[-NTU(1-C_r)\right]$$

Solving for ε finally results in

$$\varepsilon = \frac{1-\exp\left[-NTU\left(1-C_r\right)\right]}{1-C\exp\left[-NTU\left(1-C_r\right)\right]} \tag{12.38}$$

12.3.3 Other Types of Heat Exchangers

The effectiveness relations for some other types of heat exchangers are given in Table 12.2. These are based on the classical work by Kays and London (1964, 1984, 2018) and have become the standard in most heat transfer reference books. The effectiveness relations are shown as functions of two non-dimensional parameters: NTU and C_r

$$\varepsilon = f(NTU, C_r) \tag{12.39}$$

Alternative relations to Eq. (12.39) are given in Table 12.3 in the form $NTU = f(\varepsilon, C_r)$, which can be used to determine NTU following the calculation of ε from its definition.

Plots of Eq. (12.39) for various heat exchangers are shown in Figures 12.18 to 12.27.

Regarding shell-and-tube heat exchanger relations in Table 12.2, note should be taken of the following:

1) The single- and multi-pass shell-and-tube exchangers are assumed to have an overall counterflow arrangement
2) The fluids are "mixed" between passes
3) It is assumed that all passes in shell-and-tube heat exchangers are identical
4) NTU is equally distributed between passes of the same basic arrangement. The two-pass exchanger will have twice the NTU of the single-pass exchanger, three-pass exchanger three times the NTU, and so on
5) The effectiveness ε_1 in item (2.2) is for the single shell pass as given in item (2.1) in Table 12.2
6) The equation in item (2.1) can be used for 2, 4, 8, ... tube passes
7) n in the equation in item (2.2) in Table 12.2 is the number of shell passes: 1, 2, 3, ...

12.3.4 Effectiveness-NTU Thermal Design Procedure

If it is required to determine the exchanger heat transfer area A for a newly designed exchanger for the given data U, T_1, t_1, T_2, t_1, $C_h = \dot{m}_h c_{ph}$, $C_c = \dot{m}_c c_{pc}$, the following solution procedure can be used:

Table 12.2 Heat exchanger effectiveness relations $\varepsilon = f(NTU, C_r)$.

Flow pattern	
(1) Double-tube	
(1.1) Parallel-flow	$\varepsilon = \dfrac{1-\exp\left[-NTU\left(1+C_r\right)\right]}{1+C_r}$
(1.2) Counter-flow	$\varepsilon = \dfrac{1-\exp\left[-NTU\left(1-C_r\right)\right]}{1-C_r\exp\left[-NTU\left(1-C_r\right)\right]}$ $(C_r<1)$
	$\varepsilon = \dfrac{NTU}{1+NTU}$ $(C_r=1)$
(2) Shell-and-tube	
(2.1) One shell pass 2, 4, 6, tube passes	$\varepsilon_1 = \dfrac{2}{1+C_r+\left(1+C_r^2\right)^{1/2}\left(\dfrac{1+K}{1-K}\right)}$ $K=\exp\left[-(NTU)_1(1+C_r^2)^{1/2}\right]$
(2.2) n shell passes $2n$, $4n$, $6n$ tube passes	$\varepsilon = \dfrac{\left(\dfrac{1-\varepsilon_1 C_r}{1-\varepsilon_1}\right)^n - 1}{\left(\dfrac{1-\varepsilon_1 C_r}{1-\varepsilon_1}\right)^n - C_r}$
(3) Cross-flow (single-pass)	
(3.1) Both fluids unmixed	$\varepsilon = 1-\exp\left[\left(\dfrac{1}{C_r}\right)(NTU)^{0.22}\left\{\exp\left[-C_r(NTU)^{0.78}\right]-1\right\}\right]$
(3.2) C_{max}(mixed), C_{min}(unmixed)	$\varepsilon = C_r^{-1}\left(1-\exp\left\{-C_r\left[1-\exp(-NTU)\right]\right\}\right)$
(3.3) C_{min}(mixed), C_{max}(unmixed)	$\varepsilon = 1-\exp\left(-C_r^{-1}\left\{1-\exp\left[(-C_r NTU)\right]\right\}\right)$
(4) All exchangers ($C_r = 0$)	$\varepsilon = 1-\exp(-NTU)$

Table 12.3 Heat exchanger effectiveness relations $NTU = f(\varepsilon, C_r)$.

Flow pattern	
(1) Double-tube	
(1.1) Parallel-flow	$NTU = -\dfrac{\ln\left[1-\varepsilon\left(1+C_r\right)\right]}{1+C_r}$
(1.2) Counter-flow	$NTU = \dfrac{1}{C_r-1}\ln\dfrac{(\varepsilon-1)}{(\varepsilon C_r-1)}$
(2) Shell-and-tube	
One shell pass 2, 4, 6, tube passes	$NTU = -\dfrac{1}{\left(1+C_r^2\right)^{1/2}}\ln\left[\dfrac{\dfrac{2}{\varepsilon}-1-C_r-(1+C_r^2)^{1/2}}{\dfrac{2}{\varepsilon}-1-C_r+(1+C_r^2)^{1/2}}\right]$

Table 12.3 (Continued)

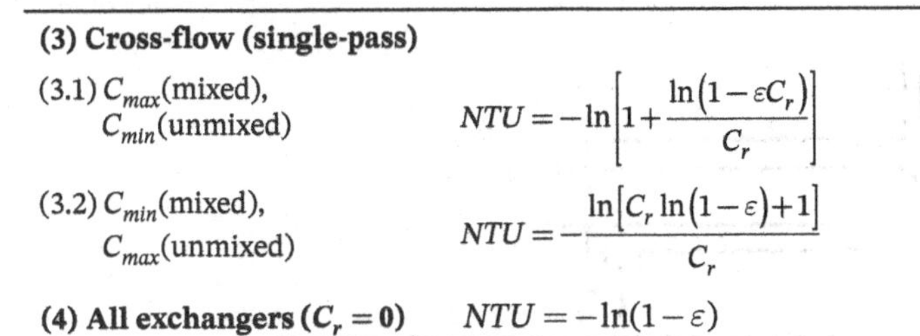

(3) Cross-flow (single-pass)	
(3.1) C_{max}(mixed), C_{min}(unmixed)	$NTU = -\ln\left[1 + \frac{\ln(1-\varepsilon C_r)}{C_r}\right]$
(3.2) C_{min}(mixed), C_{max}(unmixed)	$NTU = -\frac{\ln\left[C_r \ln(1-\varepsilon)+1\right]}{C_r}$
(4) All exchangers ($C_r = 0$)	$NTU = -\ln(1-\varepsilon)$

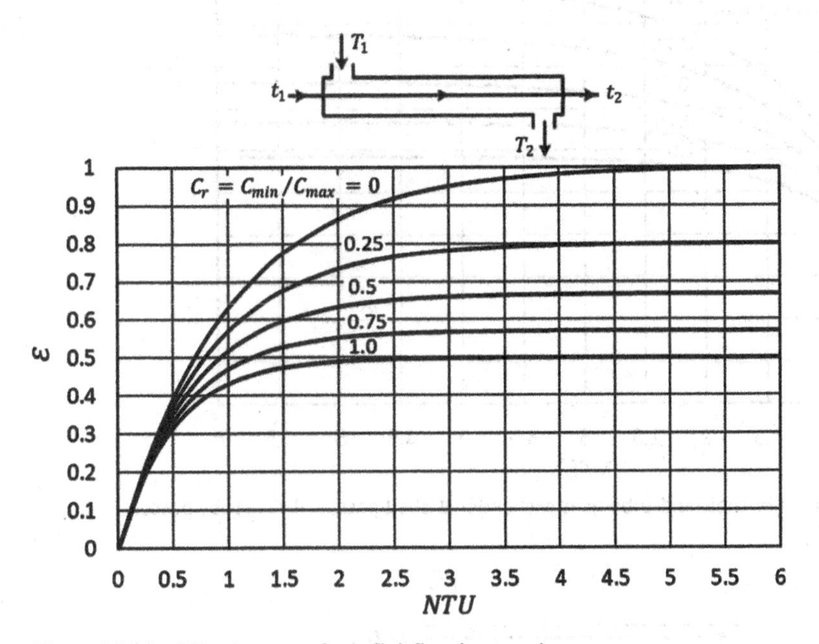

Figure 12.18 Effectiveness of parallel-flow heat exchanger.

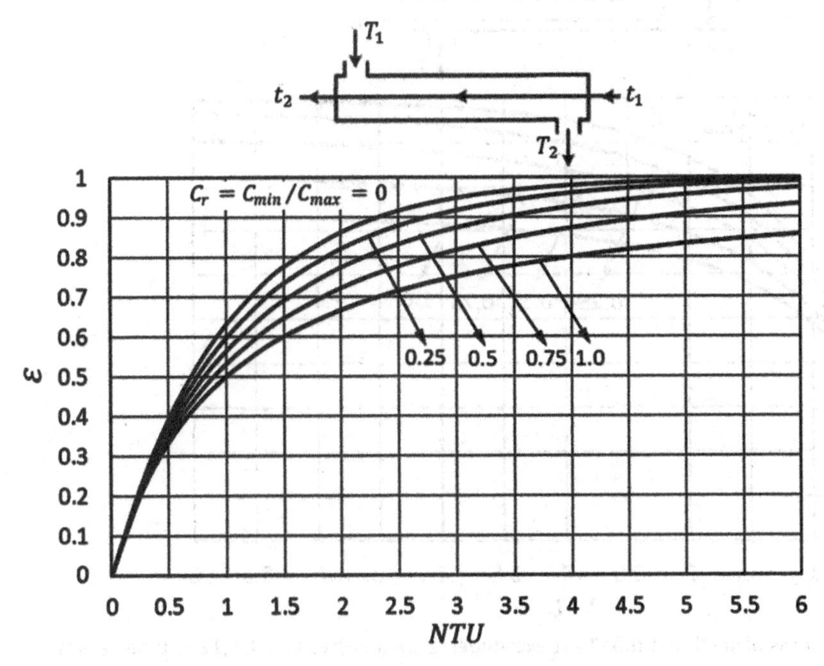

Figure 12.19 Effectiveness of counter-flow heat exchanger.

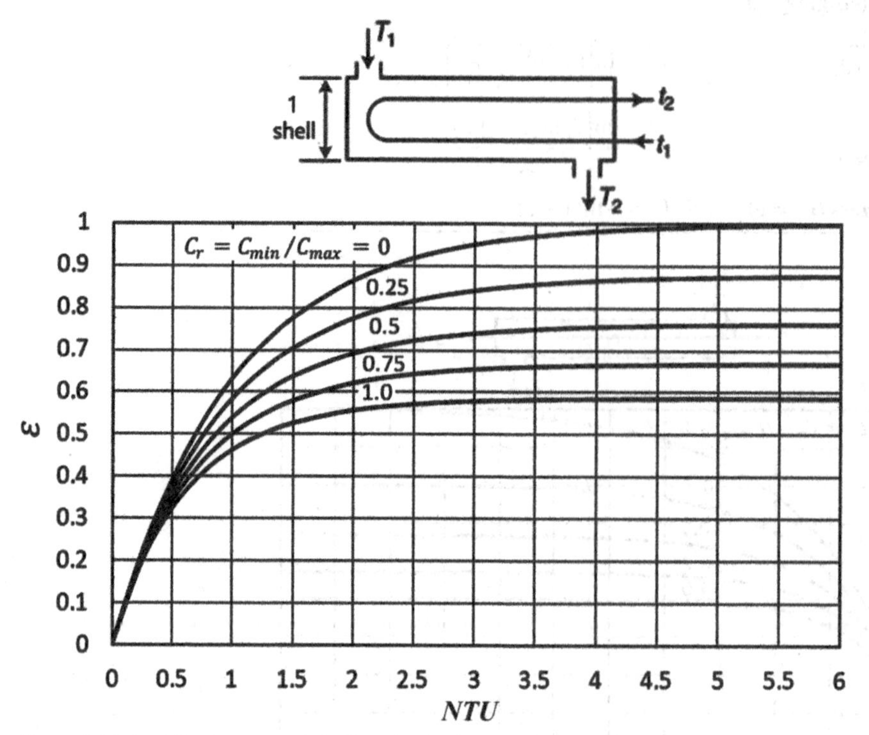

Figure 12.20 Effectiveness of shell-and-tube heat exchanger: 1 shell pass; 2, 4, 6, ... tube passes.

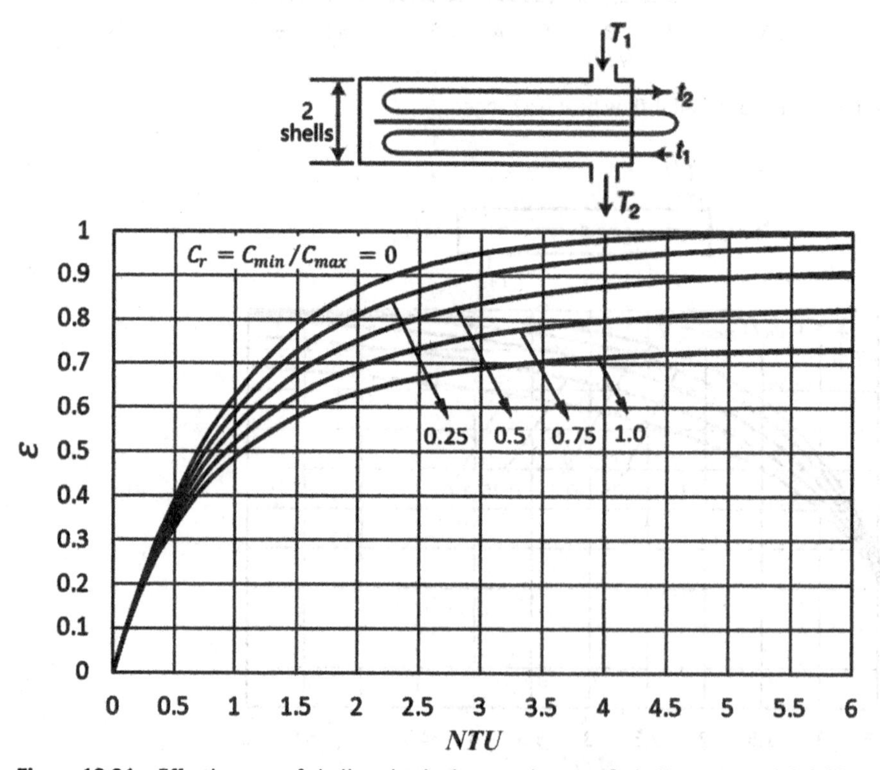

Figure 12.21 Effectiveness of shell-and-tube heat exchanger: 2 shell passes and 4, 8,12, ... tube passes.

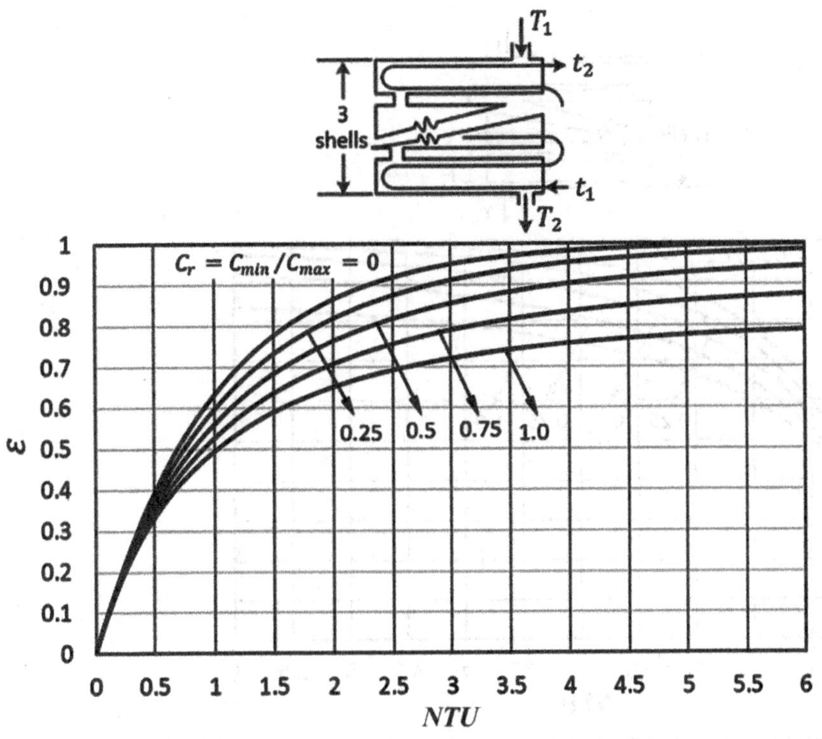

Figure 12.22 Effectiveness of shell-and-tube heat exchanger: 3 shell passes and 6, 12, 24, ... tube passes.

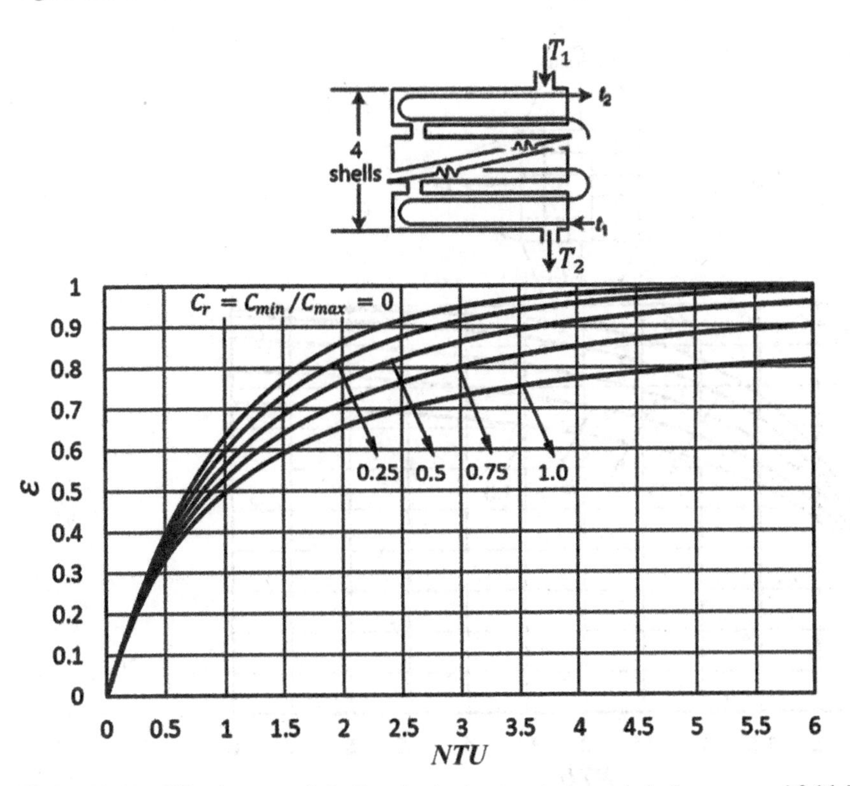

Figure 12.23 Effectiveness of shell-and-tube heat exchanger: 4 shell passes and 8,16, 32, ... tube passes.

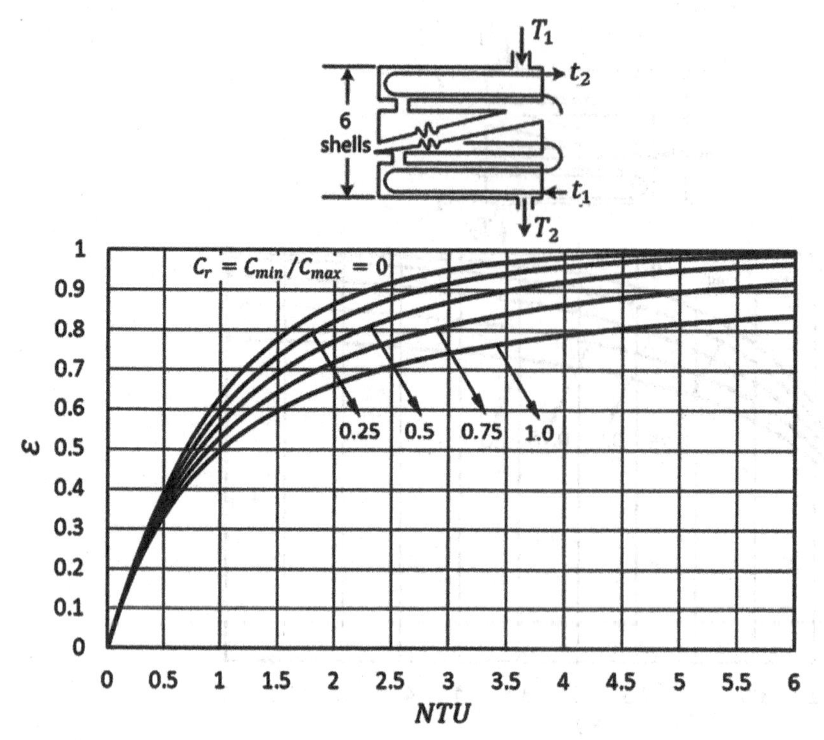

Figure 12.24 Effectiveness of shell-and-tube heat exchanger: 6 shell passes and 12, 24, 36, ... tube passes.

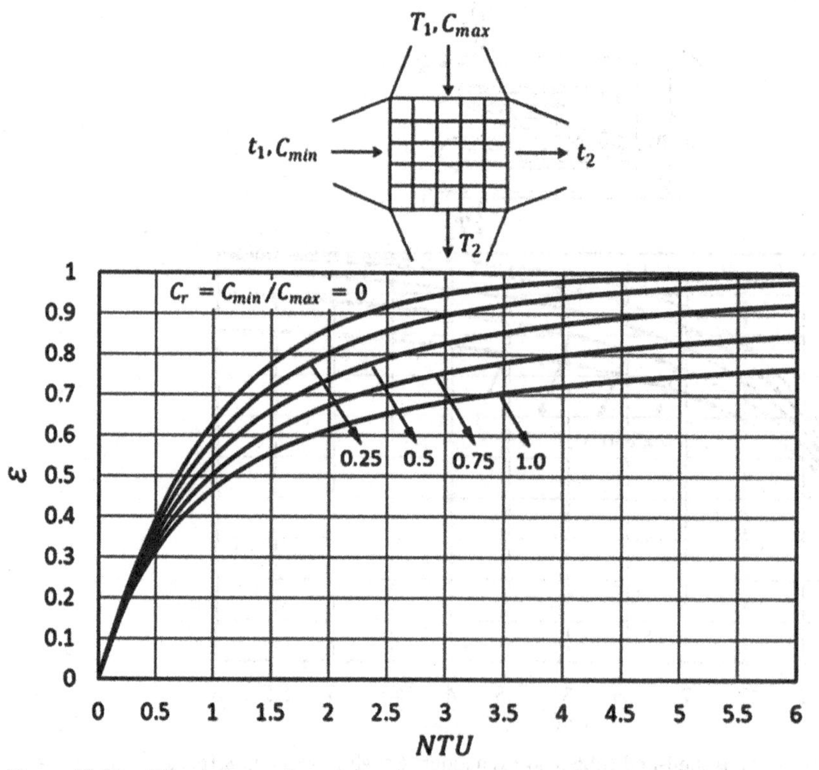

Figure 12.25 Effectiveness of one-pass, cross-flow heat exchanger: both fluids unmixed.

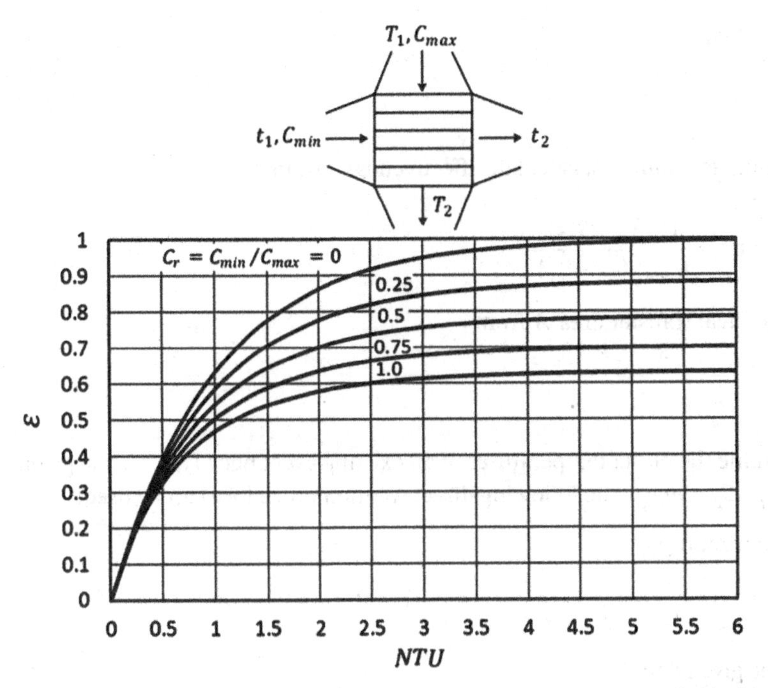

Figure 12.26 Effectiveness of one-pass cross-flow heat exchanger: fluid with C_{max} mixed; fluid with C_{min} unmixed.

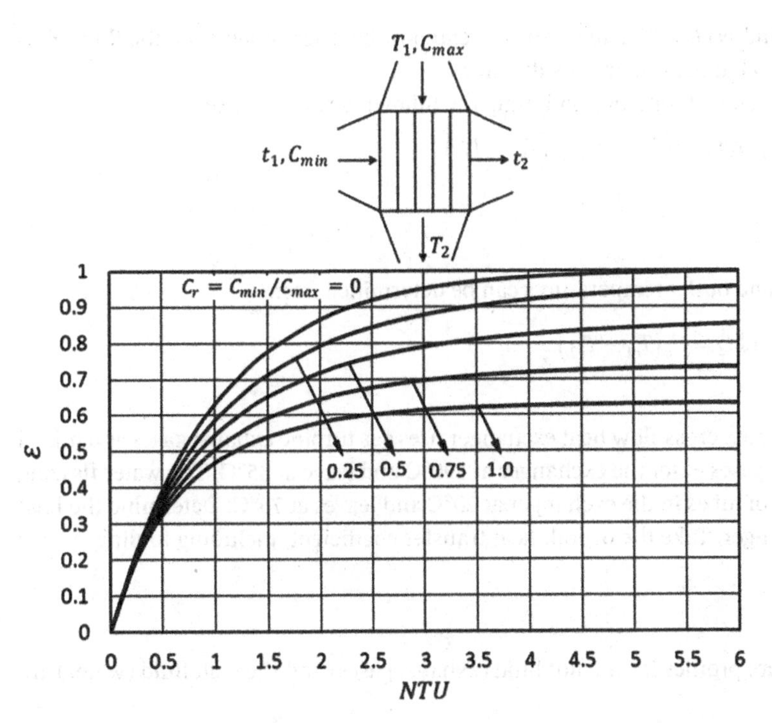

Figure 12.27 Effectiveness of one-pass cross-flow heat exchanger: fluid with C_{max} unmixed; fluid with C_{min} mixed.

1) Calculate capacitance ratio C_r

$$C_r = \frac{C_{min}}{C_{max}}$$

2) Depending on which fluid is minimum, calculate effectiveness ε from

$$\varepsilon = \frac{C_h\left(T_{h1} - T_{h2}\right)}{C_{min}\left(T_{h1} - T_{c1}\right)} \text{ or } \varepsilon = \frac{C_c\left(T_{c1} - T_{c2}\right)}{C_{min}\left(T_{h1} - T_{c1}\right)}$$

3) Calculate the exchanger heat transfer area A from

$$A = \frac{NTUC_{min}}{U}.$$

If it is required to determine the outlet temperatures of an existing exchanger T_2, t_2 for the given data $A, U, T_1, t_1, C_h = \dot{m}_h c_{ph}, C_c = \dot{m}_c c_{pc}$, the following direct solution procedure can be used:

1) Calculate the capacitance ratio C_r

 i) $C_r = \dfrac{C_{min}}{C_{max}}$

2) Calculate NTU from the given data

 i) $NTU = \dfrac{AU}{C_{min}}$

 Knowing C_{min} / C_{max} and NTU, the effectiveness ε can now be determined from the flow chart (or equation) for the exchanger under consideration

3) The heat transfer rate can be found by combining the following two equations:

 i) $\dot{Q} = C_{min}\left(T_{h1} - T_{h2}\right)$ and $\varepsilon = \left(T_{h1} - T_{h2}\right) / \left(T_{h1} - T_{c2}\right)$

 resulting in

 ii) $\dot{Q} = \varepsilon C_{min}\left(T_{h1} - T_{c2}\right)$

 Knowing $\dot{Q}$, the inlet and outlet temperatures can be determined from

$$\dot{Q} = C_h\left(T_{h1} - T_{h2}\right) \text{ and } \dot{Q} = C_c\left(T_{c1} - T_{c2}\right)$$

Example 12.5 A single-pass cross-flow heat exchanger uses gas turbine exhaust gases as a mixed stream to heat water. The gases enter the exchanger at 190°C and leave at 85°C. The water flowing at 1.5 kg / s enters a bank of tubes in the exchanger at 20°C and leaves at 70°C. Determine the heat transfer area of the exchanger. Take the overall heat transfer coefficient, including fouling, as 250 $W / m^2.K$.

Solution

The applicable temperature profiles for the hot fluid (exhaust gases) and the cold fluid (water) are shown in Figure E12.5.

$$C_w = \dot{m}_w c_{pw} = 1.5 \times 4180 = 6270\,W / K$$

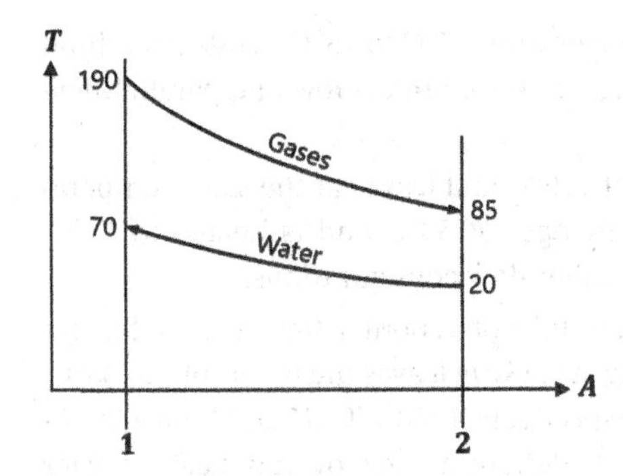

Figure E12.5 Temperature profiles for the hot and cold fluids.

$$\dot{Q} = C_g\left(T_{h1} - T_{h2}\right) = C_w\left(T_{c1} - T_{c2}\right)$$

$$C_g(190 - 85) = 6270(70 - 20)$$

$$\therefore C_g = \frac{6270 \times 50}{105} = 2985$$

Since $C_g < C_w$, the hot gas stream is the minimum stream

$$\frac{C_{min}}{C_{max}} = \frac{2985}{6270} = 0.476$$

The effectiveness of the heat exchanger in terms of the hot stream (gases) ($C_g = C_{min}$) is

$$\varepsilon = \frac{C_g\left(T_{h,i} - T_{h,o}\right)}{C_{min}\left(T_{h,i} - T_{c,i}\right)} = \frac{(190 - 85)}{(190 - 20)} = 0.618$$

or, in terms of the cold stream (water) ($C_w \neq C_{min}$)

$$\varepsilon = \frac{C_w\left(T_{c,o} - T_{c,i}\right)}{C_{min}\left(T_{h,i} - T_{c,i}\right)} = \frac{6270(70 - 20)}{2985(190 - 20)} = 0.618$$

Knowing $C_{min}/C_{max} = 0.476$ and the effectiveness $\varepsilon = 0.618$, we find from Figure 12.26 that $NTU \cong 1.3$

$$A = \frac{NTU\,C_{min}}{U} = \frac{1.3 \times 2985}{250} = 15.5\ m^2$$

Problems

12.1 Water at 80°C flows at a mass flow rate of 0.5 kg/s through a steel pipe having inner diameter $D_i = 30\ mm$ and outer diameter $D_o = 31.5\ mm$. Air flows on the outer surface at heat transfer coefficient $h_o = 30{,}000\ W/m^2.K$. Take the thermal conductivity of the steel pipe as $k_s = 60\ W/m.K$. Determine the overall heat transfer coefficient through the pipe wall.

12.2 A cold fluid is heated in a double-pipe heat exchanger from 15°C to 75°C, while a hot fluid is cooled from 100°C to 90°C. Determine the LMTD for the counter-flow and parallel-flow configurations.

12.3 A hot fluid enters a double-pipe heat exchanger at 110°C and leaves at the same temperature as it condenses. A cold fluid enters the exchanger at 50°C and is heated to 75°C. Determine the LMTD for the counter-flow and parallel-flow configurations.

12.4 A hot fluid stream flowing at 1.5 kg/s enters the outer tube of a counter-flow heat exchanger at 70°C and leaves at 30°C. The cold stream flowing at 2.2 kg/s leaves the inner tube at 30°C. The specific heats of the hot and cold fluids are, respectively, 1200 $J/kg.K$ and 1600 $J/kg.K$. If the effective area of the exchanger is 1.5 m^2, determine the overall heat transfer coefficient.

12.5 Consider the counter-flow heat exchanger and data shown in Figure P12.5. Determine the length of heat exchanger required to heat the water to 80°C if the overall heat transfer coefficient is 640 $W/m^2.K$.

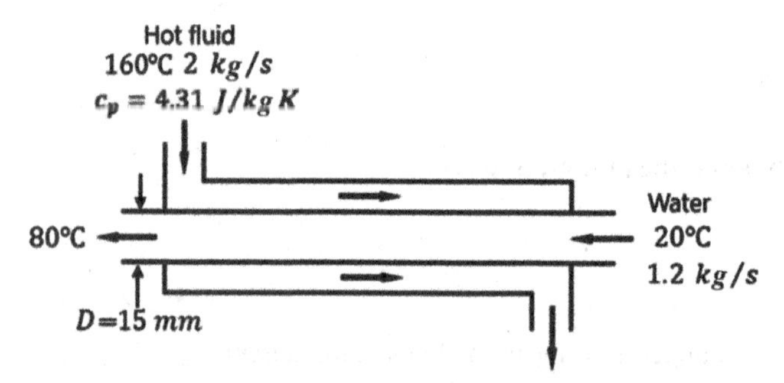

Figure P12.5 Counterflow heat exchanger schematic for Problem 12.5.

12.6 Repeat Problem 12.2 using the effectiveness-NTU method in lieu of the LMTD method.

12.7 Water flowing at the rate of 0.7 kg/s is to be heated in a counter-flow double-pipe heat exchanger at temperatures of 35°C to 90°C using hot oil flowing at the rate of 0.95 kg/s. The oil with specific heat of 2100 $J/kg.K$ enters the heat exchanger at a temperature of 175°C. For an overall heat transfer coefficient of 425 $W/m^2.K$, calculate

a) the area of the heat exchanger

b) the effectiveness of the heat exchanger.

12.8 Repeat Problem 12.7 using the effectiveness-NTU method in lieu of the LMTD method.

12.9 Air at 207 kPa flowing at a speed of 6 m/s enters a steel tube at 473K. The tube is 3m long with an inner diameter of 25 mm and thickness of 0.8 mm. Atmospheric air at 1.0 atm and temperature of 20°C flows normal to the outside of the tube with a free-stream velocity of 12 m/s. Calculate the air temperature at the exit from the tube.

12.10 Benzene having a flow rate of 0.03 kg/s is to be cooled in a counter-flow heat exchanger from 90°C to 35°C using water at 18°C flowing at 0.02 kg/s. The inner tube outside diameter is 20 mm and the overall heat transfer coefficient, based on outside area, is 650 $W/m^2.K$. The specific heats of benzene and water can be taken as 1880 and 4175 $J/kg.s$, respectively. Determine the required length of the exchanger.

12.11 Saturated steam at 0.11 *MPa* flowing at the rate of 1.8 kg/s is used in an industrial heating process requiring heating from 20°C to 70°C. Determine the flow rate of water required for the process.

12.12 A hot fluid flowing at the rate of 4 kg/s is to be cooled by water flowing at 3 kg/s in a counter-flow heat exchanger. The inlet temperature of the hot fluid is 125°C. and that of water is 30°C. The heat transfer area of the exchanger is 30 m^2 and the overall heat transfer coefficient is 820 $W/m^2.K$. The specific heat of the hot fluid can be taken as 2500 $J/kg.K$. Determine the product stream outlet temperature.

12.13 Consider the condensing shell-and-tube heat exchanger shown in Figure P12.13. The heat exchanger has a single shell and 30,000 tubes in a single bundle executing two passes. The tubes are thin-walled of 25 mm diameter, and the heat transfer coefficient of the steam condensing on the outer surface of the tubes is 11,000 $W/m^2.K$. The rate of heat transferred by the cooling water flowing at 30,000 kg/s is 2000 MW. The water enters the exchanger at 20°C and the steam condenses at 50°C. Determine:

a) the temperature of the cooling water exit temperature from the condenser

b) the required tube length per pass.

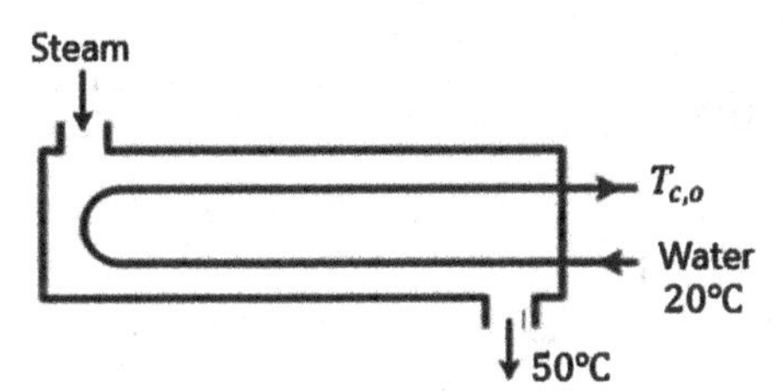

Figure P12.13 Schematic diagram of condensing 1:2 shell and tube heat exchanger.

12.14 A one-shell-pass and two-tube-pass heat exchanger is used to cool water flowing through the shell from 20°C to 6°C. The cooling stream enters the tubes at −1.0°C and leaves at 3°C. If the overall heat transfer coefficient is 1000 $W/m^2.K$ and the heat load is 6000 W, calculate the tube area.

12.15 A shell-and-tube heat exchanger with one shell pass and four tube passes is to be designed to heat water from 15°C to 65°C by hot oil entering the exchanger shell at 160°C and leaving at 85°C. The water flows through 10 metal tubes with a total flow rate of 2.5 kg/s. The overall heat transfer coefficient for the tube wall is 300 $W/m^2.K$. Calculate the rate of heat transfer to the water and the total heat transfer area of the tubes. Take the specific heat of water as $c_{pw} = 4180\ J/kg.K$.

12.16 A two-shell-pass and four-tube-pass heat exchanger is used to cool oil flowing through the tube's shell from 135°C to 50°C (Figure P12.16). The cooling stream enters the shell at 15°C and leaves at 30°C. Calculate the heat transfer rate on the basis of the following data:
Oil: $h_i = 270\ W/m^2.K$; water: $h_o = 965\ W/m^2.K$
Scale (fouling) on water side: $h_{f,o} = 2840\ W/m^2.K$
Number of tubes per pass: 120
Dimensions of each tube: $d_i = 22.1\ mm$, $d_o = 25.4\ mm$, $L = 2m$
The tube-wall thermal resistance can be ignored

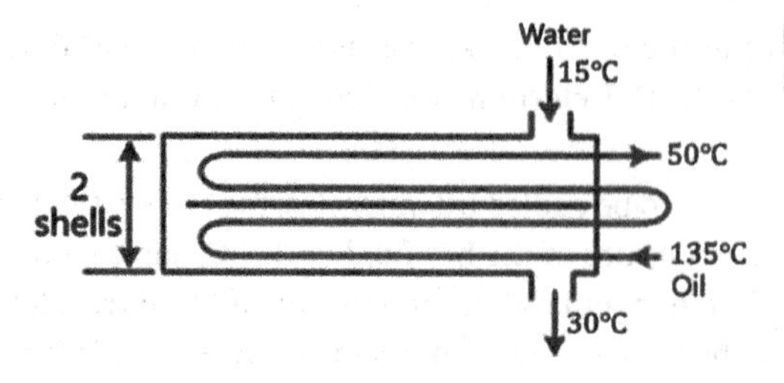

Figure P12.16 Schematic diagram of 2:4 shell and tube heat exchanger.

12.17 A single-pass cross-flow heat exchanger uses gas turbine exhaust gases as a mixed stream to heat water flowing in a bank of tubes (Figure P12.17). The gases enter the exchanger at 200°C and leave at 95°C. The water flowing at 1.8 kg/s enters the tubes in the exchanger at 18°C and leaves at 65°C. Determine the heat transfer area of the exchanger. Take the overall heat transfer coefficient, including fouling, as 260 $W/m^2.K$ and the specific heat of water as 4800 $J/kg.K$.

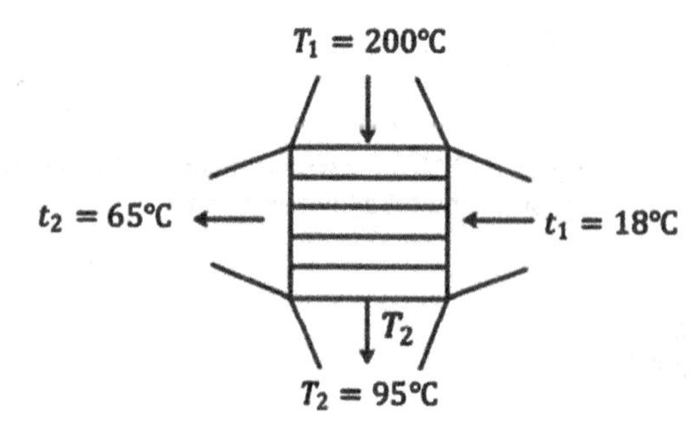

Figure P12.17 Schematic diagram of single-pass cross-flow heat exchanger.

12.18 Hot products of combustion flowing through an array of thin-walled tubes are used to boil water flowing over the tubes. At the time of installation, the overall heat transfer coefficient was 400 $W/m^2.K$. After a period of operation, the inner and outer tube surfaces are fouled, with corresponding fouling factors of 0.0015 and 0.0005, respectively. Determine the percentage decrease in the overall heat transfer coefficient over that period.

12.19 A shell-and-tube heat exchanger is used to heat 2.5 kg/s of water from 15°C to 85°C. Hot engine oil at 160°C flows through the shell side of the exchanger, providing an average convection coefficient of 400 $W/m^2.K$ on the outside of the tubes. The oil leaves the exchanger at 100°C. Water flows through ten thin-walled tubes (25 mm diameter each) making eight passes through the shell. Determine:
a)The flow rate of the oil
b)The length of the tubes.

12.20 A cross-flow heat exchanger is used to cool hot oil from 120°C to 95°C. The oil flows inside the tubes and the cooling water flows over the tubes, entering the exchanger at 20°C and leaving at 50°C. The overall heat transfer coefficient is 55 $W/m^2.K$. What size exchanger will be required to cool 3700 kg/h of oil? The specific heat of the oil is 1900 $J/kg.K$.

13

Heat Transfer With Phase Change

In this book so far, all the fluids involved in convection heat transfer processes in a thermal system were in a single-phase state and remained spatially and temporally unchanged within the control volume of the system. However, many important applications of heat transfer in engineering practice can involve processes of boiling (evaporation) and condensation. These processes usually occur at solid–liquid interfaces and are accompanied with a distinctive phase change that significantly increases the heat rate and coefficient of heat transfer.

The applications of phase change involving heat transfer include, to mention a few, evaporators and condensers in refrigeration systems, boilers and condensers in steam power plants, and evaporators and condensers in heat pipes. In this chapter, condensation and boiling heat transfer processes will be discussed in some detail.

13.1 Heat Transfer in Condensing Vapours

A simple example of condensation in a counter-flow, double-pipe heat exchanger is shown in Figure 13.1. Saturated vapour is admitted into the shell side of the exchanger and is condensed at approximately constant pressure and temperature T_h as it comes into contact with the relatively cooler outer surface of the tube. The consequent loss of thermal energy causes the vapour to condense on the surface and heat is transferred through the tube wall to the cold fluid flowing in the tube, causing its temperature to increase from T_{c1} to T_{c2}.

The condensate on the outer tube surface could be in the form of a continuous liquid film (filmwise condensation) or in the form of individual droplets (dropwise condensation), or a combination of both. To minimize the thermal resistance to heat transfer from the vapour to the cold fluid, the condensate must be removed. This is achieved naturally by gravity and could be assisted by the forced flow of the vapour along the tube.

Filmwise condensation and dropwise condensation modes will now be considered.

13.1.1 Filmwise Condensation

When a saturated vapor at temperature T_{sat} comes into contact with a surface at a lower temperature, condensation occurs. The classical solution of this problem was presented by Nusselt as far back as in 1916 in the form of continuous flow of the condensate down a vertical wall of constant

Heat Transfer Basics: A Concise Approach to Problem Solving, First Edition. Jamil Ghojel.

Companion website: www.wiley.com/go/ghojel/heat_transfer

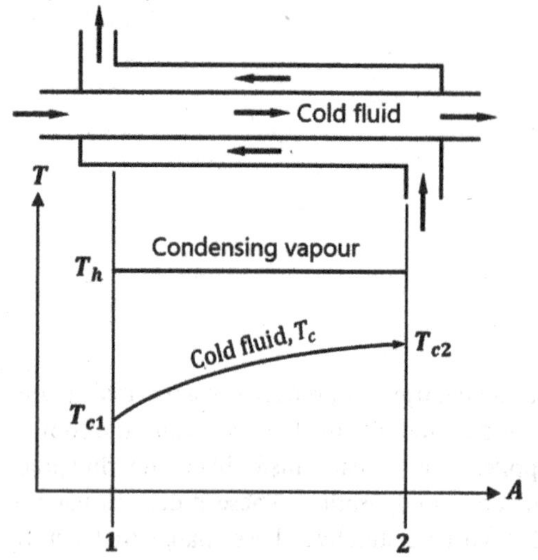

Figure 13.1 Condensing double-pipe heat exchanger.

surface temperature T_w ($T_w < T_{sat}$) under the influence of gravity forming a boundary layer, as depicted in Figure 13.2.

This model is based on the following assumptions:

1) The fluid properties are constant
2) The thickness of the condensing film increases gradually from zero at the origin $x = 0$ as pure vapour at uniform temperature T_v condenses
3) The flow of the liquid film is laminar. and the condensate temperature profile is linear
4) The vapour is at a uniform temperature equal to the saturated temperature T_{sat}
5) Heat transfer at the liquid–vapour interface is by convection only
6) Heat transfer to the wall surface across the liquid film is by conduction only
7) Velocity gradient at the liquid–vapour interface is zero
8) Shear stress at the liquid–vapor interface is negligible
9) The energy lost by the condensate as it cools below the saturation temperature is ignored (sub-cooling is ignored).

Referring to Figure 13.2, consider the hatched strip of width δ and height dx with unit depth perpendicular to the plane of the paper. The force balance on the highlighted differential element $dx.dy.1$, accounting for pressure change across the element in the x-direction (not shown in the figure), can be written as

$$\rho_l g dx dy + \left[\left(\tau + \frac{d\tau}{dy}dy\right)\right]dx + p dy = \tau dx + \left[p + \frac{dp}{dx}dx\right]dy$$

Dividing by $dxdy$ and rearranging

$$\frac{d\tau}{dy} = \frac{dp}{dx} - \rho_l g \tag{13.1}$$

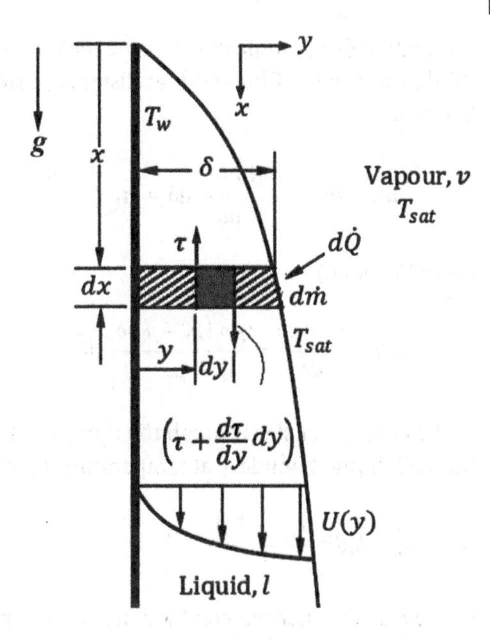

Figure 13.2 Filmwise condensation of vapour on a vertical surface.

At the liquid–vapour interface $d\tau / dy = 0$ and replacing the density ρ_l in Eq. (13.1) by the vapour density ρ_v yields

$$\frac{dp}{dx} = \rho_v g \tag{13.2}$$

Combining Eqs (13.1) and (13.2)

$$\frac{d\tau}{dy} = g\left(\rho_v - \rho_l\right) \tag{13.3}$$

but

$$\tau = \mu_l \left(\frac{dU}{dy}\right)$$

hence,

$$\left(\frac{d^2U}{dy^2}\right) = \frac{g\left(\rho_v - \rho_l\right)}{\mu_l} \tag{13.4}$$

At $y = 0$, $U = 0$ and at $y = \delta$, $dU / dy = 0$

Integrating Eq. (13.4) twice between these limits, we obtain

$$U\left(y\right) = \frac{g\left(\rho_l - \rho_v\right)\delta^2}{\mu_l}\left[\frac{y}{\delta} - \frac{1}{2}\left(\frac{y}{\delta}\right)^2\right] \tag{13.5}$$

The mass flow rate through a cross-section of the film strip per unit width at x is

$$\dot{m}\left(x\right) = \int_0^{\delta} \rho_l U dy = \frac{g\rho_l\left(\rho_l - \rho_v\right)\delta^3}{3\mu_l} \tag{13.6}$$

Equation (13.6) indicates that the local mass flow rate of the condensate is proportional to the third power of the film width at distance x from the origin. The amount of condensing film over height dx is

$$d\dot{m} = \dot{m}(x) + \frac{d\dot{m}(x)}{d\delta} d\delta - \dot{m}(x) = \frac{d\dot{m}(x)}{d\delta}$$

From Eq. (13.6)

$$d\dot{m} = \frac{d\dot{m}(x)}{d\delta} = \frac{g\rho_l(\rho_l - \rho_v)\delta^2}{\mu_l} d\delta \tag{13.7}$$

The energy transfer through the film from the temperature at the vapour–liquid interface T_{sat} to the wall–liquid boundary at temperature T_w by conduction is

$$\dot{Q} = k_l dx \frac{(T_{sat} - T_w)}{\delta} \tag{13.8}$$

where k_l is the thermal conductivity of the fluid film. This energy is equal to the latent enthalpy given up by the condensing vapour mass $d\dot{m}$

$$\dot{Q} = h_{fg} d\dot{m} = h_{fg} \frac{g\rho_l(\rho_l - \rho_v)\delta^2}{\mu_l} d\delta \tag{13.9}$$

where h_{fg} is the latent enthalpy of vaporization.

Equating Eqs (13.8) and (13.9), we obtain

$$k_l dx \frac{(T_{sat} - T_w)}{\delta} = \frac{h_{fg} g\rho_l(\rho_l - \rho_v)\delta^2}{\mu_l} d\delta$$

$$\delta^3 d\delta = \frac{\mu_l k(T_{sat} - T_w)}{h_{fg} g\rho_l(\rho_l - \rho_v)} dx \tag{13.10}$$

Integrating Eq. (13.10) thus

$$\int_0^\delta \delta^3 d\delta = \frac{\mu_l k_l(T_{sat} - T_w)}{h_{fg} g\rho_l(\rho_l - \rho_v)} \int_0^x dx$$

finally results in

$$\delta(x) = \left[\frac{4\mu_l k_l(T_{sat} - T_w)}{h_{fg} g\rho_l(\rho_l - \rho_v)} x\right]^{1/4} \tag{13.11}$$

The local heat transfer coefficient is defined as the heat flux across the film (equal to the heat flux into the wall) divided by the temperature difference across the film

$$h_x = \frac{k_l\left(\frac{T_{sat} - T_w}{\delta}\right)}{(T_{sat} - T_w)}$$

hence,

$$h_x = \frac{k_l}{\delta} \tag{13.12}$$

From Eqs (3.11) and (3.12), h_x can be rewritten as

$$h_x = \left[\frac{h_{fg} g \rho_l k_l^3 (\rho_l - \rho_v)}{4\mu_l (T_{sat} - T_w) x}\right]^{1/4} \tag{13.13}$$

The average heat transfer coefficient is

$$\bar{h}_L = \frac{1}{L}\int_0^L h_x dx = 0.943\left[\frac{\rho_l(\rho_l - \rho_v) g h_{fg} k_l^3}{\mu_l L (T_{sat} - T_w)}\right]^{1/4} \tag{13.14}$$

The latent heat of vaporization is evaluated at T_{sat} and all other properties at the mean film temperature of $T_f = (T_{sat} + T_w)/2$.

Eq. (13.14) is also valid for condensation inside or outside a vertical tube with a diameter that is much larger than the thickness of the condensate film.

If the effect of subcooling is accounted for, the latent enthalpy of vaporization h_{fg} in Eq. (13.14) is replaced by

$$h'_{fg} = h_{fg} + 0.68 c_{pl} (T_{sat} - T_w) \tag{13.15}$$

The total rate of heat transfer to wall surface area A_w $(= L \times 1)$ is

$$\dot{Q} = \bar{h}_L A_w (T_{sat} - T_w)$$

and the total condensation rate is determined from

$$\dot{m}_t = \frac{\dot{Q}}{h'_{fg}} = \frac{\bar{h}_L A_w (T_{sat} - T_w)}{h'_{fg}} = \frac{\bar{h}_L A_w (T_{sat} - T_w)}{h_{fg} + 0.68 c_{pl} (T_{sat} - T_w)} \tag{13.16}$$

The condensation rate $\dot{m}_t$ can also be determined by integrating Eq. (13.7) and multiplying by the width of the wall in a plane perpendicular to the paper. Integrating Eq. (13.7) results in

$$\dot{m} = \frac{g \rho_l (\rho_l - \rho_v) \delta^3}{3\mu_l} \tag{13.17}$$

and

$$\dot{m}_t = \dot{m} W = \frac{g \rho_l (\rho_l - \rho_v) \delta^3}{3\mu_l} \; W \tag{13.18}$$

Example 13.1 Saturated steam at $T_{sat} = 37°C$ condenses on the surface of wall (60 cm high and 40 cm wide) that is maintained at $T_w = 17°C$. Determine the average heat transfer coefficient and the amount of condensate in one hour.

Solution

Properties at the film temperature $T_f = (310 + 290)/2 = 300\ K$:

$\rho_l = 997\ kg/m^3$, $\rho_v = 0.0255\ kg/m^3$, $k_l = 0.613\ W/m.K$, $\mu_l = 8.55 \times 10^{-4}\ N.s/m^2$, $c_{pl} = 4197\ J/kg.K$, $h_{fg} = 2414\ kJ/kg\ (\text{at } T_{sat} = 310\ K)$

$$h'_{fg} = h_{fg} + 0.68 c_{pl}\left(T_{sat} - T_w\right) = 2414 + 0.68 \times 4.197(310 - 290) = 2471\ kJ/kg$$

From Eq. (13.14)

$$\bar{h}_L = 0.943\left[\frac{h_{fg} g \rho_l k_l^3 (\rho_l - \rho_v)}{\mu_l (T_{sat} - T_w) L}\right]^{1/4}$$

$$\bar{h}_L = 0.943\left[\frac{2.471 \times 10^6 \times 9.81 \times 997 \times 0.613^3 (997 - 0.0255)}{8.55 \times 10^{-4}(310 - 290) \times 0.6}\right]^{1/4} = 4548\ W/m^2.K$$

The total condensation rate can be calculated using Eq. (13.16)

$$\dot{m} = \frac{\dot{Q}}{h'_{fg}} = \frac{\bar{h}_L A (T_{sat} - T_w)}{h'_{fg}}$$

where

$$\dot{Q} = \bar{h}_L A (T_{sat} - T_w) = 4548 \times 0.6 \times 0.4 \times 20 = 21830\ W/s$$

$$\therefore \dot{m} = \frac{21830}{2471 \times 1000} = 8.83 \times 10^{-3}\ kg/s \qquad (31.79\ kg/h)$$

13.1.2 Flow Regimes of the Condensate Film

The Nusselt model discussed above assumed that the flow of the condensate film was laminar. However, experiments show that there could be two more flow regimes, namely, laminar wavy flow in which ripples form on the film, and turbulent flow (Figure 13.3).

The film Reynolds number for characterization of condensate film flow regime is defined as

$$Re = \frac{\rho_l U D_H}{\mu_l} = \frac{\rho_l U (4A/P)}{\mu_l} \tag{13.19}$$

where U is the average flow velocity, D_H is the hydraulic diameter (characteristic dimension), A is the flow area $(= \delta \times 1)$, P is the wetted perimeter, and ρ_l and μ_l are the density and dynamic viscosity of the condensate, respectively. The only wetted area for a unit width of film is the wall, hence $P = 1$ and Reynold's number can be written as

$$Re = \frac{4(\rho_l \delta U)}{\mu_l} = \frac{4\dot{m}}{\mu_l} \tag{13.20}$$

From Eqs (13.17) and (13.20)

$$Re = \frac{4 g \rho_l (\rho_l - \rho_v) \delta^3}{3\mu_l^2} \tag{13.21}$$

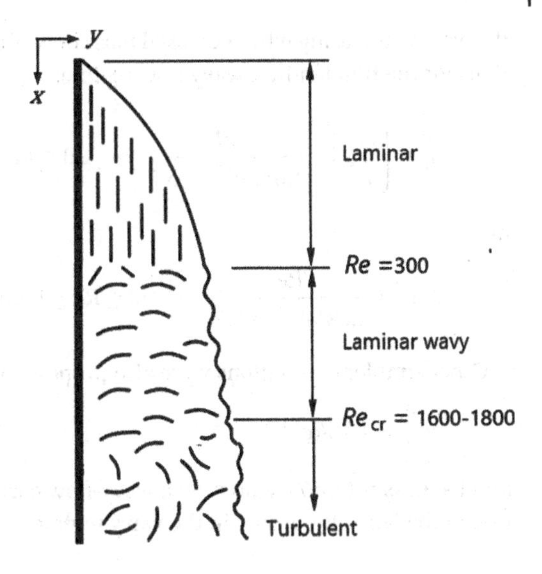

Figure 13.3 Flow regimes in film condensing on a vertical surface.

it can be assumed that $\rho_l \gg \rho_v$ (except near the critical point $T_{cr} = 647K$, $p_{cr} = 221 bar$); hence, $(\rho_l - \rho_v) \cong \rho_l$. Combining Eqs (13.11), (13.14), and (13.21), the relation between the average heat transfer coefficient and Reynold number is obtained

$$\bar{h}_L = 1.47 k_l \left(\frac{g \rho_l^2}{\mu_l^2} \right)^{1/3} Re^{-1/3} \qquad Re \le 30 \tag{13.22}$$

The inverse of the grouping in brackets in Eq. (13.22) has the dimension of length

$$\left(\frac{\mu_l^2}{g \rho_l^2} \right)^{1/3} = \left(\frac{\nu_l^2}{g} \right)^{1/3} \rightarrow \left[\frac{\left(L^2 T^{-1} \right)^2}{L T^{-2}} \right]^{1/3} \rightarrow \left(L^3 \right)^{1/3} \rightarrow L$$

and is used to define the average Nusselt number for condensation.

$$\overline{Nu}_L = \frac{\bar{h}_L \left(\nu_l^2 / g \right)^{1/3}}{k_l} \tag{13.23}$$

13.1.2.1 Laminar Flow Regime

At Reynolds numbers from zero to 30, the flow is laminar with relatively smooth condensate film surface and vertical streamlines. By combining Eqs (13.22) and (13.23), the following simple correlation for the laminar flow regime can be obtained

$$\overline{Nu}_L = 1.47 Re^{-1/3} \qquad Re \le 30 \tag{13.24}$$

13.1.2.2 Laminar Wavy Regime

At Reynolds numbers above 30 and below 1600–1800, ripples develop on the film surface forming waves. Experiments by Kapitsa (1949) showed that wavy flow can be chaotic with the amplitude of the waves increasing with increased mass flow. Kutateladze (1963) proposed the following correlations for the film laminar wavy flow regime:

$$\bar{h}_L = \left(\frac{g\rho_l^2}{\mu_l^2}\right)^{1/3} \frac{k_l Re}{1.08Re^{1.22} - 5.2} \qquad 30 \leq Re \leq 1800 \tag{13.25}$$

or

$$\overline{Nu}_L = \frac{Re}{1.08Re^{1.22} - 5.2} \qquad 30 \leq Re \leq 1800 \tag{13.26}$$

Other simpler correlations were also proposed for this regime by Whitaker (1977)

$$\overline{Nu}_L = 1.76Re^{-1/3} \tag{13.27}$$

and Labuntsov (1957), who used a wavy-flow correction factor with Eq. (13.24) for the determination of the Nusselt number in the wavy regime

$$\overline{Nu}_L == 1.47Re^{-1/3}(Re/4)^{0.04} = 1.39Re^{-0.293} \tag{13.28}$$

Figure 13.4 shows plots of Eqs (13.26), (13.27), and (13.28). The plots of the last two correlations deviate significantly from the plot of Eq. (13.26).

A correlation that agrees well with Eq. (13.26) and can be used for both the laminar and wavy laminar regimes, is obtained by fitting the experimental data in Mikheyev and Mikheyeva (1977) for a number of fluids including, in addition to water, aceton, ethanol, and ammonia

$$\overline{Nu}_L = 1.08Re^{-0.24} \tag{13.29}$$

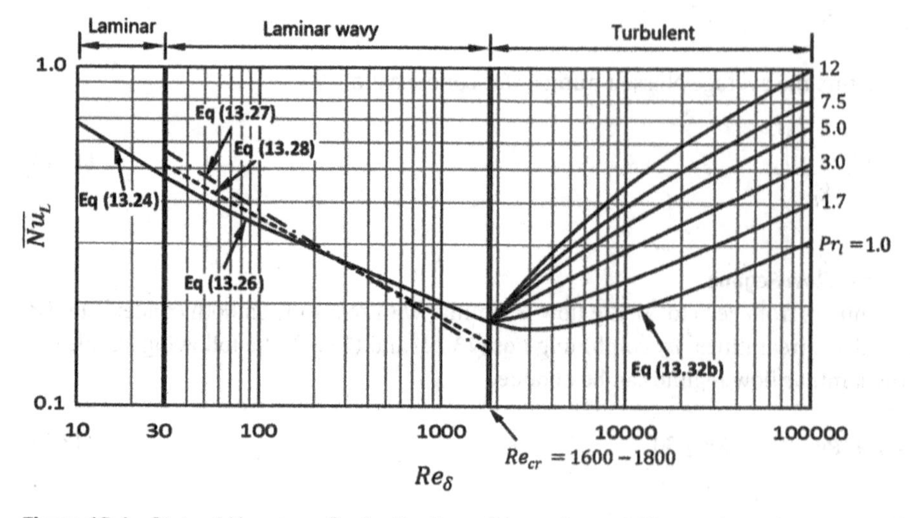

Figure 13.4 Plots of Nu_L versus Re_δ for the three flow regimes of film condensation on a vertical surface.

13.1.2.3 Turbulent Flow Regime

Elevated condensate flow rates can arise from the combined effect of significant temperature gradient ($T_{sat} - T_w$) and height of a vertical surface that can cause the flow to become turbulent. The critical Reynolds number Re_{cr}, at which transition from wavy flow regime to turbulent flow regime occurs is between 1600 and 1800.

Labuntsov (1957) proposed the following correlations for the flow in the turbulent regime:

$$\bar{h}_L = \left(\frac{g\rho_l^2}{\mu_l^2}\right)^{1/3} \frac{k_l Re}{8750 + 58Pr_l^{-0.5}\left(Re^{0.75} - 253\right)} \quad Re \geq 1800, Pr \geq 1 \tag{13.30}$$

or

$$\overline{Nu_L} = \frac{Re}{8750 + 58Pr_l^{-0.5}\left(Re^{0.75} - 253\right)} \quad Re \geq 1800, Pr \geq 1 \tag{13.31}$$

The following correlation for the turbulent flow regime from Mikheyev and Mikheyeva (1977) gives results close to Eq. (13.29)

$$\overline{Nu_L} = \frac{0.173Pr^{0.5}\left(Re / Re_{cr}\right)}{Pr^{0.5} + 1.6\left[\left(Re / Re_{cr}\right)^{0.75} - 1\right]} \quad Re \geq 1600, Pr \geq 1 \tag{13.32a}$$

If the multiplying constant in the numerator is changed from 0.173 to 1.178 and the critical Reynolds number is taken as $Re_{cr} = 1800$, Eq. (13.32a) becomes

$$\overline{Nu_L} = \frac{0.178Pr^{0.5}\left(Re / Re_{cr}\right)}{Pr^{0.5} + 1.6\left[\left(Re / Re_{cr}\right)^{0.75} - 1\right]} \quad Re \geq 1800, Pr \geq 1 \tag{13.32b}$$

Equation (13.32b) gives seamless continuation of Eqs (13.24) and (13.26), as can be seen in Figure 13.4.

If the Reynolds number is unknown, the Nusselt number or the heat transfer coefficient can be rewritten as $\overline{Nu_L} = f(Z)$ for laminar and wavy laminar flows and $\overline{Nu_L} = f(Z, Pr)$ for turbulent flows, where Z is a non-dimensional parameter defined below. Combining Eqs (13.16), (13.20), and (13.23) results in

$$Re = 4Z\overline{Nu_L} \tag{13.33}$$

where

$$Z = \frac{k_l L\left(T_{sat} - T_w\right)}{\mu_l h'_{fg}\left(\mu_l^2 / g\rho_l^2\right)^{1/3}} = \frac{k_l L\left(T_{sat} - T_w\right)}{\mu_l h'_{fg}\left(\nu_l^2 / g\right)^{1/3}} \tag{13.34}$$

Substituting for Re in Eqs (13.24), (13.26), (13.29), (13.31), and (13.32b), we obtain, respectively

$$\overline{Nu_L} = 0.943Z^{-1/4} \quad Z \leq 15.8 \tag{13.35}$$

$$\overline{Nu_L} = 0.813Z^{-0.193} \quad Z \leq 15.8 \tag{13.36}$$

$$\overline{Nu_L} = \frac{1}{Z}(0.68Z + 0.89)^{0.82} \qquad 15.8 \le Z \le 2530 \tag{13.37}$$

$$\overline{Nu_L} = \frac{1}{Z}\left[Pr_l^{0.5}(0.024Z - 53) + 89\right]^{4/3} \qquad Z \ge 2530, Pr_l \ge 1.0 \tag{13.38}$$

$$\overline{Nu_L} = \frac{240}{Z}\left[Pr_l^{0.5}\left(3.96 \times 10^{-4} Z - 1\right) + 1.6\right]^{4/3} \qquad Z \ge 2530, Pr_l \ge 1.0 \tag{13.39}$$

A useful chart for estimating the average heat transfer coefficient of condensing films on vertical surfaces can be plotted, as shown in Figure 13.5, on the basis of the following function.

$$\bar{h}_L = f\left(L\Delta T, T_{sat}\right) \tag{13.40}$$

where $\Delta T = T_{sat} - T_w$, T_{sat} is the saturation temperature, T_w is the wall (surface) temperature, and L is the height of the vertical surface.

In constructing this chart, it is assumed that

1) the laminar and wavy laminar flows are represented by Eq. (13.36) and the turbulent flow by Eq. (13.39)
2) all thermophysical properties vary with temperature and are evaluated at the saturation temperature of water
3) subcooling is ignored and the latent heat of vaporization h'_{fg} in Eq. (13.34) is replaced by h_{fg} which is read off the saturated water property tables directly.

Figure 13.5 can be used for quick estimate of the average heat transfer coefficient of film condensation of water on a vertical surface (plate or tube) for given saturation temperature, temperature gradient in the condensate film, and the height of the vertical surface. An enlarged version of the chart is given in Figure M.1 in Appendix M (can be accessed online at "www.wiley.com/go/Ghojel/heat transfer").

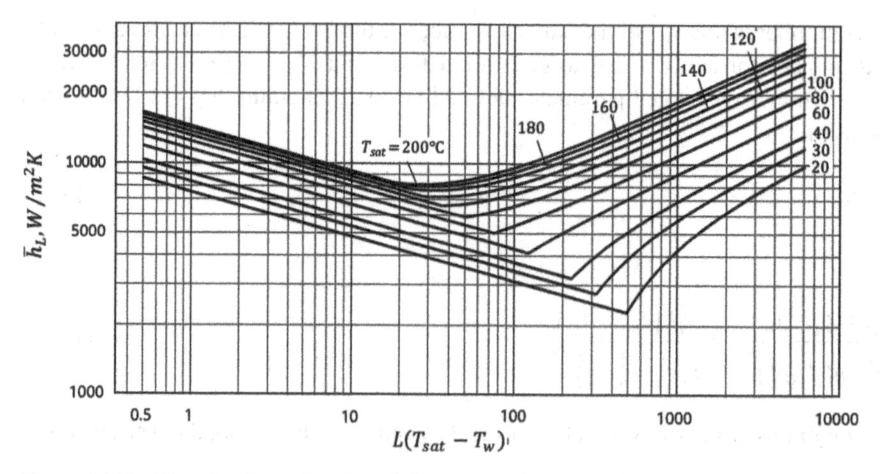

Figure 13.5 Chart for the estimation of the average heat transfer coefficient of condensing film on a vertical surface.

Example 13.2 Estimate the average heat transfer coefficient for the data given in Example 13.1,

Solution

$T_{sat} = 37°C, T_w = 17°C, L = 0.l\,m$

$\Delta T = T_{sat} - T_w = 37 - 17 = 20°C$

$L\Delta T = 0.6 \times 20 = 12$

At $L\Delta T = 10,\ T_{sat} = 40°C$, we find from Figure 13.5 that $\bar{h}_L \cong 4500\,W/m^2.K$ ($4548\,W/m^2.K$ in Example 13.1)

13.1.3 Film Condensation Outside Horizontal Tubes

Condensation on the outside of horizontal tubes is encountered in shell-and-tube heat exchangers where the vapour flows in the shell and the cooling fluid in banks of tubes, as shown in Figure 13.6a.

Nusselt's solution for the problem of laminar condensate flow down a vertical wall can also be applied to film condensation on the outer surface of a horizontal tube. Figure 13.7 shows a schematic diagram of a horizontal tube with a force balance on an annular fluid element at angle θ measured from the top of the tube

$$\mu_l \frac{d^2U}{dy^2} + (\rho_l - \rho_v) g \sin\theta = 0 \tag{13.41}$$

Solving Eq. (13.41) as before results in the Nusselt number for a single tube

$$\overline{Nu}_{D1} = C\left[\frac{\rho_l(\rho_l - \rho_v) g h'_{fg} D^3}{\mu_l k_l (T_{sat} - T_w)}\right]^{1/4} \tag{13.42}$$

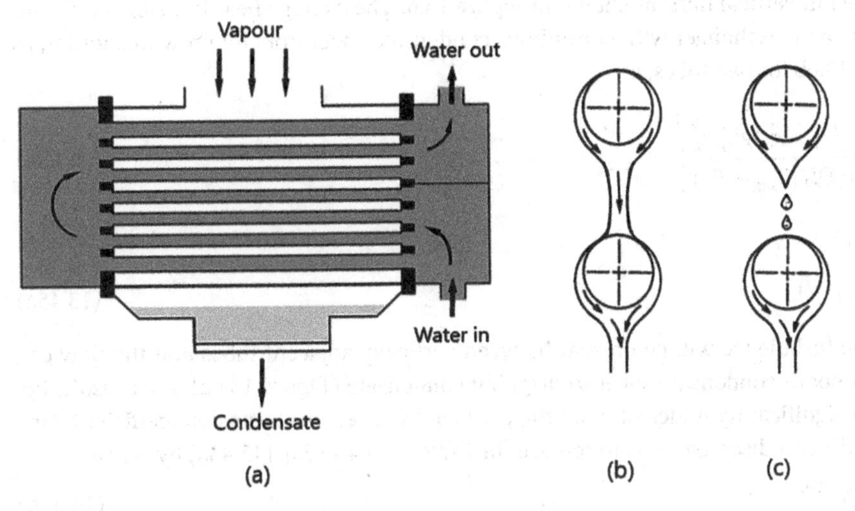

Figure 13.6 (a) Condenser shell-and-tube heat exchanger; (b) film condensation on two in-line tubes with continuous condensate sheet; (c) dripping condensate.

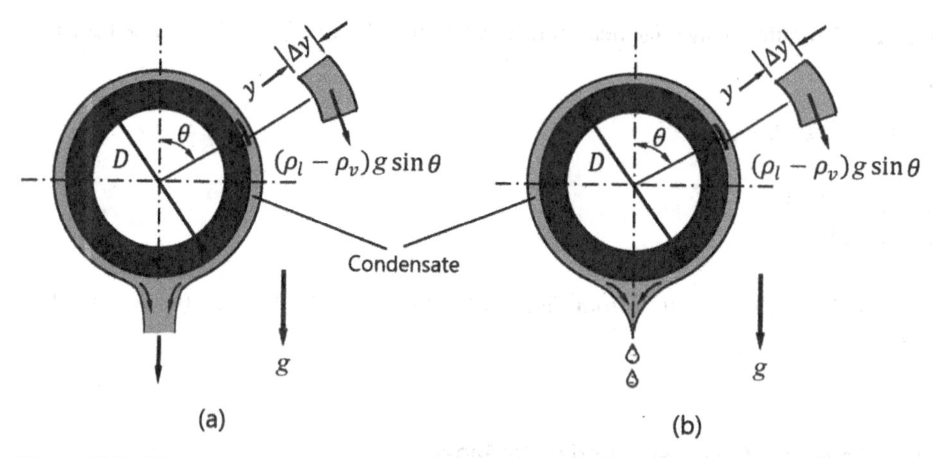

Figure 13.7 Film condensation on a single horizontal tube: (a) continuous condensate sheet; (b) dripping condensate.

where C is a constant ranging in value from 0.725 to 0.729 in various references.

Since $\overline{Nu}_{D1} = \overline{h}_{D1} D / k_l$, Eq. (13.42) can be rewritten with $\overline{h}_L$ as the dependent variable

$$\overline{h}_{D1} = C\left[\frac{\rho_l(\rho_l - \rho_v)gh'_{fg}k_l^3}{\mu_l D(T_{sat} - T_w)}\right]^{1/4} \tag{13.43}$$

The liquid properties are evaluated at the mean film temperature $T_f = (T_{sat} + T_w)/2$ and those of the vapour and latent heat of vaporization at the saturation temperature T_{sat}.

Equations (13.42) and (13.43) are valid for film condensation with continuous condensate sheet (Figure 13.7a) and dripping condensate (Figure 13.7b). The equations are essentially the same as Eq. (13.14), but with a different multiplying constant and with h'_{fg} replacing h_{fg} and the tube diameter D replacing the height L of the vertical surface

Condenser heat exchangers of the shell-and-tube type normally involve in-line tube banks, with the tubes arranged in vertical tiers, as shown in Figure 13.6. The average heat transfer coefficient for N tubes in such an exchanger with continuous condensate sheet from one row to another, as shown in Figure 13.6b for two tubes, is

$$\overline{h}_{DN} = C\left[\frac{\rho_l(\rho_l - \rho_v)gh'_{fg}k_l^3}{\mu_l DN(T_{sat} - T_w)}\right]^{1/4} \tag{13.44}$$

or, more generally,

$$\overline{h}_{DN} = \overline{h}_{D1} N^{-1/4} \tag{13.45a}$$

In reality, some turbulence will be present between vertically adjacent tubes and the flow can transition from smooth condensate sheet to dripping condensate (Figure 13.6c). As a result, Eq. (13.44) will often significantly underestimate the predicted average heat transfer coefficient. One suggestion to rectify this discrepancy is to replace the index $-1/4$ in Eq. (13.45a) by $-1/6$

$$\overline{h}_{DN} = \overline{h}_{D1} N^{-1/6} \tag{13.45b}$$

Chen (1961) proposed the following correlation that accounted for the increase of condensate thickness as a result of continuing condensation from tube to tube

$$\bar{h}_{DN} = C\left[1 + 0.2\frac{c_{pl}\left(T_{sat} - T_w\right)}{h_{fg}}(N-1)\right]N^{-1/4}\left[\frac{\rho_l\left(\rho_l - \rho_v\right)gh'_{fg}k_l^3}{\mu_l D\left(T_{sat} - T_w\right)}\right]^{1/4} \quad (13.46)$$

or

$$\bar{h}_{DN} = C\left[1 + 0.2Ja(N-1)\right]N^{-1/4}\left[\frac{\rho_l\left(\rho_l - \rho_v\right)gh'_{fg}k_l^3}{\mu_l D\left(T_{sat} - T_w\right)}\right]^{1/4} \quad (13.47)$$

where Ja is the Jacob number defined as

$$Ja = \frac{c_{pl}\left(T_{sat} - T_w\right)}{h_{fg}} \quad (13.48)$$

More generally, Eq. (13.47) can be written as

$$\bar{h}_{DN} = \bar{h}_{D1}K \quad (13.49)$$

where $\bar{h}_{D1}$ is the heat transfer coefficient for the first tube at the top given by Eq. (13.43) and K is given by

$$K = \left[1 + 0.2Ja(N-1)\right]N^{-1/4} \quad (13.50)$$

Example 13.3 compares the results from the different equations discussed above.

Example 13.3 Steam condenses on an in-line 42×42 tube bank, each of diameter $D = 6$ mm. Saturated steam enters at the top of the exchanger at $T_{sat} = 54°C$ and the outside surface of the tubes is maintained at $T_w = 25°C$ by a circulating a cold fluid. Assuming laminar condensate flow, determine

a) the heat transfer coefficient of the top tube
b) the heat transfer coefficient of a tier of 42 vertical tubes.

Solution

$$h'_{fg} = h_{fg} + 0.68c_{pl}\left(T_{sat} - T_w\right) = 2373 + 0.68 \times 0.4178 \times (327 - 298) = 2455\ kJ/kg$$

Assume $C = 0.725$ throughout.

a)

From Eq. (13.44)

$$\bar{h}_{D1} = 0.725\left[\frac{\rho_l\left(\rho_l - \rho_v\right)gh'_{fg}k_l^3}{\mu_l D\left(T_{sat} - T_w\right)}\right]^{1/4}$$

$$\bar{h}_{D1} = 0.725\left[\frac{992 \times (992 - 0.098) \times 9.81 \times 2455 \times 10^3 \times 0.631^3}{663 \times 10^{-6} \times 6 \times 10^{-3} \times 29}\right]^{1/4}$$

$$\bar{h}_{D1} = 10972\ W/m^2.K$$

b)

For $N = 42$, from Eq. (13.45a)

$$\bar{h}_{D42} = \bar{h}_{D1}N^{-1/4} = 10972 \times 42^{-1/4} = 10972 \times 0.393 = 4312\ W/m^2.K$$

and from Eq. (13.45b)

$$\bar{h}_{D42} = \bar{h}_{D1} N^{-1/6} = 10972 \times 42^{-1/6} = 10972 \times 0.536 = 5881\ W/m^2.K$$

from Eq. (13.48)

$$Ja = \frac{c_{pl}\left(T_{sat} - T_w\right)}{h_{fg}} = \frac{4178 \times (327 - 298)}{2373 \times 10^3} = 0.051$$

from Eq. (13.50)

$$K = \left[1 + 0.2Ja(N-1)\right]N^{-1/4} = 1 + 0.2 \times 0.051 \times (42-1) \times 42^{-1/4} = 0.577$$

finally, from Eq. (13.49)

$$\bar{h}_{D42} = 10972 \times 0.557 = 6111\ W/m^2.K$$

The heat transfer coefficient obtained using Eq. (13.45a) is lower than the values predicted by Eqs (13.49) and (13.45b) by about 29% and 27%, respectively.

13.1.4 Film Condensation Inside Horizontal Tubes

In the previous discussion of condensation on vertical plain surfaces and tubes and on the outside surfaces of horizontal tubes, the vapour was assumed either stationary or had negligible velocity. However, there are cases in which this assumption is no longer valid, such as in air-conditioning and refrigeration systems where condensation of the vapour moving at significant velocity occurs inside horizontal tubes.

This heat transfer mechanism in film condensation inside smooth horizontal tubes is very similar to film condensation on vertical surfaces, with some major differences. Whereas the flow on vertical surfaces is due mainly to gravity with negligible shear forces, the flow inside tubes is controlled by both gravity and vapour shear. The flow regime in tubes therefore depends on whether gravity or vapour sheer predominate. The criterion for determining the flow regime is the vapour Reynolds number at the inlet defined as

$$Re_{v,i} = \frac{\rho_v U_{m,v} D}{\mu_v} \tag{13.51}$$

13.1.4.1 Laminar Flow

During condensation inside horizontal, smooth tubes at low vapour velocities ($Re_{v,i} < 35000$), gravitational forces, which tend to pull condensate down the tube wall, are much stronger than vapour shear forces, which tend to pull the condensate in the direction of the mean flow. Thus, a condensate film forms on the top of the tube and grows in thickness as it flows around the circumference. The bottom portion of the tube is filled with the condensate forming a liquid stream that moves along the tube in the axial direction driven mainly by the vapour flow (Figure 13.8). The heat transfer coefficient varies over the circumference of the tube with the highest value at the top and the lowest at the bottom.

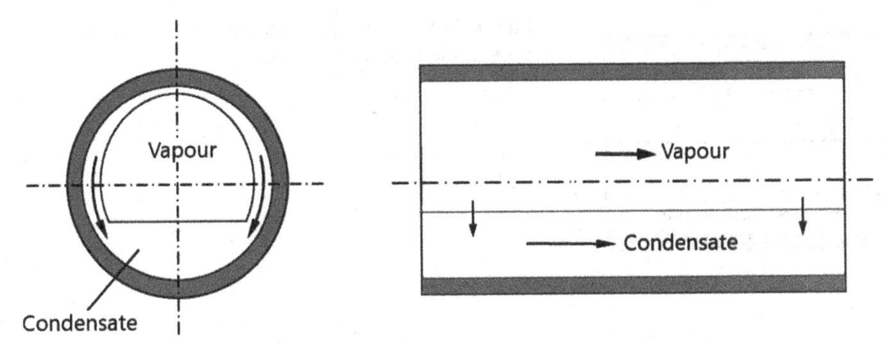

Figure 13.8 Film condensation inside horizontal tube at low vapour velocities.

Based on analytical and experimental results for R-113, Chato (1962) recommended the following equation for the local heat transfer coefficient for $Re_{v,i} < 35000$

$$h = 0.555\left[\frac{\rho_l(\rho_l - \rho_v)gh'_{fg}D^3}{\mu_l k_l(T_{sat} - T_w)}\right]^{1/4} \tag{13.52}$$

Equation (13.52) is essentially Eq. (13.42), with a lower value of the constant C and a definition of h'_{fg} that differs from Eq. (13.15), thus

$$h'_{fg} = h_{fg} + 0.375c_{pl}(T_{sat} - T_w) \tag{13.53}$$

13.1.4.2 Turbulent Flow

At vapour Reynolds numbers above 35,000, shear forces at the interface between the condensate and vapour streams become the main driving force and the condensate acquires the form of an annular flow, with vapour forming in the core which decreases in diameter as the thickness of the condenser layer increases (Figure 13.9). The condensate motion becomes turbulent as a result of alternate entrainment of liquid droplets back and forth by the vapour core and annular liquid film.

Akers et al. (1959) proposed the following correlation for the prediction of the local heat transfer coefficient for this flow regime

$$Nu_D = \frac{hD}{k_l} = 0.026Pr_l^{1/3}Re_{mix}^{0.8} \tag{13.54}$$

In this equation, the mixture Reynolds number Re_{mix} is defined as

$$Re_{mix} = \frac{D}{\mu_f}\left[G_l + G_v\left(\frac{\rho_l}{\rho_v}\right)\right] \tag{13.55}$$

G_l and G_v are the mass velocities of the condensate and vapour, respectively ($G = \dot{m}\,/\,A$ *in* $kg/s.m^2$). G_l and G_v are calculated as if each fluid occupies the entire flow area.

Dobson and Chato (1998) developed the following annular flow correlation for forced-convective condensation in a smooth horizontal tube for the prediction of the local heat transfer coefficient

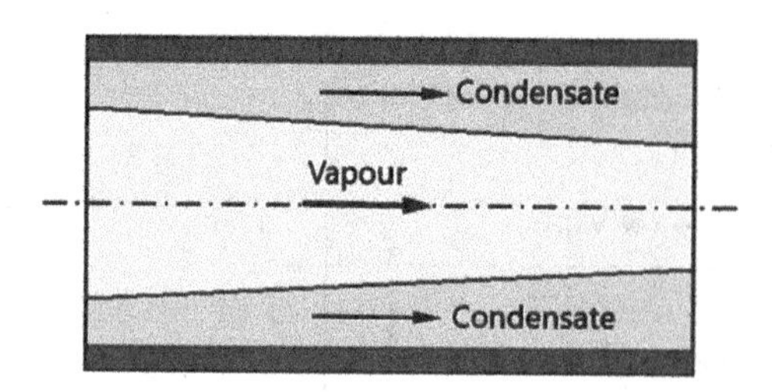

Figure 13.9 Film condensation inside horizontal tube at high vapour velocities.

$$Nu_D = \frac{hD}{k_l} = 0.023 \quad Re_l^{0.8} Pr_l^{0.4}\left[1 + \frac{2.22}{X_{tt}^{0.89}}\right] \tag{13.56}$$

In this equation, the liquid Reynolds number Re_l, vapour quality x, and turbulent-turbulent Lockhart–Martinelli parameter X_{tt} are, respectively

$$Re_l = \frac{4\dot{m}(1-X)}{\pi\mu_l D}$$

$$x = \frac{\dot{m}_v}{\dot{m}}$$

$$X_{tt} = \left(\frac{1-x}{x}\right)^{0.9}\left(\frac{\rho_l}{\rho_v}\right)^{0.5}\left(\frac{\mu_l}{\mu_v}\right)^{0.1}$$

Boiko and Kruzhilin (1966) proposed the following correlation for the estimation of the average heat transfer coefficient in a horizontal tube:

$$\frac{\bar{h}_D}{\bar{h}_{fc}} = \frac{1}{2}\left[\left(\frac{\rho_l}{\rho_m}\right)_i^{1/2} + \left(\frac{\rho_l}{\rho_m}\right)_o^{1/2}\right] \tag{13.57}$$

In this equation, $\bar{h}_{fc}$ is the average heat transfer coefficient for forced turbulent flow in a tube, which can be determined from the following correlation that was given as Eq. (8.17) in Chapter 8

$$\overline{Nu}_D = 0.021 Re_D^{0.8} Pr^{0.43}\left(\frac{Pr_b}{Pr_s}\right)^{0.25} \epsilon_L$$

$$\frac{\rho_l}{\rho_m} = 1 + \frac{\rho_l - \rho_v}{\rho_v} x$$

The density terms ρ_l, ρ_v, and ρ_m are liquid, vapour, and mean mixture densities, respectively, and x is the vapour quality. The correction factor ϵ_L can be determined from Figure 8.8 in Chapter 8 on forced convection. All properties in Eq. (13.57) are evaluated at the saturation temperature.

13.1.5 Dropwise Condensation

The presence of condensate film on a surface provides thermal resistance, which increases with film thickness. One way to improve upon the already high heat transfer coefficients provided by film condensation is to apply coatings to the heat transfer surface that inhibits the formation of a continuous condensate film and, instead, a very large number of individual droplets of varying diameters are formed on the condensing surface. The larger droplets are formed by coalescence of smaller ones as condensation continues and when they grow to a certain size, they roll down the heat transfer surface under the influence of gravity. A film of condensate remains between the droplets, albeit with a thickness much smaller than in filmwise condensation. With dropwise condensation, the heat transfer coefficient can be ten times as high as in filmwise condensation. The coatings that can be used to inhibit surface wetting include silicones, teflon, gold, waxes, metallic nonoxides, and fatty acids. The disadvantages of implementing dropwise condensation in heat exchangers include durability of the surface coating and some of the more effective coatings are too expensive for industrial use.

The dropwise condensation process is too complex to describe using mathematical models and the estimation of the heat transfer coefficient is limited to simple relations such as those by Griffith and Hewitt (1990) for condensation of steam on coated copper surface

$$\bar{h} = 51104 + 2044 t_{sat} \qquad 22°C \le t_{sat} \le 100°C \tag{13.58}$$

$$\bar{h} = 255510 \qquad t_{sat} > 100°C \tag{13.59}$$

or correlations based on experimental data such as those proposed in Isachenko et al. (1969) for dropwise condensation of stagnant saturated steam on a vertical plate

$$\overline{Nu} = \frac{\bar{h}R_{cr}}{k_l} = 3.2\times10^{-4} Re^{-0.84} \Pi_{cr}^{1.16} Pr^{1/3} \qquad 8\times10^{-4} \le Re \le 3.3\times10^{-3} \tag{13.60}$$

$$\overline{Nu} = \frac{\bar{h}R_{cr}}{k_l} = 5.0\times10^{-6} Re^{-1.57} \Pi_{cr}^{1.16} Pr^{1/3} \qquad 3.3\times10^{-3} \le Re \le 1.8\times10^{-2} \tag{13.61}$$

$$R_{cr} = \frac{2\sigma T_s}{h'_{fg}\rho_l (T_{sat} - T_w)}$$

$$Re = \frac{k_l (T_{sat} - T_w)}{\nu_l h'_{fg} \rho_l}, Pr = \frac{\mu_l c_{pl}}{k_l}$$

$$\Pi_{cr} = \frac{2\xi\sigma^2 T_s}{h'_{fg}\rho_l^2 \nu_l^2}$$

R_{cr} is the critical radius of a spherical drop, σ is the liquid–vapour surface tension, and ξ is the temperature coefficient of surface tension. The non-dimensional parameter Π_{cr} accounts for the effect of thermocapillary motion on condensation rate and heat transfer rate from the condensate.

The properties in the above equations are evaluated at the saturation temperature. Equations (13.60) and (13.61) are used to construct the chart shown in Figure 13.10, for the prediction of heat

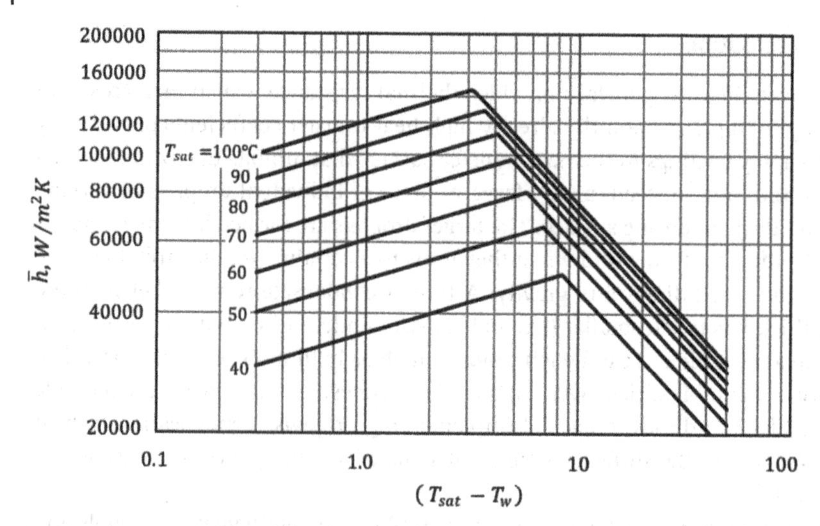

Figure 13.10 Average heat transfer coefficient during dropwise condensation of saturated steam on a vertical surface.

transfer coefficient in dropwise condensation of stagnant saturated steam. The mean heat transfer coefficient $\bar{h}$ in this chart is plotted versus temperature difference $(T_{sat} - T_w)$ for various constant saturation temperatures T_{sat}. An enlarged version of this chart is given in Figure M.2 in Appendix M (can be accessed online at "www.wiley.com/go/Ghojel/heat transfer").

13.2 Boiling Heat Transfer

In the last section, it was shown that heat transfer can be considerably increased by condensing vapour into liquid. Similar results can also be obtained if a liquid is converted to vapour by boiling. When a liquid is boiled, a comparatively large quantity of latent heat is absorbed and convection currents of increasing intensity are set in motion and high heat transfer coefficients are obtained with phase change. This process is known as boiling or evaporation heat transfer that can be categorized as pool boiling or forced-flow boiling. Applications of boiling process can be found in fossil and nuclear power cycles and of evaporation process in evaporators in vapour-compression refrigeration cycles.

13.2.1 Pool Boiling

Pool boiling or bulk boiling is a process in which a large volume of liquid is heated by a submerged heating surface. The process is shown schematically in Figure 13.10 in the form of a log-log plot of the input heat flux versus the difference between the wall temperature and the saturation temperature. Four periods can be identified in the process:

1) Free convection (A–B): The liquid behaves as a single-phase fluid and the heat flux is too low to cause boiling $(T_w - T_{sat}) \leq 5°C$.
2) Nucleate boiling (B–C): During this period, as a result of increased heat flux, a large number of bubbles form at the hot surface which travel across the liquid to the surface causing agitation

that peaks at the critical heat flux. The critical heat flux corresponds to the maximum heat transfer coefficient attainable and the bulk liquid temperature reaches the saturated temperature $5°C < (T_w - T_{sat}) < 30°C$

3) Transition boiling (C–D): After reaching peak heat flux at C, the surface suddenly becomes insulated by a continuous vapour blanket causing a sudden jump in wall temperature at E. This temperature can be more than 1000 K for water and could approach or exceed the melting point of the wall. If the wall metal can withstand high temperatures without burn-out, the stable film boiling can be attained at high heat flux values as shown by the dashed line C-E. For high heat-transfer coefficients, it is preferable to operate near maximum heat flux if burn-out can be avoided.
4) If the input heat is regulated by controlling the surface temperature instead of flux, a semi-stable regime is formed between points C and D, which transitions to a stable film boiling regime between D and E $(T_w - T_{sat}) > 150°C$ (Figure 13.11).

The coefficient of heat transfer in the region of nucleate boiling is affected by pressure, thermal conductivity, and viscosity of the liquid. Intensity of heat transfer increases with increasing thermal conductivity and decreases with increasing viscosity. Based on experimental data on nucleate boiling on a horizontal plate facing upwards in pool of liquid, Rohsenow (1952) proposed a correlation to predict the local heat transfer coefficient as a function of the temperature difference $(T_w - T_{sat})$

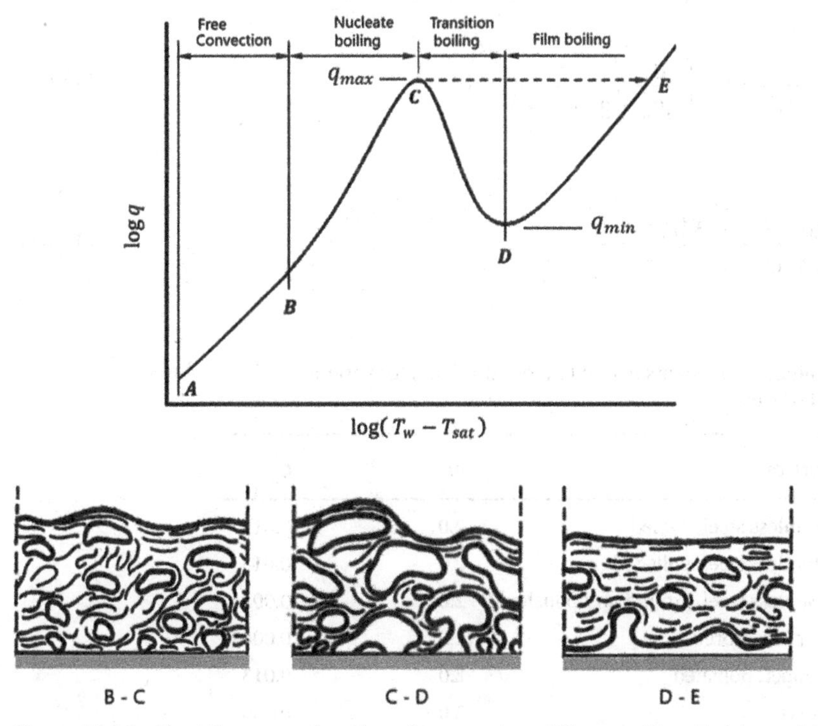

Figure 13.11 Heat flux q as a function of temperature difference $(T_w - T_{sat})$ in pool boiling by submerged heating surface.

$$Nu = \frac{hL_c}{k_l} = \frac{1}{C_{nb}^3 Pr_l^m}\left[\frac{c_{pl}\left(T_w - T_{sat}\right)}{h_{fg}}\right]^2 \tag{13.62}$$

or

$$Nu = \frac{hL_c}{k_l} = \frac{Ja^2}{C_{nb}^3 Pr_l^m} \tag{13.63}$$

where Ja is the Jacob number

$$Ja = \frac{c_{pl}\left(T_w - T_{sat}\right)}{h_{fg}} \tag{13.64}$$

The characteristic length L_c is given by

$$L_c = \left[\frac{\sigma}{g\left(\rho_l - \rho_v\right)}\right]^{1/2} \tag{13.65}$$

Equation (13.62) shows that for a given liquid the heat transfer coefficient is simply a function of ΔT^2.

Selected values of the constants m and C_{nb} for Eq. (13.62) are shown in Table 13.1. More data on these constants can be found in Kreith et al. (2011) and Thomas (1999).

Rohsenow's equation is also expressed in terms of the heat flux as

$$\frac{c_{pl}\left(T_w - T_{sat}\right)}{h_{fg} Pr_l^n} = \frac{Ja}{Pr_l^n} = C_{nb}\left[\frac{q}{\mu_l h_{fg}}\sqrt{\frac{\sigma}{g\left(\rho_l - \rho_v\right)}}\right]^{1/3} \tag{13.66}$$

or

$$q = \mu_l h_{fg}\left[\frac{c_{pl}\left(T_w - T_{sat}\right)}{h_{fg} Pr_l^n C_{nb}}\right]^3\left[\frac{g\left(\rho_l - \rho_v\right)}{\sigma}\right]^{1/2} \tag{13.67}$$

Table 13.1 Selected values of constants m and C_{nb} for Eq. (13.62) for various Liquid/solid-surface interfaces.

Liquid	Surface	m	C_{nb}
Water	Stainless steel, etched	2.0	0.013
Water	Stainless steel, polished	2.0	0.013
Water	Stainless steel, ground and polished	2.0	0.008
Water	Copper, scored	2.0	0.0068
Water	Copper, polished	2.0	0.013
Water	Brass	2.0	0.006
Benzene	Chromium	4.1	0.010
Ethanol	Chromium	4.1	0.0027
n-Butyl alcohol	Copper	4.1	0.00305

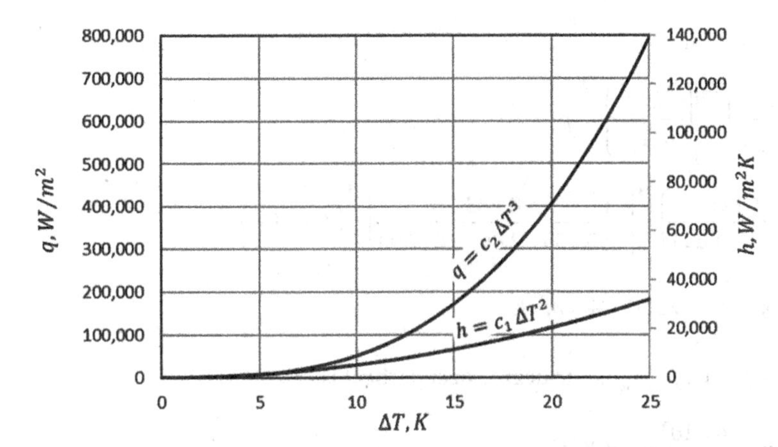

Figure 13.12 Heat flux and heat transfer coefficient versus degree of superheat during nucleate boiling of water.

The constant exponent of the Prandtl number n in Eqs (13.66) and (13.67) is equal to 1.0 for water and 1.7 for other fluids, and the constant C_{nb} is the same as in Eq. (13.62).

Mikheyev and Mikheyeva (1977) presented the following correlations for the heat transfer coefficient in terms of heat flux in developed nucleate boiling and for the maximum heat flux value

$$h = 0.075\left[1+10\left(\frac{\rho_v}{\rho_l-\rho_v}\right)^{2/3}\right]\left(\frac{k_l^2}{\sigma\nu_l T_{sat}}\right)^{1/3} q^{2/3} \tag{13.68}$$

Replacing q in Eq. (13.68) with $h(T_w - T_{sat})$, we obtain

$$h = 4.218\times10^{-4}\left[1+10\left(\frac{\rho_v}{\rho_l-\rho_v}\right)^{2/3}\right]^3\left(\frac{k_l^2}{\sigma\nu_l T_{sat}}\right)\left(T_w - T_{sat}\right)^2 \tag{13.69}$$

i.e., $h = f(\Delta T^2)$

$$q_{max} = 0.14 h_{fg}\rho_v^{0.5}\left[\sigma g\left(\rho_l-\rho_v\right)\right]^{1/4} \tag{13.70}$$

Equations (13.68) to (13.70) were experimentally validated for pool and forced-flow boiling of different fluids.

The thermal properties in Eqs (13.62) to (13.70) are evaluated at the saturation temperature T_{sat}. The plots for $h = f(\Delta T)$ and $q = f(\Delta T)$ as per Eqs (13.68) and (13.69) are shown in Figure 13.12. The heat transfer coefficient is proportional to ΔT^2, and the heat flux to ΔT^3.

Example 13.4 Determine the heat transfer coefficient due to nucleate pool boiling and the degree of superheat when water at $p_{sat} = 198.53\ kPa$ is heated on a stainless-steel surface if the applied heat flux is $4\times10^5\ W/m^2$.

Solution

Properties at $p_{sat} = 198.53\ kPa$ $T_{sat} = 120°C$: $\rho_l = 943.4\ kg/m^3$, $\rho_v = 1.121\ kg/m^3$, $h_{fg} = 2.203\times10^6\ J/kg$, $c_{pl} = 4.244\times10^3\ J/kg.K$, $k_l = 0.683\ W/m.K$, $\mu_l = 2.32\times10^{-4}\ kg/m.s$, $\nu_l = 2.46\times10^{-7}\ m^2/s$, $\sigma = 0.055$, $Pr_l = 1.44$

From Eq. (13.68)

$$h = 0.075\left[1 + 10\left(\frac{\rho_v}{\rho_l - \rho_v}\right)^{2/3}\right]\left(\frac{k_l^2}{\sigma \nu_l T_{sat}}\right)^{1/3} q^{2/3}$$

$$= 0.075 \times \left[1 + 10\left(\frac{1.121}{943.4 - 1.121}\right)^{2/3}\right] \times \left(\frac{0.683^2}{0.055 \times 2.46 \times 10^{-7} 393}\right)^{1/3} \left(4.0 \times 10^5\right)^{1/3}$$

$$= 0.075 \times 1.11 \times 44.4 \times 5429 = 20{,}067\ W/m^2.K$$

Assuming the heat flux is $q = h(T_w - T_{sat})$, the degree of superheat is

$$\Delta T = T_w - T_{sat} = \frac{q}{h} = \frac{4 \times 10^5}{2.0067 \times 10^4} \approx 20°C$$

13.2.2 Film Boiling

The boiling curve in Figure 13.11 shows a stable film boiling regime between points D and E at temperatures above 150°C. A correlation for Nusselt number by Bromley (1950) is usually used for horizontal cylinders and for spheres in this regime.

$$Nu = \frac{\bar{h}D}{k_v} = C\left[\frac{g\rho_v(\rho_l - \rho_v)h'_{fg}D^3}{\mu_v k_v (T_w - T_{sat})}\right]^{1/4} \tag{13.71}$$

where

$$h'_{fg} = h_{fg} + 0.4c_{pv}(T_w - T_{sat}) \tag{13.72}$$

D is the diameter of the sphere or outer diameter of the cylinder.
$C = 0.62$ for horizontal cylinders, $C = 0.67$ for spheres.

For vertical cylinders, Lobunstev proposed the following correlation for the average heat transfer coefficient:

$$\bar{h} = 0.25\left[\frac{\rho_v C_{pv} g k_v^2 (\rho_l - \rho_v)}{\mu_v}\right]^{1/3} \tag{13.73}$$

The liquid properties and the latent heat of vaporization h_{fg} are evaluated at the saturation temperature and the vapour properties are evaluated at the film temperature of $T_f = (T_{sat} + T_w)/2$

At wall (surface) temperatures above 300°C, radiation heat transfer across the vapour film can be significant. Bromley (1950) proposed the following correlation for the total heat transfer coefficient for film pool boiling:

$$\bar{h}^{4/3} = \bar{h}_{conv}^{4/3} + \bar{h}_{rad} + \bar{h}_{rad}\bar{h}^{1/3} \tag{13.74}$$

For $h_{conv} > \bar{h}_{rad}$, a simpler relationship can be used

$$\bar{h} = \bar{h}_{conv} + \frac{3}{4}\bar{h}_{rad} \tag{13.75}$$

The effective radiation heat transfer coefficient is expressed as

$$\bar{h}_{rad} = \frac{\varepsilon\sigma\left(T_w^4 - T_{sat}^4\right)}{T_w - T_{sat}} \tag{13.76}$$

where ε is the emissivity of the wall.

13.2.3 Forced-Convection Boiling

In many engineering applications, boiling occurs as the liquid is forced to flow outside or inside horizontal or vertical tubes with heated surfaces. The same boiling regimes involved in pool boiling are also present in this case, but there are some distinctive features that are specific to convection boiling. For example, the proportion of steam bubbles in the liquid increases as it is carried along with the flow leading to decreasing density and increasing flow velocity. The regimes associated with convective boiling inside a heated vertical tube are shown in Figure 13.13. The fluid enters at the bottom in subcooled liquid form and heat is transferred by forced convection. As the fluid ascends, vapour bubbles coalesce into large volumes of vapour (slugs) with nucleation sites on the wall. During this period, the heat transfer coefficient starts rising above the forced convection value. Further evaporation causes the flow to change to annular flow with the liquid forming an annular

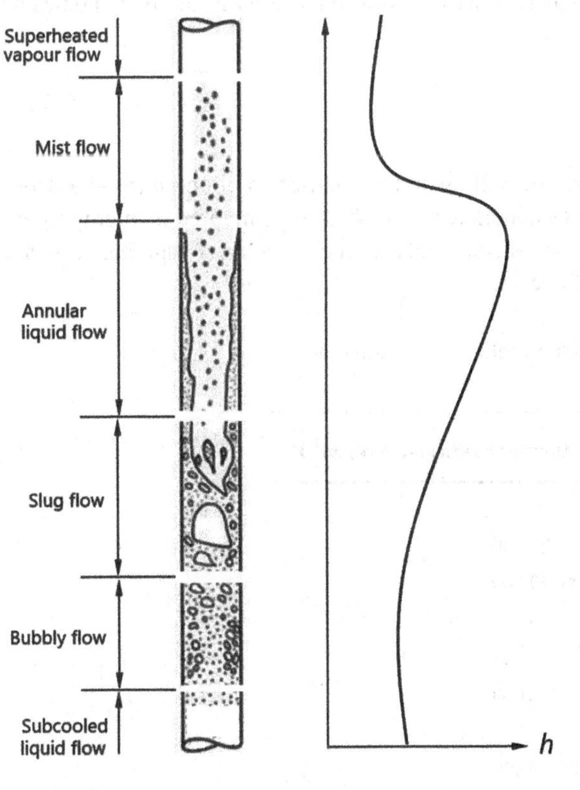

Figure 13.13 Forced convection boiling stages in once-through vertical tube.

layer and fine bubbles forming in the core. The annular liquid layer gradually thins and the flow transitions into mist flow as the liquid droplets evaporate. At this stage, the heat transfer coefficient reaches its peak and starts dropping sharply as the inner surface of the tube becomes completely dry. The drop in convection coefficient is accompanied by sharp increase in surface temperature, which could lead to tube failure if constant heat flux is applied along its length. Tube failure can be avoided if the tube wall temperature is kept constant during the process. Mist flow is followed by saturated vapor flow, which becomes superheated with further heating.

Not all the regimes of boiling discussed above are observed in actual boiling heat exchangers that tend to have shorter tubes. Figure 13.13 shows that two-phase boiling heat transfer is a very complex phenomenon encompassing several flow regimes and the simplest way to estimate the heat transfer coefficient for the whole process is to use the additive formula

$$q = q_{nb} + q_{fc} \tag{13.77}$$

where q_{nb} is the nucleate pool boiling heat flux and $q_{fc} = h_{fc}\left(T_w - T_l\right)$ is the single-phase forced convection heat flux to the liquid.

For constant temperature difference $\Delta T = T_w - T_l$, Eq. (13.77) can be rewritten in terms of the heat transfer coefficients

$$h = h_{nb} + h_{fc} \tag{13.78}$$

h_{nb} can be determined from Eq. (13.68) or (13.69) and h_{fc} can be calculated from the Dittus–Boelter correlation (see Eq. (8.15)), which can be rewritten after replacing the coefficient 0.023 by 0.019 as

$$\bar{h}_{fc} = 0.019 Re_D^{0.8} Pr_l^n \left(\frac{k_l}{D}\right) \tag{13.79}$$

This chapter has provided an overview of the boiling and condensation phenomena at a depth appropriate for an introductory text. It can be concluded that when implemented in energy-intensive engineering applications, such as nuclear reactors and rocket engines, very high heat transfer coefficients can be achieved, as shown in Table 13.2.

Table 13.2 Comparison of indicative heat transfer coefficients in different convection heat transfer modes.

Process	Heat transfer coefficient, h $W/m^2.K$
Forced convection	
Gases	25–250
Liquids	100–20,000
Free convection	
Gases	2–25
Liquids	50–1000
Convection with phase change	
Condensation	9000–24,000
Boiling	2500–100,000

Problems

13.1 Steam at a pressure of 198.5 kPa and temperature of 150°C is in contact with a plate that is 1 m high and 0.6 m wide, which is maintained at 110°C. Assuming filmwise condensation with laminar flow, determine the film thickness and local heat transfer coefficient half-way down the plate and the bottom of the plate.

13.2 For the data in Problem 13.1, determine the Reynolds number at the bottom of the plate, the average convection coefficient, and the rate of heat transfer.

13.3 Saturated steam at saturation temperature of $T_{sat} = 100°C$ condenses on a vertical plate that is maintained at 90°C by circulating cooling water on the other side. The plate is 3 m high and 5 m wide. Calculate the average heat transfer coefficient and the rate of heat transfer to the water. Compare the calculated heat transfer coefficient with the value obtained from the chart in Figure 13.5.

13.4 Saturated steam at a saturation temperature of $T_{sat} = 100°C$ condenses on a vertical plate that is maintained at 90°C by circulating cooling water on the other side. The plate is 3 m high and 5 m wide. Assuming filmwise condensation and laminar flow, tabulate and plot the film thickness and local heat transfer coefficient in vertical increments of 0.5 m.

13.5 Saturated steam condenses on the outside of a vertical tube of 0.05 m diameter and 0.8 m height. The saturation temperature of the steam is $T_{sat} = 310\ K$ and the temperature of the pipe wall is maintained constant at $T_w = 296\ K$. Calculate

(a) the film thickness at the bottom of the tube

(b) the average heat transfer coefficient, compare with the value obtained from the chart in Figure 13.5

(c) the amount of steam condensed in one hour.

13.6 Rework Problem 13.5 assuming laminar flow condensation.

13.7 Rework Problem 13.6 with the tube mounted horizontally and compare the results.

13.8 Steam at $T_{sat} = 40°C\ (p_{sat} = 7.384\ kPa)$ condenses on the outer surface of thin-walled horizontal tube of diameter $D = 30\ mm$. Cooling water flows inside the tube maintaining a constant surface temperature $T_w = 30°C$. Determine

(a) the rate of heat transfer to the water

(b) the rate of condensation of steam per unit length.

13.9 A square 25 by 25 array of 12 mm diameter thin-walled horizontal tubes is to be used to condense 10,000 kg/h of a refrigerant at 37.8°C by flowing water inside the tubes that are maintained at 32.2°C. Determine the required length of the tubes. The heat of vaporization of the refrigerant at 37.8°C is $h'_{fg} \cong h_{fg} = 130\ kJ/kg$.

The properties of the refrigerant at the film temperature $T_f = (37.8 + 32.2)/2 = 35°C$ are: $\rho_l = 1168\ kg/m^3$; $\rho_v = 43.41 kg/m^3$ $c_{pl} = 1.471 \times 10^3\ J/kg.K$, $k_l = 0.0783\ W/m.K$; and $\mu_l = 1.772 \times 10^{-4}$.

13.10 The tube bank of a steam condenser consists of a square array of 20×20 tubes, each 6 mm in diameter in an arrangement similar to Figure P13.10. The tubes through which cooling water is circulating are exposed to saturated steam at a pressure of $p_s = 15\ kPa$ and the tube surface is maintained at $T_w = 25°C$. Calculate the rate at which steam is condensed per unit length of the tube.

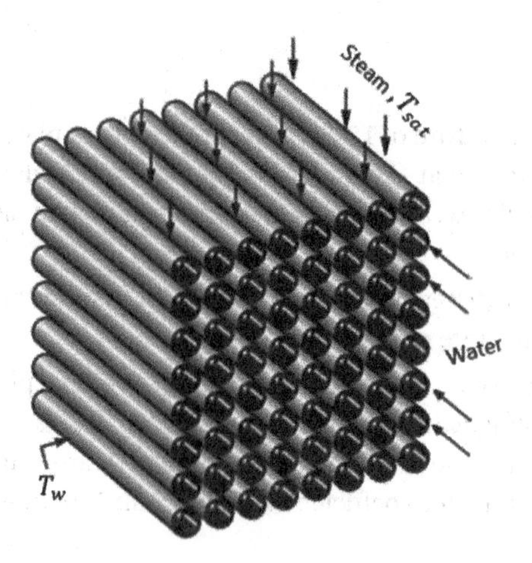

Figure P13.10 Schematic representation of the tube bank in Problem 13.10.

13.11 Saturated stagnant steam at $T_{sat} = 90°C$ is condensed on a Teflon-coated vertical plate maintained at temperature $T_w = 4°C$. Assuming dropwise condensation, estimate the average heat transfer coefficient for the process. What is the effect of increasing the plate temperature to $T_w = 10°C$?

13.12 Water is boiled at atmospheric pressure in a polished stainless-steel kettle by means of a gas burner. The kettle has a circular bottom plate of 250 *mm* diameter. The inner surface of the bottom of the kettle is kept at a constant temperature of 106°C. Determine:

(a) the heat transfer coefficient of the boiling process

(b) the rate of heat transfer to the water

(c) the rate of evaporation of water.

13.13 Horizontal tube of 12 *mm* diameter is submerged in water at atmospheric pressure. The surface temperature of the tube is maintained at 500°C. Calculate the heat transfer coefficient and the heat flux from the surface of the tube. Assume film boiling of the water on the tube.

13.14 Rework Problem 13.3 with the tube in vertical position.

13.15 Water is boiled in a polished copper pan of 350 *mm* diameter on a gas stove. The water is at atmospheric pressure and the pan is maintained at 115°C. Determine:

(a) the power input by the gas burner

(b) the rate of evaporation of water

(c) the pan burnout flux.

13.16 A nickel wire of 1 *mm* diameter and 400 *mm* long, carrying a current is used to boil water in a water bath open to atmospheric air. Calculate the voltage at the burnout point if at this point the wire carries a current of 190 *A*.

13.17 An electric wire of 1.5 mm diameter and 20 cm long is laid horizontally and submerged in water at atmospheric pressure. The current flowing through the wire is 40 A, while voltage drop is 16 V. Calculate the heat flux, heat transfer coefficient and excess temperature.

13.18 A cylindrical heating element of 10 *mm* diameter and 0.92 emissivity is immersed horizontally in a water bath exposed to atmospheric pressure. Under steady-state conditions, the surface of the element is at a temperature of 260°C. Calculate the total heat transfer coefficient of the process and the power dissipation by the element per unit length.

13.19 A pan having a diameter $D = 150\ mm$ is used to boil water on a gas stove. Under steady-state conditions, the surface temperature of the pan in contact with the water is at a temperature of $T_w = 120°C$. Calculate:

(a) the power input to the water causing the water to boil

(b) the evaporation rate of water

(c) the critical heat flux.

14

Mass Transfer

Convective heat transfer is accompanied by transfer (movement) of the fluid, either on a macroscopic scale with a pump or a blower as a motive force in forced convection, or on a microscopic scale at the level of molecules moving under the effect of buoyancy forces within the fluid in free convection. However, there is another mode of mass transfer that is encountered in processes such as condensation, evaporation, distillation, drying, humidification, and so on. This mode is known as mass transfer and is driven by the concentration difference. If the concentration of molecules of a particular species is higher in one region of a system than another, these molecules tend to slowly move towards regions where the concentration is lower. This process is known as diffusion mass transfer or, simply, mass diffusion. Examples of mass diffusion include, for example, the spread of aroma of freshly made morning coffee in a room, and the dissolution of sugar added to a cup of that coffee even without stirring. On an industrial scale, mass diffusion is observed as smoke rises from a tall chimney and disperses into the atmosphere or as water vaper is transferred into dry air in a dryer.

Mass could also be transferred between a surface and a moving fluid under the combined effect of mass diffusion and bulk fluid motion. Examples of this mode include air flowing over the water surface of a lake picking up water vapour or a stream of steam condensing on a cold surface.

14.1 Species Concentrations

Concentration means the amount of a substance in a given volume that can be quantified on a mass or molar basis. The mass concentration of a species is the density which is the mass per unit volume. The total density of a mixture of mass m and volume V is

$$\rho = \frac{m}{V} \quad kg/m^3$$

The partial density of component i of mass m_i in a multi-component mixture of total density ρ is

$$\rho_i = \frac{m_i}{V} \quad kg/m^3$$

Heat Transfer Basics: A Concise Approach to Problem Solving, First Edition. Jamil Ghojel.

Companion website: www.wiley.com/go/ghojel/heat_transfer

Hence,

$$\rho = \sum \frac{m_i}{V} = \sum \rho_i$$

The mass fraction of species i in the same mixture as above is

$$w_i = \frac{m_i}{m} = \frac{m_i}{V}\frac{V}{m} = \frac{\rho_i}{\rho} \tag{14.1}$$

and $\sum w_i = 1, \sum m_i = m$

Molar concentration or molar density is the number of moles of a species per unit volume. The total molar density of a mixture of N kmol and volume V is

$$C = \frac{N}{V} \quad kmol/m^3$$

The partial molar density of component i of N_i kmol is

$$C_i = \frac{N_i}{V} \quad kmol/m^3$$

hence,

$$C = \sum \frac{N_i}{V} = \sum C_i$$

The molar fraction of a species i

$$y_i = \frac{N_i}{N} = \frac{N_i}{V}\frac{V}{N} = \frac{C_i}{C} \tag{14.2}$$

and $\sum y_i = 1, \sum N_i = N$

The mass of 1 kmol of a substance is μ kg (μ is known as the molar mass or molecular weight). The partial molar density is thus

$$C_i = \frac{N_i}{V} = \frac{N_i \mu_i}{V \mu_i} = \frac{m_i}{V \mu_i} = \frac{\rho_i}{\mu_i}$$

and the mixture molar density

$$C = \frac{\rho}{\mu}$$

Also,

$$\mu = \frac{m}{N} = \frac{\sum N_i \mu_i}{N} = \sum \left(\frac{N_i}{N}\right)\mu_i = \sum y_i \mu_i \tag{14.3}$$

The relationship between the mass fraction and molar fraction can be determined by combining Eqs (14.1), (14.2), and (14.3), making use of the relation $C = \rho / \mu$, and rearranging

$$w_i = \left(\frac{\mu_i}{\mu}\right) y_i = \frac{y_i \mu_i}{\sum y_i \mu_i} \tag{14.4}$$

The same result can be obtained as follows:

$$w_i = \frac{m_i}{m} = \frac{m_i}{\sum m_i} = \frac{N_i \mu_i}{\sum N_i \mu_i} = \frac{\left(\frac{N_i}{N}\right)\mu_i}{\frac{\sum N_i \mu_i}{N}} = \frac{y_i \mu_i}{\sum y_i \mu_i}$$

A gas mixture can be treated as an ideal gas at low pressures, and the equation of state for species i and for the whole mixture in terms of mols of substance will then be respectively

$$p_i V = N_i \bar{R} T,\ pV = N \bar{R} T$$

where p_i is the partial pressure of the species (the pressure which the species would exert if it existed alone at the mixture temperature and volume) and $\bar{R}$ is the universal gas constant ($8.314\,kJ/kmol\,K$).

Dividing the first equation by the second, we obtain

$$\frac{p_i}{p} = \frac{N_i}{N} = y_i$$

$$p_i = y_i p \tag{14.5a}$$

The total pressure p of a mixture of gases is the sum of the partial pressures of the constituent species

$$p = \sum y_i p \tag{14.5b}$$

For a binary mixture of components A and B, the following relations apply:

$$\rho_A + \rho_B = \rho$$

$$C_A + C_B = C$$

$$w_A + w_B = 1$$

$$y_A + y_B = 1$$

$$\frac{w_A}{\mu_A} + \frac{w_B}{\mu_B} = \frac{1}{\mu}$$

$$p_A + p_B = p$$

The mass average velocity U for a binary mixture of components A and B is defined as:

$$U = \frac{\rho_A}{\rho} U_A + \frac{\rho_B}{\rho} U_B = w_A U_A + w U_B \tag{14.6}$$

$$U = \frac{C_A}{C} U_A + \frac{C_B}{C} U_B = y_A U_A + y U_B \tag{14.7}$$

Diffusive flux on mass and molar basis are defined respectively as:

$$J = \frac{\dot{m}}{A} = \rho U \quad kg/m^2 s$$

$$\bar{J} = \frac{\dot{M}}{A} = CU \quad kmol/m^2 s$$

Example 14.1 A mixture of oxygen (O_2) and carbon dioxide (CO_2) is at 288 K and 150 kPa. If the molar fraction of O_2 is 40%, determine:

a) the molar fraction of CO_2
b) the molecular mass of the mixture
c) molar concentration of the two components.

Solution

a) $y_{CO_2} + y_{O_2} = 1$

$$\therefore y_{CO_2} = 1 - 0.4 = 0.6 \ (60\%)$$

b) From Eq. (14.3)

$$\mu = y_{CO_2}\mu_{CO_2} + y_{O_2}\mu_{O_2} = 0.6 \times 44 + 0.4 \times 32 = 39.2$$

c) Partial pressures of the components in the mixture of pressure $p = 150\,kPa$

$$p_{CO_2} = y_{CO_2} p = 0.6 \times 150 = 90\,kPa$$

$$p_{O_2} = y_{O_2} p = 0.4 \times 150 = 60\,kPa$$

Assuming perfect gas, $pV = N\bar{R}T$

$$\therefore C = \frac{N}{V} = \frac{P}{\bar{R}T}$$

$$C_{CO_2} = \frac{p_{CO_2}}{8.314 \times 288} = \frac{90}{2394} = 0.0376\,kmol/m^3$$

$$C_{O_2} = \frac{p_{O_2}}{8.314 \times 288} = \frac{60}{2394} = 0.0250\,kmol/m^3$$

14.2 Diffusion Mass Transfer

Mass transfer occurs at the molecular level as a result of diffusion of matter from a region of high concentration to a region of low concentration, as shown schematically for a single species in Figure 14.1.

If the molecular concentration of a gaseous or liquid species is greater on the left side of the container than on the right side, more molecules will move from left to right than in the opposite direction with a net mass transfer to the right. The diffusion rate for this process is given by Fick's

Figure 14.1 Concentration profile in mass transfer process.

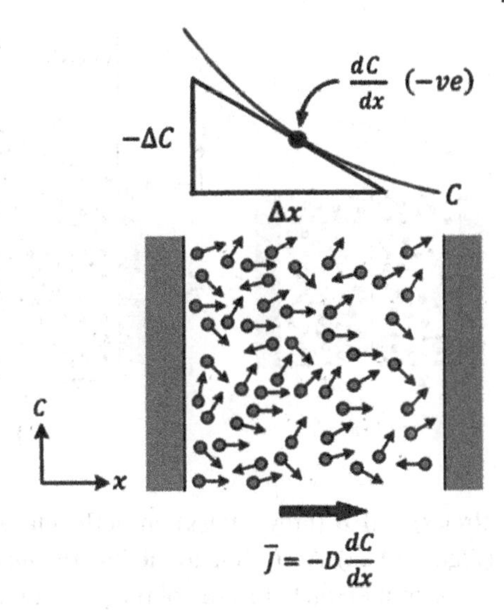

law of diffusion, which states that the mass flux $J\ (\dot{M}/A)$ of a constituent per unit area is proportional to the concentration gradient

$$\bar{J} = \frac{\dot{M}}{A} = -D\frac{dC}{dx} \quad kmole/m^2/s \tag{14.8}$$

and the molar flow rate

$$\dot{M} = -DA\frac{dC}{dx} \quad kmole/s \tag{14.9}$$

where
$\bar{J} = \dot{M}/A -$ diffusive molar flux, $kmole/m^2.s$
$\dot{M} -$ molar flow rate, $kmole/s$
$D -$ diffusion coefficient, m^2/s
$A -$ area normal to flow direction, m^2
$C -$ concentration of the species per unit volume, $kmole/m^3$
The negative sign in Flick's law is introduced to obtain a positive mass flux when the concentration gradient is negative (decreasing concentration in the x direction). Fick's law describes the diffusion of mass down a species concentration gradient in a manner similar to Fourier's law, which describes heat conduction down a temperature gradient.

Fick's law can also be written in terms of mass concentration as

$$\dot{m}_A = -DA\frac{d\rho}{dx} \quad kg/s \tag{14.10}$$

where is ρ the density of the species (mass concentration of the species per unit volume).

Generally, mass transfer occurs only in mixtures of two substances (species) or more with at least one species within the mixture moving from a region of high concentration to a region of low concentration. Consider the binary system shown in Figure 14.2 comprising species A (black

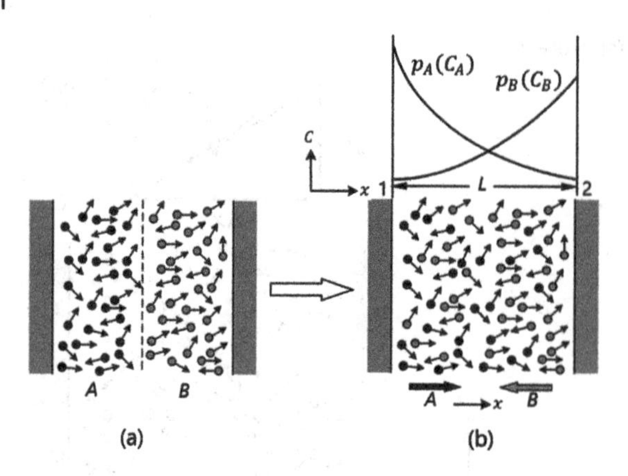

Figure 14.2 Schematic diagram of mass transfer in a binary gaseous mixture.

circles) and B (grey circles) on both sides of a partition in the middle of a chamber of width L (Figure 14.2a). According to the kinetic theory of gases, individual molecules are in a state of continual motion on both sides of the partition moving towards a lower concentration. If the partition is removed, both fluids will tend to diffuse with the net transfer of species A to the right (black arrow) and that of species B to the left (grey arrow) (Figure 14.2b). The process continues until an equilibrium state is reached with no further diffusion of either species.

As stated above, the molar concentration C_A of species A is defined as the number of moles of A per unit volume of the mixture and the molar concentration C_B for that species B as the number of moles of B per unit volume of the mixture. For an ideal gas, the molar densities for the two species are

$$C_A = \frac{N_A}{V} = \frac{p_A}{\bar{R}T},\ C_B = \frac{N_B}{V} = \frac{p_B}{\bar{R}T} \quad kmole/m^3$$

and the mass densities

$$\rho_A = \frac{\mu_A p_A}{\bar{R}T} \quad \rho_B = \frac{\mu_B p_B}{\bar{R}T} \quad kg/m^3$$

where p_A and p_B are the partial pressure and μ_A and μ_B are the molecular mass of species A and B.

Fick's law on a molar basis applied to the process in Figure 14.2, can be written in any of the following forms:

$$\dot{M}_A = -D_{AB}A\frac{dC_A}{dx} \tag{14.11a}$$

$$\dot{M}_A = -D_{AB}\frac{A}{\bar{R}T}\frac{dp_A}{dx} \tag{14.11b}$$

$$\dot{M}_A = -CD_{AB}A\frac{d(C_A/C)}{dx} \tag{14.11c}$$

$$\dot{M}_A = -CD_{AB}A\frac{dy_A}{dx} \tag{14.11d}$$

Similarly, on mass basis

$$\dot{m}_A = -D_{AB}A\frac{d\rho_A}{dx} \tag{14.12a}$$

$$\dot{m}_A = -D_{AB}\frac{M_A A}{\bar{R}T}\frac{dp_A}{dx} \tag{14.12b}$$

$$\dot{m}_A = -\rho D_{AB} A\frac{d(\rho_A/\rho)}{dx} \tag{14.12c}$$

$$\dot{m}_A = -\rho D_{AB} A\frac{dw_A}{dx} \tag{14.12d}$$

Similar equations can be written for species B, also with the subscript to the diffusion coefficient D reversed: D_{BA} instead of D_{AB}. The condition for the process to take place without s density gradient is that D_{AB} be equal to D_{BA}.

Table 14.1 lists selected values of binary diffusion coefficients at 1 atm and 298 K.

For temperatures and/or pressures other than the standard values of $T_{st} = 198K$ and $p_{st} = 1atm$, the following equation can be used:

$$\frac{D}{D_{st}} = \left(\frac{T}{T_{st}}\right)^{3/2}\left(\frac{p_{st}}{p}\right) \tag{14.13a}$$

For mixtures of water vapour and air, the following equations from Marrero and Mason (1972) are widely used:

$$D_{H_2O-air} = 1.87\times10^{-10}\frac{T^{2.072}}{p}\ m^2/s \qquad 280K < T < 450K \tag{14.13b}$$

$$D_{H_2O-air} = 2.75\times10^{-9}\frac{T^{1.632}}{p}\ m^2/s \qquad 450K < T < 1070K \tag{14.13c}$$

Table 14.1 Diffusion coefficients for various binary mixtures at pressure of 1 atm and temperature of 25 °C.

Species *A*	Species *B*	D_{AB}, m^2/s
Carbon dioxide	Air	1.6×10^{-5}
Chlorine	Air	1.2×10^{-5}
Helium (He)	Air	7.2×10^{-5}
Oxygen (O_2)	Air	2.1×10^{-5}
Carbon dioxide	Water vapour	1.6×10^{-5}
Oxygen (O_2)	Water vapour	2.5×10^{-5}
Water vapour	Air	2.5×10^{-5}
Carbon dioxide	Water	2.0×10^{-9}
Carbon dioxide	Natural rubber	1.1×10^{-10}
Hydrogen	Iron	2.6×10^{-13}
Oxygen	Pyrex glass	6.19×10^{-8}
Hydrogen	Nickel	1.2×10^{-12}
Carbon dioxide	Natural rubber	1.1×10^{-10}
Carbon dioxide	Water	2.0×10^{-9}

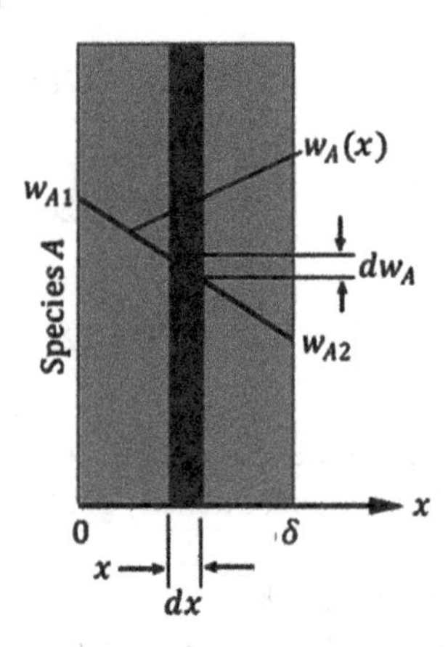

Figure 14.3 Steady mass transfer through a plane wall.

14.3 Steady Mass Diffusion Through a Plane Wall

Referring to Figure 14.3, the steady diffusion rate (kg/s) of species A through a nonreacting plane wall in terms of mass fractions can be found by integrating Eq. (14.12d) for the limits $x = 0$ to $x = \delta$, $w = w_{A1}$ to $w = w_{A2}$

$$\dot{m}_A \int_{\delta}^{0} dx = -\rho D_{AB} A \int_{w_{A2}}^{w_{A1}} dw_A$$

$$\dot{m}_A = -\frac{\rho D_{AB} A}{\delta}\left(w_{A2} - w_{A1}\right) \tag{14.14}$$

or, using Eq. (14.1)

$$\dot{m}_A = -\frac{D_{AB} A}{\delta}\left(\rho_{A2} - \rho_{A1}\right)$$

Rearranging Eq. (14.14)

$$\dot{m}_A = \frac{w_{A1} - w_{A2}}{\delta / \rho D_{AB} A} = \frac{w_{A1} - w_{A2}}{R_{diff}} \tag{14.15}$$

$R_{diff} = \left(\delta / \rho D_{AB} A\right)$ is the diffusion resistance of the wall in units of s/kg.

If analysis is conducted on molar basis

$$\dot{M}_A = -\frac{C D_{AB} A}{\delta}\left(y_{A2} - y_{A1}\right) \tag{14.16}$$

or, using Eq. (14.2)

$$\dot{M}_A = -\frac{D_{AB} A}{\delta}\left(C_{A2} - C_{A1}\right)$$

Rearranging Eq. (14.16)

$$\dot{M}_A = -\frac{y_{A2} - y_{A1}}{\delta / C D_{AB} A} = \frac{y_{A2} - y_{A1}}{R_{diff}} \tag{14.17}$$

$R_{diff} = \left(\delta / C D_{AB} A\right)$ is the molar diffusion resistance in units of $s/kmol$.

Equation (14.16) can be rewritten in terms of partial pressures

$$\dot{M}_A = A\frac{D_{AB}}{\bar{R}T}\frac{p_{A1} - p_{A2}}{x_2 - x_1} \tag{14.18}$$

where p_{A1} and p_{A2} are the partial pressures of species A at $x = 0$ and $x = \delta$, $\bar{R}$ is the universal gas constant ($8.314 kJ/kmol.K$), and T is the temperature of the mixture (K). Distances are in metres (m).

The process depicted in Figure 14.3 is comparable to a gas, such as helium, diffusing through a plane plastic membrane or leaking from spherical silica container. Another example is oxygen gas being absorbed in a rubber pipe through which it is flowing. The expressions for a fluid in contact with non-reacting cylindrical and spherical walls are:

Cylindrical tube: length L, internal radius r_1, external radius r_2

$$\dot{M}_A = 2\pi L\frac{D_{AB}}{\bar{R}T}\frac{p_{A1} - p_{A2}}{\ln(r_2/r_1)} \tag{14.19}$$

Spherical shell: internal radius r_1, external radius r_2

$$\dot{M}_A = 4\pi r_1 r_2\frac{D_{AB}}{\bar{R}T}\frac{p_{A1} - p_{A2}}{r_2 - r_1} \tag{14.20}$$

Example 14.2 Hydrogen gas is stored in a spherical shell made of nickel which has an outer diameter 4.0 m and thickness 50 mm. The molar concentration of hydrogen at the inner surface is 0.087 $kmol/m^3$ and at the outer surface is negligible. Determine the mass flow rate of hydrogen by diffusion through the nickel container wall. Take the diffusion coefficient for hydrogen–nickel interface as $1.2\times10^{-12}\ m^2/s$.

Solution

Equation (14.15) can be rewritten in terms of concentrations as

$$\dot{M}_A = 4\pi r_1 r_2 D_{AB}\frac{C_{A1} - C_{A2}}{r_2 - r_1}$$

$$\dot{M}_A = 4\pi\times1.95\times2.0\times1.2\times10^{-12}\frac{0.078 - 0.0}{0.05} = 9.174\times10^{-11} kmol/s$$

14.4 Diffusion of Vapour Through a Stationary Gas

Evaporation and condensation take place in the presence of a stationary gas in some engineering applications. Figure 14.4 is a schematic representation of a simplified isothermal evaporation process of liquid A (which could be water) and diffusion into stationary layer of gas B, which is assumed insoluble in liquid A (B could be air). The two gases are assumed to be perfect gases. Liquid A at constant pressure and temperature is open to the surroundings and the whole system is in steady state that is maintained by a slight movement of gas B at the top of the container, which removes the vapour diffusing to that point.

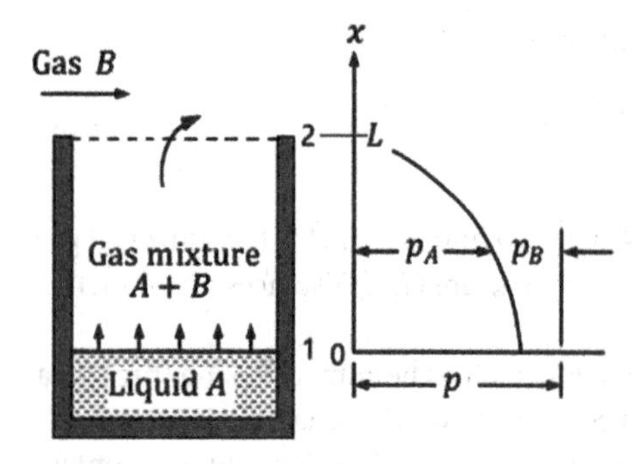

Figure 14.4 Diffusion of vapour from liquid A into stationary gas B.

The evaporating liquid A diffuses upwards through gas B and its motion is balanced by downward diffusion of B, which maintains the concentration at any x position constant. As B is insoluble in liquid A, there can be no net mass movement of B downwards; hence, there must be a bulk mass movement upwards which balances the diffusion of B downwards. It is apparent that the process is not just pure mass diffusion but rather mixed diffusion with additional mass flux driven by bulk fluid motion. With the total pressure being constant, the relative values of the partial pressures of species A and B change with x, as shown by the graph in Figure 14.4.

The mass diffusion of B downward from Eq. (14.12b)

$$\dot{m}_B = -D_{BA}\frac{\mu_B A}{\bar{R}T}\frac{dp_B}{dx} \tag{14.21}$$

The balancing bulk mass transfer rate upward at average bulk velocity U is

$$\dot{m}_B = -\rho_B AU = -\frac{\mu_B p_B}{\bar{R}T}AU \tag{14.22}$$

A in these equations is the cross-sectional area of the container.

Equating Eqs. (14.21) and (14.22) yields

$$U = \frac{D_{BA}}{p_B}\frac{dp_B}{dx} \tag{14.23}$$

The mass diffusion of vapour A from the liquid–gas interface upward is

$$\dot{m}_{A1} = -D_{AB}\frac{\mu_A A}{\bar{R}T}\frac{dp_A}{dx} \tag{14.24}$$

Making use of Eq. (14.23), the bulk mass transfer rate of vapour A moving upwards at average velocity U can be written as

$$\dot{m}_{A2} = \rho_A AU = \frac{\mu_A p_A}{\bar{R}T}AU = \frac{\mu_A p_A}{\bar{R}T}A\frac{D_{BA}}{p_B}\frac{dp_B}{dx} \tag{14.25}$$

The total vapour mass transport $\dot{m}_A$ is obtained by summing the mass transfer rates in Eqs. (14.24) and (14.25), bearing in mind that $D_{AB} = D_{BA} = D$ for a binary system with constant total concentration

$$\dot{m}_A = \dot{m}_{A1} + \dot{m}_{A2} = -D\frac{\mu_A A}{\bar{R}T}\frac{dp_A}{dx} + \frac{\mu_A p_A}{\bar{R}T}A\frac{D}{p_B}\frac{dp_B}{dx}$$

since, $p_A + p_B = p$

or

$$dp_A / dx = -dp_B / dx$$

$$\dot{m}_A = -\frac{D\mu_A A}{\bar{R}T}\frac{p}{p - p_A}\frac{dp_A}{dx} \tag{14.26}$$

This relation is known as Stefan's law, which after integration yields

$$\dot{m}_A = \frac{Dp\mu_A A}{\bar{R}T(x_2 - x_1)} \ln\left(\frac{p - p_{A2}}{p - p_{A1}}\right) = \frac{Dp\mu_A A}{\bar{R}T(x_2 - x_1)} \ln\left(\frac{p_{B2}}{p_{B1}}\right) \quad kg/s \tag{14.27}$$

In Figure 14.4, $x_1 = 0$ at level 1 and $x_2 = L$ at level 2.

The evaporation rate of species A can also be expressed as

$$\dot{m}_A = \frac{Dp\mu_A A}{\bar{R}T(x_2 - x_1)} \ln\left(\frac{p - p_{A2}}{p - p_{A1}}\right) = \frac{Dp\mu_A A}{\bar{R}T(x_2 - x_1)} \ln\left(\frac{p_{B2}}{p_{B1}}\right) \quad kg/s \tag{14.28}$$

Example 14.3 A 20-mm diameter tube is partially filled with water at 20°C. The distance of the water surface from the open end of the tube is 300 mm. Dry air at 20°C and 100 kPa is blowing over the open end of the tube so that steady state of the process is maintained. Determine the amount of water that will evaporate in a week.

Solution

Equation (14.27) for water is

$$\dot{m}_A = \frac{Dp\mu_w A}{\bar{R}T(x_2 - x_1)} \ln\left(\frac{p - p_{w2}}{p - p_{w2}}\right)$$

The partial pressure of water vapour changes from maximum at the water–vapour interface to a minimum at the open end of the tube. Since air at the top of the tube at 20°C is dry, we can assume that $p_{w2} \cong p_{w1} = 2.339 kPa$ (from saturated water table in Appendix N), $D = 2.5 \times 10^{-5}$ (from Table 14.1), $\bar{R} = 8.314 kJ/kmol.K$.

$$\dot{m}_A = \frac{2.5 \times 10^{-5} \times 100 \times 18 \times \left(\pi \times 0.02^2 / 4\right)}{8.314 \times 293 \times 0.3} \ln\left[\frac{100}{100 - 2.339}\right] = 4.5785 \times 10^{-10} kg/s$$

$$\dot{m}_A = 4.5785 \times 10^{-10} \times 7 \times 24 \times 3600 \times 1000 = 0.277 g/week$$

14.5 Steady-State Equimolar Counter Diffusion

In some engineering applications, the pressure in a gas storage tank or in a pipe transporting a gas needs to be maintained constantly by venting of the gas. To determine the rate of venting of the gas in such cases, the equimolar counter-diffusion model can be used.

Consider two chambers connected by a passage of cross-sectional area A and length L, as shown in Figure 14.5. The chambers and the passage contain a binary mixture of gaseous species A and B at uniform total pressure p and temperature T and the species concentrations are maintained constant in each chamber. The left chamber has a larger concentration of species A than B ($y_A > y_B$) and the right chamber has a larger concentration of species B than A ($y_B > y_A$). Under steady-state conditions, isothermal diffusion takes place in which equal molecules of species A and B are exchanged, with species A diffusing to the right (black arrow) and species B diffusing to the left (grey arrow). According to Flicks law, the molar fluxes for species A and B from Eqs (14.11a) and (14.11b) are, respectively

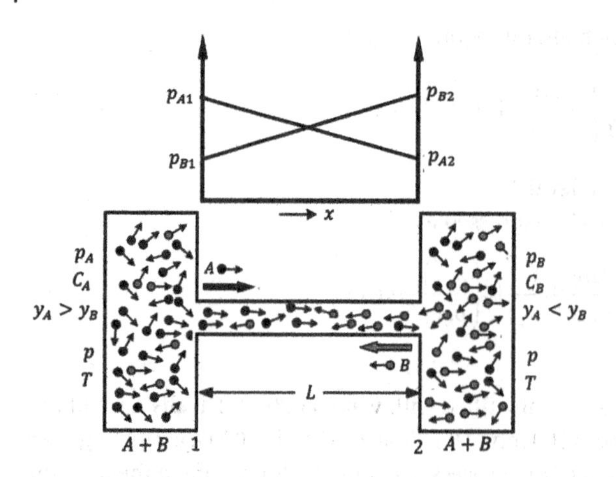

Figure 14.5 Equimolar isothermal counter-diffusion of a binary gas system.

$$\frac{\dot{M}_A}{A} = -D_{AB}\frac{dC_A}{dx} = -\frac{D_{AB}}{\bar{R}T}\frac{dp_A}{dx} \tag{14.29}$$

$$\frac{\dot{M}_B}{A} = -D_{BA}\frac{dC_B}{dx} = -\frac{D_{BA}}{\bar{R}T}\frac{dp_B}{dx} \tag{14.30}$$

The total pressure $p = p_A + p_B = const$; hence,

$$\frac{dp_A}{dx} = -\frac{dp_B}{dx}$$

Substituting for dp_A / dx and dp_B / dx in Eqs (14.29) and (14.30), and noting that $D_{AB} = D_{BA}$, shows that the molar fluxes for the two species are equal but opposite in direction $\dot{M}_A = -\dot{M}_B$.

Integrating Eqs (14.29) and (14.30) over the length L of the connecting tube, we get the molar fluxes of the two species

$$\frac{\dot{M}_A}{A} = -D_{AB}\left(\frac{C_{A2} - C_{A1}}{L}\right) = -\frac{D_{AB}}{\bar{R}T}\left(\frac{p_{A2} - p_{A1}}{L}\right) \tag{14.31}$$

$$\frac{\dot{M}_B}{A} = -D_{BA}\left(\frac{C_{B2} - C_{B1}}{L}\right) = -\frac{D_{BA}}{\bar{R}T}\left(\frac{p_{B2} - p_{B1}}{L}\right) \tag{14.32}$$

The fluxes on mass basis are

$$\frac{\dot{m}_A}{A} = -D_{AB}\mu_A\left(\frac{C_{A2} - C_{A1}}{L}\right) = -\frac{D_{AB}\mu_A}{\bar{R}T}\left(\frac{p_{A2} - p_{A1}}{L}\right) \tag{14.33}$$

$$\frac{\dot{m}_B}{A} = -D_{BA}\mu_B\left(\frac{C_{B2} - C_{B1}}{L}\right) = -\frac{D_{BA}\mu_B}{\bar{R}T}\left(\frac{p_{B2} - p_{B1}}{L}\right) \tag{14.34}$$

The net molar flow is zero, but the net mass flow is not and is equal to

$$\dot{m} = \dot{m}_A + \dot{m}_B = \dot{M}_A\mu_A + \dot{M}_B\mu_B = \dot{M}_A(\mu_A - \mu_B) \tag{14.35}$$

Figure E14.4 Representation of large tank vented to the atmosphere through a tube.

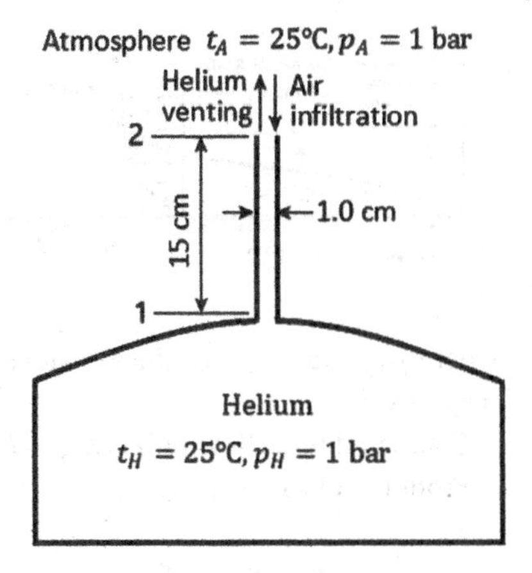

Example 14.4 A large tank of helium at 1 bar and 25°C vents to the atmosphere, which is also at 1 bar and 25°C (Figure 14.4). Venting is through a tube of 1.5 cm diameter and 15 cm long (Figure E14.4). Determine the rate of helium venting into the atmosphere and air infiltration into the tank.

Solution

From Table 14.1, $D_{AB} = 7.2\times10^{-5}$. From Eq. (14.34), the rate of helium venting, noting that the pressure of helium is 100 kPa at position 1 and 0 at position 2

$$\dot{m}_H = \frac{D\mu_H A}{\bar{R}T}\left(\frac{p_{H1}-p_{H2}}{L}\right)$$

$$= \frac{7.2\times10^{-5}\times4\times0.25\times\pi\times0.01^2}{8.314\times298}\left(\frac{100-0}{0.15}\right) = 1.522\times10^{-9}\,kg/s$$

The system is equimolar counter-diffusion in which $\dot{M}_A = -\dot{M}_H$, hence

$$\dot{m}_A = \dot{M}_H\mu_A = \frac{1.522\times10^{-9}\times29}{4} = 11.03\times10^{-9}\ kg/s$$

14.6 Mass Convection

Mass convection or mass transfer by convection is the process of mass transfer between a moving fluid and a surface, or between two moving fluids that do not have natural tendency to dissolve in each other at any concentration. Mass is transferred as a result of both mass diffusion induced by concentration gradient and bulk fluid motion induced by density gradient (free convection) or by an external source (forced convection). Additionally, the flowing fluid could be laminar, turbulent, internal, or external. In external flow (Figure 14.6), a boundary layer of the concentration of species A is formed on the surface of a plate with a density gradient between the surface and the free flow stream.

When mass transfer rates are low, there is a simple analogy between heat transfer and mass transfer that can be used to solve engineering problems. The equivalent to Newton's law of cooling of a flat plate in terms of mass concentrations (densities) is then

$$\frac{\dot{m}_A}{A_s} = h_m\left(\rho_{As} - \rho_{A\infty}\right) \tag{14.36}$$

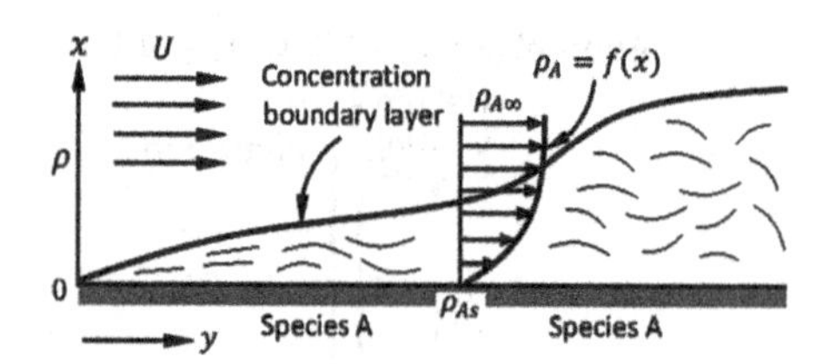

Figure 14.6 Concentration boundary layer of a moving fluid on a flat plate.

where ρ_{As} and $\rho_{A\infty}$ are the densities of species A at the surface (area A_s) and the free stream, respectively.

Compared with Newton's law $\dot{Q} = h_c A_s (T_s - T_\infty)$, $\dot{m}_A \equiv \dot{Q}$, $h_m \equiv h_c$, and $(\rho_{As} - \rho_{A\infty}) \equiv (T_s - T_\infty)$. From Eq. (14.1), $\rho_{As} = \rho w_{As}$, $\rho_{A\infty} = \rho w_{A\infty}$; hence,

$$\frac{\dot{m}_A}{A} = h_m \rho (w_{As} - w_{A\infty}) \tag{14.37}$$

where w_{As} and $w_{A\infty}$ are the mass concentrations of species A per unit volume at the surface and at the free stream; h_m is the mass transfer coefficient.

From Fick's law expressed in terms of mass concentrations (Eq. (14.12a)),

$$\frac{\dot{m}_A}{A} = D\rho \frac{w_{As} - w_{A\infty}}{\Delta x}$$

Equating the two equations yields

$$h_m = \frac{D}{\Delta x} m/s$$

Substituting h_m back into Eq. (14.37) and rearranging

$$\frac{\dot{m}_A}{A} = \left(\frac{D\rho}{\Delta x}\right)(w_{As} - w_{A\infty}) \tag{14.38}$$

The term $(D\rho / \Delta x)$ is the mass transfer conductance in $kg / m^2.s$.

Both mass and molar fluxes can also be expressed in terms of partial pressures as follows:

Mass flow from Eq. (14.12b)

$$\frac{\dot{m}_A}{A} = -D \frac{\mu_A}{\overline{R}T} \frac{\Delta p_A}{\Delta x} = -h_m \left(\frac{\mu_A}{\overline{R}T}\right) \Delta p_A \tag{14.39}$$

Molar flow from Eq. (14.11b)

$$\frac{\dot{M}_A}{A} = -\frac{D}{\overline{R}T} \frac{\Delta p_A}{\Delta x} = -h_m \left(\frac{1}{\overline{R}T}\right) \Delta p_A \tag{14.40}$$

14.6.1 Forced Mass Convection Correlations

It was shown in Chapter 6 that dimensional analysis can be used for the solution of heat transfer by forced convection by expressing heat transfer coefficient by the functional relationship

$$h = f(U, L, \rho, \mu, k, c_p)$$

from which

$$\frac{hL}{k} = f\left(\frac{U\rho L}{\mu}, \frac{c_p\mu}{k}\right) \tag{14.41}$$

or

$$Nu = f(Re, Pr) \tag{14.42}$$

Corresponding formulations for mass transfer are

$$h_m = f(D, U, \rho, L, \mu)$$

from which

$$\frac{h_m L}{D} = f\left(\frac{U\rho L}{\mu}, \frac{\mu}{\rho D}\right) \tag{14.43}$$

or

$$Sh = f(Re, Sc) \tag{14.44}$$

The non-dimensional parameter $Sh = h_m L / D$ is the Sherwood number, and $Sc = \mu / \rho D$ is the Schmidt number. The Sherwood and Schmidt numbers in mass transfer are the counterparts to the Nusselt and Prandtl numbers in forced convection heat transfer. This similarity with forced convection heat transfer in laminar or turbulent flow regimes can be used to provide a solution to corresponding mass transfer problems. For example, the correlation for laminar forced convection over a flat plat, as given in Eq. (7.15b) is

$$\overline{Nu}_L = \frac{\bar{h}L}{k} = 0.664 Re_L^{1/2} Pr^{1/3} \quad \text{for} \quad 0.6 < Pr < 10, Re_L < 5 \times 10^5 \tag{14.45a}$$

and the corresponding correlation for convective mass transfer is

$$\overline{Sh}_L = \frac{\bar{h}_m L}{D} = 0.664 Re_L^{1/2} Sc^{1/3} \quad \text{for} \quad Sc > 0.5 \tag{14.45b}$$

Presented below are pairs of correlations for different configurations for convective and mass transfer processes.

Turbulent forced flow over a flat plate:

$$\overline{Nu}_L = \frac{\bar{h}L}{k} = 0.037 Re_L^{4/5} Pr^{1/3} \quad \text{for} 5 \times 10^5 < Re_L < 10^7 \tag{14.46a}$$

$$\overline{Sh}_L = \frac{\bar{h}_m L}{D} = 0.037 Re_L^{4/5} Sc^{1/3} \quad \text{for} \quad Sc > 0.5 \tag{14.46b}$$

Fully developed laminar flow in a smooth pipe:

$$\overline{Nu}_L = 3.66 \tag{14.47a}$$

$$\overline{Sh}_L = 3.66 \tag{14.47b}$$

Turbulent forced flow in a smooth pipe:

$$\overline{Nu}_L = \frac{\bar{h}L}{k} = 0.023Re_L^{4/5}Pr^{2/5} \text{ for } Re_L > 10^4, 0.7 < Pr < 160 \tag{14.48a}$$

$$\overline{Sh}_L = \frac{\bar{h}_m L}{D} = 0.023Re_L^{4/5}Sc^{2/5} \text{ for } 0.7 < Sc < 160 \tag{14.48b}$$

14.6.2 Natural (Free) Mass Convection Correlations

The functional relationship for the heat transfer coefficient in free convection can be written as

$$h = f(k,\ c_p, \rho, \mu, l, \beta g \Delta T)$$

from which

$$Nu = f(Gr, Pr)$$

where $Gr = \beta\rho^2 L^3 g\Delta T / \mu^2$ is the Grashof number and $Pr = c_p\mu / k$ is the Prandtl number.

Similarly, the functional relationship for the mass transfer coefficient for natural mass convection can be expressed as

$$h_m = f(D, \rho, L, \mu, g\Delta\rho)$$

which results in

$$\frac{h_m L}{D} = f\left(\frac{\rho L^3 g \Delta\rho}{\mu^2}, \frac{\mu}{\rho D}\right) \tag{14.49}$$

or

$$Sh = f(Gr_m, Sc) \tag{14.50}$$

The Grashof number for mass transfer $Gr_m = \rho L^3 g\Delta\rho / \mu^2$ and the Schmidt number $Sc = \mu / \rho D$ are, respectively, the counterparts to the Grashof and Prandtl numbers in natural convection heat transfer.

Presented below are pairs of correlations for natural heat and mass convection processes.

Natural convection over a vertical plate:

$$\overline{Nu}_L = 0.62(Gr_L Pr)^{1/4} \text{ for laminar region } 10^5 < Gr_L Pr < 10^9$$

$$\overline{Sh}_L = 0.62(Gr_m Pr)^{1/4} \text{ for laminar region } 10^5 < Gr_m Sc < 10^9 \tag{14.51}$$

$$\overline{Nu}_L = 0.12(Gr_L Pr)^{1/3} \text{ for turbulent region } 10^9 < Gr_L Pr < 10^{13}$$

$$\overline{Sh}_L = 0.12(Gr_m Pr)^{1/3} \text{ for turbulent region } 10^9 < Gr_m Sc < 10^{13} \tag{14.52}$$

Natural convection over a horizontal plate with hot surface up and cold surface down:

$$\overline{Nu}_L = 0.54(Gr_L Pr)^{1/4} \text{ for} 10^4 \le Gr_L Pr \le 10^7$$

$$\overline{Sh}_L = 0.54(Gr_m Sc)^{1/4} \text{ for} 10^4 \le Gr_m Sc \le 10^7 \tag{14.53}$$

$$\overline{Nu}_L = 0.154\left(Gr_L Pr\right)^{1/3} \text{ for} 10^7 \le Gr_L Pr \le 10^{11}$$

$$\overline{Sh}_L = 0.154\left(Gr_m Sc\right)^{1/3} \text{ for} 10^7 \le Gr_m Sc \le 10^{11} \tag{14.54}$$

Natural convection over a horizontal plate with hot surface down and cold surface up:

$$\overline{Nu}_L = 0.52\left(Gr_L Pr\right)^{1/5} \text{ for } Gr_L Pr \le 10^5.$$

$$\overline{Sh}_L = 0.52\left(Gr_m Sc\right)^{1/5} \text{ for } Gr_m Sc \le 10^5 \tag{14.55}$$

$$\overline{Nu}_L = 0.27\left(Gr_L Pr\right)^{1/4} \text{ for } 10^5 \le Gr_L Pr \le 10^{10}$$

$$\overline{Sh}_L = 0.27\left(Gr_m Sc\right)^{1/4} \text{ for } 10^5 \le Gr_m Sc \le 10^{10} \tag{14.56}$$

The forced and free heat and mass convection correlations above are examples of the analogy between heat and mass transfer, where the Sherwood number Sh replaces the Nusselt number Nu, the Schmidt number Sc replaces the Prandtl number Pr, and the Grashof number for mass transfer Gr_m replaces the Grashof number for heat transfer Gr.

As a special case, consider Eqs (14.41) to (14.44). If $Sc = Pr$, then $Sh = Nu$.

hence,

$$\frac{h_m L}{D} = \frac{hL}{k}$$

from which

$$h_m = \frac{hD}{k} \tag{14.57}$$

Since $Sc = Pr$

$$\frac{\mu}{\rho D} = Pr = \frac{c_p \mu}{k} \tag{14.58}$$

combining Eqs (14.57) and (14.58) yields

$$h_m = \frac{h}{\rho c_p} \tag{14.59}$$

Equation (14.59) is the Lewis relation which can be used for air–water mixtures with reasonable accuracy.

Example 14.5 A flat plate of 5 m length and 6 m width is covered with a layer of water as a result of condensation. The temperature of the plate is maintained at 25°C. To dry the plate, a stream of air at 25°C and 1 atm is forced to flow over the plate at a velocity of $4 m/s$. If the moisture content in the air $w_{A\infty} = 0.006 kg/m^3$, determine the mass convection coefficient and the rate of evaporation of the water layer.

Solution

Thermophysical properties of air at 25°C
$\rho = 1.184\, kg/m^3$, $\mu = 1.849 \times 10^{-5}\, kg/m.s$
Air–water mixture: From Table 14.1, $D = 2.5 \times 10^{-5}$

$$Re = \frac{U\rho L}{\mu} = \frac{4 \times 1.184 \times 5}{1.849 \times 10^{-5}} = 1.28 \times 10^6$$

$$Sc = \frac{\mu}{\rho D} = \frac{1.849 \times 10^{-5}}{1.184 \times 2.5 \times 10^{-5}} = 0.62$$

Equation (14.46b) for turbulent flow can be used to determine the average Sherwood number

$$\overline{Sh}_L = \frac{h_m L}{D} = 0.037 Re_L^{4/5} Sc^{1/3} = 0.037 \times \left(1.28 \times 10^6\right)^{4/5} \times 0.62^{1/3} = 2432$$

$$h_m = \frac{2432 \times 2.5 \times 10^{-5}}{5} = 0.01216$$

From Eq. (14.37) and making use of Eq. (14.1)

$$\dot{m}_A = h_m A\left(\rho_{As} - \rho_{A\infty}\right)$$

Assuming that the water at the water–air interface is saturated, the density of the vapour at 25°C is $\rho_{As} = 0.0231\, kg/m^3$, and the rate of evaporation of the water is then

$$\dot{m}_A = h_m A\left(\rho_{As} - \rho_{A\infty}\right) = 0.01216 \times 5 \times 6 \times \left(0.0231 - 0.006\right) = 6.24 \times 10^{-3}\, kg/s$$

14.7 Simultaneous Mass and Heat Transfer

The convective mass transfer model discussed so far was based on the assumption of small temperature differences between a wet surface and a moving gas (isothermal process) and low concentrations of the diffusing substances (low partial-pressure gradient). Under such flow conditions, analogy between heat and mass flow is applicable and mass transfer can proceed independently (Figure 14.7a). However, there is some heat transfer involved in the described process, which is due to the latent heat used in the phase change (evaporation) and provided entirely by the wetted surface resulting in decrease of temperature of the water layer.

Consider the model shown in Figure 14.7b in which air is flowing past a wetted surface. If there is a difference in partial pressure between the air p_{sa} and wetted surface p_{sw}, there will be transfer of mass (water vapour), and if there is an appreciable temperature difference between the air T_a and the wetted surface T_w, there will be heat transfer by convection (radiation heat transfer can be ignored). In addition to the modes of heat transfer shown in Figure 14.7b, the transfer of mass will cause thermal energy transfer. This additional heat transfer is due to the removal of latent heat of condensation when vapour condenses in the air or addition of latent heat of vaporization as some

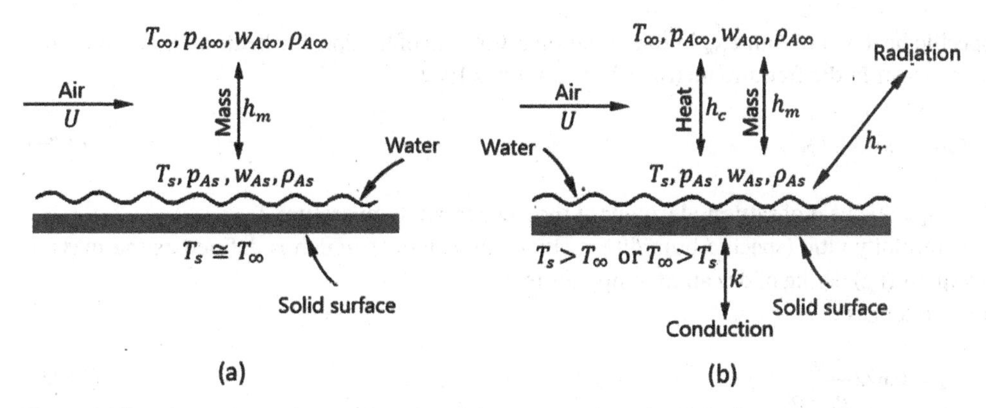

Figure 14.7 Air passing over wetted surface: (a) mass transfer only; (b) both mass and heat transfer.

water evaporates at the wetted surface, which creates a temperature gradient between the air and water that drives some heat transfer by convection from or to the air. Also, though usually ignored, additional source of heat transfer by conduction and radiation from the solid surface supporting the water can be present. The sign of the net heat flux will depend on whether the air is hotter or colder than the wetted surface. Assuming $T_w > T_a$, the rate of sensible heat transferred from the water through surface A by convection is

$$\dot{Q}_c = h_c A(T_s - T_\infty) \tag{14.60}$$

where
h_c – is the convective heat transfer coefficient
T_s, T_∞ – is temperatures of the water surface and free stream, respectively.
The rate of mass transfer from the wetted surface to the air is

$$\dot{m}_A = h_m A(\rho_{As} - \rho_{A\infty}) \tag{14.61}$$

where
h_m – is the mass transfer coefficient.
$\rho_{As}, \rho_{A\infty}$ – is density of species A (water vapour) at the water surface and that of the free stream, respectively.

The mass transferred from the water to the air causes transfer of latent heat $\dot{Q}_L$ of vaporization

$$\dot{Q}_L = h_m A(\rho_{As} - \rho_{A\infty})h_{fg} \tag{14.62}$$

where h_{fg} is the latent heat of vaporization of water at T_s.

If the heat transfer coefficient h_c is known, the mass transfer coefficient h_m can be estimated from Eq. (14.59) expressed as

$$h_m = \frac{h_c}{\rho c_{pm}} \tag{14.63}$$

The specific heat of moist air c_{pm} in this equation is the sum of the specific heats of dry air and that of water vapour in the free stream (Stoecker and Jones 1982)

$$c_{pm} = c_p + \omega_\infty c_{pv,\infty} \tag{14.64}$$

$c_{pv,\infty}$ – specific heat of saturated vapour at the free-stream temperature T_∞

ω_∞ – humidity ratio (specific humidity, or moisture content), which is defined as the mass of water vapour (kg) per kg of dry air at temperature T_∞.

For a perfect gas

$$\omega_\infty = 0.622 \frac{p_v}{p_t - p_v} \tag{14.65}$$

In this equation, p_v is the pressure of water vapour in the air and p_t is the atmospheric pressure, usually taken as $101.3kPa$. If the relative humidity at a specified temperature is known, the pressure p_v in Eq. (14.65) is calculated from

$$p_v = \phi p_s \tag{14.66}$$

where p_s is the saturation pressure of water at the specified temperature and ϕ is the relative humidity ϕ of air at the specified temperature.

Relative humidity ϕ is frequently used in air conditioning practice when dealing with air–water vapour mixtures and is defined as

$$\phi = \frac{\text{mass fraction of vapour in a given volume of mixture}}{\text{mass fraction of vapour in saturated mixture of the same volume}}$$

or

$$\phi = \frac{\text{mole fraction of vapour in the mixture}}{\text{mole fraction of vapour in saturated mixture at same temperature and total pressure}}$$

or

$$\phi = \frac{\text{partial pressure of vapour in mixture}}{\text{saturation pressure of vapour at the same temperature and total pressure}}$$

Combining Eqs (14.62) and (14.63)

$$\dot{Q}_L = \frac{h_c}{\rho c_{pm}} A(\rho_{As} - \rho_{A\infty}) h_{fg} \tag{14.67}$$

Total rate of heat transfer is the sum of the rates in Eqs (14.60) and (14.62)

$$\dot{Q}_t = \dot{Q}_c + \dot{Q}_L = h_c A(T_s - T_\infty) + h_m A(\rho_{As} - \rho_{A\infty}) h_{fg} \tag{14.68}$$

or, substituting for h_m from Eq. (14.63)

$$\dot{Q}_t = h_c A(T_s - T_\infty) + \frac{h_c}{\rho_a c_{pm}} A(\rho_{As} - \rho_{A\infty}) h_{fg} \tag{14.69}$$

Example 14.6 Calculate the humidity ratio and specific heat of moist air at 20°C and relative humidity $\phi = 50\%$.

Solution

From property tables, the saturation pressure of water at 20°C is $p_s = 2.339 kPa$, the specific heat of water vapour in saturated water is $c_{pv,20°C} = 1.867 kJ / kg.K$, and the specific heat of dry air is $c_p = 1.007 kJ / kg.K$

We first calculate the water-vapour pressure in the air

$$p_v = \phi p_s = 0.5 \times 2.339 = 1.17 kPa$$

$$\therefore \omega_\infty = 0.622 \frac{p_v}{p_t - p_v} = \frac{0.622 \times 1.17}{101.3 - 1.17} = 7.27 \times 10^{-3}$$

The specific heat of the moist air is then

$$c_{pm} = c_p + \omega_\infty c_{pv,20°C} = 1.007 + 7.27 \times 10^{-3} \times 1.867 = 1.0205 kJ / kg\,K$$

Example 14.7 Air at temperature $T_\infty = 30°C$ and 50% relative humidity is passing over a $0.4 m \times 0.5 m$ flat wetted surface at temperature $T_\infty = 30°C$. If the heat transfer coefficient is $h_c = 30 W / m^2.K$, determine the total rate of heat transfer in the process.

Solution

Properties of saturated water at 30°C: $\rho_{v\infty} = 0.0304 kg / m^3$, $p_{sat} = 4.246 kPa$, $c_{p(sat,\infty)} = 1875 J / kg.K$, $h_{fg} = 2.431 \times 10^6 J / kg$

Properties of saturated water at 20°C: $\rho_{vs} = 0.0173 kg / m^3$

Properties of dry air at 30°C: $\rho = 1.164 kg / m^3$

Convection heat transfer rate

$$\dot{Q}_c = h_c A (T_s - T_\infty) = 30 \times 0.4 \times 0.5 \times (30 - 20) = 60 W$$

Evaporation heat transfer rate

$$\dot{Q}_L = h_m A (\rho_{vs} - \rho_{v\infty}) h_{fg}$$

From Eq. (14.65)

$$\omega_\infty = 0.622 \frac{p_v}{p_t - p_v} = \frac{0.622 \times (0.5 \times 4.246)}{101.3 - 0.5 \times 4.246} = 0.0133$$

From Eq. (14.64)

$$c_{pm} = c_{pa} + \omega_\infty c_{p(v,\infty)} = 1.004 + 0.0133 \times 1.875 = 1.029 kJ / kg.K$$

From Eq. (14.63)

$$h_m = \frac{h_c}{\rho c_{pm}} = \frac{30}{1.164 \times 1029} = 0.025 m/s$$

hence,

$$\dot{Q}_L = h_m A(\rho_{vs} - \rho_{v\infty}) h_{fg} = 0.025 \times 0.4 \times 0.5 \times (0.0173 - 0.5 \times 0.0304) \times 2431 \times 10^3$$

$$\dot{Q}_L = 25.5 W$$

Total heat transfer rate

$$\dot{Q}_t = 60 + 25.5 = 85.5 W$$

Problems

14.1 Oxygen gas stored in a tank is at a pressure of 0.6 MPa and temperature of 300 K. Determine the mass and molar densities of oxygen in the tank.

14.2 The small amounts of constituents in air are often ignored and its composition is often taken as 79% nitrogen and 21% oxygen on a molar basis. What is the composition of air on a mass basis?

14.3 A gas mixture has the following molar (volume) compositions:

$CO_2 - 12.0\%$

$O_2 - 4.0\%$

$N_2 - 82.0\%$

$CO - 2.0\%$

Determine the mass composition of the mixture.

14.4 A gas mixture has the following mass composition

$CO_2 - 17.55\%$

$O_2 - 4.26\%$

$N_2 - 76.33\%$

$CO - 1.86\%$

Determine the molar composition of the mixture.

14.5 A tank containing a mixture of 80%N_2 and 20%CO_2 is connected to another tank containing 80%CO_2 and 20%N_2 by a 1.5 m long duct having a 0.1 m^2 cross-sectional area (Figure P14.5). Both tanks are kept at a pressure of 1 bar and temperature of 298 K. If the diffusion coefficient is $1.7\times10^{-5}\ m^2/s$, determine the diffusion rates of CO_2 and N_2.

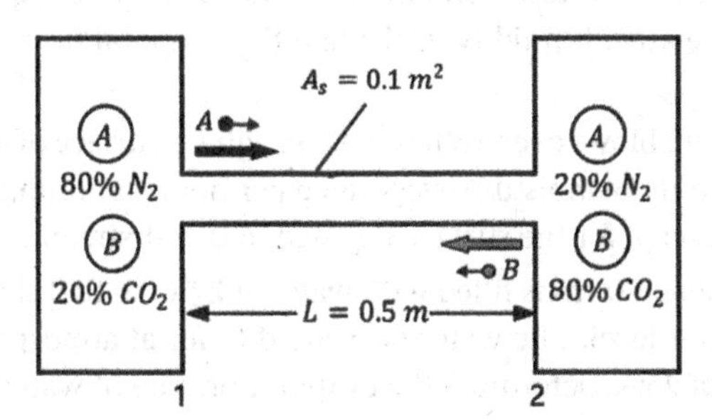

Figure P14.5 Representation of counter-diffusion of a binary gas system.

14.6 A Stefan tube of 10 mm diameter and 180 mm long is used to measure the binary diffusion coefficient of water vapour in dry atmospheric air at 25°C. The tube is partially filled with water to a depth of 30 mm and dry air is blown over the open end of the tube to maintain steady conditions. If the amount of water that has evaporated in an hour is 1.13×10^{-3}g, what is the diffusion coefficient of water?

14.7 A large tank of helium at 1 bar and 30°C vents to the atmosphere, which is also at 1 bar and 30°C (Figure P14.7). Venting is through a tube of 15 mm internal diameter and 300 mm long. Determine:

(a) the rate at which the helium is venting from the tank
(b) the rate at which the air in the atmosphere is infiltrating into the tank.

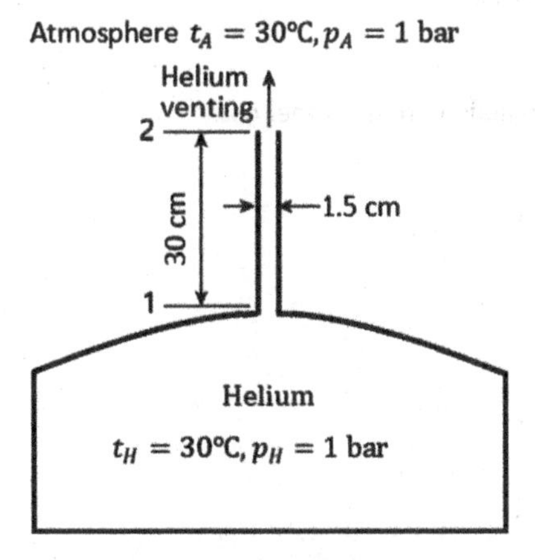

Figure P14.7 Representation of helium in a large tank venting to the atmosphere.

14.8 Air at a velocity of $U = 2.5m/s$, temperature $T_\infty = 40°C$, and relative humidity $\phi = 20\%$ is flowing over a horizontal wetted surface. The temperature of water on the surface is $T_s = 20°C$. The length of the plate is $L = 0.4m$ and width $W = 0.5m$. Determine the amount of water evaporated in a day. Take the atmospheric pressure as 101.3 kPa.

14.9 A hot water bath is kept at 50°C in a manufacturing facility. The top surface (1 m long and 3.5m wide) is exposed to moist air at 25°C, 101.3kPa, and 50% relative humidity and heat is lost to the air by free convection at $h_c = 5.5 W/m^2.K$. Estimate the ratio of the heat lost by mass convection to heat lost by convection.

14.10 An open pan 200 mm in diameter and 75 mm deep contains water at 25°C and is exposed to atmospheric air at 25°C and 50% relative humidity. Calculate the evaporation rate of water in grams per hour.

14.11 Dry air at 25°C and atmospheric pressure blows over a 30 $cm \times 30 cm$ square surface of ice at a velocity of 1.5 m/s. Estimate the amount of moisture evaporated per hour, assuming that the block of ice is perfectly insulated except for the surface exposed to the airstream.

14.12 An open cylindrical container of diameter 20 cm is filled with water at 27°C so that the top of the container is 8 cm above the water level. The water is exposed to air at atmospheric pressure 27°C and relative humidity of 25%. Determine the evaporation rate of water due to mass diffusion from stagnant liquid.

14.13 Repeat Problem 14.10 assuming mass diffusion from fluid with bulk motion.

14.14 A thin plastic membrane 0.1mm thick separates helium from a gas stream, as shown in Figure P14.14. The concentrations of the helium at the left and right surfaces of the membrane are, respectively, 0.02 and 0.005 $kmol/m^3$. If the helium plastic coefficient of diffusion is $10^{-9} m^2/s$, estimate the diffusion flux.

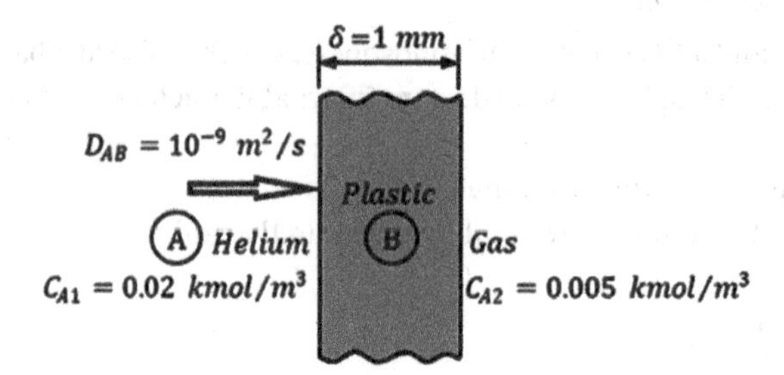

Figure P14.14 Thin plastic membrane separating helium from another gas.

Appendix B

Table B.1 Bessel functions.

Zero- and first-order functions of the first kind: $J_n(x)$								
x	$J_0(x)$	$J_1(x)$	x	$J_0(x)$	$J_1(x)$	x	$J_0(x)$	$J_1(x)$
0	1	0	4.1	−0.3887	−0.1033	8.2	0.1222	0.258
0.1	0.9975	0.0499	4.2	−0.3766	−0.1386	8.3	0.096	0.2657
0.2	0.99	0.0995	4.3	−0.361	−0.1719	8.4	0.0692	0.2708
0.3	0.9776	0.1483	4.4	−0.3423	−0.2028	8.5	0.0419	0.2731
0.4	0.9604	0.196	4.5	−0.3205	−0.2311	8.6	0.0146	0.2728
0.5	0.9385	0.2423	4.6	−0.2961	−0.2566	8.7	−0.0125	0.2697
0.6	0.912	0.2867	4.7	−0.2693	−0.2791	8.8	−0.0392	0.2641
0.7	0.8812	0.329	4.8	−0.2404	−0.2985	8.9	−0.0652	0.2559
0.8	0.8463	0.3688	4.9	−0.2097	−0.3147	9	−0.0903	0.2453
0.9	0.8075	0.4059	5	−0.1776	−0.3276	9.1	−0.1142	0.2324
1	0.7652	0.44	5.1	−0.1443	−0.3371	9.2	−0.1368	0.2174
1.1	0.7196	0.4709	5.2	−0.1103	−0.3432	9.3	−0.1577	0.2004
1.2	0.6711	0.4983	5.3	−0.0758	−0.346	9.4	−0.1768	0.1816
1.3	0.6201	0.522	5.4	−0.0412	−0.3453	9.5	−0.1939	0.1613
1.4	0.5669	0.5419	5.5	−0.0068	−0.3414	9.6	−0.209	0.1395
1.5	0.5118	0.5579	5.6	0.027	−0.3343	9.7	−0.2218	0.1166
1.6	0.4554	0.5699	5.7	0.0599	−0.3241	9.8	−0.2323	0.0928
1.7	0.398	0.5778	5.8	0.0917	−0.311	9.9	−0.2403	0.0684
1.8	0.34	0.5815	5.9	0.122	−0.2951	10	−0.2459	0.0435
1.9	0.2818	0.5812	6	0.1506	−0.2767	10.1	−0.249	0.0184
2	0.2239	0.5767	6.1	0.1773	−0.2559	10.2	−0.2496	−0.0066

(*Continued*)

Heat Transfer Basics: A Concise Approach to Problem Solving, First Edition. Jamil Ghojel.

Companion website: www.wiley.com/go/ghojel/heat_transfer

Table B.1 (Continued)

Zero- and first-order functions of the first kind: $J_n(x)$								
x	$J_0(x)$	$J_1(x)$	x	$J_0(x)$	$J_1(x)$	x	$J_0(x)$	$J_1(x)$
2.1	0.1666	0.5683	6.2	0.2017	−0.2329	10.3	−0.2477	−0.0313
2.2	0.1104	0.556	6.3	0.2238	−0.2081	10.4	−0.2434	−0.0555
2.3	0.0555	0.5399	6.4	0.2433	−0.1816	10.5	−0.2366	−0.0788
2.4	0.0025	0.5202	6.5	0.2601	−0.1538	10.6	−0.2276	−0.1012
2.5	−0.0484	0.4971	6.6	0.274	−0.125	10.7	−0.2164	−0.1224
2.6	−0.0968	0.4708	6.7	0.2851	−0.0953	10.8	−0.2032	−0.1422
2.7	−0.1424	0.4416	6.8	0.2931	−0.0652	10.9	−0.1881	−0.1604
2.8	−0.185	0.4097	6.9	0.2981	−0.0349	11	−0.1712	−0.1768
2.9	−0.2243	0.3754	7	0.3001	−0.0047	11.1	−0.1528	−0.1913
3	−0.26	0.3391	7.1	0.2991	0.0252	11.2	−0.133	−0.2028
3.1	−0.2921	0.3009	7.2	0.2951	0.0543	11.3	−0.1121	−0.2143
3.2	−0.3202	0.2613	7.3	0.2882	0.0826	11.4	−0.0902	−0.2224
3.3	−0.3443	0.2207	7.4	0.2786	0.1096	11.5	−0.0677	−0.2284
3.4	−0.3643	0.1792	7.5	0.2663	0.1352	12.0	0.0477	−0.2234
3.5	−0.3801	0.1374	7.6	0.2516	0.1592	12.5	0.1469	−0.1655
3.6	−0.3918	0.0955	7.7	0.2346	0.1813	13.0	0.2069	−0.0703
3.7	−0.3992	0.0538	7.8	0.2154	0.2014	13.5	0.215	0.038
3.8	−0.4026	0.0128	7.9	0.1944	0.2192	14.0	0.1711	0.1334
3.9	−0.4018	−0.0272	8	0.1716	0.2346	14.5	0.0875	0.1934
4	−0.3971	−0.066	8.1	0.1475	0.2476	15.0	0.0142	0.2051

Table B.2 Modified Bessel functions.

Zero-, first-, and second-order functions of the first and second kinds: $I_n(x)$ and $K_n(x)$							Fractional order functions: $I_\vartheta(x)$	
x	$I_0(x)$	$K_0(x)$	$I_1(x)$	$K_1(x)$	$I_2(x)$	$K_2(x)$	$I_{-1/3}(x)$	$I_{2/3}(x)$
0	1		0					0
0.2	1.00998	1.75266	0.10052	4.77598	0.00476	49.51250	1.61496	0.24008
0.4	1.04040	1.11454	0.20408	2.18437	0.01999	12.03640	1.33947	0.38799
0.6	1.09200	0.77750	0.31377	1.30282	0.04610	5.12025	1.25614	0.52368
0.8	1.16641	0.56535	0.43287	0.86177	0.08424	2.71977	1.25461	0.66085
1	1.26590	0.42104	0.56513	0.60189	0.13564	1.62481	1.30635	0.80752
1.2	1.39379	0.31851	0.71462	0.43459	0.20275	1.04283	1.40180	0.97010

Table B.2 (Continued)

	Zero-, first-, and second-order functions of the first and second kinds: $I_n(x)$ and $K_n(x)$						Fractional order functions: $I_\vartheta(x)$	
x	$I_0(x)$	$K_0(x)$	$I_1(x)$	$K_1(x)$	$I_2(x)$	$K_2(x)$	$I_{-1/3}(x)$	$I_{2/3}(x)$
1.4	1.55355	0.24366	0.88557	0.32082	0.28844	0.70198	1.53859	1.15472
1.6	1.74991	0.18795	1.08394	0.24064	0.39498	0.48875	1.71808	1.36782
1.8	1.98973	0.14593	1.31621	0.18262	0.52727	0.34884	1.94422	1.61651
2	2.27952	0.11390	1.58995	0.13987	0.68958	0.25377	2.22303	1.90894
2.2	2.62899	0.08926	1.91401	0.10790	0.88898	0.18735	2.56265	2.25463
2.4	3.04901	0.07022	2.29884	0.08372	1.13331	0.13999	2.97341	2.66478
2.6	3.55308	0.05540	2.75690	0.06529	1.43239	0.10562	3.46810	3.15275
2.8	4.15721	0.04382	3.30308	0.05111	1.79787	0.08033	4.06241	3.73446
3	4.88079	0.03474	3.95535	0.04016	2.24389	0.06151	4.77542	4.42900
3.2	5.74797	0.02760	4.73544	0.03164	2.78832	0.04737	5.63022	5.25929
3.4	6.78387	0.02196	5.66969	0.02500	3.44876	0.03666	6.65475	6.25286
3.6	8.02599	0.01750	6.79007	0.01980	4.25373	0.02850	7.88271	7.44286
3.8	9.51688	0.01397	8.13526	0.01571	5.23516	0.02223	9.35479	8.86919
4	11.30182	0.01116	9.75212	0.01248	6.42576	0.01740	11.12006	10.57993
4.2	13.44396	0.00893	11.69715	0.00994	7.87389	0.01366	13.23771	12.63304
4.4	16.01324	0.00715	14.03842	0.00792	9.63214	0.01075	15.77910	15.09844
4.6	19.09104	0.00573	16.85779	0.00632	11.76157	0.00848	18.83037	18.06042
4.8	22.79535	0.00460	20.25352	0.00505	14.35639	0.00670	23.52173	21.62077
5	27.23381	0.00369	24.34367	0.00404	17.49635	0.00531	26.89960	25.90231
5.2	32.57462	0.00297	29.27010	0.00324	21.31689	0.00421	32.19429	31.05334
5.4	39.01181	0.00238	35.20366	0.00260	25.97342	0.00335	38.56219	37.25299
5.6	46.72968	0.00192	42.35061	0.00208	31.60446	0.00266	46.22395	44.71763
5.8	56.01881	0.00154	50.96071	0.00167	38.44615	0.00212	55.44605	53.70870
6	67.21124	0.00124	61.33722	0.00134	46.76550	0.00169	66.55046	64.54216
6.4	96.95724	0.00081	88.94675	0.00087	69.16138	0.00108	96.04344	93.34381
6.8	140.15396	0.00053	129.20963	0.00056	102.15113	0.00069	138.89476	135.24028
7.2	202.92376	0.00034	188.08243	0.00037	150.67864	0.00044	201.23259	196.25540
7.6	294.33426	0.00022	274.25445	0.00024	222.16203	0.00029	292.02239	285.20756
8	427.46938	0.00015	400.21833	0.00016	327.41479	0.00019	424.38958	415.01453
8.4	621.69993	0.00010	583.91721	0.00010	482.67202	0.00012	617.56223	604.61070
8.8	905.57431	0.00006	851.85088	0.00007	711.97183	0.00008	899.72550	881.76523
9.2	1320.27702	0.00004	1245.01657	0.00004	1049.62124	0.00005	1312.21774	1287.22624
9.6	1926.80399	0.00003	1826.41938	0.00003	1546.29996	0.00003	1915.70113	1880.81728
10	2814.98233	0.00002	2670.72985	0.00002	2280.83636	0.00002	2799.23961	2750.40904

Table B.3 Correlations for modified Bessel functions of integer order ($x = 0 - 10$).

Correlation	$e^{-x}I_n(x) = \sum_{i=0}^{10} a_i(x)^i$ $(x = 0 - 10)$		$e^x K_n(x) = \sum_{i=0}^{5} a_n(\ln x)^i$ $(x = 0 - 10)$	
a_i	Zero-order first kind $n = 0$	First-order first kind $n = 1$	Zero-order second kind $n = 0$	First-order second kind $n = 1$
a_0	0.999801441	0.000168597	1.144453570	1.635459896
a_1	−0.992501458	0.493689474	−0.491587622	−1.139375364
a_2	0.716056275	−0.471337602	0.080521859	0.491906983
a_3	−0.355722032	0.260669695	−0.001607407	−0.171829096
a_4	0.123286713	−0.095260997	−0.001309298	0.045734142
a_5	−0.029758140	0.023687671	0.000132341	−0.006156882
a_6	0.004941450	−0.004005973	–	–
a_7	−0.000550372	0.000451603	–	–
a_8	0.000039127	−0.000032379	–	–
a_9	−0.000001600	0.000001333	–	–
a_{10}	0.000000029	−0.000000024	–	–

Table B.4 Correlations for modified Bessel functions of second order.

Correlation	$e^{-x}I_n(x) = \sum_{i=0}^{10} a_i(x)^i$	$e^x K_n(x) = \sum_{i=0}^{5} \frac{a_i}{(x)^i}$
a_i	Second-order first kind $n = 2$	Second-order second kind $n = 2$
a_0	0.007078043	0.391001052
a_1	−0.028177387	1.460168295
a_2	0.158252286	3.891981982
a_3	−0.133323271	−0.975716742
a_4	0.059665646	0.353955360
a_5	−0.016607171	−0.027018559
a_6	0.002999993	–
a_7	−0.000351638	–
a_8	0.000025787	–
a_9	−0.000001074	–
a_{10}	0.000000019	–

Table B.5 Correlations for Bessel functions.

Correlation a_i	$J_0(x)=\sum_{i=0}^{10} a_i (x)^i$	$J_1(x)=\sum_{i=0}^{5} \frac{a_i}{(x)^i}$
a_0	1.015522111	−1.61E-02
a_1	−0.143629185	0.658015764
a_2	3.65E-02	−0.354221118
a_3	−0.226802014	0.271755995
a_4	0.10353567	−0.165508293
a_5	−1.74E-02	5.01E-02
a_6	1.02E-03	−8.17E-03
a_7	4.58E-05	7.67E-04
a_8	−9.13E-06	−4.15E-05
a_9	4.43E-07	1.21E-06
a_{10}	−7.41E-09	−1.47E-08

Table B.6 Correlations for modified Bessel functions of fractional orders −1/3 and 2/3.

Correlation	$I_{-1/3}(x)= =\sum_{i=0}^{5} a_i (\ln x)^i$	$I_{-1/3}(x)=\sum_{i=0}^{10} a_i (x)^i$	$e^{-x} I_{2/3}(x)=\sum_{i=0}^{5} a_i (\ln x)^i$	$e^{-x} I_{2/3}(x) =\sum_{i=0}^{5} a_i (\ln x)^i$
a_i	$x=0-0.7$	$x=0.7-10$	$x=0-1$	$x=1-10$
a_0	1.258926187	4.351454602	0.296981026	0.296938660
a_1	0.139819580	−11.958740340	−0.019330653	−0.009372975
a_2	0.296427739	18.493613026	−0.090564653	−0.091374989
a_3	0.058327605	−15.457527486	−0.031312208	0.037609113
a_4	0.008376241	8.025186701	−0.004299994	−0.003268258
a_5	0.000316607	−2.665270221	−0.000214113	−0.000516593
a_6		0.581890141		
a_7		−0.082892660		
a_8		0.007473041		
a_9		−0.000388397		
a_{10}		0.000009093		

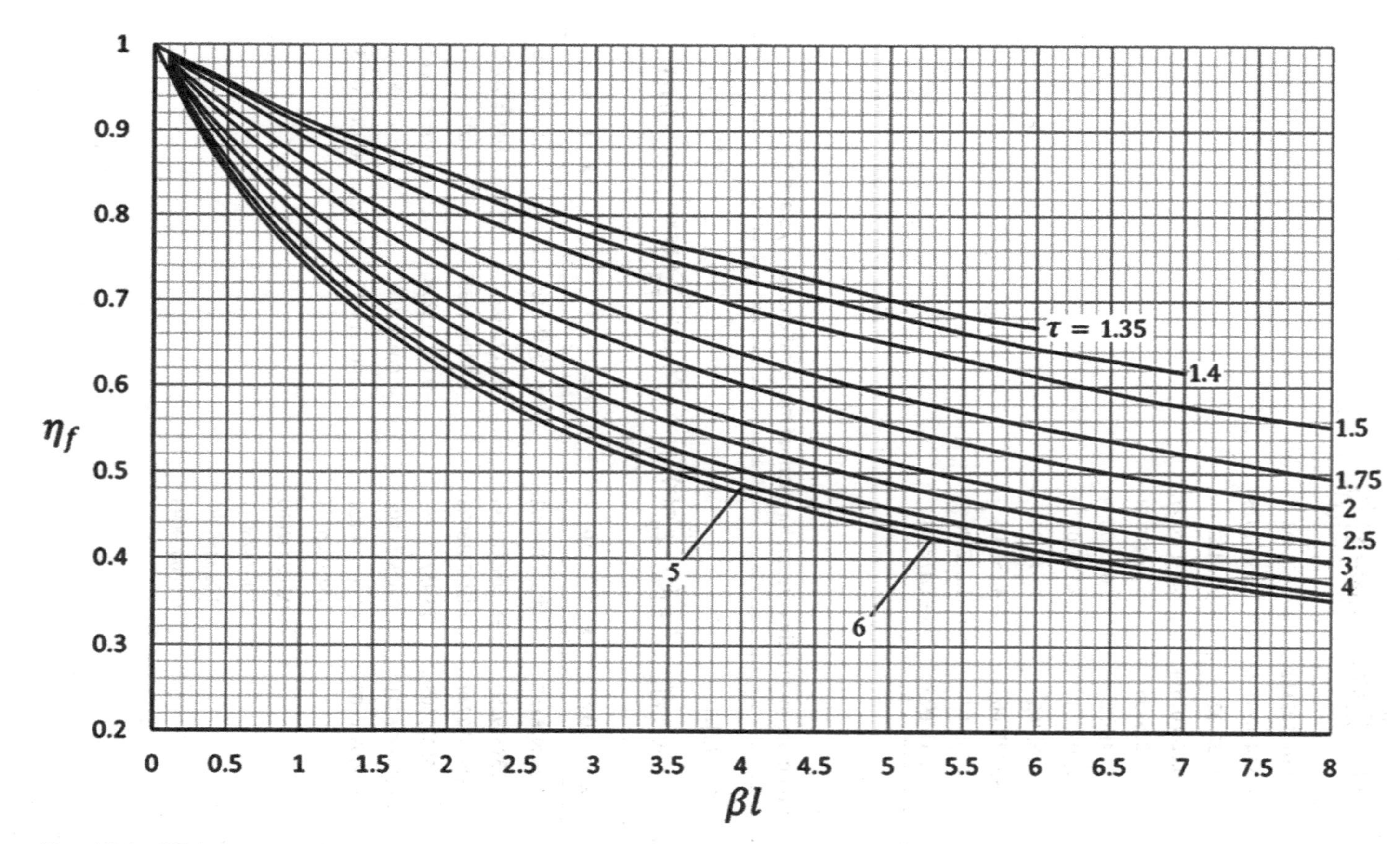

Figure B.1 Efficiency chart of straight fin of trapezoidal profile $\beta L = 0.2 - 8.0$.

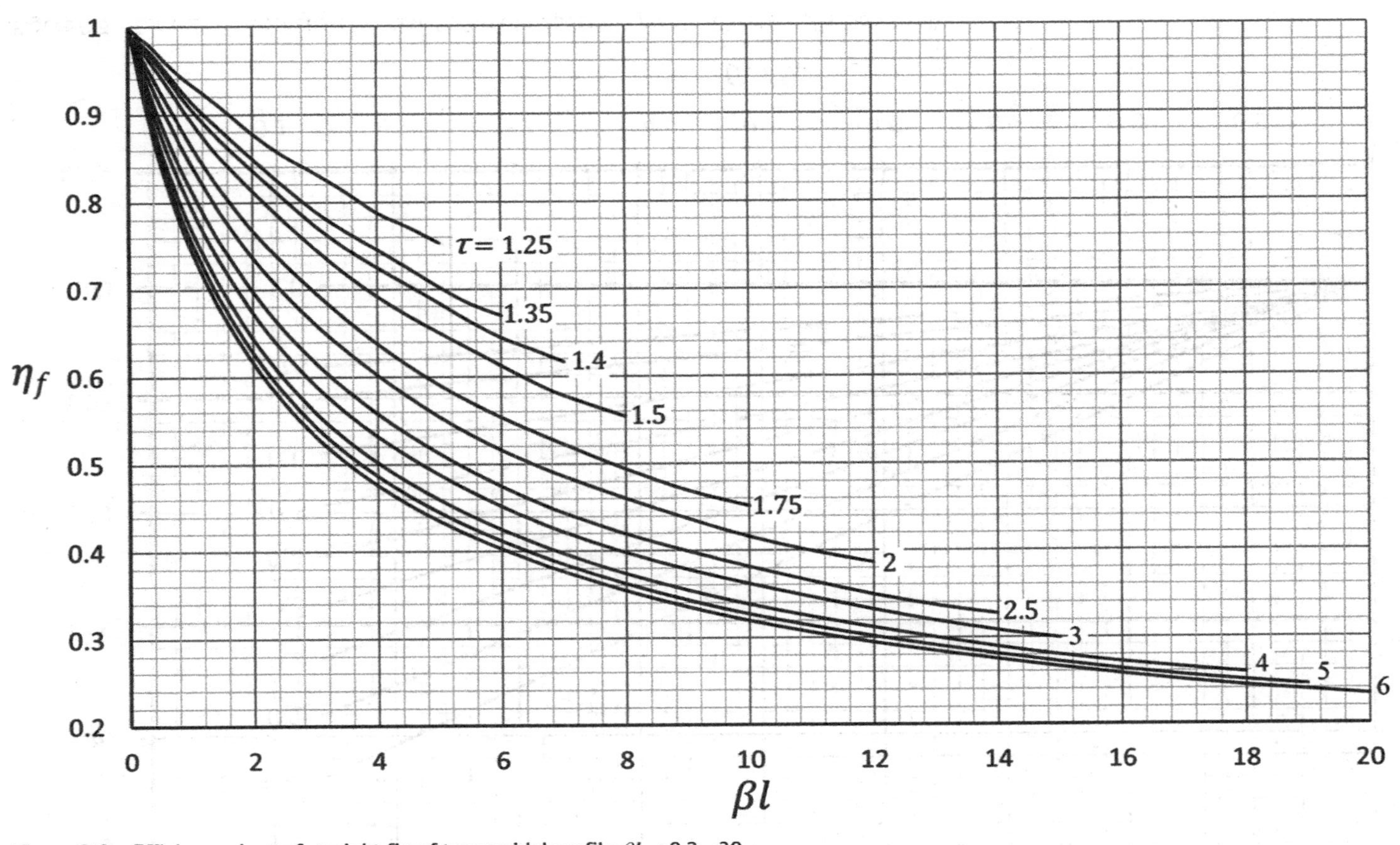

Figure B.2 Efficiency chart of straight fin of trapezoidal profile $\beta L = 0.2 - 20$.

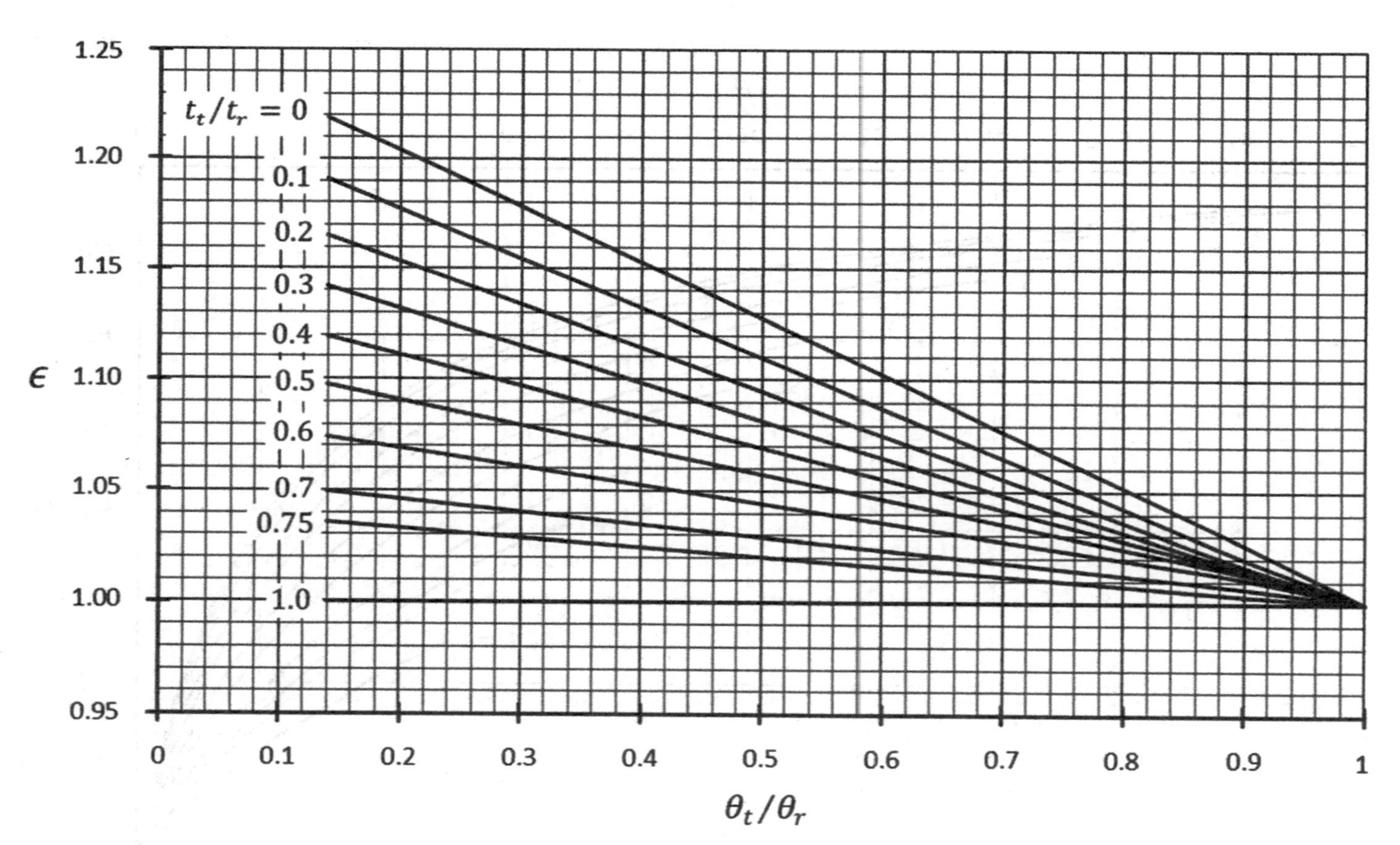

Figure B.3 Correction factor chart for straight fins of trapezoidal profile (Chapter 3, Section 3.4.4).

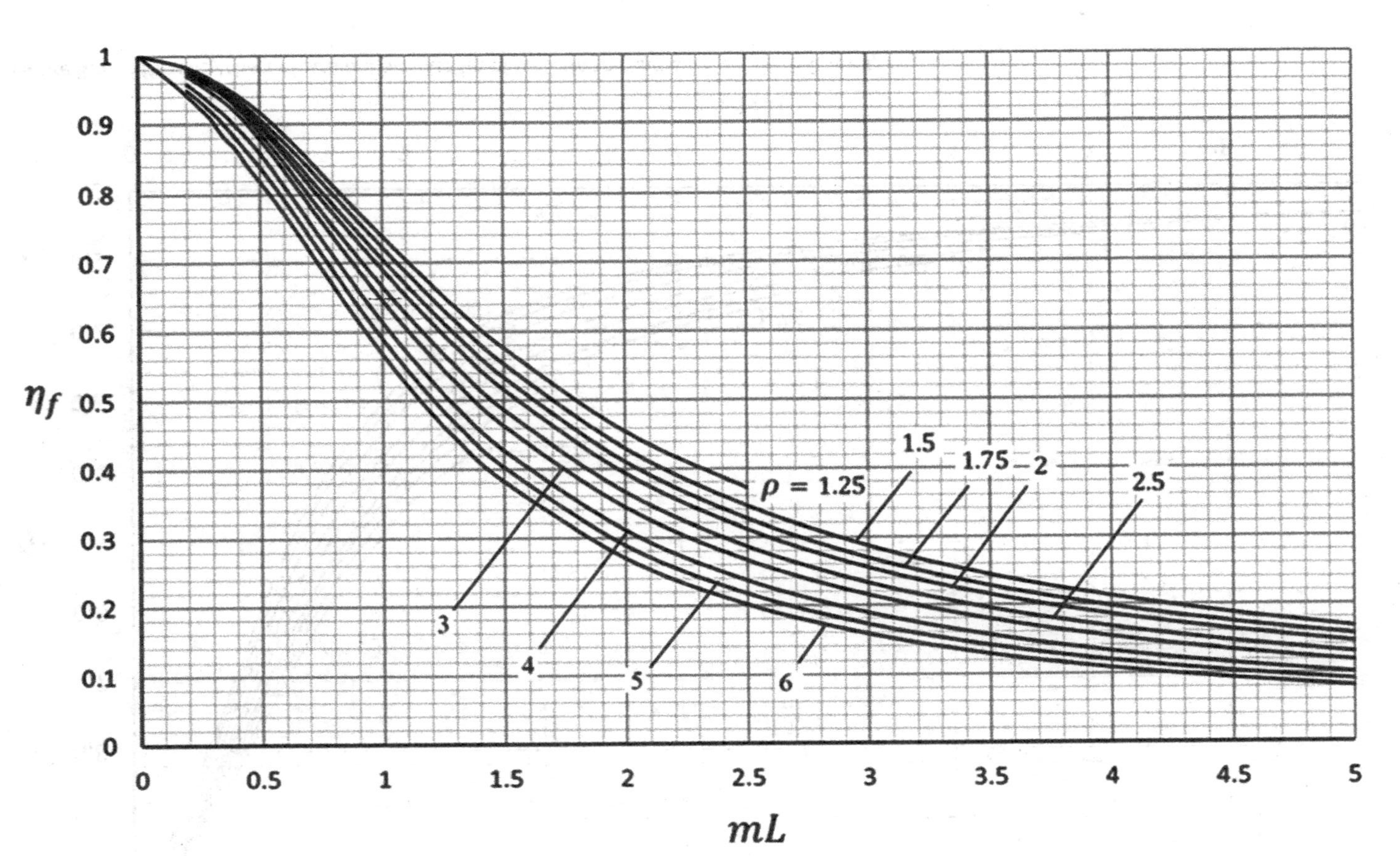

Figure B.4 Efficiency chart of straight annular fin of uniform thickness $mL = 0.2 - 5.0, \rho = t_t / t_r$.

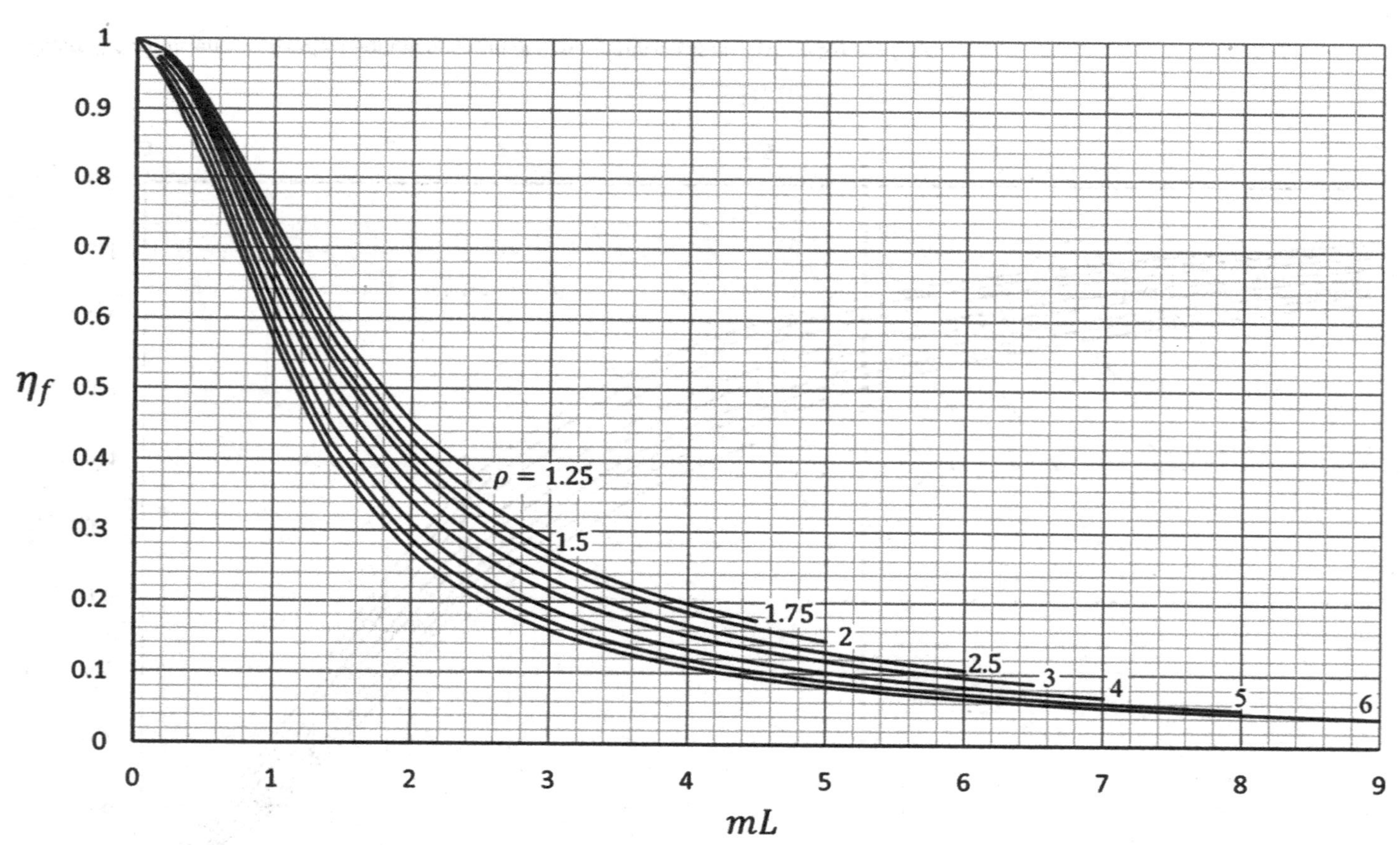

Figure B.5 Efficiency chart of straight annular fin of uniform thickness $mL = 0.2 - 9.0, \rho = t_t / t_r$.

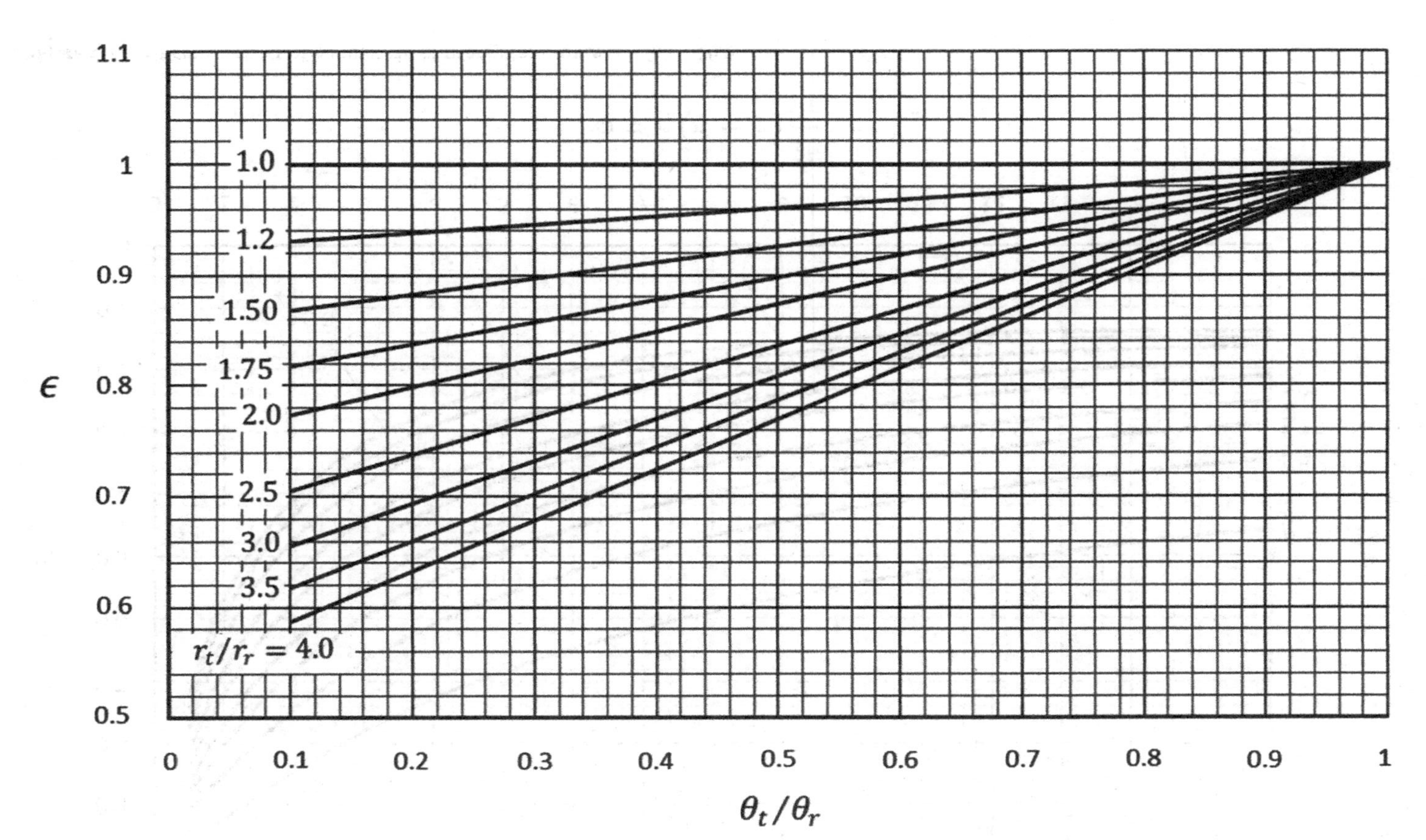

Figure B.6 Correction factor chart for straight annular fin of uniform thickness (Chapter 3, Section 3.5.3).

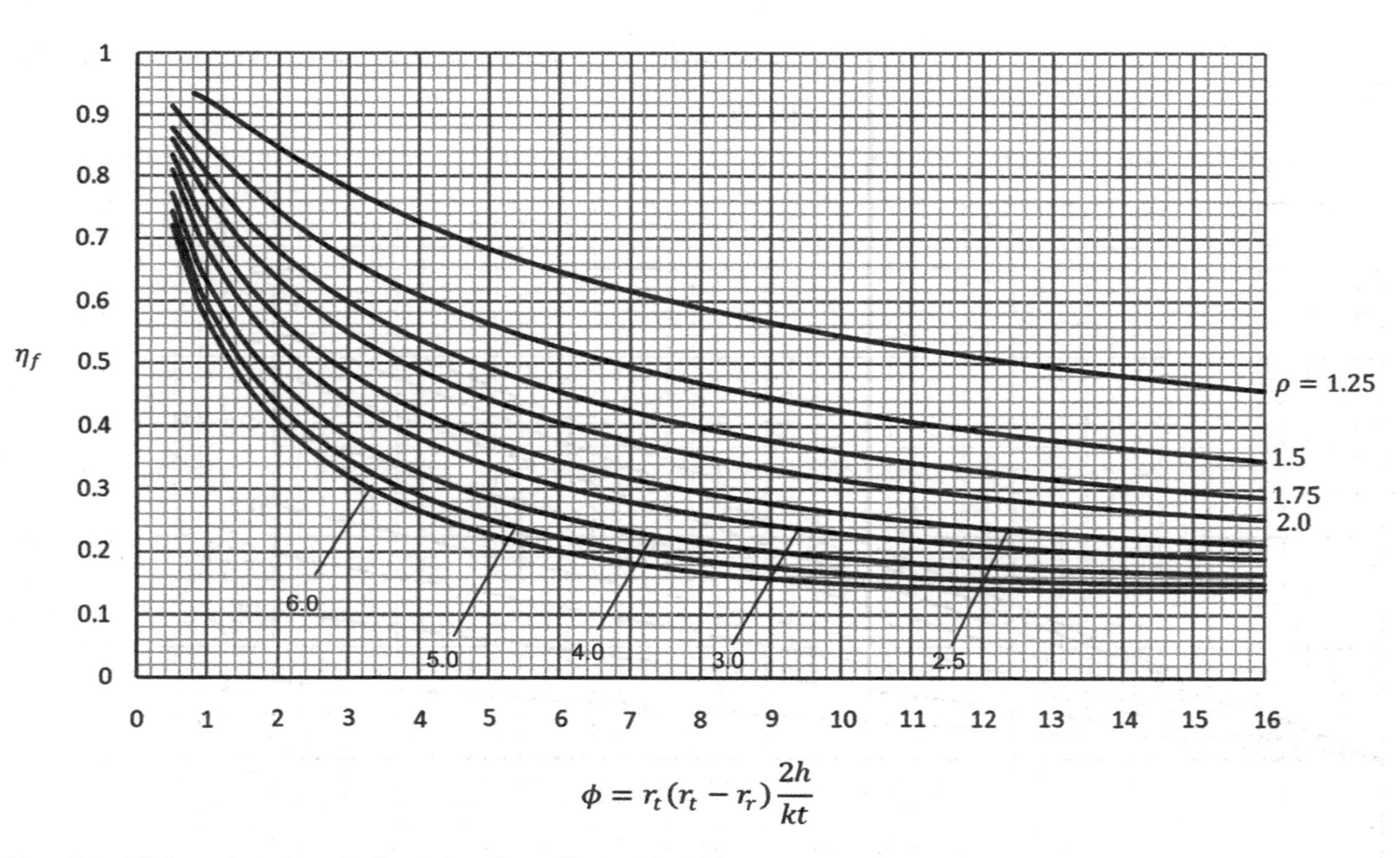

Figure B.7 Efficiency chart of annular fin of triangular profile $\phi = 0.5 - 16.0$.

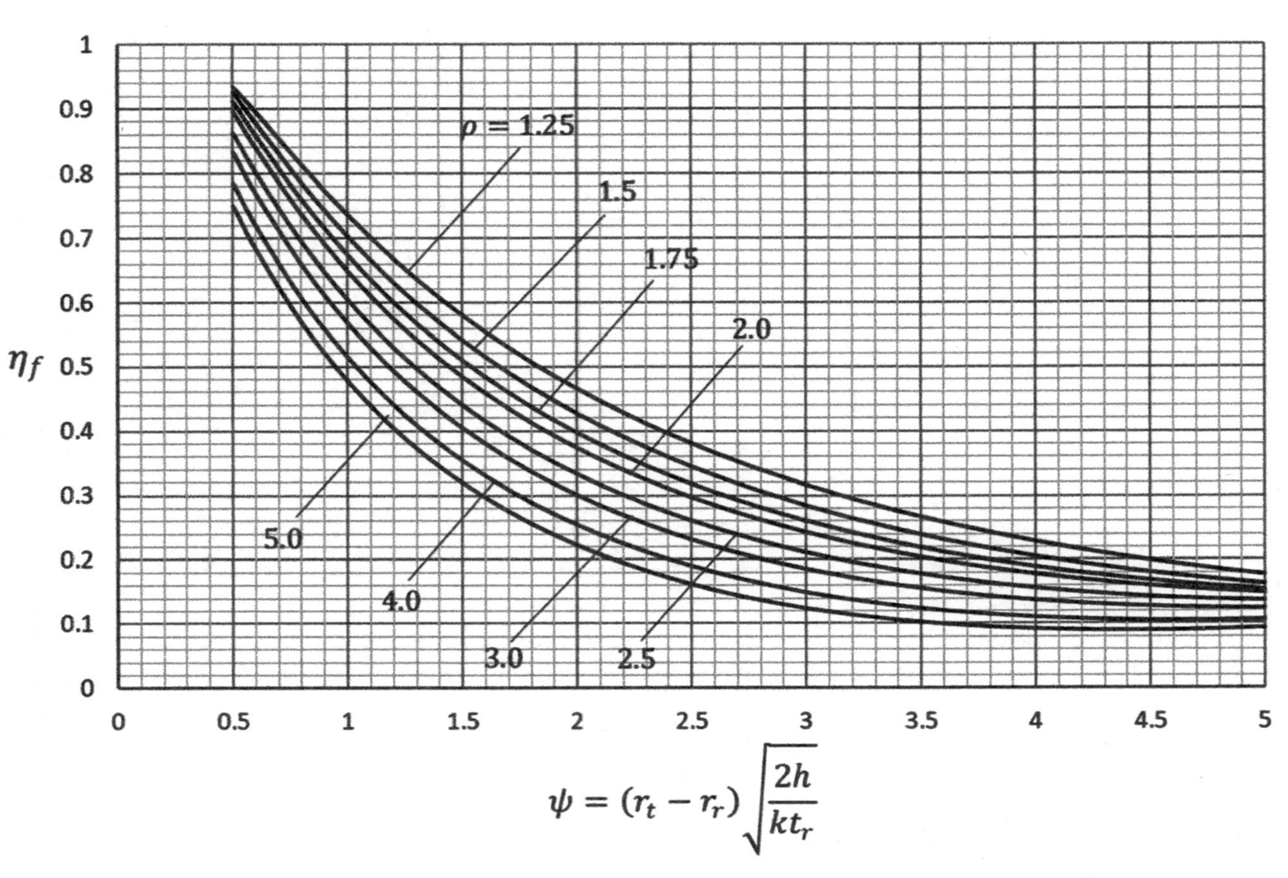

Figure B.8 Efficiency chart of annular fin of hyperbolic profile $\psi = 0.5 - 16.0$.

Appendix C

Spreadsheet implementations of methods of solving simultaneous equations of two-dimensional conduction models.

C.1 Matrix Inversion Method

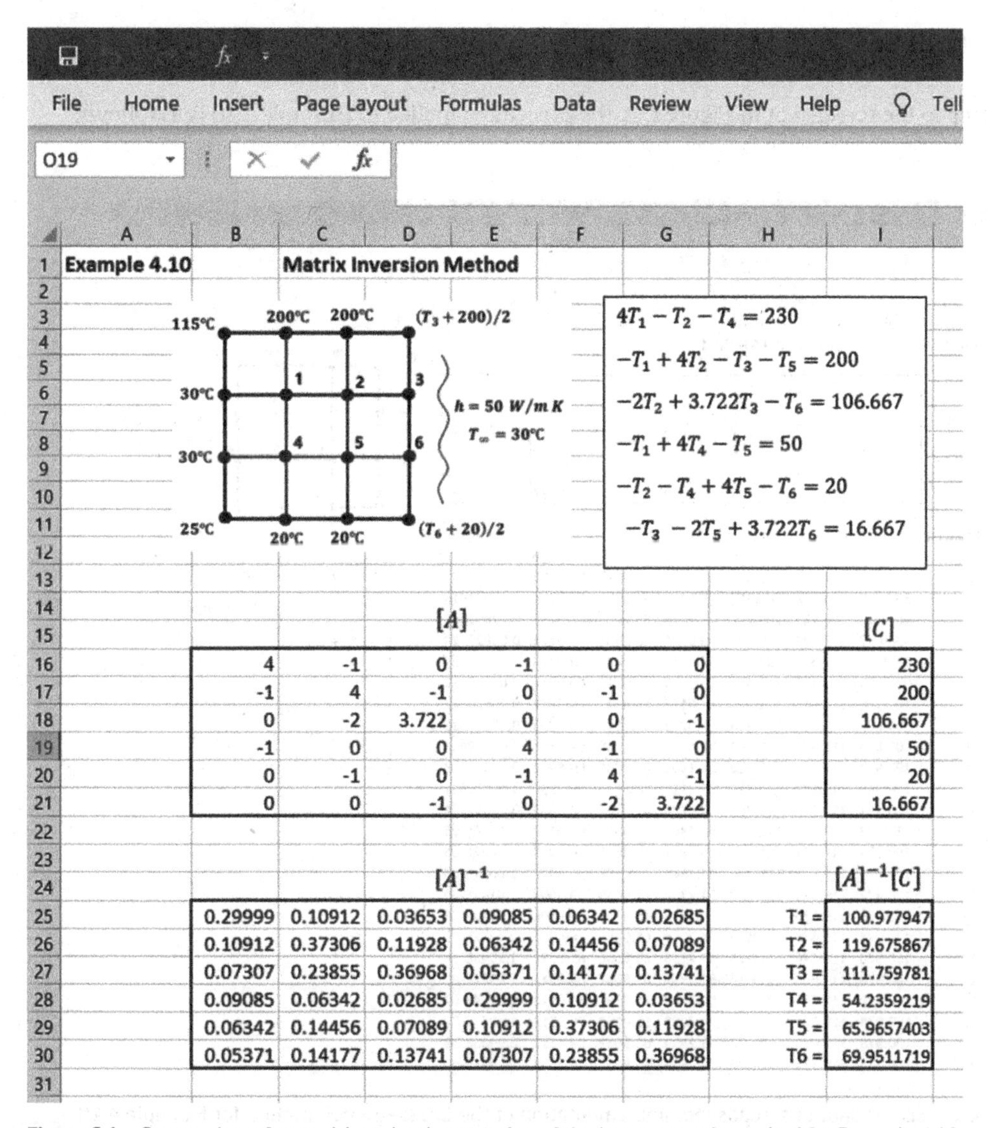

Figure C.1 Screen shot of spreadsheet implementation of the inverse matrix method for Example 4.10.

Companion website: www.wiley.com/go/ghojel/heat_transfer

Referring to the screenshot above, the procedure applied to example 4.10 is as follows:

1) Enter the coefficients a_{ij} of the unknown temperatures into cells B16:G21 to form matrix $[A]$
2) Enter the Constants C_i in cells I16:I21 to form matrix $[C]$
3) Highlight cells B25:G30
4) In the Formula Bar, type MINVERSE(B16:B21) and press **Ctrl + Shift + Enter**
   The inverse matrix $[A]^{-1}$ appears in cells B25:G30)
5) Highlight cells I25:I30
6) In the Formula Bar, type MMULT(B25:G30,I16:I21) and press **Ctrl + Shift + Enter**
7) The calculated temperatures $\{[A]^{-1}[C]\}$ appear in cells I25:I30

C.2 Gauss–Siedel Method

Referring to the screenshot in Figure C.2, the procedure applied to Example 4.10 is as follows:

Example 4.10 — **Gauss-Seidel Method**

a_{ij}						C_i
4	-1	0	-1	0	0	230
-1	4	-1	0	-1	0	200
0	-2	3.722	0	0	-1	106.667
-1	0	0	4	-1	0	50
0	-1	0	-1	4	-1	20
0	0	-1	0	-2	3.722	16.667

No of Iterations *k*	T1	T2	T3	T4	T5	T6	% error (T1)
0	0	0	0	0	0	0	
1	57.5	64.375	63.2501	26.875	27.8125	36.4165	
2	80.3125	92.8438	88.3318	39.5313	47.1979	53.5719	0.39673913
3	90.5938	106.531	100.296	46.9479	56.7627	61.9259	0.12801567
4	95.8697	113.232	106.141	50.6581	61.454	66.0172	0.05823729
5	98.4725	116.517	109.005	52.4816	63.7539	68.0226	0.0271497
6	99.7496	118.127	110.409	53.3759	64.8814	69.0057	0.01296912
7	100.376	118.917	111.098	53.8143	65.4342	69.4877	0.00627723
8	100.683	119.304	111.435	54.0292	65.7052	69.724	0.00305816
9	100.833	119.493	111.601	54.1346	65.838	69.8398	0.00149464
10	100.907	119.586	111.682	54.1862	65.9031	69.8966	0.00073163
11	100.943	119.632	111.722	54.2116	65.935	69.9244	0.00035841
12	100.961	119.654	111.741	54.224	65.9507	69.938	0.00017564
13	100.97	119.665	111.751	54.2301	65.9584	69.9447	8.6089E-05
14	100.974	119.671	111.755	54.2331	65.9621	69.948	4.22E-05
15	100.976	119.673	111.758	54.2345	65.964	69.9496	2.0687E-05
16	100.977	119.675	111.759	54.2352	65.9649	69.9504	1.0141E-05
17	100.977	119.675	111.759	54.2356	65.9653	69.9508	4.9715E-06
18	100.978	119.676	111.76	54.2358	65.9655	69.951	2.4372E-06
19	100.978	119.676	111.76	54.2358	65.9656	69.9511	1.1948E-06
20	100.978	119.676	111.76	54.2359	65.9657	69.9511	5.8572E-07
21	100.978	119.676	111.76	54.2359	65.9657	69.9512	2.8714E-07
22	100.978	119.676	111.76	54.2359	65.9657	69.9512	1.4077E-07
23	100.978	119.676	111.76	54.2359	65.9657	69.9512	6.9008E-08
24	100.978	119.676	111.76	54.2359	65.9657	69.9512	3.383E-08

115°C, 200°C, 200°C, $(T_3 + 200)/2$, 30°C, 30°C, 25°C, 20°C, 20°C, $(T_6 + 20)/2$, $h = 50\ W/m\,K$, $T_\infty = 30°C$

$$4T_1 - T_2 - T_4 = 230$$
$$-T_1 + 4T_2 - T_3 - T_5 = 200$$
$$-2T_2 + 3.722T_3 - T_6 = 106.667$$
$$-T_1 + 4T_4 - T_5 = 50$$
$$-T_2 - T_4 + 4T_5 - T_6 = 20$$
$$-T_3 - 2T_5 + 3.722T_6 = 16.667$$

Figure C.2 Screen shot of spreadsheet implementation of the Gauss–Siedel method for Example 4.10.

1) The temperature coefficients a_{ij} and the constants C_i are entered into cells B3:H8
2) Initial values of zero for temperatures are entered into cells B11 to G11
3) The equations below are entered consecutively into cells B12 to G12
   =(H$3-C$3*C11-E$3*E11)/4
   =(H$4-B$4*B12-D$4*D11-F$4*F11)/4
   =(H$5-C$5*C12-G$5*G11)/3.722
   =(H$6-B$6*B12-F$6*F11)/4
   =(H$7-C$7*C12-E$7*E12-G$7*G11)/4
   =(H$8-D$8*D12-F$8*F12)/3.722
4) Cells B12:G12 are copied and pasted to the cells blow corresponding to iterations $k = 2, 3, 4, \ldots$ until satisfactory convergence is achieved. For the current example this occurs after 23 iterations. Alternatively, cells B12:G12 can be highlighted then dragged downwards with the mouse right button pressed until convergence is achieved.

Appendix D

Table D.1 Positive roots of the transcendental equation $\mu_n \tan \mu_n = Bi$.

Bi	μ_1	μ_2	μ_3	μ_4	μ_5	μ_6
0	0	3.1416	6.2832	9.4248	12.5664	15.708
0.001	0.0316	3.1419	6.2833	9.4249	12.5665	15.708
0.002	0.0447	3.1422	6.2835	9.425	12.5665	15.7081
0.004	0.0632	3.1429	6.2838	9.4252	12.5667	15.7082
0.006	0.0632	3.1429	6.2838	9.4254	12.5668	15.7083
0.008	0.0893	3.1441	6.2845	9.4256	12.567	15.7085
0.01	0.0998	3.1448	6.2848	9.4258	12.5672	15.7086
0.02	0.141	3.1479	6.2864	9.4269	12.568	15.7092
0.04	0.1987	3.1543	6.2895	9.429	12.5696	15.7105
0.06	0.2425	3.1606	6.2927	9.4311	12.5711	15.7118
0.08	0.2791	3.1668	6.2959	9.4333	12.5727	15.7131
0.1	0.3111	3.1731	6.2991	9.4354	12.5743	15.7143
0.2	0.4328	3.2039	6.3148	9.4459	12.5823	15.7207
0.3	0.5218	3.2341	6.3305	9.4565	12.5902	15.727
0.4	0.5932	3.2636	6.3461	9.467	12.5981	15.7334
0.5	0.6533	3.2923	6.3616	9.4775	12.606	15.7397
0.6	0.7051	3.3204	6.377	9.4879	12.6139	15.746
0.7	0.7506	3.3477	6.3923	9.4983	12.6218	15.7524
0.8	0.791	3.3744	6.4074	9.5087	12.6296	15.7587
0.9	0.8274	3.4003	6.4224	9.519	12.6375	15.765
1	0.8603	3.4256	6.4373	9.5293	12.6453	15.7713

(Continued)

Heat Transfer Basics: A Concise Approach to Problem Solving, First Edition. Jamil Ghojel.

Companion website: www.wiley.com/go/ghojel/heat_transfer

Table D.1 (Continued)

Bi	μ_1	μ_2	μ_3	μ_4	μ_5	μ_6
1.5	0.9882	3.5422	6.5097	9.5801	12.6841	15.8026
2	1.0769	3.6436	6.5783	9.6296	12.7223	15.8336
3	1.1925	3.8088	6.704	9.724	12.7966	15.8945
4	1.2646	3.9352	6.814	9.8119	12.8678	15.9536
5	1.3138	4.0336	6.9096	9.8928	12.9352	16.0107
6	1.3496	4.1116	6.9924	9.9667	12.9988	16.0654
7	1.3766	4.1746	7.064	10.0339	13.0584	16.1177
8	1.3978	4.2264	7.1263	10.0949	13.1141	16.1675
9	1.4149	4.2694	7.1806	10.1502	13.166	16.2147
10	1.4289	4.3058	7.2281	10.2003	13.2142	16.2594
15	1.4729	4.4255	7.3959	10.3898	13.4078	16.4474
20	1.4961	4.4915	7.4954	10.5117	13.542	16.5864
30	1.5202	4.5615	7.6057	10.6543	13.7085	16.7691
40	1.5325	4.5979	7.6647	10.7334	13.8048	16.8794
50	1.54	4.6202	7.7012	10.7832	13.8666	16.9519
60	1.5451	4.6353	7.7259	10.8172	13.9094	17.0026
80	1.5514	4.6543	7.7573	10.8606	13.9644	17.0686
100	1.5552	4.6658	7.7764	10.8871	13.9981	17.1093
∞	1.5708	4.7124	7.854	10.9956	14.13782	17.2788

Table D.2 Positive roots of the transcendental equation $\mu_n J_1(\mu_n) - Bi J_0(\mu_n) = 0$.

Bi	μ_1	μ_2	μ_3	μ_4	μ_5	μ_6
0	0	3.8317	7.0156	10.1735	13.3237	16.4706
0.01	0.1412	3.8343	7.017	10.1745	13.3244	16.4712
0.02	0.1995	3.8369	7.0184	10.1754	13.3252	16.4718
0.04	0.2814	3.8421	7.0213	10.1774	13.3267	16.4731
0.06	0.3438	3.8473	7.0241	10.1794	13.3282	16.4743
0.08	0.396	3.825	7.027	10.1813	13.3297	16.4755
0.1	0.4417	3.8577	7.0298	10.1833	13.3312	16.4767
0.15	0.5376	3.8706	7.0369	10.1882	13.3349	16.4797
0.2	0.617	3.8835	7.044	10.1931	13.3387	16.4828
0.3	0.7465	3.9091	7.0582	10.2029	13.3462	16.4888
0.4	0.8516	3.9344	7.0723	10.2127	13.3537	16.4949
0.5	0.9408	3.9594	7.0864	10.2225	13.3611	16.501

Table D.2 (Continued)

Bi	μ_1	μ_2	μ_3	μ_4	μ_5	μ_6
0.6	1.0184	3.9841	7.1004	10.2322	13.3686	16.507
0.7	1.0873	4.0085	7.1143	10.2419	13.3761	16.5131
0.8	1.149	4.0325	7.1282	10.2516	13.3835	16.5191
0.9	1.2048	4.0562	17.1421	10.2613	13.391	16.5251
1	1.2558	4.0795	7.1558	10.271	13.3984	16.5312
1.5	1.4569	4.1902	7.2233	10.3188	13.4353	15.5612
2	1.5994	4.291	7.2884	10.3658	13.4719	16.591
3	1.7887	4.4634	7.4103	10.4566	13.5334	16.6499
4	1.9081	4.6018	7.5201	10.5423	13.6125	16.7073
5	1.9898	4.7131	7.6177	10.6223	13.6786	16.763
6	2.049	4.8033	7.7039	10.6964	13.7414	16.8168
7	2.0937	4.8772	7.7797	10.7646	13.8008	16.8684
8	2.1286	4.9384	7.8464	10.8271	13.8566	16.9179
9	2.1566	4.9897	7.9051	10.8842	13.909	16.965
10	2.1795	5.0332	7.9569	10.9363	13.958	17.0099
15	2.2509	5.1773	8.1422	11.1367	14.1576	17.2008
20	2.288	5.2568	8.2534	11.2677	14.2983	17.3442
30	2.3261	5.341	8.3771	11.4221	14.4748	17.5348
40	2.3455	5.3846	8.4432	11.5081	14.5774	17.6508
50	2.3572	5.4112	8.484	11.5621	14.6433	17.7272
60	2.3651	5.4291	8.5116	11.599	14.6889	17.7807
80	2.375	5.4516	8.5466	11.6461	14.7475	17.8502
100	2.3809	5.4652	8.5678	11.6747	14.7834	17.8931
∞	2.4048	5.5201	8.6537	11.7915	14.9309	18.0711

Table D.3 Positive roots of the transcendental equation $Bi = 1 - \mu_n \cot \mu_n$.

Bi	μ_1	μ_2	μ_3	μ_4	μ_5	μ_6
0	0	4.4934	7.7253	10.9041	14.0662	17.2208
0.001	0.0548	4.4936	7.7254	10.9042	14.0662	17.2208
0.002	0.0774	4.4939	7.7255	10.9043	14.0663	17.2209
0.003	0.0948	4.4941	7.7256	10.9044	14.0664	17.2209
0.004	0.1095	4.4943	7.7258	10.9045	14.0665	17.221
0.005	0.1224	4.4945	7.7259	10.9046	14.0666	17.2211
0.006	0.1341	4.4947	7.726	10.9047	14.0666	17.2211
0.007	0.1448	495	7.7262	10.9048	14.0667	17.2212

(Continued)

Table D.3 (Continued)

Bi	μ_1	μ_2	μ_3	μ_4	μ_5	μ_6
0.008	0.1548	4.4952	7.7263	10.9049	14.0668	17.2212
0.009	0.1642	4.4954	7.7264	10.9049	14.0668	17.2213
0.01	0.173	4.4956	7.7265	10.905	14.0669	17.2213
0.02	0.2445	4.4979	7.7287	10.906	14.0669	17.2219
0.03	0.2991	4.5001	7.7291	10.9069	14.0676	17.225
0.04	0.345	4.5023	7.7304	10.9078	14.0683	17.2231
0.05	0.3854	4.5045	7.7317	10.9087	14.0697	17.2237
0.06	0.4217	4.5068	7.733	10.9096	14.0705	17.2242
0.07	0.4551	4.509	7.7343	10.9105	14.0712	17.2248
0.08	0.486	4.5112	7.7356	10.9115	14.0719	17.2254
0.09	0.515	4.5134	7.7369	10.9124	14.0726	17.226
0.1	0.5423	4.5157	7.7382	10.9133	14.0733	17.2266
0.2	0.7593	4.5379	7.7511	10.9225	14.0804	17.2324
0.3	0.9208	4.5601	7.7641	10.9316	14.0875	17.2382
0.4	1.0528	4.5822	7.777	10.9408	14.0946	17.244
0.5	1.1656	4.6042	7.7899	10.9499	14.1017	17.2498
0.6	1.2644	4.6261	7.8028	10.9591	14.088	17.2556
0.7	1.3525	4.6479	7.8156	10.9682	14.1159	17.2614
0.8	1.432	4.6696	7.8284	10.9774	14.123	17.2672
0.9	1.5044	4.6911	7.8412	10.9865	14.1301	17.273
1	1.5708	4.7124	7.854	10.9956	14.1372	17.2788
1.5	1.8366	4.8158	7.9171	11.0409	14.1724	17.3076
2	2.0288	4.9132	7.9787	11.0856	14.2075	17.3364
3	2.2889	5.087	8.0962	11.1727	14.2764	17.3922
4	2.4557	5.2329	8.2045	11.256	14.3434	17.449
5	2.5704	5.354	8.3029	11.3349	14.408	17.5034
6	2.6537	5.4544	8.3914	11.4086	14.4699	17.5562
7	2.7165	5.5378	8.4703	11.4773	14.5288	17.6072
8	2.8363	5.6078	8.5406	11.5408	14.5847	17.6562
9	2.8044	5.6669	8.6031	11.5994	14.6374	17.7032
10	2.8363	5.7172	8.6587	11.6532	14.687	17.7481
20	2.9857	5.9783	8.9831	12.0029	15.0384	18.0887
30	3.0372	6.0766	9.1201	12.1691	15.2245	18.2869
40	3.0632	6.1273	9.1933	12.2618	15.3334	18.4085
50	3.0788	6.1582	9.2384	12.32	15.4034	18.4888
60	3.0893	6.1788	9.269	12.3599	15.4518	18.545
70	3.0967	6.1937	9.2909	12.3887	15.4872	18.5864
80	3.1023	6.2048	9.3075	12.4106	15.514	18.6181
90	3.1067	6.2135	9.3204	12.4276	15.5352	18.6431
100	3.1102	6.2204	9.3308	12.4414	15.5521	18.6632

Example 5.2 Full solution

k	40	
ρ	8000	
c_p	500	
α	0.00001	
h	1000	
L_c	0.03	
Bio	0.75	
T_∞	773	K
T_i	300	K
R	0.03	
Ts	573	K
Fo	0.530637	
t	47.75737	

$$\frac{T_s - T_\infty}{T_i - T_\infty} = \frac{2J_1(\mu_1)J_0(\mu_1)}{\mu_1[J_0^2(\mu_1) + J_1^2(\mu_1)]} e^{-(\mu_1^2 Fo)}$$

$$Fo = \frac{\alpha t}{R^2}$$

J_1

$$J_1(x) = \sum_{i=0}^{10} a_i(x)^i$$

J_0

$$J_0(x) = \sum_{i=0}^{10} a_i(x)^i$$

	J_1	J_0
a_0	-0.01611	1.015522
a_1	0.658016	-0.14363
a_2	-0.35422	0.0365
a_3	0.271756	-0.2268
a_4	-0.16551	0.103536
a_5	0.050099	-0.01741
a_6	-0.00817	0.001016
a_7	0.000767	4.58E-05
a_8	-4.2E-05	-9.1E-06
a_9	1.21E-06	4.43E-07
a_{10}	-1.5E-08	-7.4E-09

$$T = T_\infty + (T_i - T_\infty)\sum_{n=1}^{\infty} \frac{2J_1(\mu_n)}{\mu_n[J_0^2(\mu_n) + J_1^2(\mu_n)]} J_0\left(\mu_n \frac{r}{R}\right) e^{-(\mu_n^2 Fo)}$$

μ_n	$J_1(\mu_n)$	$J_1^2(\mu_n)$	$J_0(\mu_n)$	$J_0^2(\mu_n)$	$\mu_n \frac{r}{R}$	$J_0\left(\mu_n \frac{r}{R}\right)$	$\frac{2J_1(\mu_n)}{\mu_n[J_0^2(\mu_n) + J_1^2(\mu_n)]}$	$J_0\left(\mu_n \frac{r}{R}\right)$	$e^{-(\mu_n^2 Fo)}$	$\frac{T - T_\infty}{T_i - T_\infty}$	$\sum_{n=1}^{n} X$	Y, K
1.11815	0.471145	0.221977	0.716965	0.514039	0	1.015522	1.144977	1.015522	0.515079	0.598908	0.598908	489.7165
4.0205	-0.07767	0.006033	-0.39334	0.154719	0	1.015522	-0.24036	1.015522	0.000188	-4.6E-05	0.598862	489.7382
7.12125	0.032828	0.001078	0.291084	0.08473	0	1.015522	0.107446	1.015522	2.06E-12	2.24E-13	0.598862	489.7382
10.17885	0.001811	3.28E-06	-0.25293	0.063974	0	1.015522	0.005563	1.015522	1.33E-24	7.5E-27	0.598862	489.7382
13.3798	0.014064	0.000198	0.219539	0.048197	0	1.015522	0.04344	1.015522	5.55E-42	2.45E-43	0.598862	489.7382
16.5161	-0.16067	0.025814	-1.46159	2.136231	0	1.015522	-0.009	1.015522	1.37E-63	-1.3E-65	0.598862	489.7382

$$\mu_n J_1(\mu_n) - Bi\, J_0(\mu_n) = 0$$

Bi	μ_1	μ_2	μ_3	μ_4	μ_5	μ_6
0						
0.01	0.1412	3.8343	7.017	10.1745	13.3244	16.4712
0.02	0.1995	3.8369	7.0184	10.1754	13.3252	16.4718
0.04	0.2814	3.8421	7.0213	10.1774	13.3267	16.4731
0.06	0.3438	3.8473	7.0241	10.1794	13.3282	16.4743
0.08	0.396	3.825	7.027	10.1813	13.3297	16.4755
0.1	0.4417	3.8577	7.0298	10.1833	13.3312	16.4767
0.15	0.5376	3.8706	7.0369	10.1882	13.3349	16.4797
0.2	0.617	3.8835	7.044	10.1931	13.3387	16.4828
0.3	0.7465	3.9091	7.0582	10.2029	13.3462	16.4888
0.4	0.8516	3.9344	7.0723	10.2127	13.3537	16.4949
0.5	0.9408	3.9594	7.0864	10.2225	13.3611	16.501
0.6	1.0184	3.9841	7.1004	10.2322	13.3686	16.507
0.7	1.0873	4.0085	7.1143	10.2419	13.3761	16.5131
0.75	1.11815	4.0205	7.12125	10.17885	13.3798	16.5161
0.8	1.149	4.0325	7.1282	10.2516	13.3835	16.5191
0.9	1.2048	4.0562	17.1421	10.2613	13.391	16.5251
1	1.2558	4.0795	7.1558	10.271	13.3984	16.5312
1.5	1.4569	4.1902	7.2233	10.3188	13.4353	15.5612
2	1.5994	4.291	7.2884	10.3658	13.4719	16.591
3	1.7887	4.4634	7.4103	10.4566	13.5334	16.6499
4	1.9081	4.6018	7.5201	10.5423	13.6125	16.7073
5	1.9898	4.7131	7.6177	10.6223	13.6786	16.763
6	2.049	4.8033	7.7039	10.6964	13.7414	16.8168
7	2.0937	4.8772	7.7797	10.7646	13.8008	16.8684
8	2.1286	4.9384	7.8464	10.8271	13.8566	16.9179
9	2.1566	4.9897	7.9051	10.8842	13.909	16.965
10	2.1795	5.0332	7.9569	10.9363	13.958	17.0099
15	2.2509	5.1773	8.1422	11.1367	14.1576	17.2008
20	2.288	5.2568	8.2534	11.2677	14.2983	17.3442
30	2.3261	5.341	8.3771	11.4221	14.4748	17.5348

Figure D.1 Screen shot of the solution of Example 5.2.

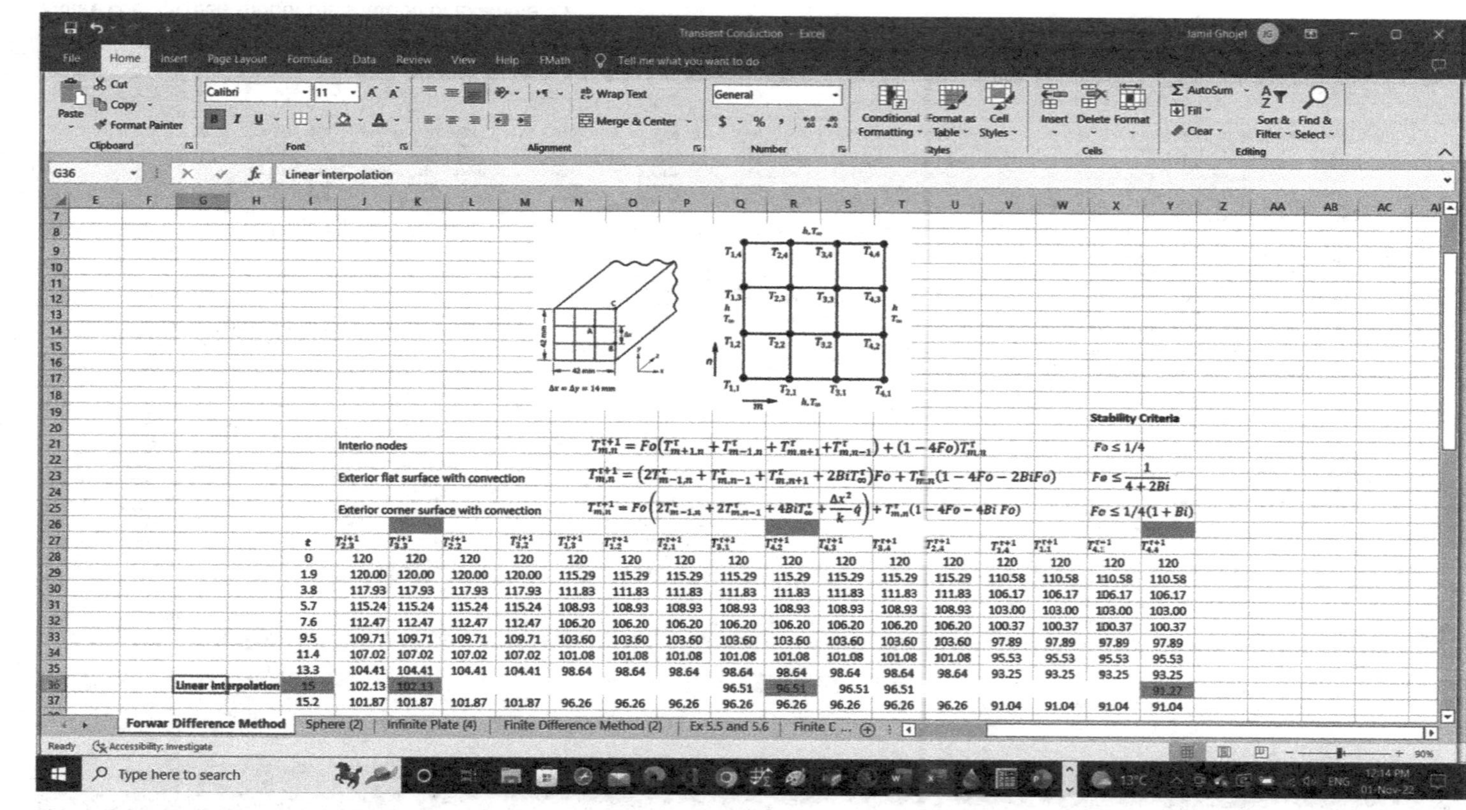

t	$T_{2,3}^{\tau+1}$	$T_{3,3}^{\tau+1}$	$T_{2,2}^{\tau+1}$	$T_{3,2}^{\tau+1}$	$T_{1,3}^{\tau+1}$	$T_{1,2}^{\tau+1}$	$T_{2,1}^{\tau+1}$	$T_{3,1}^{\tau+1}$	$T_{4,2}^{\tau+1}$	$T_{4,3}^{\tau+1}$	$T_{3,4}^{\tau+1}$	$T_{2,4}^{\tau+1}$	$T_{1,4}^{\tau+1}$	$T_{1,1}^{\tau+1}$	$T_{4,1}^{\tau+1}$	$T_{4,4}^{\tau+1}$
0	120	120	120	120	120	120	120	120	120	120	120	120	120	120	120	120
1.9	120.00	120.00	120.00	120.00	115.29	115.29	115.29	115.29	115.29	115.29	115.29	115.29	110.58	110.58	110.58	110.58
3.8	117.93	117.93	117.93	117.93	111.83	111.83	111.83	111.83	111.83	111.83	111.83	111.83	106.17	106.17	106.17	106.17
5.7	115.24	115.24	115.24	115.24	108.93	108.93	108.93	108.93	108.93	108.93	108.93	108.93	103.00	103.00	103.00	103.00
7.6	112.47	112.47	112.47	112.47	106.20	106.20	106.20	106.20	106.20	106.20	106.20	106.20	100.37	100.37	100.37	100.37
9.5	109.71	109.71	109.71	109.71	103.60	103.60	103.60	103.60	103.60	103.60	103.60	103.60	97.89	97.89	97.89	97.89
11.4	107.02	107.02	107.02	107.02	101.08	101.08	101.08	101.08	101.08	101.08	101.08	101.08	95.53	95.53	95.53	95.53
13.3	104.41	104.41	104.41	104.41	98.64	98.64	98.64	98.64	98.64	98.64	98.64	98.64	93.25	93.25	93.25	93.25
Linear interpolation 15	102.13	102.13						96.51	96.51	96.51	96.51					91.27
15.2	101.87	101.87	101.87	101.87	96.26	96.26	96.26	96.26	96.26	96.26	96.26	96.26	91.04	91.04	91.04	91.04

Figure D.2 Screen shot of the solution of Example 5.8.

Appendix N

Thermophysical Properties of Saturated Water and Air

Heat Transfer Basics: A Concise Approach to Problem Solving, First Edition. Jamil Ghojel.

Companion website: www.wiley.com/go/ghojel/heat_transfer

Table N.1 Thermophysical properties of saturated water.

			Density kg/m^3		Specific heat $c_p, J/kg.K$		Thermal conductivity $k, W/m.K$		Dynamic viscosity $\mu, kg/m.s$		Prandtl number	
									$\mu \times 10^3$	$\mu \times 10^5$		
Temp $t, °C$	Saturation pressure p_{sat}, kPa	Enthalpy of vaporization $h_{fg}, kJ/kg$	Liquid	Vapour	Liquid	Vapour	Liquid	Vapour	Liquid	Vapour	Liquid	Vapour
0	0.61	2500.9	999.8	0.0048	4220	1888	0.5678	0.0165	1.791	0.8945	13.308	1.024
0.01	0.61	2500.9	999.8	0.0048	4220	1871	0.5678	0.0165	1.790	0.8946	13.304	1.014
5	0.88	2489.1	999.9	0.0068	4205	1875	0.5770	0.0169	1.523	0.909	11.099	1.008
10	1.23	2477.2	999.7	0.0094	4194	1879	0.5857	0.0172	1.312	0.9238	9.392	1.009
15	1.71	2465.4	999.1	0.0128	4189	1882	0.5941	0.0176	1.142	0.939	8.055	1.004
20	2.34	2453.5	998.0	0.0173	4182	1887	0.6021	0.018	1.005	0.9544	6.981	1.000
25	3.17	2441.7	996.5	0.0232	4180	1891	0.6097	0.0183	0.892	0.9701	6.116	1.002
30	4.25	2429.8	994.9	0.0307	4178	1897	0.6169	0.0187	0.798	0.986	5.404	1.000
35	5.63	2417.9	993.2	0.0400	4178	1903	0.6237	0.0191	0.719	1.002	4.816	0.998
40	7.39	2406	991.4	0.0515	4179	1909	0.6302	0.0195	0.652	1.018	4.324	0.996
45	9.59	2394	989.4	0.0657	4180	1916	0.6362	0.02	0.595	1.035	3.907	0.992
50	12.35	2381.9	987.3	0.0831	4181	1924	0.6419	0.0203	0.545	1.052	3.552	0.997
55	15.76	2369.8	985.0	0.1043	4183	1933	0.6472	0.0207	0.502	1.068	3.247	0.997
60	19.94	2357.7	982.7	0.1300	4185	1941	0.6522	0.0211	0.465	1.085	2.983	0.998
65	25.03	2345.4	980.2	0.1610	4188	1951	0.6568	0.0215	0.432	1.102	2.754	1.000
70	31.18	2333	977.6	0.1979	4191	1962	0.6611	0.022	0.403	1.119	2.554	0.998
75	38.57	2320.6	974.8	0.2416	4194	1974	0.6650	0.0224	0.377	1.137	2.378	1.002
80	47.38	2308	971.9	0.2931	4198	1987	0.6685	0.0229	0.354	1.154	2.222	1.001
85	57.82	2295.3	968.9	0.3534	4202	2002	0.6717	0.0233	0.333	1.171	2.084	1.006

90	70.12	2282.5	965.8	0.4234	4207	2018	0.6746	0.0238	0.314	1.189	1.961	1.008
95	84.54	2269.6	962.5	0.5045	4212	2035	0.6772	0.0243	0.298	1.206	1.851	1.010
100	101.33	2256.4	959.1	0.5978	4218	2053	0.6794	0.0248	0.282	1.223	1.752	1.012
110	143.26	2229.6	952.0	0.8266	4231	2096	0.6829	0.0259	0.256	1.258	1.583	1.018
120	198.52	2202.1	944.3	1.122	4247	2147	0.6851	0.027	0.233	1.293	1.445	1.028
130	270.09	2173.7	936.1	1.496	4265	2207	0.6861	0.0282	0.214	1.327	1.331	1.039
140	361.29	2144.3	927.3	1.965	4285	2278	0.6859	0.0294	0.198	1.366	1.235	1.058
150	475.85	2113.7	918.1	2.545	4310	2355	0.6845	0.0308	0.184	1.396	1.156	1.067
160	617.84	2082	908.3	3.256	4337	2456	0.6820	0.0322	0.171	1.43	1.089	1.091
170	791.68	2048.8	898.0	4.117	4369	2545	0.6784	0.0338	0.160	1.464	1.032	1.102
180	1002.15	2014.2	887.3	5.153	4405	2653	0.6737	0.0354	0.151	1.499	0.985	1.123
190	1254.37	1977.9	875.9	6.388	4447	2778	0.6681	0.0372	0.142	1.533	0.946	1.145
200	1553.81	1939.7	864.1	7.852	4494	2915	0.6614	0.039	0.134	1.567	0.913	1.171
210	1906.23	1899.7	851.8	9.577	4549	3150	0.6538	0.0411	0.127	1.601	0.887	1.227
220	2317.75	1857.4	838.9	11.60	4612	3230	0.6453	0.0433	0.121	1.635	0.866	1.220
230	2794.80	1812.8	825.5	13.97	4685	3520	0.6358	0.0457	0.115	1.67	0.851	1.286
240	3344.14	1765.4	811.6	16.73	4769	3660	0.6255	0.0483	0.110	1.706	0.841	1.293
250	3972.87	1715.1	797.2	19.94	4868	4012	0.6144	0.0512	0.105	1.743	0.836	1.366
260	4688.42	1661.6	782.3	23.68	4984	4289	0.6025	0.0543	0.101	1.781	0.836	1.407

(Continued)

Table N.1 (Continued)

Temp	Saturation pressure	Enthalpy of vaporization	Density kg/m^3		Specific heat $c_p, J/kg.K$		Thermal conductivity $k, W/m.K$		Dynamic viscosity $\mu, kg/m.s$		Prandtl number	
									$\mu \times 10^3$	$\mu \times 10^5$		
$t, °C$	P_{sat}, kPa	$h_{fg}, kJ/kg$	Liquid	Vapour	Liquid	Vapour	Liquid	Vapour	Liquid	Vapour	Liquid	Vapour
270	5498.61	1604.3	766.9	28.04	5124	4650	0.5899	0.0578	0.097	1.821	0.843	1.465
280	6411.63	1543	750.9	33.13	5292	4980	0.5765	0.062	0.093	1.863	0.856	1.496
290	7436.08	1477.2	734.4	39.09	5499	5580	0.5625	0.0664	0.090	1.908	0.877	1.603
300	8581.04	1404.6	717.4	46.11	5760	6102	0.5450	0.072	0.086	1.958	0.914	1.659
310	9856.04	1329.9	699.9	54.43	6088	6740	0.5265	0.0782	0.082	2.013	0.949	1.735
320	11271.17	1238.4	667.0	64.38	6541	7730	0.5065	0.0864	0.078	2.077	1.011	1.858
330	12837.11	1125	640.0	77.00	7189	8840	0.4848	0.097	0.074	2.153	1.104	1.962
340	14565.18	1027.3	611.0	92.70	8217	10840	0.4614	0.112	0.074	2.248	1.325	2.176
350	16467.42	910.8	575.0	113.6	10100	15500	0.4365	0.135	0.066	2.374	1.523	2.726
360	18556.67	719.8	528.0	144.0	14870	20820	0.4119	0.177	0.060	2.564	2.178	3.016

Table N.2 Thermophysical properties of air at atmospheric pressure.

Temp t°C	Density ρ kg/m^3	Specific heat $c_p \times 10^{-3}$ $J/kg.K$	Dynamic viscosity $\mu \times 10^7$ $kg/m.s$	Kinematic viscosity $\nu \times 10^6$ m^2/s	Thermal conductivity $k \times 10^3$ $W/m.K$	Thermal diffusivity $\alpha \times 10^6$ m^2/s	Prandtl number Pr
−100	2.0211	1.011	117.3	5.80	15.36	7.52	0.772
−50	1.5644	1.004	145.5	9.30	19.53	12.43	0.748
−40	1.4969	1.004	150.9	10.08	20.38	13.57	0.743
−30	1.4350	1.003	156.2	10.88	21.24	14.75	0.738
−20	1.3781	1.003	161.3	11.71	22.09	15.98	0.733
−10	1.3255	1.003	166.4	12.56	22.94	17.26	0.727
0	1.2768	1.003	171.5	13.43	23.80	18.58	0.723
5	1.2538	1.003	174.0	13.87	24.22	19.26	0.720
10	1.2316	1.003	176.4	14.33	24.65	19.95	0.718
15	1.2101	1.003	178.9	14.78	25.07	20.65	0.716
20	1.1894	1.003	181.3	15.24	25.49	21.36	0.714
25	1.1694	1.004	183.7	15.71	25.91	22.08	0.712
30	1.1501	1.004	186.1	16.18	26.33	22.81	0.710
35	1.1314	1.004	188.5	16.66	26.75	23.54	0.708
40	1.1133	1.005	190.9	17.14	27.17	24.29	0.706
45	1.0957	1.005	193.2	17.63	27.58	25.05	0.704
50	1.0787	1.005	195.5	18.13	28.00	25.82	0.702
55	1.0623	1.006	197.9	18.63	28.41	26.59	0.700
60	1.0463	1.006	200.2	19.13	28.82	27.38	0.699
65	1.0308	1.007	202.4	19.64	29.23	28.17	0.697
70	1.0157	1.007	204.7	20.15	29.63	28.97	0.696
75	1.0011	1.008	207.0	20.67	30.04	29.78	0.694
80	0.9869	1.008	209.2	21.20	30.44	30.59	0.693
85	0.9731	1.009	211.5	21.73	30.84	31.42	0.692
90	0.9597	1.009	213.7	22.26	31.24	32.25	0.690
95	0.9466	1.010	215.9	22.80	31.63	33.09	0.689
100	0.9339	1.011	218.1	23.35	32.02	33.93	0.688
110	0.9095	1.012	222.4	24.45	32.80	35.64	0.686
120	0.8863	1.013	226.7	25.58	33.57	37.38	0.684
130	0.8643	1.015	230.9	26.72	34.33	39.14	0.683
140	0.8433	1.016	235.1	27.88	35.08	40.93	0.681
150	0.8234	1.018	239.3	29.06	35.81	42.73	0.680
160	0.8043	1.020	243.4	30.26	36.54	44.56	0.679
170	0.7861	1.021	247.5	31.48	37.26	46.40	0.678
180	0.7688	1.023	251.5	32.71	37.96	48.26	0.678
190	0.7521	1.025	255.5	33.97	38.65	50.14	0.677

(*Continued*)

Table N.2 (Continued)

Temp $t°C$	Density ρ kg/m^3	Specific heat $c_p \times 10^{-3}$ $J/kg.K$	Dynamic viscosity $\mu \times 10^7$ $kg/m.s$	Kinematic viscosity $\nu \times 10^6$ m^2/s	Thermal conductivity $k \times 10^3$ $W/m.K$	Thermal diffusivity $\alpha \times 10^6$ m^2/s	Prandtl number Pr
200	0.7362	1.027	259.4	35.24	39.33	52.03	0.677
210	0.7210	1.029	263.3	36.52	40.01	53.94	0.677
220	0.7063	1.031	267.2	37.83	40.67	55.86	0.677
230	0.6923	1.033	271.0	39.15	41.32	57.79	0.677
240	0.6788	1.035	274.8	40.49	41.95	59.74	0.678
250	0.6658	1.037	278.6	41.84	42.58	61.69	0.678
260	0.6533	1.039	282.3	43.21	43.20	63.66	0.679
270	0.6412	1.041	286.0	44.60	43.81	65.63	0.680
280	0.6296	1.043	289.6	46.00	44.40	67.61	0.680
290	0.6184	1.045	293.2	47.42	44.99	69.61	0.681
300	0.6076	1.047	296.8	48.85	45.57	71.61	0.682
310	0.5972	1.050	300.4	50.29	46.14	73.61	0.683
320	0.5871	1.052	303.9	51.76	46.70	75.63	0.684
330	0.5774	1.054	307.3	53.23	47.25	77.65	0.686
340	0.5679	1.056	310.8	54.72	47.79	79.68	0.687
350	0.5588	1.058	314.2	56.23	48.33	81.72	0.688
360	0.5500	1.061	317.6	57.75	48.86	83.76	0.689
370	0.5414	1.063	320.9	59.28	49.38	85.81	0.691
380	0.5331	1.065	324.3	60.82	49.89	87.87	0.692
390	0.5251	1.067	327.6	62.38	50.40	89.94	0.694
400	0.5173	1.069	330.8	63.95	50.90	92.01	0.695
410	0.5097	1.072	334.0	65.54	51.40	94.09	0.697
420	0.5024	1.074	337.2	67.13	51.89	96.18	0.698
430	0.4952	1.076	340.4	68.74	52.37	98.28	0.699
440	0.4883	1.078	343.5	70.36	52.86	100.38	0.701
450	0.4815	1.081	346.7	71.99	53.33	102.50	0.702
460	0.4749	1.083	349.7	73.64	53.81	104.62	0.704
470	0.4686	1.085	352.8	75.29	54.28	106.76	0.705
480	0.4623	1.087	355.8	76.96	54.74	108.91	0.707
490	0.4563	1.089	358.8	78.64	55.21	111.07	0.708
500	0.4504	1.092	361.8	80.33	55.67	113.24	0.709
520	0.4390	1.096	367.6	83.74	56.59	117.62	0.712
540	0.4282	1.100	373.4	87.19	57.51	122.05	0.714
560	0.4180	1.105	379.1	90.69	58.42	126.55	0.717
580	0.4082	1.109	384.6	94.23	59.34	131.12	0.719
600	0.3988	1.113	390.1	97.82	60.26	135.76	0.721
620	0.3899	1.117	395.5	101.44	61.19	140.48	0.722

Table N.2 (Continued)

Temp $t°C$	Density ρ kg/m^3	Specific heat $c_p \times 10^{-3}$ $J/kg.K$	Dynamic viscosity $\mu \times 10^7$ $kg/m.s$	Kinematic viscosity $\nu \times 10^6$ m^2/s	Thermal conductivity $k \times 10^3$ $W/m.K$	Thermal diffusivity $\alpha \times 10^6$ m^2/s	Prandtl number Pr
640	0.3814	1.121	400.9	105.11	62.13	145.28	0.724
660	0.3732	1.125	406.2	108.83	63.08	150.18	0.725
680	0.3654	1.129	411.4	112.58	64.04	155.18	0.726
700	0.3579	1.133	416.6	116.39	65.02	160.28	0.726
750	0.3405	1.143	429.3	126.11	67.54	173.53	0.727
800	0.3246	1.153	442.0	136.14	70.19	187.55	0.726
850	0.3102	1.162	454.6	146.53	72.98	202.40	0.724
900	0.2970	1.172	467.2	157.31	75.91	218.15	0.721
950	0.2849	1.181	480.1	168.50	78.98	234.82	0.718
1000	0.2738	1.189	493.1	180.13	82.19	252.41	0.714
1050	0.2634	1.198	506.4	192.23	85.52	270.93	0.710
1100	0.2539	1.207	519.9	204.82	88.95	290.34	0.705
1150	0.2450	1.216	533.7	217.88	92.48	310.59	0.701
1200	0.2367	1.224	547.7	231.42	96.07	331.64	0.698
1250	0.2289	1.233	561.7	245.41	99.72	353.42	0.694
1300	0.2216	1.241	575.8	259.83	103.40	375.85	0.691
1350	0.2148	1.250	589.9	274.64	107.11	398.87	0.689
1400	0.2083	1.259	603.8	289.79	110.85	422.43	0.686
1450	0.2023	1.269	617.5	305.26	114.61	446.49	0.684
1500	0.1966	1.279	631.0	321.01	118.40	471.02	0.682
1550	0.1911	1.290	644.1	337.02	122.27	496.04	0.679
1600	0.1860	1.301	657.1	353.27	126.23	521.60	0.677
1650	0.1811	1.314	669.8	369.79	130.34	547.77	0.675
1700	0.1765	1.328	682.3	386.60	134.67	574.69	0.673
1750	0.1721	1.344	694.7	403.76	139.29	602.53	0.670
1800	0.1679	1.361	707.2	421.31	144.30	631.51	0.667
1850	0.1638	1.381	719.8	439.32	149.80	661.87	0.664
1900	0.1600	1.405	732.6	457.86	155.93	693.91	0.660
1950	0.1567	1.431	745.6	475.78	162.82	726.11	0.655
2000	0.1533	1.461	758.9	495.14	170.62	761.91	0.650

References

Abramowitz, M. and Stegan, I.A. (1972). *Handbook of Mathematical Functions*. National Bureau of Standards. Applied mathematics Series, 55.

Akers, W.W., Deans, H.A., and Crosser, O.K. (1959). Condensation heat transfer within horizontal tubes. *Chemical Engineering Progress Symposium Series* 55 (29): 171–176.

Alberti, M., Weber, R., and Mancini, M. (2015). Re-creating Hottel's emissivity charts for carbon dioxide and extending them to 40 bar pressure using HITEMP-2010 data base. *Combustion and Flame* 162 (3): 597–612.

Alberti, M., Weber, R., and Mancini, M. (2016). Re-creating Hottel's emissivity charts for water vapor and extending them to 40 bar pressure using HITEMP-2010 data base. *Combustion and Flame* 169: 141–153.

Alberti, M., Weber, R., and Mancini, M. (2018). Gray gas emissivities for H_2O-CO_2-CO-N_2 mixtures. *Journal of Quantitative Spectroscopy and Radiative Transfer* 219: 274–291.

Alberti, M., Weber, R., and Mancini, M. (2020). New formulae for gray gas absorptivities of H_2O, CO_2, and CO. *Journal of Quantitative Spectroscopy and Radiative Transfer* 255 (15): 107227.

Arpaci, V.S. (1966). *Conduction Heat Transfer*. Reading, MA: Addison-Wesley.

Bejan, A. (2013). *Convection Heat Transfer*, 4e. Wiley.

Bejan, A. and Kraus, A.D. (2003). *Heat Transfer Handbook*. Wiley.

Bergman, T.L., Lavine, A.S., Incropera, F.P., and Dewitt, D.P. (2011). *Fundamentals of Heat and Mass Transfer*, 7e. Wiley.

Bert, C.W. (1963). Nonsymmetric temperature distributions in varying thickness circular fins. *Journal of Heat Transfer, ASME Transactions*, 85, Series C (1).

Billo, E.J. (2007). *Excel For Scientists and Engineers: Numerical Methods*. Wiley.

Boiko, L.D. and Kruzhilin, G.H. (1966). Heat transfer in vapour condensation in a tube, news of the USSR academy of science. *Energy and Transport* 5: 113–128 (in Russian).

Bowman, R.A., Mueller, A.C., and Nagle, W.M. (1940). Mean temperature differences in heat exchanger design. *Transactions of the ASME*. 62: 283–294.

Brewster, M.Q. (1992). *Thermal Radiative Transfer and Properties*. Wiley.

Bromley, L.A. (1950). Heat transfer in stable film boiling. *Chemical Engineering Progress* 46: 221–227.

Carslaw, H.S. and Jaeger, J.G. (1959). *Conduction of Heat in Solids*, 2e. Oxford University Press.

Catton, I. (1978). Natural convection in enclosures. In: *Advances in Heat Transfer, Proceedings of the 6th International Heat Transfer Conference*, vol. 6, 13–31. Toronto, Canada: Begel House.

Çengel, Y.A. and Ghajar, A.J. (2011). *Heat and Mass Transfer: Fundamentals and Applications*, 4e. McGraw Hill.

Heat Transfer Basics: A Concise Approach to Problem Solving, First Edition. Jamil Ghojel.

Companion website: www.wiley.com/go/ghojel/heat_transfer

Chato, J.C. (1962). Laminar condensation inside horizontal and inclined tubes. *ASHRAE Journal* 4: 52–60.

Chen, M.M. (1961). An analytical study of laminar film condensation. Part 1: "Flat plates," and Part 2: "Single and multiple horizontal tubes." *Transactions of the ASME*, Series C, 83: 48–60.

Churchill, S.W. (1983). Free Convection around Immersed Bodies. In: *Heat Exchanger Design Handbook* (ed. E.U. Schlünder), Section 2,5,7. 500. Hemispheres Publishing.

Churchill, S.W. and Bernstein, M. (1977). A correlating equation for forced convection from gases and liquids to a circular cylinder in crossflow. *Journal of Heat Transfer* 99: 300–306.

Churchill, S.W. and Chu, H.H.S. (1975a). Correlating equations for laminar and turbulent free convection from a vertical plate. *International Journal of Heat and Mass Transfer* 18: 1323–1329.

Churchill, S.W. and Chu, H.H.S. (1975b). Correlating equations for laminar and turbulent free convection from a horizontal cylinder. *International Journal of Heat and Mass Transfer* 18: 1049–1053.

Dobson, M.K. and Chato, J.C.J. (1998). Condensation in smooth horizontal tubes. *Journal of Heat Transfer* 120: 193–213.

Docherty, P. (1982) Prediction of gas emissivity for a wide range of process conditions. Paper R5, *Proceedings of the Seventh International Heat Transfer Conference*, vol. 2, 481–485, Munich, Germany.

Ede, A.J. (1967). *An Introduction to Heat Transfer: Principles and Calculations*. Pergamon Press.

Elenbaas, W. (1942). Heat dissipation of parallel planes by free convection. *Physica* 9 (1): 1–28.

El Sherbiny, S.M., Raithby, G.D., and Hollands, K.G.T. (1982). Heat transfer by natural convection across vertical and inclined air layers. *Journal of Heat Transfer* 104: 96–102.

Farag, I.H. (1982). Non-luminous gas radiation: approximate emissivity models. *Proceedings of the 7th International Heat Transfer Conference*, vol. R6, 487–492. Munich, Germany.

Forsberg, C.H. (2021). *Heat Transfer Principles and Applications*. Academic Press.

Gardner, K.A. (1945). Efficiency of extended surfaces. *ASME Transactions* 67 (8): 621–631.

Ghojel, J. (2004). Experimental and analytical technique for estimating interface thermal conductance in composite structural elements under simulated fire conditions. *Experimental Thermal and Fluid Science* 28 (2004): 347–354.

Globe, S. and Dropkin, D. (1959). Natural convection heat transfer in liquids confined by two horizontal plates and heated from below. *Journal Heat Transfer* 81C: 24–28.

Gnielinski, V. (1976). New equations for heat and mass transfer in in turbulent pipe and channel flow. *International Chemical Engineering* 16 (2): 359–368.

Gnielinski, V. (2010). Heat transfer in pipe flow. In: *VDI Atlas*, 2e, 693–700. Springer.

Gnielinski, V. (2013). On heat transfer in tubes. *International Journal on Heat and Mass Transfer* 63: 134–140.

Griffith, P. and Hewitt, G.F. (eds.) (1990). *Heat Exchanger Design Handbook*, Section 2.6.5, Hemisphere Publishing.

Hahn, D.W. and Ozisik, M.N. (2012). *Heat Conduction*, 3e. John Wiley & Sons, Inc.

Hahne, E. and Grigull, U. (1975). Shape factor and shape resistance for steady multidimensional heat conduction (Formfaktor Und Formwiderstand Der Stationren Mehrdimensionalen Warmeleitung). *International Journal of Heat and Mass Transfer* 18: 751–767. Pergamon Press.

Heisler, M.P. (1947). Temperature charts for induction and constant temperature heating. *Transactions of the ASME* 69: 227–236.

Hilpert, R. (1933). Warmeabgabe von geheizen Drahten und Rohren, Forsch. *Geb. Ingenieurwes.* 4: 220.

Holman, J.P. (2010). *Heat Transfer*, 10e. McGraw Hill.

Holman, J.P. and Holman, B.K. (2018). *What Every Engineer Should Know About Excel*, 2e. CRC Press.

Hottel, H.C. (1954). Radiant heat transmission. Chapter 4 In: *Heat Transmission*, 3e (ed. W.H. McAdams). McGraw-Hill.

Hottel, H.C. and Sarofim, A.F. (1967). *Radiative Transfer*. McGraw Hill.

Howell, J.R. (1982). *A Catalog of Radiation Configuration Factors*. McGraw Hill.

INTEMP. (2020). Inverse heat transfer software by David Trujillo. *truinverse@gmail.com*.

Isachenko, V., Osipova, V., and Sukomel, A. (1969). *Heat Transfer*. Moscow: Mir Publishers.

Jakob, M. (1962). *Heat Transfer*, vol. 1. Wiley.

Kapitsa, P.L. (1949). Experimental investigation of wavy flow regime. *Journal of Experimental and Theoretical Physics* 19 (2): 105–120 (in Russian).

Kays, W.M. and London, A.L. (1964). *Compact Heat Exchangers*, 2e. New York: McGraw Hill.

Kays, W.M. and London, A.L. (1984). *Compact Heat Exchangers*, 3e. New York: McGraw Hill.

Kays, W.M. and London, W.M. (2018). *Compact Heat Exchangers*. Reprint of the Third edition, 1984. Medtech (Scientific International).

Knusden, J.D. and Katz, D.L. (1958). *Fluid Dynamics and Heat Transfer*. McGraw Hill.

Kraus, A.D., Aziz, A., and Welty, J. (2001). *Extended Surface Heat Transfer*. Wiley.

Kreith, F., Manglik, R.M., Bohn, M.S., and Tiwari, S. (2011). *Principles of Heat Transfer*, 7e. (SI), Cengage Learning.

Kutateladze, S.S. (1963). *Fundamentals of Heat Transfer*. New York: Academic Press.

Labuntsov, D.A. (1957). Heat transfer in film condensation of pure vapours on vertical surfaces and horizontal tubes. *Teploenergetika* (7): 72–80 (in Russian).

Labuntsov, D.A. (1963). Computation of heat transfer in film boiling over vertical heated surfaces. *Teploenergetika* (5): 60–61 (in Russian).

Langston, L.S. (1982). Heat transfer from multidimensional objects using one-dimensional solutions for heat loss. *International Journal of Heat and Mass Transfer* 25: 149–150.

Latif, M.J. (2009). *Heat Conduction*, 3e. Springer.

Leckner, B. (1972). Spectral and total emissivity of water vapor and carbon dioxide. *Combustion and Flame* 19: 33–48.

Marrero, T.R. and Mason, E.A. (1972). Gaseous diffusion coefficients. *Journal of Physical and Chemical Reference Data* 1: 3–118.

McAdams, W.H. (1954). *Heat Transmission*, 3e. McGraw-Hill.

Mikheyev, M.A. (1964). *Fundamentals of Heat Transfer*. Moscow: MIR Publishers.

Mikheyev, M.A. and Mikheyeva, E.M. (1977). *Fundamentals of Heat Transfer*, 2e. Energiya (in Russian).

Mills, A.F. (1999). *Heat Transfer*, 2e. Prentice Hall.

Modest, M.F. (2013). *Radiative Heat Transfer*, 3e. Academic Press.

Morgan, V.T. (1975). The Overall Convective Heat Transfer from Smooth Circular Cylinders. In: *Advances in Heat Transfer*, Vol. 11 (ed. Irvine and Hartnett), 199–264. Academic Press.

National Bureau of Standards. (1949). *Tables of Bessel Functions of Fractional Order: Vol. 2 Computation Laboratory of the National Applied Mathematics Laboratory*. Columbia University Press.

Ozisik, M.N. (1979). *Heat Conduction*. New York: Wiley.

Penner, S. (1959). *Quantitative Molecular Spectroscopy and Gas Emissivities*. Addison-Wesley Publishing Company.

Petekhov, B.S. (1970). Heat transfer and friction in turbulent pipe flow with variable physical properties. In: *Advances in Heat Transfer* (ed. J.P. Hartnett and T.F. Irvine), 540–564. Academic Press.

Peterson, G.P. and Fletcher, L.S. (1987). Evaluation of the thermal contact conductance between substrate and mould compound materials. In: *Fundamentals of Conduction and Recent Developments in Contact Resistance* (ed. M. Imber, G.P. Teterson, and M.M. Yovanovitch), vol. 69, 99–105. ASME HTD.

Raithby, G.D. and Hollands, K.G.T. (1975). A general method of obtaining approximate solutions to laminar and turbulent free convection problems. In: *Advances in Heat Transfer* (ed. T.F. Irvine and J.P. Harnett), vol. 11, 265–315. Academic Press.

Ranz, W.E. and Marshall, W.R. (1952). Evaporation from drops. *Chemical Engineering Progress* 48: 141–146, 173–180.

Rathore, M.M. (2015). *Engineering Heat and Mass Transfer*, 3e. India: University Science Press.

Rogers, G. and Mayhew, Y. (1992). *Engineering Thermodynamics: Work & Heat Transfer*, 4e. Longman Scientific & Technical.

Rohsenow, W.M. (1952). A method of correlating heat transfer data for surface boiling of liquids. *Transactions of the ASME* 74: 969–976.

Rohsenow, W.M., Hartnett, J.P., and Cho, Y.I. (eds.) (1998). *Handbook of Heat Transfer*, 3e. McGraw Hill.

Sieder, E.N. and Tate, C.E. (1936). Heat transfer and pressure drop of liquids in tubes. *Industrial & Engineering Chemistry* 28: 1429.

Siegel, R. and Howell, J.R. (1992). *Thermal Radiation Heat Transfer*, 3e. Hemisphere Publishing Corporation.

Smith, P.J. and Sucec, J. (1969). Efficiency of circular fins of triangular profile. *Journal of Heat Transfer*, ASME Transactions 91: 181.

Stoecker, W.F. and Jones, J.W. (1982). *Refrigeration and Air Conditioning*. McGraw Hill International editions.

Taler, J. and Duda, P. (2006). *Solving Direct and Inverse Heat Conduction Problems*. Springer-Verlag.

TEMA. (2019). *Standards of Tubular Exchange Manufacturers Association*, 10e.

Theodore, L. (2011). *Heat Transfer Applications for the Practicing Engineer*. Wiley.

Thomas, L.C. (1999). *Heat Transfer – Professional Version*, 2e. Capstone.

Ullmann, A. and Kalman, H. (1989). Efficiency and optimized dimensions of annular fins of different cross-section shapes. *International Journal of Heat and Mass Transfer* 32 (6): 1105–1110.

Underwood, A. (1934). The calculation of the mean temperature difference in multipass heat exchangers. *Journal of the Institute of Petroleum Technology* 20: 145–158.

VanSant, J.H. (1983). *Conduction Heat Solutions, Livermore National Laboratory, National Technical Information Service, US Department of Commerce.*

Welty, J.R., Rorrer, G.L., and Foster, D.G. (2013). *Fundamentals of Momentum, Heat, and Mass Transfer*, 6e. Wiley.

Whitaker, S. (1977). *Fundamental Principles of Heat Transfer*. Pergamon Press.

Wong, H.Y. (1977). *Handbook of Essential Formulae and Data on Heat Transfer for Engineers*. Longman.

Yovanovitch, M.M., Culham, J.R., and Teerstra, P. (1997). Calculating interface resistance. *Electronics Cooling* 3 (2): 24–29.

Yurenev, V.N. and Lebedev, P.D. (eds.) (1976). *Thermal Engineering Handbook*, vol. 2. Energiya, 192 (in Russian).

Zukauskas, A. and Ulinskas, R. (1988). *Heat Transfer in Tube Banks in Cross Flow*. Hemisphere Publishing Corporation (Translation from the Russian by Jamil Ghojel).

Zukauskas, A., Makerevicius, V., and Schlanciauskus, A. (1968). *Heat Transfer in Banks of Tubes in Crossflow of Liquid*. Vilnius: Mintis (in Russian).

Index

Heat Transfer Basics: A Concise Approach to Problem Solving, First Edition. Jamil Ghojel.

Companion website: www.wiley.com/go/ghojel/heat_transfer

g

h

i

j

k

l

m

t

u

v